More information about this series at http://www.springer.com/series/7407

Lecture Notes in Computer Science 12290

Zhao Zhang · Wei Li · Ding-Zhu Du (Eds.)

Algorithmic Aspects in Information and Management

14th International Conference, AAIM 2020
Jinhua, China, August 10–12, 2020
Proceedings

 Springer

Editors
Zhao Zhang
Zhejiang Normal University
Jinhua, China

Wei Li (iD)
Texas Southern University
Houston, TX, USA

Ding-Zhu Du
University of Texas at Dallas
Richardson, TX, USA

ISSN 0302-9743   ISSN 1611-3349   (electronic)
Lecture Notes in Computer Science
ISBN 978-3-030-57601-1   ISBN 978-3-030-57602-8   (eBook)
https://doi.org/10.1007/978-3-030-57602-8

LNCS Sublibrary: SL1 – Theoretical Computer Science and General Issues

This Springer imprint is published by the registered company Springer Nature Switzerland AG
The registered company address is: Gewerbestrasse 11, 6330 Cham, Switzerland

Preface

The 14th International Conference on Algorithmic Aspects in Information and Management (AAIM 2020), was held online during August 10–12, 2020. The conference was held virtually due to the COVID-19 pandemic.

The AAIM conference series, which started in 2005 in Xi'an, China, aims to stimulate various fields for which algorithmics has become a crucial enabler, and to strengthen the ties of various research communities of algorithmics and applications. AAIM 2020 seeks to address emerging and important algorithmic problems by focusing on the fundamental background, theoretical technological development, and real-world applications associated with information and management analysis, modeling and data mining. Special considerations are given to algorithmic research that was motivated by real-world applications.

We would like to thank the two eminent keynote speakers, Rolf H. Möhring from Technische Universität Berlin, Germany, and Chandra Chekuri from University of Illinois, Urbana-Champaign, USA, for their contributions to the conference.

We would like to express our appreciation to all members of the Program Committee and the external referees whose efforts enabled us to achieve a high scientific standard for the proceedings. We would also like to thank all members of the Organizing Committee for their assistance and contribution which attributed to the success of the conference. Particularly, we would like to thank Alfred Hofmann, Celine Lanlan Chang, Anna Kramer, and their colleagues at Springer for meticulously supporting us in the timely production of this volume. Last but not least, our special thanks go to all the authors and participants for their contributions to the success of this event.

July 2020

Zhao Zhang
Wei Li
Ding-Zhu Du

Organization

General Chair

Wei Li   Texas Southern University, USA

Program Chairs

Ding-Zhu Du   The University of Texas at Dallas, USA
Zhao Zhang   Zhejiang Normal University, China

Program Committee Members

Mansoor Alam	Northern Illinois University, USA
Xujin Chen	Chinese Academy of Sciences, China
Yongxi Cheng	Xi'an Jiaotong University, China
Donglei Du	University of New Brunswick, Canada
Xiaofeng Gao	Shanghai Jiaotong University, China
Xin Han	Dalian Institute of Technology, China
Mohsin Jamali	The University of Texas of the Permian Basin, USA
Donghyun Kim	Kennesaw State University, USA
Minming Li	City University of Hong Kong, Hong Kong, China
Quanlin Li	Beijing University of Technology, China
Xianyue Li	Lanzhou University, China
Guohui Lin	University of Alberta, Canada
Wenlong Ni	Jiangxi Normal University, China
Xiaoming Sun	Institute of Computing Technology, Chinese Academy of Science, China
Guochun Tang	Shanghai Polytechnic University, China
Shaojie Tang	The University of Texas at Dallas, USA
My Thai	University of Florida, USA
Jinting Wang	Central University of Finance and Economics, China
Dachuan Xu	Beijing University of Technology, China
Boting Yang	University of Regina, Canada
Guochuan Zhang	Zhejiang University, China
Jialin Zhang	Institute of Computing Technology, Chinese Academy of Science, China
Liwei Zhong	Shanghai General Hospital, China
Peng Zhang	Shandong University, China
Binhai Zhu	Montana State University, USA
Sheng Zhu	Henan Polytechnic University, China

Contents

Polynomial-Time Algorithms for the Touring Rays and Related Problems

Xuehou Tan$^{(\boxtimes)}$ (iD)

Tokai University, 4-1-1 Kitakaname, Hiratsuka 259-1292, Japan
`tan@wing.ncc.u-tokai.ac.jp`

Abstract. The touring rays problem, which is also known as the traveling salesman problem for rays in the plane, asks to compute the shortest (closed) route that tours or intersects n given rays. We show that it can be reduced to the problem of computing a shortest route that intersects a set of ray-segments, inside a circle; at least one endpoint of every ray-segment is on the circle. Moreover, computing the shortest route intersecting all ray-segments in the circle is related to the solution of the well-known watchman route problem. Our method is further extended to solve the minimum-perimeter intersecting polygon problem, which asks for a (convex) polygon P of minimum perimeter such that P contains at least one point of every given line segment. Both of our algorithms run in $O(n^5)$ time, and they solve two long-standing open problems in computational geometry.

1 Introduction

Shortest paths are of fundamental importance in computational geometry, robotics and autonomous navigation [6]. For a given set of points, the Euclidean Traveling Salesman Problem (TSP) asks for a shortest route (closed curve) that visits each given point. An interesting variant of the problem, called the Traveling Salesman Problem with neighborhoods (TSPN), deals with a given set of the (possibly disconnected) regions; the tour is asked to visit at least one point of each region. Since the Euclidean TSP is NP-hard, TSPN is NP-hard, too.

Although TSPN is NP-hard, an interesting result is that the Traveling Salesman Problem for n lines in the plane can be solved in polynomial time. Jonsson [7] showed that it can be reduced to the watchman route problem, which asks for a shortest route that can see all points of a simple polygon P and can be solved in $O(n^4)$ time [10]. A data structure, called the *last step shortest path maps*, has been developed to solve the watchman route and related problems [3,10].

The polynomial-time result on the TSP for lines is due to the fact that all neighborhoods (lines) are unbounded. It is then natural to ask whether the Traveling Salesman Problem for n rays can be solved in polynomial time [4,7]. A ray (half-line) is described by a source point (or an origin) and a direction,

---

This work was partially supported by JSPS KAKENHI Grant Number 15K00023 and 20K11683.

Z. Zhang et al. (Eds.): AAIM 2020, LNCS 12290, pp. 1–12, 2020.
https://doi.org/10.1007/978-3-030-57602-8_1

from which the ray emanates. This problem stands open for a long time, while a linear-time 1.28-approximation algorithm has been reported [4]. As in [3], a route that intersects n rays is called a *touring route* for given rays, see Fig. 1.

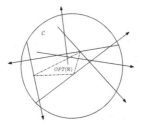

Fig. 1. A shortest touring tour for six rays.

A related problem, called the *minimum-perimeter intersecting polygon* problem (MPIP), has also been studied in the literature [5,8]. For a given set S of line segments in the plane, it asks for a polygon P of minimum perimeter such that P contains at least one point of every line segment in S. Rappaport [8] has given an $O(n \log n)$ time algorithm for MPIP when the input segments are restricted to have a constant number of orientations, but the running time becomes exponential when segments are arbitrarily oriented.

In this paper, we present the polynomial-time algorithms for the TSP for rays in the plane and the problem MPIP. Our results solve two long-standing open problems in computational geometry. Section 2 first reduces the TSP for rays in the plane to that of computing a shortest route that intersects a set of ray-segments inside a circle; at least one endpoint of every ray-segment is on the circle, and then shows that computing the shortest route visiting all ray-segments in the circle is related to the solution of the watchman route problem or the TSP for lines in the plane. Section 3 presents an $O(n^5)$ time algorithm for computing a shortest route that intersects all ray-segments inside the circle, by employing a simplified version of the last step shortest path maps. Section 4 further describes an $O(n^5)$ time solution to the problem MPIP.

2   Shortest Touring Routes for Rays in the Plane

Let $\mathcal{R}$ be the set of n given rays in the plane. Denote by $OPT(\mathcal{R})$ a shortest route that intersects or visits all rays.

Lemma 1. *(see [7]) $OPT(\mathcal{R})$ consists of straight line segments, and it is convex.*

Assume that the given rays are *not* all parallel; otherwise, $OPT(\mathcal{R})$ can be found in $O(n \log n)$ time (see the proof of **Theorem 3**). Denote by $\mathcal{C}$ the smallest enclosing circle of the origins of all rays and the intersection points among the lines through given rays, see Fig. 1. Then, $OPT(\mathcal{R})$ is contained in $\mathcal{C}$.

Lemma 2. *Suppose that the given rays are not all parallel. Then, $OPT(\mathcal{R})$ is contained in $\mathcal{C}$.*

Proof. Omitted in this extended abstract (see also [7, Lemma 2.3]).   □

Denote by $\mathcal{S}$ the set of the line segments obtained by cutting off the portions of all rays, which are outside of $\mathcal{C}$. Every segment of $\mathcal{S}$ clearly has at least one endpoint on the boundary of $\mathcal{C}$, and is thus called a *ray-segment* of $\mathcal{C}$. Under the assumption made in Lemma 2, $OPT(\mathcal{R})$ is identical to $OPT(\mathcal{S})$. In the following, we focus on how to compute $OPT(\mathcal{S})$ in $\mathcal{C}$.

2.1   A Restricted Traveling Salesman Problem for Rays

Let us consider a restricted version of the TSP for rays, in which the touring route starts at a given point s on $\mathcal{C}$. For a ray-segment $R \in \mathcal{S}$, denote by B the line segment containing R, whose two endpoints are on $\mathcal{C}$. (If the origin of R is also on $\mathcal{C}$, then R is identical to B.) Then, segment B partitions the interior of $\mathcal{C}$ into two regions, each including B itself. We call the region *not* containing s, the *pocket* of ray-segment R. For the instance of Fig. 2, pocket P_8 is shaded.[1]

Let $OPT_s(\mathcal{S})$ be an optimum route that starts at point s and visits all ray-segments in $\mathcal{C}$. Denote by $P_1, P_2, \dots P_n$ the sequence of pockets, which are encountered by a clockwise scan of $\mathcal{C}$, starting at s. Denote by $R(P_i)$ the ray-segment defining P_i, and $B(P_i)$ ($\supseteq R(P_i)$) the bounding line segment of P_i.[2] A pocket P_i is *visited* by $OPT_s(\mathcal{S})$ if at least one point of P_i is on route $OPT_s(\mathcal{S})$. We say $OPT_s(\mathcal{S})$ visits first P_i and then P_j, $1 \le i < j \le n$, if the portion of $OPT_s(\mathcal{S})$ from some visiting point of P_i to the *ending point* s goes through P_j. For the instance of Fig. 2, route $OPT_s(\mathcal{S})$ visits $P_1, P_2, \dots, P_8$ in order.

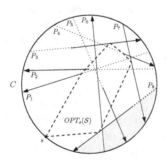

Fig. 2. An instance of $OPT_s(\mathcal{S})$.

Lemma 3. *$OPT_s(\mathcal{S})$ visits $P_1, P_2, \dots P_n$ in this order.*

---

[1] The circle $\mathcal{C}$ in Figs. 2, 4 and 5 is not drawn exactly, as it only shows an efficiently large region that contains $OPT_s(\mathcal{S})$.

[2] Although the order of pockets is defined, no order of ray-segments is specified.

Proof. It follows from the definitions of $OPT_s(\mathcal{S})$ and pockets [3,10]. □

We say that $OPT_s(\mathcal{S})$ makes a *reflection* contact with a ray-segment R if it never enters the interior of the pocket of R; otherwise, $OPT_s(\mathcal{S})$ makes a *crossing* contact with R. As noted in [3], the local optimality of $OPT_s(\mathcal{S})$ is equivalent to its global optimality. That is, a route is optimum $(OPT_s(\mathcal{S}))$ if and only if its contact point c with *every* ray-segment R is locally optimal. We can further distinguish the following contacts of $OPT_s(\mathcal{S})$ with R at point c.

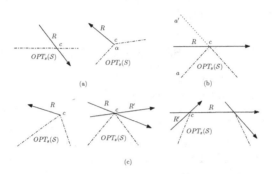

Fig. 3. Types of contact of $OPT_s(\mathcal{S})$ with a ray-segment R.

1. *Crossing contacts*: Except for the requirement that $OPT_s(\mathcal{S})$ enter the pocket of R, a line segment of $OPT_s(\mathcal{S})$ passes R through an *interior* point c, or two line segments of $OPT_s(\mathcal{S})$ have the origin c of R as their common endpoint and the interior angle α between them, which does not enclose R, is less than π. See Fig. 3(a).
2. *Perfect reflection contacts*: $OPT_s(\mathcal{S})$ reflects on R at point c such that the incoming angle of $OPT_s(\mathcal{S})$ with R is equal to the outgoing angle. See Fig. 3(b). Generally, c is not the origin of R nor an intersection point of R with some other ray-segment.
3. *Bending contacts*: It is a special case that $OPT_s(\mathcal{S})$ makes an *imperfect* reflection on R at its origin c or the intersection point c of R with another ray-segment R'. See Fig. 3(c). (From the optimality of $OPT_s(\mathcal{S})$, a slight movement or change of c on R or R' makes a longer touring route.) $OPT_s(\mathcal{S})$ may even overlap with a portion of R, because bending contacts occur at two endpoints of the overlapped portion of R, see Fig. 3(c).

Note that a route visiting all pockets $P_1, P_2, \ldots, P_n$ may *not* intersect all ray-segments, because it may just pass through the segments *extended from* some ray-segments. How to overcome this difficulty is a major issue of our work.

3   The TSP for Rays in the Plane

In this section, we first present an $O(n^4)$ time algorithm for the restricted TSP for n rays. A known data structure, called the *last step shortest path maps* [3], is used to give a solution to the restricted TSP for rays. Next, we show how to remove the restriction of a given starting point; it requires to increase the time complexity of the algorithm by a factor of n.

3.1 A Solution of the Restricted TSP for Rays

A touring route $\mathcal{W}$ is said to be *adjustable* on a ray-segment R if the contact point of $\mathcal{W}$ with R can be moved on R to get a shorter touring route. If $\mathcal{W}$ is adjustable on only one ray-segment, we say that $\mathcal{W}$ is *one-place-adjustable*; in this case, the only possible adjustment to $\mathcal{W}$ directs to the portion of R, with which optimum touring routes make contacts. Similar to the previous work on the watchman route problem [9], we have the following result.

Lemma 4. *The shortest touring route, which passes through s and intersects n given ray-segments in C, is unique.*

Proof. Omitted in this extended abstract (and see also [9]).   □

The structure of *last step shortest path maps* was originally presented to solve the *touring polygons problem* for a sequence of possibly intersecting convex polygons in the plane (and the watchman route problem as well) [3]. As pointed out in Sect. 2, the shortest route visiting all convex regions (pockets) $P_1, P_2, \ldots, P_n$ may not intersect all ray-segments.

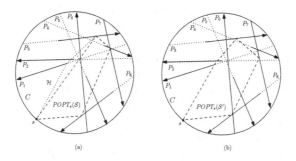

Fig. 4. Routes $POPT_s(\mathcal{S})$, $\mathcal{H}$ and $POPT_s(\mathcal{S}')$.

The above difficulty can be overcome as follows: First, we require that if the shortest touring route does not enter the *interior* of a pocket P_i, then it should reflect on its ray-segment $R(P_i)$. By placing this restriction in the construction of the last step shortest path maps, we can find a new type of shortest touring route for $P_1, P_2, \ldots, P_n$, denoted by $POPT_s(\mathcal{S})$. Still, some ray-segments, whose pockets are crossed by $POPT_s(\mathcal{S})$, may not be visited by $POPT_s(\mathcal{S})$. We call $POPT_s(\mathcal{S})$ the *shortest pseudo-touring route* for all ray-segments, see Fig. 4(a). By an argument similar to that given for **Lemma 4**, we also have:

Lemma 5. *Route $POPT_s(\mathcal{S})$ is unique.*

Consider now the situation in which some ray-segments are not visited by $POPT_s(S)$. Denote by $\mathcal{H}$ the convex hull of the vertices of polygon $POPT_s(S)$ and the origins whose ray-segments are not visited by $POPT_s(S)$. So, $\mathcal{H}$ is a touring route for all ray-segments, see Fig. 4(a).

Lemma 6. *Let R be a ray-segment, with origin p on $\mathcal{H}$, such that it is not visited by $POPT_s(S)$. Then, $OPT_s(S)$ makes a crossing contact with R.*

Proof. Let S_1 be the set of the ray-segments whose endpoints are on the arc of C from s to the endpoint of R on C clockwise, and let S_2 be the set of rest ray-segments. Assume that R does not belong to S_1 nor S_2. Then, one can consider the following two subproblems: One is asked to start at s, end at p and visit all ray-segments of S_1. And, the other starts at p, ends at s and visits all ray-segments of S_2. As shown in the proof of **Lemma 4**, the solution to either subproblem is unique. Denote by $OPT_{s,p}(S_1)$ and $OPT_{p,s}(S_2)$ the optimum touring paths obtained for two subproblems, respectively. Let $\mathcal{U}$ be the union of $OPT_{s,p}(S_1)$ and $OPT_{p,s}(S_2)$. Then, $\mathcal{U}$ is a touring route for all ray-segments.

We claim that $\mathcal{U}$ makes a crossing contact with R at point p. Assume by contradiction that $\mathcal{U}$ makes a reflection contact with R. Then, $OPT_{s,p}(S_i)$ does not enter the pocket of R. Thus, no ray-segment of S_i, $i = 1$ or 2, is wholly contained in the pocket of R. So, we can compute the shortest pseudo-touring path $POPT_{s,p}(S_i)$ for all ray-segments of S_i. Clearly, $POPT_{s,p}(S_i)$ needn't enter the pocket of R. Let $\mathcal{PU}_p$ be the union of $POPT_{s,p}(S_1)$ and $POPT_{p,s}(S_2)$. Then, $\mathcal{PU}_p$ makes a reflection contact with R and is a pseudo-touring route for all ray-segments of S. Clearly, $\mathcal{PU}_p$ is either non-adjustable or one-place-adjustable. In the former case, $\mathcal{PU}_p$ gives the other shortest pseudo-touring route starting at s, a contradiction (**Lemma 5**). In the latter case, there is an unique point q on R ($q \neq p$) such that the route $\mathcal{PU}_q$ reflecting on R at point q is optimum (**Lemma 5**). Again, a contradiction occurs. Our claim is thus proved.

From the above claim, if the interior angle of the polygon, bounded by $\mathcal{U}$, at vertex p is no more than π, the crossing contact of $\mathcal{U}$ with R at point p is locally optimal. From the optimality of $OPT_{s,p}(S_1)$ and $OPT_{p,s}(S_2)$, route $\mathcal{U}$ is just $OPT_s(S)$. Thus, the lemma follows. If the interior angle at vertex p is strictly larger than π, then $\mathcal{U}$ is adjustable only on R, i.e., $\mathcal{U}$ is one-place-adjustable. This only possible adjustment to $\mathcal{U}$ implies that $OPT_s(S)$ makes a crossing contact with R. The lemma follows, too.   □

Lemma 6 tells us that $OPT_s(S)$ does not pass through the line segment sp. So, $OPT_s(S)$ can be computed in the difference region between C and sp (i.e., $C - sp$). To obtain $OPT_s(S)$, it may require to repeatedly perform the procedure of finding the origin p a few times.

3.2   The Algorithm

A simplified version of the last step shortest path maps can be used to compute route $POPT_s(S)$. We call a path from s to a point of $R(P_i)$ ($1 \leq i \leq n$), a *partial touring path*, if it visits in order each of $P_1, \ldots, P_{i-1}$, and reflects only on ray-segments. The intersection points between all pairs of ray-segments, which can be found in $O(n^2)$ time, are needed in computing the shortest partial touring paths to the points on $R(P_i)$. We call two endpoints of a ray-segment R as well as the intersection points of R with other ray-segments, the *artificial vertices* of R. The portion of R between two consecutive vertices is called a *fragment* of R.

The last step shortest path map of P_i or $B(P_i)$, denoted by $\mathcal{M}_i$, is defined as a subdivision of circle $\mathcal{C}$ into the regions such that the *last steps* of shortest partial touring paths to all points (which are points of $B(P_{i+1})$ in the following use) in a region are combinatorially equivalent, i.e., they make the same types of contacts with $B(P_i)$ [3]. From the definition of crossing contacts, the crossing region of $B(P_i)$ is just its pocket P_i. For a fragment f of $R(P_i)$, its *reflection* region contains all the points to which the shortest partial touring paths make the last reflection contact with f. The reflection region of f is unbounded and three-sided. The bending region of a vertex v of $R(P_i)$ contains all the points to which the shortest partial touring paths lastly bend at v. The bending region of v is a cone or triangular region with the apex at v.

Lemma 7. *A data structure of $O(n^2)$ size can be built in time $O(n^3)$ such that $POPT_s(\mathcal{S})$ can be reported in time $O(n^2)$.*

Proof. Omitted in this extended abstract (and see also [10]).   □

We can now give the first result of this paper.

Theorem 1. *The restricted TSP problem for n rays can be solved in $O(n^4)$ time.*

Proof. First, compute in $O(n^3)$ time route $POPT_s(\mathcal{S})$ (**Lemma 7**). If $POPT_s(\mathcal{S})$ happens to visit all ray-segments, then it gives $OPT_s(\mathcal{S})$ and we are done. Otherwise, compute the convex hull $\mathcal{H}$, as described in Sect. 3.1. Let $p_1, p_2, \ldots, p_k$ be the origins on $\mathcal{H}$, whose ray-segments are not visited by $POPT_s(\mathcal{S})$. Delete from $\mathcal{S}$ the ray-segments having origins $p_1, p_2, \ldots, p_k$, and denote by $\mathcal{S}'$ the resulting set of ray-segments. Also, delete from $\mathcal{C}$ the convex region with vertices s, $p_1, p_2, \ldots, p_k$, and denote by $\mathcal{C}'$ the resulting region. ($\mathcal{C}'$ can still be considered as a simple polygon.) Since $OPT_s(\mathcal{S})$ makes crossing contacts with the ray-segments having origins $p_1, p_2, \ldots, p_k$ (**Lemma 6**), it is outside of the convex region with vertices s, $p_1, p_2, \ldots, p_k$. Hence, $OPT_s(\mathcal{S})$ can be obtained by computing $OPT_s(\mathcal{S}')$ inside $\mathcal{C}'$ (Fig. 4(b)). This procedure can repeatedly be performed, until $OPT_s(\mathcal{S})$ is obtained. Since the total number of the ray-segments, which can be deleted from $\mathcal{S}$, is less than n, the theorem follows.   □

3.3 Removing the Restriction of a Given Starting Point

The idea of removing the restriction of a given starting point follows from that for solving the watchman route problem without giving any starting point [9]. First, compute a special route $OPT_s(\mathcal{S})$ such that s is an artificial vertex of some ray-segment A, whose pocket does not wholly contain any other ray-segment. Since a slight movement of s to one of its incident fragments on A may produce a shorter touring route, the adjustment to $OPT_s(\mathcal{S})$ is then defined. Next, we show that $OPT(\mathcal{S})$ is unique, except for very special cases where there is an infinite number of shortest touring routes of equal length. Finally, by noticing the fact that $OPT_s(\mathcal{S})$ is adjustable only on A, $OPT(\mathcal{S})$ can be obtained by computing and adjusting routes $OPT_s(\mathcal{S})$ at most $O(n)$ times [9]. (Note that A may change to some other ray-segments in the whole process.)

Assume below that the origins of some ray-segments are *not* on $\mathcal{C}$; otherwise, the solution for $OPT(\mathcal{S})$ has been known, see [7,9]. First, we find the convex hull $\mathcal{K}$ of the ray-segments' origins, which are not on $\mathcal{C}$. If $\mathcal{K}$ happens to make crossing contacts with *all* ray-segments, then all crossing contacts are locally optimal. Hence, $\mathcal{K}$ is just $OPT(\mathcal{S})$, and we are done.

Two other situations are the followings: $\mathcal{K}$ visits *all* ray-segments but makes a reflection contact with at least one ray-segment, or some ray-segments are *wholly* outside of $\mathcal{K}$ and thus their two endpoints are on $\mathcal{C}$. In either case, there is a ray-segment A such that its pocket does not wholly contain any other ray-segment. Consider now the touring routes that are forced to *reflect* on A at an artificial vertex s. So, $OPT_s(\mathcal{S})$ is contained in the difference region between $\mathcal{C}$ and the pocket of A. Still, $OPT_s(\mathcal{S})$ can be computed using the algorithm described in Sect. 3.2. Note that if point s is allowed to float on A, then a shorter touring route can be obtained.

Figure 5 shows four types of adjustments to $OPT_s(\mathcal{S})$ on ray-segment A. The incoming angle of $OPT_s(\mathcal{S})$ with A is assumed to be smaller than the outgoing angle. A possible shorter touring route R_{new} is also shown in Fig. 5. The adjustment to $OPT_s(\mathcal{S})$ involves a change of point s to a *starting fragment* incident to s, with which the new route R_{new} makes a reflection contact. It takes $O(n)$ time to perform such an adjustment to $OPT_s(\mathcal{S})$ [1,2,9].

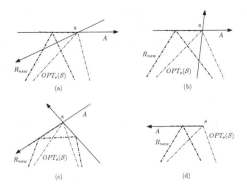

Fig. 5. Four types of adjustments to $OPT_s(\mathcal{S})$ on ray-segment A.

Theorem 2. *There is a unique non-adjustable touring route for n given rays, except for very special cases where there is an infinite number of non-adjustable routes of equal length.*

Proof. If $\mathcal{K}$ happens to give $OPT(\mathcal{S})$, then the theorem follows. Otherwise, $OPT(\mathcal{S})$ has to make the reflection contacts with some ray-segments A such that the pocket of A does not wholly contain any other ray-segment. As in [9, Lemma 4], one can show that the length function of shortest touring routes reflecting on A is monotone. The theorem can then be proved by an argument similar to that given in the proof of [9, Theorem 1]. □

We can now give our algorithm for computing a shortest touring route for n rays in the plane. If $\mathcal{K}$ happens to give $OPT(\mathcal{S})$, then we are done. Otherwise, find a ray-segment A as described above. Let $OPT_s(\mathcal{S})$ be the shortest touring route forced to reflect on A, at a vertex s of A. If $OPT_s(\mathcal{S})$ is not adjustable at point s, then it is $OPT(\mathcal{S})$ (**Theorem 2**). Otherwise, an adjustment shown in Fig. 5 is made to $OPT_s(\mathcal{S})$. If the obtained route R_{new} is non-adjustable, then it is the overall shortest touring route (**Theorem 2**) and we are done. Otherwise, R_{new} is adjustable on at least one ray-segment B at a vertex p such that the pocket of B (it may be differ from A) does not wholly contain any other ray-segment. Next, we compute the shortest touring route $OPT_p(\mathcal{S})$, which reflects on B at point p. The overall shortest touring route for all ray-segments can be obtained by performing this process repeatedly, until a non-adjustable route is found. As shown in [9], the total number of required adjustments is $O(n)$. So, the number of shortest (fixed) touring routes, which are computed in the whole process, is $O(n)$. The time taken to compute $OPT(\mathcal{S})$ is thus $O(n^5)$.

Theorem 3. *The TSP for n rays in the plane can be solved in $O(n^5)$ time.*

Proof. If the given rays are all parallel and a line segment perpendicular to the rays intersects all rays, then $OPT(\mathcal{R})$ can be computed in linear time. Otherwise, assume that all rays are vertical. Let $[y_1, y_2]$ be the *minimum* y-interval such that all rays have a non-empty intersection with $[y_1, y_2]$. Denote by x_1 and x_2 the minimal and maximal x-coordinates among all rays, respectively. It is then easy to see that $OPT(\mathcal{R})$ is contained in the rectangle with vertices (x_1, y_1), (x_1, y_2), (x_2, y_1) and (x_2, y_2). Thus, $OPT(\mathcal{R})$ can be computed in $O(n \log n)$ time [8].

In the case that the given rays are not all parallel, as described above, an overall shortest touring route can be computed in $O(n^5)$ time. □

4 An Application to the Problem MPIP

Our algorithm for the TSP for rays in the plane can be used to solve the problem MPIP. For a given set $\mathcal{S}$ of n target (line) segments in the plane, a polygon P is an *intersecting polygon* of S if every segment in $\mathcal{S}$ intersects the interior or the boundary of P, or equivalently, P contains at least one point of every target segment. The *minimum-perimeter intersecting polygon* problem then asks for an intersecting (convex) polygon $MPIP(\mathcal{S})$ of minimum perimeter.

Denote by $C(S)$ the convex hull of *all* endpoints of target segments of S. Assume that both endpoints of some segments lie in the interior of $C(S)$; otherwise, the algorithm presented in Sect. 3 can be used to give a solution.

Observation 1. *For a given set S of target segments in the plane, polygon $MPIP(S)$ is contained in $C(S)$.*

As in Sect. 3, we first consider a restricted minimum-perimeter intersecting polygon of S, which has a starting vertex s on $C(S)$. For a segment $T \in S$, denote by ET the line segment containing T, whose two endpoints are on $C(S)$. Then, ET partitions the interior of $C(S)$ into two regions. Also, we call the region not containing s the *pocket* of T. Denote by $P_1, P_2, \ldots P_n$ the sequence of pockets, which are encountered by a clockwise scan of $C(S)$, starting at s. Denote by $P_s(S)$ an optimum solution to the restricted problem MPIP. Again, $P_s(S)$ visits all pockets in the order they appear on the boundary of $C(S)$, and it is unique.

To compute $P_s(S)$, we first find the *minimum-perimeter pseudo-intersecting polygon* $PP_s(S)$ such that it intersects all pockets, and if $PP_s(S)$ makes a reflection contact with the bounding segment of pocket P_i $(1 \leq i \leq n)$, then it reflects on the target segment defining P_i.

Lemma 8. *For a point s on $C(S)$, polygon $PP_s(S)$ can be found in $O(n^3)$ time.*

Proof. As in Sect. 3, the last step shortest path map can be used to compute $PP_s(S)$. Since both endpoints of a target segment T may be in the interior of $C(S)$, the last step shortest path map for the pocket of T may have two bending regions at its endpoints. This minor difference between the last step shortest path maps used for the TSP for rays and for the problem MPIP can easily be dealt with. Hence, $PP_s(S)$ can be computed in $O(n^3)$ time, too.   □

Again, some target segments may wholly be outside of $PP_s(S)$, because $PP_s(S)$ is allowed to pass through their pockets by going across the segments extended from them (see Fig. 6(a)). Denote by T the set of target segments, which are outside of the region bounded by PP_s. For the instance of Fig. 6, T consists of four target segments, whose pockets are P_1, P_2, P_4 and P_5. Assume also that T is not empty; otherwise, $PP_s(S)$ is just $P_s(S)$.

We are going to construct an initial intersecting polygon of S, which plays the same role as $\mathcal{H}$ in Sect. 3.1. Let $\mathcal{V}$ be the set of segments' endpoints, which are *not* on $C(S)$ and whose segments belong to T. Denote by $\mathcal{L}$ the convex hull of all vertices of $PP_s(S)$ and all points of $\mathcal{V}$, see Fig. 6(b).

Lemma 9. *No target segment T in T can be an edge of $\mathcal{L}$.*

Proof. Omitted in this extended abstract.   □

From the above lemma, there may exist a segment $T \in T$ such that one endpoint a of T is on $\mathcal{L}$ and the other endpoint lies in the interior of $\mathcal{L}$. If such an endpoint a exists, we delete a from $\mathcal{V}$ and recompute the new convex hull

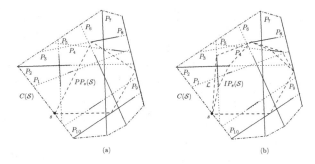

Fig. 6. Illustrating $PP_s(\mathcal{S})$, $\mathcal{L}$ and $IP_s(\mathcal{S})$.

of $PP_s(\mathcal{S})$ and $\mathcal{V}$. The operation of deleting a from $\mathcal{V}$ is repeatedly performed, until no endpoint a exists. Denote by $IP_s(\mathcal{S})$ the resulting convex hull. Clearly, $IP_s(\mathcal{S})$ is an intersecting polygon of $\mathcal{S}$, see Fig. 6(b). Since only one endpoint of a target segment in $\mathcal{T}$ needs be deleted from $\mathcal{V}$, polygon $IP_s(\mathcal{S})$ can be computed, say, simply in $O(n^2 \log n)$ time. Then, we have the following results.

Lemma 10. *Suppose that segment T is wholly outside of $PP_s(\mathcal{S})$ and an endpoint p of T is on $IP_s(\mathcal{S})$. Then, $P_s(\mathcal{S})$ makes a crossing contact with T or wholly contains T.*

Proof. From the construction of $IP_s(\mathcal{S})$, the other endpoint of T is not contained in $IP_s(\mathcal{S})$. Let Q be the polygonal region bounded by segment sp and the portion of $IP_s(\mathcal{S})$ from s to p. Denote by $\mathcal{S}_1$ the set of target segments, which are intersected by region Q, including T itself. Let $\mathcal{S}_2$ be the set of rest segments. (So, $\mathcal{S}_1 \cap \mathcal{S}_2 = \phi$ and $\mathcal{S}_1 \cup \mathcal{S}_2 = \mathcal{S}$.) By an argument analogous to that given in the proof of **Lemma 6**, we can then show that $P_s(\mathcal{S})$ makes a crossing contact with T or wholly contains T; the containment of T in $P_s(\mathcal{S})$ may occur in the case that the interior angle of the polygon bounded by the route $\mathcal{U}$, as described in the proof of **Lemma 6**, at vertex p is strictly larger than π.   □

Lemma 11. *The polygon $P_s(\mathcal{S})$ can be computed in $O(n^4)$ time.*

Proof. If no segment T described in **Lemma 10** exists, then $IP_s(\mathcal{S})$ is just $P_s(\mathcal{S})$. Otherwise, the lemma follows from **Lemmas 8** and **10**; the case of the containment of segments in $P_s(\mathcal{S})$ does not affect our algorithm at all.   □

As in Sect. 3.3, a general solution to the problem MPIP can be obtained by removing the restriction of a given starting point. Denote by $\mathcal{K}$ the convex hull of the segments' endpoints, which are *not* on $C(\mathcal{S})$. If $\mathcal{K}$ contains at least one point of every target segment of S and the contacts with the target segments, excluding those which are wholly contained in $\mathcal{K}$, are all crossing, then it is just the minimum-perimeter intersecting polygon of S and we are done.

Again, two other situations are that $\mathcal{K}$ makes a reflection contact with at least one target segment, or some segments are wholly outside of $\mathcal{K}$ and thus their two endpoints are on $C(\mathcal{S})$. In either case, there is a target segment A such that the pocket of A does not wholly contain any other segment. As in Sect. 3.3, we then consider the intersecting polygons of S such that they are forced to reflect on A at an artificial vertex s. So, $P_s(\mathcal{S})$ is computed in the difference region between $C(\mathcal{S})$ and the pocket of A. Polygon $P_s(\mathcal{S})$ is said to be *adjustable* if the starting point s can be moved on A so as to get a smaller-perimeter intersecting polygon of $\mathcal{S}$. Also, four types of adjustments on segment A can be defined.

Theorem 4 *There is a unique non-adjustable minimum-perimeter intersecting polygon of $\mathcal{S}$, except for very special cases where there is an infinite number of polygons $MPIP(\mathcal{S})$ of equal length.*

Proof. By an argument similar to that for **Theorem 2**, the theorem follows.□

Theorem 5. *The minimum-perimeter intersecting polygon of a given set of n target segments can be computed in $O(n^5)$ time.*

Proof. By an argument similar to that for **Theorem 3**, the theorem follows. □

References

1. Chin, W.P., Ntafos, S.: Optimum watchman routes. Inform. Process. Lett. **28**, 39–44 (1988)
2. Czyzowicz J., et al.: The aquarium keeper's problem, In Proc. SODA'91, pp. 459–464, 1991
3. Dror M., Efrat A., Lubiw A., Mitchell J. S. B.: Touring a sequence of polygons, In Proc. STOC'03, pp. 473–482, 2003
4. Dumitrescu A.: The traveling salesman problem for lines and rays in the plane, Discrete Mathematics, Algorithms and Applications, Vol. 4 (No. 4) 1250044 (12 pages), 2012
5. Dumitrescu, A., Jiang, M.: Minimum-perimeter intersecting polygons. Algorithmica **63**(3), 602–615 (2012)
6. Mitchell J. S. B., Geometric shortest paths and network optimization, In Sack J.-R., Urrutia J., (eds), Handbook of Computational Geometry, pp. 633–701, Elsevier Science, 2000
7. Jonsson, H.: The traveling salesman problem for lines in the plane. Inform. Process. Lett. **82**, 137–142 (2002)
8. Rappaport, D.: Minimum polygon traversals of line segments. Int. J. Comput. Geom. Appl. **5**(3), 243–265 (1995)
9. Tan, X.: Fast computation of shortest watchman routes in simple polygons. Inform. Process. Lett. **77**, 27–33 (2001)
10. Tan X., Jiang B.: Efficient algorithms for touring a sequence of convex polygons and related problems, in Proc. TAMC'2017, Lect. Notes Compt. Sci. 10185, 614–627, 2017. doi: 10.1007/978-3-319-55911-7_44

Polyhedral Circuits and Their Applications

Bin Fu[1], Pengfei Gu[1](✉), and Yuming Zhao[2]

[1] Department of Computer Science, University of Texas Rio Grande Valley,
Edinburg, TX 78539, USA
{bin.fu,pengfei.gu01}@utrgv.edu

[2] School of Computer Science, Zhaoqing University, Zhaoqing 526061,
Guangdong, People's Republic of China
ymzhao@zqu.edu.cn

Abstract. To better compute the volume and count the lattice points in geometric objects, we propose polyhedral circuits. Each polyhedral circuit characterizes a geometric region in $\mathbb{R}^d$. They can be applied to represent a rich class of geometric objects, which include all polyhedra and the union of a finite number of polyhedron. They can be also used to approximate a large class of d-dimensional manifolds in $\mathbb{R}^d$. Barvinok [3] developed polynomial time algorithms to compute the volume of a rational polyhedron, and to count the number of lattice points in a rational polyhedron in $\mathbb{R}^d$ with a fixed dimensional number d. Let d be a fixed dimensional number, $T_V(d, n)$ be polynomial time in n to compute the volume of a rational polyhedron, $T_L(d, n)$ be polynomial time in n to count the number of lattice points in a rational polyhedron, where n is the total number of linear inequalities from input polyhedra, and $T_I(d, n)$ be polynomial time in n to solve integer linear programming problem with n be the total number of input linear inequalities. We develop algorithms to count the number of lattice points in geometric region determined by a polyhedral circuit in $O\left(nd \cdot r_d(n) \cdot T_V(d, n)\right)$ time and to compute the volume of geometric region determined by a polyhedral circuit in $O\left(n \cdot r_d(n) \cdot T_I(d, n) + r_d(n)T_L(d, n)\right)$ time, where $r_d(n)$ is the maximum number of atomic regions that n hyperplanes partition $\mathbb{R}^d$. The applications to continuous polyhedra maximum coverage problem, polyhedra maximum lattice coverage problem, polyhedra $(1 - \beta)$-lattice set cover problem, and $(1 - \beta)$-continuous polyhedra set cover problem are discussed. We also show the NP-hardness of the geometric version of maximum coverage problem and set cover problem when each set is represented as union of polyhedra.

Keywords: Lattice points · Volume · Polyhedral circuits · Union

1 Introduction

Polyhedra are important topics in mathematics, and have close connection to theoretical computer science. There are two natural topics in polyhedra, computing the volume and counting the number of lattice points.

© Springer Nature Switzerland AG 2020
Z. Zhang et al. (Eds.): AAIM 2020, LNCS 12290, pp. 13–24, 2020.
https://doi.org/10.1007/978-3-030-57602-8_2

The problem of counting the number of lattice points in a polyhedron has a wide variety of applications in many areas, for example, number theory, combinatorics, representation theory, discrete optimization, and cryptography. It is related to the following problem: given a polyhedron P, which is given by a list of its vertices or by a list of linear inequalities, the goal is to compute the number $|P \cap \mathbb{Z}^d|$ of lattice points in P. Researchers have paid much attention to this problem. Ehrhart [16] introduced Ehrhart polynomials that were a higher-dimensional generalization of Pick's theorem in the Euclidean plane. Dyer [13] found polynomial time algorithms to count the number of lattice points in polyhedra when the dimensional number d is 3 or 4. Barvinok [4] designed a polynomial time algorithm for counting the number of lattice points in rational polyhedra when the dimension is fixed. The main ideas of the algorithm were using exponential sums [7,8] and decomposition of rational cones to primitive cones [29] of the polyhedra. Dyer and Kannan [14] simplified Barvinok's polynomial time algorithm and showed that only very elementary properties of exponential sums were needed to develop a polynomial time algorithm. De Loera et al. [12] described the first implementation of Barvinok's algorithm called LattE to count the number of lattice points in a rational polyhedron. Some other algebraic-analytic algorithms have been proposed by many authors (for example, see [2,5,6,22,26,27].)

Computing exactly the volume of a polytope is a basic problem that has drawn lots of researchers' attentions [1,3,11,20,23,30]. It is known that this problem is #P-complete if the polytope is given by its vertices or by its facets [15, 19]. Cohen and Hickey [11] and Von Hohenbalken [30] proposed to compute the volume of a polytope by triangulating and summing the polytope. Allgower and Schmidt [1] triangulated the boundary of the polytope to compute the volume of polytope. Lasserre [20] developed a recursive method to compute the volume of polytope. Lawrence [23] computed the volume of polytope based on Gram's relation for polytope. Barvinok [3] developed a polynomial time algorithm to compute the volume of polytope by using the exponential integral.

Our work is close to arrangements of hyperplanes. Arrangements of hyperplanes are basic problems and have a long history [10,21,24,31,32,34,36,37]. There are two natural topics in arrangements of hyperplanes, computing the number of cells of the arrangements of hyperplanes and constructing arrangements of hyperplanes. In 1826, Steiner [24] perhaps was the first researcher working on obtaining bounds on the number of arrangements of lines and circles on the planes and spheres in $\mathbb{R}^3$. Roberts [10], Alexanderson and Wetzel [37], and Wetzel [21] extended Steiner's results in other ways. Heintz et al. [17] proved that the number of d-dimensional cells in an arrangement of n hyperplanes is $(nb)^{O(d)}$, where b is some constant. Pollack and Roy [35] generalized Warren's result [33] on the number of arrangements of n hyperplanes from $O((nb/d)^d)$ to $(O(nb)/d)^d$. Edelsbrunner et al. [38] proposed an incremental algorithm to construct arrangements of n hyperplanes in $\mathbb{R}^d$ in $O(n^d)$ time with $O(n^d)$ space complexity. Clarkson and Shor [39] presented an $O(n \log n + k)$ time algorithm for constructing the arrangement of n line segments in the plane, where k is the number of vertices. Later, Chazellen and Edelsbrunner [41] developed a deterministic algorithm to construct the arrangement of n line segments in $O(n \log n + k)$ time using $O(n+k)$ storage. Balaban [40] improved the space complexity from $O(n+k)$ to $O(n)$ without affecting the asymptotic running time.

Motivation. The existing algorithms related to counting the number of lattice points in a polyhedron and to computing the volume of a polyhedron can only deal with one polyhedron, and these algorithms can not be applied to count the number of lattice points and to compute the volume of a complex geometric object. In order to have broader applications, it is essential to develop algorithms to deal with geometric objects that can be generated by a list of polyhedra via unions, intersections, and complementations. In this paper, we propose polyhedral circuits. Polyhedral circuits can be used to represent a large class of geometric objects. We propose algorithms to compute the volume of the geometric objects generated by polyhedral circuits, and to count the number of lattice points in the geometric objects generated by polyhedral circuits.

Contributions. We have the following contributions to polyhedra. 1. We introduce polyhedral circuits. Each polyhedral circuit characterizes a geometric region in $\mathbb{R}^d$. They can be applied to represent a rich class of geometric objects, which include all polyhedra and the union of a finite number of polyhedra. They can be also used to approximate a large class of d-dimensional manifolds in $\mathbb{R}^d$. 2. We develop an algorithm to compute the volume of geometric region determined by a polyhedral circuit. We also develop an algorithm to count the number of lattice points in geometric region determined by a polyhedral circuit. Our method is based on Barvinok's algorithm, which is only suitable for a rational polyhedron. 3. We apply the methods for polyhedral circuits to support new greedy algorithms for the maximum coverage problem and set cover problem, which involve more complex geometric objects. The existing research results about the geometric maximum coverage problem and set cover problem only handle simple objects such as balls and rectangular shapes. All of our algorithms run in polynomial time in $\mathbb{R}^d$ with a fixed d.

Organization. The rest of paper is organized as follows. In Sect. 2, we introduce some basic definitions and some important theorems. In Sect. 3, we propose polyhedral circuits and develop polynomial time algorithms to compute the volume of geometric region determined by a polyhedral circuit and to count the number of lattice points in geometric region determined by a polyhedral circuit. In Sect. 4, we present the applications of the algorithms, including polyhedra maximum coverage problem, polyhedra maximum lattice coverage problem, polyhedra $(1 - \beta)$-lattice set cover problem, and $(1 - \beta)$-continuous polyhedra set cover problem. Section 5 shows the NP-hardness of the geometric version of maximum coverage problem and set cover problem when each set is represented as union of polyhedra. Section 6 gives the conclusions.

2   Preliminaries

In this section, we introduce some definitions and some important theorems, which play important role in our method. We assume that $\mathbb{R}^d$ is the Euclidean d-

space and the polyhedra that we deal with are rational polyhedra throughout the paper. In this paper, computing the volume of a polyhedron means computing the volume of a polytope since the volume of a polyhedron is infinity.

Definition 1 ([28]). *A rational polyhedron $P \in \mathbb{R}^d$ is a set defined by finitely many linear inequalities:*

$$P = \left\{ x : \sum_{j=1}^{d} a_{ij}\xi_j \leq b_i \text{ for } x = (\xi_1, \cdots, \xi_d) \text{ and } i \in I \right\}$$

for some finite (possibly empty) set I, where a_{ij} and b_i are integers.

Theorem 1 ([3]). *Let us fix $d \in \mathbb{N}$. Then there exists a polynomial time algorithm for computing the volume of one rational polytope in $\mathbb{R}^d$.*

Theorem 2 ([4]). *Let us fix $d \in \mathbb{N}$. Then there exists a polynomial time algorithm for counting the number of lattice points in one rational polyhedron in $\mathbb{R}^d$.*

Definition 2 ([28]). *Integer Linear Programming problem is a constrained optimization problem of the form: $\max \{cx \mid Ax \leq b;\ x\ integral\}$, where A is a given $n \times d$ matrix, b is a n-dimensional vector and c is a d-dimensional vector, and the entries of A, b and c are rational numbers.*

Theorem 3 ([9,25]). *Let us fix $d \in \mathbb{N}$. Then there exists a polynomial time Las Vegas algorithm to solve the integer linear programming problem when the dimension d is small.*

Definition 3. *Let $d \in \mathbb{N}$ be a fixed dimensional number. Define $T_V(d, n)$ as the polynomial time in n to compute the volume of a rational polytope by Theorem 1, $T_L(d, n)$ as the polynomial time in n to count the number of lattice points in a rational polyhedron by Theorem 2, where n is the total number of linear inequalities from input polyhedra, and $T_I(d, n)$ as the polynomial time in n to solve integer linear programming problem by Theorem 3, where n is the total number of input linear inequalities.*

3   Algorithms About Polyhedral Circuits

In this section, we propose the definition of polyhedral circuits and develop algorithms to compute the volume of a geometric region determined by a polyhedral circuit and to count the number of lattice points in a geometric region determined by a polyhedral circuit. We assume that the linear inequalities that we deal with are in $\mathbb{R}^d$ with a fixed d and the coefficients of each linear inequalities are integers throughout the paper.

Definition 4. *A hyperplane in $\mathbb{R}^d$ can be defined by a linear equation of the form $a_1x_1 + a_2x_2 + \cdots + a_dx_d = b$, where $a_1, \cdots, a_d$ and b are constants.*

Definition 5. *For a set of linear inequalities in $\mathbb{R}^d$, a region is atomic if it is formed by a subset of linear inequalities via intersections, and does not contain any proper subregion that can be formed by another subset of linear inequalities via intersections (see Fig. 1).*

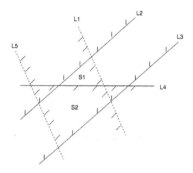

Fig. 1. Example of atomic regions. S_1 and S_2 are atomic regions, but $S_1 \cup S_2$ is not atomic region, although it is formed by a subset linear inequalities via intersections.

Definition 6. *Define $r_d(n)$ to be the maximum number of atomic regions that n hyperplanes partition $\mathbb{R}^d$.*

The range of $r_d(n)$ has been well studied in existing research papers [31,32]. Lemma 1, which has a simple proof, gives a lower bound and upper bound for $r_d(n)$ that are sufficient for its applications to the algorithms of this paper.

Lemma 1. *We have $r_d(n)$ satisfying $\left(\lfloor \frac{n}{d} \rfloor\right)^d \leq r_d(n) \leq \frac{n^d}{d!} + (n+1)^{d-1}$.*

Definition 7. *A polyhedral circuit is a circuit that consists of multiple layers of gates:*

1. *Each input gate are linear inequalities of format $\sum_{i=1}^{d} a_i x_i \leq b$ or $\sum_{i=1}^{d} a_i x_i < b$ (represent a half space in $\mathbb{R}^d$), where $a_1, \cdots, a_d$ and b are integers.*
2. *Each internal gate is either union or intersection operation.*
3. *The only output gate is on the top of the circuit.*

Example 1. $((-x < -1) \wedge (x \leq 3)) \vee ((-x \leq -2) \wedge (x \leq 5))$ is a polyhedral circuit in $\mathbb{R}^1$ (see Fig. 2). We have three layers of gates in the polyhedral circuit, the input gates are linear inequalities $-x < -1$, $x \leq 3$, $-x \leq -2$ and $x \leq 5$, the internal gates are intersection, and the output gate is union. Then polyhedral circuit $((-x < -1) \wedge (x \leq 3)) \vee ((-x \leq -2) \wedge (x \leq 5))$ is used to express the union of $(1, 3]$ and $[2, 5]$. Its output is the region $(1, 5]$ in $\mathbb{R}^1$.

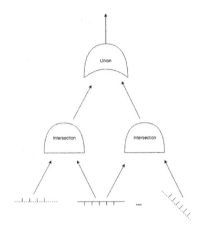

Fig. 2. Example of a polyhedral circuit.

Remark. We do not consider the negation operation in our circuit gates, since a negation operation in a gate can be handled by taking the negation of the input linear inequalities via De Morgan's laws. The polyhedral circuit can construct a complex geometric region in $\mathbb{R}^d$, since the union of multiple polyhedra can produce very complicate shape in a d-dimensional space for $d \geq 2$.

Lemma 2 shows that the output of a polyhedral circuit is a disjoint union of atomic regions.

Lemma 2. *The region generated by a polyhedral circuit is a disjoint union of atomic regions via the input linear inequalities.*

Lemma 3 shows that the volume of rational polytope formed by $Ax < b$ equals the volume of rational polytope formed by $Ax \leq b$.

Lemma 3. *The volume of atomic region formed by $Ax < b$ equals the volume of atomic region formed by $Ax \leq b$, where A is a given $n \times d$ matrix, and b is an n-dimensional vector.*

Lemma 4 shows that the number of lattice points in rational polyhedron formed by $Ax < b$ equals the number of lattice points in rational polyhedron formed by $Ax \leq b-I$, where all the elements of $n \times d$ matrix A and n-dimensional vector b are integers, and I is n-dimensional vector with all 1s.

Lemma 4. *The number of lattice points in atomic region formed by $Ax < b$ equals the number of lattice points in atomic region formed by $Ax \leq c$, where A is a given $n \times d$ matrix with integer elements, b is an n-dimensional vector with integer elements and c is an n-dimensional vector with $c = b - I$ for n-dimensional vector I whose elements are all 1s.*

The constructing arrangements of hyperplanes has been well studied in existing research works [38–41]. Lemma 5 and Lemma 6 not only present constructing arrangements of hyperplanes but also show generating interior points and integer points (if exists) from each of these atomic regions, which are essential for computing the volume of a geometric region determined by a polyhedral circuit and counting the number of lattice points in a geometric region determined by a polyhedral circuit in this paper.

Lemma 5. *There is a $O\left(nd \cdot r_d(n) \cdot T_V(d, n)\right)$ time algorithm such that given a list of n linear inequalities in $\mathbb{R}^d$, it produces all of the atomic regions with one interior point from each of them.*

Lemma 6. *There is a $O\left(n \cdot r_d(n) \cdot T_I(d, n)\right)$ time algorithm such that given a list of n linear inequalities in $\mathbb{R}^d$, it produces all of the atomic regions, and one integer point from each of them if exists.*

Theorem 4 provides an algorithm, which applies Lemma 5, to compute the volume of a geometric region by a polyhedral circuit.

Theorem 4. *There is a $O\left(nd \cdot r_d(n) \cdot T_V(d, n)\right)$ time algorithm such that given a polyhedral circuit in $\mathbb{R}^d$, the algorithm computes the volume of the region determined by the input polyhedral circuit.*

For a special case, some atomic regions have no integer point p_i, the algorithm still works by ignoring those regions that do not have integer point.

Theorem 5 presents an algorithm, which applies Lemma 6, to count the number of lattice points in the geometric region determined by a polyhedral circuit.

Theorem 5. *There is a $O\left(n \cdot r_d(n) \cdot T_I(d, n) + r_d(n)T_L(d, n)\right)$ time algorithm such that given a polyhedral circuit in $\mathbb{R}^d$, the algorithm counts the number of lattice points in the region determined by the input polyhedral circuit.*

An immediate application of the algorithm is to compute the volume of the union of a list of polyhedra, and to count the number of lattice points in the union of a list of polyhedra.

Corollary 1. *There is a $O\left(nd \cdot r_d(n) \cdot T_V(d, n)\right)$ time algorithm to compute the volume of the union of polyhedra when given a list of polyhedra.*

Corollary 2. *There is a $O\left(n \cdot r_d(n) \cdot T_I(d, n) + r_d(n)T_L(d, n)\right)$ time algorithm to count the number of lattice points in the union of polyhedra when given a list of polyhedra.*

Definition 8. *Let $d \in \mathbb{N}$ be a fixed dimensional number. Define $T_{U, L}(d, n)$ as the running time to count the number of lattice points in the union of rational polyhedra by Corollary 2 and $T_{U, V}(d, n)$ as the running time to compute the volume of the union of rational polytopes by Corollary 1 when given a list of polyhedra with n is the total number of linear inequalities from input polyhedra.*

4 Applications

In this section, we present a sample of applications of the algorithms.

4.1 Application in Continuous Polyhedra Maximum Coverage Problem

In this section, we show how to apply the method proposed at Sect. 3 to continuous polyhedra maximum coverage problem. Before presenting the algorithm, we give some definitions about the continuous polyhedra maximum coverage problem.

Definition 9. *Continuous Maximum Coverage Problem: Given an integer k, a set of regions $S_1, \cdots, S_m$, select k regions $S_{i_1}, \cdots, S_{i_k}$ such that $S_{i_1} \cup \cdots \cup S_{i_k}$ has the maximum volume of $S_1 \cup \cdots \cup S_m$.*

Definition 10. *Continuous Polyhedra Maximum Coverage Problem: It is a continuous maximum coverage problem when $S_1, \cdots, S_m$ are all polyhedra.*

Theorem 6. *There is a $O\left(km \cdot T_{U,V}(d, n)\right)$ time approximation algorithm for the continuous polyhedra maximum coverage problem with approximation ratio $\left(1 - \frac{1}{e}\right)$.*

4.2 Application in Polyhedra Maximum Lattice Coverage Problem

In this section, we show how to apply the method presented at Sect. 3 to polyhedra maximum lattice coverage problem. Before presenting the algorithm, we give some definitions about polyhedra maximum lattice coverage problem.

Definition 11. *Maximum Lattice Coverage Problem: Given an integer k, a set of regions $S_1, \cdots, S_m$, select k regions $S_{i_1}, \cdots, S_{i_k}$ such that $S_{i_1} \cup \cdots \cup S_{i_k}$ has the maximum number of lattice points of $S_1 \cup \cdots \cup S_m$.*

Definition 12. *Polyhedra Maximum Lattice Coverage Problem: It is a maximum lattice coverage problem when $S_1, \cdots, S_m$ are all polyhedra.*

Theorem 7. *There is a $O\left(km \cdot T_{U,L}(d, n)\right)$ time approximation algorithm for the polyhedra maximum lattice coverage problem with approximation ratio $\left(1 - \frac{1}{e}\right)$.*

4.3 Application in Polyhedra $(1 - \beta)$-Lattice Set Cover Problem

In this section, we show how to apply the method developed at Sect. 3 to polyhedra $(1 - \beta)$-lattice set cover problem. Before presenting the algorithm, we give some definitions about polyhedra $(1 - \beta)$-lattice set cover problem.

Definition 13. *$(1 - \beta)$-Lattice Set Cover Problem: For a real $\beta \in [0, 1)$, and a set of regions $S_1, \cdots, S_m$, select k regions $S_{i_1}, \cdots, S_{i_k}$ such that $L(S_{i_1} \cup \cdots \cup S_{i_k}) \geq (1 - \beta)L(S_1 \cup \cdots \cup S_m)$ with $L(P)$ denotes the number of lattice points in region P.*

Definition 14. *Polyhedra $(1 - \beta)$-Lattice Set Cover Problem: It is a $(1 - \beta)$-lattice set cover problem when $S_1, \cdots, S_m$ are all polyhedra.*

Theorem 8. *For reals $\alpha \in (0, 1)$ and $\beta \in [0, 1)$, there is a $O\left(m^2 \cdot T_{U, L}(d, n)\right)$ time approximation algorithm for the polyhedra $(1 - \beta)$-lattice set cover problem with the output k regions $S_{i_1}, \cdots, S_{i_k}$ satisfying $L\left(S_{i_1} \cup \cdots \cup S_{i_k}\right) \geq (1 - \alpha)$ $(1 - \beta) L\left(S_1 \cup \cdots \cup S_m\right)$ and $k \leq \left(1 + \ln \frac{1}{\alpha}\right) H$, where H is the number of sets in an optimal solution for the polyhedra $(1 - \beta)$-lattice set cover problem.*

4.4 Application in $(1 - \beta)$-Continuous Polyhedra Set Cover Problem

In this section, we show how to apply the method proposed at Sect. 3 to $(1 - \beta)$-continuous polyhedra set cover problem. Before presenting the algorithm, we give some definitions about $(1 - \beta)$-continuous polyhedra set cover problem.

Definition 15. *$(1 - \beta)$-Continuous Set Cover Problem: For a real $\beta \in [0, 1)$, and a set of regions $S_1, \cdots, S_m$, select k regions $S_{i_1}, \cdots, S_{i_k}$ such that $vol(S_{i_1} \cup \cdots \cup S_{i_k}) \geq (1 - \beta)vol(S_1 \cup \cdots \cup S_m)$ with $vol(P)$ denotes the volume of region P.*

Definition 16. *$(1 - \beta)$-Continuous Polyhedra Set Cover Problem: It is a $(1 - \beta)$-continuous set cover problem when $S_1, \cdots, S_m$ are all polyhedra.*

Theorem 9. *For reals $\alpha \in (0, 1)$ and $\beta \in [0, 1)$, there is a $O\left(m^2 \cdot T_{U, V}(d, n)\right)$ time approximation algorithm for the $(1 - \beta)$-continuous polyhedra set cover problem with the output k regions $S_{i_1}, \cdots, S_{i_k}$ satisfying $vol\left(S_{i_1} \cup \cdots \cup S_{i_k}\right) \geq$ $(1 - \alpha)(1 - \beta) vol\left(S_1 \cup \cdots \cup S_m\right)$ and $k \leq \left(1 + \ln \frac{1}{\alpha}\right) H$, where H is the number of sets in an optimal solution for the $(1 - \beta)$-continuous set cover problem.*

5 NP-Hardness and Inapproximation

In this section, we show the NP-hardness of the geometric version of maximum coverage problem and set cover problem when each set is represented as union of polyhedra.

Definition 17. *The maximum region coverage problem is that: given a list $A_1, \cdots, A_m$ of regions in $\mathbb{R}^d$, and an integer $k \geq 1$, to find k regions with the largest volume for their union.*

Definition 18. *Let $A_1, \cdots, A_k$ be a list of axis parallel rectangles. Let A be the region to be the union $A_1 \cup \cdots \cup A_k$. Then A is called an axis parallel rectangle union region.*

Definition 19. *The maximum region coverage problem with each set to be axis parallel rectangle union region is called axis parallel rectangle union maximum region coverage problem.*

Definition 20. *The lattice set cover problem with each set to be all lattice points in axis parallel rectangle union region is called lattice axis parallel rectangle union region maximum coverage problem.*

Theorem 10. *The following statements are true:*

1. *The axis parallel rectangle union maximum region coverage problem is NP-hard. Furthermore, if there is a polynomial time c-approximation for axis parallel rectangle union maximum region coverage problem, then there is a polynomial time c-approximation for classical maximum coverage problem.*
2. *The axis parallel rectangle union set problem is NP-hard. Furthermore, if there is a polynomial time c-approximation for lattice axis parallel rectangle union region set cover problem, then there is a polynomial time c-approximation for classical set cover problem.*

Using reasonable hypothesis of complexity theory, Feige [18] showed that $1 - \frac{1}{e}$ is the best possible polynomial time approximation ratio for the maximum coverage problem, and $\ln n$ is the best possible polynomial time approximation ratio for the set cover problem, where n is the number of elements in the universe set for the classical set cover. Our Theorem 10 shows the connection of inapproximation between the classical problems and their corresponding geometric versions.

6   Conclusions

In this paper, we introduce a concept of polyhedral circuits to study two problems, computing the volume of a rational polytope and counting the number of lattice points in a rational polyhedron in $\mathbb{R}^d$ with a fixed dimensional number d. An $O\left(n \cdot r_d(n) \cdot T_I(d, n) + r_d(n)T_L(d, n)\right)$ time algorithm is developed to compute the volume of the geometric region determined by a polyhedral circuit, and an $O\left(nd \cdot r_d(n) \cdot T_V(d, n)\right)$ time algorithm is presented to count the number of lattice points in the geometric region determined by a polyhedral circuit. As applications of the algorithms, we develop algorithms to continuous polyhedra maximum coverage problem, polyhedra maximum lattice coverage problem, polyhedra $(1 - \beta)$-lattice set cover problem and $(1 - \beta)$-continuous polyhedra set cover problem. We show the NP-hardness and inapproximation of the geometric version of continuous maximum coverage problem and set cover problem when each set is represented as union of polyhedra. Due to the limitation of the pages, we refer the readers to https://arxiv.org/abs/1806.05797 for the full paper.

Acknowledgement. This research is supported in part by National Science Foundation Early Career Award 0845376, Bensten Fellowship of the University of Texas Rio Grande Valley, and National Natural Science Foundation of China 61772179.

References

1. Allgower, E.L., Schmidt, P.H.: Computing volumes of polyhedra. Math. Comput. **46**(173), 171–174 (1986)

2. Baldoni-Silva, W., Vergne, M.: Residues formulae for volumes and ehrhart polynomials of convex polytopes. arXiv preprint math/0103097 (2001)
3. Barvinok, A.I.: Computing the volume, counting integral points, and exponential sums. Discrete Comput. Geom. **10**(2), 123–141 (1993). https://doi.org/10.1007/BF02573970
4. Barvinok, A.I.: A polynomial time algorithm for counting integral points in polyhedra when the dimension is fixed. Math. Oper. Res. **19**(4), 769–779 (1994)
5. Beck, M.: Counting lattice points by means of the residue theorem. Ramanujan J. **4**(3), 299–310 (2000)
6. Beck, M., Pixton, D.: The ehrhart polynomial of the birkhoff polytope. Discrete Comput. Geom. **30**(4), 623–637 (2003)
7. Brion, M.: Points entiers dans les polyedres convexes. Ann. Sci. l'Ecole Norm. Super. **21**, 653–663 (1988)
8. Brion, M.: Polyedres et réseaux. Enseign. Math. **38**, 71–88 (1992)
9. Clarkson, K.L.: Las vegas algorithms for linear and integer programming when the dimension is small. J. ACM (JACM) **42**(2), 488–499 (1995)
10. Roberts, S.: On the figures formed by the intercepts of a system of straight lines in a, plane, and on analogous relations in space of three dimensions. Proc. Lon. Math. Soc. **1**(1), 405–422 (1887)
11. Cohen, J., Hickey, T.: Two algorithms for determining volumes of convex polyhedra. J. ACM (JACM) **26**(3), 401–414 (1979)
12. De Loera, J.A., Hemmecke, R., Tauzer, J., Yoshida, R.: Effective lattice point counting in rational convex polytopes. J. Symb. Comput. **38**(4), 1273–1302 (2004)
13. Dyer, M.: On counting lattice points in polyhedra. SIAM J. Comput. **20**(4), 695–707 (1991)
14. Dyer, M., Kannan, R.: On barvinok's algorithm for counting lattice points in fixed dimension. Math. Oper. Res. **22**(3), 545–549 (1997)
15. Dyer, M.E., Frieze, A.M.: On the complexity of computing the volume of a polyhedron. SIAM J. Comput. **17**(5), 967–974 (1988)
16. Erhart, E.: Sur un probleme de géométrie diophantienne linéaire. I. IJ Reine Angew. Math **226**, 1–29 (1967)
17. Heintz, J., Roy, M.-F., Solern, P.: On the complexity of semialgebraic sets. In: Proceedings of the IFIP San Francisco, pp. 293–298 (1989)
18. Feige, U.: A threshold of ln n for approximating set cover. J. ACM (JACM) **45**(4), 634–652 (1998)
19. Leonid Genrikhovich Khachiyan: The problem of calculating the volume of a polyhedron is enumerably hard. Russ. Math. Surv. **44**(3), 199 (1989)
20. Lasserre, J.B.: An analytical expression and an algorithm for the volume of a convex polyhedron in R^n. J. Optim. Theor. Appl. **39**(3), 363–377 (1983)
21. Wetzel, J.E.: On the division of the plane by lines. Am. Math. Mon. **85**(8), 647–656 (1978)
22. Lasserre, J.B., Zeron, E.S.: Solving the knapsack problem via z-transform. Oper. Res. Lett. **30**(6), 394–400 (2002)
23. Lawrence, J.: Polytope volume computation. Math. Comput. **57**(195), 259–271 (1991)
24. Steiner, J.: Einige gesetze über die theilung der ebene und des raumes. Journal für die reine und angewandte Mathematik **1826**(1), 349–364 (1826)
25. Lenstra Jr., H.W.: Integer programming with a fixed number of variables. Math. Oper. Res. **8**(4), 538–548 (1983)
26. MacMahon, P.A.: Combinatorial Analysis. Chelsea, New York (1960). 1915

27. Pemantle, R., Wilson, M.C.: Asymptotics of multivariate sequences: I. Smooth points of the singular variety. J. Comb. Theor. Ser. A **97**(1), 129–161 (2002)
28. Schrijver, A.: Theory of Linear and Integer Programming. Wiley, New York (1998)
29. Stanley, R.P.: Cambridge studies in advanced mathematics (1997)
30. Von Hohenbalken, B.: Finding simplicial subdivisions of polytopes. Math. Program. **21**(1), 233–234 (1981)
31. Agarwal, P.K., Sharir, M.: Arrangements and their applications. In: Sack, J.-R., Urrutia, J. (eds.) Handbook of Computational Geometry, pp. 49–119. North-Holland, Amsterdam (2000)
32. Robert Creighton Buck: Partition of space. Am. Math. Monthly **50**(9), 541–544 (1943)
33. Warren, H.E.: Lower bound for approximation by nonlinear manifolds. Trans. Amer. Math. Soc. **133**, 167–178 (1968)
34. Grünbaum, B.: Arrangements of hyperplanes. In: Kaibel, V., Klee, V., Ziegler, G.M. (eds.) Convex Polytopes, vol. 221, pp. 432–454. Springer, New York (2003)
35. Pollack, R., Roy, M.F.: On the number of cells defined by a set of polynomials. C. R. Acad. Sci. Paris **316**, 573–577 (1993)
36. Zaslavsky, T.: Facing up to arrangements: Face-count formulas for partitions of space by hyperplanes. Mem. Am. Math. Soc. **154**, 1–95 (1977)
37. Alexanderson, G.L., Wetzel, J.E.: Arrangements of planes in space. Discrete Math. **34**(3), 219–240 (1981)
38. Edelsbrunner, H., O'Rourke, J., Seidel, R.: Constructing arrangements of lines and hyperplanes with applications. SIAM J. Comput. **15**(2), 341–363 (1986)
39. Clarkson, K.L., Shor, P.W.: Applications of random sampling in computational geometry, II. Discrete Comput. Geom. **4**(5), 387–421 (1989). https://doi.org/10.1007/BF02187740
40. Balaban, I.J.: An optimal algorithm for finding segments intersections. In: Proceedings of the Eleventh Annual Symposium on Computational Geometry, pp. 211–219 (1995)
41. Chazelle, B., Edelsbrunner, H.: An optimal algorithm for intersecting line segments in the plane. J. ACM (JACM) **39**(1), 1–54 (1992)

Online Bicriteria Algorithms to Balance Coverage and Cost in Team Formation

Yijing Wang[1], Dachuan Xu[1], Donglei Du[2], and Ran Ma[3(✉)]

[1] Department of Operations Research and Information Engineering,
Beijing University of Technology, Beijing 100124, People's Republic of China
`yjwang@emails.bjut.edu.cn, xudc@bjut.edu.cn`
[2] Faculty of Management, University of New Brunswick,
Fredericton E3B 5A3, Canada
`ddu@unb.ca`
[3] School of Management Engineering, Qingdao University of Technology,
Qingdao 266525, People's Republic of China
`sungirlmr@163.com`

Abstract. In this paper, we investigate the team formation problem to balance the coverage gained and the cost incurred. This problem can be formulated as maximizing the difference of two set functions $f - l$, where f is non-negative monotone approximately submodular function, and l is non-negative linear function. We propose three online bicriteria algorithms. The first two handle the cases where the function f is γ-weakly submodular, and strictly γ-weakly submodular, respectively. The last algorithm integrates the first two with more parameters introduced.

Keywords: Approximately submodular · Linear function · Balance · Bicriteria algorithm

1 Introduction

Team formation plays an important role in the labor market. According to different application backgrounds and solving skills, team formation can be expressed in various ways [8,13]. In this work, we characterize team formation in two criteria, including the coverage gained and the cost incurred. The problem of balancing the coverage and cost in team formation attracts much attention recently [1,2,12]. Ene [5] firstly introduces the team formation problem in the following way. For a given job, there is a set of experts N available. The employer would like to hire some experts $S \subseteq N$ to complete the given job while maximizing his total profit, i.e., $\max_{S \subseteq N} f(S) - l(S)$, where function $f : 2^N \to \mathbb{R}_{\geq 0}$ is non-negative monotone submodular, and function $l : 2^N \to \mathbb{R}_{\geq 0}$ is non-negative linear. Functions f and l characterize the coverage ability to the given task and the cost incurred.

Evidently, the objective function $g(S) = f(S) - l(S)$ is submodular but can be negative and non-monotone. Extant literature on submodular maximization

© Springer Nature Switzerland AG 2020
Z. Zhang et al. (Eds.): AAIM 2020, LNCS 12290, pp. 25–36, 2020.
https://doi.org/10.1007/978-3-030-57602-8_3

has been focusing on non-negative and/or monotone submodular functions. Ene shows that it is NP-hard to determine whether the optimum value of an arbitrary submodular function $g(S)$ is positive or not. Indeed, it is inapproximable to maximize a potentially negative submodular function. It is also possible that there is no multiplicative factor approximation. We focus instead on a weaker notion of approximation, i.e., bifactor approximation: an algorithm has an (α, β)-bifactor if the solution S output by the algorithm satisfies that $f(S) - l(S) \geq \alpha \cdot f(OPT) - \beta l(OPT)$, where OPT is an optimal solution to the problem and $0 \leq \alpha \leq 1$.

We investigate the model discussed above with or without constraints. With matroid constraints, Sviridenko et al. [15] reduce the problem $\max_{S \in \mathcal{S}} f(S) - l(S)$ to the one of maximizing $f(S)$ subject to both a knapsack constraint $w(S) \leq B$ and the matroid constraint $\mathcal{S}$. They first substitute the knapsack budget B with the value of $l(OPT)$. They next run a variant of the continuous greedy algorithm on the distorted model. They finally obtain a solution S such that $f(S) - l(S) \geq (1 - \frac{1}{e}) \cdot f(OPT) - l(OPT)$. Unfortunately, their algorithm involves a step guessing the value of $l(OPT)$ approximately, which significantly affects algorithm's time complexity.

Feldman [6] shows that the guessing step can be avoided and presents a distorted objective function by involving a weight vector $w(t)$ for every time $t \in [0, 1)$: $e^{t-1}F(x) - \langle w, x \rangle$, where F is the multilinear extension of f and $\langle \cdot, \cdot \rangle$ denotes the dot product. He obtains the same approximation guarantee as that of Sviridenko et al. by running a modified continuous greedy algorithm.

Ene [5] provides a comprehensive study on this type of problems subject to matroid and cardinality constraints, under offline, online and streaming models, respectively. For the offline model with matroid constraints, Ene obtains a solution set S satisfying $f(S) - l(S) \geq \frac{1}{2}f(OPT) - l(OPT)$. For the online model without any constraint, the solution set retains the same weak approximation ratio. For the streaming model with cardinality constraints, Ene returns a solution such that $f(S) - l(S) \geq \frac{1}{2}(3 - \sqrt{5})f(OPT) - l(OPT)$.

Although submodularity has been a ubiquitous property arising in many fields, there are many non-submodular functions in real-life applications, such as boosting information spread and k-center clustering in social networks. Extending submodular function to non-submodular in a parametric manner is a common practice in the literature [3,7,9,11].

For the problem subject to a cardinality constraint, Harshaw et al. [10] provide a much more efficient algorithm compared to those by Feldman [6] and Sviridenko et al. [15], where function f is γ-weakly submodular (cf. Definition 1). Instead of using the continuous greedy algorithm, they combine the time-varying distortion technique with the standard greedy algorithm to obtain an $(1 - e^{-\gamma} - \epsilon)$ weak approximation ratio, where $\gamma \in (0, 1]$. Qian [14] considers the problem of maximizing the difference between a non-negative monotone approximately submodular function and a non-negative modular function subject to a cardinality constraint, and gives an algorithm with a polynomial time approximation guarantee of $1 - e^{-1}$.

With the concept of (α, β)-bifactor, existing results are summarized in Table 1.

Table 1. The bifactors (α, β) of previous related work.

Work	Model	α	β
Sviridenko et al. [15]	Offline	$1 - e^{-1}$	1
Feldman [6]	Offline	$1 - e^{-1}$	1
Ene [5]	Online	$1/2$	1
Harshaw et al. [10]	Offline	$\left(1 - e^{-\gamma} - \epsilon\right)$	1
Qian [14]	Offline	$1 - e^{-1}$	1

Our main contribution is to propose a flexible bicriteria [4] online algorithm to balance the coverage and the cost in the team formation problem, where we want to maximize the difference of a non-negative normalized monotone γ-weakly submodular function (cf. Definition 1) and a non-negative linear function. We propose three online bicriteria algorithms. The first two handle the cases where the function f is γ-weakly submodular, and strictly γ-weakly submodular, respectively. The last algorithm integrates the first two with more parameters introduced. When $\gamma = 1$, i.e., function f is submodular, our result coincides with that of Ene [5].

The rest of this work is organized as follows. Section 2 describes the preliminaries. Section 3 presents three algorithms along with their analyses for the online model without any constraint, including both special and general cases. Finally, we present the concluding remarks in Sect. 4.

2   Preliminaries

We introduce the necessary notations and definitions. We are given a ground set $N = \{e_1, e_2, \ldots, e_n\}$ and a family of subsets $\mathcal{S} \subseteq 2^N$. A set function $f : 2^N \to \mathbb{R}$ is monotone if $f(S) \leq f(T)$, for any subsets $S \subseteq T \subseteq N$; non-negative if $f(S) \geq 0$, for any subset $S \subseteq N$; and normalized if $f(\emptyset) = 0$, respectively.

A set function $f : 2^N \to \mathbb{R}$ is submodular if $f(S) + f(T) \geq f(S \cap T) + f(S \cup T)$, for any subsets $S, T \subseteq N$; or equivalently $f(e|T) \leq f(e|S)$, for any subsets $S \subseteq T \subset N$, $e \in N \backslash T$, where $f(e|S) := f(S \cup \{e\}) - f(S)$ represents the marginal gain of f when adding any element e to any subset $S \subseteq N$. Another equivalent definition about submodular function f is $\sum_{e \in T \backslash S} f(e|S) \geq f(S \cup T) - f(S)$, for any subset $S, T \subseteq N$. Harshaw et al. [10] modify this definition and generalize submodular functions to γ-weakly submodular functions:

Definition 1. (γ-weakly submodular function)
A monotone set function f is γ-weakly submodular for $\gamma \in (0, 1]$ if $\sum_{e \in T \backslash S} f(e|S) \geq \gamma(f(S \cup T) - f(S))$, for all subsets $S, T \subseteq N$.

Function f is submodular when it is γ-weakly submodular and $\gamma = 1$. Based on the definition of γ-weakly submodular function, we call function f strictly γ-weakly submodular if $\gamma \in (0, 1)$.

From Sect. 1, we know that the goal of balancing the coverage and cost in team formation is to find a subset $S = \arg\max_{S \in \mathcal{S}}(f(S) - l(S))$, where f is non-negative normalized monotone γ-weakly submodular function, and l is non-negative linear function, such that $l(S) = \sum_{e_i \in S} l(e_i)$ and $l(T \setminus S) = l(S \cup T) - l(S)$.

Definition 2. ((α, β)-bicriteria algorithm)
An algorithm is (α, β)-bicriteria if it outputs a solution S to the problem $\max_{S \in \mathcal{S}}[f(S) - l(S)]$ such that $f(S) - l(S) \geq \alpha \cdot f(OPT) - \beta \cdot l(OPT)$, where OPT is an optimal solution to the problem and $\alpha \leq 1$.

From the computational complexity perspective, we assume that there are pre-specified oracles to evaluate the function f and l, and the number of oracle calls represents the computational complexity of an algorithm.

3   Online Algorithms for Maximizing the Difference of γ-Weakly Submodular and Linear Function

In this section, we present three online bicriteria algorithms. The first two handle the cases where the function f is γ-weakly submodular, and strictly γ-weakly submodular, respectively. The last algorithm integrates the first two with more parameters introduced. When $\gamma = 1$, i.e., function f is submodular, our result coincides with that of Ene [5].

The main idea of obtaining a good bicriteria factor (α, β) is to construct a surrogate objective function $\hat{g}$.

We focus on online models in this work, where elements arrive one at a time. When an element arrives, we decide irrevocably whether or not to add it to the solution.

3.1   An Online Algorithm for Maximizing γ-Weakly Submodular Minus Linear Function (γS-L)

In this section, we consider the model $\max_{S \in 2^V}[f(S) - l(S)]$ in the online model, where the function l is non-negative linear and f is non-negative normalized monotone γ-weakly submodular, $\gamma \in (0, 1]$. When $\gamma = 1$, function f is submodular. We consider a scaled objective $\hat{g}(S) = (1 - \frac{\gamma}{t})f(S) - l(S)$, where $\gamma \in (0, 1]$ and $t \geq 1$. Note that the function $\hat{g}$ can be negative and non-monotone. Let S be the current solution set. We add a newly arrived element e to S whenever the inequality $(t - \gamma)/t\gamma \cdot f(e|S) - l(e) > 0$ is satisfied. The detailed description is shown in Algorithm 1.

Algorithm 1. An online algorithm for γS-L

1: **Input:**
   A non-negative normalized monotone γ-weakly submodular function f, and a non-negative linear function l, $\gamma \in (0, 1]$.
2: **Output:**
   A solution set S.
3: **Process:**
4: Initially set $S := \emptyset$
   For each arriving element e
5:   if $\frac{t-\gamma}{t\gamma} \cdot f(e|S) - l(e) > 0$
6:     $S := S \cup \{e\}$
7: Return S

Theorem 1. *Algorithm 1 is a $\left(\frac{t\gamma - t + \gamma}{t\gamma}, \frac{t\gamma - t + \gamma}{\gamma(t-\gamma)}\right)$-approximation algorithm, where $\gamma \in (0, 1]$ and $t \geq 1$.*

Proof. Assume $S^* = \{s_1, s_2, \ldots, s_p\}$ is an optimal solution set to the online model, where $p = |S^*|$, and $S = \{e_1, e_2, \ldots, e_q\}$ is the solution output by Algorithm 1, where $q = |S|$. For every item $s \in S^* \setminus S$, by the element selection rule in Algorithm 1 (Line 5), we have $\frac{1}{\gamma}\left(1 - \frac{\gamma}{t}\right) f(s|S) - l(s) \leq 0$.

Let $\bar{S} = S^* \setminus S$ and $\{s_1, s_2, \ldots, s_{|\bar{S}|}\}$ be an arbitrary ordering of $\bar{S}$. Denote $\bar{S}^{(i)} = \{s_1, s_2, \ldots, s_i\}$. We have

$$
\begin{aligned}
\hat{g}(S \cup \bar{S}) - \hat{g}(S) &= \left(1 - \frac{\gamma}{t}\right) f(S \cup \bar{S}) - l(S \cup \bar{S}) - \left((1 - \frac{\gamma}{t})f(S) - l(S)\right) \\
&= \left(1 - \frac{\gamma}{t}\right)\left(f(S \cup \bar{S}) - f(S)\right) - \left(l(S \cup \bar{S}) - l(S)\right) \\
&= \left(1 - \frac{\gamma}{t}\right)\left(f(S \cup \bar{S}) - f(S)\right) - l(\bar{S}) \\
&\leq \left(1 - \frac{\gamma}{t}\right) \cdot \frac{1}{\gamma} \sum_{s \in \bar{S}} f(s|S) - \sum_{s \in \bar{S}} l(s) \\
&= \sum_{s \in \bar{S}} \left(\left(1 - \frac{\gamma}{t}\right) \cdot \frac{1}{\gamma} \cdot f(s|S) - l(s)\right) \\
&= \sum_{s \in \bar{S}} \left(\frac{t-\gamma}{t\gamma} \cdot f(s|S) - l(s)\right) \leq 0.
\end{aligned}
\tag{1}
$$

The first inequality follows by the definition of the γ-weakly submodular of function f, and the second inequality holds by the element selection rule into the solution S in the algorithm.

By the definition of $\hat{g}$, the monotonicity of f and the property of l, we have

$$
\begin{aligned}
\hat{g}(S \cup \bar{S}) - \hat{g}(S) &= \left(1 - \frac{\gamma}{t}\right)\left(f(S \cup \bar{S}) - f(S)\right) - l(\bar{S}) \\
&\geq \left(1 - \frac{\gamma}{t}\right)\left(f(S^*) - f(S)\right) - l(S^*).
\end{aligned}
\tag{2}
$$

Combining (1) and (2), we have

$$\left(1 - \frac{\gamma}{t}\right)\left(f(S^*) - f(S)\right) - l(S^*) \le 0,$$

i.e.,

$$f(S) \ge f(S^*) - \frac{t}{t - \gamma}l(S^*).$$

Denote $S^{(i)} = \{e_1, e_2, \ldots, e_i\}$. From Line 5 of Algorithm 1, for each element $e_i \in S$, we have $\frac{t-\gamma}{t\gamma}f(e_i|S^{(i-1)}) > l(e_i)$. Summing up all elements $e \in S$, we have

$$\sum_{i=1}^{|S|} \frac{t - \gamma}{t\gamma} \cdot f(e_i|S^{(i-1)})$$

$$= \frac{t - \gamma}{t\gamma} \sum_{i=1}^{|S|} f(e_i|S^{(i-1)})$$

$$= \frac{t - \gamma}{t\gamma}\left(f(S) - f(\emptyset)\right) = \frac{t - \gamma}{t\gamma}f(S)$$

$$> \sum_{i=1}^{|S|} l(e_i) = l(S).$$

With this upper bound on $l(S)$ in terms of $f(S)$, we have

$$f(S) - l(S) > f(S) - \frac{t - \gamma}{t\gamma}f(S) = \left(1 - \frac{t - \gamma}{t\gamma}\right)f(S)$$

$$\ge \frac{t\gamma - t + \gamma}{t\gamma} \cdot \left(f(S^*) - \frac{t}{t - \gamma}l(S^*)\right)$$

$$= \frac{t\gamma - t + \gamma}{t\gamma}f(S^*) - \frac{t\gamma - t + \gamma}{\gamma(t - \gamma)}l(S^*).$$

Thus Algorithm 1 is a $\left(\frac{t\gamma - t + \gamma}{t\gamma}, \frac{t\gamma - t + \gamma}{\gamma(t-\gamma)}\right)$-approximation algorithm. $\qquad\square$

When function f is submodular (i.e., $\gamma = 1$) and $t = 2$, the bicriteria factor is $(\frac{1}{2}, 1)$, and we obtain the result in Ene [5]. Table 2 shows further bifactors for varying parameters of t. From Table 2, note that, when $t = 1$ and $\gamma = 1$, the β's value is meaningless. At this point, the scaled function $\hat{g}$ becomes $-l(S)$, which is no longer the team formation problem. In view of this situation, we propose a new algorithm, as shown in the following part.

3.2 An Online Algorithm for Maximizing Strictly γ-Weakly Submodular Minus Linear Function (γSS-L)

In this section, we consider the same online model $\max_{S \in 2^V}[f(S) - l(S)]$, except that the function f is strictly γ-weakly submodular compared to that in Sect. 3.1,

Table 2. The bifactors (α, β) of Algorithm 1 with respect to varying t.

t	$\alpha = \frac{t\gamma - t + \gamma}{t\gamma}$	$\beta = \frac{t\gamma - t + \gamma}{(t-\gamma)\gamma}$
1	$\frac{2\gamma - 1}{\gamma}$	$\frac{2\gamma - 1}{\gamma(1-\gamma)}$
2	$\frac{3\gamma - 2}{2\gamma}$	$\frac{3\gamma - 2}{\gamma(2 - \gamma)}$
3	$\frac{4\gamma - 3}{3\gamma}$	$\frac{4\gamma - 3}{\gamma(3 - \gamma)}$
5	$\frac{6\gamma - 5}{5\gamma}$	$\frac{6\gamma - 5}{\gamma(5 - \gamma)}$
8	$\frac{9\gamma - 8}{8\gamma}$	$\frac{9\gamma - 8}{\gamma(8 - \gamma)}$
10	$\frac{11\gamma - 10}{10\gamma}$	$\frac{11\gamma - 10}{\gamma(10 - \gamma)}$
30	$\frac{31\gamma - 30}{30\gamma}$	$\frac{31\gamma - 30}{\gamma(30 - \gamma)}$
50	$\frac{51\gamma - 50}{50\gamma}$	$\frac{51\gamma - 50}{\gamma(50 - \gamma)}$
100	$\frac{101\gamma - 100}{100\gamma}$	$\frac{101\gamma - 100}{\gamma(100-\gamma)}$
500	$\frac{501\gamma - 500}{500\gamma}$	$\frac{501\gamma-500}{\gamma(500 - \gamma)}$
1000	$\frac{1001\gamma - 1000}{1000\gamma}$	$\frac{1001\gamma - 1000}{\gamma(1000 - \gamma)}$

i.e., $\gamma \in (0,1)$. Due to this restriction, we construct a different scaled objective $\hat{g}(S) = (1 - \gamma)f(S) - l(S)$. Similar to the selection rule in Algorithm 1, we only select the elements whose marginal gains are positive with respect to the scaled objective. The detail is as follows.

Algorithm 2. An online algorithm for γSS-L

1: **Input:**
   A non-negative normalized monotone strictly γ-weakly submodular function f, and a non-negative linear function l, $\gamma \in (0,1)$.
2: **Output:**
   A solution set S.
3: **Process:**
4: Initially set $S := \emptyset$
   For each arriving element e
5:    if $(1 - \gamma)f(e|S) - l(e) > 0$
6:        $S := S \cup \{e\}$
7: Return S

Theorem 2 *Algorithm 2 is a* $\left(\gamma, \frac{1}{1-\gamma}\right)$-*approximation algorithm, where* $\gamma \in (0,1)$.

Proof. Assume $S^* = \{s_1, s_2, \ldots, s_p\}$ is an optimal solution set, where $p = |S^*|$. For every item $s \in S^* \setminus S$, we have $\hat{g}(s|S) = (1 - \gamma)f(s|S) - l(s) \leq 0$. i.e., $(1 - \gamma)f(s|S) \leq l(s)$.

Let $\bar{S} = S^* \setminus S$ and $\{s_1, s_2, \ldots, s_{|S^* \setminus S|}\}$ be an arbitrary ordering of $\bar{S}$. Summing up all elements $s \in S^* \setminus S$, we have

$$\sum_{s \in \bar{S}} (1 - \gamma) f(s|S) \leq \sum_{s \in \bar{S}} l(s) = l(\bar{S}).$$

As function f is γ-weakly submodular, we have

$$\sum_{s \in \bar{S}} (1 - \gamma) f(s|S) \geq (1 - \gamma) \cdot \gamma \big(f(S \cup \bar{S}) - f(S) \big).$$

Combining the properties of f and l, we obtain the following inequalities

$$\gamma(1 - \gamma) \big(f(S \cup \bar{S}) - f(S) \big) \leq l(\bar{S}) \leq l(S^*),$$

and

$$f(S^*) - f(S) \leq f(S \cup \bar{S}) - f(S) \leq \frac{l(S^*)}{\gamma(1 - \gamma)}.$$

Rearranging the two inequalities, we have

$$f(S^*) - \frac{l(S^*)}{\gamma(1 - \gamma)} \leq f(S).$$

By the element selection rule in Line 5 of Algorithm 2, we have $\hat{g}(S) > 0$. In fact, let $S = \{e_1, e_2, \ldots, e_{|S|}\}$ where the elements are ordered in which they were added. Denote $S^{(i)} = \{e_1, e_2, \ldots, e_i\}$. We have

$$
\begin{aligned}
\hat{g}(S) - \hat{g}(\emptyset) &= \sum_{i=1}^{|S|} \hat{g}(e_i | S^{(i-1)}) \\
&= \sum_{i=1}^{|S|} \big((1 - \gamma) f(e_i | S^{(i-1)}) - l(e_i | S^{(i-1)}) \big) \\
&= \sum_{i=1}^{|S|} \big((1 - \gamma) f(e_i | S^{(i-1)}) - l(e_i) \big) \\
&= (1 - \gamma) \sum_{i=1}^{|S|} f(e_i | S^{(i-1)}) - \sum_{i=1}^{|S|} l(e_i) \\
&= (1 - \gamma)(f(S) - f(\emptyset)) - l(S) \\
&= (1 - \gamma) f(S) - l(S) > 0.
\end{aligned}
$$

With this upper bound $l(S) < (1 - \gamma) f(S)$ in terms of f, we have

$$
\begin{aligned}
f(S) - l(S) &> f(S) - (1 - \gamma) f(S) = \gamma f(S) \\
&\geq \gamma \cdot \left(f(S^*) - \frac{l(S^*)}{\gamma(1 - \gamma)} \right) \\
&= \gamma f(S^*) - \frac{1}{1 - \gamma} l(S^*).
\end{aligned}
$$

Table 3. The bifactors (α, β) of Algorithm 2 with respect to varying γ.

γ	$\alpha = \gamma$	$\beta = 1/(1-\gamma)$
0.01	0.01	1.0101
0.05	0.05	1.0526
0.10	0.10	1.1111
0.15	0.15	1.1765
0.20	0.20	1.2500
0.25	0.25	1.3333
0.30	0.30	1.4286
0.35	0.35	1.5385
0.40	0.40	1.6667
0.45	0.45	1.8182
0.50	0.50	2.0000
0.55	0.55	2.2222
0.60	0.60	2.5000
0.65	0.65	2.8571
0.70	0.70	3.3333
0.75	0.75	4.0000
0.80	0.80	5.0000
0.85	0.85	6.6667
0.90	0.90	10.0000
0.95	0.95	20.0000
0.99	0.99	100.0000

Thus the bicriteria factor of Algorithm 2 is $\left(\gamma, \frac{1}{1-\gamma}\right)$. $\qquad\square$

Table 3 provides further bifactors for varying parameters of γ, showing that the second factor increases with γ.

3.3 A General Online Algorithm for Maximizing γ-Weakly Submodular Minus Linear Function (γGS-L)

Based on the above two sections, we consider a general scaled objective function $\hat{g}(S) = \mu \cdot f(S) - \nu \cdot l(S)$, for any $\mu, \nu > 0$. We provide a general algorithm to analyze the bicriteria approximation ratio of the actual objective function $g(S) = f(S) - l(S)$ starting with the surrogate objective function $\hat{g}(S)$. The general algorithm is shown in Algorithm 3.

Theorem 3 *Algorithm 3 is a* $\left(\frac{\nu\gamma - \mu}{\nu\gamma}, \frac{\nu\gamma - \mu}{\mu\gamma}\right)$-*approximation algorithm, where* $\mu > 0, \nu > 0$ *and* $\gamma \in (0, 1]$.

Algorithm 3. An general algorithm for γGS-L

1: **Input:**
   A non-negative normalized monotone γ-weakly submodular function f, a non-negative linear function l, and $\gamma \in (0, 1]$.
2: **Output:**
   A solution set S.
3: **Process:**
4: Initially set $S := \emptyset$
   For each arriving element e
5:    if $\frac{\mu}{\gamma} \cdot f(e|S) - \nu \cdot l(e) > 0$
6:      $S := S \cup \{e\}$
7: Return S

Proof Assume $S^* = \{s_1, s_2, \dots, s_p\}$ is an optimal solution to the problem, where $p = |S^*|$, and $S = \{e_1, e_2, \dots, e_q\}$ is the solution output by Algorithm 3, where $q = |S|$. Denote $S^{(i)} = \{e_1, e_2, \dots, e_i\}$. Let $\bar{S} = S^* \setminus S$ and $\{s_1, s_2, \dots, s_{|\bar{S}|}\}$ be an arbitrary ordering of $\bar{S}$. Denote $\bar{S}^{(i)} = \{s_1, s_2, \dots, s_i\}, i \leq |S^* \setminus S|$. By the definition of $\hat{g}$, we have

$$
\begin{aligned}
\hat{g}(S \cup \bar{S}) - \hat{g}(S) &= \mu \cdot f(S \cup \bar{S}) - \nu \cdot l(S \cup \bar{S}) - \left(\mu \cdot f(S) - \nu \cdot l(S)\right) \\
&= \mu \cdot \left(f(S \cup \bar{S}) - f(S)\right) - \nu \cdot \left(l(S \cup \bar{S}) - l(S)\right) \\
&= \mu \cdot \left(f(S \cup \bar{S}) - f(S)\right) - \nu \cdot l(\bar{S}) \\
&\leq \mu \cdot \frac{1}{\gamma} \sum_{s \in \bar{S}} f(s|S) - \nu \cdot \sum_{s \in \bar{S}} l(s) \\
&= \sum_{s \in \bar{S}} \left(\mu \cdot \frac{1}{\gamma} \cdot f(s|S) - \nu \cdot l(s)\right) \\
&\leq 0.
\end{aligned}
\tag{3}
$$

The first inequality follows from the definition of γ-weakly submodular of the function f, and the second inequality holds by the element selection rule in the algorithm.

On the other hand, we have

$$
\begin{aligned}
\hat{g}(S \cup \bar{S}) - \hat{g}(S) &= \mu \cdot \left(f(S \cup \bar{S}) - f(S)\right) - \nu \cdot l(\bar{S}) \\
&\geq \mu \cdot \left(f(S^*) - f(S)\right) - \nu \cdot l(\bar{S}) \\
&\geq \mu \cdot \left(f(S^*) - f(S)\right) - \nu \cdot l(S^*).
\end{aligned}
\tag{4}
$$

The first inequality follows by the monotonicity of f and the second inequality holds by the property of l.

Combining inequalities (3) and (4), we obtain

$$
f(S) \geq f(S^*) - \frac{\nu}{\mu} \cdot l(S^*).
$$

By Line 5 of Algorithm 3, we know that each element e_i satisfying $\frac{\mu}{\gamma} \cdot f(e_i|S^{(i-1)}) > \nu \cdot l(e_i)$ will be added into the solution set $S^{(i-1)}$. Summing up all elements $e \in S$, we have

$$\sum_{i=1}^{|S|} \mu \cdot \frac{1}{\gamma} \cdot f(e_i|S^{(i-1)})$$

$$= \mu \cdot \frac{1}{\gamma} \sum_{i=1}^{|S|} f(e_i|S^{(i-1)})$$

$$= \frac{\mu}{\gamma}(f(S) - f(\emptyset))$$

$$> \sum_{i=1}^{|S|} \nu \cdot l(e_i) = \nu \cdot \sum_{i=1}^{|S|} l(e_i) = \nu \cdot l(S).$$

With this upper bound on $l(S) < \frac{\mu}{\nu\gamma} f(S)$ in terms of f, we have

$$f(S) - l(S) > f(S) - \frac{\mu}{\nu\gamma} f(S)$$

$$= \frac{\nu\gamma - \mu}{\nu\gamma} \cdot f(S)$$

$$\geq \frac{\nu\gamma - \mu}{\nu\gamma} \cdot \left(f(S^*) - \frac{\nu}{\mu} \cdot l(S^*) \right)$$

$$= \frac{\nu\gamma - \mu}{\nu\gamma} \cdot f(S^*) - \frac{\nu\gamma - \mu}{\mu\gamma} \cdot l(S^*).$$

Thus we complete the proof. □

4   Conclusions

In this paper, we study the team formation problem with the aim to balance the coverage and the cost incurred. This problem can be formalized as maximizing the difference of a non-negative normalized monotone γ-weakly submodular function and a non-negative linear function. We propose three online bicriteria algorithms. The first two handle the cases where the function f is γ-weakly submodular, and strictly γ-weakly submodular, respectively. The last algorithm integrates the first two with more parameters introduced. When function f is submodular, our result coincides with that of Ene [5]. For the future, the problems with cardinality constraints or matroid constraints under streaming model are worth considering.

Acknowledgements. The first and second authors are supported by Natural Science Foundation of China (Nos. 11871081, 11531014). The third author is supported by the Natural Sciences and Engineering Research Council of Canada (NSERC) grant 06446, and Natural Science Foundation of China (Nos. 11771386, 11728104). The fourth author is supported by National Natural Science Foundation of China (Nos. 11501171, 11771251).

References

1. Anagnostopoulos, A., Becchetti, L., Castillo, C., Gionis, A., Leonardi, S.: Online team formation in social networks. In: Proceedings of the 21st International Conference on World Wide Web, pp. 839–848 (2012)
2. Anagnostopoulos, A., Castillo, C., Fazzone, A., Leonardi, S., Terzi, E.: Algorithms for hiring and outsourcing in the online labor market. In: Proceedings of the 24th ACM SIGKDD International Conference on Knowledge Discovery and Data Mining, pp. 1109–1118 (2018)
3. Bian, A. A., Buhmann, J. M., Krause, A., Tschiatschek, S.: Guarantees for greedy maximization of non-submodular functions with applications. In: Proceedings of the 34th International Conference on Machine Learning, pp. 498–507 (2017)
4. Du, D., Li, Y., Xiu, N., Xu, D.: Simultaneous approximation of multi-criteria submodular functions maximization. J. Oper. Res. Soc. China **2**(3), 271–290 (2014). https://doi.org/10.1007/s40305-014-0053-z
5. Ene, A.: A note on maximizing the difference between a monotone submodular function and a linear function. arXiv preprint arXiv:2002.07782 (2020)
6. Feldman, M.: Guess free maximization of submodular and linear sums. In: Friggstad, Z., Sack, J.-R., Salavatipour, M.R. (eds.) WADS 2019. LNCS, vol. 11646, pp. 380–394. Springer, Cham (2019). https://doi.org/10.1007/978-3-030-24766-9_28
7. Friedrich, T., Göbel, A., Neumann, F., Quinzan, F., Rothenberger, R.: Greedy maximization of functions with bounded curvature under partition matroid constraints. In: Proceedings of the AAAI Conference on Artificial Intelligence, pp. 2272–2279 (2019)
8. Golshan, B., Lappas, T., Terzi, E.: Profit-maximizing cluster hires. In: Proceedings of the 20th ACM SIGKDD International Conference on Knowledge Discovery and Data Mining, pp. 1196–1205 (2014)
9. Gong, S., Nong, Q., Liu, W., Fang, Q.: Parametric monotone function maximization with matroid constraints. J. Global Optimiz. **75**(3), 833–849 (2019). https://doi.org/10.1007/s10898-019-00800-2
10. Harshaw, C., Feldman, M., Ward, J., Karbasi, A.: Submodular maximization beyond non-negativity: guarantees, fast algorithms, and applications. In: Proceedings of the 36th International Conference on Machine Learning, pp. 2634–2643 (2019)
11. Kuhnle, A., Smith, J.D., Crawford, V.G., Thai, M.T.: Fast maximization of non-submodular, monotonic functions on the integer lattice. In: Proceedings of the 35th International Conference on Machine Learning, pp. 2791–2800 (2018)
12. Lappas, T., Liu, K., Terzi, E.: Finding a team of experts in social networks. In: Proceedings of the 15th ACM SIGKDD International Conference on Knowledge Discovery and Data Mining, pp. 467–476 (2009)
13. Liu, S., Poon, C.K.: A simple greedy algorithm for the profit-aware social team formation problem. In: Proceedings of the 11th International Conference on Combinatorial Optimization and Applications, pp. 379–393 (2017)
14. Qian, C.: Multi-objective evolutionary algorithms are still good: maximizing monotone approximately submodular minus modular functions. arXiv preprint arXiv:1910.05492 (2019)
15. Sviridenko, M., Vondrák, J., Ward, J.: Optimal approximation for submodular and supermodular optimization with bounded curvature. Math. Oper. Res. **42**(4), 1197–1218 (2017)

Approximation Algorithm for Stochastic Set Cover Problem

Haiyun Sheng[1], Donglei Du[2], Yuefang Sun[3], Jian Sun[4], and Xiaoyan Zhang[1(✉)]

[1] School of Mathematical Science and Institute of Mathematics,
Nanjing Normal University, Nanjing 210023, Jiangsu, People's Republic of China
`zhangxiaoyan@njnu.edu.cn`
[2] Faculty of Management, University of New Brunswick,
Fredericton, NB E3B 5A3, Canada
[3] School of Mathematical Information, Shaoxing University,
Shaoxing 312000, Zhejiang, People's Republic of China
[4] Department of Operations Research and Information Engineering,
Beijing University of Technology, Beijing 100124, People's Republic of China

Abstract. Cover problem is a typical NP-hard problem, which has comprehensive application background and is a hot topic in recent years. In this paper, we study two stage, finite scenarios stochastic versions of set cover problem with submodular penalties which is the generalization of the stochastic vertex cover problem with submodular penalties. The goal is to minimize the sum of the first stage cost, the expected second stage cost and the expected penalty cost. By doing some research on the structural properties of submodular function, we present a primal-dual 2η-approximation algorithm for the stochastic set cover problem with submodular penalties (4-approximation algorithm for the stochastic vertex cover problem with submodular penalties when $\eta = 2$), where η is the maximum frequency of the element in the family of subsets.

Keywords: Stochastic set cover · Primal-dual · Approximation algorithm

1 Introduction

Cover problem is a typical example of NP-hard optimization problems, which is proved by Karp [13] unless P = NP. That is to say, we may not get the optimal solution in polynomial time. In computer science, the cover problem has been widely used in various fields including information retrieval, data mining and web host analysis. Furthermore, it is worth noting that the vertex cover problem is fixed-parameter tractable and is a central problem in parameterized complexity theory.

Set cover problem is defined in a given finite element set E. Formally, given a set of elements $\{e_1, e_2, ..., e_n\}$ (called the universe) and a collection $\mathbb{S}$ of t sets whose union equals the universe, each set of $\mathbb{S}$ has a cost. The set cover problem is to find the smallest cost subcollection of $\mathbb{S}$ whose union equals the universe. Obviously, the vertex cover problem is a special case of the set cover problem. Iwata and Nagano [11] also propose a relevant variant of the set cover, which is called

© Springer Nature Switzerland AG 2020
Z. Zhang et al. (Eds.): AAIM 2020, LNCS 12290, pp. 37–48, 2020.
https://doi.org/10.1007/978-3-030-57602-8_4

submodular set cover problem. Specifically, given a set of elements $\{e_1, e_2, ..., e_n\}$ (called the universe), a collection $\mathbb{S}$ of t sets whose union equals the universe and a nonnegative submodular function $f : 2^{\mathbb{S}} \to \mathbb{R}_+$. Namely, each subset of $\mathbb{S}$ has a submodular cost. The goal is to choose a minimum submodular cost set cover. Another variant called set cover problem with submodular penalties is also given a set of elements $\{e_1, e_2, ..., e_n\}$ (called the universe), a collection $\mathbb{S}$ of t sets whose union equals the universe, each set of $\mathbb{S}$ has a cost and each subset $T \subseteq E$ has a penalty cost $h(T)$, where $h(\cdot)$ is a submodular function, which is defined in the power set of E. The goal is to choose the smallest cost subcollection of $\mathbb{S}$ to cover some elements of E and penalize the uncovered elements of E.

The stochastic set cover problem is a relevant variant of the set cover which can be constructed as a two stage stochastic model as follows. In the first stage, the information of the clients (elements) is unknown. The possible scenarios and their probability distribution are given until in the second stage. The cost of sets in different stages and scenarios are distinct. We can purchase the sets in the first stage which can serve all clients (elements) in all scenarios and the sets in the second stage in each scenario can only serve the clients in that scenario. The goal is to insure that each client in each scenario should be served by purchased sets either in the scenario of the second stage or in the first stage and minimize the total expected set purchasing cost.

In this paper, we consider the stochastic set cover problem with submodular penalties and present a primal-dual 2η-approximation algorithm for the stochastic set cover problem with submodular penalties, where η is the maximum frequency of the element in the family of subsets.

2   Related Work

In recent years, the vertex cover problem has attracted significant attention, many relevant variants of the vertex cover problem have been extensively studied by many scholars [10]. There is no doubt that the relevant approximation algorithms made great progress such as greedy algorithm, primal-dual algorithm and linear programming rounding. For the vertex cover problem, Bar-Yehuda and Even [2] proposed a 2-approximation algorithm by primal-dual method. Hochbaum [9] firstly presented constant approximation algorithm by applying the linear programming rounding skill. Dinur et al. [4] proved that there doesn't exist an approximation algorithm whose approximation ratio is less than 1.3606 unless $P = NP$. However, this problem has an approximation algorithm whose approximation ratio can achieve $2 - o(1)$ according to the number of vertices and the maximum degree in a graph [8].

The relevant variants of the vertex cover problem also have many results. We only introduce a part of results which are relevant to the paper. Firstly, Hochbaum [9] introduced the generalized vertex cover problem, it is also be called prize-collecting vertex cover problem [10], and present a 2-approximation algorithm by linear programming rounding. Iwata and Nagano [11] proposed the submodular vertex cover problem and designed a 2-approximation algorithm by the convex programming. For the stochastic vertex cover problem, Gupta et *al.* [7] proposed a 8-approximation algorithm by the boosted sampling skill in 2004 and improved the approximation ratio to 4 in 2008. Ravi and Sinha [16] proposed a 2-approximation algorithm by primal-dual method for the stochastic vertex cover problem.

Set cover and its relevant variants have also been a hot topic for research [9]. There are many approximation algorithms for this problem. In 1974, the deterministic version of set cover was among the earliest NP-hard problems to be approximated with an $O(logn)$-approximation, which was provided by Johnson [12]. In 1993, Bellare [3] pointed out the fact that there doesn't exist any constant approximation algorithm unless $P = NP$. In 1997, Raz and Safra [17] presented that there doesn't exist any approximation algorithm whose approximation ratio is less than $c \cdot lnn$ for some constant c unless $P = NP$. The problem was also shown to be NP-hard to approximate better than a factor of $\Omega(logn)$ by Arora and Sudan [1] in 2003. For many relevant variants of set cover including the partial set cover [6], the prize-collecting set cover problem [14] and the generalized set cover problem [14], they are studied by relaxing request, which needs to cover all the elements. Slavik [18] has presented an $H(\Delta)$-approximation algorithm in polynomial time for the partial set cover problem, where Δ is the module of maximum set, and Ganhhi [6] proposed an f-approximation algorithm in polynomial time, where f is the maximum frequency of the element of E in the family of subsets. For the stochastic set cover problem, Ravi and Sinha [16] proved that it can be reduced to the classical set cover. In 2016, Li *et al.* [15] presented a $2(lnn + 1)$-approximation algorithm, where n is the number of elements.

The remainder of this paper is organized as follows. In Sect. 3, we introduce the stochastic set cover problem. Furthermore, applying the primal-dual method, we propose an approximation algorithm for the problem and prove that the ratio is 2η. We conclude in Sect. 4.

3   Stochastic Set Cover Problem with Submodular Penalties

In this section, we will introduce the stochastic set cover problem with submodular penalties and present the relevant programs specifically. In fact, this problem is a generalization of the stochastic vertex cover problem with submodular penalties. In the stochastic vertex cover problem with submodular penalties, each client-scenario (edge-scenario) can be covered at most two times. In this paper, we use η to denote the maximum frequency of the element in the family of subsets in stochastic set cover problem with submodular penalties.

Naturally, the stochastic set cover problems with submodular penalties can be described as follows. There are two stages in this problem: given a potential set family in the first stage, it is allowed to purchase some sets for serving any possible client (element) in advance. The purchasing cost of the set S in the first stage is c_S^0. In the second stage, all possible scenarios and the associated probabilities become known, where we only consider the case of polynomial scenarios. That is to say, the number m of the scenarios is polynomial with respect to the input of the problem. For a scenario $k \in \{1, 2, ..., m\}$ all in the second stage, the probability of scenario k is p_k, the client (element) set is denoted as E_k, the penalty cost for the unserved client set $T_k \subseteq E_k$ is $h_k(T_k)$ which is a monotone submodular function, and c_S^k denotes the purchasing cost of set S in scenario k. The client (element) in scenario k in the second stage can be covered by a set purchased in the first stage or in the second stage with respect to the scenario; otherwise, the client (element) is unserved. For convenience, we present the following notations. The notation (S, k) (for $S \in \mathbb{S}$, $k = 0, 1, ..., m$) is called a set-scenario pair and denotes the set-scenario pair set as $\mathcal{S}$. Similarly, each (e, k) (for $k = 0, 1, ..., m, e \in E_k$) represents a client-scenario (element-scenario) pair which is active in the k-th scenario. Let $\mathcal{C}$ denote the client-scenario (element-scenario) pair set, in scenario k, the set of the client-scenario (element-scenario) pair (e, k) is denoted as $\mathcal{C}_k$ for $e \in E_k$. Clearly, $\bigcup_{k=1}^{m} \mathcal{C}_k = \mathcal{C}$. The task is to determine the set of set-client pairs $\widehat{\mathbb{S}}_0$ and $\widehat{\mathbb{S}}_k$ to be purchased respectively in the first stage and in the k-th scenario in the second stage ($k = 1, ..., m$), and the set of clients $\widehat{T}_k$ ($k = 1, ..., m$) that will incur penalties. Finally, the aim is to minimize the sum of the expected set purchasing cost $\sum_{S \in \widehat{\mathbb{S}}_0} c_S^0 + \sum_{k=1}^{m} \sum_{S \in \widehat{\mathbb{S}}_k} c_S^k$, and the expected penalty cost $\sum_{k=1}^{m} p_k h_k(\widehat{T}_k)$.

The stochastic set cover problem with submodular penalties can be formulated as the following linear integer program, in which $p_0 = 1$ and $\sum_{k=1}^{m} p_k = 1$.

$$\text{min} \quad \sum_{(S,k)\in\mathcal{S}} p_k c_S^k x_S^k + \sum_{k=1}^{m} \sum_{T_k \subseteq E_k} p_k h_k(T_k) Z_{T_k}$$

$$(IP) \quad \text{s.t.} \quad \sum_{S\in\mathbb{S}:e\in S} x_S^0 + \sum_{S\in\mathbb{S}:e\in S} x_S^k + \sum_{T_k \subseteq E_k: e \subseteq T_k} Z_{T_k} \geq 1, \qquad \forall (e, k) \in \mathcal{C},$$

$$x_S^0, x_S^k, Z_{T_k} \in \{0, 1\}, \qquad \forall S \in \mathbb{S}, T_k \subseteq E_k,$$

All the variables are binary in the above formulation, x_S^0 indicates that whether set S is purchased in the first stage. If $x_S^0 = 1$, set $S \in \mathbb{S}$ is purchased in the first stage (i.e. set-scenario pair $(S, 0)$ is purchased); otherwise, $x_S^0 = 0$. Similarly, x_S^k represents whether set S is purchased in the k-th scenario of the second stage. If $x_S^k = 1$, set $S \in \mathbb{S}$ is purchased in the second stage for the k-th scenario (i.e. set-scenario pair (S, k) is purchased); otherwise, $x_S^k = 0$. Z_{T_k} indicates whether a set of clients (elements) $T_k \subseteq E_k$ incurs penalties. The first constraint models that each client-scenario (element-scenario) pair (e, k) is either covered by a set or incurs penalty. The set should be purchased either in the first stage or the corresponding scenario in the second stage.

By relaxing the integrality constraints, we obtain the LP relaxation and the corresponding dual linear program as follows.

$$\min \sum_{(S,k)\in\mathcal{S}} p_k c_S^k x_S^k + \sum_{k=1}^{m} \sum_{T_k\subseteq E_k} p_k h_k(T_k) z_{T_k}$$

(LP) s.t. $\sum_{S\in\mathbb{S}:e\in S} x_S^0 + \sum_{S\in\mathbb{S}:e\in S} x_S^k + \sum_{T_k\subseteq E_k:e\subseteq T_k} z_{T_k} \geq 1,$ $\forall(e,k)\in\mathcal{C},$

$$x_S^0, x_S^k, z_{T_k} \geq 0,   \forall S\in\mathbb{S}, T_k\subseteq E_k.$$

$$\max \sum_{(e,k)\in\mathcal{C}} y_e^k$$

(DP) s.t. $\sum_{(e,k)\in\mathcal{C}} y_e^k \leq c_S^0,$ $\forall S\in\mathbb{S},$

$\sum_{e\in E_k} y_e^k \leq p_k c_S^k,$ $\forall S\in\mathbb{S}, k=1,2,...,m,$

$\sum_{e\in T_k} y_e^k \leq p_k h_k(T_k),$ $\forall T_k\subseteq E_k, k=1,2,...,m,$

$y_e^k \geq 0,$ $\forall(e,k)\in\mathcal{C}.$

The variable y_e^k can be understood as the budget of client-scenario (element-scenario) pair (e,k) in the above dual formulation.

3.1   The Primal-Dual Algorithm

In this section, we will present a primal-dual algorithm for the stochastic set cover problem with submodular penalties. Note that we need to find the element set to be penalized in our algorithm according to the property of the submodular function.

In fact, our algorithm can be interpreted as a procedure of the dual ascent. In our algorithm, we carefully deal with the condition of penalty restriction (the third restriction in the dual programming) in order to satisfy the feasibility of duality. Therefore, by the construction of y obtained by Step 2 of Algorithm 1, it is obviously a dual feasible solution. We further give some intuitively explanation of Step 2 as follows: Step 2.1 corresponds to the event of purchasing a new set in the first stage; Step 2.2 corresponds to the event of purchasing a new set in the second stage; Step 2.3 corresponds to the event in which some clients (elements) are added to the rejected client set. In Step 2.4, $S(e,k)$ is the purchased set in which (e,k) is severed by the purchased set.

For convenience, we define the following notations.

$\widehat{T}_k(k=1,...,m)$ denotes the penalty client (element) set in the k-th scenario.

$\widehat{\mathbb{S}}_0$ represents the purchased set in the first stage.

$\widehat{\mathbb{S}}_k$ corresponds to the purchased set in the second stage with respect to k-th scenario.

$\overline{\mathcal{C}}_k$ represents the frozen client-scenario (element-scenario) pair set in the k-th scenario.

Let $\overline{\mathcal{C}} := \bigcup_{k=1}^{m} \overline{\mathcal{C}}_k$ be all the frozen client-scenario (element-scenario) pairs set.

$N(S,k)$ $((S,k)\in\mathcal{S})$ denotes the client-scenario (element-scenario) pair set in which each client-scenario (element-scenario) pair has a positive contribution to the set-scenario pair (S,k).

In the initial state, all the dual variables are zero. All the client-scenario (element-scenario) pairs are unfrozen. All the clients (elements) are not punished. All the sets are not purchased.

Algorithm 1. Primal-dual algorithm

---

begin: :

1: Initialization $y_e^k := 0$, $\widehat{T}_k := \emptyset$ $(k = 1, ..., m)$, $\widehat{S}_k := \emptyset$ $(k = 0, 1, ..., m)$, $\overline{C}_k := \emptyset$ $(k = 0, 1, ..., m)$, and $N(S, k) := \emptyset$ $((S, k) \in \mathcal{S})$.

2: Obtain a dual feasible solution

2.1: For each $S \in \mathbb{S} \backslash \widehat{S}_0$, assume that $\tau(S, 0)$ is the root of the following equation with respect to τ,

$$\sum_{(e,k) \in \overline{C}} y_e^k + \sum_{(e,k) \in \mathcal{C} \backslash \overline{C}} \tau = c_S^0.$$

Set $\tau_1 := \tau_1(S_1^*, 0) := \min_{S \in \mathbb{S} \backslash \widehat{S}_0} \tau(S, 0)$, $N(S_1^*, 0) := N(S_1^*, 0) \bigcup \{(e, k) \in \mathcal{C} \backslash \overline{C} : \tau_1 > 0\}$.

2.2: For each $S \in \mathbb{S} \backslash \widehat{S}_k$, $\tau(S, k)$ is the root of the following equation with respect to τ,

$$\sum_{(e,k) \in \overline{C}_k} y_e^k + \sum_{(e,k) \in \mathcal{C}_k \backslash \overline{C}_k} \tau = p_k c_S^k.$$

Set $\tau_2 := \tau_2(S_2^*, k_2^*) := \min_{k=1,...,m} \min_{S \in \mathbb{S} \backslash \widehat{S}_k} \tau(S, k)$ and $N(S_2^*, k_2^*) := N(S_2^*, k_2^*) \bigcup \{(e, k_2^*) \in \mathcal{C} \backslash \overline{C} : \tau_2 > 0\}$.

2.3: For each scenario k $(k = 1, 2, ..., m)$, calculate

$$\tau_{3,k} := \min_{T_k \subseteq E_k} \frac{p_k h(T_k) - \sum_{e \in T_k \cap \{e : (e,k) \in \overline{C}_k\}} y_e^k}{|T_k \backslash \{e : (e, k) \in \overline{C}_k\}|}.$$

Set

$$\tau_3 := \tau_{3,k_3^*} = \min_{k=1,...,m} \tau_{3,k},$$

and let $T_{k_3^*}$ be the optimal solution of the above formula for $k = k_3^*$.

2.4: Set $\tau^* = \min\{\tau_1, \tau_2, \tau_3\}$

Case 1. If $\tau^* = \tau_1$, set $S(e, k) := S_1^*$, $y_e^k := \tau_1$ for each client-scenario (element-scenario) pair $(e, k) \in N(S_1^*, 0) \backslash \overline{C}$ and each set S. Update $\widehat{S}_0 := \widehat{S}_0 \bigcup \{S_1^*\}$ and $\overline{C}_k := \overline{C}_k \bigcup N(S_1^*, 0)$ for each scenario $k = 0, 1, 2, ..., m$.

Case 2. If $\tau^* = \tau_2$, set $S(e, k_2^*) := S_2^*$, $y_e^{k_2^*} := \tau_2$ for each client-scenario (element-scenario) pair $(e, k_2^*) \in N(S_2^*, k_2^*) \backslash \overline{C}_{k_2^*}$ and each set S. Update $\widehat{S}_{k_2^*} := \widehat{S}_{k_2^*} \bigcup \{S_2^*\}$ and $\overline{C}_{k_2^*} := \overline{C}_{k_2^*} \bigcup N(S_2^*, k_2^*)$ for each scenario.

Case 3. If $\tau^* = \tau_3$, set $y_e^{k_3^*} := \tau_3$ for each client (element) $e \in T_{k_3^*} \backslash \{e : (e, k_3^*) \in \overline{C}_{k_3^*}\}$ and each set S. Update $\widehat{T}_{k_3^*} := \widehat{T}_{k_3^*} \bigcup T_{k_3^*}$ and $\overline{C}_{k_3^*} := \overline{C}_{k_3^*} \bigcup \{(e, k_3^*) \in T_{k_3^*}\}$. Note that if three cases occur simultaneously, execute one of three cases arbitrarily.

2.5: Set $\overline{C} := \bigcup_{k=1}^{m} \overline{C}_k$. If $\overline{C} \bigcup \overline{C}_0 := \mathcal{C} \bigcup \mathcal{C}_0$, go to Step 3; otherwise, go to Step 2.1.

3: Obtain an integer primal feasible solution.

3.1: The final selected set to purchase in the first stage is $\widehat{S}_0$, and the final selected set to purchase in the second stage with respect to the k-th scenario is $\widehat{S}_k$ for each $k = 1, 2, ..., m$. If $S_k^{(i)} \in \widehat{S}_k$, $S_k^{(j)} \in \widehat{S}_k$, and $S_k^{(i)} \subseteq S_k^{(j)}$, then we delete $S_k^{(i)}$ for $k = 0, 1, ..., m$.

3.2: Let $\widehat{T}_k$ be the set of unserved clients in the k-th scenario for each $k = 1, 2, ..., m$. Serve each client-scenario pair in $\mathcal{C}_k \backslash \{(e, k) : e \in \widehat{T}_k, k = 1, 2, ..., m\}$ in the first stage or the corresponding scenario in the second stage.

end

3.2   The Analysis of the Algorithm

In this section, we will analyze the approximation ratio of Algorithm 1. In the analysis, we still consider purchasing cost of the set obtained by the dual feasible solution in the algorithm. At the same time, we also consider the penalty cost in the algorithm. For convenience, we introduce the follow notations.

Let SOL be the solution obtained from Algorithm 1, the total cost consists of the set purchasing cost $\mathbb{S}_{SOL}$ and the penalty cost P_{SOL}. Obviously, Algorithm 1 is a polynomial time combinatorial algorithm, and the corresponding proof is presented in Lemma 1. To obtain the approximation ratio, we next bound $\mathbb{S}_{SOL}$ and P_{SOL} using the following lemmas. It's worth noting that $N(S,k) \bigcap N(S',k') \neq \emptyset$, that is, there are repeated clients which are served by S and S', where $k, k' = 0, 1, ..., m$, $S \in \widehat{\mathbb{S}}_0, S' \in \widehat{\mathbb{S}}_k$. Next, we will give the following lemmas, which are prepared for the approximation ratio.

Lemma 1. *Let τ_0 be an arbitrary time of incurring one of three cases in Step 2.4 of Algorithm 1, then we can always find the next time τ^* in polynomial time such that one of the three cases occurs again.*

Proof. Let $\widetilde{E}$ be the frozen element-scenarios, we compute the next time τ^* of incurring one of three cases in Algorithm 1.

If Case 1 incurs in the next time τ_1, then at any time of time interval (τ_0, τ_1), we have

$$\sum_{(e,k)\in\mathcal{C}\cap\widetilde{E}} y_e^k + \sum_{(e,k)\in\mathcal{C}\backslash\widetilde{E}} \tau \leq c_S^0, \quad \forall S \in \mathbb{S},$$

and then

$$\tau \leq \frac{c_S^0 - \sum\limits_{(e,k)\in\mathcal{C}\cap\widetilde{E}} y_e^k}{\sum\limits_{(e,k)\in\mathcal{C}\backslash\widetilde{E}} 1}, \quad \forall S \in \mathbb{S}, \ \mathcal{C}\backslash\widetilde{E} \neq \emptyset.$$

Suppose

$$\mu_k = \sum_{(e,k)\in\mathcal{C}\backslash\widetilde{E}} 1,$$

$$\mu'_k = \sum_{(e,k)\in\mathcal{C}\cap\widetilde{E}} y_e^k.$$

Since both μ_k and μ'_k are modular functions, we can compute

$$\tau_1 = \min_{S\in\mathbb{S}:\mathcal{C}\backslash\widetilde{E}\neq\emptyset} \frac{c_S^0 - \mu'_k}{\mu_k}$$

in polynomial time.

If Case 2 incurs in the next time τ_2, then at any time of time interval (τ_0, τ_2), we have

$$\sum_{e\in E_k\cap\widetilde{E}} y_e^k + \sum_{e\in E_k\backslash\widetilde{E}} \tau \leq p_k c_S^k, \quad \forall S \in \mathbb{S},$$

and then

$$\tau \le \frac{p_k c_S^k - \sum\limits_{e \in E_k \cap \widetilde{E}} y_e^k}{\sum\limits_{e \in E_k \setminus \widetilde{E}} 1}, \quad \forall S \in \mathbb{S}, \quad E_k \setminus \widetilde{E} \neq \emptyset.$$

Suppose

$$\psi_k = \sum_{e \in E_k \setminus \widetilde{E}} 1,$$

$$\psi_k' = \sum_{e \in E_k \cap \widetilde{E}} y_e^k.$$

Clearly, both ψ_k and ψ_k' are modular functions. Therefore we can compute

$$\tau_2 = \min_{S \in \mathbb{S}: E_k \setminus \widetilde{E} \neq \emptyset} \frac{p_k c_S^k - \psi_k'}{\psi_k}$$

in polynomial time.

If Case 3 incurs in the next time τ_3, then at any time of time interval (τ_0, τ_3), we have

$$\sum_{e \in T_k \cap \widetilde{E}} y_e^k + \sum_{e \in T_k \setminus \widetilde{E}} \tau \le p_k h_k(T_k), \quad \forall T_k \subseteq E_k,$$

and then

$$\tau \le \frac{p_k h_k(T_k) - \sum\limits_{e \in T_k \cap \widetilde{E}} y_e^k}{\sum\limits_{e \in T_k \setminus \widetilde{E}} 1}, \quad \forall T_k \subseteq E_k, \quad T_k \setminus \widetilde{E} \neq \emptyset.$$

Similar to the proof above, we have

$$\tau_3 = \min_{T_k \subseteq E_k} \frac{p_k h_k(T_k) - \sum\limits_{e \in T_k \cap \widetilde{E}} y_e^k}{\sum\limits_{e \in T_k \setminus \widetilde{E}} 1}.$$

In fact, we still deal with the last problem, which is the minimization of a ratio of a submodular function and modular function, and this can be solved in polynomial time by a combinatorial algorithm [5]. Therefore, we can also compute the above inequations in polynomial time.

In summary, $\tau^* = \min\{\tau_1, \tau_2, \tau_3\}$ is the next time of incurring cases, and thus the lemma holds. □

Lemma 2.

$$\mathbb{S}_{SOL} = \sum_{S \in \widehat{\mathbb{S}}_0} \sum_{(e,k) \in N(S,0)} y_e^k + \sum_{k=1}^{m} \sum_{S \in \widehat{\mathbb{S}}_k} \sum_{(e,k) \in N(S,k)} y_e^k.$$

Proof. Clearly, if each $S \in \widehat{\mathbb{S}}_0$, then the first cost constraint of the dual programming is tight. We have

$$c_S^0 = \sum_{(e,k) \in N(S,0)} y_e^k,$$

then

$$\sum_{S \in \widehat{\mathbb{S}}_0} c_S^0 = \sum_{S \in \widehat{\mathbb{S}}_0} \sum_{(e,k) \in N(S,0)} y_e^k.$$

This is the purchasing cost of the sets in the first stage.

Analogously, the second cost constraint is tight for each $S \in \widehat{\mathbb{S}}_k$ in the scenario k, where $k = 1, 2, .., m$. We have

$$c_S^k = \sum_{(e,k) \in N(S,k)} y_e^k,$$

then

$$\sum_{k=1}^{m} \sum_{S \in \widehat{\mathbb{S}}_k} c_S^k = \sum_{k=1}^{m} \sum_{S \in \widehat{\mathbb{S}}_k} \sum_{(e,k) \in N(S,k)} y_e^k.$$

This is the purchasing cost of the sets in the second stage. Then the above lemma holds. □

Analogously, we give the following notations. Let $\widehat{\mathcal{S}}_k := \{(S,k) | S \in \widehat{\mathbb{S}}_k\}$ and $\mathcal{C}_{\widehat{\mathbb{S}}_k} := \bigcup_{S \in \widehat{\mathbb{S}}_k} N(S,k)$ for each $k = 0, 1, 2, ..., m$, and partition the client-scenario (element-scenario) pair into two types which are denoted as $\mathcal{C}^{(1)} := \mathcal{C}_{\widehat{\mathbb{S}}_0} \bigcup (\bigcup_{k=1}^{m} \mathcal{C}_{\widehat{\mathbb{S}}_k})$ and $\mathcal{C}^{(2)} := \bigcup_{k=1}^{m} \{(e,k) : e \in \widehat{T}_k\}$ respectively. Obviously, $\mathcal{C}^{(1)}$ is the client-scenario (element-scenario) pair set in which each client-scenario (element-scenario) pair (e, k) has a positive contribution to the set-scenario pair $(S, 0) \in \widehat{\mathcal{S}}_0$ or $(S, k) \in \widehat{\mathcal{S}}_k$. $\mathcal{C}^{(2)}$ is the client-scenario (element-scenario) pair set in which each client-scenario (element-scenario) pair (e, k) is rejected in the k-th scenario. Note that $\mathcal{C}^{(1)}$ and $\mathcal{C}^{(2)}$ may be joint, and a partition of $\mathcal{C}$ can be constructed as follows: $\{\mathcal{C}^{(1)} \backslash \mathcal{C}^{(2)}, \mathcal{C}^{(2)}\}$.

The following lemma with respect to the penalty cost of *SOL* is important for the proof of deterministic set cover problem with submodular penalties. For completeness, we present the proof for the stochastic set cover problem with submodular penalties almost the same as in [19].

Lemma 3.

$$P_{SOL} = \sum_{(e,k) \in \mathcal{C}^{(2)}} y_e^k = \sum_{k=1}^{m} \sum_{e \in \widehat{T}_k} y_e^k.$$

Proof. For any iteration in which we get $\tau^* = \tau_3$ in Step 2.4. Let T_k^1 be the penalty set before this iteration and T_k^2 be the set $T_{k_3^*}$ given in this iteration. We have

$$\sum_{e \in T_k^1} y_e^k = p_k h_k(T_k^1),$$

$$\sum_{e \in T_k^2} y_e^k = p_k h_k(T_k^2).$$

We only need to show that for any $k = 1, 2, ..., m$,

$$\sum_{e \in T_k^1 \bigcup T_k^2} y_e^k = p_k h_k(T_k^1 \bigcup T_k^2).$$

From the submodularity of $h_k(\cdot)$ and Algorithm 1, we have

$$\sum_{e \in T_k^1 \bigcup T_k^2} y_e^k + \sum_{e \in T_k^1 \bigcap T_k^2} y_e^k = \sum_{e \in T_k^1} y_e^k + \sum_{e \in T_k^2 \setminus T_k^1} y_e^k + \sum_{e \in T_k^1 \bigcap T_k^2} y_e^k$$

$$= \sum_{e \in T_k^1} y_e^k + \sum_{e \in T_k^2} y_e^k$$

$$= p_k h_k(T_k^1) + p_k h_k(T_k^2)$$

$$\geq p_k h_k(T_k^1 \bigcup T_k^2) + p_k h_k(T_k^1 \bigcap T_k^2).$$

Since the following inequations always hold in Algorithm 1, we can get

$$\sum_{e \in T_k^1 \bigcup T_k^2} y_e^k \leq p_k h_k(T_k^1 \bigcup T_k^2),$$

$$\sum_{e \in T_k^1 \bigcap T_k^2} y_e^k \leq p_k h_k(T_k^1 \bigcap T_k^2).$$

By the above formulas, we can obtain

$$\sum_{e \in T_k^1 \bigcup T_k^2} y_e^k + \sum_{e \in T_k^1 \bigcap T_k^2} y_e^k = p_k h_k(T_k^1 \bigcup T_k^2) + p_k h_k(T_k^1 \bigcap T_k^2),$$

which implies that

$$\sum_{e \in T_k^1 \bigcup T_k^2} y_e^k = p_k h_k(T_k^1 \bigcup T_k^2).$$

Then the lemma is proved. $\square$

Next, we present our main result in the paper.

Theorem 1. *Algorithm 1 is a 2η-approximation algorithm for the stochastic set cover problem with submodular penalties.*

Proof. We have presented the bound on the total cost of SOL produced by Algorithm 1, and let OPT be the optimal solution of the stochastic set cover problem with submodular penalties. Note that each element appears η times at most, that is to say, each client (element) can be covered by η times at most in the corresponding scenario. Clearly, $\mathcal{C}^{(1)}, \mathcal{C}^{(2)}$ may be joint. By Lemma 2 and Lemma 3, the cost of SOL is at most

$$
\begin{aligned}
cost(SOL) &= \mathbb{S}_{SOL} + P_{SOL} \\
&= \sum_{S \in \widehat{\mathbb{S}}_0} \sum_{(e,k) \in N(S,0)} y_e^k + \sum_{k=1}^{m} \sum_{S \in \widehat{\mathbb{S}}_k} \sum_{(e,k) \in N(S,k)} y_e^k + \sum_{k=1}^{m} \sum_{e \in \widehat{T}_k} y_e^k \\
&\leq \eta \Big(\sum_{(e,k) \in \mathcal{C}_1} y_e^k + \sum_{(e,k) \in \mathcal{C}_2} y_e^k \Big) \\
&\leq 2\eta \Big(\sum_{(e,k) \in \mathcal{C}} y_e^k \Big) \\
&\leq 2\eta OPT.
\end{aligned}
$$

Therefore, we conclude the proof. □

As explained previously, the stochastic vertex cover problem with submodular penalties is a particular case of the stochastic set cover problem with submodular penalties. Specifically, each client-scenario (edge-scenario) can be covered at most two times in the stochastic vertex cover problem with submodular penalties. This is just the case of $\eta = 2$ in the stochastic set cover problem with submodular penalties. Therefore, we have the following corollary.

Corollary 1. *There exists a 4-approximation algorithm for the stochastic vertex cover problem with submodular penalties.*

4 Conclusions

Considering the stochastic set cover problem, we present a primal-dual 2η-approximation algorithm. Many researchers made substantial contribution in stochastic problems by considering the use of sampling, cost sharing function, and primal-dual, etc. In the future, we believe that there will be more substantial progress in approximation algorithms for the stochastic optimization problems, and it will be interesting to improve the approximation ratios of these problems.

Acknowledgements. This research is supported or partially supported by the National Natural Science Foundation of China (Grant Nos. 11871280, 11371001, 11771386, 11401389 and 11728104), the Natural Sciences and Engineering Re-search Council of Canada (NSERC) Grant 06446, Zhejiang Provincial Natural Science Foundation (No. LY20A010013) and Qinglan Project.

References

1. Arora, A., Sudan, M.: Improved low degree testing and applications. Combinatorica **23**(3), 365–426 (2003). https://doi.org/10.1007/s00493-003-0025-0
2. Bar-Yehuda, R., Even, S.: A linear-time approximation algorithm for the weighted vertex cover problem. J. Algorithms **2**(2), 198–203 (1981)
3. Bellare, M., Goldwass, S., Lund, C., Russell, A.: Efficient probabilistically checkable proofs and applications to approximations. In: Proceedings of the 26th Annual ACM Symposium on Theory of Computing (STOC), pp. 23–25. ACM, Canada (1994)
4. Dinur, I., Safra, S.: On the hardness of approximating minimum vertex cover. Ann. Math. **162**(2), 439–485 (2005)
5. Fleischer, L., Iwata, S.: A push-relabel framework for submodular function minimization and applications to parametric optimization. Discrete Appl. Math. **131**(2), 311–322 (2003)
6. Gandhi, R., Khuller, S., Srinivasan, A.: Approximation algorithms for partial covering problems. J. Algorithms **53**(1), 55–84 (2004)
7. Gupta, A., Pál, M., Ravi, R., Sinha, A.: Sampling and cost-sharing: approximation algorithms for stochastic optimization problems. SIAM J. Comput. **40**(5), 1361–1401 (2011)
8. Halperin, E.: Improved approximation algorithms for the vertex cover problem in graphs and hypergraphs. SIAM J. Comput. **31**(5), 1608–1623 (2002)
9. Hochbaum, D.-S.: Approximation algorithm for set covering and vertex cover problems. SIAM J. Comput. **11**(3), 555–556 (1982)
10. Hochbaum, D.-S.: Approximation algorithms for NP-hard problems. ACM SIGACT News **28**(2), 40–52 (1997)
11. Iwata, S., Nagano, K.: Submodular function minimization under covering constraints. In: Proceedings of the 50th Annual IEEE Symposium on Foundations of Computer Science (FOCS), pp. 671–680. IEEE, USA (2009)
12. Johnson, D.-S.: Approximation algorithms for combinatorial problems. J. Comput. Syst. Sci. **9**(3), 256–278 (1974)
13. Karp, R.-M.: Reducibility among combinatorial problems. In: Miller, R.E., Thatcher, J.W., Bohlinger, J.D. (eds.) Complexity of Computer Computations. The IBM Research Symposia Series, pp. 85–103. Springer, Boston (1972). https://doi.org/10.1007/978-1-4684-2001-2_9
14. Könemann, J., Parekh, O., Segev, D.: A unified approach to approximating partial covering problems. Algorithmica **59**(4), 489–509 (2011)
15. Li, J., Liu, Y.: Approximation algorithms for stochastic combinatorial optimization problems. J. Oper. Res. Soc. China **4**(1), 1–47 (2016)
16. Ravic, R., Sinhac, A.: Hedging uncertainty: approximation algorithms for stochastic optimization problems. Math. Program. **108**(1), 97–114 (2006)
17. Raz, R., Safra, S.: A sub-constant error-probability low-degree test, and a sub-constant error-probability PCP characterization of NP. In: Proceedings of the 29th Annual ACM Symposium on the Theory of Computing (STOC), pp. 4–6. ACM, USA (1997)
18. Slavik, P.: Improved performance of greedy algorithm for partial over. Inf. Process. Lett. **64**(5), 251–254 (1997)
19. Xu, D., Gao, D., Wu, C.: A primal-dual 3-approximation algorithm for the stochastic facility location problem with submodular penalties. Acta Mathematicae Applicatae Sinica English **64**(3), 617–626 (2015)

On Approximations for Constructing 1-Line Minimum Rectilinear Steiner Trees in the Euclidean Plane $\mathbb{R}^2$

Junran Lichen[1,2], Jianping Li[1(✉)], Wencheng Wang[1], Jean Yeh[3],
YeongNan Yeh[4], Xingxing Yu[5], and Yujie Zheng[1]

[1] Department of Mathematics, Yunnan University,
Kunming, People's Republic of China
junranlichen@hotmail.com, jianping@ynu.edu.cn, wencheng2018a@163.com,
zyj15837625043@163.com
[2] Institute of Applied Mathematics, Academy of Mathematics and Systems Science,
Beijing, People's Republic of China
[3] Department of Mathematics, National Kaohsiung Normal University,
Kaohsiung, Taiwan
chunchen.yeh@gmail.com
[4] Institute of Mathematics, Academia Sinica, Taipei, Taiwan
mayeh@math.sinica.edu.tw
[5] School of Mathematics, Georgia Institute of Technology, Atlanta, USA
yu@math.gatech.edu

Abstract. In this paper, we consider the 1-line minimum rectilinear Steiner tree (1L-MRST) problem, which is defined as follows. Given n points in the Euclidean plane $\mathbb{R}^2$, we are asked to find the location of a line l and a Steiner tree $T(l)$, which consists of vertical and horizontal line segments plus the line l, to interconnect these n points and at least one point on the line l, the objective is to minimize total weight of $T(l)$, *i.e.*, $\min\{\sum_{uv \in T(l)} w(u,v) \mid T(l)$ is a Steiner tree as mentioned-above$\}$, where weight $w(u,v) = 0$ if two endpoints u, v of an edge $uv \in T(l)$ is located on the line l and weight $w(u,v)$ as the rectilinear distance between u and v otherwise. Given a line l as an input, we refer to this problem as the 1-line-fixed minimum rectilinear Steiner tree (1LF-MRST) problem; In addition, when Steiner points of $T(l)$ are all located on the line l, we refer to this problem problem as the constrained minimum rectilinear Steiner tree (CMRST) problem.

We obtain three main results as follows. (1) We design an exact algorithm in time $O(n \log n)$ to solve the CMRST problem; (2) We show that the same algorithm in (1) is a 1.5-approximation algorithm to solve the 1LF-MRST problem; (3) Using a combination of the algorithm in (1)

This paper is supported by Project of the National Natural Science Foundation of China [Nos. 11861075, 11801498], Project for Innovation Team (Cultivation) of Yunnan Province, Joint Key Project of Yunnan Provincial Science and Technology Department and Yunnan University [No. 2018FY001014] and IRTSTYN. In addition, J.R. Lichen is also supported by Project of Doctorial Fellow Award of Yunnan Province [No. 2018010514].

Z. Zhang et al. (Eds.): AAIM 2020, LNCS 12290, pp. 49–61, 2020.
https://doi.org/10.1007/978-3-030-57602-8_5

for many times and a key lemma proved by using some techniques of computational geometry, we provide a 1.5-approximation algorithm in time $O(n^3 \log n)$ to solve the 1L-MRST problem.

Keywords: 1-line minimum rectilinear Steiner tree · Constrained minimum rectilinear Steiner tree · Approximation algorithms · Complexity

1   Introduction

The minimum spanning tree problem is one of important combinatorial optimization problems, and this problem has many applications in our reality life. There are many polynomial-time exact algorithms to solve this minimization problem, for example, the Kruskal algorithm [11] and the Prim algorithm [13].

The Euclidean minimum spanning tree problem is a special version of the minimum spanning tree problem, and it is defined as follows. Given n points in the Euclidean plane $\mathbb{R}^2$, the objective is to construct a tree to span these n points with the minimum total length, where the length of each edge in such a spanning tree is the Euclidean distance between its two end-points. The exact algorithms as mentioned-above can be applied to solve the Euclidean minimum spanning tree problem, each running time of which is at least $O(n^2)$. However, using the Voronoi diagram, Shamos and Hoey [16] presented an exact algorithm in time $O(n \log n)$ to solve the Euclidean minimum spanning tree problem.

The Euclidean minimum rectilinear spanning tree problem is similarly defined as follows. Given n points in $\mathbb{R}^2$, the objective is to find a tree, which consists only of vertical and horizontal line segments, to span these n points with the minimum total length, where the length of each edge in such a spanning tree is the rectilinear distance (sometimes, referred as the Manhattan distance) between its two end-points. Usually, the exact algorithms as mentioned-above for general graphs can be applied to solve this Euclidean minimum rectilinear spanning tree problem, each running time of which is at least $O(n^2)$. Constructing a Delaunay triangulation and a Voronoi diagram as in [16], Hwang [8] in 1979 presented an exact algorithm in time $O(n \log n)$ to solve the Euclidean minimum rectilinear spanning tree problem. However, Delaunay triangulation is not easily defined in the rectilinear distance. Establishing a framework for minimum spanning tree construction which is based on a general concept of spanning graphs and not necessarily on a Delaunay triangulation, Zhou et al. [19] in 2002 designed a sweep-line algorithm in time $O(n \log n)$ to construct a Euclidean minimum rectilinear spanning tree without using Delaunay triangulation.

The minimum Steiner tree problem is one of the fundamental combinatorial optimization problems, and it has many wide applications in our reality life. Bern and Plassmann [2] showed that the minimum Steiner tree problem is *max-SNP*-hard, even for unit weights. Since the 1990s, the minimum Steiner tree problem and its variations have been studied extensively, and some good approximation algorithms to solve these *NP*-hard problems have been found in [9,18].

The Euclidean minimum Steiner tree (EMST) problem is a new version of the minimum Steiner tree problem, and it is defined as follows. Given n points in $\mathbb{R}^2$, it is asked to construct a tree to interconnect these n points with the minimum total length, where the length of each edge in such a tree is the Euclidean distance between its two end-points. This tree may contain some extra points, called as Steiner points, different from these n points so that the total length of such a tree is minimized. Garey, Graham and Johnson [4] in 1977 showed the EMST problem is NP-hard. There are some good approximation algorithms to solve the EMST problem [9, 17].

Garey and Johnson [5] in 1977 also reconsidered the Euclidean minimum rectilinear Steiner tree (MRST) problem, which is modelled as follows. Given n points in $\mathbb{R}^2$, it is asked to find a tree, which consists only of vertical and horizontal line segments, to interconnect these n points with the shortest possible length, where the length of each edge in such a tree is the rectilinear distance between its two end-points. The MRST problem has received many attentions, and it has potential applications to wire layout for printed circuit boards. Though Garey and Johnson [5] showed that the MRST problem is NP-complete, this problem has many applications in making electric wire connections on a control panel in very large scale integrated layout designs (VLSI layout designs) [14, 15] and communication networks [10, 14]. Several heuristics for the MRST problem have been proposed [9, 14]. Many applications of Steiner trees, especially rectilinear Steiner trees, in industries can be found in the book by Cheng and Du [3].

Finding the addition of a "Steiner line" into Steiner points whose weight is not counted in the resulting network, Holby [6] in 2017 considered a variation of the Euclidean Steiner tree problem, which we refer as the 1-line minimum Steiner tree (1L-MST) problem. Given a set $P = \{r_1, r_2, \ldots, r_n\}$ of n points in $\mathbb{R}^2$, it is asked to find the location of a line l and a Steiner tree $T(l)$ on the set P such that at least one Steiner point is located at the line l, the objective is to minimize total weight of such a Steiner tree $T(l)$, i.e., $\min\{\sum_{uv \in T(l)} w(u, v) \mid T(l)$ is a Steiner tree as mentioned-above$\}$, where weight $w(u, v) = 0$ if two end-points u, v of edge $uv \in T(l)$ are located on the line l and weight $w(u, v)$ as the Euclidean distance between u and v otherwise. Holby [6] only discussed a heuristic algorithm to produce a feasible solution for the 1L-MST problem on larger sets and then presented some related properties.

Motivated by the problems as mentioned-above and some applications in making electric wire connections on a control panel in VLSI layout designs, we address the 1-line minimum rectilinear Steiner tree (1L-MRST) problem, which is defined as follows. Given a set $P = \{r_1, r_2, \ldots, r_n\}$ of n points in $\mathbb{R}^2$, we are asked to find the location of a line l and a Steiner tree $T(l)$, which consists of vertical and horizontal line segments plus the line l, to interconnect these n points in P and at least one point located at the line l, the objective is to minimize total weight of such a Steiner tree $T(l)$, i.e., $\min\{\sum_{uv \in T(l)} w(u, v) \mid T(l)$ is a Steiner tree as mentioned-above$\}$, where weight $w(u, v) = 0$ if two endpoints u, v of edge $uv \in T(l)$ are located on the line l and weight $w(u, v)$ as the rectilinear distance between u and v otherwise. Given a fixed line l as an input in $\mathbb{R}^2$, we refer to this problem as the 1-line-fixed minimum rectilinear Steiner tree (1LF-MRST) problem. In addition, when Steiner points of $T(l)$ are all located on the line l, we refer to this problem as the constrained minimum rectilinear Steiner tree (CMRST) problem.

The 1LF-MRST problem and the 1L-MRST problem have many applications in our reality life, respectively, such as transportation, communication, or VLSI layout designs. In the sequel, we hope to design an exact algorithm to solve the CMRST problem, and then to approximate the 1LF-MRST problem. We finally hope to provide an approximation algorithm to solve the 1L-MRST problem.

Our paper is well organized as follows. In Sect. 2, we present a few notations, terminologies and fundamental lemmas; In Sect. 3, using an exact algorithm to find a constrained minimum rectilinear Steiner tree, we design a 1.5-approximation algorithm to solve the 1LF-MRST problem; In Sect. 4, using the algorithm designed-above for many times, a technique of finding linear facility location and a key Lemma 9 proved by using some techniques of computational geometry, we provide a 1.5-approximation algorithm to solve the 1L-MRST problem; In Sect. 5, we give our conclusion and further research.

2   Terminologies and Fundamental Lemmas

We present some notations, terminologies and fundamental lemmas in order to prove our results in the following sections. And other materials not defined in this paper can be found in the references [1,15,17].

Consider a set S of some points in $\mathbb{R}^2$ where the distance may be either the rectilinear distance (sometimes, referred as the Manhattan distance) or the Euclidean distance. More concretely, if two points p and q are identified by their Cartesian coordinates, *i.e.*, $p = (x_p, y_p)$ and $q = (x_q, y_q)$, we denote by $[p, q]_r$, simply $[p, q]$, the rectilinear segment between two points p and q, and we define by $rd(p, q)$ the rectilinear distance between two points p and q as follows

$$rd(p, q) = |x_p - x_q| + |y_p - y_q| \qquad (1)$$

Given a point $p = (x_p, y_p)$ in S and a fixed line l, whose equation satisfies the following: $y = kx + b$, in $\mathbb{R}^2$, we define the rectilinear distance $rd(p, l)$ between this point p and that fixed line l as follows

$$rd(p, l) = \min\{rd(p, q) \mid q \text{ is a point at this fixed line } l\} \qquad (2)$$

For convenience, we define the Euclidean distance between two points. Given two points $p = (x_p, y_p)$ and $q = (x_q, y_q)$ in $\mathbb{R}^2$, we denote by $\overline{pq}$ the line segment, simply segment, between two points p and q, and we define by $d(p, q)$ the Euclidean distance between two points p and q as follows

$$d(p, q) = \sqrt{(x_p - x_q)^2 + (y_p - y_q)^2} \qquad (3)$$

Given a point $p = (x_p, y_p)$ and a fixed line l in $\mathbb{R}^2$, we denote by $\overline{pl}$ the perpendicular segment from this point p to that fixed line l, and at the same time, we denote by l_p the vertical foot from this point p to that fixed line l, *i.e.*, l_p is the sole point as the intersection of this fixed line l and the vertical line passing through the point p. Similarly, we may define by $d(p, l)$ the Euclidean distance between this point p and that fixed line l as follows

$$d(p, l) = \min\{d(p, q) \mid q \text{ is a point at this fixed line } l\} \tag{4}$$

By the vertical foot l_p and the Euclidean distance $d(p, l)$ as mentioned-above, we immediately obtain the fact $d(p, l) = d(p, l_p)$.

In addition, if this fixed line l satisfies the equation: $y = kx + b$ in $\mathbb{R}^2$, using some techniques of Computational Geometry [1], we obtain the Euclidean distance $d(p, l)$ between this point p and that fixed line l as follows

$$d(p, l) = \frac{|y_p - kx_p - b|}{\sqrt{k^2 + 1}} \tag{5}$$

For convenience, given a point $p = (x_p, y_p)$ and a fixed line l, we denote by l_{pX} (l_{pY}, respectively) the intersection point of that fixed line l and the x-parallel line (y-parallel line, respectively) which passes through the point $p = (x_p, y_p)$, where we also denote $rd(p, l_{pX}) \ (= d(p, l_{pX})) = +\infty$ or $rd(p, l_{pY}) \ (= d(p, l_{pY})) = +\infty$ if this point l_{pX} or l_{pY} does not exist, respectively. We denote this point l_{pX} (l_{pY}, respectively) as the x-parallel point (the y-parallel point, respectively) located at the line l corresponding to the point p, simply, the x-parallel point l_{pX} (the y-parallel point l_{pY}, respectively) if no confusion. In addition, this line segment $\overline{pl_{pX}}$ or $\overline{pl_{pY}}$ is called as an x-parallel segment or an y-parallel segment, respectively.

Using the definition of rectilinear distance $rd(p, l)$, we obtain the following lemmas in turn, and we omit the proofs in details.

Lemma 1. *Given a point $p = (x_p, y_p)$ and a fixed line l, whose equation satisfies the form: $y = kx + b$, in $\mathbb{R}^2$, we can obtain the following*

$$rd(p, l) = \min\left\{rd(p, l_{pX}), rd(p, l_{pY})\right\} \tag{6}$$
$$= \min\left\{d(p, l_{pX}), d(p, l_{pY})\right\}$$

In addition, (i) only when $|k| = 1$, this minimum value $rd(p, l)$ is attained at any point on the line segment $\overline{l_{pX}l_{pY}}$, (ii) only when $|k| > 1$, this minimum value $rd(p, l)$ is attained at the point l_{pX}, and (iii) only when $|k| < 1$, this minimum value $rd(p, l)$ is attained at the point l_{pY}, where l_{pX} and l_{pY} are defined as mentioned-above.

Lemma 2. *Given a point $p = (x_p, y_p)$ and a fixed line l, whose equation satisfies the form: $y = kx + b$, in $\mathbb{R}^2$, we obtain the following*

$$rd(p, l) = \min\left\{\frac{d(p, l_p)}{|\cos\alpha|}, \frac{d(p, l_p)}{|\sin\alpha|}\right\} \tag{7}$$
$$= d(p, l_p) \cdot \min\left\{\frac{1}{|\cos\alpha|}, \frac{1}{|\sin\alpha|}\right\}$$

where l_p is the vertical foot from this point p to that fixed line l and α $(0 \leq \alpha \leq \pi)$ is the inclined angle between that fixed line l and the x-axis, satisfying $\tan \alpha = k$.

Lemma 3. [11] *The minimum spanning tree problem can be optimally solved by the Kruskal algorithm in time $O(m \log n)$, where $n = |V(G)|$ and $m = |E(G)|$.*

Lemma 4. [8,19] *A minimum rectilinear spanning tree on n points in $\mathbb{R}^2$ can be constructed by a polynomial-time exact algorithm, denoted by the MRspanT algorithm, in time $O(n \log n)$.*

We denote the rectilinear Steiner ratio as the minimum upper-bound for the ratio between total lengths of a minimum rectilinear spanning tree and a minimum rectilinear Steiner tree for the same set P of n points in $\mathbb{R}^2$.

Lemma 5. [7] *Given a set P of n points in $\mathbb{R}^2$, denote by $L_M(P)$ the total lengths of a minimum rectilinear spanning tree and by $L_S(P)$ the total lengths of a minimum rectilinear Steiner tree on the same set P, respectively. Then we have the fact $L_M(P) \leq 3/2 \cdot L_S(P)$.*

Given a set $P = \{r_1, r_2, \ldots, r_n\}$ of n points and a fixed line l, having its equation as the form $y = kx + b$ (where $|k| \geq 1$), in $\mathbb{R}^2$ for the CMRST problem, for each bipartition $\{P_1, P_2\}$ of P, we denote $\delta_l^*(P_1)$ to be the set $E_1 \cup E_2$, where $E_1 = \{[r_i, r_j] \mid r_i \in P_1 \text{ and } r_j \in P_2 \text{ are both located at same side of the line } l\}$ and $E_2 = \{[r_i, l_{r_i X}] \mid r_i \in P_1, \text{ and } l_{r_i X} \text{ is the } x\text{-parallel point located at the line } l \text{ corresponding to } r_i\}$. In this case, we have properties $[r_i, l_{r_i X}] = \overline{r_i l_{r_i X}}$ for each point $r_i \in P_1$. Similarly, when linear equation is as the form $y = kx + b$ (where $|k| < 1$), we still denote $\delta_l^*(P_1)$ to be the set $E_1 \cup E_2$, where E_1 and E_2 are defined as before, except substituting $l_{r_i Y}$ for $l_{r_i X}$ in E_2 for each $r_i \in P_1$, then we have properties $[r_i, l_{r_i Y}] = \overline{r_i l_{r_i Y}}$ for each point $r_i \in P_1$. Finally, we can obtain the following

Lemma 6. *Given a set $P = \{r_1, r_2, \ldots, r_n\}$ of n points and a fixed line l, having its equation in the form $y = kx + b$ (where $|k| \geq 1$), in $\mathbb{R}^2$ for the CMRST problem, then $T(l)$ is a constrained minimum rectilinear Steiner tree if and only if, for every edge $uv \in E(T(l))$ which satisfies $w(u,v) \neq 0$, i.e., at least one of two end-points u and v of this edge uv is not at the line l, there exists a subset P_1 $(\subseteq P)$ consisting of all points in a connected component of $T - \{uv\}$ such that $[u, v]$ is a shortest rectilinear segment of $\delta_l^*(P_1)$.*

3   The 1LF-MRST Problem

In this section, we consider the 1-line-fixed minimum rectilinear Steiner tree (1LF-MRST) problem. We hope to design a polynomial-time exact algorithm to optimally solve the constrained minimum rectilinear Steiner tree (CMRST) problem. We may reminder that the CMRST problem is a special version of the 1LF-MRST problem, where Steiner points are all located on the fixed line l.

The following two results play important roles to find an optimal solution for solving the CMRST problem and the 1LF-MRST problem, respectively, whose proofs are simple and clear, and we can omitted the proofs in details.

Lemma 7. *Given a set P of n points and a fixed line l, whose equation satisfies the form: $y = kx + b$, in $\mathbb{R}^2$, then we obtain the following*

(1) *If $r_i s_i$ is a rectilinear segment in a constrained minimum rectilinear Steiner tree T for the CMRST problem, where r_i is a point in P and s_i is a point on the line l, then this point s_i may be chosen as either the x-parallel point $l_{r_i X}$ (for the case $|k| \geq 1$) or the y-parallel point $l_{r_i Y}$ (for the case $|k| < 1$) on the line l.*

(2) *If $r_i s_i$ is a rectilinear segment in a 1-line-fixed minimum rectilinear Steiner tree T for the 1LF-MRST problem, where r_i is either a point in P or a Steiner point of T, which is not on the line l, and s_i is a point on the line l, then this point s_i may be chosen as either the x-parallel point $l_{r_i X}$ (for the case $|k| \geq 1$) or the y-parallel point $l_{r_i Y}$ (for the case $|k| < 1$) on the line l.*

For the case where the line l has its equation in the form $y = kx + b$ to satisfy $|k| \geq 1$, using Lemma 7, we can design an algorithm to solve the CMRST problem in the following strategies. (1) Use the MRspanT algorithm [8,19] to produce a minimum rectilinear spanning tree $T = (P, E_T)$ on the set $P = \{r_1, r_2, \ldots, r_n\}$; (2) Construct a weighted graph $G = (P \cup \{r_0\}, E, w)$, where r_0 is a new vertex to represent the fixed line l and $E = \{r_i r_0 \mid r_i \in P\} \cup \{r_i r_j \mid [r_i, r_j] \in E_T$ is vertex-disjoint from that line $l\}$, and we denote $w(r_i, r_j) = rd(r_i, r_j)$ for each $[r_i, r_j] \in E_T$ and $w(r_i, r_0) = d(r_i, l_{r_i X})$ for each $r_i \in P$ using Lemmas 1 and 7, where the point $l_{r_i X}$ is the x-parallel point located at the line l; (3) Use the Kruskal algorithm [11] to find a minimum spanning tree T_G in G equipped with a weighted function $w(\cdot)$, then construct a constrained rectilinear Steiner tree from this minimum spanning tree T_G of G.

For the case where the line l has its equation in the form $y = kx + b$ to satisfy $|k| < 1$, we only denote $w(r_i, r_0) = d(r_i, l_{r_i Y})$ for each $r_i \in P$ in the strategy (2) using Lemmas 1 and 7, where the point $l_{r_i Y}$ is the y-parallel point located at the line l, and other steps need not be changed. For convenience, we may mention the same weighted graph $G = (P \cup \{r_0\}, E, w)$ in both cases.

Our algorithm $\mathcal{A}_{CMRST}$ to solve the CMRST problem is described as follows.

Algorithm: $\mathcal{A}_{CMRST}$

Input:	A fixed line l with its equation in form: $y = kx + b$, and a set $P = \{r_1, r_2, \dots, r_n\}$ of n points in $\mathbb{R}^2$.

Output: a constrained minimum rectilinear Steiner tree $T(l)$.

Begin

Step 1 Denote $T(l) = (V^*, E^*)$, where $V^* = P$ and $E^* = \emptyset$.

Step 2 Use the MRspanT algorithm [8,19] to find a minimum rectilinear spanning tree $T = (P, E_T)$ on the set P.

Step 3 Construct a weighted graph $G = (P \cup \{r_0\}, E, w)$ as mentioned-above.

Step 4 Use the Kruskal algorithm [11] to find a minimum spanning tree T_G of G.

Step 5 For each edge $e \in E(T_G)$ do
  If $(e = r_i r_j \in E(T_G)$, where $i \geq 1$ and $j \geq 1)$ then
    we denote $E^* = E^* \cup \{[r_i, r_j]\}$
  If $(e = r_i r_0 \in E(T_G)$, where $i \geq 1)$ then
    we denote $E^* = E^* \cup \{[r_i, s_i]\}$ and $V^* = V^* \cup \{s_i\}$, where s_i is either $l_{r_i X}$ (if $|k| \geq 1$) or $l_{r_i Y}$ (if $|k| < 1$) on the fixed line l.

Step 6 Output $T(l) = (V^*, E^* \cup E^{**})$, where $E^{**} = \{\overline{s_i s_j} \mid s_i$ and s_j in V^* are successive vertices on the line $l\}$.

End

Using the algorithm $\mathcal{A}_{CMRST}$, we obtain the following

Theorem 1. *The algorithm $\mathcal{A}_{CMRST}$ optimally solves the CMRST problem in time $O(n \log n)$, where an instance of the CMRST problem consists of a fixed line l and a set $P = \{r_1, r_2, \dots, r_n\}$ of n points in $\mathbb{R}^2$.*

Proof. Given a set P of n points and a fixed line l to satisfy its equation: $y = kx + b$, in $\mathbb{R}^2$, we may assume, without loss of generality, that $|k| \geq 1$ holds, implying that, for each $r_i \in P$, we construct an edge $r_i r_0$ in such a weighted graph $G = (P \cup \{r_0\}, E, w)$ to represent the x-parallel segment $r_i l_{r_i X}$, having $w(r_i r_0) = rd(r_i, l) = d(r_i, l_{r_i X})$ using Lemma 7, where $l_{r_i X}$ is the x-parallel point located on the line l of the point r_i in the set P.

By Lemma 6, we obtain the fact that $T(l)$ is an optimal solution if and only if, for every edge $uv \in E(T(l))$ which satisfies $w(uv) \neq 0$, *i.e.*, at least one of two end-points u, v of this edge uv is not at the line l, $[u, v]$ is a shortest rectilinear segment of $\delta_l^*(P_1)$, where P_1 is the vertex set of a connected component of $T - uv$ which contains no point on the line l.

Suppose, to the contrary, that we may assume that there exists one edge $e = [u, v] \in E(T(l))$, which satisfies $w(u, v) \neq 0$, such that $[u, v]$ is not the shortest rectilinear segment in $\delta_l^*(P_1)$. For the choice of P_1 and the fact that there exists some rectilinear segment in $\delta_l^*(P_1)$ to connect two components of $T(l) - \{[u, v]\}$, we choose $[p, q]$ to be a shortest rectilinear segment in $\delta_l^*(P_1)$, which satisfies $w(pq) \neq 0$, then we obtain $rd(p, q) < rd(u, v)$.

If the rectilinear segment $[p, q] = [r_i, r_j]$ for two points $r_i, r_j \in P$, then $[r_i, r_j]$ is a shortest rectilinear segment in $\delta_l^*(P_1)$. Since T is a minimum rectilinear spanning tree on P by Step 2 in the $\mathcal{A}_{CMRST}$ algorithm (in fact, the MRspanT algorithm), using Lemma 6, we can assume that $[r_i, r_j]$ belongs to E_T, implying $[r_i, r_j] \in E$. Then we have $w(r_i, r_j) < w(u, v)$, and we obtain the fact that $T' = (T - [u, v]) + [r_i, r_j]$ is another rectilinear spanning tree on P with $w(T') = w(T) + w(r_i, r_j) - w(u, v) < w(T)$, contradicting the fact that T is a minimum rectilinear spanning tree on P.

If the rectilinear segment $[p, q] = [r_i, l_{r_i X}]$, where $r_i \in P$ and $l_{r_i X}$ is the x-parallel point at the line l of r_i in P, we obtain $r_i r_0 \in E$. Using Lemma 6 and the similar arguments as mentioned-above, we can reach a contradiction that T_G is a minimum spanning tree in the weighted graph G.

The complexity of the Algorithm $\mathcal{A}_{CMRST}$ can be determined as follows. (1) By Lemma 4, a minimum rectilinear spanning tree $T = (P, E_T)$ on these n points in $\mathbb{R}^2$ can be determined in time $O(n \log n)$ [8,19]. (2) Step 3 can be implemented in $O(n)$ time. (3) Since this spanning tree $T = (P, E_T)$ has $n - 1$ edges, the graph G has exactly $n + 1$ vertices and at most $2n - 1$ edges, then the Kruskal algorithm can determine a minimum spanning tree of the weighted graph G in time $O(n \log n)$ in this case. (4) Since there are exactly n edges in $E(T_G)$, Step 5 can be implemented to run in $O(n)$. Thus, the whole algorithm $\mathcal{A}_{CMRST}$ can be implemented to run in $O(n \log n)$.

This completes the proof of this theorem.   ∎

Using Lemma 5 for many times, we obtain the following

Theorem 2. *The algorithm $\mathcal{A}_{CMRST}$ is a 1.5-approximation algorithm in time $O(n \log n)$ to solve the 1LF-MRST problem.*

Proof. Given a fixed line l and a set P of n points for the 1LF-MRST problem, we may suppose that an optimal 1-line-fixed minimum rectilinear Steiner tree $T_S^*(l)$ has exactly q Steiner points located successively at that fixed line l, saying $s_1, s_2, \ldots, s_q$. (At the same time, we may permit that there are some other Steiner points (if any) out of the fixed line l).

We should consider all maximal rectilinear Steiner subtrees in the reduced graph $T_S^*(l) - E(l)$, i.e., the reduced graph of $T_S^*(l)$ removing all edges at the fixed line l. Given each $i = 1, 2, \ldots, q$, we may assume that $T_{s_i}^*(l)$ is a maximal rectilinear Steiner subtree in $T_S^*(l) - E(l)$ to contain a sole point s_i located on the line l. Now, we can partition the vertex set of each maximal rectilinear Steiner subtree $T_{s_i}^*(l)$ into three parts: (1) a subset $P_{s_i} = V(T_{s_i}^*(l)) \cap (P - \{s_i\})$, (2) a subset $\{s_i\}$, and (3) a subset $S_{s_i} = V(T_{s_i}^*(l)) - (P_{s_i} \cup \{s_i\})$.

We can denote by $T_{s_i}^{**}(l)$ a minimum rectilinear spanning tree on the set $P_{s_i} \cup \{s_i\}$ for each $i = 1, 2, \ldots, q$. Using Lemma 5 on the set $P_{s_i} \cup \{s_i\}$, we can obtain $L_M(P_{s_i} \cup \{s_i\}) \leq 1.5 \cdot L_S(P_{s_i} \cup \{s_i\})$, i.e., $w(T_{s_i}^{**}(l)) \leq 1.5 \cdot w(T_{s_i}^*(l))$. Using the definition of weight $w(s_{i-1} s_i) = 0$ for each $i = 2, 3, \ldots, q$, we can have $w(T_S^*(l)) = \sum_{i=1}^{q} L_S(P_{s_i} \cup \{s_i\}) = \sum_{i=1}^{q} w(T_{s_i}^*(l))$.

At the same time, we can construct a 1-line-fixed rectilinear spanning tree $T = \cup_{i=1}^{q} T_{s_i}^{**}(l) \cup \{s_{i-1} s_i \mid i = 2, 3, \ldots, q\}$ on the set $P \cup \{s_1, s_2, \ldots, s_q\}$. It is easy to see that this spanning tree T is actually a constrained rectilinear Steiner tree to contain all points in P and q Steiner points $s_1, s_2, \ldots, s_q$, where these

q Steiner points are all successively located on that fixed line l and there are no other Steiner points out of that fixed line l. By Theorem 1, the algorithm $\mathcal{A}_{CMRST}$ produces a constrained minimum rectilinear Steiner tree $T(l)$, then we obtain $w(T(l)) \leq w(T)$.

Finally, we obtain the following

$$
\begin{aligned}
w(T(l)) &\leq w(T) \\
&= w(\cup_{i=1}^{q} T_{s_i}^{**}(l)) + \sum_{i=2}^{q} w(s_{i-1}s_i) \\
&= \sum_{i=1}^{q} w(T_{s_i}^{**}(l)) + \sum_{i=2}^{q} w(s_{i-1}s_i) \\
&\leq 1.5 \cdot \sum_{i=1}^{q} w(T_{s_i}^{*}(l)) + 1.5 \cdot \sum_{i=2}^{q} w(s_{i-1}s_i) \\
&= 1.5 \cdot w(T_S^{*}(l))
\end{aligned}
\tag{8}
$$

All arguments mentioned-above show that the algorithm $\mathcal{A}_{CMRST}$ is a 1.5-approximation algorithm to solve the 1LF-MRST problem.

The proof Theorem 1 indeed shows that the algorithm $\mathcal{A}_{CMRST}$ runs in time $O(n \log n)$.

Thus, we can obtain the conclusion of this theorem.   ∎

4   The 1L-MRST Problem

In this section, we consider the 1-line minimum rectilinear Steiner tree (1L-MRST) problem. We introduce an optimization problem and some fundamental lemmas, and using the algorithm $\mathcal{A}_{CMRST}$ for many times, we hope to design an approximation algorithm to solve the 1L-MRST problem.

Definition 1. *[12] Given m "demand" points $Q = \{q_1, q_2, \ldots, q_m\}$ in $\mathbb{R}^2$ with coordinates $q_i = (x_i, y_i)$ and weights c_i, the linear facility can be described by a line whose equation is $y = k_q x + b_q$, where (k_q, b_q) is an optimal solution to solve the following optimization problem*

$$
\min_{q \in Q} \ f_q(k, b) = \sum_{i=1}^{m} c_i \cdot \frac{|y_i - kx_i - b|}{\sqrt{k^2 + 1}}
\tag{9}
$$

i.e., Formula 9 is equivalent to find a line in $\mathbb{R}^2$, whose equation is $y = k_q x + b_q$, such that the sum of Euclidean distances from these m points $q_1, q_2, \ldots, q_m$ to this line is minimized.

We find the following lemma due to Morris and Norback [12] to solve the optimization problem in Formula 9.

Lemma 8. [12] *Given m "demand" points $Q = \{q_1, q_2, \ldots, q_m\}$ in $\mathbb{R}^2$ with coordinates $q_i = (x_i, y_i)$ and weights c_i, an optimal solution to the optimization problem as mentioned-above (seeing Formula 9) exists which is a line satisfying equation $y = k_q x + b_q$ to pass through at least two "demand" points in Q.*

We have the following result similar to Lemma 8, which plays an important role to find an optimal solution for the 1L-MRST problem.

Lemma 9. *Given a set P of n points in $\mathbb{R}^2$ as an instance of the 1L-MRST problem, there exists an optimum solution in which we can choose a line l to pass through at least two points in the set P.*

Using Lemma 9, we design an algorithm, referred as the algorithm $\mathcal{A}_{1L-MRST}$, to solve the 1L-MRST problem.

Algorithm: $\mathcal{A}_{1L-MRST}$

Input: A set $P = \{r_1, r_2, \ldots, r_n\}$ of n points in $\mathbb{R}^2$.
Output: a line l and a Steiner tree T.

Begin
Step 1 For $i = 1$ to n do:
        For $j = 1$ to n $(j \neq i)$ do:
           Choose a line l_{ij} to pass through the two points r_i and r_j;
           Use the algorithm $\mathcal{A}_{CMRST}$ on the set $P - \{r_i, r_j\}$ to construct a constrained minimum rectilinear Steiner tree $T(l_{ij})$.
Step 2 Choose two points r_{i_0} and r_{j_0} to satisfy
    $w(T(l_{i_0 j_0})) = \min\{w(T(l_{ij})) | 1 \leq i \leq n,\ 1 \leq j \leq n \text{ and } j \neq i\}$.
Step 3 Output "the line $l_{i_0 j_0}$ and the 1-line rectilinear Steiner tree $T(l_{i_0 j_0})$".
End

Using the algorithm $\mathcal{A}_{1L-MRST}$, we obtain the following

Theorem 3. *The algorithm $\mathcal{A}_{1L-MRST}$ is a 1.5-approximation algorithm for the 1L-MRST problem, and its time complexity is $O(n^3 \log n)$, where n is the number of points in $\mathbb{R}^2$.*

Proof. We may suppose, for a set $P = \{r_1, r_2, \ldots, r_n\}$ of n points in $\mathbb{R}^2$, that $T(l^*)$ is an optimal minimum rectilinear Steiner tree in $\mathbb{R}^2$ and $T(l_{i_0 j_0})$ is a 1-line rectilinear Steiner tree produced by the algorithm $\mathcal{A}_{1L-MRST}$.

By Lemma 9, we may choose an optimal 1-line minimum rectilinear Steiner tree $T(l^*)$, where this optimal solution contains a line l^* to pass through at least two points in P, denoted by r_i and r_j.

We denote by $T_{r_i r_j}$ the constrained minimum rectilinear Steiner tree for the line l^*. For the fixed line l^*, using Theorem 2, we obtain the fact $w(T_{r_i r_j}) \leq 1.5 \cdot w(T(l^*))$. Since the algorithm $\mathcal{A}_{1L-MRST}$ enumerates all possibilities of lines to pass any two points, and the algorithm $\mathcal{A}_{1L-MRST}$ produces a 1-line rectilinear Steiner tree $T(l_{i_0 j_0})$ with minimum weight, then we obtain the following

$$w(T(l_{i_0 j_0})) \leq w(T_{r_i r_j}) \leq 1.5 \cdot w(T(l^*)) \tag{10}$$

implying that the algorithm $\mathcal{A}_{1L-MRST}$ is a 1.5-approximation algorithm for the 1L-MRST problem.

Since Theorem 1 shows that the algorithm $\mathcal{A}_{CMRST}$ runs in time $O(n \log n)$, then we obtain that Step 1 can be implemented in time $O(n^3 \log n)$, and other steps run in time at most $O(n^2)$. Thus, the whole algorithm $\mathcal{A}_{1L-MRST}$ can be implemented in $O(n^3 \log n)$ time.

Thus, we can obtain the conclusion of this theorem.   ∎

5   Conclusion and Further Research

In this paper, we consider the 1-line minimum rectilinear Steiner tree problem, and we obtain the following three main results.

(1) We design a polynomial-time exact algorithm in time $O(n \log n)$ to solve the constrained minimum rectilinear Steiner tree (CMRST) problem;
(2) We show that the algorithm designed in (1) is a 1.5-approximation algorithm to solve the 1-line-fixed minimum rectilinear Steiner tree (1LF-MRST) problem;
(3) Using the algorithm in (1) for many times and Lemma 9, we provide a 1.5-approximation algorithm in time $O(n^3 \log n)$ to solve the 1-line minimum rectilinear Steiner tree (1L-MRST) problem.

A challenging task for further research is to design some approximation algorithms to solve the 1L-MRST problem either with smaller performance ratios or in lower time complexity.

References

1. Berg, M.d., Cheong, O., Kreveld, Mv., Overmars, M.: Computational Geometry: Algorithms and Applications. Springer, New York (2008)
2. Bern, M., Plassmann, P.: The Steiner problem with edge lengths 1 and 2. Inf. Process. Lett. **32**(4), 171–176 (1989)
3. Cheng, X.Z., Du, D.Z.: Steiner Trees in Industry. Combinatorial Optimization 11. Kluwer Academic Publishers, Dordrecht, The Netherlands (2001)
4. Garey, M.R., Graham, R.L., Johnson, D.S.: The complexity of computing Steiner minimal trees. SIAM J. Appl. Math. **32**(4), 835–859 (1977)
5. Garey, M.R., Johnson, D.S.: The rectilinear Steiner tree problem is NP-complete. SIAM J. Appl. Math. **32**(4), 826–834 (1977)
6. Holby, J.: Variations on the Euclidean Steiner tree problem and algorithms. Rose-Hulman Undergrad. Math. J. **18**(1), 123–155 (2017)
7. Hwang, F.K.: On Steiner minimal trees with rectilinear distance. SIAM J. Appl. Math. **30**(1), 104–114 (1976)
8. Hwang, F.K.: An $O(n \log n)$ algorithm for rectilinear minimal spanning trees. J. ACM **26**(2), 177–182 (1979)
9. Hwang, F.K., Richards, D.S.: Steiner tree problems. Networks **22**(1), 55–89 (1992)

10. Imase, M., Waxman, B.M.: Dynamic Steiner tree problem, Technical Report No. WUCS-89-11, Department of Computer Science, Washington University, St. Louis, MO (1989)
11. Kruskal, J.B.: On the shortest spanning subtree of a graph and the traveling salesman problem. Proc. Am. Math. Soc. **7**(1), 48–50 (1956)
12. Morris, J.G., Norback, J.P: A simple approach to linear facility location. Transp. Sci. **14**(1), 1–8 (1980)
13. Prim, R.C.: Shortest connection networks and some generalizations. Bell Syst. Tech. J. **36**, 1389–1401 (1957)
14. Riehards, D.: Fast heuristic algorithms for rectilinear Steiner trees. Algorithmica **4**, 191–207 (1989)
15. Schrijver, A.: Combinatorial Optimization: Polyhedra and Efficiency. Springer, Berlin (2003)
16. Shamos, M.I., Hoey, D.: Closest-point problems. In: 16th Annual Symposium on Foundations of Computer Science, pp. 151–162. IEEE Computer Society (1975)
17. Vazirani, V.V.: Approximation Algorithms. Springer, Heidelberg (2003). https://doi.org/10.1007/978-3-662-04565-7_30
18. Williamson, D.P., Shmoys, D.B.: The Design of Approximation Algorithms. Cambridge University Press, Cambridge (2011)
19. Zhou, H., Shenoy, N., Nicholls, W.: Efficient minimum spanning tree construction without Delaunay triangulation. Inf. Process. Lett. **81**, 271–276 (2002)

Minimum Diameter Vertex-Weighted Steiner Tree

Wei Ding[1][(✉)] [iD] and Ke Qiu[2]

[1] Zhejiang University of Water Resources and Electric Power,
Hangzhou 310018, Zhejiang, China
dingweicumt@163.com
[2] Department of Computer Science, Brock University,
St. Catharines, Canada
kqiu@brocku.ca

Abstract. Let $G = (V, E, w, \rho, \mathcal{T})$ be a weighted connected graph, where V is the vertex set, E is the edge set, $\mathcal{T} \subseteq V$ is a *terminal* subset, $w : E \to \mathbb{R}^+$ is an edge-weight function and $\rho : V \to \mathbb{R}^+$ is a vertex-weight function. The *weighted diameter* of a Steiner tree T in G spanning $\mathcal{T}$ is referred to as the longest weighted tree distance on T between terminals. The objective of the **Minimum Diameter Vertex-Weighted Steiner Tree Problem (MDWSTP)** is to construct a Steiner tree in G spanning $\mathcal{T}$ to minimize the weighted diameter.

In this paper, we study the MDWSTP in two classes of *parameterized graphs*, $\langle \mathcal{T}, \mu \rangle$-PG and $(\mathcal{T}, \lambda)$-PG, which are introduced from the perspective of the parameterized upper bound on the ratio of two vertex-weights, and a weaker version of the parameterized triangle inequality, respectively, and achieve simple approximation algorithms. For the MDWSTP in edge-weighted $\langle \mathcal{T}, \mu \rangle$-PG, we obtain a $\frac{\mu+1}{2}$-factor approximation algorithm where $\frac{\mu+1}{2}$ is tight. For the MDWSTP in vertex-weighted $(\mathcal{T}, \lambda)$-PG, we first obtain a λ-factor approximation algorithm where λ is tight, and then develop a slightly improved approximation algorithm.

Keywords: Steiner tree · Diameter · Vertex-weighted · Approximation

1 Introduction

Let $G = (V, E, w, \mathcal{T})$ be a weighted connected graph, where V is the set of n vertices, E is the set of m edges, $\mathcal{T} \subseteq V$ is a subset of p *terminals*, and $w : E \to \mathbb{R}^+$ is an edge-weight function. A *Steiner tree* is an acyclic connected subgraph of G spanning $\mathcal{T}$. The **Steiner Minimum Tree Problem (SMTP)** in G seeks a minimum cost connected subgraph of G spanning $\mathcal{T}$. It is one of the well-known combinatorial optimization problems and has many applications in a variety of fields [7,12], such as communication networks and computational biology. It has been proved to be NP-hard in the strong sense [8].

© Springer Nature Switzerland AG 2020
Z. Zhang et al. (Eds.): AAIM 2020, LNCS 12290, pp. 62–72, 2020.
https://doi.org/10.1007/978-3-030-57602-8_6

1.1   Related Works

Let T be a tree and $\mathcal{L}(T)$ denote the subset of leaves of T. For any pair of vertices of T, v and u, the unique v-u path on T is denoted by $\pi_T(v, u)$, and the length of $\pi_T(v, u)$ is defined as the v-u *tree distance* on T and denoted by $d_T(v, u)$. The longest tree distance on T between vertices must be one between leaves. It is called the *diameter* of T and denoted by diam(T), i.e.,

$$\text{diam}(T) = \max_{v,u \in \mathcal{L}(T)} d_T(v, u). \tag{1}$$

The problems involving the diameter of tree have been widely studied in the past decades, including the **Minimum Diameter Spanning Tree Problem (MDTP)** and the **Minimum Diameter Steiner Tree Problem (MDSTP)**. Given a set of n points in the Euclidean plane, Ho *et al.* [11] first studied the geometrical MDSTP and MDTP. They proved that the geometrical MDSTP is reducible to the *minimum enclosing circle problem* and admits a linear-time algorithm. For the geometrical MDTP (equal to MDTP in the *Euclidean graph* induced by the n points), they proved that a minimum diameter spanning tree is *monopolar* or *dipolar* and shown a $\Theta(n^3)$-time algorithm. In general, this result applies to MDTP in complete graphs with a *metric* edge-weight function. Chan [2] proposed an $o(n^3)$-time algorithm which is the first sub-cubic time algorithm for the problem. Gudmundsson *et al.* [9] presented a 1.2-approximation algorithm with a time complexity of $O(n^2 \log n)$ and a $(1 + \epsilon)$-approximation algorithm with a time complexity of $O(\epsilon^{-5} + n)$, for any $\epsilon > 0$, and later Spriggs *et al.* [16] presented a $(1 + \epsilon)$-approximation algorithm with a lower time complexity of $O(\epsilon^{-3} + n)$. Furthermore, Ihler *et al.* [13] considered the *minimum diameter spanning tree problem with classes* and obtained an $O(n^3)$-time algorithm by generalizing the results of Ho *et al.* [11].

In weighted connected graphs, MDSTP is equal to MDTP when $\mathcal{T} = V$. For MDTP in $G = (V, E, w)$ where the edge-weight function $w(\cdot)$ is not necessary metric, Hassin and Tamir [10] proved that MDTP is equivalent to the *absolute 1-center problem*, and gave an $O(mn + n^2 \log n)$-time exact algorithm. Later, Bui *et al.* [1] developed a distributed algorithm with a time complexity of $O(n)$ and a message complexity of $O(mn)$. In the last few years, we have worked on MDSTP and some related problems in graphs. In [5], we studied MDSTP in $G = (V, E, w, \mathcal{T})$, which asks for a Steiner tree in G spanning $\mathcal{T}$ to minimize the diameter, and gave an $O(mp + np \log p)$-time exact algorithm. Also, we considered the restricted version of MDSTP, called the **Minimum Diameter Terminal Steiner Tree Problem (MDTSTP)**, where each terminal appears as a leaf of tree [5]. We designed an $O(p(n-p)^2)$-time exact algorithm for the *metric* version of MDTSTP, as well as an $O((n - p) \log p)$-time 2-approximation algorithm and an $O(np(n - p))$-time exact algorithm for the *nonmetric* version of MDTSTP. In [6], we proposed the **Minimum Diameter k-Steiner Forest Problem (MDkSFP)**, which seeks a k-Steiner forest, i.e., a collection of k disjoint Steiner trees, such that the maximum diameter of the k Steiner trees is minimized. We established the relationship between MDkSFP and the *absolute k-Steiner center* *problem*, and first achieved a 2-approximation algorithm and further developed a slightly improved approximation algorithm by perturbing facilities and re-clustering terminals. Besides, some *bicriteria* Steiner tree problems involving the cost and diameter of tree in double-weighted connected graphs have been studied [3,4,15].

1.2 Our Results

Let $G = (V, E, w, \rho, \mathcal{T})$ be a weighted connected graph, where V is the vertex set, E is the edge set, $\mathcal{T} \subseteq V$ is a *terminal* subset, $w : E \to \mathbb{R}^+$ is an edge-weight function and $\rho : V \to \mathbb{R}^+$ is a vertex-weight function, and let T be a Steiner tree in G spanning $\mathcal{T}$. The longest weighted tree distance on T between terminals is called the *weighted diameter* of T. This paper deals with the **Minimum Diameter Vertex-Weighted Steiner Tree Problem (MDWSTP)**, the goal of which is to construct a Steiner tree in G spanning $\mathcal{T}$ to minimize the weighted diameter.

In this paper, we propose two classes of *parameterized graphs (PG)*, $\langle \mathcal{T}, \mu \rangle$-PG and $(\mathcal{T}, \lambda)$-PG, from the angle of the parameterized upper bound on the ratio of two vertex-weights, and a weaker version of the parameterized triangle inequality, respectively. This paper focuses on the MDWSTP in these parameterized graphs and achieves approximation algorithms. For the MDWSTP in edge-weighted $\langle \mathcal{T}, \mu \rangle$-PG, we obtain a $\frac{\mu+1}{2}$-factor approximation algorithm where $\frac{\mu+1}{2}$ is tight. For the MDWSTP in vertex-weighted $(\mathcal{T}, \lambda)$-PG, we first obtain a λ-factor simple approximation algorithm where λ is tight, and then develop a slightly improved approximation algorithm.

Organization. The rest of this paper is organized as follows. In Sect. 2, we define some notations used frequently and MDWSTP formally. In Sect. 3, we introduce two classes of parameterized graphs (PG), $\langle \mathcal{T}, \mu \rangle$-PG and $(\mathcal{T}, \lambda)$-PG. In Sect. 4, for the MDWSTP in edge-weighted $\langle \mathcal{T}, \mu \rangle$-PG, we achieve a $\frac{\mu+1}{2}$-factor approximation algorithm. In Sect. 5, for the MDWSTP in vertex-weighted $(\mathcal{T}, \lambda)$-PG, we achieve a λ-factor approximation algorithm where λ is tight as well as a slightly improved approximation algorithm. In Sect. 6, we present some concluding remarks.

2   Preliminaries

2.1   Notations

Let $G = (V, E, w, \rho, \mathcal{T})$ be a weighted undirected connected graph, where V is the vertex set, E is the edge set, $\mathcal{T} \subseteq V$ is a terminal subset, $w : E \to \mathbb{R}^+$ is an edge-weight function and $\rho : V \to \mathbb{R}^+$ is a vertex-weight function. All the vertices in V are numbered in sequence by $1, 2, \ldots, |V|$. For any vertex subset, $U \subseteq V$, we let $I(U)$ denote the *index set* of U. Clearly, $I(V) = \{1, 2, \ldots, |V|\}$. For each $i \in I(V)$, we let v_i denote the vertex with index i. Specifically, we let t_i denote the terminal with index i, for each $i \in I(\mathcal{T})$.

For any index pair, $i, j \in \{1, 2, \ldots, |V|\}$, we let $\pi^*(v_i, v_j)$ denote a *shortest path* in G between v_i and v_j, abbreviated as a v_i-v_j shortest path. The length of $\pi^*(v_i, v_j)$ is called the *shortest path distance (SPD)* in G between v_i and v_j, abbreviated as the v_i-v_j SPD and denoted by $d(v_i, v_j)$. Let $p(e)$ be the set of all the *continuum* points on e, for any edge $e \in E$, and $p(G)$ be the set of all the continuum points on the edges of G. Likewise, for any point pair, $x, y \in p(G)$, we also use $\pi^*(x, y)$ and $d(x, y)$ to denote the x-y shortest path and SPD in G, respectively. All the shortest paths from a point x to terminals form a tree, called the *shortest path tree (SPT)* and denoted by T_x, which roots at x and spans all the terminals in T. So, T_x must be a Steiner tree. Specifically, we let T_k denote the SPT rooted at $v_k, k \in I(V)$.

For any point $x \in p(G)$, the maximum distance from x to T, $r(x, T)$, is called the x-to-T *radius*, and a point of G, x^*, minimizing the x-to-T radius is called an *absolute 1-Steiner center (A1SC)* of G. We have

$$r(x, T) = \max_{i \in I(T)} d(x, t_i), \tag{2}$$

and

$$r(x^*, T) = \min_{x \in p(G)} r(x, T). \tag{3}$$

Moreover, we define the x-to-v_i *weighted distance* as $\rho(v_i) \cdot d(x, v_i)$, for any $x \in p(G)$ and $i \in I(V)$. The maximum weighted distance from x to T, $r^\rho(x, T)$, is called the x-to-T *weighted radius*, and a point of G, x^*, minimizing the x-to-T weighted radius is called a *weighted absolute 1-Steiner center (WA1SC)*. We have

$$r^\rho(x, T) = \max_{i \in I(T)} \{\rho(t_i) \cdot d(x, t_i)\}, \tag{4}$$

and

$$r^\rho(x^*, T) = \min_{x \in p(G)} r^\rho(x, T). \tag{5}$$

2.2   Problem Definition

Let $T = (V', E')$ be a Steiner tree of $G = (V, E, w, \rho, T)$ spanning T. For any index pair, $i, j \in I(V')$, the v_i-to-v_j *weighted tree distance* on T is defined as $\rho(v_j) \cdot d_T(v_i, v_j)$. The maximum weighted tree distance on T between terminals is called the *weighted diameter* of T and denoted by $\text{diam}^\rho(T)$, and accordingly a path on T between terminals with the weighted tree distance equal to $\text{diam}^\rho(T)$ is called a *longest weighted path* on T. Let $\mathbf{T}^*$ denote a Steiner tree of minimum weighted diameter. We have

$$\text{diam}^\rho(T) = \max_{i, j \in I(T)} \{\rho(t_i) \cdot d_T(t_i, t_j)\}, \tag{6}$$

and

$$\text{diam}^\rho(\mathbf{T}^*) = \min_T \{\text{diam}^\rho(T)\}. \tag{7}$$

For ease of presentation, we let INPUT represent an input as follows: an undirected connected graph, $G = (V, E, w, \rho, \mathcal{T})$, where V is the vertex set, E is the edge set, $\mathcal{T} \subseteq V$ is a terminal subset, $w : E \rightarrow \mathbb{R}^+$ is an edge-weight function and $\rho : V \rightarrow \mathbb{R}^+$ is a vertex-weight function.

Problem 1. Given an INPUT, the goal of **Minimum Diameter Vertex-Weighted Steiner Tree Problem (MDWSTP)** is to construct a Steiner tree, $\mathbf{T}^*$, spanning $\mathcal{T}$ of minimum weighted diameter.

3   Parameterized Graphs

In this section, we will introduce two classes of *parameterized graphs (PG)*, both of which arise from the real-world problems.

First, we consider such instances where the given vertex-weighted graph satisfies that the ratio of any two vertex weights is always no more than a fixed parameter. This inspires us to introduce the following vertex-weight parameterized graph.

Definition 1. *Given a vertex-weighted undirected connected graph, $G = (V, E, \rho)$, where $\rho : V \rightarrow \mathbb{R}^+$ is a vertex-weight function, a vertex subset $U \subseteq V$, and a real number, $\mu > 1$, if G satisfies that $\frac{\rho(v_i)}{\rho(v_j)} \leq \mu$, for any $i, j \in I(U), i \neq j$, then G is called a $\langle U, \mu \rangle$-parameterized graph and abbreviated as $\langle U, \mu \rangle$-PG.*

Given an edge-weighted connected graph, $G = (V, E, w)$, the SPD function $d : V \times V \rightarrow \mathbb{R}^+ \cup \{0\}$ in G is a *metric*, i.e., $d(v_i, v_k) + d(v_k, v_j) \geq d(v_i, v_j), \forall i, j, k \in I(V)$ and $d(v_i, v_i) = 0$. Let $G^c = (V, E, d)$ denote the *shortest path graph (SPG)* induced by G. Furthermore, we consider a class of graphs whose SPG's also satisfy that $d(v_i, v_k) + d(v_k, v_j) \leq \lambda \cdot d(v_i, v_j), \forall i, j, k \in I(V)$, where $\lambda > 1$ is a fixed parameter. This inequality is called a *parameterized triangle inequality*. Using a weaker version of this inequality, we define a class of edge-weight parameterized graphs as follows.

Definition 2. *Given an edge-weighted undirected connected graph, $G = (V, E, w)$, where $w : E \rightarrow \mathbb{R}^+$ is an edge-weight function, a vertex subset $U \subset V$, and a real number, $\lambda > 1$, if the SPG, $G^c = (V, E, d)$, induced by G satisfies that $d(v_i, v_k) + d(v_k, v_j) \leq \lambda \cdot d(v_i, v_j)$, for any $i, j \in I(U), i \neq j$ and $k \in I(V \setminus U)$, then G is called a (U, λ)-parameterized graph and abbreviated as (U, λ)-PG.*

4   Approximating MDWSTP in $\langle \mathcal{T}, \mu \rangle$-PG

In this section, we suppose that the input graph in INPUT of MDWSTP is an edge-weighted $\langle \mathcal{T}, \mu \rangle$-PG, denoted as $G_{\langle \mathcal{T}, \mu \rangle} = (V, E, w, \rho, \mathcal{T}, \mu)$, where $\mu > 1$ is a real number and $\mathcal{T}$ is the terminal set in INPUT.

Recall that x^* is a WA1SC of G and T_{x^*} is the SPT in G with x^* as the origin spanning $\mathcal{T}$. In this section, we will prove that T_{x^*} is a good approximation to MDWSTP in $\langle \mathcal{T}, \mu \rangle$-PG, and accordingly propose an approximation algorithm.

4.1   Preliminary Lemmas

In this subsection, we show three preliminary lemmas which will play an important role in design and analysis of our approximation algorithm for MDWSTP in $\langle \mathcal{T}, \mu \rangle$-PG.

For a Steiner tree T of $G = (V, E, w, \rho, \mathcal{T})$ spanning $\mathcal{T}$, we let $r_T^\rho(x, \mathcal{T})$ denote the x-to-$\mathcal{T}$ weighted radius in T, and $x_T^\star$ denote a WA1SC of T, likewise. We have

$$r_T^\rho(x, \mathcal{T}) = \max_{i \in I(\mathcal{T})} \{\rho(t_i) \cdot d_T(x, t_i)\}, \tag{8}$$

and

$$r_T^\rho(x^\star, \mathcal{T}) = \min_{x \in p(T)} r_T^\rho(x, \mathcal{T}). \tag{9}$$

Lemma 1. $x^\star$ is also a WA1SC of $T_{x^\star}$.

Lemma 2. In $T_{x^\star}$, there exist at least two terminal indices, $i, j \in I(\mathcal{T})$, such that

(1) $\rho(t_i) \cdot d(x^\star, t_i) = \rho(t_j) \cdot d(x^\star, t_j) = r^\rho(x^\star, \mathcal{T})$;
(2) $\pi_{T_{x^\star}}(t_i, t_j)$ passes through $x^\star$.

The following lemma shows a *lower bound* on the diameter of an optimal solution tree to MDWSTP.

Lemma 3. $\text{diam}^\rho(\mathbf{T}^*) \geq 2 \cdot r^\rho(x^\star, \mathcal{T})$.

For general edge-weighted trees, the following lemma always holds.

Lemma 4. Given an edge-weighted tree, $T = (V_T, E_T, w_T)$, it always holds that

$$d_T(v_1, v_2) \leq d_T(v_1, x) + d_T(x, v_2), \quad \forall v_1, v_2 \in V_T; x \in p(T), \tag{10}$$

and "=" holds iff $\pi_T(v_1, v_2)$ passes through x.

4.2   An Approximation Algorithm

In this subsection, we present an approximation algorithm, called Approx-VPG, for MDWSTP in $\langle \mathcal{T}, \mu \rangle$-PG. The main idea of it is described as follows: we first compute a WA1SC, $x^\star$, of $\langle \mathcal{T}, \mu \rangle$-PG, and then output the SPT, $T_{x^\star}$, as the solution tree. Note that Kariv and Hakimi's exact algorithm [14] can be easily adapted to compute $x^\star$ by restricting computation to all the terminals instead of all the vertices, whose time cost is certainly not beyond the original Kariv and Hakimi's algorithm, i.e., no more than $O(mn \log n)$ when the distance matrix is known as well as $O(mn \log n + n^3)$ when the distance matrix is unknown. Moreover, we use Dijkstra's algorithm to compute $T_{x^\star}$ and it terminates once all $x^\star$-to-terminal SPD's are obtained.

Approx-VPG($G_{\langle T,\mu \rangle}$): Approximation in $\langle T,\mu \rangle$-PG.
Input: $G_{\langle T,\mu \rangle} = (V, E, w, \rho, T, \mu)$.
Output: a solution tree T^A.
1: Adapt Kariv and Hakimi's exact algorithm [15] to compute a WA1SC, $x^\star$, of $G_{\langle T,\mu \rangle}$;
2: Use Dijkstra's algorithm to compute the SPT, $T_{x^\star}$, in $G_{\langle T,\mu \rangle}$ spanning T; Return $T_{x^\star}$;

Theorem 1. *Given $G_{\langle T,\mu \rangle} = (V, E, w, \rho, T, \mu)$ with n vertices and m edges, where $\mu > 1$ and $T \subseteq V$ is a subset of s terminals, Approx-VPG can produce a $\frac{1+\mu}{2}$-approximation to MDWSTP in $G_{\langle T,\mu \rangle}$, where the factor of $\frac{1+\mu}{2}$ is tight, within $O(mn \log n)$ time when the distance matrix is known and $O(mn \log n + n^3)$ time when the distance matrix is unknown.*

For the special case of $\mu = 1$, $G_{\langle T,1 \rangle} = (V, E, w, \rho, T, 1)$ is equivalent to a *vertex-unweighted* graph, $G = (V, E, w, T)$, in essence. Accordingly, MDWSTP (*resp.* WA1SC) in $G_{\langle T,1 \rangle}$ is equal to MDSTP (*resp.* A1SC) in G, and so Approx-VPG is just an exact algorithm for the classic MDSTP.

5   Approximating MDWSTP in (T, λ)-PG

In this section, we suppose that the input graph G in INPUT of MDWSTP is a vertex-weighted (T, λ)-PG, denoted as $G_{(T,\lambda)} = (V, E, w, \rho, T, \lambda)$, where $\lambda > 1$ is a real number and T is the terminal set in INPUT.

Recall that x^* is an A1SC of G and T_{x^*} is the SPT in G with x^* as the origin spanning T. In this section, we will discover that the SPT (including T_{x^*}) with any vertex as the origin is always an approximation to MDWSTP in (T, λ)-PG with a uniform performance factor guarantee, and accordingly design two approximation algorithms.

5.1   A Simple Approximation Algorithm

In this subsection, we design a simple approximation algorithm for MDWSTP in (T, λ)-PG. First, we show two important lemmas.

Lemma 5. $\text{diam}^\rho(\mathbf{T}^*) \geq \max\limits_{i,j \in I(T)} \{\rho(t_i) \cdot d(t_i, t_j)\}$.

Lemma 6. *In $G_{(T,\lambda)}$, it always holds that, for each $k \in I(V)$,*

$$\text{diam}^\rho(T_k) \leq \lambda \cdot \max\limits_{i,j \in I(T)} \{\rho(t_i) \cdot d(t_i, t_j)\}. \tag{11}$$

Essentially, Lemma 5 shows a *lower bound* on the diameter of an optimal solution tree to MDWSTP, and Lemma 6 shows an *upper bound* on the diameter of each SPT, $T_k, k \in I(V)$. The combination of them inspires us to design a simple approximation algorithm, called Approx1-EPG, for MDWSTP in (T, λ)-PG, whose performance analysis is presented in Theorem 2. Let T^{A_1} denote the output tree of Approx1-EPG. Its main idea is to select a vertex index $k^\circ \in I(V)$ arbitrarily and then take the SPT, T_{k°, as the solution tree.

Approx1-EPG($G_{(\mathcal{T},\lambda)}$): Approximation in $(\mathcal{T},\lambda)$-PG.
Input: $G_{(\mathcal{T},\lambda)} = (V, E, w, \rho, \mathcal{T}, \lambda)$.
Output: a solution tree T^{A_1}.
1: Select an index $k° \in I(V)$ arbitrarily;
2: Use Dijkstra's algorithm to compute the SPT, $T_{k°}$,   in $G_{(\mathcal{T},\lambda)}$ spanning $\mathcal{T}$; Return $T_{k°}$;

Theorem 2. *Given* $G_{(\mathcal{T},\lambda)} = (V, E, w, \rho, \mathcal{T}, \lambda)$ *with* n *vertices and* m *edges, where* $\lambda > 1$ *and* $\mathcal{T} \subseteq V$ *is a subset of* s *terminals,* Approx1-EPG *can produce a* λ*-approximation to MDWSTP in* $G_{(\mathcal{T},\lambda)}$ *within* $O(m + n \log n)$ *time, where the factor of* λ *is tight.*

5.2 A Slightly Improved Approximation Algorithm

In this subsection, based on Approx1-EPG, we select a minimum weighted diameter SPT starting at a vertex as a solution tree. As a result, we obtain a slightly improved algorithm, called Approx2-EPG, whose performance analysis is shown in Theorem 3.

Let $T = (V', E')$ be a Steiner tree of $G = (V, E, w, \rho, \mathcal{T})$ spanning $\mathcal{T}$. Since each leaf of T is a terminal and a terminal may be a nonleaf of T, all the weighted tree distances between terminals can be divided into two types: *terminal-to-nonleaf* and *terminal-to-leaf*. In order to compute the weighted diameter of T, we need to compute all the weighted tree distances on T between terminals. The following lemma shows a property that helps to reduce the amount of computation to some extent.

Lemma 7. *The weighted diameter of a Steiner tree,* $T = (V', E')$, *of* $G = (V, E, w, \rho, \mathcal{T})$ *spanning* $\mathcal{T}$ *cannot be a terminal-to-nonleaf weighted tree distance on* T.

It is implied by Lemma 7 that the weighted diameter of a tree must be a terminal-to-leaf weighted tree distance. As a result, we can rewrite Eq. (6) to be

$$\text{diam}^\rho(T) = \max_{i \in I(\mathcal{T})} \left\{ \rho(t_i) \cdot \max_{j \in I(\mathcal{L}(T))} d_T(t_i, t_j) \right\}. \tag{12}$$

According to Eq. (12), we design a procedure, called DIAM, to compute the weighted diameter of a Steiner tree. The input of DIAM includes a Steiner tree T and a terminal subset $\mathcal{T}$. The idea of DIAM can be described as follows: we use DFS (*depth first search*) to traverse T with t_i as the origin, for each terminal $t_i, i \in I(\mathcal{T})$. Note that only terminal-to-leaf weighted tree distances,

$d_T(t_i, t_j), j \in I(\mathcal{L}(T))$, are computed and recorded during the traversing, and DFS terminates once all the leaves are visited. We use M_i to record the maximum of all $d_T(t_i, t_j), j \in I(\mathcal{L}(T))$. After DFS runs $|I(T)|$ times, the maximum of all the products, $\rho(t_i) \cdot M_i, i \in I(T)$, is determined and output as the weighted diameter of T.

DIAM(T, T): Compute the weighted diameter.
Input: $T = (V_T, E_T, w_T)$ and $T \subseteq V_T$.   **Output:** the diameter of T.
1: **for** each $i \in I(T)$ **do**   2:   Use DFS to traverse T with t_i as the origin to   compute all $d_T(t_i, t_j), j \in I(\mathcal{L}(T))$;   3: $M_i \leftarrow \max\limits_{j \in I(\mathcal{L}(T))} d_T(t_i, t_j)$;   4: **endfor**   5: diam$^\rho(T) \leftarrow \max\limits_{i \in I(T)} \rho(t_i) \cdot M_i$; Return diam$^\rho(T)$;

Applying DIAM as a sub-procedure, we present a slightly improved approximation algorithm, called Approx2-EPG, for MDWSTP in (T, λ)-PG. Its main body can be described as follows: we first use Dijkstra's algorithm to compute the SPT, T_k, in $G_{(T,\lambda)}$ spanning T, and then call DIAM(T_k, T) to compute the diameter of T_k, for each vertex index $k \in I(V)$. Finally, the SPT having a minimum weighted diameter is output as a solution tree of Approx2-EPG, denoted by T^{A_2}. In addition, we let

$$LB = \min_{k \in I(V)} \text{diam}^\rho(T_k), \qquad UB = \max_{k \in I(V)} \text{diam}^\rho(T_k). \qquad (13)$$

Approx2-EPG$(G_{(T,\lambda)})$: Improved Approximation in (T, λ)-PG.
Input: $G_{(T,\lambda)} = (V, E, w, \rho, T, \lambda)$.   **Output:** a solution tree T^{A_2}.
1: **for** each $k \in I(V)$ **do**   2:   Use Dijkstra's algorithm to compute the SPT, T_k, in $G_{(T,\lambda)}$ spanning T;   3:   Call DIAM(T_k, T) to compute the weighted diameter of T_k;   4: **endfor**   5: $k^* \leftarrow \arg\min_{k \in I(V)} \text{diam}^\rho(T_k)$; Return T_{k^*};

Theorem 3. *Given* $G_{(T,\lambda)} = (V, E, w, \rho, T, \lambda)$ *with* n *vertices and* m *edges, where* $\lambda > 1$ *and* $T \subseteq V$ *is a subset of* s *terminals, Approx2-EPG can produce a new approximation to MDWSTP in* $G_{(T,\lambda)}$ *within* $O(mn + n^2(\log n + s))$ *time, which improves the output of Approx1-EPG by a factor of* $\lambda \Delta$, *where* Δ *is a real number satisfying that*

$$0 \leq \Delta \leq 1 - \frac{LB}{UB}. \qquad (14)$$

6   Conclusions

This paper studied the minimum diameter vertex-weighted Steiner tree problem (MDWSTP) in weighted connected graphs. We proposed two classes of parameterized graphs, $\langle \mathcal{T}, \mu \rangle$-PG and $(\mathcal{T}, \lambda)$-PG, from the angle of the parameterized upper bound on the ratio of two vertex-weights, and a weaker version of the parameterized triangle inequality, respectively. For the MDWSTP in edge-weighted $\langle \mathcal{T}, \mu \rangle$-PG, we presented an approximation algorithm with a tight performance factor of $\frac{\mu+1}{2}$. For the MDWSTP in vertex-weighted $(\mathcal{T}, \lambda)$-PG, we first designed a λ-factor simple approximation algorithm where λ is tight, and then developed a slightly improved approximation algorithm. As for these parameterized graphs, our approximation algorithms can be adapted to the **Minimum Diameter Vertex-Weighted Spanning Tree Problem (MDWTP)**. The intractability of MDWSTP remains open. We conjecture that it is NP-complete and MAX SNP-hard.

References

1. Bui, M., Butelle, F., Lavault, C.: A distributed algorithm for constructing a minimum diameter spanning tree. J. Parallel Distrib. Comput. **64**(5), 571–577 (2004)
2. Chan, T.M.: Semi-online maintenance of geometric optima and measures. In: Proceedings of 13th ACM-SIAM Symposium on Discrete Algorithms (SODA 2002), pp. 474–483 (2002)
3. Ding, W., Lin, G., Xue, G.: Diameter-constrained steiner trees. Discrete Math. Algorithms Appl. **3**(4), 491–502 (2011)
4. Ding, W., Xue, G.: Minimum diameter cost-constrained Steiner trees. J. Comb. Optim. **27**(1), 32–48 (2013). https://doi.org/10.1007/s10878-013-9611-2
5. Ding, W., Qiu, K.: Algorithms for the minimum diameter terminal Steiner tree problem. J. Comb. Optim. **28**(4), 837–853 (2014)
6. Ding, W., Qiu, K.: A 2-approximation algorithm and beyond for the minimum diameter k-Steiner forest problem. Theor. Comput. Sci. (2020). https://doi.org/10.1016/j.tcs.2019.12.012
7. Du, D.Z., Hu, X.: Steiner Tree Problems in Computer Communication Networks. World Scientific Publishing Co., Pte. Ltd., Singapore (2008)
8. Garey, M.R., Johnson, D.S.: Computers and Intractability: A Guide to the Theory of NP-Completeness. Freeman, San Francisco (1979)
9. Gudmundsson, J., Haverkort, H., Park, S.M., Shin, C.S., Wolff, A.: Approximating the geometric minimum-diameter spanning tree. In: Proceedings of 18th European Workshop on Computational Geometry (EWCG 2002), pp. 41–45 (2002)
10. Hassin, R., Tamir, A.: On the minimum diameter spanning tree problem. Inf. Proc. Lett. **53**, 109–111 (1995)
11. Ho, J.M., Lee, D.T., Chang, C.H., Wong, C.K.: Minimum diameter spanning trees and related problems. SIAM J. Comput. **20**, 987–997 (1991)
12. Hwang, F.K., Richards, D.S., Winter, P.: The Steiner Tree Problem. Annals of Discrete Mathematics, vol. 53. North-Holland, Amsterdam (1992)
13. Ihler, E., Reich, G., Widmayer, P.: On shortest networks for classes of points in the plane. In: Bieri, H., Noltemeier, H. (eds.) CG 1991. LNCS, vol. 553, pp. 103–111. Springer, Heidelberg (1991). https://doi.org/10.1007/3-540-54891-2_8

14. Kariv, O., Hakimi, S.L.: An algorithmic approach to network location problems. I: the p-centers. SIAM J. Appl. Math. **37**(3), 513–538 (1979)
15. Marathe, M.V., Ravi, R., Sundaram, R., Ravi, S.S., Rosenkrantz, D.J., Hunt III, H.B.: Bicriteria network design problems. J. Algorithms **28**(1), 142–171 (1998)
16. Spriggs, M.J., Keil, J.M., Bespamyatnikh, S., Segal, M., Snoeyink, J.: Computing a $(1 + \epsilon)$-approximate geometric minimum-diameter spanning tree. Algorithmica **38**(4), 577–589 (2004)

Community-Based Rumor Blocking Maximization in Social Networks

Qiufen Ni[1,2] , Jianxiong Guo[3] , Chuanhe Huang[1,2(✉)], and Weili Wu[3]

[1] School of Computer Science, Wuhan University, Wuhan 430072, China
{niqiufen,huangch}@whu.edu.cn
[2] Collaborative Innovation Center of Geospatial Technology, Wuhan 430072, China
[3] Department of Computer Science, The University of Texas at Dallas, Richardson,
TX 75080, USA
{jianxiong.guo,weiliwu}@utdallas.edu

Abstract. Even though the widespread use of social networks brings a lot of convenience to people's life, it also cause a lot of negative effects. The spread of misinformation in social networks would lead to public panic and even serious economic or political crisis. We study the community-based rumor blocking problem to select b seed users as protectors such that expected number of users eventually not being influenced by rumor sources is maximized, called Community-based Rumor Blocking Maximization Problem (CRBMP). We consider the community structure in the social network and solve our problem in two stages, in the first stage, we allocate budget b for all the communities with the technique of submodular function maximization on an integer lattice, which is different from most of the existing work with the submodular function over a set function. We prove that the objective function for the budget allocation problem is monotone and DR-submodular, then a greedy algorithm is devised to get a $1 - 1/e$ approximation ratio; then we solve the Protector Seed Selection (PSS) problem in the second stage after we obtained the budget allocation vector for communities, we greedily choose protectors for each communities with the budget constraints to achieve the maximization of the influence of protectors. The greedy algorithm for PSS problem can achieve a $\frac{1}{2}$-approximation guarantee. At last, we verified the effectiveness and superiority of our algorithms on three real world datasets .

Keywords: Social network · Community structure · Influence maximization · Rumor blocking

1 Introduction

Social networks are booming rapidly in recent years, it becomes an important platform for people to communicate with each other and generate a lot of information at any time. For example, there are at most 2.5 billion monthly active

This work is supported by the National Natural Science Foundation of China (No. 61772385, No. 61572370) and supported partially by NSF 1907472.

ⓒ Springer Nature Switzerland AG 2020
Z. Zhang et al. (Eds.): AAIM 2020, LNCS 12290, pp. 73–84, 2020.
https://doi.org/10.1007/978-3-030-57602-8_7

users and more than 1 billion registered accounts in the most popular network Facebook according to the latest statistics in January 2020. The photo-sharing app Instagram which ranks sixth has 1 billion monthly active accounts [1]. Information spreads very fast in social network, it brings a lot of convenience to our life, for example, businesses advertise products through social networks in viral marketing; we can exchanging information and ideas in online social network. On the other hand, it also bring us many negative influence, such as rumors propagate quickly in social networks, which may cause public panic and economic or politic crisis. For example, during the coronavirus (COVID-19) pandemic, there are many rumors related with it spreading in the social network recently. On March, 2020, someone in China posts a rumor say that the outbreak of the COVID-19 may trigger a global grain crisis, we should hoard grain for at least three month as china is a big importer of grain, which caused panic among public. Thousands of people in china went to the supermarket and scrambled for grain after the rumor got out [2]. In England, also on March 2020, some people said that 5G is accelerating the spread of the new coronavirus. Some criminals in England went to damage the phone masts and telecoms engineers abused their power, which happen apparently as inspired by the rumor circulating online [3]. Therefore, effective methods to restrain the spread of rumors have been a popular topic.

Information spreads in social networks through diffusion cascade. There are two classical cascades in social networks: Independent Cascade (IC) and Linear Threshold (LT), which are proposed by Kemp *et al.* in [4]. The influence propagation process in social networks is also firstly formulated In [4]. They propose the Influence Maximization (IM) problem which aims at selecting an initial set of seed users to maximize the expected influence propagation in the given diffusion model. The rumor blocking problem which is proposed by Budak *et al.* in [5]. Existing works on rumor blocking in social networks mainly focusing on the goal of minimizing the influence of rumors or maximizing the influence of positive information [6,7]. Usually they would construct a monotone and submodular objective function which can get a 1-1/e approximation ratio with greedy method directly [8,9]. For a non-submodular objective function, it is more challenging to solve it. Let X with $|X| = n$ be a *ground set*. A set function on X is a function $h: 2^X \to R$. A set function $h: 2^X \to R$ is submodular if for any $A \subseteq B \subseteq X$ and $u \in X \backslash B$, we have $h(A \cup \{u\}) - h(A) \geq h(B \cup \{u\}) - h(B)$. There is another equivalent definition for submodularity, that is $h(A \cap B) - h(A \cup B) \leq h(A) + h(B)$. For supermodularity, the inequality is reversed to submodularity.

In this paper, we consider a strategy that initiates a set of users whose number of elements is b as protectors to spread the opposite view in social networks, we use IC diffusion model as the fundamental model, so both protectors and rumors spread in IC model, and we call the IC with two competing diffusion models in our problem as competitive Independent Cascade (CIC) model.

As we know that the social network has the feature of community structure [10]. An important property of the community structure is that the nodes within a community is dense and sparse with nodes in other communities [11]. According, the influence propagates much faster within a community than spreading out its own community [12]. We partition a social network into disjoint communities based on influence before solving our problem, then we study the Community-based Rumor Blocking Maximization Problem (CRBMP), which aims at selecting a set of protectors with budget b to maximize the rumor blocking.

We summarize the main contributions in this paper as follows:

- We formulate the Community-based Rumor Blocking Maximization Problem (CRBMP) into two subproblems under the CIC model in social networks: 1. Allocation problem; 2. Protector Seed Selection (PSS) problem.
- We consider the general situation where rumors are distributed in different communities, we get the community budget allocation vector for all the communities with the method of submodular function maximization on an interger lattice in the first stage. We prove that the objective function for the budget allocation problem is monotone and DR-submodular, then a greedy algorithm is devised to get a $1 - 1/e$ approximation ratio. Then we propose a greedy algorithm for PSS problem to select protectors within each community to maximize the blocked nodes by protectors, which obtains a $\frac{1}{2}$ approximation ratio.
- We evaluate our algorithms on three real-world datasets. The results show the effectiveness of the proposed algorithms.

The result of the paper is organized as follows. Section 2 devoted to the existing work. In Sect. 3, we construct network model and formulate the Community-based Rumor Blocking Problem as two subproblems: Allocation problem and Protector Selection Seed Selection (PSS) problem. In Sect. 4 we solve our problems in two stages under general situation, and give properties of the objective function. Section 5 presents the simulation results, while finally, the conclusion is presented in Sect. 6.

2   Related Work

In this paper, we focus on community-based rumor blocking problem in social networks. Below we discuss recent related work on the related topics and classified them to three categories, there is some community-based related work in each category.

Removing a set of the most influential nodes to minimize the spread of rumors. We should notice that when we remove a node from the network, the relationship that this node with its incoming node and outgoing nodes are broken. If a node has very high degree, it would cause a dramatically change for the network structure. Moreover, removing nodes from a social network takes high cost and not easy to operate. In [13], J. Zheng *et al.* study the Least Cost Rumor Community Blocking Optimization (LCRCBO) problem. It aims at select a minimal set of nodes to remove from the social network such that rumors within the rumor community can be blocked, their method can also ensure that the expected number of nodes infected by rumors is less than K. They propose a Minimum Vertex Cover Based Greedy (MVCBG) algorithm to solve the problem.

Removing a certain number of edges to restrain rumors' propagation. Removing an edge needs to broke the relationship between two nodes, which also changes the structure of the network, but this method is easier to deal with than deleting nodes. J. Guo *et al.* [14] study a target protection maximization (TPM) problem which aims at protecting the expected ration of nodes in targeted set influenced by rumors is at most β through blocking the least edges. They propose a simple Greedy algorithm which gets a theoretical bound and General-TIM algorithm which reduces the running time. They also consider the community structure in social networks which can help them improved the performance by removing some unrelated communities when running the proposed two algorithms.

Spreading positive information to compete with rumors such that the positive influence is maximized. Most of the existing works consider that there are just one misinformation cascade and one positive cascade, ln [15], they propose a reverse-tuple based randomized (RBR) algorithm for the rumor blocking problem, in which one rumor is detected, the target is to select k positive seed users to maximize the influence of positive information. They show that the proposed algorithm is superior the existing algorithms in running time. X. Chen *et al.* [16] study a centralized rumor blocking problem which the objective function is proved to be non-decreasing and submodular, which can obtain a $1 - 1/e$ approximation with a greedy algorithm. They further propose a decentralized rumor blocking problem with two protectors who compete each other to limit the spread of rumors to maximize their own profits.

3   Network Model and Problem Formulation

3.1   The Network Model

A social network is modeled as a directed graph $G(V, E)$, where each vertex $v \in V$ represents a user, and each edge $(u, v) \in E$ is the relationship between user u and v. There are two cascades evolving simultaneously in the network: R (for "rumor") and P (for "protector"). Each node has three states: infected by rumor nodes, protected by protector nodes, inactive. There is a rumor node set T exists in the social network for cascade R and initial protector seed D for cascade P. Supposing that the rumor spreads in cascade R is detected with a delay r, then the protector seeds are selected and begin to spread. Protector nodes and rumor nodes spread simultaneously in the network, when R and P reach a node v at the same time, P has the priority to activate v. Once a node v is protected or infected, it stays the status forever. In the social network, the incoming neighbors set and the outgoing neighbors set of a node v is denoted as $N^-(v)$ and $N^+(v)$, respectively. Next we extend the IC model for our problem.

Competitive Independent Cascade (CIC) model : Each edge $(u, v) \in E$ is associated with one probability $p_{uv} \in [0, 1]$. No matter a protector or a rumor reach a node v, the probability that they activate v is the same. If there are two or more nodes trying to activate v at the same time, at most one of them can succeed. Each newly influenced user u in step t only has a single chance to influence his uninfluenced outgoing neighbors in step $t+1$. After time step $t+1$, u could not activate any of its outgoing neighbors. The diffusion completes until there is no new user can be influenced.

3.2 Problem Formulation

In this section, we introduce our problem based on aforementioned CIC model. As we all know, social networks have the feature of containing community structure, and the influence propagation is rarely across different communities. We take into account this advantage of social networks which have the community structure. Ni *et al.* [17] proposed a influence-based community partition method under IC model, and we adopt it to partition our network preliminarily.

A social network $G = (V, E)$ with m disjoint communities $S = \{S_1, S_2, \cdots, S_m\}$ that satisfying $\bigcup_{i=1}^{m} V(S_i) = V$, where $V(S_i)$ denotes node set in community S_i. This community is partitioned by using influence propagation as the metric. Then, our Community-based Rumor Blocking Maximization Problem (CRBMP) is formulated, which can be divided into two stages.

Problem 1 ((CRBMP)). *Given a graph $G = (V, E, S)$ and budget b, in the first stage, we need to allocate this budget into each community, thus we can get an allocation $\boldsymbol{x} = (x_1, \cdots, x_m)$ where $\sum x_i = b$. In the second stage, we aim to select an optimal nodes set D such that protect nodes from being infected by rumor as much as possible, where we have $|D \cap S_i| \leq x_i$ for all $S_i \in S$.*

Thus, our CRBMP can be divided into two subproblems, Allocation problem in the first stage and Protector Seed Selection (PSS) problem in the second stage.

Define $\theta_{S,T}(\emptyset)$ as the expected number of nodes that are infected by rumor seed set T in the social network when there is not any protector. Let $\theta_{S,T}(D)$ be the expected number of nodes that are infected by rumor seed set T when the protector seed set D coexists with rumor set T. Let $f(D)$ be the expected protected nodes when protector seed nodes are selected as D. So

$$f(D) = \theta_{S,T}(\emptyset) - \theta_{S,T}(D)$$

Our PSS problem in the second stage can be formulated as follows: Given a social network $G = (V, E, S)$, given a rumor node set T, and allocation vector $\boldsymbol{x} = (x_1, \cdots, x_m)$ where $\sum x_i = b$ for the protectors. The PSS problem aims at selecting a set of protectors under the allocation vector to protect the most nodes from being infected by rumor nodes. I.e., find the optimal allocation D^* which maximizes $f(D)$, where $|D|$ is the cardinality of protector seed set D and $\{|D \cap S_i| \leq x_i, D \subseteq \{V(G) \backslash T\}\}$. In [9], they proved that the community-based rumor blocking problem can be reduced to a set cover problem which is a NP-hard problem proved in [18]. So our community-based rumor blocking problem is NP-hard.

4 Solution for CRBMP in Two Stages

We assume that the rumor nodes have already detected by us. As we already partition communities for the social network, we can easily detect the communities which contain the rumor nodes.

In this section, we solve the CRBMP in two stages. In the first stage, we use the technique of submodular function on integer lattice [19] to find a budget allocation for each community. Most of the existing work about the influence maximization problem consider the submodular function over a set-submodular function which selecting a subset of seed users and return the value of influence propagation. We consider the submodular function maximization over a multiset integer lattice in this problem. In the second stage, we select protectors in each community based on the protector seed allocation constraint which is obtained in the first step.

4.1 Allocation Problem

In the first stage, we begin to find a budget allocation for each community. We define a protector allocation vector for the m communities as $\boldsymbol{x} = (x_1, \cdots, x_m)$ where $\sum x_i = b$, i.e., if we allocate x_i to community S_i, at most x_i protectors could be selected from community S_i. Let $\phi_{S,T}(\boldsymbol{x})$ be the expected infected nodes after we choose $\boldsymbol{x}$ as protector allocation vector, where it selects x_i nodes from $S_i \backslash T$ uniformly and randomly. Let $g(\boldsymbol{x})$ be the expected protected nodes given the allocation vector $\boldsymbol{x}$. So

$$g(\boldsymbol{x}) = \phi_{S,T}(\boldsymbol{0}) - \phi_{S,T}(\boldsymbol{x})$$

where $\phi_{S,T}(\boldsymbol{0})$ denotes the expected influence of rumor nodes when the protector allocation vector is empty. Here, $g(\boldsymbol{x})$ is defined on integer lattice. We can not use the submodularity property for $g(\boldsymbol{x})$ and the greedy algorithm for this problem is also cannot be used directly. For a vector function over integer lattice, it has a similar condition like submodularity: the diminishing return submodular (DR-submodular). A function $h \colon \mathbb{Z}^m \to \mathbb{R}$ is diminishing return submodular (DR-submodular) if $h(\boldsymbol{x} + \boldsymbol{e}_i) - h(\boldsymbol{x}) \geq h(\boldsymbol{y} + \boldsymbol{e}_i) - h(\boldsymbol{y})$ for any $\boldsymbol{x} \leq \boldsymbol{y} \in \mathbb{Z}^m$, where $\boldsymbol{e}_i$ is the i-th unit vector and $i \in \{1, 2, \cdots, m\}$. Another important property is monotonicity for function h. Function h is monotone if $h(\boldsymbol{x}) \leq h(\boldsymbol{y})$ for any $\boldsymbol{x} \leq \boldsymbol{y} \in \mathbb{Z}^m$. If our objective function $g(\boldsymbol{x})$ is monotone and DR-submodular, the greedy algorithm can be applied to get a good performance [20]. Then we begin to prove the properties of $g(\boldsymbol{x})$.

Let $\mathcal{D}(\boldsymbol{x}) \subseteq 2^V$ be the collection that contains all set D satisfying $|D \cap S_i| = x_i$. Let $\hat{G}$ be a realization which is a subgraph of $G = (V, E)$, for each edge $(u, v) \in E$, the probability that this edge appears in realization $\hat{G}$ is p_{uv} and the edges which appear in realization $\hat{G}$ are live-edges. From [4], we can know the expected number of infected nodes given the allocation vector $\boldsymbol{x}$ is

$$\phi_{S,T}(\boldsymbol{x}) = \sum_{\hat{G}} \Pr[\hat{G}] \sum_{D \in \mathcal{D}(\boldsymbol{x})} \Pr[D] \cdot \theta_{S,T}(\hat{G}, D)$$

Algorithm 1. Greedy Community Budget Allocation Algorithm

Input: Graph $G = (V, E, S)$, m, and g.
Output: x
1: Initialize $x \leftarrow \mathbf{0}$
2: **while** j=1 to b **do**
3:     $i \in \arg\max_{\{k=1,\cdots,m\}} g(x + e_k) - g(x)$
4:     $x = x + e_i$
5: **end while**
6: **return** community budget allocation vector x

---

where $\sum_{\hat{G}} \Pr[\hat{G}]$ is all the possible live-edge subgraphs $\hat{G}$ of G, $\Pr[D]$ is the probability that set D is selected as protector set according to allocation x, $\theta_{S,T}(\hat{G}, D)$ is the number of nodes infected by rumors in graph $\hat{G}$ at the end of the CIC propagation process when the protector set is D. We rewrite $\phi_{S,T}(x)$ as $\phi_{S,T}(x) = \sum_{\hat{G}} \Pr[\hat{G}] \cdot \phi_{S,T}(\hat{G}, x)$, where $\Pr[\hat{G}]$ is the probability of sampling $\hat{G}$, and $\phi_{S,T}(\hat{G}, x) = \sum_{D \in \mathcal{D}(x)} \Pr[D] \cdot \theta_{S,T}(\hat{G}, D)$.

Theorem 1. *The objective function $g(x)$ for the problem of protector allocation for each community is monotone non-decreasing and DR-submodular.*

Proof. We omitted the proof due to the limitation of conference pages.

Since the objective function is monotone non-decreasing and DR-submodular, then we can devise our greedy algorithm. The pseudo-code is in Algorithm 1. We increase budget for the community which can bring the maximum marginal ˙ gain for the objective function g until the total budget b is exhausted. Next, we analyze the performance and time complexity of the Algorithm 1, which is shown as follows:

Lemma 1. *Let $y = \{y_i, \cdots, y_m\}$ where $y_i \in \mathbb{Z}^*$ and $\sum_i y_i = b$, then we have $g(x + y) - g(x) \leq \sum_i y_i(g(x + e_i) - g(x))$.*

Proof. We omitted the proof due to the limitation of conference pages.

Theorem 2. *Algorithm 1 returns a $(1 - \frac{1}{e})$-approximation (in expectation) for the problem of maximizing $g(x)$ subject to $\sum_i x_i \leq b$.*

Proof. We omitted the proof due to the limitation of conference pages.

Theorem 3. *The complexity of Algorithm 1 is upper bounded by $O(bn|\beta||$ $\eta||E|\pi)$.*

Proof. In [21], T. Soma *et al.* prove that for a monotone submodular function maximization over integer lattice, if f is the DR-submodular, and the cost is uniform, then we can get the solution in $O(bn\alpha)$ time, where α is the running time for evaluating f. In our problem, we prove that f is DR-submodular and the cost is uniform. Then we need to get the value of α. As the protectors in each

---

Algorithm 2. Greedy Protector Selection Algorithm

Input: Graph $G = (V, E, S)$, x, and f.
Output: D
1: Initialize $D \leftarrow \emptyset$, $D = \bigcup_{i \in \{1, \cdots, m\}} D_i$
2: **while** $b > 0$ **do**
3:     $I = \{i \in \{1, 2, \cdots, m\} : |D_i| \leq x_i\}$
4:     select $(v, i) = \arg \max_{v \in S_i \setminus (D_i \cup T), i \in [I]} f(D \cup (v, i)) - f(D)$
5:     $D \leftarrow D \cup (v, i)$; $b \leftarrow b - 1$
6: **end while**
7: **return** protector seed selection set D.

---

communities are selected uniformly at random, we estimate $\theta_{S,T}(\mathbf{x})$ with the sample average approximation (SAA) method which is an approach for solving the stochastic optimization problems by Monte Carlo simulation [22]. Let $\eta \subset \mathcal{D}(\mathbf{x})$ be a sample set from all possible protector selection sets and π be the number of Monte Carlo simulation. $\theta_{S,T}(\hat{G}, D)$ is estimated with the method proposed by Kempe $et\ al.$ [4]: sampling from the set of live-edge graphs. For each node $v \in V$, we generates a live-edge graph by selecting at most one of its incoming edges u with the probability p_{uv}. We use β to denote this sample set. It takes $O(|\beta||\eta||E|\pi)$ time to estimate $\theta_{S,T}(\mathbf{x})$. So the time complexity for Algorithm 1 is $O(bn|\beta||\eta||E|\pi)$.

4.2 Protector Seed Selection (PSS) Problem

After we obtained the protector allocation vector $\mathbf{x} = (x_1, x_2, \cdots, x_m)$, where $\sum_i x_i = b$, we begin to select protector seed for each community. In each community S_k, the number of selected protector seed is constrained by x_k. We assume the protector seed set $D = \{D_1, D_2, \cdots, D_m\}$, where D_i is the protector seed set in community S_i and $|D_i| \leq x_i$. Here, we select the protector seed node for each community based on that which node can bring the maximum marginal gain to f. The iteration stops until the budget b is reduced to 0. The pseudo-code of proposed greedy algorithm for the second step of PSS problem is shown in Algorithm 2.

Theorem 4. *Given the allocation $\mathbf{x}$, Algorithm 2 returns a $\frac{1}{2}$-approximation (in expectation) for the problem of maximizing $f(D)$ subject to $|D \cap S_i| \leq x_i$ for any $i \in \{1, \cdots, m\}$.*

Proof. We omitted the proof due to the limitation of conference pages.

5 Experimental Results

In this section, we verify the efficiency and effectiveness of the proposed algorithms on three real world datasets, we also compare our algorithms with three baseline algorithms.

5.1   Experimental Setup

Datasets: The first two datasets we used in the experiments are from net-workrepository.com [23], which is an online repository including different kinds of networks. The dataset 1 is a small co-authorship network, namely coauthorship of scientists in the field of network theory and experiment. It has 379 nodes and 914 edges. The dataset 2 is a medium Wiki-vote network, which is Wikipedia who-votes-on-whom network. It includes 914 nodes and 2914 edges. Dataset 3 is from [24], which is a large bitcoin alpha trust weighted signed network of people who trade using Bitcoin on a platform called Bitcoin Alpha. It has 3783 nodes and 24186 edges.

Experiment setup: In this paper, we adopt the IC model as the influence model. The propagation probability of each directed edge e is assigned as $p(e) = 1/|N^-(v)|$. This setting method of $p(e)$ is widely used in previous literatures [25]. We set the number of community as $m = 3$, the number of rumor node as $|T| = 25$, who have the maximum out-degree in given network. The budget changes from 1 to 25. The number of Monte Carlo simulation is set as 100 to estimate the number of expected influenced nodes and the running time, where the running time is measured in second.

Baseline Methods: Random. It randomly selects nodes as protectors, which is a classical baseline algorithm.

Proximity. It selects protector seed node set which are the out-neighbours with the maximum out-degree of the rumor seed nodes. We select these neighbors with high out-degree in priority.

5.2   Experimental Results

In this section we evaluate the performance of our proposed algorithms.   In Fig. 1, the three sub-figures show the changing of the expected nodes that are protected by protectors when we vary the budget b. In sub-figure (a), it shows the result in dataset 1, we can observe that our greedy method is superior to the proximity method and random method. The expected protected nodes increases

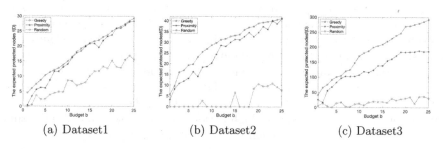

| (a) Dataset1 | (b) Dataset2 | (c) Dataset3 |

Fig. 1. Performance comparision achieved by different algorithms with the changing of budget under 3 datasets

with the increasing of the budget b with our greedy method since we can select more protectors, and we can also find that the growth rate of the greedy method curve is gradually getting smaller which makes the curve flatter, this phenomenon verifies the submodularity of the objective function $f(D)$, the marginal gain will be smaller and smaller with the increasing of protectors in the social networks. The above results is also can be found in sub-figure (2) and (3). Comparing the three sub-figures, we can also find that the advantages of our algorithm are more and more obvious with the growth of dataset size. In a small dataset 1, our algorithm just a little more than the proximity method, when the dataset changes to a medium dataset 2, the performance gap between our greedy algorithm and the proximity is big which shows in subfigure (b), the gap becomes bigger when the dataset changes to a large dataset 3. We can observe from the three subfigures in Fig. 1 that the larger the dataset, the greater the expected protected nodes, and the random method has very bad performance in all the three datasets.

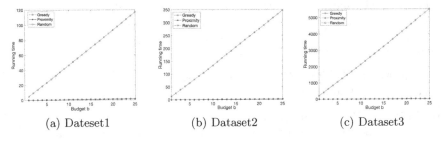

(a) Dateset1	(b) Dataset2	(c) Dataset3

Fig. 2. Running time comparision used by different algorithms with the changing of budget under 3 datasets

In Fig. 2, the three sub-figures show the changing of the running time that takes by different methods when we vary the budget from 1 to 25. In the three sub-figures, the running time of proximity method and random method is very small since these two methods select protectors trivially, they do not need much computational cost. In sub-figure (a) of Fig. 2, it shows the running time in dataset 1, we can observe that our greedy method takes much more time than the random method and proximity method, and running time linearly increases with the increasing of budget b since it takes much more time to choose more protectors. The same results are shown in sub-figure (b) and (c). Comparing the results in sub-figure (a) with (b) and (c), we can get that when the dataset becomes larger, the running time of our greedy method also becomes larger, it is because that in a larger dataset, the candidate set becomes larger for our algorithm, which increases the computational cost. From the experimental results, we can conclude that although the high performance advantage of our algorithm is obvious, its disadvantages of high time complexity is also obvious.

6 Conclusion

In this paper, we have studied the rumor blocking problem based on the community structure of social networks. We first partition the social network into communities based on the influence, then we aim to selecting the number of protectors with budget b to protect the most nodes in social networks from being influenced by rumors. We first propose budget allocation algorithm with the technique of submodular function maximization on an integer lattice, which obtains a $1 - 1/e$ performance guarantee since the objective function is monotone and DR-submodular; then we devise a greedy algorithm to solve the PSS problem which can achieve a $\frac{1}{2}$ performance guarantee. Finally, experiments show the effectiveness and efficiency of the rumor blocking effect of the proposed algorithms.

Acknowledgment. This work is supported by the National Natural Science Foundation of China (No.61772385, No.61572370) and supported partially by NSF 1907472.

References

1. https://www.statista.com/statistics/272014/global-social-networks-ranked-by-number-of-users/
2. https://www.zhihu.com/question/383352936/answer/1122919261
3. https://www.fema.gov/coronavirus-rumor-control
4. Kempe, D., Kleinberg, J., Tardos, É.: Maximizing the spread of influence through a social network. In: Proceedings of the Ninth ACM SIGKDD International Conference on Knowledge Discovery and Data Mining, pp. 137–146. ACM (2003)
5. Budak, C., Agrawal, D., El Abbadi, A.: Limiting the spread of misinformation in social networks. In: Proceedings of the 20th International Conference on World Wide Web, pp. 665–67 (2011)
6. Wang, B., Chen, G., Luoyi, F., Song, L., Wang, X.: Drimux: dynamic rumor influence minimization with user experience in social networks. IEEE Trans. Knowl. Data Eng. **29**(10), 2168–2181 (2017)
7. Ding, L., Hu, P., Guan, Z.-H., Li, T.: An efficient hybrid control strategy for restraining rumor spreading. IEEE Trans. Syst. Man Cybern. Syst. 1–13 (2020)
8. He, X., Song, G., Chen, W., Jiang, Q.: Influence blocking maximization in social networks under the competitive linear threshold model. In: Proceedings of the 2012 SIAM International Conference on Data Mining, pp. 463–474. SIAM (2012)
9. Fan, L., Lu, Z., Wu, W., Thuraisingham, B., Ma, H., Bi, Y.: Least cost rumor blocking in social networks. In: 2013 IEEE 33rd International Conference on Distributed Computing Systems, pp. 540–549. IEEE (2013)
10. Nedioui, M.A., Moussaoui, A., Saoud, B.: Detecting communities in social networks based on cliques. Phys. A Stat. Mech. Appl. **551**, 124100 (2020)
11. Kim, A.C.H., Newman, J.I., Kwon, W.: Developing community structure on the sidelines: a social network analysis of youth sport league parents. Soc. Sci. J. **57**, 1–1 (2020)
12. Ma, T., Liu, Q., Cao, J., Tian, Y., Al-Dhelaan, A., Al-Rodhaan, M.: Lgiem: global and local node influence based community detection. Future Gener. Comput. Syst. **105**, 533–546 (2020)

13. Zheng, J., Pan, L.: Least cost rumor community blocking optimization in social networks. In: 2018 Third International Conference on Security of Smart Cities, Industrial Control System and Communications (SSIC), pp. 1–5. IEEE (2018)
14. Guo, J., Li, Y., Wu, W.: Targeted protection maximization in social networks. IEEE Trans. Netw. Sci. Eng. **PP**(99), 1 (2019)
15. Tong, G., Wu, W., Guo, L., Li, D., Liu, C., Liu, B., Du, D.-Z.: An efficient randomized algorithm for rumor blocking in online social networks. IEEE Trans. Netw. Sci. Eng. **7**, 845–854 (2017)
16. Chen, X., Nong, Q., Feng, Y., Cao, Y., Gong, S., Fang, Q., Ko, K.-I.: Centralized and decentralized rumor blocking problems. J. Comb. Optim. **34**(1), 314–329 (2016). https://doi.org/10.1007/s10878-016-0067-z
17. Ni, Q., Guo, J., Huang, C., Wu, W.: Influence-based community partition with sandwich method for social networks. arXiv preprint arXiv:2003.10439 (2020)
18. Feige, U.: A threshold of ln n for approximating set cover. J. ACM (JACM) **45**(4), 634–652 (1998)
19. Soma, T., Yoshida, Y.: Maximizing monotone submodular functions over the integer lattice. Math. Program. **172**(1–2), 539–563 (2018)
20. Guo, J., Chen, T., Wu, W.: Continuous activity maximization in online social networks. IEEE Trans. Netw. Sci. Eng. 1 (2020)
21. Soma, T., Kakimura, N., Inaba, K., Kawarabayashi, K.: Optimal budget allocation: theoretical guarantee and efficient algorithm. In: International Conference on Machine Learning, pp. 351–359 (2014)
22. Kleywegt, A.J., Shapiro, A., Homem-de Mello, T.: The sample average approximation method for stochastic discrete optimization. SIAM J. Optim. **12**(2), 479–502 (2002)
23. Rossi, R., Ahmed, N.: The network data repository with interactive graph analytics and visualization. In: Twenty-Ninth AAAI Conference on Artificial Intelligence (2015)
24. Kumar, S., Hooi, B., Makhija, D., Kumar, M., Faloutsos, C., Subrahmanian, V.S.: Rev2: fraudulent user prediction in rating platforms. In: Proceedings of the Eleventh ACM International Conference on Web Search and Data Mining, pp. 333–341. ACM (2018)
25. Yang, Y., Mao, X., Pei, J., He, X.: Continuous influence maximization: what discounts should we offer to social network users? In: Proceedings of the 2016 International Conference on Management of Data, pp. 727–741. ACM (2016)

Improved Hardness and Approximation Results for Single Allocation Hub Location

Xing Wang[1], Guangting Chen[2], Yong Chen[1], Guohui Lin[3(✉)] ⓘ,
Yonghao Wang[4], and An Zhang[1(✉)] ⓘ

[1] Department of Mathematics, Hangzhou Dianzi University, Hangzhou, China
{wx198491,chenyong,anzhang}@hdu.edu.cn
[2] Taizhou University, Taizhou, China
gtchen@hdu.edu.cn
[3] Department of Computing Science, University of Alberta, Edmonton, Canada
guohui@ualberta.ca
[4] Computer Science and Technology, Hangzhou Dianzi University, Hangzhou, China
764574072@qq.com

Abstract. Given a metric graph $G = (V, E, w)$ and an integer k, we aim to find a single allocation k-hub location, which is a spanning subgraph consisting of a clique of size k and an independent set of size $|V| - k$, such that each node in the independent set is adjacent to exactly one node in the clique. For various optimization objective functions studied in the literature, we present improved hardness and approximation results.

Keywords: Hub location · NP-hard · Inapproximability · Approximation algorithm

1 Introduction

We study the *single allocation hub location* (SAHL) problem on metric graphs. A metric graph denoted by $G = (V, E, w)$ is an edge-weighted complete graph, where the edge weight function $w(\cdot)$ satisfies the triangle inequality. Mostly due to the important application background in transportation and telecommunication systems, in this context graphs are often referred to as *networks* and the vertices of V are referred to as *nodes*.

Given a metric graph $G = (V, E, w)$ and an integer $0 < k < |V|$, a spanning subgraph H of G is called a *single allocation k-hub location* (or simply k-hub location) if it contains a clique of k nodes so that all the nodes outside the clique are pairwise non-adjacent (i.e., form an independent set) and each of them is adjacent to exactly one node inside the clique [7,9]. In other words, a k-hub location is a *clique-star* with the clique size k. With respect to a k-hub location H in G, the nodes in the clique C are called *hub nodes* or *hub facilities*, the nodes in $V \setminus C$ are called *client nodes* or *demand nodes*, each is *hung on* C or *hung on*

Z. Zhang et al. (Eds.): AAIM 2020, LNCS 12290, pp. 85–96, 2020.
https://doi.org/10.1007/978-3-030-57602-8_8

a hub node in C, and the fact that each demand node is hung on (i.e., adjacent) to exactly a hub node is referred to as the *single allocation* rule.

Hub location is one of the novel and thriving research problems in facility location theory. It has many practical applications in airline, transportation and telecommunication systems [3,10,12]. The SAHL problem concerns how to locate a set of hub facilities and allocate client demands to the facilities so as to project a service path between each origin-destination pair. Many variants of the SAHL problem have been proposed and widely studied, see for example surveys [1,6,12,14] and the references therein. Among them, single allocation is a simple but important construction rule.

Several different optimization objective functions have been proposed for the SAHL problem. Some of them aim to minimize the total transportation distance/cost (i.e., min-sum) while others are to minimize the maximum transportation distance/cost (i.e., min-max), between pairs of nodes in the entire network or in a certain part of the network.

The SAHL problem with the min-sum objective is often recognized as the *k-hub median problem*, for which the first linear integer program was formulated by Compbell [4]. Compbell also proposed two heuristics for the problem [5]. The k-hub median problem is NP-hard when $k \geq 3$, even the hub nodes are fixed [16,19]; most of the research is on (exponential time) exact algorithms or heuristic algorithms, see for example [1,6,12,14].

Approximation algorithms with theoretical quality guarantee have been designed, but limited to subproblems. For example, Iwasa et al. [15] studied a subproblem of the k-hub median problem in *hub-and-spoke* networks where the hub nodes are fixed, and the required amount of flow and the transportation cost per unit flow between each pair of nodes are given; in particular, they presented a deterministic 3-approximation algorithm and a randomized 2-approximation algorithm, and when the number of hub nodes is equal to three, the problem remains NP-hard [20] and the authors provided a better 5/4-approximation algorithm. Chen et al. [9] studied a variant called the *at most k*-hub median problem on a class of generalized so-called Δ_β-*metric* graphs [1], in which the solution subgraph contains less than or equal to k hub nodes; they presented a 2β-approximation algorithm for any $\beta \geq 1/2$ and showed that the problem is already strongly NP-hard when the edge weights are from $\{1, 2, 3\}$ (the so-called $\{1, 2, 3\}$-weighted metric graphs).

In an optimal solution to the k-hub median problem, the maximum distance/cost between two nodes can be arbitrarily large. Compbell [4] also introduced the SAHL problem with the min-max objective, also known as the *k-hub center problem* [1,12]. Kara and Tansel [17] gave a combinatorial formulation of the k-hub center problem too, for which there are several exact and heuristic methods [2,11,18].

From the approximation algorithm perspective, Chen et al. [7] studied the SAHL problem to minimize the diameter of the k-hub location. They first

---

[1] The Δ_β metric uses the *parameterized* triangle inequality $w(v_1, v_2) \leq \beta(w(v_1, u) + w(u, v_2))$, for all nodes $v_1, v_2, u \in V$.

showed that, for any $\epsilon > 0$, the problem cannot be approximated within the factor $4/3 - \epsilon$ unless $P = NP$, even on $\{1, 2, 3\}$-weighted metric graphs; then designed a $5/3$-approximation algorithm for any metric graphs. In [8], Chen et al. extended both the inapproximability and approximation results to the more general Δ_β-metric graphs. They showed that the problem admits an $r(\beta)$-approximation algorithm while cannot be approximated within the factor $g(\beta) - \epsilon$, where $r(\beta)$ and $g(\beta)$ are functions in β; in particular, for the classic metric graphs, $r(1) = 2$ and $g(1) = 3/2$.

In this paper, we study the SAHL problem with both the min-max and min-sum objectives, on the classic metric graphs. The input consists of a metric graph $G = (V, E, w)$ and a positive integer $k < |V|$. For a k-hub location H, let $d_H(u, v)$ denote the shortest distance between a pair of nodes u and v in H; the following defines six different objective functions for minimization:

$$D(H) = \max_{u,v \in V} d_H(u, v), \tag{1}$$

$$D(H \setminus C) = \max_{u,v \in V \setminus C} d_H(u, v), \tag{2}$$

$$w_{\max}(H) = \max_{(u,v) \in E(H)} w(u, v), \tag{3}$$

$$dis(H) = \sum_{u,v \in V} d_H(u, v), \tag{4}$$

$$dis(H \setminus C) = \sum_{u,v \in V \setminus C} d_H(u, v), \tag{5}$$

$$w(H) = \sum_{(u,v) \in E(H)} w(u, v); \tag{6}$$

and we have correspondingly six SAHL problems. We show that it is NP-hard to approximate the SAHL problem to minimize $D(H)$ ($D(H \setminus C)$, $w_{\max}(H)$, respectively) within the factor $4/3 - \epsilon$ ($4/3 - \epsilon$, $2 - \epsilon$, respectively), for any $\epsilon > 0$; and we propose 2-approximation algorithms for the SAHL problem to minimize $D(H \setminus C)$ and $w_{\max}(H)$, respectively. We lastly prove that the SAHL problem to minimize any of the three min-sum objectives is strongly NP-hard. These and the previously best hardness and approximation results on the SAHL problems are summarized in Table 1, in which our improved results are marked with *.

2   Hardness Results

In this section, we prove several complexity and inapproximability results through polynomial reductions from the EXACT COVER BY 3-SETS (X3C), which is known strongly NP-complete [13]. Recall that the SAHL problem to minimize $D(H)$ has been shown not approximable within the factor $4/3 - \epsilon$ on $\{1, 2, 3\}$-weighted metric graphs, for any $\epsilon > 0$, unless $P = NP$ [7], and the *at most* k-hub median problem has been shown strongly NP-hard on $\{1, 2, 3\}$-weighted metric graphs [9]. Our new results in the second column of Table 1 are

Table 1. The state-of-the-art hardness and approximation results on SAHL. A '*' indicates a result achieved in this paper; a '-' suggests no known result, to the best of our knowledge; a '+' suggests an implied result from another cell in the table; 's. NP-hard' means 'strongly NP-hard'.

Objective	$\{1,2\}$-weighted metric graphs		General graphs	
$D(H)$	$(4/3 - \epsilon)$-inapprox*	4/3-approx*	$(3/2 - \epsilon)$-inapprox [8]	5/3-approx [7]
$D(H \setminus C)$	$(4/3 - \epsilon)$-inapprox*	4/3-approx*	+	2-approx*
$w_{\max}(H)$	$(2 - \epsilon)$-inapprox*	+	+	2-approx*
$dis(H)$	s. NP-hard*	+	+	2-approx [9]
$dis(H \setminus C)$	s. NP-hard*	2-approx*	+	-
$w(H)$	s. NP-hard*	2-approx*	+	-

all on $\{1,2\}$-weighted graphs (which are surely metric and are special cases of $\{1,2,3\}$-weighted metric graphs), and thus they are improvements over the prior results.

Definition 1. EXACT COVER BY 3-SETS (X3C)

Input: A universe $\mathcal{U}$ of $3q$ elements and a collection $\mathcal{S}$ of m 3-subsets of $\mathcal{U}$.
Query: Is there a sub-collection $\mathcal{S}^* \subset \mathcal{S}$ with $|\mathcal{S}^*| = q$ such that $\mathcal{S}^*$ covers all elements, i.e., $\bigcup_{S \in \mathcal{S}^*} S = \mathcal{U}$?

Theorem 1. For any $\epsilon > 0$, it is NP-hard to approximate the SAHL problem to minimize $D(H)$ or $D(H \setminus C)$ within the factor $4/3 - \epsilon$, even restricted on $\{1,2\}$-weighted graphs.

Proof. Given an instance I of X3C (without loss of generality, we suppose $m \geq q \geq 4$), we construct the following instance I' of SAHL.

Let $k = q$ and let $G = (V, E, w)$ be a complete graph, where V consists of all the elements and subsets in I, i.e., $V = \mathcal{U} \cup \mathcal{S}$. Each edge connecting two elements has a weight 2. Each edge connecting two subsets has a weight 1 if they do not intersect each other or 2 otherwise. If an element is included by a subset, then their edge has a weight 1; otherwise, their edge has a weight 2.

On one hand, if the X3C instance I admits a sub-collection $\mathcal{S}^* \subset \mathcal{S}$ with $|\mathcal{S}^*| = q$ such that $\mathcal{S}^*$ covers all elements, then a solution to the SAHL instance I', i.e., a spanning subgraph H of G with $D(H) = D(H \setminus C) = 3$ can be constructed as follows. Let $\mathcal{S}^*$ be the set of hub nodes. Each element is hung on the subset in $\mathcal{S}^*$ that covers it, and each subset outside $\mathcal{S}^*$ is hung on one of the subsets in $\mathcal{S}^*$ that has no intersection with it. Since all elements are covered by $\mathcal{S}^*$ and notice that at most three subsets in $\mathcal{S}^*$ might have intersection with a subset outside $\mathcal{S}^*$ (hence, at least one subset in $\mathcal{S}^*$ does not intersect with it due to $q \geq 4$), the spanning subgraph H is a k-hub location. Moreover, each edge of H must have a weight 1 by the construction of I'. Hence we obtain $D(H) = D(H \setminus C) = 3$.

On the other hand, if the SAHL instance I' has a k-hub location H with $D(H) = 3$ or $D(H \backslash C) = 3$, then we claim the clique C of hub nodes must consist of q subsets in S and cover all elements. In fact, if there are two elements each of which is either hung on a subset that does not cover it or on an element in C, then their distance must be greater than 3. Moreover, if C covers all elements but one, then it must consist entirely of subsets in S. Consider the subset that hangs the element not covered by C, at most two other elements are hung on it. Thus there must be an element which is not hung on this subset. The distance between this element and the element not covered by C must be greater than 3 too. Hence, we can deduce that all elements must be covered by C. That is, C gives a YES answer to the X3C instance I.

By the above argument, the X3C instance I answers YES if and only if the SAHL instance I' has a k-hub location H with $D(H) = 3$ or $D(H \backslash C) = 3$. Hence, for any $\epsilon > 0$, if there is a $(4/3 - \epsilon)$-approximation algorithm for the SAHL problem to minimize $D(H)$ or $D(H \backslash C)$, then it can answer whether the corresponding instance of X3C answers YES or not, which cannot be true unless $P = NP$. This proves the theorem.   □

By the same reduction, we have the following two results.

Theorem 2. *For any $\epsilon > 0$, it is NP-hard to approximate the* SAHL *problem to minimize $w_{\max}(H)$ within the factor $2 - \epsilon$, even restricted on $\{1, 2\}$-weighted graphs.*

Theorem 3. *It is strongly NP-hard to approximate the* SAHL *problem to minimize $w(H)$, even restricted on $\{1, 2\}$-weighted graphs.*

Theorem 4. *It is strongly NP-hard to approximate the* SAHL *problem to minimize $dis(H)$, even restricted on $\{1, 2\}$-weighted graphs.*

Proof. Given an instance I of X3C, we first add N new elements to $\mathcal{U}$ and add the set $\mathcal{N}$ containing all these N elements to S, where $N \gg mq$ (precisely, $N = m^2 + (9q + 2)m + 13q^2 + 7q + 8$). For convenience, the set $\mathcal{N}$ is referred to as a 'subset' too. Now I contains $N + 3q$ elements and $m + 1$ subsets. We construct the following instance I' of SAHL.

Let $k = q + 1$ and let $G = (V, E, w)$ be a complete graph, where V consists of all the elements and subsets in I, i.e., $V = \mathcal{U} \cup \mathcal{N} \cup S \cup \{\mathcal{N}\}$. The weights of edges are set in the same way as in the proof of Theorem 1. Denote $Y = N^2 + (2m + 9q)N + m^2 + 9qm + 10q^2 - 4q$.

On one hand, if the X3C instance I admits a sub-collection $S^* \subset S$ with $|S^*| = q$ such that S^* covers all elements in $\mathcal{U}$, then a solution to the SAHL instance I', i.e., a k-hub location H with $dis(H) \leq Y$, can be constructed as follows. Let $C = S^* \cup \{\mathcal{N}\}$ be the set of hub nodes. Each element is hung on the subset in C that covers it, and each subset outside C is hung on $\mathcal{N}$. Accordingly, we see that

$$\sum_{u,v \in C} d_H(u, v) = \frac{q(q + 1)}{2},$$

$$\sum_{u \in V \setminus C, v \in C} d_H(u,v) = (N+m-q+3q)(2q+1),$$

and

$$\sum_{u,v \in V \setminus C} d_H(u,v) = 6q + (N+m-q)(N+m-q-1) + 3 \times \left(\frac{9q(q-1)}{2} + 3q(N+m-q) \right).$$

Hence,

$$dis(H) = \sum_{u,v \in C} d_H(u,v) + \sum_{u \in V \setminus C, v \in C} d_H(u,v) + \sum_{u,v \in V \setminus C} d_H(u,v)$$
$$= N^2 + (2m+9q)N + m^2 + 9qm + 10q^2 - 4q = Y.$$

On the other hand, if the SAHL instance I' has a k-hub location H with $dis(H) \leq Y$, then we have the following four claims.

Claim 1. $\mathcal{N} \in C$.

If $\mathcal{N} \notin C$, then the distance of any two elements in $\mathcal{N} \setminus C$ must be at least 4. We thus derive a contradiction:

$$dis(H) > \sum_{u,v \in \mathcal{N} \setminus C} d_H(u,v) \geq 2(N-q-1)(N-q-2) > Y.$$

Claim 2. Any element of $\mathcal{N}$ must be connected to the subset $\mathcal{N}$.

If an element of $\mathcal{N}$ is not connected to the subset $\mathcal{N}$, then it must be hung on a node in $C \setminus \{\mathcal{N}\}$. Hence it is at least 4 away from all elements in $\mathcal{N} \setminus C$ and all elements but possibly one (which hangs it) in $\mathcal{N} \cap C$. Let N_0 denote the number of elements in $\mathcal{N}$ that are not connected to $\mathcal{N}$. Let $N_{in} = |\mathcal{N} \cap C|$ and let N_{out} be the number of elements in $\mathcal{N}$ which are hung on $\mathcal{N}$. Thus we have $N = N_0 + N_{out} + N_{in}$, implying that $N_0 + N_{out} \geq N - q$. Suppose that $N_0 \geq 1$, then we have

$$\sum_{u,v \in \mathcal{N}} d_H(u,v) \geq (N-N_0)(N-N_0-1) + 4N_0(N-1) \geq N^2 + N - 2,$$

$$\sum_{u=\mathcal{N}, v \in \mathcal{N}} d_H(u,v) \geq N_{out} + N_{in} + 3N_0 \geq N + 2,$$

and

$$\sum_{u \in \mathcal{U} \cup \mathcal{S}, v \in \mathcal{N}} d_H(u,v) \geq 3N_0(|\mathcal{U}|+|\mathcal{S}|-1) + 2N_0 + 2N_{in}(|\mathcal{U}|+|\mathcal{S}|) + 2N_{out}|\mathcal{S}| + 3N_{out}|\mathcal{U}|$$
$$= 2N(|\mathcal{U}|+|\mathcal{S}|-1) + (N_0+N_{out})|\mathcal{U}| + 2N_0 + 2N_{in} + 2N_{out}$$
$$\geq 2N(m+3q-1) + 3q(N-q) + 2N = (2m+9q)N - 9q^2.$$

Hence we again derive a contradiction:

$$dis(H) > \sum_{u,v \in \mathcal{N}} d_H(u,v) + \sum_{u=\mathcal{N}, v \in \mathcal{N}} d_H(u,v) + \sum_{u \in \mathcal{U} \cup \mathcal{S}, v \in \mathcal{N}} d_H(u,v)$$

$$\geq N^2 + (2m + 9q + 2)N - 9q^2 > Y.$$

Claim 3. All subsets in $\mathcal{S} \setminus C$ must be hung on $\mathcal{N}$, and all elements in $\mathcal{U} \setminus C$ must be hung either on a subset of $\mathcal{S} \cap C$ that covers it, or on $\mathcal{N}$.

By Claim 2, we know that $N_0 = 0$ and hence $N_{out} = |\mathcal{N} \setminus C| = N - N_{in} \geq N - q$. By a similar calculation, we have

$$\sum_{u,v \in \mathcal{N}} d_H(u,v) = N(N-1), \quad \sum_{u=\mathcal{N}, v \in \mathcal{N}} d_H(u,v) = N.$$

Note that a subset of $\mathcal{S} \setminus C$ not hung on $\mathcal{N}$ must be at least 3 away from each element in $\mathcal{N} \setminus C$, and an element of $\mathcal{U} \setminus C$ that is not hung on a subset of $\mathcal{S} \cap C$ that covers it or on $\mathcal{N}$ must be at least 4 away from each element in $\mathcal{N} \setminus C$. Therefore, if the claim is false, then we must have

$$\sum_{u \in \mathcal{U} \cup \mathcal{S}, v \in \mathcal{N}} d_H(u,v) \geq 2N_{in}(|\mathcal{U}| + |\mathcal{S}|) + 2N_{out}|\mathcal{S}| + 3N_{out}|\mathcal{U}| + N_{out}$$

$$= 2N(|\mathcal{U}| + |\mathcal{S}|) + N_{out}(|\mathcal{U}| + 1)$$
$$\geq 2N(m + 3q) + (N - q)(3q + 1)$$
$$= (2m + 9q + 1)N - 3q^2 - q.$$

Hence we derive a contradiction:

$$dis(H) > \sum_{u,v \in \mathcal{N}} d_H(u,v) + \sum_{u=\mathcal{N}, v \in \mathcal{N}} d_H(u,v) + \sum_{u \in \mathcal{U} \cup \mathcal{S}, v \in \mathcal{N}} d_H(u,v)$$
$$\geq N^2 + (2m + 9q + 1)N - 3q^2 - q > Y.$$

Denote $S = |\mathcal{S}| = m$ and $U = |\mathcal{U}| = 3q$. Let $S_{in} = |\mathcal{S} \cap C|$ and $S_{out} = |\mathcal{S} \setminus C|$, then $S_{in} + S_{out} = S$. Let $U_{in} = |\mathcal{U} \cap C|$ and $U_{out} = |\mathcal{U} \setminus C|$, then $U_{in} + U_{out} = U$ and $U_{in} + S_{in} + N_{in} = q$. Moreover, suppose there are U_{out}^1 elements of $\mathcal{U} \setminus C$ hung on $\mathcal{S} \cap C$ and U_{out}^2 elements of $\mathcal{U} \setminus C$ hung on $\mathcal{N}$. Thus $U_{out} = U_{out}^1 + U_{out}^2$. By Claim 1–3, the connection between nodes must be as shown in Fig. 1. Then we have another claim.

Claim 4. $U_{out}^2 = 0$, $N_{in} = 0$ and $U_{in} = 0$.

By a more careful calculation, we have

$$\sum_{u \in \mathcal{N}, v \in V} d_H(u,v) = N + S + 2U, \quad \sum_{u,v \in \mathcal{N}} d_H(u,v) = N(N-1), \quad \sum_{u \in \mathcal{S}, v \in \mathcal{N}} d_H(u,v) = 2NS,$$

$$\sum_{u \in \mathcal{U}, v \in \mathcal{N}} d_H(u,v) = 3NU_{out} + 3N_{out}U_{in} + 2N_{in}U_{in}$$

$$= 3N(U - U_{in}) + 3(N - N_{in})U_{in} + 2N_{in}U_{in}$$
$$= 3NU - N_{in}U_{in},$$

and

$$\sum_{u,v \in \mathcal{S}} d_H(u,v) \geq S(S-1) - \frac{1}{2}S_{in}(S_{in} - 1),$$

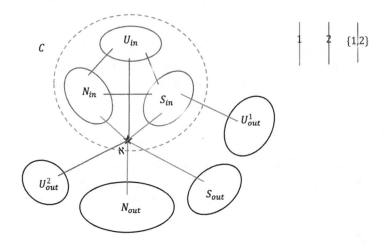

Fig. 1. An illustration of the k-hub location H meeting Claims 1–3.

$$\sum_{u \in \mathcal{U}, v \in \mathcal{S}} d_H(u,v) \geq 3SU_{out}^2 + 3(U_{in} + U_{out}^1)S_{out} + (U_{in} + U_{out}^1)[1 + 2(S_{in} - 1)]$$

$$= 3SU_{out}^2 + (U_{in} + U_{out}^1)S_{out} + 2S(U_{in} + U_{out}^1) - (U_{in} + U_{out}^1)$$
$$= 3SU - (U_{in} + U_{out}^1)(S - S_{out} + 1)$$
$$= 3SU - (U - U_{out}^2)(S_{in} + 1).$$

Note that any subset of $\mathcal{S}$ covers at most three elements of $\mathcal{U}$, each of the U_{out}^1 elements that are hung on a subset of $\mathcal{S} \cap C$ must be at least 3 away from at least $U_{out}^1 - 3$ elements of them, and at least 2 away from the others of them. Hence, we obtain

$$\sum_{u,v \in \mathcal{U}} d_H(u,v) \geq 2U_{out}^2(U_{out}^2 - 1) + 4U_{out}^2(U_{in} + U_{out}^1) + U_{in}(U_{in} - 1)$$

$$+ 2U_{in}U_{out}^1 + \frac{1}{2}U_{out}^1[4 + 3(U_{out}^1 - 3)]$$
$$= 2U_{out}^2(U - 1) + 2U_{out}^2(U_{in} + U_{out}^1)$$
$$+ (U_{in} + U_{out}^1)(U_{in} + U_{out}^1 - 1) + \frac{1}{2}U_{out}^1(U_{out}^1 - 3)$$
$$= 2U_{out}^2(U - 1) + (U_{in} + U_{out}^1)(U + U_{out}^2 - 1) + \frac{1}{2}U_{out}^1(U_{out}^1 - 3)$$
$$= U(U - 1) + U_{out}^2(2U - U_{out}^2 - 1) + \frac{1}{2}U_{out}^1(U_{out}^1 - 3).$$

Summing up the above, we have

$$dis(H) \geq N^2 + (2S + 3U)N + S^2 + 3SU + U^2 + \Delta$$

$$= N^2 + (2m + 3q)N + m^2 + 9qm + 9q^2 + \Delta,$$

where

$$\Delta = U - N_{in}U_{in} - \frac{1}{2}S_{in}(S_{in} - 1) - (U - U_{out}^2)(S_{in} + 1) + U_{out}^2(2U - U_{out}^2 - 1)$$

$$+ \frac{1}{2}U_{out}^1(U_{out}^1 - 3)$$

$$= U_{out}^2(2U + S_{in} - U_{out}^2) + \frac{1}{2}U_{out}^1(U_{out}^1 - 3) - \frac{1}{2}S_{in}(S_{in} - 1) - US_{in} - N_{in}U_{in}$$

$$= U_{out}^2(2U + S_{in} - U_{out}^2) + \frac{1}{2}(U - U_{out}^2 - U_{in})(U - U_{out}^2 - U_{in} - 3)$$

$$- \frac{1}{2}(q - N_{in} - U_{in})(q - N_{in} - U_{in} - 1) - U(q - N_{in} - U_{in}) - N_{in}U_{in}$$

$$= \frac{U(U - 3)}{2} - \frac{q(q - 1)}{2} - Uq + U_{out}^2(U + U_{in} + S_{in} - \frac{U_{out}^2}{2} + \frac{3}{2})$$

$$+ \frac{1}{2}(N_{in} + U_{in})(2U + 2q - N_{in} - U_{in} - 1) - UU_{in} + \frac{U_{in}(U_{in} + 3)}{2} - N_{in}U_{in}$$

$$\geq q^2 - 4q + \frac{U_{out}^2(U_{out}^2 + 3)}{2} + U(N_{in} + U_{in}) - UU_{in} + \frac{U_{in}(U_{in} + 3)}{2} - N_{in}U_{in}$$

$$\geq q^2 - 4q + \frac{U_{out}^2(U_{out}^2 + 3)}{2} + 2qN_{in} + \frac{U_{in}(U_{in} + 3)}{2}.$$

Therefore,

$$dis(H) \geq Y + \frac{U_{out}^2(U_{out}^2 + 3)}{2} + 2qN_{in} + \frac{U_{in}(U_{in} + 3)}{2},$$

which leads to $U_{out}^2 = 0, N_{in} = 0$ and $U_{in} = 0$ by the assumption $dis(H) \leq Y$.

By Claims 1–4, we conclude that $S \cap C$ covers all elements of $\mathcal{U}$. Hence it gives a "YES" answer to the X3C instance. This proves the theorem.   □

The same reduction also gives the following result.

Theorem 5. *It is strongly NP-hard to approximate the SAHL problem to minimize* $dis(H \setminus C)$, *even restricted on* $\{1, 2\}$-*weighted graphs.*

3   Approximation Algorithms

Chen et al. [7] have introduced several ideas for constructing a k-hub location H to minimize $D(H)$, among which, we are most interested in the one that hangs all demand nodes on a common hub node. We call such a solution a *star-like* k-hub location below.

Let H^* be an optimal k-hub location and C^* be the set of hub nodes in H^*. Firstly, if G is a $\{1, 2\}$-weighted graph, then whether $D(H^*) = 2$ (resp. $D(H^* \setminus C^*) = 2$) or not can be determined in polynomial time. In fact, $D(H^*) = 2$ (resp. $D(H^* \setminus C^*) = 2$) if and only if there is a node that is unit weight away from all the other $n - 1$ nodes (resp. at least $n - k$ other nodes) of V. Thus we only need to check the star-like k-hub locations, which can be done in $O(n^2)$ time (see APPROX-1). If $D(H^*) \geq 3$ (resp. $D(H^* \setminus C^*) \geq 3$), then any star-like k-hub location H must have $D(H) \leq 4 \leq \frac{4}{3}D(H^*)$ (resp. $D(H \setminus C) \leq \frac{4}{3}D(H^* \setminus C^*)$).

Theorem 6. *Given a $\{1,2\}$-weighted graph G, the algorithm* APPROX-1 *outputs a star-like k-hub location H with $D(H) \leq \frac{4}{3}D(H^*)$ (resp. $D(H \setminus C) \leq \frac{4}{3}D(H^* \setminus C^*))$ (Fig. 2).*

APPROX-1 for SAHL to minimize $D(H)$ (resp. $D(H \setminus C)$):

1. For any node $v \in V$, construct a star-like k-hub location H_v by adding v and $k-1$ furthest neighbors of v to C and hanging all nodes in $V \setminus C$ on v.
2. Return the best star-like k-hub location H from $\{H_v \mid v \in V\}$.

Fig. 2. A description of APPROX-1.

On general metric graphs, any star-like k-hub location is a 2-approximation for minimizing $D(H)$ [7]. However, it can be extremely bad if the objective is to minimize $D(H \setminus C)$. We show next that APPROX-1 remains a 2-approximation for minimizing $D(H \setminus C)$.

Theorem 7. *The algorithm* APPROX-1 *is a 2-approximation algorithm for the SAHL problem on general metric graphs to minimize $D(H \setminus C)$.*

Proof. Assume H^* is an optimal k-hub location in which C^* is the set of hub nodes. Let v^* be a hub node in C^* which has at least one leaf $u^* \in V \setminus C^*$, and consider the star-like k-hub location H_{v^*}. Suppose that u is the k-th furthest neighbor of v^*, then $u \in V \setminus C$ and $D(H_{v^*} \setminus C) \leq 2w(u, v^*)$. If $w(u^*, v^*) \geq w(u, v^*)$, then

$$D(H \setminus C) \leq D(H_{v^*} \setminus C) \leq 2w(u, v^*) \leq 2w(u^*, v^*) \leq 2D(H^* \setminus C^*).$$

Otherwise, $u^* \in V \setminus C$. Since $v^* \in C^*$ and $|C \setminus \{v^*\} \cup \{u\}| = k$, at least one node of $C \setminus \{v^*\} \cup \{u\}$ must be outside C^*. Denote one such node by v, then $d_{H^*}(v, v^*) \geq w(v, v^*) \geq w(u, v^*)$. Thus the distance between the two leaves, v and u^* of H^*, must satisfy

$$d_{H^*}(v, u^*) \geq d_{H^*}(v, v^*) + w(u^*, v^*) > w(u, v^*),$$

which leads to $D(H \setminus C) \leq D(H_{v^*} \setminus C) \leq 2w(u, v^*) < 2d_{H^*}(v, u^*) \leq 2D(H^* \setminus C^*)$. This proves the theorem. $\square$

For minimizing $w_{\max}(H) = \max_{(u,v) \in E(H)} w(u, v)$, any feasible k-hub location is a trivial 2-approximation on $\{1,2\}$-weighted graphs. Our next algorithm APPROX-2 on general metric graphs also picks a star-like k-hub location, but in an alternative way (Fig. 3).

Approx-2 for SAHL to minimize $w_{\max}(H)$:

1. For any node $v \in V$, construct a star-like k-hub location H_v by adding v and $k-1$ nearest neighbors of v to C and hanging all nodes in $V \setminus C$ on v.
2. Return the best star-like k-hub location H from $\{H_v \mid v \in V\}$.

Fig. 3. A description of Approx-2.

Theorem 8. *The algorithm* Approx-2 *is a 2-approximation algorithm for the* SAHL *problem on general metric graphs to minimize* $w_{\max}(H)$.

Proof. Assume H^* is an optimal k-hub location in which C^* is the set of hub nodes. For each $v \in V \setminus C^*$, let $f^*(v)$ denote its unique neighbor in C^*. Let $v^* = \arg\max_{v \in V \setminus C^*} w(f^*(v), v)$ and $\ell = w(f^*(v^*), v^*)$. Let $\gamma = \max_{u \in C^*}\{w(v^*, u)\}$. Then $w_{\max}(H^*) \geq \max\{\ell, \gamma\}$. For any $v \in V \setminus C^*$, $w(v, v^*) \leq w(v, f^*(v)) + w(v^*, f^*(v)) \leq \ell + \gamma$.

Consider the star-like k-hub location H_{v^*}. Clearly, $w_{\max}(H) \leq w_{\max}(H_{v^*})$. Since Approx-2 adds the $k-1$ nearest neighbors of v to C, we have $w(v^*, u) \leq \gamma$ for any $u \in C$, and $w(v, v^*) \leq \ell + \gamma$ for any $v \in V \setminus C$. Moreover, for any $u_1, u_2 \in C \setminus \{v^*\}$, $w(u_1, u_2) \leq w(v^*, u_1) + w(v^*, u_2) \leq 2\gamma$. Hence, we can conclude that $w_{\max}(H_{v^*}) \leq \max\{\ell + \gamma, 2\gamma\} \leq 2w_{\max}(H^*)$. This proves the theorem. $\square$

For minimizing $dis(H \setminus C)$ or $w(H)$, any star-like k-hub location is a 2-approximation for the SAHL problem on $\{1, 2\}$-weighted graphs; we leave it open on how to approximate the problem on general metric graphs.

4 Conclusions

We studied the single allocation k-hub location problem on metric graphs, and obtained several improved negative hardness results and several positive approximation algorithms. We showed that the SAHL problem is either strongly NP-hard or APX-hard for all six objectives, even restricted on $\{1, 2\}$-weighted graphs; some of the same conclusions were previously established on the larger class of $\{1, 2, 3\}$-weighted metric graphs [7,9]. The presented approximation algorithms aim to find the best star-like k-hub location, and they were proved to have solid performance. With respect to Table 1, the SAHL problem to minimize $w_{\max}(H)$ has now been completely solved; but for the other five objectives there are gaps, some bigger than the other, between the lower bound of inapproximability and the state-of-the-art approximation ratio, each of which deserves further research.

Acknowledgements. XW, GC, YC and AZ are supported by the NSFC Grants 11771114 and 11971139; YC and AZ are supported by the CSC Grants 201508330054 and 201908330090, respectively. GL is supported by the NSERC Canada.

References

1. Alumur, S.A., Kara, B.Y.: Network hub location problems: The state of the art. Eur. J. Oper. Res. **190**, 1–21 (2008)
2. Brimberg, J., Mladenović, N., Todosijević, R., Urošević, D.: General variable neighborhood search for the uncapacitated single allocation p-hub center problem. Optim. Lett. **11**, 377–388 (2017)
3. Bryan, D.L., O'Kelly, M.E.: Hub-and-spoke networks in air transportation: an analytical review. J. Reg. Sci. **39**, 275–295 (1999)
4. Campbell, J.F.: Integer programming formulations of discrete hub location problems. Eur. J. Oper. Res. **72**, 387–405 (1994)
5. Campbell, J.F.: Hub locaiton and the p-hub median problem. Oper. Res. **44**, 1–13 (1996)
6. Campbell, J.F., O'Kelly, M.E.: Twenty-five years of hub location research. Transp. Sci. **46**, 153–169 (2012)
7. Chen, L.-H., Cheng, D.-W., Hsieh, S.-Y., Hung, L.-J., Lee, C.-W., Wu, B.Y.: Approximation algorithms for single allocation k-hub center problem. In: Proceedings of the 33rd Workshop on Combinatorial Mathematics and Computation Theory (CMCT 2016), pp. 13–18 (2016)
8. Chen, L.-H., Hsieh, S.-Y., Hung, L.-J., Klasing, R.: The approximability of the p-hub center problem with parameterized triangle inequality. In: Proceedings of the 23rd Annual International Computing and Combinatorics Conference (COCOON 2017), pp. 112–123 (2017)
9. Chen, L.-H., Hsieh, S.-Y., Hung, L.-J., Klasing, R.: Approximation algorithms for the p-hub center routing problem in parameterized metric graphs. Theor. Comput. Sci. **806**, 271–280 (2020)
10. Chung, S.H., Myung, Y.S., Tcha, D.W.: Optimal design of a distributed network with a two-level hierarchical structure. Eur. J. Oper. Res. **62**, 105–115 (1992)
11. Ernst, A.T., Hamacher, H., Jiang, H., Krishnamoorthy, M., Woeginger, G.: Uncapacitated single and multiple allocation p-hub center problems. Comput. Oper. Res. **36**, 2230–2241 (2009)
12. Farahani, R.Z., Hekmatfar, M., Arabani, A.B., Nikbakhsh, E.: Hub location problems: A review of models, classification, solution techniques, and applications. Comput. Ind. Eng. **64**, 1096–1109 (2013)
13. Garey, M.R., Johnson, D.S.: Computers and Intractability: A Guide to the Theory of NP-Completeness. W. H. Freeman & Co., San Francisco (1979)
14. Hsieh, S.Y., Kao, S.S.: A survey of hub location problems. J. Interconnection Netw. **19**, 1940005 (2019)
15. Iwasa, M., Satio, H., Matsui, T.: Approximation algorithms for the single allocation problem in hub-and-spoke networks and related metric labeling problems. Discrete Appl. Math. **157**, 2078–2088 (2009)
16. Kara, B.Y.: Modeling and analysis of issues in hub location problems. Ph.D. thesis, Bilkent University Industrial Engineering Department (1999)
17. Kara, B.Y., Tansel, B.C.: On the single-assignment k-hub center problem. Eur. J. Oper. Res. **125**, 648–655 (2000)
18. Meyer, T., Ernst, A.T., Krishnamoorthy, M.: A 2-phase algorithm for solving the single allocation p-hub center problems. Comput. Oper. Res. **36**, 3143–3151 (2009)
19. Sohn, J., Park, S.: A linear program for the two-hub location problem. Eur. J. Oper. Res. **100**, 617–622 (1997)
20. Sohn, J., Park, S.: The single-allocation problem in the interacting three-hub network. Networks **35**, 17–25 (2000)

Approximation Algorithm
for the Balanced 2-correlation Clustering
Problem on Well-Proportional Graphs

Sai Ji[1], Dachuan Xu[1], Donglei Du[2], and Ling Gai[3(✉)]

[1] Department of Operations Research and Information Engineering,
Beijing University of Technology, Beijing 100124, People's Republic of China
`jisai@emails.bjut.edu.cn, xudc@bjut.edu.cn`
[2] Faculty of Management, University of New Brunswick, Fredericton,
NB E3B 9Y2, Canada
`ddu@unb.ca`
[3] Glorious Sun School of Business and Management, Donghua University,
Shanghai 200051, People's Republic of China
`lgai@dhu.edu.cn`

Abstract. In this paper, we consider the balanced 2-correlation clustering problem on well-proportional graphs, which has applications in protein interaction networks, cross-lingual link detection, communication networks, among many others. Given a complete graph $G = (V, E)$ with each edge $(u, v) \in E$ labeled by $+$ or $-$, the goal is to partition the vertices into two clusters of equal size to minimize the number of positive edges whose endpoints lie in different clusters plus the number of negative edges whose endpoints lie in the same cluster. We provide a $(3, \max\{4(M + 1), 16\})$-balanced approximation algorithm for the balanced 2-correlation clustering problem on M-proportional graphs. Namely, the cost of the vertex partition $\{V_1, V_2\}$ returned by the algorithm is at most $\max\{4(M + 1), 16\}$ times the optimum solution, and $\min\{|V_1|, |V_2|\} \leq 3 \max\{|V_1|, |V_2|\}$.

Keywords: Balanced · k-correlation clustering · Well-proportional graphs · Approximation algorithm

1 Introduction

Clustering problems arise in many applications such as machine learning, computer vision, data mining and data compression. These problems have been widely studied in the literature [2,7,11,12,22,23].

The correlation clustering problem, introduced by Bansal et al. [8], has applications in protein interaction networks, cross-lingual link detection, communication networks, and so on. This problem has two versions: minimizing disagreement and maximizing agreement. In the minimizing disagreement version, we are given a complete graph $G = (V, E)$, where each edge $(u, v) \in E$ is labeled by

© Springer Nature Switzerland AG 2020
Z. Zhang et al. (Eds.): AAIM 2020, LNCS 12290, pp. 97–107, 2020.
https://doi.org/10.1007/978-3-030-57602-8_9

+ or − depending on whether vertex u and vertex v are similar or different. The positive edges whose endpoints lie in different clusters and the negative edges whose endpoints lie in the same cluster are called the error edges. The goal of this version is to partition the vertex set V into disjoint clusters so as to minimize the total number of error edges. In other words, the goal is to partition the vertices into clusters such that the vertices lie in the same cluster are as similar as possible and the vertices between the different clusters are as dissimilar as possible. In the maximizing agreement version, the negative edges whose endpoints are in different clusters and the positive edges whose endpoints are within the same cluster are called the right edges. The goal of this version is to partition the vertex set V into disjoint clusters so as to maximize the total number of right edges.

Bansal et al. [8] prove that the correlation clustering problem is NP-hard, that is, one cannot obtain an optimal solution in polynomial time unless $P = NP$. There are many existing work on approximation algorithms for this problem [1, 10, 15, 17, 26].

For the minimizing disagreement version, Bansal et al. [8] give the first constant-factor approximation algorithm. Demaine et al. [15] design an $O(\log n)$-approximation algorithm for general graphs. Charikar et al. [13] prove that the minimizing disagreement problem is APX-hard. They provide a 4-approximation algorithm on complete graphs as well as an $O(\log n)$-approximation algorithm for general graphs based on the LP-rounding technique. Ailon et al. [5] provide a randomized 3-approximation algorithm and further improve the ratio to 2. Chawla et al. [14] provide a 2.06-approximation algorithm based on the LP-rounding technique, which is the best deterministic approximation algorithm for this problem. When the graph is a complete bipartite graph, Amit [6] provides an LP-rounding 11-approximation algorithm. Ailon et al. [4] give a deterministic LP-rounding 4-approximation algorithm as well as a randomized LP-rounding 4-approximation algorithm based on [5].

For the maximizing agreement version, Bansal et al. [8] give a polynomial time approximation scheme. Charikar et al. [13] provide a 0.766-approximation algorithm by the semidefinite programming technique. Swamy [30] gives a 0.75-approximation algorithm for the weighted maximizing agreement problem based on the technique of [18], and the number of clusters returned by this algorithm is at most 4. Furthermore, he improves the approximation ratio from 0.75 to 0.766 by the technique of [16], and the maximum number of clusters returned by the latter algorithm is 6.

There are many variants of the correlation clustering problem [3, 17, 20, 25–28].

One particular problem that is relevant to our problem in this paper is the correlation clustering problem with a fixed number of clusters [8], where the maximum number of clusters that we are allowed to use is k. We use Max-Agree[k] and MinDisAgree[k] to denote the minimizing disagreement version and the maximizing agreement version of the correlation clustering with a fixed number of clusters, respectively. Shamir et al. [29] prove both the MaxAgree[k] and MinDisAgree[k] are NP-hard for every $k \geq 2$ by using a rather complicated reduction. Giotis and Guruswami [17] provide a much simpler NP-hardness proof and prove that both MaxAgree[k] and MinDisAgree[k] admit a polynomial time approximation scheme for every $k \geq 2$.

Balanced clustering arise in many application such as wireless sensor networks, routing and resource allocation problem. It emphasizes the concept of the average and fairness. The balanced clustering problems have been widely studied in the literature [9,19,21,24,31]. There is no requirement for the number of vertices in each cluster in the previous study of correlation clustering problems. For this reason, we introduce the balanced concept into the correlation clustering problem.

In this paper, we study the minimizing disagreement version of the balanced k-correlation clustering problem. In this problem, the number of clusters equals k and each cluster contains the same number of vertices. In particular, we consider the balanced 2-correlation clustering problem on M-proportional graphs and give a polynomial time $(3, \max\{4(M+1), 16\})$-balanced approximation algorithm for this problem. Namely, the solution $\mathcal{C} = \{V_1, V_2\}$ returned by the algorithm has the cost that is at most $\max\{4(M+1), 16\}$ times the optimum solution and $\min\{|V_1|, |V_2|\} \leq 3 \max\{|V_1|, |V_2|\}$.

The rest of this paper is organized as follows. In Sect. 2, we provide some definitions as well as the formulation of the balanced k-correlation clustering problem. The algorithm is provided in Sect. 3 and the theoretical analysis is presented in Sect. 4. Some discussions are given in Sect. 5.

2   Preliminaries

In this section, we provide the definitions of the minimization version of the k-correlation clustering problem and the minimizing disagreement version of the balanced k-correlation clustering problem. Furthermore, we give the formulation for the minimizing disagreement version of the balanced k-correlation clustering problem. The problem we studied in this paper are all for the minimizing disagreement version.

Definition 1 (k-correlation clustering problem). *In this problem, we are given a labeled complete graph $G = (V, E)$ together with a positive integer k, and each edge $(u, v) \in E$ is labeled by $+$ or $-$. We use set E^+ and set E^- to denote the sets of positive and negative edges, respectively. The goal of this problem is to partition the vertices into k clusters so as to minimize the number of positive edges whose endpoints lie in different clusters plus the number of negative edges whose endpoints lie in the same cluster.*

Definition 2 (Balanced k-correlation clustering problem). *In this problem, we are given a labeled complete graph $G = (V, E)$ together with a positive integer k, and each edge $(u, v) \in E$ is labeled by $+$ or $-$. We use set E^+ and set E^- to denote the set of positive edges and the set of negative edges, respectively.*

The goal of this problem is to partition the vertices into k clusters so as to min-imize the number of positive edges whose endpoints lie in different clusters plus the number of negative edges whose endpoints lie in the same cluster. Moreover, each cluster contains the same number of vertices.

Definition 3 ((t,r)-balanced approximation algorithm). *Algorithm is a (t,r)-balanced approximation algorithm for the balanced k-correlation clustering problem if for any instance, it returns a solution $\mathcal{C} = \{V_1, V_2, \ldots, V_k\}$ whose cost is at most r times the optimum solution and $\min\{|V_1|, |V_2|, \ldots, |V_k|\} \leq t \max\{|V_1|, |V_2|, \ldots, |V_k|\}$.*

For each vertex v, denote

$$P_v := \{(v,t) \in E^+ : t \in V\},$$
$$N_v := \{(v,t) \in E^- : t \in V\}.$$

Definition 4 (Well-proportional graph). *A graph is a well-proportional graph if there exists a constant M satisfying*

$$\min\{|P_v|, |N_v||\} \leq M \max\{|P_v|, |N_v||\}, \forall v \in V. \tag{1}$$

A well-proportional graph is M-proportional if M is the minimum constant that satisfies (1).

Let even number N be the number of vertices. For each edge $(u,v) \in E$, we introduce a binary decision variable x_{uv} to denote whether the vertices u and v are in the same cluster. Variable $x_{uv} = 0$ if u and v are in the same cluster, and $x_{uv} = 1$ otherwise. Then the balanced k-correlation clustering problem can be formulated as follows:

$$\min \sum_{(u,v)\in E^+} x_{uv} + \sum_{(u,v)\in E^-} (1 - x_{uv})$$

$$\text{s. t. } x_{uv} + x_{vw} \geq x_{uw}, \qquad \forall u, v, w \in V,$$

$$\sum_{v \in V} (1 - x_{uv}) = N/k, \qquad \forall u \in V,$$

$$x_{uu} = 0, \qquad \forall u \in V,$$

$$x_{uv} \in \{0,1\}, \qquad \forall u, v \in V. \tag{2}$$

The objective function contains two parts, the first part is the number of posi-tive edges whose endpoints lie in different clusters, while the second part is the number of negatives edges whose endpoints lie in the same cluster. The first constraint ensures that the solution returned by the formulation is a feasible solution of the correlation clustering problem. The second constraint ensures that each cluster contains the same number of vertices. The LP relaxation of (2) is given as follows:

$$\min \sum_{(u,v)\in E^+} x_{uv} + \sum_{(u,v)\in E^-} (1 - x_{uv})$$

$$\text{s. t. } x_{uv} + x_{vw} \geq x_{uw}, \qquad \forall u, v, w \in V,$$

$$\sum_{v\in V}(1 - x_{uv}) = N/k, \qquad \forall u \in V,$$

$$x_{uu} = 0, \qquad \forall u \in V,$$

$$0 \leq x_{uv} \leq 1, \qquad \forall u, v \in V. \qquad (3)$$

3   Algorithm

In this paper, we present an algorithm for the balanced 2-correlation clustering problem on well-proportional graphs.

We give a high level description of our algorithm. The algorithm consists of two main steps. The first step is a computational process. First, we solve (3) to obtain a fractional optimal solution x^*. For each vertex $v \in V$, we sort all the vertices in a non-decreasingly order of x from v. Let T_v be the set of the first $|V|/2$ vertices according to this order. Note that the value of x_{vw} is the distance between the vertices w and v. Then, we calculate the average distance Evg_v of the first $|V|/2$ vertices to vertex v. The second step is a clustering process. First, we find the vertex u with the minimum average distance Evg_u. If Evg_u is large enough, then we let T_u be a cluster and $V \backslash T_u$ be a cluster. This solution satisfies $|T_u| = |V \backslash T_u|$. It is a feasible solution of the balanced 2-correlation clustering problem. Otherwise, we put all the vertices that are no more than $1/2$ away from vertex u into one cluster and assign the rest of the vertices to another cluster. This solution may be infeasible to the balanced 2-correlation clustering problem and we need to establish the relationships between the vertices in the two clusters in order to construct a feasible solution. The detailed algorithm is shown as follows:

4   Analysis

The clusters returned by Algorithm 1 is either a type 1 cluster or a type 2 cluster. In Subsect.4.1, we analyze the upper bound for error edges of type 1 cluster. In Subsect.4.2, we analyze the upper bound for error edges of type 2 cluster.

4.1   Type 1 Cluster

In this case, the clusters V_1 and V_2 returned by the algorithm is a feasible partition of the balanced 2-correlation clustering problem. There are two types of error edges produced by the partition. As shown in Fig. 1, one is the positive edges $(v, p) \in E^+$, where $v \in V_1$, and $q \in V_2$, and the other is the negatives $(v, w) \in E^-$, where $v, w \in V_i, i = 1, 2$.

We analyze the upper bound of error edges produced by the positive edges in the following three lemmas.

Algorithm 1

Input: A labeled M-proportional complete graph $G = (V, E)$ with even number of vertices, $k = 2$.
Output: A partition of vertices V_1, V_2.
1: Initialize $V_1 = V_2 = \emptyset$.
2: Solve the LP relaxation of (2) to obtain an optimal solution x^*.
3: **for** each vertex $v \in V$ **do**
4:    Sort the vertices in V (including vertex v) in a nondecreasing order of x from v. Let T_v be the set of the first $|V|/2$ vertices in V according to this order. Denote

$$\mathrm{Evg}_v = \frac{\sum_{w \in T_v} x^*_{vw}}{|T_v|}.$$

5: **end for**
6: Choose the vertex u with the minimum Evg_u.
7: **if** $\mathrm{Evg}_u \geq \frac{1}{4}$ **then**
8:    Update $V_1 := T_u$, and $V_2 := V \backslash V_1$ (Type 1 cluster)
9: **else**
10:    Update $V_1 := \{v \in T_u : x^*_{uv} \leq \frac{1}{2}\}$, and $V_2 := V \backslash V_1$ (Type 2 cluster)
11: **end if**
12: **return** the partition V_1, V_2.

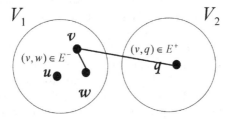

Fig. 1. Partition of V

Lemma 1. *For each $v \in V_1$, if for any $q \in V_2$, edge (v, q) is a positive edge, then the number of error edges can be bounded by*

$$4 \sum_{(v,q) \in E, q \in V_2} x^*_{vq}.$$

If there exists an edge (v, p), where $p \in V_2$ is a negative edge, then we analyze the upper bound of error edges produced by the positive edges by Lemma 2–3.

Lemma 2. *For each $v \in V_1$, if $|P_v| > |N_v|$, then there exists a set $S \subseteq P_v$ with $|S| = N/2$ and the number of error edges can be bounded by*

$$4 \sum_{(v,q) \in E, q \in S} x^*_{vq}.$$

Lemma 3. *For each $v \in V_1$, if $|P_v| < |N_v|$, then there exist a set $S \subseteq N_v$ with $|S| = N/2$ and the number of error edges can be bounded by*

$$4 \sum_{(v,w)\in E, q\in S} (1 - x^*_{vw}).$$

Corollary 1. *Recall Lemma 1–Lemma 3. In the upper bound of all the error edges produced by positive edges, each edge appears at most twice in the objective function.*

Next, we analyze the error edges produced by the negative edges. The analysis of the upper bounds of error edges in cluster V_1 and V_2 are similar. We only consider V_1.

Lemma 4. *For each $v \in V_1$, if for any $w \in V_1$, edge (v, w) is a negative edge, then the number of error edges can be bounded by*

$$4 \sum_{(v,w)\in E, w\in V_1} (1 - x^*_{vw}).$$

If there exists a positive edge (v, q) $(q \in V_1)$, then we analyze the upper bound of the error edges produced by the negative edges in Lemmas 5–6.

Lemma 5. *For each $v \in V_1$, if $|P_v| > |N_v|$, then there exists a set $S \subseteq P_v$ with $|S| = N/2$ and the number of error edges can be bounded by*

$$4 \sum_{(v,q)\in E, q\in S} x^*_{vq}.$$

Lemma 6. *For each $v \in V_1$, if $|P_v| < |N_v|$, then there exists a set $S \subseteq N_v$ with $|S| = N/2$ and the number of error edges can be bounded by*

$$4 \sum_{(v,w)\in E, w\in S} (1 - x^*_{vw}).$$

Corollary 2. *Recall Lemma 4–Lemma 6. In the upper bound of all the error edges produced by the negatives edges, each edge appears at most twice in the objective function.*

Corollary 3. *From Corollaries 1–2. In the upper bound of all the error edges, each edge appears at most four times in the objective function.*

Combining Lemmas 1–6 and Corollary 3, we obtain the following theorem.

Theorem 1. *If the clusters returned by Algorithm 1 is a type 1 cluster, then we have $|V_1| = |V_2|$ and the upper bound of the error edges is bounded by*

$$16 \left[\sum_{(u,v)\in E^+} x^*_{uv} + \sum_{(u,v)\in E^-} (1 - x^*_{uv}) \right],$$

where x^* is the optimal fractional solution of (3).

4.2   Type 2 Cluster

In this case, we have

$$\text{Evg}_u = \frac{\sum_{v \in T_u} x^*_{uv}}{|T_u|} < \frac{1}{4}. \tag{4}$$

There are two kinds of errors in this type of cluster. One is the positive edges $(v,q) \in E^+$, where $v \in V_1$, $q \in V_2$, and the other one is the negatives $(v,w) \in E^-$, where $v,w \in V_i, i = 1,2$. First, we analyze the error edges produced by the positive edges.

Lemma 7. *In any type 2 cluster, we have*

(i) $|V_1| > \frac{N}{4}$;
(ii) $|V_2| > \frac{N}{4}$.

We relabel the vertices (other than u) so that $p < w$ if $x^*_{up} < x^*_{uw}$ by breaking ties arbitrarily. Then, we have the following lemmas to analyze the upper bound of the error edges.

Lemma 8. *[13] The upper bound of the errors produced by the positive edges satisfies the following two properties:*

(1) For each vertex $q \in V_2$, if $x^*_{uq} \geq 3/4$, then for each positive edge $(v,q) \in E^+$ where $v \in V_1$, the error can be bounded by

$$\frac{1}{4} x^*_{vq};$$

(2) For each vertex $q \in V_2$, if $1/2 \leq x^*_{uq} < 3/4$, then the number of error associated with q can be bounded by

$$4 \left[\sum_{(v,q) \in E^+, v \in V_1} x^*_{vq} + \sum_{(v,q) \in E^-, v \in V_1} (1 - x^*_{vq}) \right].$$

Lemma 9. *[13] The upper bound of the errors produced by the negative edges in V_1 satisfies the following two properties:*

(1) For each negative edge (w,p) with $w,p \in V_1$, if $x^*_{uw}, x^*_{up} \leq 1/4$, then the number of error edges can be bounded by

$$2(1 - x^*_{wp});$$

(2) For each vertex $p \in T_u$, if $1/4 < x^*_{up} \leq 1/2$, then the total error produced by all the negative edges (w,p) with $w < p$ can be bounded by

$$4 \left[\sum_{(w,p) \in E^+, w < p} x^*_{wp} + \sum_{(w,p) \in E^-, w < p} (1 - x^*_{wp}) \right].$$

In the following, we use Lemma 10 to analyze the upper bound of errors produced by the negative edges in V_2. For each $q \in V_2$, we have $|P_q| = |N_q|$. Furthermore, for each $q \in V_2$, the error edges produced by the negative edges is no less than $|N_q|$, which implies that error edges produced by the negative edges is no less than $|P_q|$. Then, the next thing we need to do is to analyze the upper bound of $|P_q|, q \in V_2$.

We partition the set P_q into P_q^1 and P_q^2 as follows:

$$P_q^1 := \{(v, q) \in E^+ : v \in V_1\};$$
$$P_q^2 := \{(w, q) \in E^+ : w \in V_2\}.$$

The upper bound of $|P_q^1|$ has been analyzed by Lemma 8. Now, we use Lemma 10 to analyze the upper bound of $|P_q^2|$.

Lemma 10. *The upper bound of* $|P_q^2|$ *can be bounded by*

$$2 \sum_{w \in P_q^2} x_{wq}^*.$$

Corollary 4. *Recall that graph is a M-proportional graph. In the upper bound of all the error edges, each edge appears at most $M + 1$ times in the objective function.*

Combining Lemmas 7–10 and Corollary 4, we obtain Theorem 2.

Theorem 2. *If the clusters returned by Algorithm 1 is a type 2 cluster, then we have $\min\{|V_1|, |V_2|\} \leq 3 \max\{|V_1|, |V_2|\}$ and the upper bound of the error edges is bounded by*

$$4(M + 1) \left[\sum_{(u,v) \in E^+} x_{uv}^* + \sum_{(u,v) \in E^-} (1 - x_{uv}^*) \right],$$

where x^* is the optimal fractional solution of (3).

From Theorems 1–2, we obtain our main result.

Theorem 3. *Algorithm 1 is a $(3, \max\{4(M + 1), 16\})$-balanced approximation algorithm for the balanced 2-correlation clustering problem on M-proportional graphs.*

5   Discussions

In this paper, we introduce the balanced k-correlation clustering problem and provide a $(3, \max\{4(M + 1), 16\})$-balanced approximation algorithm for the balanced 2-correlation clustering problem on M-proportional graphs. There are several research directions for this problem in the future:

- Study the balanced 2-correlation clustering problem on general complete graphs based on the local search technique.
- Study the balanced k-correlation clustering problem on complete graphs based on the LP-rounding technique.
- Study the maximization version of the balanced k-correlation clustering problem by using semidefinite programming relaxation.

Acknowledgements. The first two authors are supported by National Natural Science Foundation of China (Nos. 11531014, 11871081). The third author is supported by the Natural Sciences and Engineering Research Council of Canada (NSERC) grant 06446, and National Natural Science Foundation of China (Nos. 11771386, 11728104). The fourth author is supported by National Natural Science Foundation of China (No. 11201333).

References

1. Achtert, E., Böhm, C., David, J., Kröger, P., Zimek, A.: Global correlation clustering based on the Hough transform. Stat. Anal. Data Mining **1**, 111–127 (2010)
2. Ahmadian, S., Norouzi-Fard, A., Svensson, O., Ward, J.: Better guarantees for k-means and Euclidean k-median by primal-dual algorithms. In: Proceedings of FOCS, pp. 61–72 (2017)
3. Ahn, K.J., Cormode, G., Guha, S., Mcgregor, A., Wirth, A.: Correlation clustering in data streams. In: Proceedings of ICML, pp. 2237–2246 (2015)
4. Ailon, N., Avigdor-Elgrabli, N., Liberty, E., Zuylen, A.V.: Improved approximation algorithms for bipartite correlation clustering. SIAM J. Comput. **41**, 1110–1121 (2012)
5. Ailon, N., Charikar, M., Newman, A.: Aggregating inconsistent information: ranking and clustering. J. ACM **55**(5) (2008). Article No. 23
6. Amit, N.: The bicluster graph editing problem. Diss, Tel Aviv University (2004)
7. Arthur, D., Vassilvitskii, S.: k-Means++: the advantages of careful seeding. In: Proceedings of SODA, pp. 1027–1035 (2007)
8. Bansal, N., Blum, A., Chawla, S.: Correlation clustering. Mach. Learn. **56**, 89–113 (2004)
9. Behsaz, B., Friggstad, Z., Salavatipour, M.R., Sivakumar, R.: Approximation algorithms for min-sum k-clustering and balanced k-median. In: Halldórsson, M.M., Iwama, K., Kobayashi, N., Speckmann, B. (eds.) ICALP 2015. LNCS, vol. 9134, pp. 116–128. Springer, Heidelberg (2015). https://doi.org/10.1007/978-3-662-47672-7_10
10. Bonchi, F.: Overlapping correlation clustering. Knowl. Inf. Syst. **35**, 1–32 (2013)
11. Braverman, V., Lang, H., Levin, K., Monemizadeh, M.: Clustering problems on sliding windows. In: Proceedings of SODA, pp. 1374–1390 (2016)
12. Byrka, J., Fleszar, K., Rybicki, B., Spoerhase, J.: Bi-factor approximation algorithms for hard capacitated k-median problems. In: Proceedings of SODA, pp. 722–736 (2015)
13. Charikar, M., Guruswami, V., Wirth, A.: Clustering with qualitative information. J. Comput. Syst. Sci. **71**, 360–383 (2005)
14. Chawla, S., Makarychev, K., Schramm, T., Yaroslavtsev, G.: Near optimal LP rounding algorithm for correlation clustering on complete and complete k-partite graphs. In: Proceedings of STOC, pp. 219–228 (2015)

15. Demaine, E., Emanuel, D., Fiat, A., Immorlica, N.: Correlation clustering in general weighted graphs. Theoret. Comput. Sci. **361**, 172–187 (2006)
16. Frieze, A., Jerrum, M.: Improved approximation algorithms for max k-cut and max bisection. Algorithmica **18**, 67–81 (1997)
17. Giotis, I., Guruswami, V.: Correlation clustering with a fixed number of clusters. In: Proceedings of SODA, pp. 1167–1176 (2006)
18. Goemans, M.X., Williamson, D.P.: Improved approximation algorithms for maximum cut and satisfiability problems using semidefinite programming. J. ACM **42**, 1115–1145 (1995)
19. Hendrickx, J.M., Tsitsiklis, J.N.: Convergence of type-symmetric and cut-balanced consensus seeking systems. IEEE Trans. Autom. Control **58**, 214–218 (2013)
20. Ji, S., Xu, D., Li, M., Wang, Y.: Approximation algorithm for the correlation clustering problem with non-uniform hard constrained cluster sizes. In: Du, D.-Z., Li, L., Sun, X., Zhang, J. (eds.) AAIM 2019. LNCS, vol. 11640, pp. 159–168. Springer, Cham (2019). https://doi.org/10.1007/978-3-030-27195-4_15
21. Kuila, P., Jana, P.K.: Approximation schemes for load balanced clustering in wireless sensor networks. J. Supercomput. **68**(1), 87–105 (2013). https://doi.org/10.1007/s11227-013-1024-6
22. Li, S.: On uniform capacitated k-median beyond the natural LP relaxation. ACM Trans. Algorithms **13**(2) (2017). Article No. 22
23. Li, M., Xu, D., Zhang, D., Zhang, T.: A streaming algorithm for k-means with approximate coreset. Asia Pacific J. Oper. Res. **36**(01), 1950006 (2019)
24. Liao, Y., Qi, H., Li, W.: Load-balanced clustering algorithm with distributed self-organization for wireless sensor networks. IEEE Sens. J. **13**, 1498–1506 (2013)
25. Mathieu, C., Sankur, O., Schudy, W.: Online correlation clustering. Comput. Stat. **21**, 211–229 (2010)
26. Mathieu, C., Schudy, W.: Correlation clustering with noisy input. In: Proceedings of SODA, pp. 712–728 (2010)
27. Puleo, G.J., Milenkovic, O.: Correlation clustering with constrained cluster sizes and extended weights bounds. SIAM J. Optim. **25**, 1857–1872 (2015)
28. Puleo, G.J., Milenkovic, O.: Correlation clustering and biclustering with locally bounded errors. IEEE Trans. Inf. Theory **64**, 4105–4119 (2018)
29. Shamir, R., Sharan, R., Tsur, D.: Cluster graph modification problems. Discrete Appl. Math. **144**, 173–182 (2004)
30. Swamy, C.: Correlation clustering: maximizing agreements via semidefinite programming. In: Proceedings of SODA, pp. 526–527 (2004)
31. Zhao, M., Yang, Y., Wang, C.: Mobile data gathering with load balanced clustering and dual data uploading in wireless sensor networks. IEEE Trans. Mob. Comput. **14**, 770–785 (2015)

2-Level Station Location for Bike Sharing

Fengmin Wang[1]($\boxtimes$)(iD), Xiaodong Hu[2,3](iD), and Chenchen Wu[4](iD)

[1] Beijing Jinghang Research Institute of Computing and Communication,
Beijing 100074, China
casic_wfm@163.com
[2] Academy of Mathematics and Systems Science, Chinese Academy of Sciences,
Beijing 100190, China
[3] University of Chinese Academy of Sciences, Beijing 100049, China
[4] College of Science, Tianjin University of Technology, No. 399, Binshui West Street,
Xiqing District, Tianjin 300384, China

Abstract. This paper is motivated by designing bike sharing systems that flourish in many large cities. Such systems should locate bike stations allowing users to pick up and drop off bicycles. Considering the interests of both station investors and bike users, we propose an optimization model for determining the locations of bike stations at two levels, and solve it in polynomial time of a randomized approximation algorithm with expected approximation ratio 3.

Keywords: Bike sharing · Facility location · Approximation algorithm · Randomized rounding.

1 Introduction

Bike sharing systems have become one of important means to solve the first or last mile problem in people transportation planning and lessen the impact of traffic pollution on environment in many large cities [7]. These systems provide short-term shared bike usage services for citizens. Bicycles are parked at various stations in the city. The registered user of the system can walk from his origin to a bike station, pick up a bicycle from the station, ride it for a short journey, drop off it at another bike station, and then walk to his destination. The user will pay for the ride depending on how much time he spends in the journey. The first bike sharing system was installed in Amsterdam in 1965. In recent years, the number of bike sharing systems is rising at a very high rate in urban cities around the world [10]. The algorithmic research on designing and managing bike sharing systems has sprung up like mushrooms in the last decade [4]. Zhang *et al.* studied the bike sharing systems expansion [13]. Quilliot and Sarbinowski [9] discussed a strategic related vehicle sharing station location model. Lin *et al.* [6] presented a survey of formulations and solution procedures for hub location

Supported in part by NNSF of China under Grant No.11901544, No. 11531014, and No.11971349.

models with performance constraints. Dell'Amico *et al.* [8] considered the bike sharing rebalancing problem. Despite the extensive research on various aspects of bike sharing, little literature has been known for their approximation algorithm design.

Locate the stations for the bike sharing system is actually an extension of classical facility location problem, which has been extensively investigated in the field of combinatorial optimization. In the facility location problem, the inputs are a facility set F, a client set C, a nonnegative facility opening cost for every facility in F, and a nonnegative service cost for connecting each pair of a facility in F and a client in C. The connection cost is often assumed to be metric. The objective is to open (locate) some facilities in F, and connect (allocate) each client in C to one of the open facilities, in such a way that the sum of opening and connection costs is minimized. The problem has two widely investigated versions. In the uncapacitated version, each facility can serve an unlimited number of clients while in the capacitated version, each facility can serve at most a certain number of clients. Shmoys *et al.* [11] developed the first constant factor approximation algorithm for the metric uncapacitated facility location problem. They used the LP-rounding technique to obtain the approximate ratio 3.16. The ratio was improved later by Chudak and Shmoys [2] who provided a randomized rounding based 1.736-approximation algorithm. For the capacitated version, Chudak and Williamson [3] obtained a 3-approximation algorithm in the *soft* case that there is no limit on the number of copies of a facility which can be opened at a location. Shmoys *et al.* [11] studied the problem of 2-level facility location, obtaining a 3.16-approximation algorithm. The approximation ratio was later improved to 3 by Aardal *et al.* [1], based on randomized rounding of an LP solution to an integer one. Furthermore, the (randomized) LP-rounding techniques have been successfully used to design several algorithms for the facility location problem and its variants (see [5,12] and reference therein).

This paper studies the problem of how bike stations (which are partitioned into two groups) in a bike sharing system are located. Our objective is to minimize the sum of investment costs for setting up stations and user costs for traveling. We propose a model of 2-level station location, and solve the cost minimization problem by a randomized approximation algorithm with approximation ratio 3 in expectation.

Our algorithm is a slight adaptation of the randomized approximation algorithm for the 2-level uncapacitated facility location problem by Aardal *et al.* [1] (referred to as ACS Algorithm henceforth), where we think of the origin-destination pair of each bike user as an "imaginary" client in the facility location problem (see Remark 2(i) in Sect. 2 and Remark 3 in Sect. 3). The major contribution of this paper is to show that the algorithm of Aardal *et al.* [1] can be easily adapted to solve the 2-level station location problem, attaining the same expected approximation ratio 3. Since the "imaginary" replacements of origin-destination pairs with clients have no direct corresponding projections from the metric space for bike sharing to the one for facility location, the analysis on service costs for the ACS Algorithm [1] cannot be applied here (see Remark 2(ii) in Sect. 2). By exploring the structural properties exclusively associated with the 2-level station location problem, we successfully get around the difficulties of directly fulfilling the metric requirement on service costs which is indispensable to the approximation ratio 3 of the ACS Algorithm [1]. This is our main technical contribution (see Remark 4 in Sect. 3).

The remainder of this paper is organized as follows. In Sect. 2, we present the mathematical model of the problem. In Sect. 3, we make use of the randomized rounding technique to design the algorithm for the problem, and analyze its correctness and performance. In Sect. 4, we conclude the paper with discussions on future works.

2   The 2-Level bike-Station Location Problem

Given a finite set N of locations, the travel costs between locations s and t, $c_{st} \geq 0$, $s, t \in N$, are symmetric and satisfy the triangle inequality. There is a *demand set* D of *origin-destination* (a.k.a. *demand*) pairs (i, j), where $i, j \in N$ are distinct, i is the origin and j is the destination. Each demand pair $(i, j) \in D$ corresponds to a bike user who wants to travel from i to j. Two levels of bike stations are to be opened to serve the user demands $(i, j) \in D$. A user will walk from his origin i to an open station at the first level where he will rent a bike. Then the user will ride the bike to an open station at the second level where he will return the bike. After that, he will walk to his destination j. It is assumed that the travel cost of walking from i directly to j is much larger than that of walking from i to any level-1 station and than that of walking from any level-2 station to j. In the 2-level bike-station location problem (2L-BSL), there are a set S_1 of locations where stations at level 1 may be opened, and a set S_2 of locations where stations at level 2 may be opened, where $S_1 \cap S_2 = \emptyset$. Each demand pair in D must be assigned to precisely one open station at each of the two levels, i.e., assigned to exactly one pair of stations in $S_1 \times S_2$. For each location $b \in S_1 \cup S_2$, the nonnegative cost of opening a bike station at b is s_b. The total connection cost incurred by assigning demand pair $(i, j) \in D$ to bike stations $k \in S_1$ and $l \in S_2$ is equal to $c_{ijkl} = c_{ik} + c_{kl} + c_{lj}$. The aim of the 2L-BSL is to minimize the total cost of setting up (opening) the bike stations and of connecting the demand pairs and their assigned open stations. The general solution structure of the 2L-BSL addressed in this study is represented in Fig. 1.

We introduce the following three decision variables: y_k, z_l and x_{ijkl}. If a station at level 1 is open at location $k \in S_1$, then $y_k = 1$, otherwise, $y_k = 0$. If a station at level 2 is open at location $l \in S_2$, then $z_l = 1$, otherwise, $z_l = 0$. If the (i, j) is assigned to stations $k \in S_1$ and $l \in S_2$, then $x_{ijkl} = 1$, otherwise, $x_{ijkl} = 0$. The 2L-BSL model is formulated as the following integer linear program.

$$\min \sum_{k \in S_1} s_k y_k + \sum_{l \in S_2} s_l z_l + \sum_{(i,j) \in D} \sum_{(k,l) \in S_1 \times S_2} c_{ijkl} x_{ijkl} \tag{1}$$

$$\text{s. t.} \sum_{k \in S_1} \sum_{l \in S_2} x_{ijkl} = 1, \quad \forall (i,j) \in D,$$

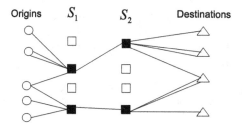

Fig. 1. The solution structure of the 2L-BSL.

$$\sum_{l \in S_2} x_{ijkl} \leq y_k, \quad \forall (i,j) \in D, k \in S_1,$$

$$\sum_{k \in S_1} x_{ijkl} \leq z_l, \quad \forall (i,j) \in D, l \in S_2,$$

$$x_{ijkl}, y_k, z_l \in \{0,1\}, \quad \forall\, (i,j) \in D, k \in S_1, l \in S_2.$$

The first set of constraints guarantees that each demand pair $(i,j) \in D$ should be assigned to exactly one pair of level-1 station and level-2 station. The second (third) set of constraints indicates that if pair (i,j) is assigned to station $k \in S_1$ ($l \in S_2$), then the station k (resp. l) must be open.

Remark 1. The essential solution to the 2L-BSL concerns only locations of open station pairs (once the open station pairs have been determined, the assignment task to minimize connection costs is trivial). This is similar to the uncapacitated facility location problem.

Remark 2. It is worth noting the close relation between the 2L-BSL and 2-level uncapacitated facility location problem (2L-UFL) that demand pairs and stations in the former problem correspond to clients and facilities in the latter problem, respectively.

(i) *The similar side*: Let us think of each demand pair (i,j) in the 2L-BSL as a client in the 2L-UFL, and connection cost c_{ijkl} of assigning (i,j) to station pair $(k,l) \in S_1 \times S_2$ in the 2L-BSL as the service cost of using level-1 facility k and level-2 facility l to serve the client in the 2L-UFL. In this way, the program (1) for the 2L-BSL is exactly the natural integer linear program presented by Aardal *et al.* [1] for the 2L-UFL.

(ii) *The different side*: To assure constant approximation ratios for the 2L-UFL, all known research (including [1]) assumes the service costs satisfy the triangle inequality. Although we have the correspondence stated in (i), the construction of service costs is not complete (i.e., not all the station pairs can serve the demand pairs), and it is unclear whether the construction can be completed such that the triangle inequality are satisfied (since it is unclear whether any two points can be connected, if there are two points can not be connected, then the triangle inequality does not exist).

In the rest of this section, we present some preliminaries which are crucial to our algorithm design and analysis for solving the 2L-BSL, i.e., program (1). Relaxing the first and the fourth sets of constraints in (1), we have the following linear programming relaxation.

$$\min \sum_{k \in S_1} s_k y_k + \sum_{l \in S_2} s_l z_l + \sum_{(i,j) \in D} \sum_{(k,l) \in S_1 \times S_2} c_{ijkl} x_{ijkl} \qquad (2)$$

$$\text{s. t.} \sum_{k \in S_1} \sum_{l \in S_2} x_{ijkl} \geq 1, \quad \forall (i,j) \in D,$$

$$\sum_{l \in S_2} x_{ijkl} \leq y_k, \quad \forall (i,j) \in D, k \in S_1,$$

$$\sum_{k \in S_1} x_{ijkl} \leq z_l, \quad \forall (i,j) \in D, l \in S_2,$$

$$x_{ijkl}, y_k, z_l \geq 0, \quad \forall (i,j) \in D, k \in S_1, l \in S_2.$$

Let α_{ij}, β_{ijk} and γ_{ijl} denote the dual variables corresponding to the primal constraints in (2), respectively. The dual program of the above linear programming relaxation (2) is as follows.

$$\max \sum_{(i,j) \in D} \alpha_{ij} \qquad (3)$$

$$\text{s. t.} \alpha_{ij} \leq c_{ijkl} + \beta_{ijk} + \gamma_{ijl}, \quad \forall (i,j) \in D, k \in S_1, l \in S_2,$$

$$\sum_{(i,j) \in D} \beta_{ijk} \leq s_k, \quad \forall k \in S_1,$$

$$\sum_{(i,j) \in D} \gamma_{ijl} \leq s_l, \quad \forall l \in S_2,$$

$$\alpha_{ij}, \beta_{ijk}, \gamma_{ijl} \geq 0, \quad \forall (i,j) \in D, k \in S_1, l \in S_2.$$

In view of Remark 2(i), we see from [1] that programs (2) and (3) can both be solved in polynomial time, and the following lemma holds.

Lemma 1 (Aardal et al. [1]). *Let $(\hat{x}, \hat{y}, \hat{z})$ and $(\hat{\alpha}, \hat{\beta}, \hat{\gamma})$ be optimal solutions to the primal and dual linear programs (2) and (3), respectively. Then $\hat{x}_{ijkl} > 0$ implies that $c_{ijkl} \leq \hat{\alpha}_{ij}$, for each $(i,j) \in D, k \in S_1, l \in S_2$.* □

3   Algorithm and Its Analysis

In this section, we present a polynomial-time randomized approximation algorithm for the 2L-BSL, and prove that the expectation of its approximation ratio is 3. Our algorithm is a direct adaptation of the ACS Algorithm, i.e., the randomized rounding algorithm for 2L-UFL by Aardal et al. [1]. We first solve the primal and dual linear programs (2) and (3) to their optimality. Then, based on the optimal primal (fractional) solution, we construct mutually disjoint clusters of stations and demand pairs such that all demand pairs are covered. Finally, we open in each cluster a pair of stations (according to a probability distribution defined in the optimal primal solution) to serve all demand pairs in the cluster. The details of our algorithm are as follows.

Algorithm 1 INPUT: A 2L-BSL instance as specified in Section 2.
OUTPUT: A feasible solution of program (1).

---

1. Solve the primal and dual linear programs (2) and (3). Denote the optimal solutions by $(\hat{x}, \hat{y}, \hat{z})$ and $(\hat{\alpha}, \hat{\beta}, \hat{\gamma})$, respectively.
2. DETERMINISTIC CLUSTERING: Let D^c denote the set of demand pairs that are selected as the centers of the clusters, $\mathcal{C}$ denote the set of clusters, U denote the set of demand pairs that have not been clustered. At the beginning of the clustering, set $D^c := \emptyset$, $\mathcal{C} := \emptyset$, $U := D$.
   - Consider each demand pair $(i, j) \in U$, let $\hat{c}_{ij} := \sum_{k \in S_1, l \in S_2} c_{ijkl} \hat{x}_{ijkl}$. Choose the demand pair $(i^c, j^c) \in U$ such that the sum $\hat{\alpha}_{i^c j^c} + \hat{c}_{i^c j^c}$ is minimum. Set

$$S^{(i^c,j^c)} := \{(k,l) \in S_1 \times S_2 : \hat{x}_{i^c j^c kl} > 0\},$$
$$S_1^{(i^c,j^c)} := \{k \in S_1 : \text{there exists } l \in S_2 \text{ with } \hat{x}_{i^c j^c kl} > 0\},$$
$$S_2^{(i^c,j^c)} := \{l \in S_2 : \text{there exists } k \in S_1 \text{ with } \hat{x}_{i^c j^c kl} > 0\},$$
$$D^{(i^c,j^c)} := \{(i,j) \in D : \hat{x}_{ijkl} > 0 \text{ with } k \in S_1^{(i^c,j^c)} \text{ or } l \in S_2^{(i^c,j^c)}\}.$$

   Define the cluster centered at (i^c, j^c) as $C^{(i^c,j^c)} := S^{(i^c,j^c)} \cup D^{(i^c,j^c)}$.
   - Update $D^c := D^c \cup \{(i^c, j^c)\}$, $\mathcal{C} := \mathcal{C} \cup \{C^{(i^c,j^c)}\}$, $U := U - D^{(i^c,j^c)}$.
   (Note that the update of U implies that the clusters $C^{(i^c,j^c)}$ are mutually disjoint.)
   - Iterate over the above clustering process, until $U = \emptyset$.
   (Note that at the end of clustering, each demand pair belongs to precisely one cluster in $\mathcal{C}$.)
3. RANDOM CONSTRUCTION: The algorithm constructs a feasible solution of program (1) based on $(\hat{x}, \hat{y}, \hat{z})$ by considering (processing) all clusters in $\mathcal{C}$ one by one.
   - Initially, all stations at both levels are unopened, and all demand pairs are unassigned, i.e., we set x_{ijkl}, y_k, z_l to be 0 for all $(i, j) \in D, k \in S_1, l \in S_2$.
   - For each cluster $C^{(i^c,j^c)} \in \mathcal{C}$, we choose a station pair $(k, l) \in S^{(i^c,j^c)}$ with probability $\hat{x}_{i^c j^c kl}$ (note from the optimality of $(\hat{x}, \hat{y}, \hat{z})$ that $\sum_{k \in S_1, l \in S_2} \hat{x}_{i^c j^c kl} = 1$). Write the chosen station pair as (k^c, l^c). We open k^c and l^c, i.e., reset $y_{k^c} := 1$ and $z_{l^c} := 1$. Then we assign all the demand pairs $(i, j) \in D^{(i^c,j^c)}$ to the open station pair (k^c, l^c), i.e., reset $x_{ijk^c l^c} := 1$ for all $(i, j) \in D^{(i^c,j^c)}$.

Since each demand pair belongs to precisely one cluster in $\mathcal{C}$, and all demand pairs in every cluster have been assigned to a pair of open stations (at levels 1 and 2 respectively) in the same cluster, we see that all demand pairs in D have been assigned to open station pairs. Furthermore, since clusters in $\mathcal{C}$ are mutually disjoint, we see that each demand pair in D is assigned to exactly one pair of open stations. The feasibility of the solution (x, y, z) follows.

Remark 3. Algorithm 1 is *in essence* the same as the ACS Algorithm [1]. On the other hand, two differences from [1] are obvious.

(i) Clients in [1] are replaced with demands pairs.
(ii) The iterative rounding steps in the ACS Algorithm [1] are replaced with independent constructions within disjoint clusters, which avoids the iterative maintenance of a feasible solution of the LP relaxation as Ardal *et al.* [1] did, and yields a more direct argument for the algorithmic correctness as presented above. Essentially, one iteration in the ACS Algorithm [1] corresponds to our processing of a cluster.

Next, we analyze the approximation ratio of Algorithm 1. We first prove two lemmas (Lemmas 2 and 3) that bound the expected station (opening) cost and the expected connection cost of the solution output by Algorithm 1, respectively. In view of "the same side" stated in Remark 2(i) and the 1-1 correspondence between iterations in the ACS Algorithm [1] and processing of clusters in our algorithm as stated in 3(ii), the following lemma has been proved by Aardal *et al.* (see inequalities (11) in [1]).

Lemma 2 (Aardal *et al.* [1]). *The expected station cost of the feasible integer solution (x, y, z) output by Algorithm 1 is no more than the station cost of the optimal fractional solution $(\hat{x}, \hat{y}, \hat{z})$.* □

Lemma 3. *The expected connection cost of the feasible integer solution (x, y, z) output by Algorithm 1 is no more than $\sum_{(i,j) \in D} (2\hat{\alpha}_{ij} + \hat{c}_{ij})$.*

Proof. According to Step 2 in Algorithm 1, D is the disjoint union of $D^{(i^c, j^c)}$ for all $(i^c, j^c) \in D^c$. We consider any demand pair $(i, j) \in D^{(i^c, j^c)}$, which has been assigned to open station pair (k^c, l^c) by Step 3 when processing cluster $C^{(i^c, j^c)}$.

If $(i, j) = (i^c, j^c)$ is the center of cluster $C^{(i^c, j^c)}$, then

$$
\mathbb{E}\left[\sum_{(k,l) \in S_1 \times S_2} c_{ijkl} x_{ijkl} \right] = \mathbb{E}\left[\sum_{(k,l) \in S^{(i^c, j^c)}} c_{i^c j^c kl} x_{i^c j^c kl} \right]
$$
$$
= \sum_{(k,l) \in S^{(i^c, j^c)}} c_{i^c j^c kl} \hat{x}_{i^c j^c kl}
$$
$$
= \hat{c}_{i^c j^c}
$$

where the second equality is implied by the random opening of the station pair in $S^{(i^c, j^c)}$ to which (i, j) is assigned, and the last equality is guaranteed by the definitions of $S^{(i^c, j^c)}$ and $\hat{c}_{i^c j^c}$.

It remains to consider the case that $(i, j) \in D^{(i^c, j^c)} - \{(i^c, j^c)\}$, for which we will prove that

$$
\Lambda := \mathbb{E}\left[\sum_{(k,l) \in S_1 \times S_2} c_{ijkl} x_{ijkl} \right] = \sum_{(k^c, l^c) \in S^{(i^c, j^c)}} c_{ijk^c l^c} \hat{x}_{i^c j^c k^c l^c}
$$

is bounded above by $2\hat{\alpha}_{ij} + \hat{c}_{ij}$. This in turn establishes the lemma. Recall that there exists $(k, l) \in S_1 \times S_2$ with $k \in S_1^{(i^c, j^c)}$ or $l \in S_2^{(i^c, j^c)}$ such that $\hat{x}_{ijkl} > 0$. It follows from Lemma 1 that $c_{ijkl} \leq \hat{\alpha}_{ij}$.

If $(k, l) = (k^c, l^c)$, then $c_{ijk^c l^c} \leq \hat{\alpha}_{ij}$, which implies

$$\Lambda \leq \sum_{(k^c, l^c) \in S^{(i^c, j^c)}} \hat{\alpha}_{ij} \hat{x}_{i^c j^c k^c l^c} = \hat{\alpha}_{ij},$$

where the equality is implied by $\sum_{(k^c, l^c) \in S^{(i^c, j^c)}} \hat{x}_{i^c j^c k^c l^c} = 1$ (recall the optimality of $\hat{x}$).

If $(k, l) \neq (k^c, l^c)$, we distinguish among a couple of cases (Case1 – Case3) to show $\Lambda \leq 2\hat{\alpha}_{ij} + \hat{c}_{ij}$. Recalling the minimality of $\hat{\alpha}_{i^c j^c} + \hat{c}_{i^c j^c}$, it suffices to establish $\Lambda \leq \hat{\alpha}_{ij} + \hat{\alpha}_{i^c j^c} + \hat{c}_{i^c j^c}$. Furthermore, as $c_{ijkl} \leq \hat{\alpha}_{ij}$, we only need to prove $\Lambda \leq c_{ijkl} + \hat{\alpha}_{i^c j^c} + \hat{c}_{i^c j^c}$. The inequality will be implied by the following claim.

Claim. $c_{ijk^c l^c} \leq c_{ijkl} + \hat{\alpha}_{i^c j^c} + c_{i^c j^c k^c l^c}$.

Indeed, assuming the claim, we will have

$$\Lambda \leq \sum_{(k^c, l^c) \in S^{(i^c, j^c)}} (c_{ijkl} + \hat{\alpha}_{i^c j^c} + c_{i^c j^c k^c l^c}) \hat{x}_{i^c j^c k^c l^c}$$

$$= (c_{ijkl} + \hat{\alpha}_{i^c j^c}) \sum_{(k^c, l^c) \in S^{(i^c, j^c)}} \hat{x}_{i^c j^c k^c l^c} + \sum_{(k^c, l^c) \in S^{(i^c, j^c)}} c_{i^c j^c k^c l^c} \hat{x}_{i^c j^c k^c l^c}$$

$$= c_{ijkl} + \hat{\alpha}_{i^c j^c} + \hat{c}_{i^c j^c}$$

where the second equality is implied by $\sum_{(k^c, l^c) \in S^{(i^c, j^c)}} \hat{x}_{i^c j^c k^c l^c} = 1$ (recall the optimality of $\hat{x}$) and $\sum_{(k^c, l^c) \in S^{(i^c, j^c)}} c_{i^c j^c k^c l^c} \hat{x}_{i^c j^c k^c l^c} = \sum_{(\bar{k}, \bar{l}) \in S_1 \times S_2} c_{i^c j^c \bar{k} \bar{l}} \hat{x}_{i^c j^c \bar{k} \bar{l}} = \hat{c}_{i^c j^c}$ (recall the definitions of $S^{(i^c, j^c)}$ and $\hat{c}_{i^c j^c}$). It remains to verify the claim when $\{k, l\} \neq \{k^c, l^c\}$, for which we distinguish among three cases according to different outcomes of $\{i, j\} \cap \{i^c, j^c\}$: $i \neq i^c$ and $j \neq j^c$, or $i = i^c$ and $j \neq j^c$, or $i \neq i^c$ and $j = j^c$.

Case 1. $i \neq i^c$ and $j \neq j^c$. We consider three subcases as shown in Fig. 2 depending on whether $\{k, l\} \cap \{k^c, l^c\}$ is empty or not and whether (k, l) belongs to $S^{(i^c, j^c)}$ or not.

Case 1.1. $\{k, l\} \cap \{k^c, l^c\} = \emptyset$ and $(k, l) \in S^{(i^c, j^c)}$ (as depicted in Fig. 2(1.1)). The claimed inequality is proved by

$$c_{ijk^c l^c} = c_{ik^c} + c_{k^c l^c} + c_{l^c j}$$
$$\leq (c_{ik} + c_{i^c k} + c_{i^c k^c}) + c_{k^c l^c} + (c_{lj} + c_{lj^c} + c_{l^c j^c})$$
$$\leq c_{ijkl} + c_{i^c j^c kl} + c_{i^c j^c k^c l^c}$$
$$\leq c_{ijkl} + \hat{\alpha}_{i^c j^c} + c_{i^c j^c k^c l^c}$$

116   F. Wang et al.

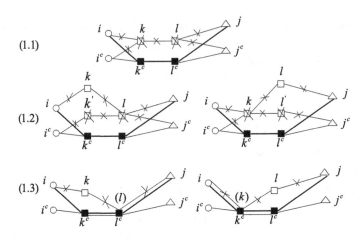

Fig. 2. The three subcases of Case 1 with $i \neq i^c$ and $j \neq j^c$.

where the first inequality is implied by the triangle inequality, the second is by the definitions of c_{ijkl}, $c_{i^c j^c kl}$ and $c_{i^c j^c k^c l^c}$, and the third is by $c_{i^c j^c kl} \leq \hat{\alpha}_{i^c j^c}$ (note from $(k, l) \in S^{(i^c, j^c)}$ that $\hat{x}_{i^c j^c kl} > 0$, which along with Lemma 1 implies $c_{i^c j^c kl} \leq \hat{\alpha}_{i^c j^c}$).

Case 1.2. $\{k, l\} \cap \{k^c, l^c\} = \emptyset$ and $(k, l) \notin S^{(i^c, j^c)}$ (as depicted in Fig. 2(1.2)). By symmetry we assume w.l.o.g. that $l \in S_2^{(i^c, j^c)}$ and $\hat{x}_{i^c j^c k' l} > 0$ for some $k' \in S_1 \setminus \{k\}$ as shown in the left part of Fig. 2(1.2). Note that possibly $k' = k^c$. By Lemma 1, $\hat{x}_{i^c j^c k' l} > 0$ implies $c_{i^c j^c k' l} \leq \hat{\alpha}_{i^c j^c}$. It follows that

$$
\begin{aligned}
&c_{ijk^c l^c} \\
&= c_{ik^c} + c_{k^c l^c} + c_{l^c j} \\
&\leq \left(c_{i^c k^c} + c_{i^c k'} + c_{ik} + c_{kl} + c_{k' l}\right) + c_{k^c l^c} + \left(c_{lj} + c_{lj^c} + c_{l^c j^c}\right) \quad \text{(triangle inequality)} \\
&= c_{ijkl} + c_{i^c j^c k' l} + c_{i^c j^c k^c l^c} \quad \text{(by the definitions of } c) \\
&\leq c_{ijkl} + \hat{\alpha}_{i^c j^c} + c_{i^c j^c k^c l^c}
\end{aligned}
$$

as claimed.

Case 1.3. $\{k, l\} \cap \{k^c, l^c\} \neq \emptyset$ (as depicted in Fig. 2(1.3)). By symmetry, we may assume w.l.o.g. that $\{k, l\} \cap \{k^c, l^c\} = \{l^c\} = \{l\}$ as in the left part of (1.3). It follows that

$$
\begin{aligned}
c_{ijk^c l^c} &= c_{ik^c} + c_{k^c l^c} + c_{l^c j} \\
&\leq \left(c_{ik} + c_{kl^c} + c_{k^c l^c}\right) + c_{k^c l^c} + c_{l^c j} \quad \text{(by the triangle inequality)} \\
&\leq c_{ijkl^c} + 2c_{i^c j^c k^c l^c} \quad \text{(by the definitions of } c) \\
&\leq c_{ijkl} + \hat{\alpha}_{i^c j^c} + c_{i^c j^c k^c l^c} \quad \text{(by } l^c = l \text{ and } c_{i^c j^c k^c l^c} \leq \hat{\alpha}_{i^c j^c})
\end{aligned}
$$

as claimed.

Due to space constraints, we omit the proof of Case 2. $i = i^c$ and $j \neq j^c$, and Case 3. $i \neq i^c$ and $j = j^c$, which are similar to the proof of Case 1.

The proof of Lemma 3 has been completed.   □

Remark 4. Our analysis tool used in Lemma 3 on the linearity of expectation and the triangle inequality are the same as those in [1] for bounding expected service cost. However, our situations are much more complicated in that each demand pair involves two different locations instead of one as its client counterpart in 2L-UFL (recall Remark 2(i)). An even more serious complication occurs when two different demand pairs intersect at a same location, which never happens for different clients in the 2L-UFL. The proof of Lemma 3 constitutes our main technical contribution, which is derived by exploring the structural properties of the 2L-BSL.

We are now ready to prove the performance of Algorithm 1. Let OPT denote the total cost of an optimal solution of the given 2L-BSL instance, i.e., the optimal objective value of the integer program (1). Instantly, OPT is lower bounded by the optimal objective value of program (2), which is the LP relaxation of (1).

Theorem 1. *The expected total cost of the feasible integer solution (x, y, z) output by Algorithm 1 is no more than 3 times of the OPT, i.e.,*

$$\mathbb{E}\left[\sum_{k \in S_1} s_k y_k + \sum_{l \in S_2} s_l z_l + \sum_{(i,j) \in D} \sum_{(k,l) \in S_1 \times S_2} c_{ijkl} x_{ijkl}\right] \leq 3OPT.$$

Proof. Recall that $(\hat{x}, \hat{y}, \hat{z})$ and $(\hat{\alpha}, \hat{\beta}, \hat{\gamma})$ are optimal solutions of primal and dual linear programs (2) and (3), respectively, whose equal objective value is a lower bound of OPT, i.e.,

$$\sum_{k \in S_1} s_k \hat{y}_k + \sum_{l \in S_2} s_l \hat{z}_l + \sum_{(i,j) \in D} \sum_{(k,l) \in S_1 \times S_2} c_{ijkl} \hat{x}_{ijkl} = \sum_{(i,j) \in D} \hat{\alpha}_{ij} \leq OPT.$$

It follows from Lemmas 2 and 3, and the definition of $\hat{c}_{ij}$ that proving the theorem. □

Remark 5. Recalling Remark 1, we might do better than Algorithm 1 by using it to randomly open station pairs and assigning each demand pair to a best available open station pair.

4   Conclusion

In this paper, we study locating bike-stations for cost-effective constructions of bike sharing systems. We propose the model of 2-level bike sharing station location, and design an approximation algorithm for the model, achieving an approximation ratio 3 in expectation. Regarding future research, it might be interesting to consider the problem with soft capacities, where each location is associated with a capacity, each open station at the location can serve demand pairs no more than the capacity, and any number of stations can be opened at the location to meet the service need. Furthermore, it would be challenging to design good approximation algorithms for the problems with hard capacities and time window constraints.

References

1. Aardal, K.I., Chudak, F., Shmoys, D.B.: A 3-approximation algorithm for the k-level uncapacitated facility location problem. Inf. Process. Lett. **72**, 161–167 (1999)
2. Chudak, F.A., Shmoys, D.B.: Improved approximation algorithms for the uncapacitated facility location problem. SIAM J. Comput. **33**, 1–25 (2003)
3. Chudak, F.A., Williamson, D.P.: Improved approximation algorithms for the capacitated facility location problem. Math. Program. **102**(2), 207–222 (2005)
4. Gavalas, D., Konstantopoulos, C., Pantziou, G.: Design and management of vehicle-sharing systems: a survey of algorithmic approaches. In: Smart Cities and Homes, pp. 261–289 (2016)
5. Li, S.: A 1.488 approximation algorithm for the uncapacitated facility location problem. Inf. Comput. **222**, 45–58 (2013)
6. Lin, J.R., Yang, T.H., Chang, Y.C.: A hub location inventory model for bicycle sharing system design: formulation and solution. Comput. Ind. Eng. **65**(1), 77–86 (2013)
7. Midgley, P.: Bicycle-sharing schemes: enhancing sustainable mobility in urban areas. In: Back ground Paper No. 8, CSD19/2011/BP8, Commission on Sustainable Development, Department of Economic and Social Affairs. United Nations (2011)
8. Dell'Amico, M., Hadjicostantinou, E., Iori, M., Novellani, S.: The bike sharing rebalancing problem: mathematical formulations and benchmark instances. Omega **45**(2), 7–19 (2014)
9. Quilliot, A., Sarbinowski, A.: Facility location models for vehicle sharing systems. In: Computer Science and Information Systems IEEE (2016)
10. Shaheen, S., Guzman, S., Zhang, H.: Bike sharing in Europe, the Americas, and Asia: past, present, and future. Transp. Res. Rec. **2143**, 159–167 (2010)
11. Shmoys, D.B., Tardös, É, Aardal, K.I.: Approximation algorithms for facility location problems. In: Proceedings of the 29th Annual ACM Symposium on Theory of Computing, pp. 265–274 (1997)
12. Wu, C., Xu, D.: An improved approximation algorithm for the k-level facility location problem with soft capacities. Acta Math. Applicatae Sinica-English Ser. **11**(33), 1015–1024 (2017)
13. Zhang, J., Pan, X., Li, M., Yu, P.S.: Bicycle-sharing systems expansion: station redeployment through crowd planning. In: ACM Sigspatial International Conference on Advances in Geographic Information Systems, vol. 2 (2016)

Approximation Algorithms for the Lower-Bounded Knapsack Median Problem

Lu Han[1], Chunlin Hao[2], Chenchen Wu[3]([✉]), and Zhenning Zhang[2]

[1] Academy of Mathematics and Systems Science, Chinese Academy of Sciences, Beijing 100190, People's Republic of China
hanlu@amss.ac.cn
[2] Department of Operations Research and Information Engineering, Beijing University of Technology, Beijing 100124, People's Republic of China
{haochl,zhangzhenning}@bjut.edu.cn
[3] College of Science, Tianjin University of Technology, Tianjin 300384, People's Republic of China
wu_chenchen_tjut@163.com

Abstract. In this paper, we introduce the lower-bounded knapsack median problem (LB knapsack median). In this problem, we are given a set of facilities, a set of clients, a budget B and a lower bound L. Every facility is associated with a weight. Every facility-client pair is associated with a connection cost. The aim is to select a subset of facilities to open and connect every client to some opened facility, such that the total weights of the selected facilities is no more than B, any opened facility is connected by at least L clients and the total connection costs is minimized.

As our main contribution, we study the LB knapsack median and present two approximation algorithms with ratios of 2730 and 1608. The first algorithm is based on reduction and the improved second algorithm is based on an intuitive observation. Additionally, we adapt these two algorithms to the lower-bounded k-median problem (LB k-median) and obtain the approximation ratios of 610 and 387.

Keywords: Knapsack median · Lower bounds · Approximation algorithm

1 Introduction

The k-median problem (k-median) has extensive applications from the context of clustering to data mining. In this problem, we are given a set of facilities, a set of clients and an integer k. Every facility-client pair is associated with a connection cost. The goal is to open at most k facilities and connect every client

© Springer Nature Switzerland AG 2020
Z. Zhang et al. (Eds.): AAIM 2020, LNCS 12290, pp. 119–130, 2020.
https://doi.org/10.1007/978-3-030-57602-8_11

to some opened facility so as to minimize the total connection costs. In general, we assume that the connection costs are non-negative, symmetric and satisfy the triangle inequality (i.e., the connection costs are metric).

As a well-known NP-hard problem, the k-median has received a lot of attention on designing approximation algorithm for it as well as its variants [2–5,7,8,11,16]. For the k-median, based on LP-rounding, Charikar et al. [5] give the first constant-factor approximation algorithm with a ratio of $6\frac{2}{3}$ and based on dependent-rounding, Byrka et al. [4] propose the current best $(2.675 + \epsilon)$-approximation algorithm. Under the assumption that $\text{NP} \not\subseteq \text{DTIME}(n^{O(\log \log n)})$, Jain et al. [7] provide the 1.736-hardness of approximation. If we generalize the cardinality constraint in the k-median to a knapsack constraint, we get the knapsack median problem (knapsack median). To be specific, in the knapsack median, we are given a budget B rather than the integer k and every facility is associated with a weight. We wish to select a subset of facilities to open and connect every client to an opened facility, such that the total weights of the selected facilities is no more than B and the total connection costs is minimized. By adding constraints to strengthen the natural linear program of the knapsack median, Kumar [10] presents the first constant approximation ratio of 2700. Through considering a preprocessing step which includes adding randomization and using sparsification, Byrka et al. [3] give the state-of-art approximation ratio of 17.46.

When every facility in the k-median has an additional opening cost, the cardinality constraint does not need to be satisfied and the objective becomes to open a subset of facilities and connect every client to an opened facility so as to minimize the total opening as well as connection costs, we obtain the uncapacitated facility location problem (UFLP). The UFLP is a well-studied NP-hard problem, which plays an important role in both the field of operations management and computer science. In general, we have the assumption that the connection costs are metric. The first constant 3.16-approximation algorithm, which is based on the technique of LP-rounding, is given by Shmoys et al. [14]. Li [12] applies the techniques of LP-rounding and dual-fitting to offer the current best 1.488-approximation algorithm. Recently, the meaningful lower-bounded generalization of the UFLP named lower-bounded facility location problem (i.e., LBFLP) has raised great attention, since the lower bound constraints is motivated from both the facility location and data privacy perspective. In the LBFLP, a set of facilities, a set of clients and a lower bound L are given. Every facility is associated with an opening cost and every facility-client is associated with a connection cost. We aim to open a subset of facilities and connect every client to an opened facility such that any opened facility is connected by at least L clients and the total opening and connection costs is minimized. Guha et al. [6] and Karger and Minkoff [9] introduce this problem simultaneously and both design an $O(1)$-bi-criteria approximation algorithm for the LBFLP. Svitkina [15] reduces the LBFLP to the capacitated facility location problem (CFLP) and gives the first true 488-approximation algorithm. Ahmadian and Swamy [1] make improvement on solving a structured LBFLP and improve the approximation ratio to 82.6. For the general case of the LBFLP where every facility is associated with a non-uniform lower bound, also based on reducing the LBFLP to the CFLP, Li [13] provide a 4000-approximation algorithm.

Despite the fact that many interesting and meaningful generalizations of the k-median have been studied, to our knowledge, very little work has been done for the natural lower-bounded generalization of the k-median or its variants. This situation stimulates us to introduce and study the lower-bounded knapsack median problem (LB knapsack median), which generalizes the lower-bounded k-median problem (LB k-median). Compared with the knapsack median, the LB knapsack median has extra lower bound constraints that need to be respected. The main contribution of this paper is to design two approximation algorithms with ratios of 2730 and 1608 for the LB knapsack median. The first algorithm is inspired by the works of Svitkina [15] and Ahmadian and Swamy [1] on the LBFLP and related to their reduction steps. The main idea behind the second algorithm relies on an observation that simply running an algorithm for the knapsack median to satisfy the knapsack constraint and then running an algorithm for the LBFLP to satisfy the lower bound constraints works adequately, rather than a reduction process. Moreover, we adapt the algorithms for the LB knapsack median to produce algorithms for the LB k-median and obtain the approximation ratios of 610 and 387.

The remainder of our paper is organized as follows. Section 2 describes the LB knapsack median and the relevant knapsack facility location problem (knapsack FLP) along with their integer programs. Section 3 presents a basic 2730-approximation algorithm for the LB knapsack median. Section 4 improves the approximation ratio to 1608. Section 5 adapts the algorithms for the LB knapsack median to the LB k-median and obtain the approximation ratios of 610 and 387. Section 6 gives some discussions. Due to space constraint, all proofs are removed but will further appear in a full version of this paper.

2   Preliminaries

In the LB knapsack median, we are given a set $\mathcal{F}$ of facilities, a set $\mathcal{D}$ of clients, a budget B and a lower bound L. Every facility $i \in \mathcal{F}$ is associated with a weight w_i. Every facility-client pair (i, j) is associated with a connection cost c_{ij} where $i \in \mathcal{F}$ and $j \in \mathcal{D}$. Assume that the connection costs are metric. The goal is to open a subset $S \subseteq \mathcal{F}$ of facilities subject to the knapsack constraint $\sum_{i \in S} w_i \leq B$, and connect every client to some opened facility, such that any opened facility is connected by at least L clients and the total connection costs is minimized.

The LB knapsack median can be formulated as the following integer program:

$$\min \sum_{i \in \mathcal{F}} \sum_{j \in \mathcal{D}} c_{ij} x_{ij} \tag{1}$$

$$\text{s. t.} \sum_{i \in \mathcal{F}} x_{ij} \geq 1, \qquad\qquad \forall j \in \mathcal{D}, \tag{2}$$

$$x_{ij} \leq y_i, \qquad\qquad \forall i \in \mathcal{F}, j \in \mathcal{D}, \tag{3}$$

$$\sum_{j \in \mathcal{D}} x_{ij} \geq L y_i, \qquad\qquad \forall i \in \mathcal{F}, \tag{4}$$

$$\sum_{i \in \mathcal{F}} w_i y_i \leq B, \tag{5}$$

$$x_{ij} \in \{0, 1\}, \qquad\qquad \forall i \in \mathcal{F}, j \in \mathcal{D}, \tag{6}$$

$$y_i \in \{0, 1\}, \qquad\qquad \forall i \in \mathcal{F}. \tag{7}$$

In program (1–7), the variables x_{ij} and y_i indicate whether client j is connected to facility i and whether facility i is opened, respectively. The objective function is the total connection costs. The first constraints state that any client must be connected to some facility. The second constraints show that any facility that is connected by some client must be opened. The third constraints say that any opened facility must be connected by at least L clients. The fourth constraint guarantee that the sum of weights of the opened facilities is no more than the budget B.

When every facility $i \in \mathcal{F}$ in the LB knapsack median is additionally associated with an opening cost f_i rather than the lower bound L and the objective is to open a subset $S \subseteq \mathcal{F}$ of facilities subject to the knapsack constraint $\sum_{i \in S} w_i \leq B$, and connect every client to some opened facility so as to minimize the total opening and connection costs, the LB knapsack median becomes the knapsack FLP. With the same variables x_{ij} and y_i as in the program (1–7), the knapsack FLP can be formulated as the following integer program:

$$\min \sum_{i \in \mathcal{F}} f_i y_i + \sum_{i \in \mathcal{F}} \sum_{j \in \mathcal{D}} c_{ij} x_{ij} \tag{8}$$

$$\text{s. t.} \sum_{i \in \mathcal{F}} x_{ij} \geq 1, \qquad\qquad \forall j \in \mathcal{D}, \tag{9}$$

$$x_{ij} \leq y_i, \qquad\qquad \forall i \in \mathcal{F}, j \in \mathcal{D}, \tag{10}$$

$$\sum_{i \in \mathcal{F}} w_i y_i \leq B, \tag{11}$$

$$x_{ij} \in \{0, 1\}, \qquad\qquad \forall i \in \mathcal{F}, j \in \mathcal{D}, \tag{12}$$

$$y_i \in \{0, 1\}, \qquad\qquad \forall i \in \mathcal{F}. \tag{13}$$

In program (8–13), the objective function is the sum of opening as well as connection costs.

We have the following lemma which deserves to be noticed.

Lemma 1. *When the program (1–7) for the LB knapsack median and the program (8–13) for the knapsack FLP have the same inputs of $\mathcal{F}$, $\mathcal{D}$ and B, it is explicit that any feasible solution for the LB knapsack median is also a feasible solution for the knapsack FLP.*

In the rest of this paper, let $\mathcal{D}_i$ be the set of closest L clients to the facility $i \in \mathcal{F}$ and we use binary (S, σ) to denote a solution of some instance of the LB knapsack median or its relevant problems, where $S \subseteq \mathcal{F}$ indicates the set of opened facilities and $\sigma : \mathcal{D} \rightarrow S$ indicates a function that maps every client $j \in \mathcal{D}$ to some facility $i \in S$. Let $\sigma(j)$ denote the facility which is connected by client j in solution (S, σ).

3   A Basic Approximation Algorithm

In this section, we present a basic 2730-approximation algorithm for the LB knapsack median. The idea behind the algorithm is similar to those for the LBFLP, namely to reduce the LB knapsack median to the CFLP.

The basic algorithm, which consists of four steps, is formally shown in Algorithm 1. First, from the instance $\mathcal{I}$ of the LB knapsack median, we pick a constant $\alpha \in (\frac{1}{2}, 1)$ and construct an instance $\mathcal{I}_1(\alpha)$ of the knapsack FLP. Second, we use the current best approximation algorithm for the knapsack FLP to solve the instance $\mathcal{I}_1(\alpha)$ and obtain a solution (S_1, σ_1) for $\mathcal{I}_1$. Based on the solution (S_1, σ_1), we construct a bi-criteria solution (S_b, σ_b), which connects at least αL clients to any opened facility $i \in S_b$, for $\mathcal{I}$. Third, based on the solution (S_b, σ_b), we convert the instance $\mathcal{I}$ to a new instance $\mathcal{I}_2(\alpha)$ of the LB knapsack median, which is also an instance of some structured LBFLP. For an instance $\mathcal{I}(\alpha)$ of the LBFLP, it is a structured LBFLP instance if it has the following structures: every client is located at the position of some facility, every facility is co-located with at least αL clients and its opening cost is zero. Last, we solve the instance $\mathcal{I}_2(\alpha)$ with the current existing approximation algorithm for the structured LBFLP in order to obtain a solution (S_2, σ_2) for $\mathcal{I}_2(\alpha)$ as well as $\mathcal{I}$.

The main result of Algorithm 1 is as follows.

Theorem 1. *For any instance $\mathcal{I}$ of the LB knapsack median, Algorithm 1 can output a feasible solution (S_2, σ_2) for $\mathcal{I}$ where the total connection costs of (S_2, σ_2) is within a factor of 2730 of the total connection costs of the optimal solution for $\mathcal{I}$.*

The feasibility of the solution (S_2, σ_2) is not hard to see, since Step 2 and Step 4 in Algorithm 1 guarantee that (S_2, σ_2) satisfies the knapsack constraint and the lower bound constraints, respectively. From now on, we will focus on analyzing the approximation ratio of Algorithm 1. For any instance $\mathcal{I}$ of the LB knapsack median, let $(S_{\mathcal{I}}^*, \sigma_{\mathcal{I}}^*)$ denote the optimal solution for $\mathcal{I}$ and let $OPT_{\mathcal{I}}$ denote the total connection costs of $(S_{\mathcal{I}}^*, \sigma_{\mathcal{I}}^*)$, i.e., $OPT_{\mathcal{I}} = \sum_{j \in \mathcal{D}} c_{\sigma_{\mathcal{I}}^*(j)j}$. Let $OPT_{\mathcal{I}_1(\alpha)}$ and $OPT_{\mathcal{I}_2(\alpha)}$ be the total opening and connection costs of the optimal solution for $\mathcal{I}_1(\alpha)$ and the total connection costs of the optimal solution for $\mathcal{I}_2(\alpha)$, respectively.

Algorithm 1

Step 1 Convert $\mathcal{I}$ to $\mathcal{I}_1(\alpha)$ of the knapsack FLP.

For the instance $\mathcal{I} = (\mathcal{F}, \mathcal{D}, B, L, \{w_i\}_{i \in \mathcal{F}}, \{c_{ij}\}_{i \in \mathcal{F}, j \in \mathcal{D}})$ of the LB knapsack median, pick a constant $\alpha \in (\frac{1}{2}, 1)$, get rid of the lower bounds L from $\mathcal{I}$ and

$$set \ f_i := \frac{2\alpha}{1 - \alpha} \sum_{j \in \mathcal{D}_i} c_{ij} \ for \ every \ i \in \mathcal{F},$$

to obtain an instance $\mathcal{I}_1(\alpha) = (\mathcal{F}, \mathcal{D}, B, \{f_i\}_{i \in \mathcal{F}}, \{w_i\}_{i \in \mathcal{F}}, \{c_{ij}\}_{i \in \mathcal{F}, j \in \mathcal{D}})$ of the knapsack FLP.

Step 2 Based on $\mathcal{I}_1(\alpha)$, obtain a bi-criteria solution (S_b, σ_b) for $\mathcal{I}$.

Step 2.1 Solve the instance $\mathcal{I}_1(\alpha)$ with the current best ρ-approximation algorithm for the knapsack FLP (see [3]) and obtain a feasible solution (S_1, σ_1) where $\rho = 17.46$.

Step 2.2 Initialize $S_b := S_1$ and $\sigma_b(j) := \sigma_1(j)$ for any $j \in \mathcal{D}$. Define $T_i := \{j \in \mathcal{D} : \sigma_b(j) = i\}$ for any $i \in \mathcal{F}$ and $S_r := \{i \in S_b : |T_i| < \alpha L\}$.

Step 2.3 While $S_r \neq \emptyset$ do

Arbitrarily choose some facility $i \in S_r$ and close it. Reconnect every client j with $\sigma_b(j) = i$ to its closest facility i' in $S_b \setminus \{i\}$ and update $\sigma_b(j) := i'$. Update $S_b := S_b \setminus \{i\}$. Update T_i for any facility $i \in \mathcal{F}$ and S_r.

end while

Obtain solution (S_b, σ_b).

Step 3 Based on (S_b, σ_b), convert $\mathcal{I}$ to $\mathcal{I}_2(\alpha)$ of the LB knapsack median.

From the instance $\mathcal{I} = (\mathcal{F}, \mathcal{D}, B, L, \{w_i\}_{i \in \mathcal{F}}, \{c_{ij}\}_{i \in \mathcal{F}, j \in \mathcal{D}})$ of the LB knapsack median and its bi-criteria solution (S_b, σ_b), construct a new instance $\mathcal{I}_2(\alpha) = (\mathcal{F}_2, \mathcal{D}, B, L, \{w_i\}_{i \in \mathcal{F}_2}, \{c'_{ij}\}_{i \in \mathcal{F}_2, j \in \mathcal{D}})$ of the LB knapsack median where $\mathcal{F}_2 = S_b$, $c'_{ij} = c_{i\sigma_b(j)}$ for any $i \in \mathcal{F}_2$ and $j \in \mathcal{D}$ (see Fig. 1). Since the instance $\mathcal{I}_2(\alpha)$ already satisfies that $\sum_{i \in S_b} w_i \leq B$, it can also be viewed as $(S_b, \mathcal{D}, L, \{f'_i\}_{i \in S_b}, \{c'_{ij}\}_{i \in S_b, j \in \mathcal{D}})$ where $f'_i = 0$ for any $i \in S_b$, which is an instance of the structured LBFLP.

Step 4 Solve $\mathcal{I}_2(\alpha)$ and obtain a solution (S_2, σ_2) for $\mathcal{I}$.

Solve the instance $\mathcal{I}_2(\alpha)$ with the current best $g(\alpha)$-approximation algorithm for the structured LBFLP (see [1]) and obtain a feasible solution (S_2, σ_2) where $g(\alpha) = \frac{2}{\alpha} + \frac{2\alpha}{2\alpha - 1} + 2\sqrt{\frac{2}{\alpha^2} + \frac{4}{2\alpha - 1}}$. Output (S_2, σ_2) as the solution for the instance $\mathcal{I}$.

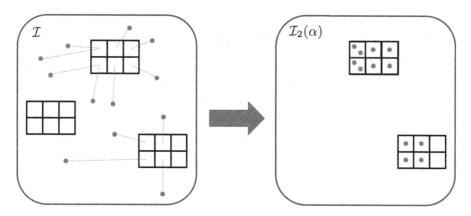

Fig. 1. An illustration of how to convert $\mathcal{I}$ to $\mathcal{I}_2(\alpha)$. The rectangles presents the facilities with lower bound $B = 6$. The black dots present the clients. For the bi-criteria solution (S_b, σ_b), the set of its opened facilities is all the facilities that are touched by some dotted line and the function of connections is presents by all the dotted lines.

Note that Step 1–2 in Algorithm 1 can be regarded as a bi-criteria algorithm for the LB knapsack median, which is a contribution in our working paper. For completeness, we present the main result of it in the following theorem.

Theorem 2. *For any instance $\mathcal{I}$ of the LB knapsack median, Step 1–2 in Algorithm 1 can give a bi-criteria solution (S_b, σ_b) for $\mathcal{I}$, which connects at least αL clients to every facility $i \in S_b$ and costs within a factor of $\frac{1+\alpha}{1-\alpha}\rho$ of $OPT_\mathcal{I}$ where $\alpha \in (\frac{1}{2}, 1)$ and $\rho = 17.46$.*

The following lemma bounds $OPT_{\mathcal{I}_2(\alpha)}$ in terms of $OPT_\mathcal{I}$.

Lemma 2. *The total connection costs of the optimal solution for $\mathcal{I}_2(\alpha)$ is within a factor of $2(\frac{1+\alpha}{1-\alpha}\rho + 1)$ of $OPT_\mathcal{I}$ where $\alpha \in (\frac{1}{2}, 1)$ and $\rho = 17.46$.*

Now we are ready to give the approximation ratio of Algorithm 1.

Lemma 3. *The $g(\alpha)$-approximation solution (S_2, σ_2) for the structured LBFLP instance $\mathcal{I}_2(\alpha)$ is a $\left((2g(\alpha) + 1)\frac{1+\alpha}{1-\alpha}\rho + 2g(\alpha)\right)$-approximation solution for the LB knapsack median instance $\mathcal{I}$, where $\alpha \in (\frac{1}{2}, 1)$, $\rho = 17.46$ and $g(\alpha) = \frac{2}{\alpha} + \frac{2\alpha}{2\alpha-1} + 2\sqrt{\frac{2}{\alpha^2} + \frac{4}{2\alpha-1}}$. When setting $\alpha = 0.64$, the approximation ratio is no more than 2730.*

4   An Improved Approximation Algorithm

In this section, we propose an intuitive and dramatically better 1608-approximation algorithm for the LB knapsack median. The idea of the algorithm based on an intuitive observation that running an algorithm for the knapsack median in order to satisfy the knapsack constraint and then an algorithm for the LBFLP in order to satisfy the lower bound constraints works well.

The improved algorithm, which comprises four steps, is formally given in Algorithm 2. First, from the instance $\mathcal{I}$ of the LB knapsack median, we construct an instance $\mathcal{I}_1'$ of the knapsack median. Second, we use the current best approximation algorithm for the knapsack median to solve the instance $\mathcal{I}_1$ and obtain a solution (S_1', σ_1') for $\mathcal{I}_1'$. Third, based on the solution (S_1', σ_1'), we convert the instance $\mathcal{I}$ to a new instance $\mathcal{I}_2'$ of the LB knapsack median, which is also an instance of some special case of the LBFLP. Last, we solve the instance $\mathcal{I}_2'$ with the state-of-art approximation algorithm for the LBFLP in order to obtain a solution (S_2', σ_2') for $\mathcal{I}_2$ as well as $\mathcal{I}$.

Algorithm 2

Step 1 Convert $\mathcal{I}$ to $\mathcal{I}_1'$ of the knapsack median.
   For the instance $\mathcal{I} = (\mathcal{F}, \mathcal{D}, B, L, \{w_i\}_{i \in \mathcal{F}}, \{c_{ij}\}_{i \in \mathcal{F}, j \in \mathcal{D}})$ of the LB knapsack median, get rid of the lower bounds L from $\mathcal{I}$ and obtain an instance $\mathcal{I}_1' = (\mathcal{F}, \mathcal{D}, B, \{w_i\}_{i \in \mathcal{F}}, \{c_{ij}\}_{i \in \mathcal{F}, j \in \mathcal{D}})$ of the knapsack median.

Step 2 Solve $\mathcal{I}_1'$ and obtain a solution (S_1', σ_1') for $\mathcal{I}_1'$.
   Solve the instance $\mathcal{I}_1'$ with the current best η-approximation algorithm for the knapsack median (see [3]) and obtain a feasible solution (S_1', σ_1') where $\eta = 17.46$.

Step 3 Based on (S_1', σ_1'), convert $\mathcal{I}$ to $\mathcal{I}_2'$ of the LB knapsack median.
   From the instance $\mathcal{I} = (\mathcal{F}, \mathcal{D}, B, L, \{w_i\}_{i \in \mathcal{F}}, \{c_{ij}\}_{i \in \mathcal{F}, j \in \mathcal{D}})$ of the LB knapsack median and the solution (S_1', σ_1'), construct a new instance $\mathcal{I}_2' = (\mathcal{F}_2', \mathcal{D}, B, L, \{w_i\}_{i \in \mathcal{F}_2'}, \{c_{ij}\}_{i \in \mathcal{F}_2', j \in \mathcal{D}})$ of the LB knapsack median where $\mathcal{F}_2' = S_1'$ (see Fig. 2). Since the instance $\mathcal{I}_2'$ already satisfies that $\sum_{i \in S_1'} w_i \leq B$, it can also be viewed as $(S_1', \mathcal{D}, L, \{f_i'\}_{i \in S_1'}, \{c_{ij}\}_{i \in S_1', j \in \mathcal{D}})$ where $f_i' = 0$ for any $i \in S_1'$, which is an instance of a special case of the LBFLP.

Step 4 Solve $\mathcal{I}_2'$ and obtain a solution (S_2', σ_2') for $\mathcal{I}$.
   Solve the instance $\mathcal{I}_2'$ with the current best θ-approximation algorithm for the LBFLP (see [1]) and obtain a feasible solution (S_2', σ_2') where $\theta = 82.6$. Output (S_2', σ_2') as the solution for the instance $\mathcal{I}$.

The main result of Algorithm 2 is as follows.

Theorem 3. *For any instance $\mathcal{I}$ of the LB knapsack median, Algorithm 2 can output a feasible solution (S_2', σ_2') for $\mathcal{I}$ where the total connection costs of (S_2', σ_2') is within a factor of 1608 of the total connection costs of the optimal solution for $\mathcal{I}$.*

The feasibility of the solution (S_2', σ_2') is easy to understand, since Step 2 and Step 4 in Algorithm 2 guarantee that (S_2', σ_2') satisfies the knapsack constraint and the lower bound constraints, respectively. The remainder of this section pays attention on analyzing the approximation ratio of Algorithm 2. Denote $OPT_{\mathcal{I}_1'}$

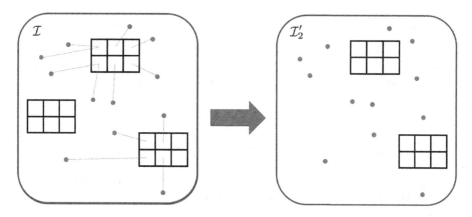

Fig. 2. An illustration of how to convert $\mathcal{I}$ to $\mathcal{I}_2'$. The rectangles presents the facilities with lower bound $B = 6$. The black dots present the clients. For the solution (S_1', σ_1'), the set of its opened facilities is all the facilities that are touched by some dotted line and the function of connections is presents by all the dotted lines.

and $OPT_{\mathcal{I}_2'}$ as the total connection costs of the optimal solution for $\mathcal{I}_1'$ and $\mathcal{I}_2'$, respectively.

The following lemma bounds $OPT_{\mathcal{I}_2'}$ in terms of $OPT_{\mathcal{I}}$.

Lemma 4. *The total connection costs of the optimal solution for $\mathcal{I}_2'$ is within a factor of $(\eta + 2)$ of $OPT_{\mathcal{I}}$ where $\eta = 17.46$.*

Now we are ready to give the approximation ratio of Algorithm 2.

Lemma 5. *The θ-approximation solution (S_2', σ_2') for the LBFLP instance $\mathcal{I}_2'$ is a $\theta(\eta + 2)$-approximation solution for the LB knapsack median instance $\mathcal{I}$, where $\eta = 17.46$ and $\theta = 82.6$.*

5   The Lower-Bounded k-median

In this section, we show that the ideas of algorithms for the LB knapsack median can be used to produce approximation algorithms for the LB k-median, which is a special case of the LB knapsack median, and obtain approximation ratios that are significantly better than 2730 and 1608.

Algorithms 1 and 2 for the LB knapsack median can be adapted to Algorithms 3 and 4 for the LB k-median, respectively. The main results of Algorithms 3 and 4 are also given.

Theorem 4. *For any instance $\mathcal{I}$ of the LB kmedian, Algorithm 3 can output a feasible solution (S_2, σ_2) for $\mathcal{I}$ where the total connection costs of (S_2, σ_2) is within a factor of 610 of the total connection costs of the optimal solution for $\mathcal{I}$.*

Theorem 5. *For any instance $\mathcal{I}$ of the LB k-median, Algorithm 4 can output a feasible solution (S_2', σ_2') for $\mathcal{I}$ where the total connection costs of (S_2', σ_2') is within a factor of 387 of the total connection costs of the optimal solution for $\mathcal{I}$.*

Algorithm 3

Step 1 Convert $\mathcal{I}$ to $\mathcal{I}_1(\alpha)$ of the k-FLP.
For the instance $\mathcal{I} = (\mathcal{F}, \mathcal{D}, k, L, \{c_{ij}\}_{i \in \mathcal{F}, j \in \mathcal{D}})$ of the LB k-median, pick a constant $\alpha \in (\frac{1}{2}, 1)$, get rid of the lower bounds L from $\mathcal{I}$ and

$$set \; f_i := \frac{2\alpha}{1 - \alpha} \sum_{j \in \mathcal{D}_i} c_{ij} \; for \; every \; i \in \mathcal{F},$$

to obtain an instance $\mathcal{I}_1(\alpha) = (\mathcal{F}, \mathcal{D}, k, \{f_i\}_{i \in \mathcal{F}}, \{c_{ij}\}_{i \in \mathcal{F}, j \in \mathcal{D}})$ of the k-FLP.

Step 2 Based on $\mathcal{I}_1(\alpha)$, obtain a bi-criteria solution (S_b, σ_b) for $\mathcal{I}$.

  Step 2.1 Solve the instance $\mathcal{I}_1(\alpha)$ with the current best ρ'-approximation algorithm for the k-FLP (see [16]) and obtain a feasible solution (S_1, σ_1) where $\rho' = 2 + \sqrt{3} + \epsilon$.

  Step 2.2 Same as Step 2.2 in Algorithm 1.

  Step 2.3 Same as Step 2.3 in Algorithm 1. At the end of this step, obtain a bi-criteria solution (S_b, σ_b).

Step 3 Based on (S_b, σ_b), convert $\mathcal{I}$ to $\mathcal{I}_2(\alpha)$ of the LB k-median.
From the instance $\mathcal{I}$ of the LB k-median and its bi-criteria solution (S_b, σ_b), construct a new instance $\mathcal{I}_2(\alpha) = (\mathcal{F}_2, \mathcal{D}, k, L, \{c'_{ij}\}_{i \in \mathcal{F}_2, j \in \mathcal{D}})$ of the LB k-median where $\mathcal{F}_2 = S_b$, $c'_{ij} = c_{i\sigma_b(j)}$ for any $i \in \mathcal{F}_2$ and $j \in \mathcal{D}$. Since the instance $\mathcal{I}_2(\alpha)$ already satisfies the cardinality constraint, it can also be viewed as $(S_b, \mathcal{D}, L, \{f'_i\}_{i \in S_b}, \{c'_{ij}\}_{i \in S_b, j \in \mathcal{D}})$ where $f'_i = 0$ for any $i \in S_b$, which is an instance of the structured LBFLP.

Step 4 Solve $\mathcal{I}_2(\alpha)$ and obtain a solution (S_2, σ_2) for $\mathcal{I}$.
Same as Step 4 in Algorithm 1. At the end of this step, output (S_2, σ_2) as the solution for the instance $\mathcal{I}$.

Algorithm 4

Step 1 Convert $\mathcal{I}$ to $\mathcal{I}'_1$ of the k-median.
For the instance $\mathcal{I} = (\mathcal{F}, \mathcal{D}, k, L, \{c_{ij}\}_{i \in \mathcal{F}, j \in \mathcal{D}})$ of the LB k-median, get rid of the lower bounds L from $\mathcal{I}$ and obtain an instance $\mathcal{I}'_1 = (\mathcal{F}, \mathcal{D}, k, \{c_{ij}\}_{i \in \mathcal{F}, j \in \mathcal{D}})$ of the k-median.

Step 2 Solve $\mathcal{I}'_1$ and obtain a solution (S'_1, σ'_1) for $\mathcal{I}'_1$.
Solve the instance $\mathcal{I}'_1$ with the current best η'-approximation algorithm for the k-median (see [4]) and obtain a feasible solution (S'_1, σ'_1) where $\eta' = 2.675 + \epsilon$.

Step 3 Based on (S'_1, σ'_1), convert $\mathcal{I}$ to $\mathcal{I}'_2$ of the LB k-median.
From the instance $\mathcal{I}$ of the LB k-median and the solution (S'_1, σ'_1), construct a new instance $\mathcal{I}'_2 = (\mathcal{F}'_2, \mathcal{D}, k, L, \{c_{ij}\}_{i \in \mathcal{F}'_2, j \in \mathcal{D}})$ of the LB k-median where $\mathcal{F}'_2 = S'_1$. Since the instance $\mathcal{I}'_2$ already satisfies the cardinality constraint, it can also be viewed as $(S'_1, \mathcal{D}, L, \{f'_i\}_{i \in S'_1}, \{c_{ij}\}_{i \in S'_1, j \in \mathcal{D}})$ where $f'_i = 0$ for any $i \in S'_1$, which is an instance of a special case of the LBFLP.

Step 4 Solve $\mathcal{I}'_2$ and obtain a solution (S'_2, σ'_2) for $\mathcal{I}$.
Same as Step 4 in Algorithm 2. At the end of this step, output (S'_2, σ'_2) as the solution for the instance $\mathcal{I}$.

6 Discussions

In this paper, we introduce the LB knapsack median, which is a natural generalization of the LB k-median, and propose approximation algorithms for the LB knapsack median with approximation ratio of 2730 and 1608. In addition, these algorithms can be adapted to produce algorithms for the LB k-median with approximation ratios of 610 and 387. Although the approximation ratios in this paper are not practical and far from the hardness of approximation, we do believe them take the first step on providing practical approximation algorithms and heuristics for the LB knapsack median.

Acknowledgements. The first author is supported by National Natural Science Foundation of China (No. 11531014). The second author is supported by National Natural Science Foundation of China (No. 11771003). The third author is supported by Natural Science Foundation of China (No. 11971349). The fourth author is supported by National Natural Science Foundation of China (No. 11871081) and the Science and Technology Program of Beijing Education Commission (No. KM201810005006).

References

1. Ahmadian, S., Swamy, C.: Improved approximation guarantees for lower-bounded facility location. In: Proceedings of the 10th International Workshop on Approximation and Online Algorithms, pp. 257–271 (2012)
2. Arya, V., Garg, N., Khandekar, R., Meyerson, A., Munagala, K., Pandit, V.: Local search heuristics for k-median and facility location problems. SIAM J. Comput. **33**, 544–562 (2004)
3. Byrka, J., Pensyl, T., Rybicki, B., Spoerhase, J., Srinivasan, A., Trinh, K.: An improved approximation algorithm for knapsack median using sparsification. Algorithmica **80**, 1093–1114 (2018)
4. Byrka, J., Pensyl, T., Rybicki, B., Srinivasan, A., Trinh, K.: An improved approximation for k-median, and positive correlation in budgeted optimization. ACM Trans. Algorithms **13**, 1–31 (2017)
5. Charikar, M., Guha, S., Tardos, É., Shmoys, D.B.: A constant-factor approximation algorithm for the k-median problem. J. Comput. Syst. Sci. **65**, 129–149 (2002)
6. Guha, S., Meyerson, A., Munagala, K.: Hierarchical placement and network design problems. In: Proceedings of the 41st Annual Symposium on Foundations of Computer Science, pp. 603–612 (2000)
7. Jain, K., Mahdian, M., Saberi, A.: A new greedy approach for facility location problems. In: Proceedings of the 34th Annual ACM Symposium on Theory of Computing, pp. 731–740 (2002)
8. Jain, K., Vazirani, V.V.: Approximation algorithms for metric facility location and k-median problems using the primal-dual schema and Lagrangian relaxation. J. ACM **48**, 274–296 (2001)
9. Karger, D.R., Minkoff, M.: Building Steiner trees with incomplete global knowledge. In: Proceedings of the 41st Annual Symposium on Foundations of Computer Science, pp. 613–623 (2000)
10. Kumar, A.: Constant factor approximation algorithm for the knapsack median problem. In: Proceedings of the 23rd Annual ACM-SIAM Symposium on Discrete Algorithms, pp. 824–832 (2012)

11. Li, S., Svensson, O.: Approximating k-median via pseudo-approximation. SIAM J. Comput. **45**, 530–547 (2016)
12. Li, S.: A 1.488 approximation algorithm for the uncapacitated facility location problem. Inf. Comput. **222**, 45–58 (2013)
13. Li, S.: On facility location with general lower bounds. In: Proceedings of the 30th Annual ACM-SIAM Symposium on Discrete Algorithms, pp. 2279–2290 (2019)
14. Shmoys, D.B., Tardos, É., Aardal, K.I.: Approximation algorithms for facility location problems. In: Proceedings of the 29th Annual ACM symposium on Theory of Computing, pp. 265–274 (1997)
15. Svitkina, Z.: Lower-bounded facility location. ACM Trans. Algorithms **6**, 1–16 (2010)
16. Zhang, P.: A new approximation algorithm for the k-facility location problem. Theoret. Comput. Sci. **384**, 126–135 (2007)

The Spherical k-means++ Algorithm via Local Search

Xiaoyun Tian[1], Dachuan Xu[1], Donglei Du[2], and Ling Gai[3(✉)]

[1] Department of Operations Research and Information Engineering,
Beijing University of Technology, Beijing 100124, People's Republic of China
`xiaoyun_txy@emails.bjut.edu.cn, xudc@bjut.edu.cn`
[2] Faculty of Management, University of New Brunswick,
Fredericton, NB E3B 9Y2, Canada
`ddu@unb.ca`
[3] Glorious Sun School of Business and Management, Donghua University,
Shanghai 200051, People's Republic of China
`lgai@dhu.edu.cn`

Abstract. We consider the spherical k-means problem (SKMP), a generalization of the k-means clustering problem (KMP). Given a data set of n points $\mathcal{P}$ in d-dimensional unit sphere $\mathbb{R}^d$, and an integer $k \leq n$, it aims to partition the data set $\mathcal{P}$ into k sets so as to minimize the sum of cosine dissimilarity measure from each data point to its closest center. We present a constant expected approximation guarantee for this problem based on integrating the k-means++ seeding algorithm for the KMP and the local search technique.

Keywords: Spherical k-means · Cosine dissimilarity · Local search · Seeding algorithm · Approximation algorithm

1 Introduction

Clustering problems arise in many different applications, including text clustering, data compression, pattern classification and machine learning. Further information on clustering and clustering algorithms can be found in [7,8,10]. One of the most popular and widely studied clustering models for points in Euclidean space is called the k-means problem (KMP), which tries to minimize the mean squared Euclidean distance from each data point to its closest center (refer to e.g. [1,2,9–11,15] for more information on KMP).

It is of significant importance to be able to mine valuable information from large-scale high-dimensional text data, which are becoming more and more common. Our main interest in this paper is the spherical k-means clustering (SKMP) problem, one of the most representative approaches to cluster text documents. The goal is to partition a given set of objects into clusters such that objects in the same cluster should be similar to each other while objects in different clusters should be less similar. Specifically, given a data set $\mathcal{P}$ in the unit sphere

Z. Zhang et al. (Eds.): AAIM 2020, LNCS 12290, pp. 131–140, 2020.
https://doi.org/10.1007/978-3-030-57602-8_12

$\mathbb{R}^d$, and an integer $k \leq n$, it aims to partition the data set $\mathcal{P}$ into k sets so as to minimize the sum of cosine dissimilarity measure from each data point to its closest center. We shall assume that the document vectors have been normalized to have a unit Euclidean norm, that is, they can be thought of as points on a high-dimensional unit sphere.

The KMP has been extensively studied in the literature. Perhaps the most famous heuristic for k-means is the well-known Lloyd's algorithm [14] (a.k.a., the k-means algorithm), which is a very powerful clustering method for KMP in practice. However one major limitation of this algorithm is that it strongly depends on the initial value, and hence does not provide any theoretical approximation guarantee. To decrease the dependence, Arthur and Vassilvitskii propose an $O(\ln k)$-approximation algorithm, called k-means++ algorithm [2], which incrementally chooses a set of k centers by sampling the first k centers with pre-specified probabilities. In practice, k-means++ is easy to implement and it is often used as a starting solution for Lloyd's algorithm. However, the approximation guarantee of the k-means++ is not a constant factor.

For the SKMP, Dhillon and Modha [3] present the primitive spherical k-means clustering algorithm with cosine similarities based on the KMP. Honik et al. [6] propose a new spherical k-means clustering algorithm which is based on [3] and [15]. It is easy to find that the algorithms in both [3] and [6] depend on the initial value. Inspired by the k-means++ algorithm [2], Endo and Miyamoto [4] present an algorithm which works well for the SKMP, called the spherical k-means++ clustering algorithm, and obtain an $O(\log k)$-approximation guarantee. Li et al. [13] then prove that the algorithm based on [2] is also an $O(\log k)$-approximation for the SKMP and has a constant approximation ratio for the SKMP with separable sets.

Zhang et al. [16] present a $(8(2 + \sqrt{3}) + \varepsilon)$-approximation algorithm by using a direct local search technique based on the work of Kanungo et al. [9] for the SKMP. In [9], they present a local improvement heuristic based on swapping centers in and out and prove that it yields a $(9 + \varepsilon)$-approximation algorithm for the KMP.

Recently, Lattanzi and Sohler [12] develop a random variant of the k-means++ seeding algorithm that achieves a constant expected approximation guarantee. They pose the question of whether there exists a simple and practical approximation algorithm for KMP. This work tries to answer this question by designing a simple local search algorithm with an expected constant approximation ratio for the SKMP. We present an approximation algorithm via integrating the k-means++ seeding algorithm and the local search technique. We obtain a constant approximation for SKMP, which is also a generalization of the k-means++ seeding approach via local search in [12].

The rest of this paper is organized as follows. In Sect. 2, we describe the formulation for the spherical k-means problem (SKMP) and provide several lemmas related to the SKMP, which lead to the proof of Theorem 1. In Sect. 3, we present a SpheMeans++ algorithm via the local search for the SKMP along with our main result. In Sect. 4, we provide the analysis of the approximation ratio.

Conclusions are given in Sect. 5. In addition, all proofs are omitted due to space constraints, and will be available in the journal version.

2   Notations, Definitions and Preliminaries

In this section, we formally define the spherical k-means problem (SKMP), as well as the notations used in the paper. Let $\mathcal{X}$ be a data set of n points in the d-dimensional space $\mathbb{R}^d$. Without loss of generality, we assume that each point in $\mathcal{X}$ is normalized as a unit vector according to the Euclidean norm, i.e, $\mathcal{X} = \{x \in \mathbb{R}^d : \|x\| = 1\}$, and hence the norms of all data are normalized as one and are located on the unit sphere.

For any two points u and v in $\mathcal{X}$, we use the distance $d(u,v)$ to denote their cosine dissimilarity: $d(u,v) = 1 - \cos(u,v) = \frac{1}{2}\|u-v\|^2$, where $\|u-v\|^2$ measures their squared Euclidean distance between u and v. So $d(u,v)$ is just a half of their squared Euclidean distance. Moreover, for any set $\mathcal{S} \subseteq \mathcal{X}$ and a point $x \in \mathcal{X}$, the cosine dissimilarity from x to S is defined as $d(\{x\}, \mathcal{S}) = \min_{s \in \mathcal{S}} d(x, s)$.

For any set $\mathcal{S} \subseteq \mathcal{X}$, the point $scen(\mathcal{S}) = \left(\sum_{s \in \mathcal{S}} s\right) / \|\sum_{s' \in \mathcal{S}} s'\|$ is called the spherical center of the mass of all points in $\mathcal{S}$. In particular, if $\sum_{s \in \mathcal{S}} s = 0$, then $scen(\mathcal{S})$ includes all d-dimensional unit vectors in $\mathcal{S}$. Analogously, given a set $\mathcal{C} \in \mathbb{R}^d$, the total sum of the cosine dissimilarity over each point of $\mathcal{X}$ to the set $\mathcal{C}$ is given by $d(\mathcal{X}, \mathcal{C}) = \sum_{x \in \mathcal{X}} d(\{x\}, \mathcal{C})$.

We now formally define the spherical k-means problem (SKMP). Suppose that we are given an integer k, as well as a data points set $\mathcal{P} = \{p_1, p_2, \cdots, p_n\}$ in $\mathcal{X}$. The goal of the SKMP is to find a set $\mathcal{C}$ of k data points in $\mathcal{X}$, called centers, to minimize the sum of cosine dissimilarity measure, or total cost for short, from each data point in $\mathcal{P}$ to its closest center. Namely, find a set $\mathcal{C} = \{c_1, c_2, \cdots, c_k\} \subseteq \mathcal{X}$ minimizing the cost function

$$cost(\mathcal{P}, \mathcal{C}) := \sum_{j=1}^{n} \min_{i \in \{1,2,\cdots,k\}} d(p_j, c_i) = \frac{1}{2} \sum_{j=1}^{n} \min_{i \in \{1,2,\cdots,k\}} \|p_j - c_i\|^2.$$

Obviously, for the SKMP, the centers set $\mathcal{C}$ partitions the input points $\mathcal{P}$ with unit length into k clusters $\{\mathcal{P}_1, \cdots, \mathcal{P}_k\}$, where each $\mathcal{P}_i := \{\arg\min_p d(p, c_i) | p \in \mathcal{P}\}$ is with respect to a cluster center c_i. In other words, the cluster $\mathcal{P}_i$ is the set of points assigned to their closest center c_i respectively in terms of cosine dissimilarity. If the closest center is not unique, then p is assigned to one of them arbitrarily. Assume that the optimal center set is $\mathcal{O} = \{o_1, \cdots, o_k\}$. Then $\mathcal{P}$ is partitioned into k disjoint subsets $\{O_1, \cdots, O_k\}$ induced by these optimal centers. Denote the cost of the optimal solution as opt_k.

The following property in [4] is a vital property of spherical centroidal solutions that are similar to the center of the mass in the k-means problem.

Lemma 2.1. *For any point $c \in \mathcal{X}$, we have*

$$cost\left(\mathcal{S}, \{c\}\right) = cost\left(\mathcal{S}, \{scen(\mathcal{S})\}\right) + \left\|\sum_{s \in \mathcal{S}} s\right\| \cdot cost\left(\{scen(\mathcal{S})\}, \{c\}\right)$$

where $cost\,(\mathcal{S}, \{scen(\mathcal{S})\}) = \min_{c \in \mathcal{X}} cost(\mathcal{S}, \{c\})$. Furthermore, the above equality is equivalent to

$$\sum_{s \in \mathcal{S}} \|s - c\|^2 = \sum_{s \in \mathcal{S}} \|s - scen(\mathcal{S})\|^2 + \|\sum_{s \in \mathcal{S}} s\| \cdot \|scen(\mathcal{S}) - c\|^2$$

Hence, by the centroidal property, it is not hard to see that $scen(\mathcal{S})$ is the unique optimal 1-means of $\mathcal{S}$. In this case, the objective function for the SKMP can also be represented by

$$\frac{1}{2} \sum_{i=1}^{k} \min_{p \in \mathcal{P}_i} \|p - scen(P_i)\|^2.$$

Based on the definition of the cosine dissimilarity, we know that the cosine dissimilarity measure always satisfies the triangle inequality within a factor of two in the following lemma.

Lemma 2.2. *For any $u, v, w \in \mathcal{X}$ we have*

$$d(u, v) \le 2d(u, w) + 2d(v, w)$$

Therefore, $\|u - v\| \le \sqrt{2}\,(\|u - w\| + \|v - w\|)$. The following lemma, a generalization of the similar result in [5], is crucial to our analysis.

Lemma 2.3. *Let $\varepsilon > 0$, $u, v \in \mathcal{X}$ and $\mathcal{C} \subseteq \mathcal{X}$ is a set of k centers. Then*

$$|\, cost(\{u\}, \mathcal{C}) - cost(\{v\}, \mathcal{C})\, | \le 2\varepsilon cost(\{u\}, \mathcal{C}) + \left(1 + \frac{1}{\varepsilon}\right) \|u - v\|^2.$$

Proof.

$$|\, cost(\{u\}, \mathcal{C}) - cost(\{v\}, \mathcal{C})\, |$$
$$=|\, d(\{u\}, \mathcal{C}) - d(\{v\}, \mathcal{C})\, |$$
$$=|\, \sqrt{d(\{u\}, \mathcal{C})} - \sqrt{d(\{v\}, \mathcal{C})}\, | \cdot \left(\sqrt{d(\{u\}, \mathcal{C})} + \sqrt{d(\{v\}, \mathcal{C})}\right)$$
$$\le \sqrt{2}\sqrt{d(u, v)} \cdot \left((1 + \sqrt{2})\sqrt{d(\{u\}, \mathcal{C})} + \sqrt{2}\sqrt{d(u, v)}\right)$$
$$= 2d(u, v) + (2 + \sqrt{2})\sqrt{d(\{u\}, \mathcal{C})} \cdot \sqrt{d(u, v)}$$
$$= \|u - v\|^2 + (2 + \sqrt{2})\sqrt{d(\{u\}, \mathcal{C})} \cdot \frac{\sqrt{2}}{2}\|u - v\|$$
$$\le \|u - v\|^2 + 2\sqrt{2}\sqrt{d(\{u\}, \mathcal{C})} \cdot \|u - v\|$$
$$= \|u - v\|^2 + 2\sqrt{2\varepsilon}\sqrt{d(\{u\}, \mathcal{C})} \cdot \frac{\|u - v\|}{\sqrt{\varepsilon}}$$
$$\le \|u - v\|^2 + 2\varepsilon d(\{u\}, \mathcal{C}) + \frac{\|u - v\|^2}{\varepsilon}$$
$$= \left(1 + \frac{1}{\varepsilon}\right) \|u - v\|^2 + 2\varepsilon d(\{u\}, \mathcal{C})$$

where the first inequality is due to the symmetry and Lemmas 2.2, and the last follows from $2uv \le u^2 + v^2$ for $u, v \in \mathcal{X}$. $\qquad\square$

Algorithm 1. SpheMeans++ with local search

Input: A data point set $\mathcal{P} \subseteq \mathcal{X}$ with n unit length points, integer k, K and $\mathcal{C} = \phi$.
Output: An approximate spherical k-means++ $\mathcal{C} \in \mathcal{X}$ for $\mathcal{P}$.
 1: Sample the first center c_1 uniformly at random from $\mathcal{P}$ and set $\mathcal{C} = \mathcal{C} \cup \{c_1\}$.
 2: **for** $i = 2$ to k **do**
 3:     Sample the center $c_i = p' \in \mathcal{P}$ with probability $\frac{d(\{p'\},\mathcal{C})}{cost(\mathcal{P},\mathcal{C})}$.
 4:     Set $\mathcal{C} := \mathcal{C} \cup \{c_i\}$.
 5: **end for**
 6: **for** $i = 1$ to k **do**
 7:     Set the cluster $\mathcal{P}_i := \{p \in \mathcal{P} : d(p, c_i) \leq d(p, c_j), \forall j \in [k], j \neq i\}$;
 8: **end for**
 9: **for** $i = 1$ to K **do**
10:     Sample the center $p' \in \mathcal{P}$ with probability $\frac{d(\{p'\},\mathcal{C})}{cost(\mathcal{P},\mathcal{C})}$.
11:     **if** $\exists\ c \in \mathcal{C}$ s.t. $cost(\mathcal{P}, \mathcal{C} \setminus \{c\} \cup \{p'\}) < cost(\mathcal{P},\mathcal{C})$ **then**
12:         $c := \arg\min_{c \in \mathcal{C}} cost(\mathcal{P}, \mathcal{C} \setminus \{c\} \cup \{p'\})$.
13:         Set $\mathcal{C} := \mathcal{C} \setminus \{c\} \cup \{p'\}$.
14:     **end if**
15:     Break the for loop if $\mathcal{C}$ does not change.
16: **end for**
17: **return** the set $\mathcal{C}$

---

3   An Improved k-means++ Approximate Algorithm via Local Search

In this section, we develop a SpheMeans++ algorithm via the local search technique with single-swap operations for the spherical k-means problem (SKMP) as follows. The main idea of our SpheMeans++ algorithm is to swap centers in a certain way. Initially, we sample k points in $\mathcal{P}$ to construct a feasible solution $\mathcal{C}$, akin to the k-means++ algorithm for the k-means problem [12]. Then we perform the single-swap local search operations to improve the current solution repeatedly. Specifically, we iteratively remove one center $c \in \mathcal{C}$ from $\mathcal{C}$ and add another point $p' \in \mathcal{P} \backslash \mathcal{C}$ into $\mathcal{C}$. The formal algorithm is presented in Algorithm 1.

From Step 1 to Step 5, the initial clustering $\mathcal{C}$ from the input data set $\mathcal{P}$ is sampled with given probabilities. We start with an empty solution $\mathcal{C}$ and choose the first cluster center c_1 uniformly at random from $\mathcal{P}$. To select an initial set of k centers, the remaining clustering centers will be iteratively sampled at random from $\mathcal{P}$ where the probability to sample a point $p' \in \mathcal{P}$ is proportional to its cost $d(\{p'\}, \mathcal{C})$ in the current clustering. We use $d(\{p'\}, \mathcal{C})$ to denote the closest distance from a data point $\mathcal{P}$ to the closest center which we have already chosen. Then one can obtain the current clustering with corresponding partition $\{\mathcal{P}_1, \cdots, \mathcal{P}_k\}$ from Step 6 to Step 8.

As mentioned above, our approach is based on two ideas. The first is to construct a feasible solution $\mathcal{C}$ similarly to the seeding algorithm, and the second is based on the local search process from Step 9 to Step 16. The local search procedure that we shall consider is a natural one. Each current solution is specified by a subset $\mathcal{C} \in \mathcal{P}$ of exactly k points and the clustering will be updated by the single-swap operations. Intuitively, we repeatedly check to see if any swap operation yields a solution of lower cost: if so, the resulting solution is our new current solution; otherwise, we discard the sampled center.

Formally, a single-swap operation $swap(c, p')$ involves swapping two points: if there exists a point $c \in \mathcal{C}$ such that $cost(\mathcal{P}, \mathcal{C} \setminus \{c\} \cup \{p'\}) < cost(\mathcal{P}, \mathcal{C})$, we find such a new point p' that reduces the cost function as much as possible from $\mathcal{P}$ and replace the old center c by this point. Afterwards, we reassign each client to its closest center. It is easy to see that the cost of the new clustering over $\mathcal{P}$ will not increase over iterations because of the local optimal property.

In practical implementations, the local search process is repeated to improve the current solution until the partition and the centers become stable. For simplicity, we will assume that the algorithm terminates when no single-swap reduces the cost function anymore.

4   Analysis

In this section, we show that the Algorithm 1 reduces the cost of the current solution by a $O(1/k)$ factor in every iteration, which means that the solution returned by the algorithm after polynomial number of iteration is a constant approximation. For the SKMP, the spherical centroid centers may be placed anywhere in the spherical space $\mathbb{R}^d$, we impose that all candidate centers are chosen from $\mathcal{X}$ to make sure the algorithm runs in polynomial time. Before proceeding to the analysis for the approximation guarantee of Algorithm 1, we present some necessary notations and propositions.

4.1   Preliminary Step

Let $\mathcal{C} = \{c_1, c_2, \cdots, c_k\}$ and $\mathcal{O} = \{o_1, o_2, \cdots, o_k\}$ denote the current feasible solution and optimal solution, respectively. For each optimal center $o \in \mathcal{O}$, let c_o denote its closest center among all centers in $\mathcal{C}$, that is $c_o = \arg\min_{c \in \mathcal{C}} d(o, c)$. We say that o is captured by c_o. Without causing confusion, we drop indices when they are irrelevant. Therefore, there are three cases to be considered.

Case 1. Each optimal center $o \in \mathcal{O}$ is captured by exactly one center from $\mathcal{C}$, and the index set of such centers is E;

Case 2. Each center $c \in \mathcal{C}$ may capture more than one optimal centers, and the index set of such centers is M;

Case 3. Some centers in $\mathcal{C}$ do not capture any of the optimal centers, called lonely centers, and its index set is L.

Our analysis is based on constructing a set of k special swap pairs $swap(c, o)$ for $\mathcal{C}$ and $\mathcal{O}$. For a cluster center with index $e \in E$, that is, the center c_e in Case 1, we generate a swap pair consisting of the center and its captured center. In this case, if c_e is far away from the center of the optimal cluster, then we are likely to sample a point near the center with high probability and displace it. For a cluster center in Case 2 with index $m \in M$, we generate swap pairs between the lone centers and the optimal centers such that each of the optimal centers involves only one swap pair and each of the lone centers involves at most two swap pairs. On the other hand, if c_l is a lone center, we move it to a different cluster and replace it by sampling a point from other clusters with high probability.

Now that we analyze the cost change caused by the swap operations. It is necessary to compute the cost of swapping the new sample point with an old center. For a center $c_e \in C$, where $e \in E$, we assign all points in the cluster of c_e that are not in the captured optimal cluster to a different center. A change in cost like this is called the reassignment cost. Formally, the reassignment cost of c_e is defined as $reas(\{c_e\}) = cost(\mathcal{P} \backslash O_e, \mathcal{C} \backslash \{c_e\}) - cost(\mathcal{P} \backslash O_e, \mathcal{C})$. For a center $c_l \in C$, where $l \in L$, the cost of assigning all points in $\mathcal{P}$ to other centers is denoted as $reas(\{c_l\}) = cost(\mathcal{P}, \mathcal{C} \backslash \{c_l\}) - cost(\mathcal{P}, \mathcal{C})$. In fact, we sample such points with high probability such that the reassignment cost is lower than the increase for this cluster.

To complete the analysis, we categorize all centers in the set C into three classes as follows, similar to [12].

Definition 4.1. (Good and Bad Center) *Consider a feasible solution C and the optimal solution O.*

- *For a center $c_e \in C$, if $3 \sum_{e \in E} cost(O_e, \mathcal{C}) > cost(\mathcal{P}, \mathcal{C})$, we have*

$$cost(O_e, \mathcal{C}) - reas(\{c_e\}) - 4cost(O_e, \{o_e\}) > \frac{1}{100k} cost(\mathcal{P}, \mathcal{C}),$$

and we call it a good center.
- *For a center $c_i \in C, i \in \{1, \cdots, k\} \backslash E$ there exists a center index $l \in L$ such that*

$$cost(O_i, \mathcal{C}) - reas(\{c_l\}) - 4cost(O_i, \{o_i\}) > \frac{1}{100k} cost(\mathcal{P}, \mathcal{C}).$$

and we also call it a good center.

Otherwise, we call it a bad center.

4.2 Main Step

We establish an upper bound on the cost of reassignment points in the following lemma.

Lemma 4.1. *For $u \in E \cup L$, we have*

$$reas(\{c_u\}) \leq \frac{21}{100} cost(\mathcal{P}_u, \mathcal{C}) + 45cost(\mathcal{P}_u, \mathcal{O}).$$

Now, we argue that we have a high probability to sample a good cluster and we have an upper bound on the reassignment cost. From the definition of a good center, we have classified all centers in $\mathcal{C}$ into three categories, and two of them are good centers. Next, we analyze the total cost of good clusters.

For $e \in E$, we estimate the cost of replacing c_e with a point close to the center of O_e by considering a cluster that reassigns the points in $\mathcal{P}_e \setminus O_e$ and assigns all points in $\mathcal{P}_e \cap O_e$ to the new center. For $v \in M \cup L$, we estimate the gain of removing l and inserting a new cluster center close to the center of $\mathcal{P}_i$. In fact, we cannot easily move the cluster center in M without affecting other clusters. Therefore, we focus on the centers $v \in M \cup L$ and use the centers in L as candidate centers for a swap, that is, swapping an arbitrary center $l \in L$ with an arbitrary point that is close to an optimal center of a cluster $\mathcal{P}_v$ for some $v \in M \cup L$.

We use the index g or b of the cluster center to indicate whether the center is good or bad. Then, the cost can be bounded as follows.

Lemma 4.2. *Let* $cost(\mathcal{P},\mathcal{C}) > 500opt_k$. *If*

$$3 \sum_{e \in E} cost(O_e,\mathcal{C}) > cost(\mathcal{P},\mathcal{C}),$$

then we have

$$75 \sum_{e_g \in E} cost(O_{e_g},\mathcal{C}) \geq cost(\mathcal{P},\mathcal{C});$$

otherwise, for $v \in L \cup M$, *we have*

$$21 \sum_{v_g \in L \cup M} cost(O_{v_g},\mathcal{C}) \geq cost(\mathcal{P},\mathcal{C}),$$

where the index e_g *and* v_g *indicate that the corresponding centers are good, respectively.*

From the above lemma, we know that the total cost of any good cluster is large. In the following part, we just need to argue that the probability of sampling a good center is high enough. Consider a (good) cluster G in the optimal solution and the set of center points $\mathcal{C}$ in the current solution.

Lemma 4.3. *Let* $G \subseteq \mathcal{X}$ *be a cluster and* $\mathcal{C} \subseteq \mathcal{X}$ *be a set of* k *centers w.r.t the current solution. If* $cost(G,\mathcal{C}) \geq \delta cost(G,\{scen(G)\})$, *then*

$$\frac{cost(R,\mathcal{C})}{cost(G,\{scen(G)\})} \geq \frac{\delta-1}{8}$$

where $\delta \geq 9$ *and* $R \subseteq G$ *such that*

$$\left\| \sum_{g \in G} g \right\| cost(R,\{scen(G)\}) \leq 2cost(G,\{scen(G)\}).$$

The above lemmas together imply the following important lemma, which again implies the main theorem.

Lemma 4.4. *Let $\mathcal{P}$ and $\mathcal{C}$ be a set of input data points from Algorithm 1 and a set of centers with $\mid C \mid = k$ and $cost(\mathcal{P},\mathcal{C}) \geq 500opt_k$. Let $\mathcal{C}'$ be obtained from the local search procedure in Algorithm 1 from Step 9 to Step 16. Then*

$$cost\,(\mathcal{P},\mathcal{C}') \leq \left(1 - \frac{1}{100k}\right) cost(\mathcal{P},\mathcal{C}) \text{ with probability } \frac{1}{1000}.$$

The main results obtained by SpheMeans++ local search in this paper can be stated as below.

Theorem 4.1. *Let $\mathcal{P}$ be a set of data points in $\mathcal{X}$, $\mathcal{C}$ is the returned set by Algorithm 1 with $K \geq 100000k \log(\log k)$. For the spherical k-means problem, Algorithm 1 has an expected constant approximation ratio*

$$E[cost(\mathcal{P},\mathcal{C})] \leq 517cost(\mathcal{P},\mathcal{O}),$$

where the set $\mathcal{O}$ contains k optimal centers. Moreover, the running time of the algorithm is $O(dnk^2 \log \log k)$.

A full proof will be shown in the journal version.

5   Discussions

In this paper, we present an expected constant approximation algorithm for the SKMP by employing the local search scheme. There are two possible future research questions to pursue. One is to study other clustering problems, such as the robust k-means problem, or the spherical k-means problem with penalties. Another is to design an algorithm to reduce the number of local search steps in Algorithm 1.

Acknowledgements. The first two authors are supported by National Natural Science Foundation of China (Nos. 11871081, 11531014). The third author is supported by the Natural Sciences and Engineering Research Council of Canada (NSERC) grant 06446, and National Natural Science Foundation of China (Nos. 11771386, 11728104). The fourth author is supported by National Natural Science Foundation of China (No. 11201333).

References

1. Ahmadian, S., Norouzi-Fard, A., Svensson, O., Ward, J.: Better guarantees for k-means and Euclidean k-median by primal-dual algorithms. In: Proceedings of FOCS, pp. 61–72 (2017)
2. Arthur, D., Vassilvitskii, S.: k-means++: the advantages of careful seeding. In: Proceedings of SODA, pp. 1027–1035 (2007)

3. Dhillon, I.S., Modha, D.S.: Concept decompositions for large sparse text data using clustering. Mach. Learn. **42**, 143–175 (2001)
4. Endo, Y., Miyamoto, S.: Spherical k-means++ clustering. In: Torra, V., Narukawa, Y. (eds.) MDAI 2015. LNCS (LNAI), vol. 9321, pp. 103–114. Springer, Cham (2015). https://doi.org/10.1007/978-3-319-23240-9_9
5. Feldman, D., Schmidt, M., Sohler, C.: Turning big data into tiny data: constant-size coresets for k-means, PCA and projective clustering. ArXiv preprint arXiv:1807.04518 (2018)
6. Hornik, K., Feinerer, I., Kober, M., Buchta, C.: Spherical k-means clustering. J. Stat. Softw. **50**, 1–22 (2012)
7. Jain, A.K., Dubes, R.C.: Algorithms for Clustering Data. Prentice Hall, New Jersey (1988)
8. Jain, A.K., Murty, M.N., Flynn, P.J.: Data clustering: a review. ACM Comput. Surv. **31**, 264–323 (1999)
9. Kanungo, T., Mount, D.M., Netanyahu, N.S., Piatko, C.D., Silverman, R., Wu, A.Y.: A local search approximation algorithm for k-means clustering. Comput. Geom.: Theory Appl. **28**, 89–112 (2004)
10. Kanungo, T., Mount, D.M., Netanyahu, N.S., Piatko, C.D., Silverman, R., Wu, A.Y.: An efficient k-means clustering algorithm: analysis and implementation. IEEE Trans. Pattern Anal. Mach. Intell. **24**, 881–892 (2002)
11. Kumar, A., Sabharwal, Y., Sen, S.: A simple linear time $(1 + \varepsilon)$-approximation algorithm for k-means clustering in any dimensions. In: Proceedings of FOCS, pp. 454–462 (2004)
12. Lattanzi, S., Sohler, C.: A better k-means++ algorithm via local search. In: Proceedings of ICML, pp. 3662–3671 (2019)
13. Li, M., Xu, D., Zhang, D., Zou, J.: The seeding algorithms for spherical k-means clustering. J. Glob. Optim. **76**, 695–708 (2020)
14. Lloyd, S.: Least squares quantization in PCM. IEEE Trans. Inf. Theory **28**, 129–137 (1982)
15. Macqueen, J.B.: Some methods for classification and analysis of multivariate observations. In: Proceedings of the 5th Berkeley Symposium on Mathematical Statistics and Probability & Statistics, pp. 281–297 (1966)
16. Zhang, Dongmei, Cheng, Yukun, Li, Min, Wang, Yishui, Xu, Dachuan: Local search approximation algorithms for the spherical k-means problem. In: Du, Ding-Zhu, Li, Lian, Sun, Xiaoming, Zhang, Jialin (eds.) AAIM 2019. LNCS, vol. 11640, pp. 341–351. Springer, Cham (2019). https://doi.org/10.1007/978-3-030-27195-4_31

Local Search Algorithm for the Spherical k-Means Problem with Outliers

Yishui Wang[1], Chenchen Wu[2], Dongmei Zhang[3(✉)], and Juan Zou[4]

[1] University of Science and Technology Beijing, 30 Xueyuan Road, Haidian District,
Beijing 100083, People's Republic of China
ys.wang1@siat.ac.cn

[2] College of Science, Tianjin University of Technology,
Tianjin 300384, People's Republic of China
wu_chenchen_tjut@163.com

[3] School of Computer Science and Technology, Shandong Jianzhu University,
Jinan 250101, People's Republic of China
zhangdongmei@sdjzu.edu.cn

[4] School of Mathematical Sciences, Qufu Normal University,
Qufu 273165, People's Republic of China
zoujuanjn@163.com

Abstract. We study the spherical k-means problem with outliers, a variant of the classical k-means problem, in which data points are on the unit sphere and a small set of points called outliers (as a constraint, the number of outliers can not be greater than a given integer) can be ignored. Using local search method, we give a constant-factor approximation algorithm that may violate slightly the constraint about the number of outliers.

Keywords: Spherical k-means · k-Means with outliers · Local search · Approximation algorithm

1 Introduction

In the classical k-means problem, given a set of n sample points in the Euclidean space, we need to select at most k points in the space as cluster centers, and assign each sample point to the nearest center according to the Euclidean distance, minimizing the sum of squares of distances of each sample point to its center. However, the Euclidean distance is not suitable to measure the dissimilarity of two points for textual data, since the Euclidean distance of a long text and a short text is great but their meanings may be similar. To address this issue, Dhillon and Modha [5] present the spherical k-means problem that uses the so-called cosine dissimilarity based on the angle between two points(vectors), or equivalently, uses the Euclidean distance of the projections of vectors onto the unit sphere.

We consider the spherical k-means problem with outliers, in which a small set of sample points can be ignored, that is, we can choose at most z sample

© Springer Nature Switzerland AG 2020
Z. Zhang et al. (Eds.): AAIM 2020, LNCS 12290, pp. 141–148, 2020.
https://doi.org/10.1007/978-3-030-57602-8_13

points as outliers, and these outliers need not to be clustered. The motivation of clustering problems with outliers is to handle the noise and error data (seen as outliers). By removing outliers, we can reduce the clustering cost dramatically and improve the quality of the clustering consequently.

As a classical clustering problem, the k-means problem has been studied widely for several decades. The most popular algorithm for the k-means problem is the well-known Lloyd's algorithm [13] (it is also called k-means algorithm in many literatures). Although Lloyd's algorithm has a good performance in practice, its approximation ratio can not be bounded [11] (we say an algorithm has an approximation ratio α if it can produce a solution with the objective value that is not worse than α times the optimum for any instance). Arthur and Vassilvitskii [2] provide the so-called k-means++ algorithm with a special random seeding method to find the initial solution, together with Lloyd's iteration, resulting in a $O(8 \ln k + 2)$-approximation. Jain and Vazirani [10] present the first constant 108-approximation algorithm for the k-means problem using the primal-dual method and Lagrangian relaxation. Kanungo et al. [11] give a $(9+\varepsilon)$-approximation using local search scheme. The best known approximation ratio of the k-means problem is 6.357 proposed by Ahmadian el al. [1].

Based on the primitive spherical k-means [5], Hornik et al. [9] present the standard and extended spherical k-means problems, and discuss some heuristic algorithms. Endo et al. [6] present a generalized problem called α-spherical k-means that extends the cosine dissimilarity to satisfy the triangle inequality. Utilizing this extended dissimilarity and based on the k-means++ algorithm, they provide the spherical k-means++ algorithm which is a $O(4 \ln k + 2)$-approximation. Li et al. [12] study a special case called spherical separable k-means problem, a special case of spherical k-means problem. They present a constant factor approximation algorithm for this problem and generalize it to the α-spherical separable k-means problem. Zhang et al. [14] use the local search method to provide a $(2(4 + \sqrt{7}) + \varepsilon)$-approximation algorithm.

The clustering problem with outliers is firstly introduced by Charikar et al. [3]. They study the outlier version of the facility location problem and k-median problem. For the former, a 3-approximation algorithm is presented. For the latter, a bi-criteria $(4(1 + 1/\varepsilon), 1 + \varepsilon)$-approximation algorithm is presented, where $4(1 + 1/\varepsilon)$ is the approximation ratio and $1 + \varepsilon$ is the factor of violating the outlier constraint, that is, there are at most $(1 + \varepsilon)z$ outliers in the solution. Chen [4] uses Lagrangian relaxation and local search technique to give the first true approximation algorithm with a large approximation ratio for the k-median problem with outliers. Gupta et al. [8] provide a bi-criteria $(274, O(k \log n\Delta))$-approximation algorithm for the k-means with outliers, where Δ is the maximal distance between sample points. Recently, Friggstad et al. [7] consider the cases where the distance is doubling metric, a generalization of fixed dimensional Euclidean metrics. They firstly show that a multi-swap local search algorithm is a PTAS for the uniform-cost facility location problem with outliers on doubling metric. Then, they extend this result to k-median and k-means problems with outliers, to obtain a bi-criteria $(1 + \varepsilon, 1 + \varepsilon)$-approximation algorithm where the first $1 + \varepsilon$ is the approximation ratio and the second is the factor of violating the k constraint. Furthermore, a $(25 + \varepsilon, 1 + \varepsilon)$-approximation algorithm is obtained for the k-means with outliers on general metrics. They also show that a natural local search algorithm satisfying both the k and outlier constraint can not yield a bounded approximation ratio, even in Euclidean metrics.

In this paper, we give a bi-criteria approximation algorithm for the spherical k-means problem with outliers, using the local search scheme inspired by the works of [8,14]. In Sect. 2, we give the formulation of the problem, the local search algorithm and the corresponding analysis. The conclusion is given in Sect. 3.

2 The Local Search Algorithm for the Spherical k-means Problem with Outliers

2.1 Problem

Given a set of sample points U in the d-dimensional unit sphere $\mathbb{S}^d = \{s \in \mathbb{R}^d | \ \|s\| = 1\}$ (we use $\| \cdot \|$ to denote the l_2-norm in this paper), the spherical k-means with outliers problem can be formulated as follows:

$$\min_{C \subseteq \mathbb{S}^d, Z \subseteq U : |C| \leq k, |Z| \leq z} \sum_{u \in U \backslash Z} d(u, C)^2, \tag{1}$$

where $d(u, v) = \|u - v\|$, $d(u, C) = \min_{v \in C} d(u, v)$. Let $outlier(C, R)$ be the set of z farthest points from $C \subseteq U \setminus R$, and let $outlier(C) = outlier(C, \emptyset)$. If the center set C is fixed, then set $Z = outlier(C)$ minimizes the cost. Thus, the formulation (1) is equivalent to

$$\min_{C \subseteq \mathbb{S}^d : |C| \leq k} \sum_{u \in U \backslash outlier(C)} d(u, C)^2.$$

2.2 Algorithm

Gupta et al. [8] present a single-swap local search algorithm for the k-means problem with outliers. Based on their work, we present a constant factor approximation algorithm for the problem of spherical version. Let $cost(C, Z)$ be the cost of the solution (C, Z) (we use (C, Z) to denote a solution with the center set C and outlier set Z). We call our algorithm as LS-Single-Swap-Outlier. The formal description of LS-Single-Swap-Outlier is given in Algorithm 1.

From an arbitrary feasible solution, the algorithm improves the cost by adding outliers and single-swap operation iteratively, until the cost can not be improved (as shown in Proposition 1). In the phase of adding outliers without swap (lines 6–8 in Algorithm 1), it adds z outliers if the cost will be reduced by a factor after this operation. In the phase of single-swap (lines 9–12 in Algorithm 1), the algorithm improves the cost by searching solution from the neighborhoods $\{C \setminus \{a\} \cup \{b\} | a \in C, b \in U\}$ of the current center set C and adding z additional outliers. We denote the swap operation $C \setminus \{a\} \cup \{b\}$ by $swap(a, b)$.

From the description of LS-Single-Swap-Outlier, we can obtain Proposition 1(1) due to the outliers adding operation and Proposition 1(2) due to the single-swap operation.

Proposition 1. *Let C be the center set and Z be the outlier set returned by LS-Single-Swap-Outlier(U, k, z, p, ε). We have*

(1) $cost(C, Z \cup outlier(C, Z)) \geq (1 - \frac{\varepsilon}{k})cost(C, Z)$,
(2) $cost(C \setminus \{a\} \cup \{b\}, Z \cup outlier(C \setminus \{a\} \cup \{b\}, Z)) \geq (1 - \frac{\varepsilon}{k})cost(C, Z)$ for any $a \in C$ and $b \in U$.

Algorithm 1. LS-Single-Swap-Outlier(U, k, z, ε)

Input: Sample set $U \subseteq \mathbb{S}^d$, positive integer k as the maximal number of centers, positive integer z as the maximal number of outliers, real number $\varepsilon > 0$.
Output: Center set $C \subseteq \mathbb{S}^d$ and outlier set $Z \subseteq U$.
1: $C \leftarrow$ an arbitrary set of k points from U
2: $Z \leftarrow outlier(C)$
3: $\alpha \leftarrow +\infty$
4: **while** $cost(C, Z) < \alpha(1 - \frac{\varepsilon}{k})$ **do**
5:    $\alpha \leftarrow cost(C, Z)$
6:    **if** $cost(C, Z \cup outlier(C, Z)) < (1 - \frac{\varepsilon}{k})cost(C, Z)$ **then**
7:        $Z \leftarrow Z \cup outlier(C, Z)$
8:    **end if**
9:    **if** $\exists a \in C$ and $b \in U$ s.t. $cost(C \setminus \{a\} \cup \{b\}, Z \cup outlier(C \setminus \{a\} \cup \{b\}, Z)) < (1 - \frac{\varepsilon}{k})cost(C, Z)$ **then**
10:       $C \leftarrow C \setminus \{a\} \cup \{b\}$
11:       $Z \leftarrow Z \cup outlier(C \setminus \{a\} \cup \{b\}, Z)$
12:    **end if**
13: **end while**
14: **return** C and Z

2.3   Analysis

We first give the time complexity of LS-Single-Swap-Outlier as shown in the following theorem.

Theorem 1. *Let δ be the maximal value of the distance between two points in U, i.e. $\delta = \max_{u,v \in U} d(u, v)$. The time complexity of LS-Single-Swap-Outlier(U, k, z, ε) is $O(\frac{k^2 n^2}{\varepsilon} \log(n\delta))$.*

Proof. The proof is similar to that in [8], with loss of generality, we give it here. From the objective function of the problem we study, the cost of any feasible solution is at most $n\delta^2$. The cost is reduced by at least $1 - \frac{\varepsilon}{k}$ factor after each iteration, so the number of iterations is at most $O(\log_{1-\varepsilon/k}(n\delta^2)) = O(\frac{k}{\varepsilon}\log(n\delta))$ under the assumption that the optimal value $OPT \leq 1$ (this assumption is easy to be satisfied by scaling the distances). In each iteration, the outlier adding operation takes time $O(1)$, and the single-swap operation takes time $O(kn)$. Thus, the total time is $O(\frac{k^2n^2}{\varepsilon}\log(n\delta))$. □

Next, we analyze the approximation ratio and the bound of the number of outliers returned by LS-Single-Swap-Outlier. To analyze the approximation ratio, we will construct some swap operations between the optimal solution denoted by (C^*, Z^*) and the solution (C, Z) returned by the algorithm, and then use Proposition 1 to get an inequality for each swap, finally combining these inequalities we can bound $cost(C, Z)$ by $cost(C^*, Z^*)$.

Here, we use the same method in [14] to construct the swap operations. Firstly we introduce the notion of *capture* defined as follows. For each center $o \in C^*$, let $\pi(o) = \operatorname{argmin}_{c \in C} d(o, c)$. We say that $\pi(o)$ captures o. Let *capture*(c) be the set of points captured by the center c. If *capture*$(c) = \emptyset$, we call c as a good center, otherwise, we call c as a bad center.

With loss of generality, assume that $|C| = |C^*|$. We partition C and C^* by the following procedure. Let $C_i^* = capture(c_i)$ and $C_i = \{c_i\}$ for each bad center c_i. Since any center $o \in C^*$ is captured by exactly one center in C, we get the partition of $C^* = \dot{\cup}_{i=1}^m C_i^*$. Put arbitrary good centers into C_i until $|C_i| = |C_i^*|$ for each $i = 1, \dots, m$. Because $|C| = |C^*|$, this procedure is feasible, and we get the partition of $C = \dot{\cup}_{i=1}^m C_i$.

For a no-outlier point v, let v^c and v^o be v's center in the solution (C, Z) and (C^*, Z^*) respectively. For a center $a \in C$, let N_a be the set of points in $U \setminus Z$ assigned to the center a, i.e.

$$N_a = \{v \in U \setminus Z | v^c = a\}.$$

Similarly, let

$$N_b^* = \{v \in U \setminus Z^* | v^o = b\}.$$

Note that the points a and b for swap in our algorithm are contained in U. Thus, to use Proposition 1, we define

$$\hat{b} = \operatorname*{argmin}_{u \in N_b^*} d(u, b).$$

Then, we construct the swap operations in each pair (C_i, C_i^*) for each $i = 1, \dots, m$. We consider the following two cases.

Case 1: $|C_i| = 1$. Let $C_i = \{a_i\}$ and $C_i^* = \{b_i\}$. We construct $swap(a_i, \hat{b}_i)$.
Case 2: $|C_i| > 1$. Let $C_i = \{a_i^1, a_i^2, \dots, a_i^{m_i}\}$ and $C_i^* = \{b_i^1, b_i^2, \dots, b_i^{m_i}\}$, where a_i^1 is the unique bad point. We construct $swap(a_i^2, \hat{b}_i^1)$, $swap(a_i^2, \hat{b}_i^2)$, $swap(a_i^3, \hat{b}_i^3)$, ..., $swap(a_i^{m_i}, \hat{b}_i^{m_i})$.

Let $\mathcal{P}$ be the set of all swap operations we construct. It is obvious that $|\mathcal{P}| = k$. Denote the cost after $swap(a,b)$ in LS-Single-Swap-Outlier by $cost(a,b)$, i.e.

$$cost(a,b) = cost(C \setminus \{a\} \cup \{b\}, Z \cup outlier(C \setminus \{a\} \cup \{b\}, Z)).$$

The following lemma shows the upper bound of the cost after a swap operation.

Lemma 1. *For each $swap(a,\hat{b})$ in $\mathcal{P}$, we have*

$$cost(a,\hat{b}) - cost(C,Z) \leq$$
$$4 \sum_{v \in N_b^*} d(v,v^o)^2 - \sum_{v \in N_b^* \setminus Z} d(v,v^c)^2 + \sum_{v \in N_a \setminus (N_b^* \cup Z^*)} \left(d(v,\pi(v^o))^2 - d(v,v^c)^2 \right).$$
$$(2)$$

Proof. The proof is deferred to the journal version.

Also, we need to the following two results to proceed the analysis.

Lemma 2. ([14]) $\sum_{v \in N_b^*} d(v,\hat{b})^2 \leq 4 \sum_{v \in N_b^*} d(v,b)^2$.

Lemma 3. ([11])

(1) $\sum_{v \in U \setminus (Z \cup Z^*)} d(v,v^c)d(v,v^o) \leq \sqrt{cost(C,Z)cost(C^*,Z^*)},$

(2) $d(v,\pi(v^o)) \leq d(v,v^c) + 2d(v,v^o), \quad \forall v \in U \setminus (Z \cup Z^*).$

Note that each $a \in C$ appears at most twice and each $b \in C^*$ appears exactly once in $\mathcal{P}$. So, summing the inequality (2) for all swaps and using Lemma 3(2), we have

$$\sum_{(a,\hat{b}) \in \mathcal{P}} \left(cost(a,\hat{b}) - cost(C,Z) \right)$$

$$\leq \sum_{(a,\hat{b}) \in \mathcal{P}} \left(4 \sum_{v \in N_b^*} d(v,v^o)^2 - \sum_{v \in N_b^* \setminus Z} d(v,v^c)^2 \right.$$

$$\left. + \sum_{v \in N_a \setminus (N_b^* \cup Z^*)} \left(d(v,\pi(v^o))^2 - d(v,v^c)^2 \right) \right)$$

$$\leq 4 \sum_{v \in U \setminus Z^*} d(v,v^o)^2 - \sum_{v \in U \setminus Z} d(v,v^c)^2 + \sum_{Z^* \setminus Z} d(v,v^c)^2$$

$$+ 2 \sum_{v \in U \setminus (Z \cup Z^*)} \left(4d(v,v^o)^2 + 4d(v,v^c)d(v,v^o) \right).$$

$$\leq 12cost(C^*,Z^*) - cost(C,Z) + \sum_{Z^* \setminus Z} d(v,v^c)^2 + 8 \sum_{v \in U \setminus (Z \cup Z^*)} d(v,v^c)d(v,v^o).$$
$$(3)$$

From Proposition 1(1), we know

$$\frac{\varepsilon}{k}cost(C, Z) \geq cost(C, Z) - cost(C, Z \cup outlier(C, Z)) = \sum_{v \in outlier(C,Z)} d(v, v^c)^2.$$

(4)

Recalling the definition of $outlier(C, Z)$, this set contains z most expensive points out of Z as additional outliers. Thus, we have

$$\sum_{Z^* \setminus Z} d(v, v^c)^2 \leq \frac{\varepsilon}{k}cost(C, Z),$$

(5)

since the inequality (4) and $|Z^* \setminus Z| \leq z$.

Combining the inequalities (3), (5), Lemma 3(1), and Proposition 1(2), we have

$$- \varepsilon \cdot cost(C, Z)$$

$$\leq \sum_{(a,\hat{b}) \in \mathcal{P}} \left(cost(a, \hat{b}) - cost(C, Z) \right)$$

$$\leq 12cost(C^*, Z^*) - (1 - \frac{\varepsilon}{k})cost(C, Z) + 8\sqrt{cost(C, Z)cost(C^*Z^*)}$$

(6)

By factorization of (6), we have

$$0 \leq \left(\sqrt{12cost(C^*, Z^*)} + \beta\sqrt{cost(C, Z)} \right) \left(\sqrt{12cost(C^*, Z^*)} - \gamma\sqrt{cost(C, Z)} \right)$$

(7)

where

$$\beta = \frac{2\sqrt{3}}{3} + \sqrt{\frac{7}{3} - \frac{\varepsilon}{k}} - \varepsilon, \gamma = -\frac{2\sqrt{3}}{3} + \sqrt{\frac{7}{3} - \frac{\varepsilon}{k}} - \varepsilon.$$

From the inequality (7) and $\sqrt{12cost(C^*, Z^*)} + \beta\sqrt{cost(C, Z)} \geq 0$, we have

$$\sqrt{12cost(C^*, Z^*)} - \gamma\sqrt{cost(C, Z)} \geq 0.$$

After a simple calculation, we get

$$cost(C, Z) \leq \frac{36}{11 - 3(\varepsilon/k + \varepsilon) - 4\sqrt{7 - 3(\varepsilon/k + \varepsilon)}} \cdot cost(C^*, Z^*).$$

Since $36/(11 - 3(\varepsilon/k + \varepsilon) - 4\sqrt{7 - 3(\varepsilon/k + \varepsilon)}) \to 4(11 + 4\sqrt{7})$ as $\varepsilon \to 0$, we get a $(4(11 + 4\sqrt{7}) + \varepsilon')$-approximation where ε' is a sufficiently small positive number.

Finally, we estimate the upper bound of the number of outliers. In each iteration of LS-Single-Swap-Outlier, it puts at most $2z$ additional outliers into Z. From the proof of Theorem 1, we know that the number of iterations is $O(\frac{zk}{\varepsilon}\log(n\delta))$. So, the number of outliers is bounded by $O(\frac{zk}{\varepsilon}\log(n\delta))$.

Through above analysis, we obtain the following theorem.

Theorem 2. *For the spherical k-means problem with outliers, LS-Single-Swap-Outlier is a bi-criteria $(4(11 + 4\sqrt{7}) + \varepsilon, O(\frac{k}{\varepsilon}\log(n\delta)))$-approximation algorithm, where $4(11 + 4\sqrt{7}) + \varepsilon$ is the approximation ratio and $O(\frac{k}{\varepsilon}\log(n\delta))$ is the factor violating the outlier constraint.*

3   Conclusion

In this paper, we study the spherical k-means problem with outliers. Based on the local search technique with single-swap, we present a bi-criteria approximation algorithm that yields an approximation ratio $4(11 + 4\sqrt{7}) + \varepsilon$ and may violate the outlier constraint by a factor $O(\frac{k}{\varepsilon}\log(n\delta))$. In the future, it has two possible improvement of the algorithm. One is extending the single-swap to multi-swap. Another is to reduce the number of outliers.

Acknowledgements. The second author is supported by National Natural Science Foundation of China (No. 11971349). The third author is supported by National Natural Science Foundation of China (No. 11871081). The fourth author is supported by National Natural Science Foundation of China (No. 11801310).

References

1. Ahmadian, S., Norouzi-Fard, A., Svensson, O., Ward, J.: Better guarantees for k-means and Euclidean k-median by primal-dual algorithms. In: Proceedings of FOCS, pp. 61–72 (2017)
2. Arthur, D., Vassilvitskii, S.: K-means++: the advantages of careful seeding. In: Proceedings of SODA, pp. 1027–1035 (2007)
3. Charikar, M., Khuller, S., Mount, D.M., Narasimhan, G.: Algorithms for facility location problems with outliers. In: Proceedings of SODA, pp. 642–651 (2001)
4. Chen, K.: A constant factor approximation algorithm for k-median clustering with outliers. In: Proceedings of SODA, pp. 826–835 (2008)
5. Dhillon, I.S., Modha, D.S.: Concept decompositions for large sparse text data using clustering. Mach. Learn. **42**, 143–175 (2001)
6. Endo, Y., Miyamoto, S.: Spherical k-Means++ clustering. In: Torra, V., Narukawa, Y. (eds.) MDAI 2015. LNCS (LNAI), vol. 9321, pp. 103–114. Springer, Cham (2015). https://doi.org/10.1007/978-3-319-23240-9_9
7. Friggstad, Z., Khodamoradi, K., Rezapour, M., Salavatipour, M.: Approximation schemes for clustering with outliers. ACM Trans. Algorithms **15**(2), 26 (2019)
8. Gupta, S., Kumar, R., Lu, K., Moseley, B., Vassilvitskii, S.: Local search methods for k-means with outliers. Proc. VLDB Endow. **10**(7), 757–768 (2017)
9. Hornik, K., Feinerer, I., Kober, M., Buchta, C.: Spherical k-means clustering. J. Stat. Softw. **50**(10), 1–22 (2012)
10. Jain, K., Vazirani, V.V.: Approximation algorithms for metric facility location and k-median problems using the primal-dual schema and Lagrangian relaxation. J. ACM **48**, 274–296 (2001)
11. Kanungo, T., Mount, D.M., Netanyahu, N.S., Piatko, C.D., Silverman, R., Wu, A.Y.: A local search approximation algorithm for k-means clustering. Comput. Geom.: Theory Appl. **28**, 89–112 (2004)
12. Li, M., Xu, D., Zhang, D., Zou, J.: The seeding algorithms for spherical k-means clustering. J. Glob. Optim. **76**, 695–708 (2020)
13. Lloyd, S.: Least squares quantization in PCM. IEEE Trans. Inf. Theory **28**, 129–137 (1982)
14. Zhang, D., Cheng, Y., Li, M., Wang, Y., Xu, D.: Approximation algorithms for spherical k-means problem using local search scheme. Theoret. Comput. Sci. https://doi.org/10.1016/j.tcs.2020.06.029

A Bi-criteria Analysis for Fuzzy C-means Problem

Yang Zhou, Jianxin Liu, Min Li, and Qian Liu$^{(\boxtimes)}$

School of Mathematics and Statistics, Shandong Normal University, Jinan 250014,
People's Republic of China
zhyg1212@163.com, 3286330436@qq.com, liminEmily@sdnu.edu.cn, lq_qsh@163.com

Abstract. Fuzzy C-means problem has a broad application prospect as a branch of clustering problem. This paper deeply explores the fuzzy C-means bi-criteria problem in two different algorithms and extends the previous known $O(k^2\ln k)$ and $O(k\ln k)$ performance guarantee. It is shown that for any constant $\beta \geq 1$, selecting βk cluster centers can achieve $O(k^2)$ and $O(k)$ approximation. Preliminary numerical experiments are proposed to support the theoretical results of the paper, in which we run these algorithms on real data sets with different parameter values.

Keywords: Fuzzy C-means problem · Bi-criteria analysis · Fuzzy C-means algorithm · Seeding algorithm

1 Introduction

Partitional cluster analysis is defined as the problem of dividing a group of objects into clusters with similar characteristics [15]. As a class of classical problems, partitional cluster analysis originates from taxonomy and is widely used in data mining, machine learning, anomaly detection and other fields. At present, there are many different clustering problems [6,18,19]. According to whether each observation point definitely belongs to one cluster, these problems can be divided into hard and fuzzy clustering problems [4,7,20].

The k-means problem is one of the most well-known hard clustering problems. Given an input observation set X and a positive integer k, the task is to select k cluster centers, and the data samples are arranged to k clusters such that the sum of distances between any points and their corresponding cluster centers is minimized.

The Lloyd's algorithm [10] is widely applied to the k-means problem because of its advantage of simplicity and rapidity. Arthur and Vassilvitskii propose a k-means++ algorithm [3] based on the Lloyd's algorithm and prove that this

---

Supported by Higher Educational Science and Technology Program of Shandong Province (No. J17KA171), Natural Science Foundation of Shandong Province (Nos. ZR2019MA032, ZR2019PA004) of China.

Z. Zhang et al. (Eds.): AAIM 2020, LNCS 12290, pp. 149–160, 2020.
https://doi.org/10.1007/978-3-030-57602-8_14

algorithm is $O(\ln k)$ approximation with optimum k-clustering. This seeding algorithm, which is also known as D^2-weighting, selects k centers according to the contribution of observation points to the potential function.

Many researchers also study the bi-criteria approximation for k-means problem [1,2,8,11]. In many cases, the clustering number k is often unknown or even difficult to predict. Thus the bi-criteria analysis is introduced and we study the solution of the problem with relaxation the restriction of the number of candidates [1,5]. In particular, Wei studies the D^l-sampling ($l \geq 2$) algorithm that k-means++ algorithm belongs to [17], which shows that a bi-criteria approximation can be maintained. It is also shown that for any k, at most 20% oversampling is required to guarantee a better approximation.

Due to the property of strict clustering of each observation point, it is hard to adjust the result of clustering. For example, a famous butterfly problem shows that the uncertainty of which cluster the point belongs to [13]. Fuzzy C-means (FCM) problem is one of the most popular fuzzy clustering problems, which is proposed based on the fuzzy set theory [21]. In the fuzzy C-means problem, the degree of membership between each point and some clusters determines the affiliation strength of sample points, so as to achieve the purpose of automatic classification. Fuzzy C-means problem can also be regarded as a generalization of the k-means problem with membership degree of 0 or 1.

Bezdek [16] proposes the standard fuzzy C-means algorithm. Compared with k-means++ algorithm, the fuzzifier parameter m and the nonzero membership degree μ_{ij} are introduced [22]. Currently, there are some research efforts on the selection of parameter and the general used value is $m = 2$ in applications [12,14]. Stetco [15] proposes the fuzzy C-means++ (FCM++) algorithm to ameliorate the FCM algorithm, in which the seeding strategy of k-means++ is used to improve the effectiveness of the standard algorithm. Then, the theoretical analysis and performance guarantee $O(k^2 \ln k)$ are provided by Liu et al. [9]. In addition, they also propose a $O(k \ln k)$-approximation seeding algorithm (NFCM) in which the centers are chosen by a probability distribution μ^2-weighting based on the contribution to the potential function of the fuzzy C-means problem.

In this paper, we consider the bi-criteria setting of the fuzzy C-means problem by invoking two improved FCM++ [15] and NFCM [9] algorithms. We select βk centers ($\beta \geq 1$, for any constant), and approximate the optimum k-clustering, which is applicable to all data sets. The contribution of this paper is that for any constant $\beta \geq 1$, selecting βk cluster centers can achieve $O(k^2)$ and $O(k)$ approximation by the two algorithms. Meanwhile, it can be concluded that in order to obtain a better approximation than $O(k^2 \ln k)$ and $O(k \ln k)$ with high probability, we only need to sample more than 20 percentage experimentally. We also propose some numerical experiments as a support on these theoretical results, in which we run these algorithms on real data sets with different parameter values.

The structure of the paper is as follows. In Sect. 2, we present the fuzzy C-means problem with its classical algorithm and two seeding algorithms, FCM++ and NFCM. Some existing lemmas related to this paper is also introduced. In Sect. 3, we prove the performance guarantee of the expected potential function. Numerical experiments about the bi-criteria approximation are proposed in Sect. 4.

2 Preliminaries

2.1 Problem Description

Given a positive integer k and a set of observation points $X = \{x_1, \ldots, x_n\}$, the fuzzy C-means problem is to find a clustering set $C = \{c_1, \ldots, c_k\}$ that minimize the potential function

$$\phi_m(X, C) = \sum_{i=1}^{n} \sum_{j=1}^{k} \mu_{ij}^m ||x_i - c_j||^2, \quad m \geq 1,$$

where m denotes the fuzzifier parameter and $\mu_{ij} \in [0, 1] (i = 1, 2, \ldots, n; j = 1, 2, \ldots, k)$ denotes the degree of membership between the i-th point and j-th center, satisfying $\sum_{j=1}^{k} \mu_{ij} = 1$ for $i = 1, 2, \ldots, n$. We can also limit the potential function to an arbitrary subset A of X,

$$\phi_m(A, C) = \sum_{x_i \in A} \sum_{j=1}^{k} \mu_{ij}^m ||x_i - c_j||^2, \quad m \geq 1.$$

The k-means problem can be seen as the special case of the fuzzy C-means problem as $m = 1$. When m is close to 1, the solution of the fuzzy C-means algorithm is similar to the one of k-means. On the contrary, when m is large, fuzziness is also large and clusters are blurred. Typical values for the parameters m are between 1 and 2. In addition, we take $m = 2$ in the following discussion. By deriving the first order optimality condition of ϕ, it is easy to obtain that $\mu_{ij} = \dfrac{1}{\sum_{l=1}^{k} \left(\frac{||x_i - c_j||}{||x_i - c_l||} \right)^2}, i = 1, 2, \ldots, n; j = 1, 2, \ldots, k$ for a given set $C = \{c_1, \ldots, c_k\}$, we can then rewrite the potential function as

$$\phi(X, C) := \phi_2(X, C) = \sum_{i=1}^{n} \frac{1}{\sum_{j=1}^{k} \frac{1}{||x_i - c_j||^2}}.$$

Throughout this paper, we denote ϕ^* as the optimum value of the potential function corresponding to the optimum clustering C^* with k cluster centers. Meanwhile, each cluster is implicitly defined as a subset of X which is closer to this center than others. The aim of this paper is to select βk cluster centers where $\beta \geq 1$ for any constant, and approximate the optimum k-clustering in expectation.

2.2 Algorithms and Main Results

In this part, we present the standard fuzzy C-means algorithm and two kinds of seeding algorithms. In the fuzzy C-means algorithm, as Algorithm 1 in the following, we randomly determine k initial centers, and compute the initial degrees of membership of each observation point x_i to clustering c_j. The centers and degrees of membership are then updated alternatively until the process is stable.

The seeding strategy to find a good initial center set is introduced to improve the performance of the above algorithm. The first seeding algorithm is fuzzy C-means++ (FCM++) algorithm, which is the seeding procedure in k-means++ algorithm. It samples the initial centers with probabilities proportional to the current k-means problem potential function [17]. The second seeding algorithm NFCM innovates in the probability distribution of selecting the initial centers by adding the membership degree as an indicator. In the other word, it proposes an effective probability which depends on the contribution of the potential function in the fuzzy C-means problem.

Hereafter we call the probability distribution used in Algorithm 1 (FCM++) as D^2-weighting, and the one in Algorithm 2 (NFCM) as μ^2-weighting. When we focus on the bi-criteria analysis, the number of centers in these algorithm becomes βk accordingly. We denote $\Phi(X, C) = \sum_{x_i \in X} \min_{c_j \in C} \| x_i - c_j \|^2$ as the potential function of the k-means problem for convenience.

Algorithm 1. Fuzzy C-means algorithm for fuzzy C-means problem

---

Input: A set of observation points $X = \{x_1, \ldots, x_n\}$, the clusters number k, the initial centers $C = \{c_1, \ldots, c_k\}$ and the initial null matrix of degree of membership $\mu_{n \times k}$.

1: **for** i from 1 to n and j from 1 to k **do**

2:     Update the degree of membership $\mu_{ij} := \dfrac{1}{\sum_{l=1}^{k} \left(\frac{\| x_i - c_j \|}{\| x_i - c_l \|} \right)^2}$;

3: **end for**

4: **for** j from 1 to k **do**

5:     Update the centers $c_j := \dfrac{\sum_{i=1}^{n} \mu_{ij}^2 x_i}{\sum_{i=1}^{n} \mu_{ij}^2}$;

6: **end for**

7: Repeat Steps 1 to 6 until ϕ changes less than 1e-5;

Output: the clustering set C and matrix of degree of membership $\mu_{n \times k}$.

Algorithm 2. Bi-criteria seeding procedure of k-means++ algorithm

---

Input: A set of observation points $X = \{x_1, \ldots, x_n\}$, the clusters number k, a constant $\beta \geq 1$ and null clustering set C.

1: $C \longleftarrow C \cup \{c_1\}$, which is selected uniformly at random from X;

2: **for** i from 2 to βk **do**

3:     Choose the i-th center $c_i \in X$ with probability $\frac{\Phi(c_i, C)}{\Phi(X, C)}$;

4:     $C \longleftarrow C \cup \{c_j\}$;

5: **end for**

Output: the clustering set C.

Algorithm 3. Novel bi-criteria seeding algorithm for fuzzy C-means problem

Input: A set of observation points $X = \{x_1, \ldots, x_n\}$, the clusters number k, a constant $\beta \geq 1$ and null clustering set C.

1: $C \longleftarrow C \cup \{c_1\}$, which is selected uniformly at random from X;
2: **for** i from 2 to βk **do**
3:     Choose the i-th center $c_i \in X$ with probability $\frac{\phi(c_i, C)}{\phi(X, C)}$;
4:     $C \longleftarrow C \cup \{c_i\}$;
5: **end for**

Output: the clustering set C.

Theorem 1. *Let ϕ be the potential value after selecting βk centers in Algorithm 2 or 3. And ϕ^* denotes the optimum value with respect to the optimum k-clustering.*

(I) If the cluster centers are selected with D^2-weighting by Algorithm 2, then

$$\frac{\mathrm{E}[\phi]}{\phi^*} \leq 16k^2(1 + \min\{\frac{\varphi(k-2)}{(\beta-1)k+\varphi}, H_{k-1}\}) - \Theta(\frac{1}{n});$$

(II) If the cluster centers are selected with μ^2-weighting by Algorithm 3, then

$$\frac{\mathrm{E}[\phi]}{\phi^*} \leq 16k(1 + \min\{\frac{\varphi(k-2)}{(\beta-1)k+\varphi}, H_{k-1}\}) - \Theta(\frac{1}{n}).$$

Here, n and H_k denote the number of points in input observation set X and the harmonic number, respectively. $\varphi = \frac{1+\sqrt{5}}{2} \approx 1.618$ is the golden section ratio.

Corollary 1. *With the same definitions as in Theorem 1, we can obtain*

$$(I) \ \frac{\mathrm{E}[\phi]}{\phi^*} \leq 16k^2(1 + \frac{\varphi}{\beta-1});$$

$$(II) \ \frac{\mathrm{E}[\phi]}{\phi^*} \leq 16k(1 + \frac{\varphi}{\beta-1}).$$

By combining Theorem 1 with Corollary 1, we can confirm that there is an approximation dependent on two input parameters k and β. Also note that, the parentheses of the bound in Theorem 1 means that we reduce the series H_{k-1} and the other term. Obviously, it improves the previous $O(k^2\ln k)$ and $O(k\ln k)$ results. As k increases, we can simplify it even more.

2.3 Existing Lemmas Relating to FCM++ and NFCM

The first lemma shows why the performance guarantee of fuzzy C-means problem is $O(k\ln k)$.

Lemma 1. *Assume A to be an arbitrary cluster, $C = \{c_1, \ldots, c_t\}$ to be a clustering set. Then*

$$\Phi(A, C) \leq t\phi(A, C).$$

The following lemmas in [9] give some results on the upper bound of the potential function in expectation through selecting exactly one center from an arbitrary optimum cluster.

Lemma 2 ([9]). *Assume that A is an arbitrary optimum cluster, and C is the clustering with only one center, which is selected uniformly at random from A. Then*

$$E[\phi(A, C)] \leq 4k\phi^*(A).$$

Lemma 3 ([9]). *Assume that A is an arbitrary optimum cluster, and $C = \{c_1, \ldots, c_t\}$ is an arbitrary clustering. C' represents the resulting clustering after we add a random center to C from A with D^2-weighting (or μ^2-weighting). Then*

$$E[\phi(A, C')] \leq 16k\phi^*(A).$$

3   Proof of Main Results

In this Section, we present our main results and demonstrate the theoretical analysis.

In fact, no matter which algorithm is considered, the potential function will not increase after initialization, so we only need to judge the performance guarantee after initialization.

We will show the most critical intermediate results below. Our analysis consists of two parts. First of all, Lemma 4 shows that the change of the function value when the number of added centers is smaller. Lemma 4 also appears as a stronger result than the original lemma [9], reducing the coefficient from $1 + H_t$ to $1 + H_{t-1}$. In the following content, we say that an optimum cluster A is uncovered if no center is selected from A.

Lemma 4. *Assume that C is an arbitrary clustering with corresponding potential ϕ, and $u > 0$ be the number of uncovered optimum clusters, X_u denotes the set of points in these clusters. Also let $X_v = X - X_u$. Now we add $t \leq u$ cluster centers to C. And C' be the resulting clustering with corresponding potential ϕ'.*

(I) If the cluster centers are selected with D^2-weighting, then

$$E[\phi' \mid \phi] \leq (1 + H_t)\Phi(X_v, C) + (1 + H_{t-1})16k^2\phi^*(X_u) + \frac{u-t}{u}\Phi(X_u, C);$$

(II) If the cluster centers are selected with μ^2-weighting, then

$$E[\phi' \mid \phi] \leq (1 + H_t)\phi(X_v, C) + (1 + H_{t-1})16k\phi^*(X_u) + \frac{u-t}{u}\phi(X_u, C).$$

Here, H_t denotes the harmonic number. Without loss of generality, assume $H_{-1} = -1, H_0 = 0$.

The following lemma is closer to the conditions of the bi-criteria analysis. When more centers are selected, we can get inspiration from Lemma 5.

Lemma 5. *Assume that C is an arbitrary clustering with corresponding potential ϕ, and $u > 0$ be the number of uncovered optimum clusters, X_u denotes the set of points in these clusters. Also let $X_v = X - X_u$. Now we add $t \geq u$ cluster centers to C. And C' be the resulting clustering with corresponding potential ϕ'.*

(I) If the cluster centers are selected with D^2-weighting, then

$$E[\phi' \mid \phi] \leq \varphi_v(t, u)\Phi(X_v, C) + \varphi_u(t, u)16k^2\phi^*(X_u);$$

(II) If the cluster centers are selected with μ^2-weighting, then

$$E[\phi' \mid \phi] \leq \varphi_v(t, u)\phi(X_v, C) + \varphi_u(t, u)16k\phi^*(X_u).$$

Where $\varphi_v(t, u) = 1 + \frac{\varphi u}{t - u + \varphi}$; $\varphi_u(t, u) = \begin{cases} 1 + \frac{\varphi(u-1)}{t-u+\varphi}, & u > 0 \\ 0, & u = 0. \end{cases}$

The main proof way is similar to Wei [17]. Here we present a simple sketch of this proof by induction. And we substitute $\phi(X_v)$, $\phi(X_u)$ and $\varrho(X_u)$ for $\phi(X_v, C)$, $\phi(X_u, C)$ and $16k\phi^*(X_u)$ for convenience. First of all, it is easy to check that the conclusion is true for $(t, 0), t \geq 0$. As for $t = u, t \geq 1$, it can be deduced from the special case of Lemma 4.

Assume that the conclusion holds for (t, u) and $(t, u + 1)$. We can complete the proof as long as proving that it also holds for $(t+1, u+1)$. This can be shown by three intermediate results. Each of these outcomes is a known condition for the next outcome.

The first result considers selecting centers from the covered clusters X_v or uncovered clusters X_u. Then the $E[\phi' \mid \phi]$ is at most:

$$\min \left\{ \frac{\varphi_v(t, u)\phi(X_u) + \varphi_v(t, u + 1)\phi(X_v)}{\phi(X_u) + \phi(X_v)}\phi(X_v) \right.$$
$$\left. + \frac{\varphi_v(t, u)\phi(X_u) + \varphi_u(t, u + 1)\phi(X_v)}{\phi(X_u) + \phi(X_v)}\varrho(X_u), \phi(X_u) + \phi(X_v) \right\}.$$

As noted above, the expected potential function is determined not only by $\varrho(X_u)$, $\phi(X_v)$, but also by the bounds of $\phi(X_u)$. Therefore, we regard the above formula as a monotone function of the variable $\phi(X_u)$, and eliminate this influence by taking the derivative and the extremum with respect to $\phi(X_u)$. Then the $E[\phi' \mid \phi]$ is at most:

$$\frac{1}{2}\varphi_v(t, u)[\phi(X_v) + \varrho(X_u)] + \frac{1}{2}\max\{\varphi_v(t, u)[\phi(X_v) + \varrho(X_u)], \sqrt{\Delta}\},$$

where

$$\Delta = [\varphi_v^2(t, u) - 4\varphi_v(t, u) + 4\varphi_v(t, u + 1)]\phi^2(X_v)$$
$$+ 2[\varphi_v^2(t, u) - 2\varphi_v(t, u) + 2\varphi_u(t, u + 1)]\phi(X_v)\varrho(X_u) + \varphi_v^2(t, u)\varrho^2(X_u).$$

However, subsequently, it turns out that this result is nonlinear. So our next goal is to bound the quadratic function Δ' by the square $[a\phi(X_v) + b\phi(X_u)]^2$. Then the $E[\phi' \mid \phi]$ is at most:

$$\frac{1}{2}[\varphi_v(t,u) + \sqrt{\varphi_v^2(t,u) + 4\max\{\varphi_v(t,u+1) - \varphi_v(t,u), 0\}}]\phi(X_v) + \varphi_v(t,u)\varrho(X_u).$$

Finally, directly replacing $\varphi_v(t,u)$ and $\varphi_u(t,u)$ in Lemma 5, and we can obtain:

$$\begin{cases} \varphi_v(t+1, u+1) \geq \frac{1}{2}[\varphi_v(t,u) + \sqrt{\varphi_v^2(t,u) + 4\max\{\varphi_v(t,u+1) - \varphi_v(t,u), 0\}}], \\ \varphi_u(t+1, u+1) \geq \varphi_v(t,u). \end{cases}$$

So far, the inductive process of Lemma 5 is proved successfully.

Proof of Theorem 1

We just prove the second result, and the first can be obtained in the same way. Let A denote the optimum cluster from which we select the first center, n denote the number of points in X and n_A denote the number of points in the optimum cluster A. By applying Lemma 2 and Lemma 5 (II) with $u = k - 1$, $t = \beta k - 1$, $X_v = A$, $X_u = X - A$, we have

$$\begin{aligned} E[\phi' \mid \phi] &\leq \varphi_v(\beta k - 1, k - 1)\phi(A) + \varphi_u(\beta k - 1, k - 1)16k\phi^*(X - A) \\ &\leq 4k\varphi_v(\beta k - 1, k - 1)\phi^*(A) + \varphi_u(\beta k - 1, k - 1)16k\phi^*(X - A). \end{aligned}$$

Further, we can still obtain

$$\begin{aligned} E[\phi'] &\leq \sum_{A \subseteq X} \frac{n_A}{n} E[\phi' \mid \phi] \\ &\leq \sum_{A \subseteq X} \frac{n_A}{n} 4k\varphi_v(\beta k - 1, k - 1)\phi^*(A) \\ &\quad + \sum_{A \subseteq X} \frac{n_A}{n} 16k\varphi_u(\beta k - 1, k - 1)\phi^*(X - A) \\ &\leq 16k\varphi_u(\beta k - 1, k - 1)\phi^* - C\sum_{A \subseteq X} \frac{n_A}{n}\phi^*(A) \\ &\leq 16k\varphi_u(\beta k - 1, k - 1)\phi^* - \frac{2C}{n}\phi^*. \end{aligned}$$

The last inequality can be obtained by

$$\sum_{A \subseteq X} \frac{n_A}{n}\phi^*(A) \geq \frac{2}{n}\phi^*.$$

By replacing the $\phi_v(t,u)$ and $\phi_u(t,u)$ in Lemma 5, we can obtain

$$\frac{E[\phi']}{\phi^*} \leq 16k(1 + \frac{\varphi(k-2)}{(\beta-1)k + \varphi}) - \frac{2C}{n},$$

where $C = 16k\varphi_u(\beta k - 1, k - 1) - 4k\varphi_v(\beta k - 1, k - 1) \geq 0$. By combining with the known bound $16k(1 + H_{k-1})$ [9] and minimizing the two terms, we now get the desired bound,

$$\frac{\mathrm{E}[\phi']}{\phi^*} \leq 16k(1 + \min\{\frac{\varphi(k-2)}{(\beta-1)k + \varphi}, H_{k-1}\}) - \Theta(\frac{1}{n}).$$

Proof of Corollary 1

As k increases, the first term in parentheses approaches $\frac{\varphi}{\beta-1}$, and the second term approaches infinity. At this point, by taking the minimum value, we can get the result. And notice that the $\frac{1}{n}$ term can be ignored when n is big enough.

4   Numerical Experiment

In this section we show the numerical experiments and evaluate the feasibility of bi-criteria approximation, as well as consider the impact of β. The experiments are implemented in Matlab 2015b.

A common data set Spambase from UC Irvine Machine Learning Repository is used in this section for numerical experiments. The complete Spambase contains information from 4,601 e-mails, each marked as spam or non-spam. Each data also has 57 features, representing the length and number of letters, the frequency of certain words, etc. The last feature indicates whether the e-mail is considered spam and is excluded from the numerical experiment.

Although we have theoretically proposed the approximation guarantee of the bi-criteria in the previous content, it is difficult to design the numerical experiment from this perspective. However, we would like to notice that both the original fuzzy C-means problem and the bi-criteria exploration are based on the optimum k-clustering. We also know that the FCM++ and NFCM algorithms can guarantee $O(k^2\ln k)$ and $O(k\ln k)$ approximation ratio respectively, in expectation. Therefore, there is a more measurable way to study this problem. The way is that we compare the potential function values when k centers are selected and βk centers are selected by the FCM++ (or NFCM) algorithm.

First we give some new notations. $F_{\beta,k}$ denotes the potential target value of βk centers obtained by applying the FCM++ algorithm, while $N_{\beta,k}$ is that of NFCM algorithm. According the theoretical analysis in this paper, it is expected that $r_{F,\beta,k} = \frac{F_{\beta,k}}{F_{1,k}}$ and $r_{N,\beta,k} = \frac{N_{\beta,k}}{N_{1,k}}$ are proportional with a ratio $O(\frac{1}{\ln k})$. For each algorithm, we record and analyze the experimental results by changing the values of parameters k and β, in which k is increased from 10 to 150 by 10 as steps, and β takes 1.2, 1.5, 2, 2.2.

Figure 1 and Fig. 2 separately plot the variation tendency of $r_{F,\beta,K}$ and $r_{N,\beta,K}$ with respect to the two parameters k, β. First it is natural to see that for fixed k, both $r_{F,\beta,k}$ and $r_{N,\beta,k}$ go smaller by increasing β, which means that the potential function will decrease by increasing the number of centers. It can also be seen from the figures that for the case $\beta = 2$ and 2.2 the $r_{F,\beta,k}$ and $r_{N,\beta,k}$ slightly decrease with the increase of k, which can be seen as a partial evidence

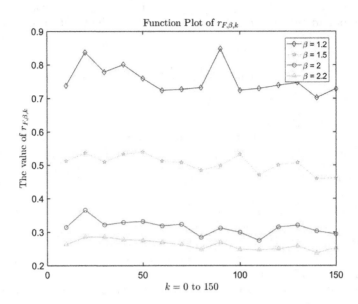

Fig. 1. Numerical results of $r_{F,\beta,k} = \frac{F_{\beta,k}}{F_{1,k}}$ using FCM++ Algorithm with Spambase data set.

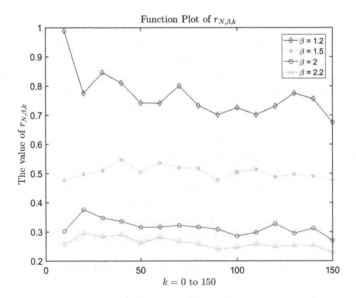

Fig. 2. Numerical results of $r_{N,\beta,k} \doteq \frac{N_{\beta,k}}{N_{1,k}}$, using NFCM Algorithm with Spambase data set.

that the two ratios are of $O(\frac{1}{\ln k})$. Therefore, we illustrate the feature of β from the perspective of analysis and numerical experiment.

5 Conclusion and Future Work

This paper shows that applying FCM++ and NFCM algorithms to optimize optimum k-clustering will produce a bi-criteria approximation ratio in expectation. The result further generalizes the previous conclusion [9,15].

It is worth studying whether we can improve the approximate ratio by reducing the coefficient in Lemma 1 to 3. We are also working on extending the results with arbitrary fuzzifier parameter m. To sum up, the content of the article needs to be further improved.

References

1. Aggarwal, A., Deshpande, A., Kannan, R.: Adaptive sampling for k-means clustering. In: Dinur, I., Jansen, K., Naor, J., Rolim, J. (eds.) APPROX/RANDOM -2009. LNCS, vol. 5687, pp. 15–28. Springer, Heidelberg (2009). https://doi.org/10.1007/978-3-642-03685-9_2
2. Ailon, N., Jaiswal, R., Monteleoni, C.: Streaming k-means approximation. In: Annual Conference on Neural Information Processing Systems, pp. 10–18 (2009)
3. Arthur, D., Vassilvitskii, S.: k-means++: the advantages of careful seeding. In: Proceedings of the Eighteenth Annual ACM-SIAM Symposium on Discrete Algorithms, pp. 1027–1035 (2007)
4. Bonis, T., Oudot, S.: A fuzzy clustering algorithm for the mode-seeking framework. Pattern Recogn. Lett. **102**, 37–43 (2018)
5. Chen, K.: On coresets for k-median and k-means clustering in metric and Euclidean spaces and their applications. SIAM J. Comput. **39**, 923–947 (2009)
6. Cornuejols, A., Wemmert, C., Gancarski, P., Bennani, Y.: Collaborative clustering: why, when, what and how. Inf. Fusion **39**, 81–95 (2018)
7. Ferreira, M., De Carvalho, F., Simoes, E.: Kernel-based hard clustering methods with kernelization of the metric and automatic weighting of the variables. Pattern Recogn. **51**, 310–321 (2016)
8. Li, M.: The bi-criteria seeding algorithms for two variants of k-means problem. J. Comb. Optim. **40**, 1–12 (2020)
9. Liu, Q., Liu, J., Li, M., Zhou, Y.: A novel initialization algorithm for fuzzy C-means problem. Theory and Applications of Models of Computation (2020, to be published)
10. Lloyd, S.: Least squares quantization in PCM. IEEE Trans. Inf. Theory **28**, 129–137 (1982)
11. Makarychev, K., Makarychev, Y., Sviridenko, M., Ward, J.: A bi-criteria approximation algorithm for k-means. In: International Workshop and International Workshop on Approximation Randomization and Combinatorial Optimization Algorithms and Techniques, pp. 1–20 (2016)
12. Memon, K.: A histogram approach for determining fuzzifier values of interval type-2 fuzzy C-means. Expert Syst. Appl. **91**, 27–35 (2018)
13. Ruspini, E.: Numerical methods for fuzzy clustering. Inf. Sci. **2**, 319–350 (1970)

14. Shen, Y., Shi, H., Zhang, J.: Improvement and optimization of a fuzzy C-means clustering algorithm. In: IEEE International Instrumentation and Measurement Technology Conference, pp. 1430–1433 (2001)
15. Stetco, A., Zeng, X., Keane, J.: Fuzzy C-means++: fuzzy C-means with effective seeding initialization. Expert Syst. Appl. **42**, 7541–7548 (2015)
16. Wang, P.: Pattern recognition with fuzzy objective function algorithms (James C. Bezdek). SIAM Rev. **25**, 442 (1983)
17. Wei, D.: A constant-factor bi-criteria approximation guarantee for k-means++. In: Annual Conference on Neural Information Processing Systems, pp. 604–612 (2016)
18. Xu, R., Wunsch, D.: Survey of clustering algorithms. IEEE Trans. Neural Netw. **16**, 645–678 (2005)
19. Xu, Y., Wu, X.: An affine subspace clustering algorithm based on ridge regression. Pattern Anal. Appl. **20**(2), 557–566 (2016). https://doi.org/10.1007/s10044-016-0564-9
20. Yang, M.: A survey of fuzzy clustering. Math. Comput. Model. **18**, 1–16 (1993)
21. Zadeh, L.: Fuzzy sets. Inf. Control **8**, 338–353 (1965)
22. Zhou, K., Yang, S., Shao, Z.: Household monthly electricity consumption pattern mining: a fuzzy clustering-based model and a case study. J. Cleaner Prod. **141**, 900–908 (2017)

Approximating Max k-Uncut via LP-rounding Plus Greed, with Applications to Densest k-Subgraph

Peng Zhang[1](✉) and Zhendong Liu[2]

[1] School of Software, Shandong University, Jinan, Shandong 250101, China
algzhang@sdu.edu.cn
[2] School of Computer Science and Technology, Shandong Jianzhu University, Jinan, Shandong 250101, China
liuzd2000@126.com

Abstract. The Max k-Uncut problem arose from the study of homophily of large-scale networks. Given an n-vertex undirected graph $G = (V, E)$ with nonnegative weights defined on edges, and a positive integer k, the Max k-Uncut problem asks to find a partition $\{V_1, V_2, \cdots, V_k\}$ of V such that the total weight of edges that are *not* cut is maximized. Max k-Uncut can also be viewed as a clustering problem with the measure being the total weight of uncut edges in the solution. This problem is the complement of the classic Min k-Cut problem, and was proved to have surprisingly rich connection to the Densest k-Subgraph problem. In this paper, we give approximation algorithms for Max k-Uncut using a non-uniform approach combining LP-rounding and the greedy strategy. With a limited violation of the constraint k, we present a good expected approximation ratio $\frac{1}{2}(1 + (\frac{n-k}{n})^2)$ for Max k-Uncut.

Keywords: Max k-Uncut · Densest k-Subgraph · Approximation algorithm · Combinatorial optimization

1 Introduction

The Max k-Uncut problem was recently proposed by Zhang et al. [18] when the authors [18] studied the *homophily* law [7, Chapter 4] of large scale networks. Being one of the most important basic laws governing the structures of large scale networks, the homophily law states that edges in a network tend to connect vertices with the same or similar attributes, just as an old proverb says, "birds of a feather flock together".

To simplify the situation, we consider the case that there is one attribute for vertices in a network. For this attribute, we may assume that the number of its values, denoted by k, is known. For example, for a paper-citation network in computer science, it can be assumed known that how many research directions there are in the network. So, a natural question is, given a network and an integer k, how to partition the network into k parts such that the number of uncut

© Springer Nature Switzerland AG 2020
Z. Zhang et al. (Eds.): AAIM 2020, LNCS 12290, pp. 161–172, 2020.
https://doi.org/10.1007/978-3-030-57602-8_15

edges is maximized? This is precisely the Max k-Uncut problem. Intuitively, the Max k-Uncut problem asks for a flat clustering of a network such that the resulting partition reflects the homophily property to the most degree, where the measure is the number of uncut edges.

Viewing attribute values as colors, we say that an edge in a network is *happy* if its two endpoints have the same color. That is, an edge is happy if and only if it is uncut. The Max k-Uncut problem can also be equivalently stated as a coloring problem: Given a graph and a color set $\{1, 2, \ldots, k\}$, how to color the vertices in k colors so that the number of happy edges in the resulting graph is maximized?

We define the Max k-Uncut problem in a more generalized weighted case.

Definition 1. The Max k-Uncut Problem [18].

(Instance) *We are given an undirected graph $G = (V, E)$ with nonnegative edge weights $\{w_e \mid e \in E\}$, and a positive integer k.*

(Goal) *The problem asks to find a partition $\{V_1, V_2, \cdots, V_k\}$ of V such that the total weight of happy edges is maximized.*

In the definition of Max k-Uncut, the k-partition $\{V_1, V_2, \cdots, V_k\}$ can also be called a k-coloring, which means a coloring scheme using *exactly* k colors, in which V_i is the set of vertices whose color is i. We will interchangeably use k-coloring and k-partition. Note that the requirement of exactly k colors is necessary, otherwise (if we allow at most k colors) we can color all vertices in one color and all edges are happy.

Some common notations and terms are listed here. Given a graph $G = (V, E)$, let n be the number of its vertices. Given an optimization problem, let OPT denote the value of its optimal solution. For simplicity, let MkU be the abbreviation of Max k-Uncut. Given k being an positive integer, the notation $[k]$ denotes the set $\{1, 2, \cdots, k\}$.

The MkU problem is NP-hard since it is the complement of the classic Min k-Cut problem [8]. Whereas Min k-Cut has been proposed for a long time (see Sect. 1.2), MkU is a new problem. There are two known approximation results for MkU. Zhang et al. [18] gave a simple randomized greedy algorithm for MkU with approximation ratio $(1 - \frac{k}{n})^2$. This ratio is good when k is not too large. For example, if $k \leq n/2$, then $(1 - \frac{k}{n})^2 \geq 1/4$. However, when k approaches $n - 1$, $(1 - \frac{k}{n})^2$ becomes worse and worse, and equals to $\frac{1}{n^2}$ finally. This observation suggests that the most difficult case of approximating Max k-Uncut should be the case when k is close to n, say, $k = n - O(\log n)$. Zhang et al. [18] also showed that for any constant $\epsilon > 0$, MkU can be approximated within $\Omega(1/n^{\frac{1}{4}+\epsilon})$ by reducing it to the Densest k-Subgraph problem (DkS for short) [3].

1.1   Our Results

In this paper, we give a bicriteria approximation algorithms for MkU. The MkU algorithm partitions the vertices of the input graph into at least $(1 - \frac{1}{e})k \geq 0.6321k$ parts in expectation, and obtains expected total weight of happy edges at least $\frac{1}{2}(1+\alpha)\text{OPT}_{\text{M}k\text{U}}$, where $\alpha = (1-\frac{k}{n})^2$ and $\text{OPT}_{\text{M}k\text{U}}$ is the value of an optimal solution to MkU. Note that the approximation ratio $\frac{1}{2}(1+\alpha)$ is always at least $1/2$. It is much better in contrast with the previously known ratios $(1-\frac{k}{n})^2$ and $\Omega(1/n^{\frac{1}{4}+\epsilon})$ [18]. So, we obtain a good ratio at least one half for MkU, at the cost of cutting down the part number k by a fraction of at most 0.3679.

Our method to approximate MkU is a non-uniform approach combining LP-rounding and greedy strategy. To round the fractional optimal solution to a natural LP-relaxation for MkU, we use the Kleinberg-Tardos rounding (KT-rounding) method [10]. The KT-rounding method has been successfully applied in approximating many combinatorial optimization problems, including the Metric Labeling problem [10], the Multiway Uncut problem [11], the Maximum Happy Edges problem, and the Maximum Happy Vertices problem [19]. Our novelty is that the KT-rounding method is applied to an LP-relaxation with a color number constraint (see (2) of (LP-U)). To the best of our knowledge, this is the first time that KT-rounding is applied in such a situation.

The approximation ratio of the solution produced by KT-rounding is $1/2$. This ratio, is further improved to $\frac{1}{2}(1+\alpha)$ by combining KT-rounding with a randomized greedy strategy for MkU (whose ratio is α). In general, the ratio $\frac{1}{2}(1+\alpha)$ is better both than $1/2$ and α. When k is near to 1, the ratio $\frac{1}{2}(1+\alpha)$ is much better than $1/2$. When k is near to n, the ratio $\frac{1}{2}(1+\alpha)$ is much better than α.

It was known that if MkU can be approximated within ρ, then DkS (recall that DkS denotes Densest k-Subgraph) can be approximated within $\rho/2$ [18]. Building upon the approximation results for MkU introduced above, we design an approximation algorithm for DkS which finds in expected polynomial time a subgraph such that its total edge weight is at least $\frac{1}{4}(1+(\frac{k-1}{n})^2)\text{OPT}_{\text{D}k\text{S}}$ in expectation, where $\text{OPT}_{\text{D}k\text{S}}$ is the optimum of DkS, and that it contains in expectation at most $\frac{1}{e}(n-k+1)$ extra vertices beyond k. We must say that this is a weak result for DkS. However, it has its own meaning for which the reasons are twofold.

First, note that $\frac{1}{4}(1+(\frac{k-1}{n})^2)$ is a good ratio since it is always at least $1/4$. One would point out that this is obtained at the cost of using too many extra vertices. We should say that when $k \geq \alpha n$ for some constant α, the number of extra vertices in our result is only $\leq \frac{1}{e}(\frac{1}{\alpha}-1)k$. When $\alpha \geq \frac{1}{e+1} \approx 0.269$, we have $\frac{1}{e}(\frac{1}{\alpha}-1) \leq 1$. Note that the case when $k \geq \alpha n$ for DkS deserves to be studied. For example, Ye and Zhang [16] studied DkS when $k = \frac{1}{2}n$. They obtained the approximation ratio 0.586 for DkS using the semidefinite programming technique. Han et al. [9] studied DkS for various values of α. Moreover, note that the number of extra vertices (i.e., $\frac{1}{e}(n-k+1)$) in our result is a worst case upper bound. The real number in the solution may be small and far away from this upper bound. At last, note that the upper bound becomes worse when k is near to one. It is known that the case when k is near to one is the most difficult case of DkS.

Second, to the best of our knowledge, our method is a new approximation approach for DkS. We hope that this new approach may develop further and open a door in the way tackling DkS, as DkS has been a notorious difficult problem in approximation algorithms for a long time.

Our algorithm for DkS (Algorithm $\mathcal{B}$ in Sect. 3) may find applications in the situations that there is no rigorous requirement on the subgraph vertex number, e.g., finding a community whose size is around some number in a social network. On the other hand, Algorithm $\mathcal{B}$ actually finds a dense core with guaranteed performance from an input graph. This can be used as a promising starting point to deal with DkS. Based on the dense core found by Algorithm $\mathcal{B}$, one can continue to design heuristics in practice for DkS, or to design further improved algorithms.

1.2 Related Work

The MkU problem has rich connection to existing problems. Max k-Uncut is just the complement of the classic Min k-Cut problem. The Min k-Cut problem asks for a k-partition such that the total weight of cut edges is minimized. The Min k-Cut problem is strongly NP-hard [8], and its current best approximation ratio is 2 [15]. Manurangsi [14] proved this is the optimal one under the Small Set Expansion hypothesis. Downey et al. [6] proved that Min k-Cut is W[1]-hard when k is used as the parameter. When k is a constant, the Min k-Cut problem can be optimally solved in polynomial time [8]. Obviously, Max k-Uncut with constant k is also polynomial time solvable.

In literature, the "uncut" problems have also been studied extensively. Besides Max k-Uncut, other examples include Min Uncut [1], Multiway Uncut [11,19], and the complement of Min Bisection [16].

We would like to indicate that the Maximum Happy Edges problem (MHE for short) [17], which is also obtained from the study of homophily of networks, is closely related to MkU. MHE can be approximated within $\frac{1}{2} + \frac{\sqrt{2}}{4} f(k) \geq 0.8535$ for some function $f(k) \geq 1$ [19]. As a new coloring problem, MHE (as well as Maximum Happy Vertices, its companion problem) attracts much attention of researchers after it was proposed (see, e.g., [2,4,12]).

A (slightly) surprising result proved in [18] states that MkU and DkS are actually equivalent in approximability. More precisely, if MkU can be approximated within a factor of ρ, then DkS can be approximated within $\rho/2$. Moreover, the converse direction is also true: If DkS can be approximated within a factor of ρ, then MkU can be approximated within $\rho/2$. The relation between DkS and MkU suggests that not like what it looks at the first glimpse, MkU is actually a difficult problem.

2   Approximating Max k-Uncut

The algorithm for MkU is based on a non-uniform approach consisting of LP-rounding and a greedy strategy. We first show the LP-rounding algorithm in Sect. 2.1, then show the final non-uniform approach for MkU in Sect. 2.2.

2.1   LP-rounding

The following linear program (LP-U) is an LP-relaxation for the MkU problem. To see this, let us consider the corresponding integer program of (LP-U), in which every variable takes value in $\{0,1\}$. Variable y_v^i indicates whether vertex v is colored in i, x_e^i indicates whether edge e is happy by color i (i.e., its two endpoints are both colored in i), and x_e indicates whether edge e is happy.

$$\max \quad \sum_{e \in E} w_e x_e \tag{LP-U}$$

$$\text{s.t.} \quad \sum_{i=1}^{k} y_v^i = 1, \qquad \forall v \tag{1}$$

$$\sum_{v \in V} y_v^i \geq 1, \qquad \forall i \tag{2}$$

$$x_e^i = \min\{y_u^i, y_v^i\}, \quad \forall i, \forall e = (u,v) \tag{3}$$

$$x_e = \sum_{i=1}^{k} x_e^i, \qquad \forall e \tag{4}$$

$$x_e, x_e^i, y_v^i \geq 0, \qquad \forall i, \forall v, \forall e$$

Constraint (1) says that each vertex has exactly one color. Constraint (2) says that every color must be used by at least one vertex, guaranteeing that the solution is a k-coloring. Constraint (3) says that edge e is happy by color i only when both its two endpoints are colored in i. Note that constraint (3) is linear since it can be replaced by two constraints $x_e^i \leq y_u^i$ and $x_e^i \leq y_v^i$. Furthermore, by constraint (1), it is impossible that an edge is simultaneously satisfied by two different colors. Finally, the objective function is to maximize the total weight of happy edges. Therefore, the integer version of (LP-U) formulates the MkU problem.

A Straightforward Rounding Strategy. Let (x,y) be an optimal fractional solution to (LP-U). It can be easily seen from constraint (1) that $\{y_v^i\}$ just constitutes a probability distribution for vertex v. Thus, a straightforward strategy to obtain an integral solution is to color vertex v in i with probability y_v^i. However, the expected value of the resulting solution can be as bad as $1/k$ times $\text{OPT}_f(\text{LP-U})$, where $\text{OPT}_f(\text{LP-U})$ denotes the fractional optimum of (LP-U).

Therefore, instead of using the straightforward randomized rounding approach described above, we use the Kleinberg-Tardos rounding technique [10] to round a fractional optimal solution to (LP-U). The algorithm is shown as Algorithm $\mathcal{R}$.

Let us call an iteration of the while loop of Algorithm $\mathcal{R}$ a *round*. Due to space limitation, some lemmas are given in the following without proofs. The proofs will be given in the full version.

Lemma 1. *Fix a round of Algorithm $\mathcal{R}$. The probability that a vertex v is colored in this round is $1/k$.*

Algorithm 2.1 (Algorithm $\mathcal{R}$ for MkU)

---
1 Solve (LP-U) to obtain an optimal solution (x, y).
2 **while** there exists some uncolored vertex **do**
3   Pick a color $i \in [k]$ uniformly at random.
4   Pick a parameter $\rho \in [0, 1]$ uniformly at random.
5   For each uncolored vertex v, if $y_v^i \geq \rho$, then color v in i.
6 **end while**

Lemma 2. *Let v be a vertex and i be a color. In Algorithm $\mathcal{R}$, the probability that vertex v is colored in i is y_v^i.*

Lemma 3. *The probability that there exists a vertex which is colored in color i is at least $1 - \frac{1}{e}$.*

Proof. By Lemma 2, the probability that vertex v is *not* colored in color i is $(1 - y_v^i)$. By Algorithm $\mathcal{R}$, the events that vertex v_1 is colored in i, vertex v_2 is colored in i, ..., and vertex v_n is colored in i are mutually independent. Therefore, the probability that there exists a vertex which is colored in color i is

$$1 - \prod_v \left(1 - y_v^i\right) = 1 - e^{\sum_v \ln\left(1 - y_v^i\right)} \geq 1 - e^{-\sum_v y_v^i} \underset{(2)}{\geq} 1 - \frac{1}{e},$$

where $\sum_v \ln\left(1 - y_v^i\right) \leq -\sum_v y_v^i$ since $\ln x \leq x - 1$ when $x \in [0, 1]$.

Lemma 4. *In expectation, the solution produced by Algorithm $\mathcal{R}$ uses at least $(1 - \frac{1}{e})k$ colors.*

Proof. Let i be a color. Define random variable X_i as follows.

$$X_i = \begin{cases} 1, & \text{there exists a vertex which is colored in color } i. \\ 0, & \text{otherwise.} \end{cases}$$

Then the random variable $X = X_1 + X_2 + \cdots + X_k$ is the number of colors used in the final solution found by Algorithm $\mathcal{R}$. By Lemma 3, for variable X we have $\mathrm{E}[X] = \sum_i \mathrm{E}[X_i] \geq (1 - \frac{1}{e})k$.

Lemma 4 indicates that Algorithm $\mathcal{R}$ may produce an infeasible solution. However, we can guarantee that the expected number of parts of the partition obtained by Algorithm $\mathcal{R}$ is at least $(1 - \frac{1}{e})k$. Note that the same thing happens to the straightforward randomized rounding strategy given after (LP-U).

Next, we will analyze the approximation ratio of the solution produced by Algorithm $\mathcal{R}$.

Lemma 5. *Let $e = (u, v)$ be an edge. Suppose that u and v are not colored before the current round. Then, the probability that in this round both u and v are colored (hence in the same color) is $\frac{1}{k} x_e$.*

Lemma 6. *Let $e = (u, v)$ be an edge. Suppose that u and v are not colored before the current round. Then, the probability that in this round u or v are colored is $\frac{1}{k}(2 - x_e)$.*

Lemma 7. *The probability that an edge e is happy in Algorithm $\mathcal{R}$ is at least $\frac{x_e}{2 - x_e}$.*

Proof. Let $e = (u, v)$ be the edge. By the coloring strategy of Algorithm $\mathcal{R}$, the probability that edge e is happy is the sum of (i) the probability that u and v are colored simultaneously in some one round and (ii) the probability that u and v are colored in the same color in two different rounds. We omit the latter probability. The former probability is

$$\sum_{r=1}^{\infty} \Pr[\text{both } u \text{ and } v \text{ are not colored before the } r\text{-th round}] \cdot$$

$$\Pr[u \text{ and } v \text{ are colored in the } r\text{-th round},$$

$$\text{conditioned on that } u \text{ and } v \text{ are not colored}].$$

By Lemma 5 and Lemma 6, this probability is equal to $\sum_{r=1}^{\infty} \left(1 - \frac{2 - x_e}{k}\right)^{r-1} \cdot \frac{1}{k} x_e = \frac{x_e}{2 - x_e}$.

So far, we actually obtain the following result.

Theorem 1. *Algorithm $\mathcal{R}$ is a randomized $(\frac{1}{2}, 1 - \frac{1}{e})$-approximation algorithm for Max k-Uncut. That is, the algorithm finds a partition of the vertices of graph G in expected polynomial time, such that the expected total weight of happy edges produced by the partition is at least $\frac{1}{2}\mathrm{OPT}_{MkU}$, and that the partition contains at least $(1 - \frac{1}{e})k$ parts in expectation.*

Proof. By Lemma 7, the expected solution value of Algorithm $\mathcal{R}$ is at least $\sum_e w_e \frac{x_e}{2 - x_e} \geq \frac{1}{2} \sum_e w_e x_e$. By Lemma 4, the solution output by Algorithm $\mathcal{R}$ using at least $(1 - \frac{1}{e})k$ colors in expectation. Finally, it is not hard to see that the expected running time of Algorithm $\mathcal{R}$ is polynomial, concluding the theorem.

In the following, we use a non-uniform approach combining Algorithm $\mathcal{R}$ and a greedy algorithm to further improve the approximation ratio of MkU.

2.2 A Non-uniform Approach

A Greedy Algorithm. Intuitively, if we can make a subgraph as large as possible, then we may get a large weight of happy edges in this subgraph. So, a simple but clever greedy strategy is to pick $k - 1$ vertices as singleton sets, and to put all the remaining vertices in a separate set, obtaining a subgraph with the largest possible size (number of vertices). This idea leads to the following greedy Algorithm $\mathcal{G}$ for MkU, which was given in [18].

Theorem 2 ([18]). *The expected approximation ratio of Algorithm $\mathcal{G}$ is $\left(1 - \frac{k}{n}\right)^2$.*

Algorithm 2.2 (Algorithm $\mathcal{G}$ for MkU [18])

---

1 Pick randomly $k - 1$ vertices from V, and color them respectively in colors 1 to $k - 1$.
2 Color all the remaining vertices in color k.

Let $\alpha = \left(1 - \frac{k}{n}\right)^2$ be the (expected) approximation ratio of Algorithm $\mathcal{G}$.

A Non-uniform Approach. The following Algorithm $\mathcal{A}$ is a non-uniform approach dealing with MkU. The algorithm just combines Algorithms $\mathcal{R}$ and $\mathcal{G}$ with appropriate probabilities. In the algorithm, λ is a parameter which will be fixed in the subsequent analysis (see Theorem 3).

Algorithm 2.3 (Algorithm $\mathcal{A}$ for MkU)

---

1 With probability λ, run Algorithm $\mathcal{R}$; with probability $1 - \lambda$, run Algorithm $\mathcal{G}$.
2 **return** the solution found by either $\mathcal{R}$ or $\mathcal{G}$.

Theorem 3. *In expected polynomial time, Algorithm $\mathcal{A}$ finds a partition of the vertices of graph G, such that the expected total weight of happy edges produced by the partition is at least $\frac{1}{2}(1 + \alpha)\mathrm{OPT}_{MkU}$, where $\alpha = \left(1 - \frac{k}{n}\right)^2$, and that the partition contains at least $(1 - \frac{1}{e})k$ parts in expectation.*

Proof. Let random variables W_A, W_R, and W_G be the total weights of the happy edges found by Algorithms $\mathcal{A}$, $\mathcal{R}$, and $\mathcal{G}$, respectively. Then we have

$$\mathrm{E}[W_A] = (1 - \lambda)\mathrm{E}[W_G] + \lambda\mathrm{E}[W_R]$$

$$= (1 - \lambda)\sum_e w_e \Pr[e \text{ is happy in } \mathcal{G}] +$$

$$\lambda\sum_e w_e \Pr[e \text{ is happy in } \mathcal{R}]. \tag{5}$$

By Theorem 2 and Lemma 7, we have

$$\text{RHS of (5)} \geq (1 - \lambda)\sum_e w_e \alpha + \lambda\sum_e w_e \frac{x_e}{2 - x_e}$$

$$= \sum_e \left[(1 - \lambda)\alpha w_e + \frac{\lambda w_e x_e}{2 - x_e}\right]$$

$$\geq \sum_{e:\, x_e > 0} \left[\frac{(1 - \lambda)\alpha}{x_e} + \frac{\lambda}{2 - x_e}\right] w_e x_e. \tag{6}$$

By calculus, function $f(x) = \frac{(1-\lambda)\alpha}{x} + \frac{\lambda}{2-x}$ is always $\geq \frac{1}{2}[\lambda + \alpha - \alpha\lambda + 2\sqrt{\alpha\lambda(1 - \lambda)}]$ when x ranges from 0 to 1. So, we get

$$\text{RHS of (6)} \geq \sum_{e:\, x_e > 0} \frac{1}{2}\left[\lambda + \alpha - \alpha\lambda + 2\sqrt{\alpha\lambda(1 - \lambda)}\right] w_e x_e. \tag{7}$$

Function $g(\lambda) = \lambda + \alpha - \alpha\lambda + 2\sqrt{\alpha\lambda(1-\lambda)}$ is a concave function of $0 \leq \lambda \leq 1$. When $\lambda = \frac{1}{\alpha+1}$, $g(\lambda)$ gets its maximum. So, we set $\lambda = \frac{1}{\alpha+1}$ in Algorithm $\mathcal{R}$. Consequently, we have

$$\text{RHS of (7)} = \sum_{e : \, x_e > 0} \frac{1}{2}(1+\alpha)w_e x_e = \frac{1}{2}(1+\alpha)\sum_e w_e x_e. \tag{8}$$

Combining (5), (6), (7), and (8) together, we get that $\mathrm{E}[W_A] \geq \frac{1}{2}(1 + \alpha)\mathrm{OPT}_{\mathsf{M}k\mathsf{U}}$.

Algorithm $\mathcal{G}$ partitions the vertex set $V(G)$ into exactly k parts, while Algorithm $\mathcal{R}$ partitions $V(G)$ into at least $(1-\frac{1}{e})k$ parts in expectation. So, no matter which algorithm is run in Algorithm $\mathcal{A}$, the solution output by $\mathcal{A}$ contains at least $(1 - \frac{1}{e})k$ parts in expectation. Finally, it is not hard to see that Algorithm $\mathcal{A}$ runs in expected polynomial time. The theorem follows.

3   Approximating Densest k-Subgraph

In this section, we show that the approximation results of MkU in Sect. 2 can be extended to the Densest k-Subgraph problem, thus giving a new result to this famous optimization problem.

Definition 2. The Densest k-Subgraph **Problem.**
(Instance) *We are given an undirected graph* $G = (V, E)$ *with nonnegative edge weights* $\{w_e \mid e \in E\}$, *and a positive integer* k.
(Goal) *The problem asks to find a* k-vertex subgraph G' *such that the total weight of edges in* $E(G')$ *is maximized.*

DkS is known as a notorious hard problem in approximation algorithms. In 2010, Bhaskara et al. [3] gave an $\Omega(n^{-(1/4+\epsilon)})$-approximation algorithm for DkS for any small constant $\epsilon > 0$. This is the current best approximation ratio of DkS. When $k = \alpha n$, where $\alpha \in (0,1)$ is a constant, Han et al. [9] proved much better approximation ratios for DkS using the semidefinite programming technique. Recently, Chen et al. [5] considered the connected DkS problem (in which the subgraph to be found should be connected) and gave an $\Omega(n^{-0.4})$-approximation algorithm for it. There are also several hardness results for DkS which are based on different complexity assumptions. Manurangsi [13] proved that assuming ETH is true, there is a constant $c > 0$ such that no polynomial time algorithm can approximate DkS within $n^{-1/(\log\log n)^c}$. Further, if one assumes Gap-ETH is true, then there is no polynomial time algorithm which can approximate DkS within $n^{-f(n)}$ for any function f satisfying $f(n) \in o(1)$.

Our Idea of Approximating DkS**.** Our overall idea of approximating DkS is to view it as a *cut* problem. It is not new to view DkS as a 2-cut problem. That is, the k-vertex subgraph forms one part of the 2-cut (i.e., bipartition) and the remaining subgraph forms another part.
Let

$$\bar{k} = n - k + 1. \tag{9}$$

Our new idea is to view DkS as a $\bar{k}$-cut problem. In our perspective, the k-vertex subgraph forms one part of the $\bar{k}$-partition of $V(G)$, and each of the remaining vertices forms a separate part. See Fig. 1 for an illustration.

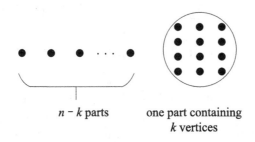

$n - k$ parts   one part containing
                     k vertices

Fig. 1. An illustration of $(n - k + 1)$-partition.

An immediate question is that, what is the merit of $\bar{k}$-cut, after all $\bar{k}$ may be much larger than two? The reasons are the following. If we view DkS as a 2-cut problem, then it is difficult to define the objective function in the 2-cut problem. This is because what we want is to maximize the total weight of edges only in one part of the 2-partition (the k-subgraph part). These edges are *uncut* edges. However, according to the 2-cut, there are still uncut edges lying in the other part. We cannot define the objective function as to maximize the total weight of uncut edges in the viewpoint of 2-cut.

If we view DkS as a $\bar{k}$-cut problem satisfying an additional size constraint that in the $\bar{k}$-partition there are one part of size k and $n-k$ parts of size one, then the edges that we want are *all* uncut edges. In this way, we reduce DkS to M$\bar{k}$U with additional size constraints. Note that M$\bar{k}$U and MkU are the same problem with different input parameters. We use the name M$\bar{k}$U to emphasize that when we reduce DkS to M$\bar{k}$U, we get an input parameter $\bar{k}$ (see (9)) different to k.

Zhang et al. [18] proved that if MkU can be approximated within ρ, then DkS can be approximated within $\rho/2$. We extend the idea in this result to the bicriteria approximation algorithm $\mathcal{A}$ for MkU, obtaining a bicriteria approximation algorithm for DkS, shown as Algorithm $\mathcal{B}$.

In step 4 of Algorithm $\mathcal{B}$, if we scatter the removed $\bar{k}' - 1$ vertices into singleton sets $V_1', \cdots, V_{\bar{k}'-1}'$, then we actually obtain a new $\bar{k}'$-partition $\mathcal{P}' = \{V_1', V_2', \cdots, V_{\bar{k}'-1}', V'\}$. Step 4 actually converts partition $\mathcal{P}$ to partition $\mathcal{P}'$.

For a vertex subset S, let $w(S)$ denote the total weight of happy edges whose two endpoints are both in S. For Algorithm $\mathcal{B}$ we have Theorem 4. Its proof will be given in the full version.

Algorithm 3.1 (Algorithm $\mathcal{B}$ for DkS)

Input: A DkS instance (G, w, k).
Output: A vertex subset $V' \subseteq V(G)$.
1 $\bar{k} \leftarrow n - k + 1$.
2 **call** Algorithm $\mathcal{A}$ on the M$\bar{k}$U instance $(G, w, \bar{k})$, obtaining a $\bar{k}'$-partition $\mathcal{P} = \{V_1, V_2, \cdots, V_{\bar{k}'}\}$ of $V(G)$ for some integer $\bar{k}'$.
3 Without loss of generality, we may assume that $w(V_1) \leq w(V_2) \leq \cdots \leq w(V_{\bar{k}'})$. Find $\ell \in [\bar{k}']$ such that $|V_1| + \cdots + |V_\ell| \geq \bar{k}' - 1$ and $|V_1| + \cdots + |V_{\ell-1}| < \bar{k}' - 1$.
4 Remove arbitrary $\bar{k}' - 1$ vertices from $V_1 \cup \cdots \cup V_\ell$. The remaining vertices in $V_1 \cup \cdots \cup V_\ell$, together with all vertices in $V_{\ell+1} \cup \cdots \cup V_{\bar{k}'}$, form a single set V'.
5 **return** V'.

Theorem 4. *In expected polynomial time, Algorithm $\mathcal{B}$ finds a vertex subset V' such that*

(i) $E[w(V')] \geq \frac{1}{4}(1 + (\frac{k-1}{n})^2)\mathrm{OPT}_{DkS}$ (that is, the expected total weight of edges in the induced subgraph $G' = G[V']$ is at least $\frac{1}{4}(1 + (\frac{k-1}{n})^2)\mathrm{OPT}_{DkS}$), and
(ii) $E[|V'|] \leq k + \frac{1}{e}(n - k + 1)$ (that is, the number of extra vertices used in G' is at most $\frac{1}{e}(n - k + 1)$).

4 Conclusions

In this paper, we give approximation algorithms for MkU using a non-uniform approach combining LP-rounding and the greedy strategy. With a limited violation of the constraint k, we present a good expected approximation ratio $\frac{1}{2}(1 + (\frac{n-k}{n})^2)$ for MkU. We also illustrate how this result extends to DkS. Hope that the methods presented in this paper could inspire new results for MkU and (especially) DkS.

Acknowledgements. Peng Zhang is supported by the National Natural Science Foundation of China (61972228, 61672323), and the Natural Science Foundation of Shandong Province (ZR2016AM28, ZR2019MF072). Zhendong Liu is supported by the National Natural Science Foundation of China (61672328).

References

1. Agarwal, A., Charikar, M., Makarychev, K., Makarychev, Y.: $O(\sqrt{\log n})$ approximation algorithms for min uncut, min 2CNF deletion, and directed cut problems. In: Proceedings of the 37th Annual ACM Symposium on Theory of Computing (STOC), pp. 573–581 (2005)
2. Agrawal, A.: On the parameterized complexity of happy vertex coloring. In: Brankovic, L., Ryan, J., Smyth, W.F. (eds.) IWOCA 2017. LNCS, vol. 10765, pp. 103–115. Springer, Cham (2018). https://doi.org/10.1007/978-3-319-78825-8_9
3. Bhaskara, A., Charikar, M., Chlamtac, E., Feige, U., Vijayaraghavan, A.: Detecting high log-densities: an $O(n^{1/4})$ approximation for densest k-subgraph. In: Proceedings of the 42nd Annual ACM Symposium on Theory of Computing (STOC), pp. 201–210 (2010)

4. Bliznets, I., Sagunov, D.: Lower bounds for the happy coloring problems. In: Proceedings of the 25th International Computing and Combinatorics Conference (COCOON), pp. 490–502 (2019)
5. Chen, X., Hu, X., Wang, C.: Finding connected k-subgraphs with high density. Inf. Comput. **256**, 160–173 (2017)
6. Downey, R.G., Estivill-Castro, V., Fellows, M.R., Prieto-Rodriguez, E., Rosamond, F.A.: Cutting up is hard to do: the parameterized complexity of k-Cut and related problems. Electron. Notes Theoret. Comput. Sci. **78**, 209–222 (2003)
7. Easley, D., Kleinberg, J.: Networks, Crowds, and Markets: Reasoning About a Highly Connected World. Cambridge University Press, Cambridge (2010)
8. Goldschmidt, O., Hochbaum, D.: A polynomial algorithm for the k-cut problem for fixed k. Math. Oper. Res. **19**(1), 24–37 (1994)
9. Han, Q., Ye, Y., Zhang, J.: An improved rounding method and semidefinite programming relaxation for graph partition. Math. Program. **92**(3), 509–535 (2002)
10. Kleinberg, J., Tardos, E.: Approximation algorithms for classification problems with pairwise relationships: metric labeling and Markov random fields. J. ACM **49**(5), 616–639 (2002)
11. Langberg, M., Rabani, Y., Swamy, C.: Approximation algorithms for graph homomorphism problems. In: Díaz, J., Jansen, K., Rolim, J.D.P., Zwick, U. (eds.) APPROX/RANDOM -2006. LNCS, vol. 4110, pp. 176–187. Springer, Heidelberg (2006). https://doi.org/10.1007/11830924_18
12. Lewis, R., Thiruvady, D., Morgan, K.: Finding happiness: an analysis of the maximum happy vertices problem. Comput. Oper. Res. **103**, 265–276 (2019)
13. Manurangsi, P.: Almost-polynomial ratio eth-hardness of approximating densest k-subgraph. In: Proceedings of the 49th Annual ACM Symposium on Theory of Computing (STOC), pp. 954–961 (2017)
14. Manurangsi, P.: Inapproximability of maximum biclique problems, minimum k-cut and densest at-least-k-subgraph from the small set expansion hypothesis. Algorithms **11**(1), 10:1–10:22 (2018)
15. Saran, H., Vazirani, V.: Finding k-cuts within twice the optimal. SIAM J. Comput. **24**, 101–108 (1995)
16. Ye, Y., Zhang, J.: Approximation of dense-$n/2$-subgraph and the complement of min-bisection. J. Glob. Optim. **25**(1), 55–73 (2003)
17. Zhang, P., Li, A.: Algorithmic aspects of homophyly of networks. Theoret. Comput. Sci. **593**, 117–131 (2015)
18. Zhang, P., Wu, C., Xu, D.: Approximation and hardness results for the max k-uncut problem. Theoret. Comput. Sci. **749**, 47–58 (2018)
19. Zhang, P., Xu, Y., Jiang, T., Li, A., Lin, G., Miyano, E.: Improved approximation algorithms for the maximum happy vertices and edges problems. Algorithmica **80**, 1412–1438 (2018)

Online BP Functions Maximization

Ling Chen[1], Zhicheng Liu[1], Hong Chang[1], Donglei Du[2],
and Xiaoyan Zhang[1(✉)]

[1] School of Mathematical Science and Institute of Mathematics,
Nanjing Normal University, Nanjing 210023, People's Republic of China
chenling8646@126.com, manlzhic@163.com, {changh,zhangxiaoyan}@njnu.edu.cn
[2] Faculty of Management, University of New Brunswick, Fredericton,
NB E3B 5A3, Canada
ddu@unb.ca

Abstract. The BP problem maximizes the sum of a suBmodular function and a suPermodular function(BP) subject to some constraints, where both functions are nonnegative and monotonic. This type of problems arises naturally in many applications in machine learning, data science and artificial intelligence. We consider two online BP problems. The first is a BP maximization problem subject to a uniform matroid constraint when the items arrive one-by-one, for which we offer an online algorithm with constant competitive ratio. The second is a BP maximization problem subject to a partition matroid constraint where items arrive in a random order, for which we present a randomized linear-time approximation algorithm with constant competitive ratio.

Keywords: Online BP maximization · Partition matroid ·
Submodular term · Supermodular term · Competitive ratio ·
Approximation algorithm

1 Introduction

Submodular function maximization has been widely studied under various constraints and models in recent years. Submodular functions play an important role in various fields, including machine learning and algorithmic game theory. In machine learning applications, maximization of submodular functions has been used for information gathering [8,9], document summarization [12,13], string alignment [14] and sensor placement [7,10,11]. In algorithmic game theory, the problems in calculating market expansion [5] and calculating the core value of certain types of games [15] can be simplified to submodular maximization.

However, some subset selection issues in data science are not just submodular. For instance, when selecting a subset of training data in a machine learning system [16], there may be both redundancies and complementarities between certain subsets of elements, where the full collective utility of these elements can be achieved only when bundled together. Submodular functions can only reduce, rather than enhance, the utility of data items in the presence of other data items, whereas supermodular functions can model this phenomena.

© Springer Nature Switzerland AG 2020
Z. Zhang et al. (Eds.): AAIM 2020, LNCS 12290, pp. 173–184, 2020.
https://doi.org/10.1007/978-3-030-57602-8_16

Given a ground set $V = \{1, \ldots, n\}$ along with its power set $2^V = \{X : X \subseteq V\}$, a normalized non-decreasing submodular function $f : 2^V \to \mathbb{R}_+$, and a normalized non-decreasing supermodular function $g : 2^V \to \mathbb{R}_+$, consider the following suBmodular+suPermodular (BP) maximization problem, first proposed in [2]

$$\max_{X \in \mathcal{I}} h(X) := f(X) + g(X), \tag{1.1}$$

where $\mathcal{I}$ is the family of the independent sets of a matroid $\mathcal{M} = (V, \mathcal{I})$. A matroid satisfies three properties: (M_0) $\emptyset \in \mathcal{I}$; (M_1) If $J' \subseteq J \in \mathcal{I}$, then $J' \in \mathcal{I}$; and (M_2) $\forall A \subseteq V$, every maximal independent subset of A has the same cardinality.

A set function $f : 2^V \to \mathbb{R}_+$ is normalized if $f(\emptyset) = 0$. It is non-decreasing if $f(S) \leq f(T), \forall S \subseteq T \subseteq V$. It is submodular if $f(S) + f(T) \geq f(S \cap T) + f(S \cup T), \forall S, T \subseteq V$. It is supermodular if its negative is submodular.

The objective function $h = f + g$ is called the suBmodular+suPermodular (BP) function in [2]. The BP maximization problem (1.1) arises naturally in the field of machine learning, data mining and artificial intelligence.

In this work, we focus on two online BP problems: (i) a BP maximization problem subject to a uniform matroid constraint; (ii) a BP maximization problem subject to a (binary) partition matroid constraint.

For both problems, we design competitive algorithms whose competitive ratios depends on the curvatures of the two functions involved. For a given set function $f : 2^V \to \mathbb{R}_+$, its curvature [4] is defined as follows

$$k(f) = 1 - \min_{v \in V} \frac{f(V) - f(V \setminus \{v\})}{f(v)}. \tag{1.2}$$

Online Problem. The first problem addresses an online version of the BP maximization problem subject to a uniform matroid constraint; namely, in (1.1), the feasible set $\mathcal{I} = \{X \subseteq V : |X| \leq k\}$ is the family of all independent sets of the uniform matroid $\mathcal{M} = (V, \mathcal{I})$ for a given cardinality parameter k which is a natural number. The elements of the ground set $V = \{1, \ldots, n\}$ arrive one-by-one over a list. Whenever an element arrives, an online algorithm must make an irrevocable decision on whether or not to include this newly arrived element into the final solution without violating the cardinality constraint. In contrast to the offline version of the problem, at any time, we can only call upon the value oracle who returns function values over the elements that have been revealed.

To measure the performance of an online algorithm, we utilize the concept of competitive ratio which is defined to be the supremum of ratio $h(ON)/h(OFF)$ over all problem instances, where $ON \in \mathcal{F}$ and $OFF \in \mathcal{F}$ are the solutions returned by the online and the optimal offline algorithms, respectively.

Our first contribution is to present a $\frac{(1-k_f)(1-k^g)^3}{(2-k^g)^2}$-competitive algorithm for this problem, where $k_f = k(f)$ is the submodular curvature as in (1.2), and $k^g = k(\hat{g})$ is the supermodular curvature as in (1.2), where $\hat{g} : 2^V \to \mathbb{R}_+$ is a derived set function from g such that $\hat{g}(X) = g(V) - g(V \setminus X), \forall X \subseteq V$.

In terms of relevant work, online submodular function maximization problems have been investigated in the literature. [1] studies an online Max-SAT problem, for which they give a $\frac{2}{3}$-competitive algorithm. [6] considers an online version of maximizing a non-negative non-monotone submodular function and proposes a double-sided myopic algorithms. [3] considers an online preemptive version of maximizing a non-negative monotone submodular function and offers a $1/e$-competitive algorithm for the unconstrained case and a $1/4$-competitive algorithm for the cardinality constrained model, respectively.

Partition Matroid Constrained Problem. The second problem is a BP maximization problem subject to a partition matroid constraint. Firstly, we consider the binary partition matroid, namely, in (1.1), the feasible set

$$\mathcal{I} = \{X \subseteq V : |X \cap P_t| \leq 1, \forall t = 1, \ldots, m\}$$

is the family of all independent sets of a binary partition matroid $\mathcal{M} = (V, \mathcal{I})$, where the ground set $V = \{a_1, b_1, a_2, b_2, \ldots, a_m, b_m\}$ consists of $n = 2m$ elements, and $P_t = \{a_t, b_t\}$ ($t = 1, \ldots, m$) is a partition of the ground set V. Secondly, we generalize the case to the $(1, \ell)$-partition matroid where P_t contains ℓ elements.

Our second contributions are to offer a $\frac{2-2k^g}{3-2k^g}$-approximation algorithm for this problem under the binary partition matroid case and offer a $\frac{\ell-\ell k^g}{2\ell-1-\ell k^g}$-approximation algorithm for this problem under the $(1, \ell)$- partition matroid case, where $k^g = k(\hat{g})$ is the supermodular curvature as in (1.2), where $\hat{g} : 2^V \to \mathbb{R}_+$ is a derived set function from g such that $g(X) = g(V) - g(V \backslash X), \forall X \subseteq V$.

In terms of relevant work, BP maximization problems have been investigated in the past. In [2], they consider the BP maximization problem (offline case) subject to two types of constraints, either a cardinality constraint or $k^g \geq 1$ matroid independence constraints.

The remainder of our paper is organized as follows. Section 2 introduces some preliminaries. Sections 3 and 4 present the algorithms along with their analysis for these two problems, respectively. Finally, we offer concluding remarks in Sect. 5.

2 Preliminaries

In this section, we introduce some results that will be used later. The first property says that the curvature of any submodular function can be computed efficiently under value oracle (hence the supermodular curvature can also be computed efficiently).

Proposition 1 ([2]). *The curvature k_f can be computed with at most $2|V| + 1$ oracle queries of f.*

The next result concerns some facts of BP functions.

Lemma 1 ([2]). *Given a BP function $h(X) = f(X) + g(X)$, where f and g are non-negative, monotonic non-decreasing submodular and supermodular functions respectively, we have*

(i) $h(v|Y) \geq (1 - k_f)h(v|X)$, $\forall X \subseteq Y \subset V$ and $v \notin Y$;
(ii) $h(v|Y) \leq \frac{1}{1-k^g}h(v|X)$, $\forall X \subseteq Y \subset V$ and $v \notin Y$;
(iii) $h(X|Y) \geq (1 - k_f)\sum_{v \in X \setminus Y} h(v|Y)$, $\forall X, Y \subseteq V$;
(iv) $h(X|Y) \leq \frac{1}{1-k^g}\sum_{v \in X \setminus Y} h(v|Y)$, $\forall X, Y \subseteq V$,

where $h(v|Y) = h(Y + v) - h(Y)$ is the marginal contribution of an element v to a set Y and $h(X|Y) = h(X \cup Y) - h(Y)$ is the marginal contribution of a set X to a set Y.

3   Online BP Maximization

We present a greedy algorithm in Sect. 3.1 and analyze its competitive ratio in Sect. 3.2.

3.1   Algorithm

The main idea of Algorithm 1 is as follows. We first choose a parameter $c > 0$. Assume the elements of V are revealed in the order of $1, \ldots, n$. Let S_{i-1} be the solution generated by the algorithm when elements $1, \ldots, i - 1$ have been revealed. When the element i arrives, if $i \in \{1, \ldots, k\}$, then include i into S_{i-1}; otherwise, swap i with an element selected greedily from S_{i-1} whenever some threshold is reached.

Algorithm 1: Greedy for online BP maximization

Let $S_0 = \emptyset$.
foreach *element u_i revealed* **do**
  if $i \leq k$ **then**
  | Let $S_i \leftarrow S_{i-1} + u_i$.
  else
  | Let u_i' be the element of S_{i-1} maximizing $h(S_{i-1} + u_i - u_i')$
  | **if** $h(S_{i-1} + u_i - u_i') - h(S_{i-1}) \geq \frac{c \cdot h(S_{i-1})}{k(1-k^g)}$ **then**
  | | Let $S_i \leftarrow S_{i-1} + u_i - u_i'$.
  | **else**
  | | Let $S_i \leftarrow S_{i-1}$.

For convenience, we define the following notations. Let $h(u|S) = h(S + u) - h(S)$ be the marginal contribution of an element u to a set S. Let $A_i = \bigcup_{j=1}^{i} S_j$ for every $0 \leq i \leq n$. In fact, A_i is the set of elements of $\{u_1, u_2, \ldots, u_i\}$ originally accepted by Algorithm 1, regardless of whether they are preempted or not.

3.2 Analysis

To analyze the competitive ratio of Algorithm 1, we need the following lemma.

Lemma 2 $\forall i = k+1, \ldots, n$, we have

$$h(S_{i-1} + u_i - u_i') - h(S_{i-1}) \geq (1 - k_f)(1 - k^g)h(u_i|A_{i-1}) - \frac{h(S_{i-1})}{(1 - k^g)k}.$$

Proof. We have

$$
\begin{aligned}
&h(S_{i-1} + u_i - u_i') - h(S_{i-1}) \\
&\geq \frac{\sum_{u_i' \in S_{i-1}} [h(S_{i-1} + u_i - u_i') - h(S_{i-1})]}{k} \\
&= \frac{\sum_{u_i' \in S_{i-1}} [h(S_{i-1} + u_i - u_i') - h(S_{i-1} + u_i) + h(S_{i-1} + u_i) - h(S_{i-1})]}{k} \\
&\geq \frac{\sum_{u_i' \in S_{i-1}} (1 - k_f)[h(S_{i-1} + u_i - u_i') - h(S_{i-1} - u_i')]}{k} \\
&\quad + \frac{\sum_{u_i' \in S_{i-1}} [h(S_{i-1} + u_i - u_i') - h(S_{i-1} + u_i)]}{k} \\
&= \underbrace{\frac{\sum_{u_i' \in S_{i-1}} (1 - k_f)h(u_i|S_{i-1} - u_i')}{k}}_{a} - \underbrace{\frac{\sum_{u_i' \in S_{i-1}} h(u_i'|S_{i-1} + u_i - u_i')}{k}}_{b},
\end{aligned}
$$

where the second inequality follows from Lemma 1 in [2].

The first term a in the last quantity can be bounded as follows. Note that $S_{i-1} - u_i' \subseteq A_{i-1}$. From Lemma 1 in [2], we have $h(u_i|S_{i-1} - u_i') \geq (1 - k^g)h(u_i|A_{i-1})$. Thus,

$$a \geq (1 - k_f)(1 - k^g)h(u_i|A_{i-1}). \tag{3.1}$$

The second term b can be bounded via Lemma 1 in [2] and the non-negativity of h as follows

$$\frac{\sum_{u_i' \in S_{i-1}} h(u_i'|S_{i-1} + u_i - u_i')}{k} \leq \frac{h(S_{i-1}) - h(\emptyset)}{(1 - k^g)k} \leq \frac{h(S_{i-1})}{(1 - k^g)k}. \tag{3.2}$$

Together, (3.1) and (3.2) imply the desired bound.

$\square$

The following lemma is related to S_i and A_i.

Lemma 3. $\forall i = 0, \ldots, n$, we have

$$h(S_i) \geq \frac{c}{c+1}(1 - k_f)(1 - k^g)h(A_i).$$

Proof. For $i = 0$, the lemma follows from $h(A_0) = h(\emptyset) = h(S_0)$. Thus, it suffices to show that for every $i \in \{1, \ldots, n\}$,

$$h(S_i) - h(S_{i-1}) \geq \frac{c}{c+1}(1 - k_f)(1 - k^g)[h(A_i) - h(A_{i-1})].$$

If the algorithm never accepts u_i, the claim is trivial since then $S_i = S_{i-1}$ and $A_i = A_{i-1}$. Hence, we only consider the case when the algorithm accepts u_i.

If $i \leq k$, then $h(S_i) - h(S_{i-1}) = h(u_i|S_{i-1}) \geq (1 - k^g)h(u_i|A_{i-1}) = (1 - k^g)[h(A_i) - h(A_{i-1})]$, where the inequality follows from Lemma 1 in [2] since $S_{i-1} \subseteq A_{i-1}$.

If $i > k$, then the quantity $h(S_i) - h(S_{i-1}) = h(S_{i-1} + u_i - u_i') - h(S_{i-1})$ have two lower bounds $(1 - k_f)(1 - k^g)h(u_i|A_{i-1}) - h(S_{i-1})/[(1 - k^g)k]$ from Lemma 2, and $c \cdot h(S_{i-1})/[k(1 - k^g)]$ from the algorithm since u_i is accepted. With both lower bounds, we have

$$h(S_i) - h(S_{i-1}) \geq \max\left\{(1 - k_f)(1 - k^g)h(u_i|A_{i-1}) - \frac{h(S_{i-1})}{(1 - k^g)k}, \frac{ch(S_{i-1})}{k(1 - k^g)}\right\}$$

$$\geq \frac{c\left((1 - k_f)(1 - k^g)h(u_i|A_{i-1}) - \frac{h(S_{i-1})}{(1-k^g)k}\right)}{c+1} + \frac{c\frac{h(S_{i-1})}{k(1-k^g)}}{c+1}$$

$$\geq \frac{c}{c+1}(1 - k_f)(1 - k^g)h(u_i|A_{i-1})$$

$$= \frac{c}{c+1}(1 - k_f)(1 - k^g)[h(A_i) - h(A_{i-1})].$$

$\square$

We are now ready to prove the competitive ratio of Algorithm 1.

Theorem 1. *The competitive ratio of Algorithm 1 is at least* $\frac{c(1-k_f)(1-k^g)^3}{(c+1)((1-k^g)^2+c)}$. *Hence, for* $c = 1 - k^g$ *the competitive ratio of Algorithm 1 is at least* $\frac{(1-k_f)(1-k^g)^3}{(2-k^g)^2}$.

Proof. Let OPT be the optimal solution. Consider an element $u_i \in OPT \setminus A_i$. Since u_i was rejected by Algorithm 1, the following inequality must hold

$$\frac{ch(S_{i-1})}{k(1 - k^g)} > h(S_i) - h(S_{i-1}) \geq (1 - k_f)(1 - k^g)h(u_i|A_{i-1}) - \frac{h(S_{i-1})}{(1 - k^g)k},$$

where the second inequality follows from Lemma 2. Rearranging the above yields

$$h(u_i|A_{i-1}) < \frac{c+1}{k(1 - k^g)^2(1 - k_f)}h(S_{i-1}) \leq \frac{c+1}{k(1 - k^g)^2(1 - k_f)}h(S_n),$$

where the second inequality uses the monotonicity of $h(s_i)$ (as a function of i). In conclusion, Lemma 1 in [2], Lemma 3 and the monotonicity of h together imply the desired result

$$h(OPT) \leq h(OPT \cup A_n)$$

$$\leq h(A_n) + \frac{1}{1 - k^g} \sum_{u \in OPT \setminus A_n} h(u|A_n)$$

$$< h(A_n) + \frac{1}{1 - k^g} \sum_{u \in OPT \setminus A_n} \left(\frac{c + 1}{k(1 - k^g)^2(1 - k_f)} h(S_n) \right)$$

$$\leq h(A_n) + \frac{c + 1}{(1 - k^g)^3(1 - k_f)} h(S_n)$$

$$\leq \left(\frac{c + 1}{c(1 - k_f)(1 - k^g)} + \frac{c + 1}{(1 - k^g)^3(1 - k_f)} \right) h(S_n)$$

$$= \frac{(c + 1)\left((1 - k^g)^2 + c\right)}{c(1 - k_f)(1 - k^g)^3} h(S_n).$$

$\square$

4   BP Maximization Under a Partition Matroid

We present a proportional selecting algorithm in Sect. 4.1 and analyze its competitive ratio in Sect. 4.2.

4.1   Algorithm

The main idea of the proportional selecting algorithm below is to consider the partition subsets of the matroid in an arbitrary order and select one element from each partition subset. The algorithm randomly selects one of the two elements in proportion to their marginal contribution, different from the natural greedy algorithm.

Algorithm 2: Proportional selecting

Input:
    A monotone submodular function $f : 2^V \to \mathbb{R}_+$
    A monotone supermodular function $g : 2^V \to \mathbb{R}_+$
    A binary partition matroid $\mathcal{M}$
Output:
    A set $S \subseteq V$ approximating the maximum of $h = f + g$ under $\mathcal{M}$
1: $S_0 \leftarrow \emptyset$
2: **for** $t \leftarrow 1$ to m **do**
3:     $w_t \leftarrow h(a_t|S_{t-1}) + h(b_t|S_{t-1})$
4:     $P_{a_t} = \frac{h(a_t|S_{t-1})}{w_t}$
5:     $P_{b_t} = \frac{h(b_t|S_{t-1})}{w_t}$
6:     Pick s_t from $\{a_t, b_t\}$ with respective probabilities (P_{a_t}, P_{b_t})
7:     $S_t \leftarrow S_{t-1} \cup \{s_t\}$
8: **end for**
9: $S \leftarrow S_m$

For convenience, we define the following notations. Let $h(a|S) = h(S + a) - h(S)$ be the marginal contribution of an element a to a set S. Let $A \subseteq V$ be an independent set of a matroid $\mathcal{M}$. Let $O_A \subseteq V$ be an optimal solution to the problem of maximizing h under $\mathcal{M}$ that satisfies $A \subseteq O_A$. Namely, $O_A = \arg\max_{T \in \mathcal{I}, A \subseteq T} h(T)$. Also, we let $OPT_A = h(O_A)$ be the value that h assigns the set O_A. Note that the value of the optimal solution that maximizes h under $\mathcal{M}$ is $OPT = OPT_\emptyset$.

4.2   Analysis

In this section, we analyze the competitive ratio of Algorithm 2.

Lemma 4. *Let $A \subseteq V$ be an independent set of a matroid $\mathcal{M}$ such that $A \cap P_t = \emptyset$. Moreover, let x_t be the element of P_t that belongs to O_A and y_t the element of P_t that does not appear in O_A. Then,*

$$OPT_A - OPT_{A+y_t} \le \frac{1}{1 - k^g} h(x_t|A).$$

Proof. We can easily obain that

$$
\begin{aligned}
OPT_A = h(O_A) &\le h(O_A + y_t) \\
&= h(O_A + y_t) - h(O_A + y_t - x_t) + h(O_A + y_t - x_t) \\
&= h(x_t|O_A + y_t - x_t) + h(O_A + y_t - x_t),
\end{aligned}
$$

where the inequality is due to the monotonicity of h. Note that $A \subseteq O_A + y_t - x_t$, and according to Lemma 1 in [2], we know that $h(x_t|O_A+y_t-x_t) \le \frac{1}{1-k^g} h(x_t|A)$. Furthermore,

$$h(O_A + y_t - x_t) \le h(O_{A+y_t}) = OPT_{A+y_t},$$

where the inequality follows as O_{A+y_t} is optimal with respect to $A + y_t \subseteq O_A - x_t + y_t$. Therefore, we conclude that $OPT_A \le \frac{1}{1-k^g} h(x_t|A) + OPT_{A+y_t}$.

We denote $L_t = OPT_{S_{t-1}} - OPT_{S_t}$ as the loss of the algorithm at Step t of the main loop. The following observation relates the sum of these losses over all steps in the main loop to the difference between the optimal value and the solution value.

Lemma 5. $\sum_{t=1}^{m} L_t = OPT - h(S)$.

Proof. Note that

$$
\begin{aligned}
\sum_{t=1}^{m} L_t &= \sum_{t=1}^{m} (OPT_{S_{t-1}} - OPT_{S_t}) = OPT_{S_0} - OPT_{S_m} \\
&= OPT_\emptyset - h(S_m) = OPT - h(S).
\end{aligned}
$$

$\square$

We are ready to prove the main result of this section.

Theorem 2.
$$\mathbb{E}[h(S)] \geq \frac{2 - 2k^g}{3 - 2k^g} OPT.$$

Proof. Consider Step t of the algorithm, where one element of the partition subset P_t is selected. Note that only one element $x_t \in P_t \cap O_{S_{t-1}}$, while the other element $y_t \in P_t$ but $y_t \notin O_{S_{t-1}}$.

We first bound the expected loss of the algorithm at Step t, given the set of elements selected before Step $t-1$. Recall that s_t is the element selected at Step t of the algorithm.

$$
\begin{aligned}
\mathbb{E}[L_t|S_{t-1}] &= P(s_t = x_t|S_{t-1})(OPT_{S_{t-1}} - OPT_{S_{t-1}+x_t}) \\
&\quad + P(s_t = y_t|S_{t-1})(OPT_{S_{t-1}} - OPT_{S_{t-1}+y_t}) \\
&= P(s_t = x_t|S_{t-1}) \cdot 0 + P(s_t = y_t|S_{t-1})(OPT_{S_{t-1}} - OPT_{S_{t-1}+y_t}) \\
&\leq P(s_t = y_t|S_{t-1}) \cdot \frac{1}{1 - k^g} h(x_t|S_{t-1}) \\
&= \frac{1}{1 - k^g} \cdot \frac{h(y_t|S_{t-1})h(x_t|S_{t-1})}{h(y_t|S_{t-1}) + h(x_t|S_{t-1})},
\end{aligned}
$$

where the inequality is due to Lemma 4, and the last equality is attained because the algorithm selects y_t with probability $h(y_t|S_{t-1})/(h(y_t|S_{t-1}) + h(x_t|S_{t-1}))$.

We now calculate the expected gain of the algorithm at Step t, given the set of elements selected up to Step $t - 1$.

$$
\begin{aligned}
\mathbb{E}[h(s_t|S_{t-1})|S_{t-1}] &= P(s_t = x_t|S_{t-1}) \cdot h(x_t|S_{t-1}) + P(s_t = y_t|S_{t-1}) \cdot h(y_t|S_{t-1}) \\
&= \frac{h(x_t|S_{t-1})}{h(y_t|S_{t-1}) + h(x_t|S_{t-1})} \cdot h(x_t|S_{t-1}) \\
&\quad + \frac{h(y_t|S_{t-1})}{h(y_t|S_{t-1}) + h(x_t|S_{t-1})} \cdot h(y_t|S_{t-1}) \\
&= \frac{h(y_t|S_{t-1})^2 + h(x_t|S_{t-1})^2}{h(y_t|S_{t-1}) + h(x_t|S_{t-1})}.
\end{aligned}
$$

Together the expected gain loss rate is

$$
\frac{\mathbb{E}[L_t|S_{t-1}]}{\mathbb{E}[h(s_t|S_{t-1})|S_{t-1}]} \leq \frac{1}{1 - k^g} \cdot \frac{h(y_t|S_{t-1})h(x_t|S_{t-1})}{h(y_t|S_{t-1})^2 + h(x_t|S_{t-1})^2} \leq \frac{1}{2(1 - k^g)},
$$

where the last inequality holds since $2ab \leq a^2 + b^2$, for any $a, b \in \mathbb{R}$. We can now bound the expected loss of the algorithm at Step t as follows

$$
\begin{aligned}
\mathbb{E}[L_t] &= \sum_{S_{t-1} \subseteq V} \mathbb{E}[L_t|S_{t-1}] \cdot P\left(\begin{array}{c}\text{the algorithm selects} \\ S_{t-1} \text{ up to Step } t - 1\end{array}\right) \\
&\leq \frac{1}{2(1 - k^g)} \sum_{S_{t-1} \subseteq V} \mathbb{E}[h(s_t|S_{t-1})|S_{t-1}] \cdot P\left(\begin{array}{c}\text{the algorithm selects} \\ S_{t-1} \text{ up to Step } t - 1\end{array}\right) \\
&= \frac{1}{2(1 - k^g)} \mathbb{E}[h(s_t|S_{t-1})].
\end{aligned}
$$

Consequently, we have

$$OPT - \mathbb{E}[h(S)] = \mathbb{E}\left[\sum_{t=1}^{m} L_t\right] = \sum_{t=1}^{m} \mathbb{E}[L_t]$$

$$\leq \frac{1}{2(1-k^g)} \sum_{t=1}^{m} \mathbb{E}[h(s_t|S_{t-1})]$$

$$= \frac{1}{2(1-k^g)} \mathbb{E}\left[\sum_{t=1}^{m} h(s_t|S_{t-1})\right]$$

$$= \frac{1}{2(1-k^g)} \mathbb{E}[h(S)].$$

Thus, $\mathbb{E}[h(S)] \geq \frac{2-2k^g}{3-2k^g} OPT$.

$\square$

4.3   Algorithm for BP Maximization Under a $(1, \ell)$ Partition Matroid

In this section, we generalize the binary matroid in problem 2 to the $(1, \ell)$ partition matroid. The following algorithm modifies Algorithm 2.

Algorithm 3: Proportional selecting

Input:
    A monotone submodular function $f : 2^V \to \mathbb{R}_+$
    A monotone supermodular function $g : 2^V \to \mathbb{R}_+$
    A binary partition matroid $\mathcal{M}$
Output:
    A set $S \subseteq V$ approximating the maximum of $h = f + g$ under $\mathcal{M}$
1: $S_0 \leftarrow \emptyset$
2: **for** $t \leftarrow 1$ to m **do**
3:     $w_t \leftarrow \sum_{i=1}^{\ell} (h(a_t^i|S_{t-1}))^{\ell-1}$
4:     $P_{a_t^i} \leftarrow \frac{(h(a_t^i|S_{t-1}))^{\ell-1}}{w_t}$ $(1 \leq i \leq \ell)$
5:     Pick s_t from $\{a_t^1, a_t^2, \ldots, a_t^\ell\}$ with respective probabilities
        $(P_{a_t^1}, P_{a_t^2}, \ldots, P_{a_t^\ell})$
6:     $S_t \leftarrow S_{t-1} \cup \{s_t\}$
7: **end for**
8: $S \leftarrow S_m$

Theorem 3. *Algorithm 3 returns a set S with*

$$\mathbb{E}[h(S)] \geq \frac{\ell - \ell k^g}{2\ell - 1 - \ell k^g} OPT.$$

The proof of this result is deferred to the full version of the paper.

5   Discussions

In this paper, we introduce two online BP maximization problems. The first is an online BP maximization problem subject to a uniform matroid constraint when the items arrive one-by-one. We present a greedy algorithm with a constant competitive ratio. The other consider an online BP maximization problem subject to a partition matroid constraint where items arrives online in a random order. We present a proportional selecting algorithm with a constant competitive ratio. As one of the future research directions, it is interesting to further improve the competitive ratios for the two problems. While extensive work has been devoted towards submodular maximization problems in the literature, only a few results exist for online BP maximization problems and we believe more research is needed for this fertile filed.

Acknowledgements. This research is supported or partially supported by the National Natural Science Foundation of China (Grant Nos. 11871280, 11971349, 11371001, 11771386 and 11728104), the Natural Sciences and Engineering Research Council of Canada (NSERC) Grant 06446 and Qinglan Project.

References

1. Azar, Y., Gamzu, I., Roth, R.: Submodular max-SAT. In: Demetrescu, C., Halldórsson, M.M. (eds.) ESA 2011. LNCS, vol. 6942, pp. 323–334. Springer, Heidelberg (2011). https://doi.org/10.1007/978-3-642-23719-5_28
2. Bai, W., Bilmes, J.A.: Greed is still good: maximizing monotone submodular+supermodular (BP) functions. In: Proceedings of ICML, pp. 304–313 (2018)
3. Buchbinder, N., Feldman, M., Schwartz, R.: Online submodular maximization with preemption. In: Proceedings of SODA, pp. 1202–1216 (2015)
4. Conforti, M., Cornuejols, G.: Submodular set functions, matroids and the greedy algorithm: tight worst-case bounds and some generalizations of the Rado-Edmonds theorem. Discrete Appl. Math. **7**(3), 251–274 (1984)
5. Dughmi, S., Roughgarden, T., Sundararajan, M.: Revenue submodularity. In: Proceedings of EC, pp. 243–252 (2009)
6. Huang, N., Borodin, A.: Bounds on double-sided myopic algorithms for unconstrained non-monotone submodular maximization. arXiv:1312.2173v2 (2014)
7. Krause, A., Guestrin, C.: Near-optimal nonmyopic value of information in graphical models. In: Proceedings of UAI, pp. 324–331 (2005)
8. Krause, A., Guestrin, C.: Near-optimal observation selection using submodular functions. In: Proceedings of AAAI, pp. 1650–1654 (2007)
9. Krause, A., Guestrin, C., Gupta, A., Kleinberg, J.: Near-optimal sensor placements: maximizing information while minimizing communication cost. In: Proceedings of IPSN, pp. 2–10 (2006)
10. Krause, A., Leskovec, J., Guestrin, C., VanBriesen, J., Faloutsos, C.: Efficient sensor placement optimization for securing large water distribution networks. J. Water Res. Plan. Man. **134**(6), 516–526 (2008)
11. Krause, A., Singh, A., Guestrin, C.: Near-optimal sensor placements in gaussian processes: theory, efficient algorithms and empirical studies. J. Mach. Learn. Res. **9**, 235–284 (2008)

12. Lin, H., Bilmes, J.: Multi-document summarization via budgeted maximization of submodular functions. In: Proceedings of NAACL, pp. 912–920 (2010)
13. Lin, H., Bilmes, J.: A class of submodular functions for document summarization. In: Proceedings of HLT, pp. 510–520 (2011)
14. Lin, H., Bilmes, J.: Word alignment via submodular maximization over matroids. In: Proceedings of ACL-HLT, pp. 170–175 (2011)
15. Schulz, A.S., Uhan, N.A.: Approximating the least core value and least core of cooperative games with supermodular costs. Discrete Optim. **10**(2), 163–180 (2013)
16. Wei, K., Iyer, R., Bilmes, J.: Submodularity in data subset selection and active learning. In: Proceedings of ICML, pp. 1954–1963 (2015)

Adaptive Robust Submodular Optimization and Beyond

Shaojie Tang[1][(✉)] and Jing Yuan[2]

[1] Naveen Jindal School of Management, University of Texas at Dallas,
Richardson, USA
shaojie.tang@utdallas.edu
[2] Department of Computer Science,
University of Texas at Dallas, Richardson, USA
csyuanjing@gmail.com

Abstract. Constrained submodular maximization has been extensively studied in the recent years. In this paper, we study adaptive robust optimization with nearly submodular structure (ARONSS). Our objective is to randomly select a subset of items that maximizes the worst case value of several reward functions simultaneously. Our work differs from existing studies in two ways: (1) we study the robust optimization problem under the adaptive setting, i.e., one needs to adaptively select items based on the feedback collected from picked items, and (2) our results apply to a broad range of reward functions characterized by ϵ-nearly submodular function. We first analyze the adaptivity gap of ARONSS and show that the gap between the best adaptive solution and the best non-adaptive solution is bounded. Then we propose an approximate solution to this problem when all reward functions are submodular. In particular, our algorithm achieves approximation ratio $(1 - 1/e)$ when considering a single matroid constraint. At last, we present two heuristics for the general case with nearly submodular functions. All proposed solutions are non-adaptive which are easy to implement.

1 Introduction

Constrained submodular maximization has attracted growing attention recently [4–6]. Most existing work on submodular maximization focuses on selecting a subset of items subject to given constraints so as to maximize a submodular objective function [13]. In this paper, we study adaptive robust optimization with nearly submodular structure (ARONSS). This study belongs to the category of robust submodular maximization. Our objective is to randomly select a subset of items that performs well over several reward functions. Although robust submodular maximization has been well studied [1,9,12,15,19], most of existing studies assume a non-adaptive setting, i.e., one has to select a subset of items all at once in advance, and submodular reward function. However, in many applications from artificial intelligence [10,16,18,20,21], the outcome of

© Springer Nature Switzerland AG 2020
Z. Zhang et al. (Eds.): AAIM 2020, LNCS 12290, pp. 185–194, 2020.
https://doi.org/10.1007/978-3-030-57602-8_17

an objective function is often uncertain, one needs to make a sequence of decisions adaptively based on the outcomes of the previous decisions. Moreover, the reward function is not necessarily submodular. This motivates us to study the adaptive robust optimization problem with general reward functions.

The main contribution of this paper is three-fold:

- We extend the previous studies on robust submodular maximization in two directions: (1) we consider the robust optimization problem under the adaptive setting, i.e., one can select one item at a time and observe the outcome of picked items, before selecting the next item, and (2) our results apply to a broad range of reward functions characterized by ϵ-nearly submodular function.
- We first analyze the adaptivity gap of ARONSS and show that the gap between the best adaptive solution and the best non-adaptive solution is bounded. This enables us to focus on designing non-adaptive solutions which are much easier to work with.
- Then we propose an approximate solution to this problem with submodular reward functions subject to many practical constraints. In particular, our algorithm achieves a $1 - 1/e$ approximation ratio when considering a single matroid constraint. We also present two algorithms that achieve bounded approximation ratios for the general case. All algorithms are non-adaptive and easy to implement.

2   Preliminaries and Problem Formulation

We first introduce some notations, then formulate our problem.

2.1   Submodular Function

A set function $h(S)$ that maps subsets of a finite ground set Ω to non-negative real numbers is said to be submodular if for every $S_1, S_2 \subseteq \Omega$ with $S_1 \subseteq S_2$ and every $v \in \Omega \backslash S_2$, we have that

$$h(S_1 \cup \{v\}) - h(S_1) \geq h(S_2 \cup \{v\}) - h(S_2)$$

A submodular function h is said to be monotone if $h(S_1) \leq h(S_2)$ whenever $S_1 \subseteq S_2$.

2.2   Items and States

Let $E = \{e_1, e_2, \ldots, e_n\}$ denote a finite set of n items, and each item is in a particular state from a set $O = \{o_1, o_2, \ldots, o_m\}$ of m possible states. Each item $e_i \in E$ is associated with a random variable $Y_i \in O$ that represents a random realization of e_i's state. We use $\mathbf{Y} = \{Y_i \mid i \in [n]\}$ to denote the collection of all variables. We assume there is a known prior probability distribution $p(i) = \{\Pr[Y_i = o_j] \mid j \in [m]\}$ over realizations O for each item $e_i \in E$. For notation simplicity, let $p_{ij} = \Pr[Y_i = o_j]$ for all $i \in [n], j \in [m]$. We further assume that the states of all items are decided independently from each other [2], i.e., $\mathbf{Y}$ is drawn randomly from the product distribution $\prod_{i \in [n]} p(i)$. Let $\mathbf{y} : E \to O$ denote a realization of item states. After picking an item e_i, we are able to observe its state $y_i \in O$.

2.3 ϵ-nearly Submodular Reward Functions

We are given a family of L reward functions $\mathcal{F} = \{f_1, f_2, \ldots, f_L\}$, where each $f_l \in \mathcal{F} : 2^{E \times O} \to \mathbb{R}_{\geq 0}$ maps a set of items and their states $X \subseteq E \times O$ to some reward $\mathbb{R}_{\geq 0}$. In this work, we assume each function f_l is monotone, i.e., $f_l(A) \leq f_l(B)$ for all $A \subseteq B$, and ϵ-*nearly submodular*, i.e., for any $f_l \in \mathcal{F}$, there is a submodular function g_l such that for any $X \subseteq E \times O$, we have $\epsilon g_l(X) \leq f_l(X) \leq \frac{1}{\epsilon} g_l(X)$ where $\epsilon \in (0, 1]$. It is clear that any submodular function is 1-nearly submodular.

2.4 Adaptive Policies

We model the adaptive strategy of picking items through a policy π [10]. Formally, a policy π is a function that specifies which item to pick next under the observations made so far: $\pi : 2^{V \times O} \to E$. Note that π can be regarded as some decision tree that specifies a rule for picking items adaptively. Assume that when the items are in state $\mathbf{Y} = \mathbf{y}$, the policy π picks a set of items (and corresponding states), which is denoted by $S(\pi, \mathbf{y}) \subseteq E \times O$, then the expected reward received π from function f_l is $\mathcal{U}(\pi, f_l) := \mathbb{E}_{\mathbf{y}}[f_l(S(\pi, \mathbf{y}))]$ where the expectation is taken over $\mathbf{y}$ with respect to $\prod_{i \in [n]} p(i)$. In the context of robust optimization, our goal is to pick a set of items (and corresponding states) that achieves high reward in the worst case over reward functions in $\mathcal{F}$. Thus, we define the utility $\mathcal{U}(\pi, \mathcal{F})$ of π as

$$\mathcal{U}(\pi, \mathcal{F}) = \min_{l \in [L]} \mathcal{U}(\pi, f_l)$$

Let $\mathcal{I}$ be a *downward-closed* family of subsets of E, i.e., a family of subsets $\mathcal{I}$ is downward-closed if for any $U \in \mathcal{I}$ and any $W \subseteq U$, we have $W \in \mathcal{I}$. We use $E(\pi, \mathbf{y})$ to refer to the subset of items picked by policy π given state $\mathbf{y}$. We say a policy π is *feasible* if for any $\mathbf{y}$, $E(\pi, \mathbf{y}) \in \mathcal{I}$. This downward-closed family generalizes many useful systems which give rise to natural constraints such as matroid and knapsack constraints. Our goal is to identify the best feasible policy that maximizes its expected utility.

$$\max_{\pi} \mathcal{U}(\pi, \mathcal{F}) \text{ subject to } E(\pi, \mathbf{y}) \in \mathcal{I} \text{ for any } \mathbf{y}.$$

3   Analysis on Adaptivity Gap

We say a policy is non-adaptive if it always picks the next item independent of the states of the picked items. Clearly adaptive polices obtain at least as much utility as non-adaptive policies. Perhaps surprisingly, building on recent advances in stochastic submodular probing [3], we show that this adaptivity gap is upper bounded by a constant (given that ϵ is a constant). Based on this result, we can focus on designing non-adaptive polices which are much easier to work with.

Theorem 1. *Given any adaptive policy π, there exists a non-adaptive algorithm σ_π such that $\mathcal{U}(\sigma_\pi, \mathcal{F}) \geq \frac{\epsilon^2}{2}\mathcal{U}(\pi, \mathcal{F})$.*

Proof: Given any adaptive policy π, we follow the idea in [11] and define a non-adaptive policy σ_π: randomly draw a state vector $\mathbf{y}$ from the product distribution $\prod_{i \in [n]} p(i)$ (this step is done virtually), pick $E(\pi, \mathbf{y}) \subseteq E$, i.e., pick all items picked by π given $\mathbf{y}$. Let $\mathbf{y}'_E$ be the state of all items drawn virtually by σ_π and $\mathbf{y}$ be the true state of all items when picked by σ_π.

Now consider any $l \in [L]$, the expected value of f_l obtained by σ_π is

$$\mathcal{U}(\sigma_\pi, f_l) = \mathbb{E}_{\mathbf{y}'_E}\left[\mathbb{E}_{\mathbf{y}}[f_l(\bigcup_{e_i \in E(\pi, \mathbf{y}'_E)} (e_i, y_i))]\right] \tag{1}$$

Because f_l is ϵ-nearly submodular, for some submodular function g_l, we have

$$\mathbb{E}_{\mathbf{y}'_E}\left[\mathbb{E}_{\mathbf{y}}[f_l(\bigcup_{e_i \in E(\pi, \mathbf{y}'_E)} (e_i, y_i))]\right] \geq \mathbb{E}_{\mathbf{y}'_E}\left[\mathbb{E}_{\mathbf{y}}[\epsilon g_l(\bigcup_{e_i \in E(\pi, \mathbf{y}'_E)} (e_i, y_i))]\right] = \epsilon\mathcal{U}(\sigma_\pi, g_l) \tag{2}$$

(1) and (2) together imply that

$$\mathcal{U}(\sigma_\pi, f_l) \geq \epsilon\mathcal{U}(\sigma_\pi, g_l) \tag{3}$$

We next analyze the utility of π. The expected value of f_l obtained by π is

$$\mathcal{U}(\pi, f_l) = \mathbb{E}_{\mathbf{y}}[f_l(S(\pi, \mathbf{y}))] \tag{4}$$

Because f_l is ϵ-nearly submodular, we have

$$\mathbb{E}_{\mathbf{y}}[f_l(S(\pi, \mathbf{y}))] \leq \mathbb{E}_{\mathbf{y}}[\frac{1}{\epsilon}g_l(S(\pi, \mathbf{y}))] = \frac{1}{\epsilon}\mathcal{U}(\pi, g_l) \tag{5}$$

(4) and (5) together imply that

$$\mathcal{U}(\pi, f_l) \leq \frac{1}{\epsilon}\mathcal{U}(\pi, g_l) \tag{6}$$

Because g_l is submodular, the ratio between $\mathcal{U}(\pi, g_l)$ and $\mathcal{U}(\sigma_\pi, g_l)$ is upper bounded by 2 [3], i.e., $\mathcal{U}(\pi, g_l) \leq 2\mathcal{U}(\sigma_\pi, g_l)$. This together with (4) and (6) implies that

$$\mathcal{U}(\sigma_\pi, f_l) \geq \frac{\epsilon^2}{2}\mathcal{U}(\pi, f_l) \tag{7}$$

It follows that

$$\mathcal{U}(\sigma_\pi, \mathcal{F}) = \min_{l \in [L]} \mathcal{U}(\sigma_\pi, f_l) \tag{8}$$

$$\geq \min_{l \in [L]} \frac{\epsilon^2}{2} \mathcal{U}(\pi, f_l) \tag{9}$$

$$= \frac{\epsilon^2}{2} \mathcal{U}(\pi, \mathcal{F}) \tag{10}$$

□

It was worth noting that Theorem 1 holds when $\mathcal{I}$ is a *prefix-closed* family of constraints, where a family of subsets $\mathcal{I}$ is prefix-closed if for any subsequence in $\mathcal{I}$, its prefix also belongs to $\mathcal{I}$.

4 Robust Continuous Greedy for Submodular Reward Function

We first focus on the case when $\epsilon = 1$, i.e., all reward functions are submodular. We propose a constant approximate non-adaptive policy to this special case subject to many practical constraints. In the rest of this paper, we use σ to denote a non-adaptive policy.

Before introducing our algorithm, we first introduce some important notations. For a independence system $\mathcal{I}$, the polytope of $\mathcal{I}$ is defined as $P(\mathcal{I}) = \text{conv}\{\mathbf{1}_I : I \in \mathcal{I}\}$ where $\mathbf{1}_I \in [0, 1]^n$ denotes the vector with entries I one and all other entries zero. Abusing notation, let $f(X)$ denote the expected utility of X for any $X \subseteq E$ over $\prod_{i \in [n]} p(i)$: $f(X) = \mathbb{E}_{\mathbf{y}}[f(\cup_{e_i \in X}(e_i, y_i))]$. Given a vector $\mathbf{x} \in [0, 1]^n$, the *multilinear extension* F of f is defined as

$$F(\mathbf{x}) = \sum_{X \subseteq E} f(X) \prod_{i \in X} x_i \prod_{i \notin X} (1 - x_i)$$

We further introduce a matrix $\mathbf{z} \in [0, 1]^{n \times m}$. Given a matrix $\mathbf{z}$, the *expanded multilinear extension* F' of f is defined as

$$F'(\mathbf{z}) = \sum_{Z \subseteq E \times O} f(Z) \prod_{(e_i, y_j) \in Z} z_{ij} \prod_{(e_i, y_j) \notin Z} (1 - z_{ij})$$

Define the marginal of (e_i, y_j) for F' as $F'((e_i, y_j) \mid \mathbf{z}) = F'(\mathbf{z} \vee \mathbf{1}_{(e_i, y_j)}) - F'(\mathbf{z})$ where $\mathbf{1}_{(e_i, y_j)}$ denotes a matrix with entry (i, j) one and all other entries zero, and $\mathbf{z} \vee \mathbf{1}_{(e_i, y_j)}$ denotes the component wise maximum.

We next propose a *Robust Continuous Greedy Policy*, denoted by σ^{rg}, that achieves a constant approximation ratio for ARONSS (with $\epsilon = 1$) subject many practical constraints such as matroid and knapsack constraints. Our design is inspired by [7], and we generalize their idea to the adaptive setting. In particular, we first compute a fractional solution $\mathbf{x}^{1/\delta} \in P(\mathcal{I})$ from Algorithm 1. Then we round it to an integer solution using some existing techniques. Note that Algorithm 1 can be viewed as a variant of the classic continuous greedy algorithm [6], and we adapt it to the robust and adaptive setting.

Algorithm 1. Robust Continuous Greedy

1: Set $\delta = \frac{1}{9(nm)^2}, t = 0; \mathbf{z}^0 = [0]^{n \times m}, \mathbf{x}^0 = [0]^n$.
2: **while** $t < 1$ **do**
3:    For each $(e_i, y_j) \in E \times O$, estimate $F'((e_i, y_j) \mid \mathbf{z}^t)$
4:    Find a feasible solution $\mathbf{x}^*$ to **P1** or terminates the algorithm and outputs a
       certificate that no feasible solution is found

5:    | **P1:** *Maximize* 0
      | **subject to:** $\mathbf{x} \in P_{\mathcal{I}}; \forall l \in [L] : \sum_{(e_i, y_j) \in E \times O} F'_l((e_i, y_j) \mid \mathbf{z}^t) p_{ij} x_i \geq \gamma;$

6:    For all $i \in [n]$, $j \in [m]$: $z_{ij}^{t+\delta} = z_{ij}^t + \delta p_{ij} x_i^*$; $x_i^{t+\delta} = x_i^t + \delta x_i^*$;
7:    Increment $t = t + \delta$;
8: **return** $\mathbf{x}^{1/\delta}$;

Lemma 1. *Given L submodular functions $f_1, f_2, \ldots, f_l$, a value γ, and an independence system $\mathcal{I}$, Algorithm 1 finds a point $\mathbf{x}^{1/\delta} \in P(\mathcal{I})$ such that $F_l(\mathbf{x}^{1/\delta}) \geq (1 - 1/e)\gamma, \forall l \in [L]$ or outputs a certificate that there is no such an adaptive policy π with $\mathcal{U}(\pi, \mathcal{F}) \geq \gamma$.*

Proof: First, if our algorithm can not find a feasible solution to **P1**, it stops and outputs a certificate that there is no adaptive policy with $\mathcal{U}(\sigma, \mathcal{F}) \geq \gamma$.

Fix a matrix $\mathbf{z}$. If there exists an adaptive policy, say π^{opt}, such that $\mathcal{U}(\pi^{opt}, f_l) \geq \gamma, \forall l \in [L]$, we have

$$\gamma \leq \mathcal{U}(\pi^{opt}, f_l) = \sum_{Z \in E \times O} \beta_Z^{opt} f_l(Z) \tag{11}$$

$$\leq \sum_{Z \in E \times O} \beta_Z^{opt} \left(F'(\mathbf{z}) + \sum_{(e_i, o_j) \in Z} F'((e_i, o_j) \mid \mathbf{z}) \right)$$

$$= F'(\mathbf{z}) + \sum_{(e_i, o_j) \in E \times O} \left(\sum_{(e_i, o_j) \in Z} \beta_Z^{opt} \right) F'((e_i, o_j) \mid \mathbf{z}) \tag{12}$$

$$= F'(\mathbf{z}) + \sum_{(e_i, o_j) \in E \times O} x_i^{opt} p_{ij} F'((e_i, o_j) \mid \mathbf{z}) \tag{13}$$

where β_Z^{opt} denotes the probability that Z is being observed and selected by running π^{opt}, and x_i^{opt} is the probability that e_i is selected by π^{opt}. Because $\mathbf{x}^{opt} \in P(\mathcal{I})$, (13) implies that for any fractional solution $\mathbf{z}$, there exits a direction $\mathbf{z}^* = \{z_{ij}^* = x_i p_{ij} \mid \mathbf{x} \in P(\mathcal{I})\}$ such that $\mathbf{z}^* \cdot \nabla F'_l(\mathbf{z}) \geq \gamma - F'_l(\mathbf{z}), \forall l \in [L]$. This direction can be found using linear program **P1**. Based on the same analysis of the classic continuous greedy algorithm [6] and the above discussion, we have

$$F'_l(\mathbf{z}^{1/\delta}) \geq (1 - 1/e)\gamma, \forall l \in [L] \tag{14}$$

through induction.

We next show that $F_l(\mathbf{x}^{1/\delta}) \geq F'_l(\mathbf{z}^{1/\delta}), \forall l \in [L]$. Note that $F'_l(\mathbf{z}^{1/\delta})$ is the expected utility of f_l when each realization $(e_i, o_j) \in E \times O$ is selected *independently* with probability $x_i^{1/\delta} p_{ij}$. On the other hand, $F_l(\mathbf{x}^{1/\delta})$ is the expected utility of f_l with respect to the following selection process: for each $i \in [n]$, we randomly select *one* realization from $\{(e_i, o_j) \cup \{\emptyset\} \mid j \in [m]\}$ such that $(e_i, o_j) \in E \times O$ is being selected with probability $x_i^{1/\delta} p_{ij}$ and $\emptyset$ is being selected with probability $1 - \sum_{j \in [m]} x_i^{1/\delta} p_{ij}$. Note that the above constraint that only one realization can selected from $\{(e_i, o_j) \cup \{\emptyset\} \mid j \in [m]\}$ can be viewed as a (basic) partition matroid constraint. Based on the above discussion and Lemma 3.7 in [6], we have

$$F_l(\mathbf{x}^{1/\delta}) \geq F_l'(\mathbf{z}^{1/\delta}), \forall l \in [L] \tag{15}$$

(14) and (15) imply that $F_l(\mathbf{x}^{1/\delta}) \geq (1 - 1/e)\gamma, \forall l \in [L]$. $\qquad \square$

Based on Lemma 1, we can perform a binary search on γ to find a $(1 - 1/e)$-approximate fractional solution. At last, depending on the type of $\mathcal{I}$, we use an appropriate technique to round the fractional solution to an integral solution. In particular, for a single matroid constraint, using swap rounding [7] achieves $\zeta = 1$. Many other useful constraints such as knapsack and the intersection of knapsack and matroid constraints admit good rounding techniques [8].

Theorem 2. σ^{rg} *returns a solution that achieves approximation ratio* $(1-1/e)\zeta$, *where* $\zeta \in [0, 1]$ *is the performance loss due to rounding.*

Remark: The runtime of Algorithm 1 is dependent on m, the number of possible states. When m is a constant or a polynomial of n, Algorithm 1is a polynomial-time algorithm. When m is exponential or even infinity, Theorem 2 implies that there exists a non-adaptive policy whose approximation ratio is bounded by $(1 - 1/e)\zeta$. This can be viewed as an enhanced bound on the adaptivity gap, as compared with the adaptivity gap (whose value is $1/2$ when $\epsilon = 1$) derived for the general prefix-closed family of constraints (Theorem 1).

5 Two Heuristics for Nearly Submodular Reward Functions

In this section, we introduce two non-adaptive policies for the general case. In the rest of this paper, assume π^{opt} is the optimal adaptive policy.

5.1 A $1/L$-approximate Solution

Algorithm 2. $1/L$-approximate Non-Adaptive Policy $\sigma^{1/L}$

1: Set $l = 1$.
2: **while** $l \leq L$ **do**
3:    $E_l \leftarrow \text{APPROX}(\max_{S \in \mathcal{I}} f_l(S))$
4:    $l \leftarrow l + 1$
5: Randomly pick an index $l \in [L]$
6: **return** E_l

The basic idea of our first $1/L$-*approximate Non-Adaptive Policy* (Algorithm 2), denoted by $\sigma^{1/L}$, is very simple: First solving $\max_\sigma \mathcal{U}(\sigma, f_l)$ for each $l \in [L]$, then randomly pick one among L outputs as solution. One can verify that solving $\max_\sigma \mathcal{U}(\sigma, f_l)$ is equivalent to solving $\max_{S \in \mathcal{I}} f_l(S)$.

To carry out these steps, $\sigma^{1/L}$ requires one oracle $\mathrm{APPROX}(\max_{S \in \mathcal{I}} f_l(S))$ which returns an approximate solution to $\max_{S \in \mathcal{I}} f_l(S)$ for each $l \in [L]$. Assume for each $l \in [L]$, the approximation ratio of $\mathrm{APPROX}(\max_{S \in \mathcal{I}} f_l(S))$ is α_l and the time complexity of $\mathrm{APPROX}(\max_{S \in \mathcal{I}} f_l(S))$ is d_l, we have

Theorem 3. *Let $\alpha = \min_{l \in [L]} \alpha_l$, our first policy $\sigma^{1/L}$ achieves $\frac{\epsilon^2 \alpha}{2L}$ approximation ratio for ARONSS, i.e., $\mathcal{U}(\sigma^{1/L}, \mathcal{F}) \geq \frac{\epsilon^2 \alpha}{2L} \mathcal{U}(\pi^{opt}, \mathcal{F})$. The time complexity of $\sigma^{1/L}$ is $O(Ld)$ where $d = \max_{l \in [L]} d_l$.*

Proof: First,

$$\max_{S \in \mathcal{I}} f_l(S) = \max_\sigma \mathcal{U}(\sigma, f_l) \geq \max_\sigma \min_{l \in [L]} \mathcal{U}(\sigma, f_l) = \max_\sigma \mathcal{U}(\sigma, \mathcal{F}) \qquad (16)$$

Recall that $\mathrm{APPROX}(\max_{S \in \mathcal{I}} f_l(S))$ is returned as the final solution with probability $1/L$. Because $\mathrm{APPROX}(\max_{S \in \mathcal{I}} f_l(S))$ achieves approximation ratio α, we have $\mathcal{U}(\sigma^{1/L}, f_l) \geq \frac{\alpha}{L} \max_{S \in \mathcal{I}} f_l(S)$, it follows that $\mathcal{U}(\sigma^{1/L}, f_l) \geq \frac{\alpha}{L} \max_\sigma \mathcal{U}(\sigma, \mathcal{F})$. Thus,

$$\mathcal{U}(\sigma^{1/L}, \mathcal{F}) = \min_{l \in [L]} \mathcal{U}(\sigma^{1/L}, f_l) \geq \frac{\alpha}{L} \max_\sigma \mathcal{U}(\sigma, \mathcal{F})$$

due to (16). Since $\max_\sigma \mathcal{U}(\sigma, \mathcal{F}) \geq \frac{\epsilon^2}{2} \mathcal{U}(\pi^{opt}, \mathcal{F})$ due to Theorem 1, we have $\mathcal{U}(\sigma^{1/L}, \mathcal{F}) \geq \frac{\epsilon^2 \alpha}{2L} \mathcal{U}(\pi^{opt}, \mathcal{F})$. This finishes the proof of the first part of this theorem. The proof of time complexity is trivial since $\sigma^{1/L}$ calls APPROX L times. $\qquad \square$

Discussion on the Value of α. We next briefly discuss how to solve $\max_{S \in \mathcal{I}} f_l(S)$. Consider a special case when all reward functions in $\mathcal{F}$ are submodular, i.e., $\epsilon = 1$, and $\mathcal{I}$ is a family of subsets that satisfies a knapsack constraint or a matroid constraint [6], there exist algorithms that achieve $1 - 1/e$ approximation ratio, i.e., $\alpha = 1 - 1/e$. For more complicated constraints such as intersection of a fixed number of knapsack and matroid constraints, [8] provide approximate solutions via the multilinear relaxation and contention resolution schemes.

5.2   Double-Oracle Algorithm

We next present a double-oracle-based solution [14] to the non-adaptive optimization problem. Without loss of generality, assume that double oracle algorithm σ^{DO} finds a β approximate solution to the non-adaptive optimization problem, i.e., $\mathcal{U}(\sigma^{DO}, \mathcal{F}) \geq \beta \max_\sigma \mathcal{U}(\sigma, \mathcal{F})$, we have $\mathcal{U}(\sigma^{DO}, \mathcal{F}) \geq \frac{\epsilon^2 \beta}{2} \mathcal{U}(\pi^{opt}, \mathcal{F})$ due to the adaptivity gap proved in Theorem 1.

Theorem 4. *Assume* σ^{DO} *finds a* β *approximate solution to the non-adaptive robust* ϵ-*nearly submodular maximization problem, then* σ^{DO} *achieves* $\frac{\epsilon^2 \beta}{2}$ *approximation ratio for ARONSS, i.e.,* $\mathcal{U}(\sigma^{\mathrm{DO}}, \mathcal{F}) \geq \frac{\epsilon^2 \beta}{2} \mathcal{U}(\pi^{opt}, \mathcal{F})$.

As compared with $\sigma^{1/L}$, we remove $1/L$ from the above approximation ratio, however, the time complexity of σ^{DO} could be exponential.

6 Conclusion

To the best of our knowledge, we are the first to systematically study the problem of adaptive robust optimization with nearly submodular structure. We analyze the adaptivity gap of ARONSS. Then we propose an approximate solution to this problem when all reward functions are submodular. In particular, our algorithm achieves a $(1 - 1/e)$ approximation ratio when considering a single matroid constraint. At last, we develop two algorithms that achieve bounded approximation ratios for the general case. In the future, we would like to relax the assumption that all items are independent, and analyze the adaptivity gap in the presence of dependent items [17].

References

1. Anari, N., Haghtalab, N., Pokutta, S., Singh, M., Torrico, A., et al.: Structured robust submodular maximization: offline and online algorithms. arXiv preprint arXiv:1710.04740 (2017)
2. Asadpour, A., Nazerzadeh, H., Saberi, A.: Stochastic submodular maximization. In: Papadimitriou, C., Zhang, S. (eds.) WINE 2008. LNCS, vol. 5385, pp. 477–489. Springer, Heidelberg (2008). https://doi.org/10.1007/978-3-540-92185-1_53
3. Bradac, D., Singla, S., Zuzic, G.: (near) optimal adaptivity gaps for stochastic multi-value probing. arXiv preprint arXiv:1902.01461 (2019)
4. Buchbinder, N., Feldman, M.: Deterministic algorithms for submodular maximization problems. ACM Trans. Algorithms (TALG) **14**(3), 32 (2018)
5. Buchbinder, N., Feldman, M., Naor, J.S., Schwartz, R.: Submodular maximization with cardinality constraints. In: Proceedings of the Twenty-Fifth Annual ACM-SIAM Symposium on Discrete Algorithms, pp. 1433–1452. Society for Industrial and Applied Mathematics (2014)
6. Calinescu, G., Chekuri, C., Pál, M., Vondrák, J.: Maximizing a monotone submodular function subject to a matroid constraint. SIAM J. Comput. **40**(6), 1740–1766 (2011)
7. Chekuri, C., Vondrak, J., Zenklusen, R.: Dependent randomized rounding via exchange properties of combinatorial structures. In: 2010 IEEE 51st Annual Symposium on Foundations of Computer Science, pp. 575–584. IEEE (2010)
8. Chekuri, C., Vondrák, J., Zenklusen, R.: Submodular function maximization via the multilinear relaxation and contention resolution schemes. SIAM J. Comput. **43**(6), 1831–1879 (2014)
9. Chen, W., Lin, T., Tan, Z., Zhao, M., Zhou, X.: Robust influence maximization. In: Proceedings of the 22nd ACM SIGKDD International Conference on Knowledge Discovery and Data Mining, pp. 795–804. ACM (2016)

10. Golovin, D., Krause, A.: Adaptive submodularity: theory and applications in active learning and stochastic optimization. J. Artif. Intell. Res. **42**, 427–486 (2011)
11. Gupta, A., Nagarajan, V., Singla, S.: Adaptivity gaps for stochastic probing: submodular and XOS functions. In: Proceedings of the Twenty-Eighth Annual ACM-SIAM Symposium on Discrete Algorithms, pp. 1688–1702. SIAM (2017)
12. Krause, A., McMahan, H.B., Guestrin, C., Gupta, A.: Robust submodular observation selection. J. Mach. Learn. Res. **9**(Dec), 2761–2801 (2008)
13. Krause, A., Singh, A., Guestrin, C.: Near-optimal sensor placements in Gaussian processes: theory, efficient algorithms and empirical studies. J. Mach. Learn. Res. **9**(Feb), 235–284 (2008)
14. McMahan, H.B., Gordon, G.J., Blum, A.: Planning in the presence of cost functions controlled by an adversary. In: ICML, pp. 536–543 (2003)
15. Orlin, J.B., Schulz, A.S., Udwani, R.: Robust monotone submodular function maximization. Math. Program. **172**, 505–537 (2018). https://doi.org/10.1007/s10107-018-1320-2
16. Tang, S.: When social advertising meets viral marketing: sequencing social advertisements for influence maximization. In: Thirty-Second AAAI Conference on Artificial Intelligence (2018)
17. Tang, S.: Price of dependence: stochastic submodular maximization with dependent items. J. Comb. Optimiz. **39**(2), 305–314 (2019). https://doi.org/10.1007/s10878-019-00470-6
18. Tang, S., Yuan, J.: Influence maximization with partial feedback. Oper. Res. Lett. **48**(1), 24–28 (2020)
19. Udwani, R.: Multi-objective maximization of monotone submodular functions with cardinality constraint. In: Advances in Neural Information Processing Systems, pp. 9493–9504 (2018)
20. Yuan, J., Tang, S.: Adaptive discount allocation in social networks. In: Proceedings of the Eighteenth ACM International Symposium on Mobile Ad Hoc Networking and Computing. ACM (2017)
21. Yuan, J., Tang, S.: No time to observe: adaptive influence maximization with partial feedback. In: Proceedings of the 26th International Joint Conference on Artificial Intelligence, pp. 3908–3914. AAAI Press (2017)

Approximation Guarantees
for Parallelized Maximization
of Monotone Non-submodular Function
with a Cardinality Constraint

Min Cui[1], Dachuan Xu[1], Longkun Guo[2(✉)], and Dan Wu[3]

[1] Department of Operations Research and Information Engineering,
Beijing University of Technology, Beijing 100124, People's Republic of China
B201840005@emails.bjut.edu.cn,xudc@bjut.edu.cn
[2] Shandong Key Laboratory of Computer Networks,
School of Computer Science and Technology,
Shandong Computer Science Center, Qilu University of Technology
(Shandong Academy of Sciences), Jinan 250353, People's Republic of China
longkun.guo@gmail.com
[3] School of Mathematics and Statistics,
Henan University of Science and Technology, Luoyang 471023, China
lywd2964@126.com

Abstract. In this paper, we present an adaptive algorithm for maximizing a monotone nonsubmodular function with a cardinality constraint. Based on the relationship between OPT and the maximum marginal gain of the elements in the ground set, the algorithm first calculates all possible values of OPT, then computes in parallel a family of sets each of which corresponding to each value of OPT, and lastly selects the set with maximum value as the desired solution. For the first, we divide the value range of OPT into finite parts and take the lower bound of each part as a possible value of OPT. As the main ingredient for the parallel computation, we use the Bernoulli distribution to independently sample elements so as to ensure further parallelism. Provided the generic submodularity ratio γ of the monotone set function, we prove the algorithm deserves an approximation ratio $1 - e^{-\gamma^2} - \varepsilon$, consumes $O(log(n/\eta)/\varepsilon^2)$ adaptive rounds, and needs $O(nlog(k)/\varepsilon^3)$ oracle queries in expectation. Moreover, if the set function is submodular (i. e. $\gamma = 1$), our algorithm can achieve an approximation guarantee $1 - 1/e - \varepsilon$ coinciding with the state-of-art result.

Keywords: Non-submodular optimization · Cardinality constraint · Submodularity ratio · Parallel algorithm

Z. Zhang et al. (Eds.): AAIM 2020, LNCS 12290, pp. 195–203, 2020.
https://doi.org/10.1007/978-3-030-57602-8_18

1   Introduction

In this paper, we devise adaptivity algorithms for the following optimization problem:

$$\max_{S \subseteq N} \{f(S) : |S| \leq k\}, \tag{1}$$

where $f : 2^N \rightarrow \mathbb{R}^+$ is a nonnegative monotone nonsubmodular set function, $N = \{1, ..., n\}$ is a ground set, and the aim is to find a subset $S \subseteq N$ with its size bounded by k and the objective value $f(S)$ maximized.

In past decades, submodularity has received considerable research interest in theoretical computer science [1], machine learning [2], and computer networks [3], because submodular function has wide applications in these fields [4,5]. There are several algorithms based on greedy approach [6,7] and local search [8], achieving constant factor approximation guarantees for maximizing a submodular function. Among them, the most famous one is designated for maximizing a non-decreasing submodular function subject to a cardinality constraint. The algorithm is with the key idea of iteratively adding the element of maximum marginal gain and achieves a $1 - 1/e$ approximation [9]. The approximation ratio is already optimal due to Feige [10].

Because the applications of bigdata impose parallelization requirement of submodular optimization methods, adaptability has been attracting research interest in recent years where an algorithm with low adaptability can be significantly speeded up by parallel computing. However, the above greedy algorithms suffer high adaptivity, and hence can not be efficiently parallelized. In the context, Balkanski and Singer [7] began to study the adaptability of submodular maximization. They discussed the adaptability of maximizing the monotone submodular function with the cardinality constraints and eventually gave an $1/3 - \varepsilon$ approximation algorithm with $O((log(n))/\varepsilon^2)$ adaptivity rounds. Later, the research on adaptive algorithms for maximizing monotone submodular function with cardinality constraints has borne fruits. In the same year, three groups Balkanski et al. [11], Ene et al. [12], Fahrbach et al. [13] used different techniques to improve the approximation ratio of $1/3 - \varepsilon$, and obtain an adaptive algorithm with the same approximation ratio of $1 - 1/e - \varepsilon$ in $O((log(n))/\varepsilon^2)$ adaptivity rounds. Later, Ene et al. [16] study the adaptive algorithm of maximizing submodular non-monotone function obtain the $1/2 - \varepsilon$-approximation in $O((log(n))^2/\varepsilon^2)$ adaptivity rounds. Recently, Chekuri et al. [14,15] and Ene et al. [16] had studied more complicated constraints in the adaptive complexity model.

Compared to submodular functions, the nonsubmodular function is sometimes more realistic and has broader applications. Bian et al. [17] combined and generalized the ideas of curvature α and submodularity ratio ξ [9], and consequently derived the first tight constant-factor approximation guarantee of $\frac{1}{\alpha}(1 - e^{-\xi\alpha})$ for maximizing a non-submodular nondecreasing set function with a cardinality constraint. Kuhnle et al. [18] provided approximation algorithms for maximizing a nonsubmodular function on the integer lattice with cardinality constraints. These are the first algorithms with polynomial query complexity, based on combining Greedy approach and Threshold-based Greedy to improve the performance ratio to $1 - e^{-\nu\xi} - \varepsilon$ for ν not smaller than ξ. Gong et al. [19] proposed a more practical measurement γ which is called generic submodularity ratio and is used to characterizes how close a nonnegative monotone set function is to be submodular. Nong et al. [20] make a systematic analysis of greedy algorithms for maximizing a monotone and normalized set function with a generic submodularity ratio γ under cardinality constraints that the approximation ratio is $1 - e^{-\gamma} - \varepsilon$ and the oracle queries is $O(nk)$. However, to the best of our knowledge, there exist no adaptive algorithms for maximizing a monotone nonsubmodular function with cardinality constraint k.

Contribution. In the paper, we present the first parallel algorithm for maximizing a monotone nonsubmodular function with a cardinality constraint. In expectation, our algorithm achieves an approximation ratio of $1 - e^{-\gamma^2} - \varepsilon$, with $O(log(n/\eta)/\varepsilon^2)$ adaptive rounds and $O(nlog(k)/\varepsilon^3)$ oracle queries. To this end, we combine the high-level idea of Fahrbach et al. [13] and the definition Generic submodularity ratio. We first present an algorithm called Lower-Bound-Sampling. For a given threshold (according to the guessed value of OPT), the algorithm returns a subset of the ground set in which each element has an expected marginal contribution at least the generic submodularity ration times the bound in $O(log(n/\eta)/\varepsilon)$ adaptive rounds. Then employing Lower-Bound-Sampling algorithm as a subroutine, we propose the Enumeration algorithm which constructs a solution by gradually reducing the threshold. Note that the subroutine Lower-Bound-Sampling can run in parallel subject to many different initial threshold values, so consequently the adaptivity complexity is reduced.

Organization. The remainder of the paper is organized as below: Sect. 2 gives preliminaries of the paper; Sect. 3 presents the subroutine Lower-Bound-Sampling algorithm and its analysis; Sect. 4 proposes the main algorithm as well as the related lemmas; Sect. 5 lastly concludes the paper. All formal proofs in the paper are given in the journal version.

2 Preliminaries

Let N be the ground set and $|N| = n$. We assume that function $f : 2^N \to \mathbb{R}^+$ is a monotone nonsubmodular set function with $f(\emptyset) = 0$ throughout the paper. We say a set function is monotone if and only if

$$T \subseteq S \subseteq N \Rightarrow f(T) \leq f(S) \tag{2}$$

The *marginal gain* of f at $T \subset N$ respect to $S \subset N$ can be defined as $\rho_T(S) = f(S \cup T) - f(S)$. Let S be a set output as a solution to the maximization problem $\max_{S \subseteq N}\{f(S) : |S| \leq k\}$, S^* be the optimum solution, and $\mathcal{U}(X, t)$ represents the uniform distribution on all subsets of X of size t.

Next, we introduce a concept to describe how close is a set function to be submodular.

Definition 1. *(Generic submodularity ratio [19]) Given a ground set N and a nondecreasing set function $f : 2^N \to \mathbb{R}^+$, the generic submodularity ratio of f is the largest scalar γ such that for any $A \subseteq T \subseteq N$ and any $x \in 2^N \setminus T$, we have:*

$$\rho_x(A) \geqslant \gamma \cdot \rho_x(T) \tag{3}$$

Proposition 1. *(Property of generic submodularity ratio [19]) For an increasing set function $f : 2^N \to R^+$ with generic submodularity ratio γ, it holds that*

a) $\gamma \in (0,1]$;
b) The function f is submodular iff $\gamma = 1$;
c) $\sum_{\omega \in T \setminus A} \rho_\omega(A) \geq \gamma \cdot \rho_T(A)$ *holds for any pair of $A, T \subseteq N$*

The adaptability of the algorithm is essentially the minimum number of consecutive rounds in which queries for the evaluation oracle of the function. In each round, the algorithm allows polynomially-many parallel queries. In our algorithm, the *evaluation oracle* of f proceeds as below: for any query $S \subseteq N$, it returns the value $f(S)$ in $O(1)$time. For the given evaluation oracle, the adaptivity of our algorithm is the minimum number of rounds, in each of which the algorithm performs polynomially-many independent queries by calling the evaluation oracle. In our algorithm, the performance of the algorithm needs to consider both the query and the adaptive complexity. For briefness, we set $\eta^{-1} = \Omega(poly(n))$.

3   Lower-Bound-Sampling Algorithm

In this section, we develop the Lower-Bound-Sampling algorithm with respect to a given bound and show that it can effectively maximize a monotone non-submodular function subject to a cardinality constraint. For the input lower bound β, the algorithm constructs the solution A by repeatedly adding a subset until either the number of elements in A attains k or the marginal gain of the remaining elements is less than β.

In the beginning of the algorithm, the solution A is an empty set and all elements in ground set N are recorded as C. In each round, the algorithm filters set C at first and keeps the elements whose marginal benefit is no less than lower bound. We denote the filtered set by H, which is essentially the set of element candidates. Then, the algorithm seeks z^*, the maximum number of the set elements which guarantees the marginal gain expectation of elements in the set is no less than $\gamma(1 - \varepsilon)\beta$ in expectation. Then the algorithm samples T from H uniformly with size z^*, adds T into A, and updates the set C with $C \setminus T$. The procedure repeats until the algorithm stops when $|A| = k$ or $H = \emptyset$.

Next, we define the distribution $\mathcal{D}_z$ that is used to sample and to estimate the maximum set size z^*. Sampling in the distribution can be simulated by calling evaluation oracle twice.

Algorithm 1. Lower-Bound-Sampling

Input: evaluation oracle $f : 2^N \rightarrow \mathbb{R}^+$, constraint k, lower bound β, generic submodularity ratio γ, error ε, failure probability η

Output: a solution A that $E[f(A)/\mid A \mid] \geq \gamma(1 - \varepsilon)\beta$

1: Set smaller error bound $\widehat{\varepsilon}_b \longleftarrow \varepsilon/3$
2: Set iteration bounds $r_b \longleftarrow \lceil log_{\gamma \cdot (1-\widehat{\varepsilon})^{-1}}(3n/\eta) \rceil$, $m_b \leftarrow \lceil log(k)/\widehat{\varepsilon} \rceil$
3: Set smaller failure probability $\widehat{\eta}_b \leftarrow 2\eta/(3r_b(m_b + 1))$
4: Initialize $A \leftarrow \emptyset$, $C \leftarrow N$
5: **for** r_b rounds **do**
6:     Filter $H \leftarrow \{x \in C : \rho_x(A) \geq \beta\}$
7:     **if** $|H| = 0$ **then**
8:         break
9:     **for** $i = 0$ to m_b **do**
10:         Set $z \leftarrow min\{\lfloor (1 + \widehat{\varepsilon})^i \rfloor, |H|\}$
11:         Set number of samples $l \leftarrow 16\lceil log(3/\widehat{\eta}_b/\widehat{\varepsilon}_b{}^2 \rceil$
12:         Sample $X_1, X_2, ..., X_l \backsim D_z$
13:         Set $\overline{\mu} \leftarrow \frac{1}{l}\sum_{j=1}^l X_i$
14:         **if** $\overline{\mu} \leqslant 1 - 1.5\widehat{\varepsilon}_b$ **then**
15:             break
16:     Set $z^* \leftarrow min\{z, k- \mid A \mid\}$
17:     Sample $T \backsim \mathcal{U}(H, z^*)$
18:     Update $A \leftarrow A \cup T$, $C \leftarrow C \setminus T$
19:     **if** $\mid A \mid = k$ **then**
20:         break
21: **return** A

Definition 2. *(Distribution [13]) Conditioned on the current state of the algorithm, consider the process where the set $T \sim \mathcal{U}(H, z - 1)$ and then the element $x \sim H \setminus T$ are drawn uniformly. Let $\mathcal{D}_z$ denote the probability distribution over the indicator random variable*

$$I_z = \mathbf{1}[\rho_x(A \cup T) \geqslant \beta] \tag{4}$$

$E[I_z]$ *can be regarded as the probability that the marginal benefit of the z-th element is at least the lower bound β.*

3.1   Analysis of Lower-Bound-Sampling Algorithm

First, we give the exact guarantee of the inner loop lines 10–15 of Algorithm 1 is a standard unbiased estimator for the mean of $\mathcal{D}_z$. Then we show after sampling $T \backsim \mathcal{U}(H, z^*)$ and adding the elements of T to A, the expected marginal gain of the randomly sampled subset T is at least $\gamma(1 - \varepsilon)\beta$. In each round, the elements in the candidate set can be filtered by $\widehat{\varepsilon}_b$ -fraction. At last, we show if the number of elements in solution is less than the constraint, the selection of number of rounds ensures that the marginal gain of unchosen elements is less than β with high probability.

Lemma 1. *For any Bernoulli distribution* $\mathcal{D}_z$, *lines 10–15 of the Lower-Bound-Sampling algorithm hold one of the following properties with a high probability of at least* $1 - \widehat{\eta}_b$:

1. *If* $\overline{\mu} \leq 1 - 1.5\widehat{\varepsilon}_b$, *then the mean of* $\mathcal{D}_z$ *is* $\mu \leq 1 - \widehat{\varepsilon}_b$;
2. *If* $\overline{\mu} > 1 - 1.5\widehat{\varepsilon}_b$, *then the mean of* $\mathcal{D}_z$ *is* $\mu \geqslant 1 - 2\widehat{\varepsilon}_b$;

Lemma 2. *In each round of the Lower-Bound-Sampling algorithm, we have*
   1. $E[I_p] \leq \frac{E[I_q]}{\gamma}$, $\forall p \geq q \geq 1$, $p = 1, 2, 3, ..., |H|$;
   2. *There are* $\widehat{\varepsilon}_b$-*fraction of candidate elements filtered with probability at least* $1 - \widehat{\eta}_b$.

Lemma 3. *If the number of elements in output A of the Lower-Bound-Sampling algorithm is less than cardinality constraint k, then with probability at least $1 - \eta$ H is empty set.*

Theorem 1. *The Lower-Bound-Sampling algorithm outputs the $A \subseteq N$ with $|A| \leq k$ in $O(log(n/\eta)/\varepsilon)$ adaptive rounds such that with probability at least $1 - \eta$ the following properties hold:*

1. *There are $O(n/\varepsilon)$ oracle queries in expectation.*
2. $E[f(A)/ | A |] \geq \gamma(1 - \varepsilon)\beta$.
3. *If $|A| \leq k$, then $\rho_x(A) < \beta$ for all $x \in N$.*

Due to length limitation, the full proofs of Lemma 1–3 and Theorem 1 are given in the appendix.

4   Enumeration Algorithm

In this section, we put the Lower-Bound-Sampling algorithm into a greedy framework for maximizing monotone non-submodular function with a cardinality constrain, namely Enumeration algorithm. First, we give a general overview of the Enumeration algorithm. In every round, given an initial bound β, the Enumeration algorithm at a decreasing bound $(1-\varepsilon)^j \beta$ repeatedly calls subroutine Lower-Bound-Sampling algorithm and obtains a candidate solution. The operation of calling the Lower-Bound-Sampling is greedy, the expected average marginal gain of the elements of the return set is at least $\gamma(1 - \varepsilon)\beta$, the individual contribution of other elements in ground set is less than β. At last, the algorithm compares the function value of all candidate solutions and chooses the maximum as the output.

To guarantee the quality of the solution, let $\lambda_0^* = \max\limits_{x \in N} \rho_x(\emptyset) = max\{f(x) : x \in N\}$ denote the upper bound for marginal contributions of all individual elements in the empty set. From the definition of generic submodularity ratio: for any $T \subseteq N$ and any $x \in 2^N \setminus T$, we have $\rho_x(T) \leq \frac{\rho_x(\emptyset)}{\gamma} \leq \frac{\lambda_0^*}{\gamma}$, so the upper bound for all marginal contributions is $\lambda^* := \lambda_0^*/\gamma$. Hence, $\lambda_0^* \leq OPT \leq k\lambda^* = k\lambda_0^*/\gamma$. The bound which the algorithm desires is $\beta^* = OPT/k$, and the expected

---

Algorithm 2. Enumeration

Input: evaluation oracle $f : 2^N \rightarrow \mathbb{R}^+$, constraint k, generic submodularity ratio γ, error ε, failure probability η
Output: the solution S
1: Set upper bounds $\lambda_0^* \leftarrow max\{f(x) : x \in N\}$
2: Set iteration bound $r \leftarrow \lceil 2log(k/\gamma)/\varepsilon \rceil, m \leftarrow \lceil log(3/\gamma)/\varepsilon \rceil$
3: Set smaller failure probability $\hat{\eta} \leftarrow \eta/(r(m+1))$
4: Initialize $S \leftarrow \emptyset$
5: **for** $i = 0$ to r in parallel **do**
6:     Set $\beta \leftarrow (1+\varepsilon)^i \lambda_0^*/k$
7:     Initialize $A \leftarrow \emptyset$
8:     **for** $j = 0$ to m **do**
9:         Set T $\leftarrow$ Lower-Bound-Sampling $(f_A, k - |A|, (1 - \varepsilon)^j \beta, \varepsilon, \hat{\eta})$
10:         Update $A \leftarrow A \cup T$
11:         **if** $| A |= k$ **then**
12:             break
13:     **if** $f(A) > f(S)$ **then**
14:         Update $S \leftarrow A$
15: **return** S

---

average marginal gain of the elements is at least $\gamma(1 - \varepsilon)\beta$. So beginning with the initial bound $(1 + \varepsilon)^i \lambda_0^*/k$, the algorithm has $O(log(k)/\varepsilon)$ rounds that can suffice to approximate OPT. We try all the computed bounds in parallel such that the adaptivity complexity of the algorithm is retained.

4.1 Analysis of Enumeration Algorithm

By constructing the average random process, we analyse the expected approximation factor of Enumeration algorithm.

Lemma 4. *For the fixed input β and the size of output set S is k, then the*

$$E[f(A)] \geq (1 - e^{-\gamma^2} - \varepsilon)OPT. \tag{5}$$

Theorem 2. *For any monotone, nonnegative non-submodular function f, Enumeration Algorithm outputs a set $S \subseteq N$ with $|S| \leq k$ in $O(log(n/\eta)/\varepsilon^2)$ adaptive rounds such that with probability at least $1 - \eta$, the following properties hold:*

1. *There are $O(nlog(k)/\varepsilon^3)$ oracle queries in expectation*
2. *$E[f(S)] \geq (1 - e^{-\gamma^2} - \varepsilon)OPT$*

5 Conclusion

In the paper, we devised the first parallel algorithm for maximizing a monotone nonsubmodular function with a cardinality constraint. The key idea of our algorithm is to divide the range that possibly contains OPT into logarithmical parts, and then uses the threshold-based greedy approach against each part in parallel. Provided the generic submodularity ratio γ for the monotone set function, we show that the algorithm deserves an approximation ratio $(1-e^{-\gamma^2}-\varepsilon)$ in expectation. In particular, the algorithm achieves an approximation ratio $(1-1/e-\varepsilon)$ when the set function is submodular, coinciding with the state-of-art approximation ratio due to Fahrbach et al. [13].

Acknowledgements. The first two authors are supported by National Natural Science Foundation of China (No. 11531014, 11871081). The third author is supported by National Natural Science Foundation of China (No. 61772005) and Natural Science Foundation of Fujian Province (No. 2017J01753). The fourth author is supported by National Natural Science Foundation of China (No. 11701150).

References

1. Bian, A., Levy, K.Y., Krause, A., Buhmann, J.M.: Continuous DR-submodular maximization: structure and algorithms. In: 31st International Proceedings Conference on Neural Information Processing Systems, pp. 486–496. Curran Associates Inc., Red Hook (2017)
2. Kulesza, A., Taskar, B.: Determinantal point processes for machine learning. Found. Trends Mach. Learn. **5**(2–3), 123–286 (2012)
3. Ito, S., Fujimaki, R.: Large-scale price optimization via network flow. In: 30th International Proceedings on Neural Information Processing Systems, pp. 3862–3870. Curran Associates Inc., Red Hook (2016)
4. Das, A., Kempe, D.: Submodular meets spectral: greedy algorithms for subset selection, sparse approximation and dictionary selection. In: 28th International Proceedings on International Conference on Machine Learning, pp. 1057–1064. Omnipress, Madison (2011)
5. Parotsidis, N., Pitoura, E., Tsaparas, P.: Centrality-aware link recommendations. In: 9th International Proceedings on Web Search and Data Mining, pp. 503–512. Association for Computing Machinery, New York (2016)
6. Buchbinder, N., Feldman, M., Naor, J., Schwartz, R.: A tight linear time (1/2)-approximation for unconstrained submodular maximization. SIAM J. Comput. **44**(5), 1384–1402 (2015)
7. Feige, U., Mirrokni, V., Vondrak, J.: Maximizing non-monotone submodular functions. SIAM J. Comput. **40**(4), 1133–1153 (2011)
8. Balkanski, E., Singer, Y.: The adaptive complexity of maximizing a submodular function. In: 50th Annual ACM SIGACT Symposium on Theory of Computing, pp. 1138–1151. Association for Computing Machinery, New York (2018)
9. Nemhauser, G.L., Wolsey, L.A., Fisher, M.L.: An analysis of approximations for maximizing submodular set functions—I. Math. Program. **14**(1), 265–294 (1978)
10. Feige, U.: A threshold of ln n for approximating set cover. J. ACM **45**(4), 314–318 (1999)
11. Balkanski, E., Rubinstein, A., Singer, Y.: An exponential speedup in parallel running time for submodular maximization without loss in approximation. In: 30th Annual ACM-SIAM Symposium on Discrete Algorithms, pp. 283–302. Society for Industrial and Applied Mathematics, USA (2019)

12. Ene, A., Nguyen, H.L.: Submodular maximization with nearly-optimal approximation and adaptivity in nearly-linear time. In: 30th Annual ACM-SIAM Symposium on Discrete Algorithms, pp. 274–282. Society for Industrial and Applied Mathematics, USA (2019)

13. Fahrbach, M., Mirrokni, V., Zadimoghaddam, M.: Submodular maximization with nearly optimal approximation, adaptivity and query complexity. In: 30th Annual ACM-SIAM Symposium on Discrete Algorithms, pp. 255–273. Society for Industrial and Applied Mathematics, USA (2019)

14. Chekuri, C., Quanrud, K.: Submodular function maximization in parallel via the multilinear relaxation. In: 30th Annual ACM-SIAM Symposium on Discrete Algorithms, pp. 303–322. Society for Industrial and Applied Mathematics, USA (2019)

15. Chekuri, C., Quanrud, K.: Parallelizing greedy for submodular set function maximization in matroids and beyond. In: 51st Annual ACM SIGACT Symposium on Theory of Computing, pp. 78–89. Association for Computing Machinery, New York (2019)

16. Ene, A., Nguyn, H.L., Vladu, A.: Submodular maximization with matroid and packing constraints in parallel. In: 51st Annual ACM SIGACT Symposium on Theory of Computing, pp. 90–101. Association for Computing Machinery, New York (2019)

17. Bian, A.A., Buhmann, J.M., Krause, A., Tschiatschek, S.: Guarantees for greedy maximization of non-submodular functions with applications. In: 34th International Proceedings on International Conference on Machine Learning - Volume 70, pp. 498–507. JMLR.org (2017)

18. Kuhnle, A., Smith, J., Crawford, V.G., Thai, M.T.: Fast maximization of non-submodular, monotonic functions on the integer lattice. In: Proceedings on International Conference on Machine Learning, pp. 2786–2795 (2018)

19. Gong, S., Nong, Q., Liu, W., Fang, Q.: Parametric monotone function maximization with matroid constraints. J. Global Optimiz. **75**(3), 833–849 (2019). https://doi.org/10.1007/s10898-019-00800-2

20. Nong, Q., Sun, T., Gong, S., Fang, Q., Du, D., Shao, X.: Maximize a monotone function with a generic submodularity ratio. In: Du, D.-Z., Li, L., Sun, X., Zhang, J. (eds.) AAIM 2019. LNCS, vol. 11640, pp. 249–260. Springer, Cham (2019). https://doi.org/10.1007/978-3-030-27195-4_23

Fast Algorithms for Maximizing Monotone Nonsubmodular Functions

Bin Liu$^{(\boxtimes)}$ⓓ and Miaomiao Huⓓ

School of Mathematical Sciences, Ocean University of China,
Qingdao 266100, China
binliu@ouc.edu.cn

Abstract. Recently, there has been much progress on improving approximation for problems of maximizing monotone (nonsubmodular) objective functions, and many interesting techniques have been developed to solve these problems. In this paper, we develop approximation algorithms for maximizing a monotone function f with generic submodularity ratio γ subject to certain constraints. Our first result is a simple algorithm that gives a $(1 - e^{-\gamma} - \epsilon)$-approximation for a cardinality constraint using $O(\frac{n}{\epsilon} log \frac{n}{\epsilon})$ queries to the function value oracle. The second result is a new variant of the continuous greedy algorithm for a matroid constraint. We combine the variant of continuous greedy method and contention resolution schemes to find a solution with approximation ratio $(\gamma^2(1 - \frac{1}{e})^2 - O(\epsilon))$, and the algorithm makes $O(rn\epsilon^{-4}log^2\frac{n}{\epsilon})$ queries to the function value oracle.

Keywords: Nonsubmodular function · Maximization · Cardinality constraint · Matroid constraint · Approximation algorithm

1 Introduction

In these years, optimization problems involving maximization of a set function have attracted much attention. Many combinatorial optimization problems can be formulated as the maximization of a set function. For example, the welfare maximization problem is a submodular function maximization problem. Although submodular functions have some good properties, such as diminishing marginal returns, and they also have important applications, many objective functions in practical problems are not submodular. In these settings, we turn to study the problem of maximizing nonsubmodular functions.

The problems of maximizing a submodular function subject to combinatorial constraints are generally NP-hard, so we turned to find approximation algorithms for solving these problems. The greedy approach is a basic technique for these problems: start from an empty set; iteratively add to the current solution

This work was supported in part by the National Natural Science Foundation of China (11971447, 11871442), and the Fundamental Research Funds for the Central Universities.

Z. Zhang et al. (Eds.): AAIM 2020, LNCS 12290, pp. 204–213, 2020.
https://doi.org/10.1007/978-3-030-57602-8_19

set one element that results in the largest marginal gain of the objective function while satisfying the constraints. Meanwhile, the continuous greedy approach is basic technique for maximizing a submodular function under a matroid constraint. We should note that a continuous greedy algorithm is always combined with some rounding methods in order to get the feasible solution, such as pipage rounding and swap rounding.

Given a ground set $N = \{1, 2, ..., n\}$, a set function f is *nonnegative* if $f(S) \geq 0$ for any $S \subseteq N$. The function f is *monotone* if $f(S) \leq f(T)$ whenever $S \subseteq T$. Moreover, f is called *submodular* if $f(S \cup \{j\}) - f(S) \geq f(T \cup \{j\}) - f(T)$ for any $S \subseteq T \subseteq N$, $j \in N \setminus T$. Without loss of generality, for any pair of $S, T \subseteq N$, denote by $f_S(T) = f(S \cup T) - f(S)$ the marginal gain of the set T in S. Specially, denote by $f_S(j) = f(S \cup \{j\}) - f(S)$ the marginal gain of the singleton set $\{j\}$ in S, for any $S \subseteq N$ and any $j \in N$. Moreover, we assume there is a value oracle for the function f.

In this paper, we deal with the following two optimization problems:

Problem (1): $\max\{f(S) : |S| \leq k, S \subseteq N\}$

Problem (2): $\max\{f(S) : S \in \mathcal{I}, S \subseteq N\}$

where $f : 2^N \to R_+$ is a monotone (nonsubmodular) function, k is a positive integer, and $(N, \mathcal{I})$ is a matroid.

In the previous studies, it is proved that, when f is nonnegative, monotone and submodular, the greedy approach yield a $(1 - \frac{1}{e})$-approximation for a cardinality constraint [14], which is also proved to be optimal [13]. After that, lots of results obtained for maximizing a submodular function subject to different constraints. But for nonsubmodular functions, there are only a few results. On the purpose of using known results or methods for maximizing submodular functions, one can define some parameters, such as submodularity ratio, to deal with the maximization of nonsubmodular functions. Das and Kempe [5] first defined the *submodularity ratio* $\hat{\gamma} = min_{S,T \subseteq N} \frac{\sum_{j \in T \setminus S} f_S(j)}{f_S(T)}$, which describes how close a function is to being submodular. Afterwards, Bian et al. [3] proved that the greedy approach gives a $\frac{1}{\alpha}(1 - e^{-\alpha\hat{\gamma}})$-approximation for maximizing a monotone nonsubmodular function with curvature α and submodularity ratio $\hat{\gamma}$ under a cardinality constraint. Recently, Nong et al. [15] proposed the *generic submodularity ratio* which is the largest scalar γ that satisfies $f_S(j) \geq \gamma f_T(j)$, for any $S \subseteq T \subseteq N$. What's more, they showed that the greedy algorithm can achieve a $(1 - e^{-\gamma})$-approximation and a $\frac{\gamma}{1+\gamma}$-approximation for maximizing a strictly monotone nonsubmodular function with generic submodularity ratio γ under a cardinality constraint and a matroid constraint respectively.

Continuous greedy algorithm is always combined with some rounding methods in solving the problem of maximizing a submodular function under a matroid constraint. In the previous studies, Vondrák [17] showed that there exists a $\frac{(1-e^{-\alpha})}{\alpha}$-approximation algorithm for any monotone submodular function with curvature α and matroid constraints, achieving by the continuous greedy approach and the pipage rounding technique [1]. Later, Badanidiyuru et al. [2] proposed an accelerated continuous greedy algorithm for maximizing a monotone submodular function under a matroid constraint using the multilinear extension and swap rounding, and they achieved a $(1 - \frac{1}{e} - \epsilon)$-approximation. The pipage rounding and swap rounding technique are effective methods, but it depends on the submodularity of the objective function. In addition, Chekuri et al. [4] proposed the contention resolution schemes, another framework for rounding, and showed a $(1 - \frac{1}{e})$-approximation for rounding a fractional solution under a matroid constraint. Recently, Gong et al. [9] combined the continuous greedy algorithm and contention resolution schemes technique to achieve a $\gamma(1 - e^{-1})(1 - e^{-\gamma} - O(1))$-approximation.

For nonsubmodular functions optimization, there are other research efforts in application-driving, such as supermodular-degree [7,8], difference of submodular functions [10,12,18], and discrete difference of convex functions [11,19].

Our Result. In this paper, our main contribution is to develop algorithms that have both theoretical approximation guarantees, and fewer queries of the function value oracle. We use simple decreasing threshold algorithm to solve the problem of maximizing a nonsubmodular function under a cardinality constraint. The following Theorem 1 implies the result in [2] for submodular functions (the case that $\gamma = 1$ in Theorem 1). Besides, we use the continuous greedy approach and contention resolution schemes to resolve the nonsubmodular maximizing problem under a matroid constraint. In Theorem 2, we improves the query times of a former result in [15], from $O(n^2)$ to $O(rn\epsilon^{-4}log^2\frac{n}{\epsilon})$. Formally, we obtain the following two theorems.

Theorem 1. *There is a $(1 - e^{-\gamma} - \epsilon)$-approximation algorithm for maximizing a monotone function with generic submodularity ratio γ subject to a cardinality constraint, using $O(\frac{n}{\epsilon}log\frac{n}{\epsilon})$ queries to the function oracle.*

Theorem 2. *There is a $(\gamma^2(1 - e^{-1})^2 - O(\epsilon))$-approximation algorithm for maximizing a monotone function with generic submodularity ratio γ subject to a matroid constraint, using $O(rn\epsilon^{-4}log^2\frac{n}{\epsilon})$ queries to the function oracle.*

2   Preliminary

In this section, we propose some definitions and properties that we will use in the following of the paper.

Definition 1 (Generic Submodularity Ratio [15]).
Given a ground set N and a monotone set function $f : 2^N \rightarrow R_+$, the generic submodularity ratio of f is the largest scalar γ such that for any $S \subseteq T \subseteq N$ and any $j \in N \setminus T, f_S(j) \geq \gamma f_T(j)$.

Definition 2 (The Multilinear Extension [2]).
For a function $f : 2^N \rightarrow R_+$, we define the multilinear extension of f is $F(\mathbf{x}) = E[f(R(\mathbf{x}))]$, where $R(\mathbf{x})$ is a random set where element i appears independently with probability $\mathbf{x}$.

Definition 3 (Matroid [6]).
A matroid $\mathcal{M} = (N, \mathcal{I})$ can be defined as a finite N and a nonempty family $\mathcal{I} \subset 2^N$, called independent set, such that:
(i) $A \subset B, B \in \mathcal{I}$, then $A \in \mathcal{I}$;
(ii) $A, B \in \mathcal{I}, |A| \leq |B|$, then there is an element $j \in B \backslash A$, $A + j \in \mathcal{I}$.

Definition 4 (Matroid Polytope [9]).
Given a matroid $(N, \mathcal{I})$, the matroid polytope is defined as $\mathcal{P}_{\mathcal{I}} = conv\{1_I : I \in \mathcal{I}\} = \{x \geq 0 : for\ any\ S \subset N; \sum_{j \in S} x_j \leq r_{\mathcal{I}}(S), where\ r_{\mathcal{I}}(S) = max\{|I| : I \subset S, I \subset \mathcal{I}\}$ is the rank function of matroid $(N, \mathcal{I})$(hereinafter called r).

Definition 5 (Contention Resolution Schemes (CR Schemes) [4]).
For any $x \in \mathcal{P}_{\mathcal{I}}$ and any subset $A \subset N$, a CR scheme π for $\mathcal{P}_{\mathcal{I}}$ is a removal procedure that returns a random set $\pi_x(A)$ such that $\pi_x(A) \subset A \cap support(x)$ where $support(x) = \{j \in N | x_j > 0\}$ and $\pi_x(A) \in (I)$ with probability 1.

Afterwards, Gong et al. [9] proved that the CR schemes have a $\gamma(1 - \frac{1}{e})$-approximation in the nonsubmodular setting, where γ is the generic submodularity ratio of the objective function.

Lemma 1 (Property of the generic submodularity ratio [9]).

(a) $\gamma \in [0, 1]$;
(b) f is submodular iff $\gamma = 1$;
(c) $\sum_{j \in T \backslash S} f_S(j) \geq \gamma f_S(T)$, for any $S, T \subseteq N$

Lemma 2 ([16]).
Let $\mathcal{M} = (N, \mathcal{I})$ be a matroid, and $B_1, B_2 \in \mathcal{B}$ be two bases. Then there is a bijection $\phi : B_1 \to B_2$ such that for every $b \in B_1$, we have $B_1 - b + \phi(b) \in \mathcal{B}$.

Lemma 3 ([2]). *(Relative + Additive Chernoff Bound) Let $X_1, X_2, ..., X_m$ be independent random variables such that for each i, $X_i \in [0, 1]$, and let $X = \frac{1}{m} \sum_{i=1}^{m} X_i$ and $\mu = E[X]$. Then*

$$Pr[X > (1 + \alpha)\mu + \beta] \leq e^{-\frac{m\alpha\beta}{3}},$$

and

$$Pr[X < (1 - \alpha)\mu - \beta] \leq e^{-\frac{m\alpha\beta}{2}}.$$

Note that the generic submodularity ratio of a strictly monotone function is greater than 0. In the following of the paper, we consider the problem of maximizing a nonnegative strictly monotone and normalized set function under certain constraints.

Algorithm 1. Simple Decreasing Threshold Algorithm

Input: $f : 2^N \to R_+, k \in \{1, 2, ..., n\}$.
Output: A set $S \subset N$ satisfying $|S| \leq k$.
1: $S \leftarrow \emptyset$;
2: $d \leftarrow max_{j \in N} f(j)$;
3: **for** $(w = \frac{d}{\gamma}; w \geq \frac{\epsilon d}{n\gamma}; w \leftarrow \frac{1-\epsilon}{\gamma} w)$ **do**
4:     **for all** $i \in N$ **do**
5:         **if** $|S \cup \{j\}| \leq k$ and $f_S(j) \geq w$ **then**
6:             $S \leftarrow S \cup \{j\}$
7:         **end if**
8:     **end for**
9: **end for**
10: return S

3 Cardinality Constraint

First we present a simple algorithm, Algorithm 1, for **Problem (1)**: $\max\{f(S) : |S| \leq k, S \subseteq N\}$, where f is a monotone function with generic submodularity ratio γ. Our goal is to develop an algorithm that have both theoretical approximation guarantee, and fewer queries of function value oracle.

Next we prove Theorem 1. Firstly, we check the number of queries of Algorithm 1. Obviously, there are two loops in Algorithm 1. Each inner loop executes n queries of value oracle. According to the termination condition of the outer loop, we get the query numbers of per outer loop is $O(\frac{1}{\epsilon} log \frac{n}{\epsilon})$. Therefore the algorithm using $O(\frac{n}{\epsilon} log \frac{n}{\epsilon})$ queries of the value oracle. For the approximation ratio, it is necessary to prove the following claim.

Claim 1. Let O be an optimal solution. Given a current solution S at the beginning of each iteration, the gain of the element added to S is at least $\frac{1-\epsilon}{k} \sum_{a \in O \backslash S} f_S(a)$.

Proof. Suppose that the next element chosen is a and the current threshold value is w. Then it implies the following inequalities

$$\begin{cases} f_S(x) \geq w, & if \ x = a; \\ f_S(x) \leq \frac{w}{1-\epsilon}, & if \ x \in O \setminus S \cup \{a\}. \end{cases}$$

The above inequalities imply that $f_S(a) \geq (1 - \epsilon)f_S(x)$ for each $x \in O \backslash (S \cup \{a\})$. Taking an average over these inequations, we have

$$f_S(a) \geq \frac{1-\epsilon}{|O \backslash S|} \sum_{x \in O \backslash S} f_S(x) \geq \frac{1-\epsilon}{k} \sum_{x \in O \backslash S} f_S(x).$$

Now we finish the proof of Claim 1. $\qquad\qquad\qquad\qquad\qquad\qquad\qquad\square$

Then it is straightforward to finish the proof of Theorem 1.

Proof. Consider on a solution $S_i = \{a_1, a_2, ..., a_i\}$. After i steps, by Claim 1, we have

$$f_{S_i}(a_{i+1}) \geq \frac{1-\epsilon}{k} \sum_{a \in O \setminus S_i} f_{S_i}(a).$$

Then

$$\sum_{a \in O \setminus S_i} f_{S_i}(a) \geq \gamma f_{S_i}(O) \geq \gamma(f(O) - f(S_i)).$$

Therefore,

$$f(S_{i+1}) - f(S_i) = f_{S_i}(a+1) \geq \frac{1-\epsilon}{k} \gamma(f(O) - f(S_i)).$$

We have

$$f(S_k) \geq (1 - (1 - \frac{(1-\epsilon)\gamma}{k})^k) f(O)$$
$$\geq (1 - e^{-\gamma(1-\epsilon)}) f(O)$$
$$\geq (1 - e^{-\gamma} - \epsilon) f(O).$$

So we complete the proof of Theorem 1. $\qquad\qquad\square$

4 Matroid Constraint

In this section we give a $(\gamma^2(1 - \frac{1}{e})^2 - O(\epsilon))$-approximation algorithm for **Problem (2)**: $\max\{f(S) : S \in \mathcal{I}, S \subseteq N\}$, where f is a monotone function with generic submodularity ratio γ, using $O(rn\epsilon^{-4}log^2 \frac{n}{\epsilon})$ queries to the value oracle. The general outline of our algorithm follows from the continuous greedy algorithm in [2]. With a fractional solution being built up gradually from $\mathbf{x} = \mathbf{0}$, and finally using the contention resolution schemes from [4] to convert the fractional solution to an integer one.

Notation. In the following, for $\mathbf{x} \in [0,1]^N$, we denote by $R(\mathbf{x})$ a random set that contains each element $i \in N$ independently with probability x_i. We denote $R(\mathbf{x} + \epsilon 1_S)$ as $R(\mathbf{x}, S)$. Before we analyse the approximation of Algorithm 2, we give and analyse a subroutine, which is used in Algorithm 2. This subroutine takes a current fractional solution $\mathbf{x}$ and adds to it an increment corresponding to an independent set B, to obtain $\mathbf{x} + \epsilon 1_B$. The way we find B in Algorithm 3 is similar to that in Algorithm 1.

Claim 2. Let O be an optimal solution. Given a fractional solution $\mathbf{x}$, Algorithm 3 produces a new fractional solution $\mathbf{x}' = \mathbf{x} + \epsilon 1_B$ such that

$$F(\mathbf{x}') - F(\mathbf{x}) \geq \epsilon(\gamma(1 - \epsilon) - \frac{2\epsilon}{\gamma}) f(O) - F(\mathbf{x}').$$

Algorithm 2. Revised Continuous Greedy Algorithm

Input: $f : 2^N \to R_+$, $\mathcal{I} \subseteq 2^N$.
Output: A set $S \subseteq N$ satisfying $S \in \mathcal{I}$.
1: $\mathbf{x} \leftarrow \mathbf{0}$;
2: **for** $(t \leftarrow \epsilon; t \leq 1; t \leftarrow t + \epsilon)$ **do**
3:     $B \leftarrow$ the output of Algorithm 3
4:     $\mathbf{x} \leftarrow \mathbf{x} + \epsilon \mathbf{1}_B$
5: **end for**
6: $S \leftarrow$ contention resolution schemes $(\mathbf{x}, \mathcal{I})$
7: **return** S

Algorithm 3. Decreasing Threshold procedure

Input: $f : 2^N \to R_+$, $\mathbf{x} \in [0,1]^N$, $\epsilon \in [0,1]$, $\mathcal{I} \subseteq 2^N$.
Output: A set $B \subset N$ satisfying $B \in \mathcal{I}$
1: $B \leftarrow \emptyset$;
2: $d \leftarrow max_{j \in N} f(j)$;
3: **for** $(w = \frac{d}{\gamma}; w \geq \frac{\epsilon d}{r\gamma}; w \leftarrow \frac{1-\epsilon}{\gamma} w)$ **do**
4:     **for all** $e \in N$ **do**
5:         $w_e(B, \mathbf{x}) \leftarrow$ estimate of $E[f_{R(\mathbf{x} + \epsilon \mathbf{1}_B)}(e)]$ by averaging $\frac{r \log n}{\epsilon^2}$ random samples.
6:         **if** $B \cup \{e\} \in \mathcal{I}$ and $w_e(B, \mathbf{x}) \geq w$ **then**
7:             $B \leftarrow B \cup \{e\}$
8:         **end if**
9:     **end for**
10: **end for**
11: **return** B

Proof. Suppose that Algorithm 3 returns r elements, $B = \{b_1, b_2, ..., b_r\}$ (indexed in the order in which they were chosen). In fact, Algorithm 3 might return fewer than r elements if the threshold w drops below $\frac{\epsilon d}{r}$ before termination. In this case, we formally add dummy elements of value 0 so that $|B| = r$.

Let $O = \{o_1, o_2, ..., o_r\}$ be an optimal solution, with $\phi(b_i) = o_i$ as specified by Lemma 2. Additionally, let B_i denote the first i elements of B, and let O_i denote the first i elements of O.

Note that by Lemma 3, we get that there is an error while using $w_e(B_i, \mathbf{x})$ to estimate $E[f_{R(\mathbf{x}, B_i)}(e)]$, with high probability we have the following inequality

$$|w_e(B_i, \mathbf{x}) - E[f_{R(\mathbf{x}, B_i)}(e)]| \leq \frac{\epsilon f(O)}{\gamma r} + \epsilon E[f_{R(\mathbf{x}, B_i)}(e)]. \tag{1}$$

When an element b_i is chosen, o_i is a candidate element which could have been chosen instead of b_i. Thus, according to Algorithm 3, and because either o_i is a potential candidate of value within a factor of $1 - \epsilon$ of the element we chose instead, or the algorithm terminated and all remaining elements have value below $\frac{\epsilon d}{\gamma r}$, we have

$$w_{b_i}(B_{i-1}, \mathbf{x}) \geq (1 - \epsilon) w_{o_i}(B_{i-1}, \mathbf{x}) - \frac{\epsilon d}{\gamma r}. \tag{2}$$

Combining (1) and (2), and the fact that $f(O) \geq d$, we have

$$E[f_{R(\mathbf{x}, B_{i-1})}(b_i)] \geq (1 - \epsilon)E[f_{R(\mathbf{x}, B_{i-1})}(o_i)] - 2\frac{\epsilon f(O)}{\gamma r}. \tag{3}$$

Then at each step in Algorithm 2:

$$F(\mathbf{x}') - F(\mathbf{x}) = F(\mathbf{x} + \epsilon \mathbf{1}_B) - F(\mathbf{x})$$

$$= \sum_{i=1}^{r}(F(\mathbf{x} + \epsilon \mathbf{1}_{B_i}) - F(\mathbf{x} + \epsilon \mathbf{1}_{B_{i-1}}))$$

$$= \sum_{i=1}^{r} \epsilon \frac{\partial F}{\partial x_{b_i}}\Big|_{\mathbf{x} + \epsilon \mathbf{1}_{B_{i-1}}}$$

$$\geq \sum_{i=1}^{r} \epsilon E[f_{\mathbf{x} + \epsilon \mathbf{1}_{B_{i-1}}}(b_i)]$$

$$\geq \sum_{i=1}^{r} \epsilon((1 - \epsilon)E[f_{\mathbf{x} + \epsilon \mathbf{1}_{B_{i-1}}}(o_i)] - 2\frac{\epsilon f(O)}{\gamma r})$$

$$\geq \sum_{i=1}^{r} \epsilon((1 - \epsilon)\gamma E[f_{R(\mathbf{x} + \epsilon \mathbf{1}_{B \cup \{o_1, o_2, \dots, o_{i-1}\}})}(o_i)] - 2\frac{\epsilon f(O)}{\gamma r})$$

$$= \epsilon((1 - \epsilon)\gamma E[f(R(\mathbf{x}') \cup O) - f(R(\mathbf{x}'))] - 2\frac{\epsilon f(O)}{\gamma})$$

$$\geq (\gamma\epsilon(1 - \epsilon) - \frac{2\epsilon^2}{\gamma})f(O) - \epsilon F(\mathbf{x}')$$

$$= \epsilon((\gamma(1 - \epsilon) - \frac{2\epsilon}{\gamma})f(O) - F(\mathbf{x}')).$$

The second inequality follows from that o_i is a candidate element when b_i is chosen. The first and last inequalities are due to monotonicity, and the third inequality is due to the definition of generic submodularity ratio γ. $\qquad \square$

Claim 3. Algorithm 3 makes $O(\frac{1}{\epsilon^3} nr\log^2 \frac{n}{\epsilon})$ queries to the function oracle.

Proof. Obviously, there are two loops in Algorithm 3. According to the termination condition of the outer loop, we get the query numbers of per outer loop is $O(\frac{1}{\epsilon}\log\frac{n}{\epsilon})$. The number of iterations in the inner loop is n, and the number of samples per evaluation of F is $\frac{1}{\epsilon^2}r\log n$ in per inner loop. Therefore, Algorithm 3 makes $O(\frac{1}{\epsilon^3} nr\log^2 \frac{n}{\epsilon})$ queries to the value oracle. $\qquad \square$

Claim 4. Algorithm 2 has an approximation ratio of $\gamma^2(1 - \frac{1}{e})^2 - O(\epsilon)$.

Proof. Define $\Omega = (\gamma(1 - \epsilon) - \frac{2\epsilon}{\gamma})f(O)$. Substituting this in the result of Claim 2, we have

$$F(\mathbf{x}(t + \epsilon)) - F(\mathbf{x}(t)) \geq \epsilon(\Omega - F(\mathbf{x}(t + \epsilon))).$$

Add $\Omega - F(\mathbf{x}(t + \epsilon))$ to the inequation we have

$$\Omega - F(\mathbf{x}(t + \epsilon)) \leq \frac{\Omega - F(\mathbf{x}(t))}{1 + \epsilon}.$$

Using induction to this inequation, we have

$$\Omega - F(\mathbf{x}(t)) \leq \frac{\Omega}{(1 + \epsilon)^{\frac{t}{\epsilon}}}.$$

Substituting $t = 1$ and rewriting the inequation, we have

$$F(\mathbf{x}(t)) \geq (1 - \frac{1}{(1 + \epsilon)^{\frac{1}{\epsilon}}})\Omega$$

$$= (1 - \frac{1}{(1 + \epsilon)^{\frac{1}{\epsilon}}})(\gamma(1 - \epsilon) - \frac{2\epsilon}{\gamma})$$

$$\geq \gamma(1 - \frac{1}{e}) - O(\epsilon).$$

Besides, when we use CR schemes to convert the fractional solution to the integer one, there also have an approximation ratio which is $\gamma(1 - \frac{1}{e})$.

Therefore, the approximation ratio of Algorithm 2 is $\gamma^2(1 - \frac{1}{e})^2 - O(\epsilon)$.   □

Claim 5. Algorithm 2 makes $O(\frac{1}{\epsilon^4} nrlog^2 \frac{n}{\epsilon})$ queries to the function oracle.

Proof. Observe that in Algorithm 2, the queries to the function oracle is only related to Algorithm 3. Therefore the total number of oracle calls to the function is equal to the number of the loop multiplied with the number of oracle calls in one iteration. So we get the queries to the function oracle are at most $O(\frac{1}{\epsilon^4} nrlog^2 \frac{n}{\epsilon})$.   □

References

1. Ageev, A.A., Sviridenko, M.I.: Pipage rounding: a new method of constructing algorithms with proven performance guarantee. J. Comb. Optimiz. **8**(3), 307–328 (2014). https://doi.org/10.1023/B:JOCO.0000038913.96607.c2
2. Badanidiyuru, A., Vondrák, J.: Fast algorithm for maximizing submodular functions. In: SODA, pp. 1497–1514 (2014)
3. Bian, A.A., Buhmann, J.M., Krause, A., Tschiatschek, S.: Guarantees for greedy maximization of nonsubmodular functions with applications. In: ICML (2017)
4. Chekuri, C., Vondrk, J., Zenklusen, R.: Submodular function maximization via the multilinear relaxation and contention resolution schemes. SIAM J. Comput. **43**(6), 1831–1879 (2014)
5. Das, A., David, K.: Submodular meets spectral: greedy algorithms for subset selection, sparse approximation and dictionary selection. In: ICML (2011)
6. Edmonds, J.: Submodular functions, matroids, and certain polyhedra. In: Jünger, M., Reinelt, G., Rinaldi, G. (eds.) Combinatorial Optimization—Eureka, You Shrink!. LNCS, vol. 2570, pp. 11–26. Springer, Heidelberg (2003). https://doi.org/10.1007/3-540-36478-1_2

7. Feldman, M., Izsak, R.: Constrained monotone function maximization and the supermodular degree. In: International Workshop and International Workshop on Approximation Randomization and Combinatorial Optimization Algorithms and Techniques, pp. 160–175 (2014)
8. Feige, U., Izsak, R.: Welfare maximization and the supermodular degree. In: Proceedings of the 4th Conference on Innovations in Theoretical Computer Science, pp. 247–256. ACM (2013)
9. Gong, S., Nong, Q., Liu, W., et al.: Parametric monotone function maximization with matroid constraints. J. Global Optimiz. **75**(3), 833–849 (2019). https://doi.org/10.1007/s10898-019-00800-2
10. Iyer, R., Bilmes, J.: Algorithms for approximate minimization of the difference between submodular functions, with applications. In: UAI, pp. 407–417 (2012)
11. Maehara, T., Murota, K.: A framework of discrete DC programming by discrete convex analysis. Math. Program. **152**(1–2), 435–466 (2015). https://doi.org/10.1007/s10107-014-0792-y
12. Narasimhan, M., Bilmes, J.A.: A submodular-supermodular procedure with applications to discriminative structure learning. arXiv:1207.1404 (2012)
13. Nemhauser, G.L., Wolsey, L.A.: Best algorithms for approximating the maximum of a submodular set functions. Math. Oper. Res. **3**(3), 177–188 (1978)
14. Nemhauser, G.L., Wolsey, L.A., Fisher, M.L.: An analysis of approximations for maximizing submodular set functions-I. Math. Program. **14**(1), 265–294 (1978). https://doi.org/10.1007/BF01588971
15. Nong, Q., Sun, T., Gong, S., et al.: Maximize a monotone function with a generic submodularity ratio. In: AAIM (2019)
16. Sviridenko, M.: A note on maximizing a submodular set function subject to a Knapsack constraint. Oper. Res. Lett. **32**(1), 41–43 (2004)
17. Vondrák, J.: Submodularity and Curvature: The Optimal Algorithm. Rims Kokyuroku Bessatsu B (2010)
18. Wu, W.-L., Zhang, Z., Du, D.-Z.: Set function optimization. J. Oper. Res. Soc. China **7**(2), 183–193 (2018). https://doi.org/10.1007/s40305-018-0233-3
19. Wu, C., et al.: Solving the degree-concentrated fault-tolerant spanning subgraph problem by DC programming. Math. Program. **169**(1), 255–275 (2018). https://doi.org/10.1007/s10107-018-1242-z

Non-Submodular Streaming Maximization with Minimum Memory and Low Adaptive Complexity

Meixia Li[1], Xueling Zhou[2], Jingjing Tan[1](✉), and Wenchao Wang[2]

[1] School of Mathematics and Information Science, Weifang University,
Weifang 261061, Shandong, China
limeixia001@163.com, tanjingjing1108@163.com
[2] College of Mathematics and Systems Science,
Shandong University of Science and Technology, Qingdao 266590, Shandong, China
zhou_xueling@yeah.net, wangwenchao928@163.com

Abstract. Extracting representative elements from a large stream of data is an important and interesting problem. Such problem can be formulated as maximizing a normalized monotone non-submodular set function subject to a cardinality constraint. In this paper, we first present an algorithm called NON-SUBMODULAR-SIEVE-STREAMING^{++} for solving this problem by utilizing the concept of diminishing-return ratio, which requires only one pass over the data and obtain tight approximation ratio and minimum memory complexity. Then, for reducing the number of adaptive complexity, we propose an algorithm called NON-SUBMODULAR-BATCH-SIEVE-STREAMING^{++} by buffering a small fraction of the stream and applying a filtering procedure. We analyze the approximation ratios of the two algorithms, which generalize the results of SIEVE-STREAMING^{++} and BATCH-SIEVE-STREAMING^{++} to the non-submodular case. Finally, we illustrate the feasibility and effectiveness of the two algorithms through a numerical example and compare the corresponding results with the existing algorithms.

Keywords: Streaming algorithm · Non-submodular functions · Cardinality constraint

1 Introduction

With the development of science and technology, there are massive data coming from social networks, sensor data, videos, etc. In such cases, the amount of input data is much larger than the main memory capacity of individual computers. It is necessary to extract useful information from the massive data. Therefore, the study on the data streaming models is meaningful (cf. [30] and references therein). That is to say, we consider the situation where each item in the ground set arrives sequentially, and we are allowed to keep only a small number of the items in memory at any point. In general, the memory is restricted to be limited

© Springer Nature Switzerland AG 2020
Z. Zhang et al. (Eds.): AAIM 2020, LNCS 12290, pp. 214–224, 2020.
https://doi.org/10.1007/978-3-030-57602-8_20

space which is sublinear with respect to the input size. Data streaming models have many applications in lots of aforementioned fields (cf. [1,13,14,16,22,23,36] and references therein). For data summarization, a systematic way is to turn the problem into selecting a subset of data elements optimizing a utility function that quantifies "respectiveness" of the selected set. These objective functions often satisfy submodularity. Hence, many data streaming models require maximizing submodular set functions subject to cardinality constraints [12,21,22,25].

Many interesting results obtained about the submodular optimization, such as [4,5,18,31] subject to a cardinality constraint, [3,15,27,33] subject to a knapsack constraint, [7,20,34] subject to a matroid, etc. On the data streaming summarization, Badanidiyuru et al. [2] provided the first efficient method called SIEVE-STREAMING for monotone submodular function maximation, subject to the constraint that at most k points are selected. It requires only a single pass over the data, in arbitrary order, and presented a $(\frac{1}{2} - \epsilon)$ approximation ratio, $O(\log k/\epsilon)$ update time and $O(k \log k/\epsilon)$ memory cost for any $\epsilon > 0$. Then, Buchbider et al. [6] proposed an improved algorithm with a $\frac{1}{4}$ approximation ratio, but the memory complexity being $\Theta(k)$. Norouzi-Fard et al. [32] introduced an algorithm for random order stream with $\frac{1}{2}$ approximation ratio. Following the results in [2], Kazemi et al. [24] presented an algorithm called SIEVE-STREAMING^{++} with $(\frac{1}{2} - \epsilon)$ approximation ratio, $O(\log k/\epsilon)$ update time and $O(k/\epsilon)$ memory cost for any $\epsilon > 0$. The memory cost is superior to the result of SIEVE-STREAMING in [2]. Furthermore, the SIEVE-STREAMING^{++} requires at least one query to the oracle for each incoming element of the stream, which increases its adaptive complexity. To reduce the adaptive complexity, Kazemi et al. [24] designed an hybrid algorithm called BATCH-SIEVE-STREAMING^{++} with $(\frac{1}{2} - \epsilon)$ approximation ratio, $O(B + k/\epsilon)$ memory complexity, and $O(N \log B \log k/(B\epsilon))$ adaptive complexity, where ϵ is a constant smaller than $\frac{1}{3}$, N is the total number of elements in the stream and B is the buffer size. The other references on the data streaming models, readers can see [8–10,19,29], etc.

However, for many applications, including experimental design and sparse Gaussian processes [28], objective function is in general not submodular [26]. To explain the good empirical performance, Das and Kempe [11] proposed the submodularity ratio, a quantity characterizing how close a set function is to be submodular. Following the idea of [2], Elenberg et al. [17] proposed the first streaming algorithm for weakly submodular case with constant approximation ratio being $\gamma_f(1 - \epsilon)(3 - e^{-\gamma_f/2} - 2\sqrt{2 - e^{-\gamma_f/2}})/2$ where $\epsilon > 0$ and γ_f is the weak submodularity ratio. Recently, Wang et al. [35] designed and analyzed four dimishing-return sieve-streaming algorithms for non-submodular function utilizing the concept of diminishing-return ratio.

In this paper, we consider the problem

$$\max_{S \subseteq V} \; f(S)$$
$$\text{s.t.} \; |S| \leq k, \tag{1}$$

where $f : 2^V \rightarrow R_+$ is a non-negative and normalized monotone nonsubmodular function with the diminishing-return ratio γ, V is a data stream and k is a positive integer.

In order to overcome (1), we propose two new algorithms which generalize the SIEVE-STREAMING^{++} algorithm and BATCH-SIEVE-STREAMING^{++}. Using the diminishing-return ratio, we analyze the approximation ratios of these algorithms and illustrate the feasibility and effectiveness of the algorithms through numerical experiment.

The rest of this paper is organized as follows. In Sect. 2, we introduce some definitions and results about submodular and non-submodular set function. In Sect. 3, we present two algorithms with the analysis of the approximation ratios. In Sect. 4, we give a numerical example to illustrate the effectiveness of the two algorithms and compare them with the corresponding algorithms. We give some conclusions in Sect. 5. In addition, the formal proofs are omitted due to the length limitation but are nevertheless given in the appendix.

2   Preliminaries

In this section, we give some definitions and results.

A set function $f : 2^V \rightarrow R_+$ on a ground set V is called submodular if it satisfies the diminishing marginal return property, i.e., for any subsets $S \subseteq T \subseteq V$ and $e \in V \setminus T$, we have

$$f(S \cup \{e\}) - f(S) \geq f(T \cup \{e\}) - f(T).$$

An equivalent definition is

$$f(X) + f(Y) \geq f(X \cap Y) + f(X \cup Y),$$

for any subsets $X, Y \subseteq V$.

The marginal gain of f with respect to e and S is defined as

$$f(\{e\}|S) = f(S \cup \{e\}) - f(S)$$

for any given $e \in V$ and $S \subseteq V$.

Similarly, for any subsets $S, T \subseteq V$, we define the marginal gain of f with respect to S and T is

$$f(S|T) = f(S \cup T) - f(T).$$

We say that f is monotone if $f(X) \leq f(Y)$ for any $X \subseteq Y$ and f is non-negative and normalized if $f(S) \geq 0$ for any $S \subseteq V$ and $f(\emptyset) = 0$.

Now we introduce the definitions of Diminishing-Return Ratio and Weak Submodularity Ratio.

Definition 1. *(Diminishing-Return Ratio)* [35] *The diminishing-return ratio of a normalized nonnegative monotone set function f is the largest scalar $\gamma \in [0,1]$ such that for any $S \subseteq T$, $e \notin T$,*

$$\gamma f(e|T) \leq f(e|S).$$

This γ can be reformulated as

$$\gamma = \min_{S \subseteq T, e \notin T} \frac{f(e|S)}{f(e|T)}.$$

Definition 2. *(Weak Submodularity Ratio)* [17] *A monotone non-negative set function* $f : 2^V \to R_+$ *is called* γ_f-*weakly submodular for an integer* k *if*

$$\gamma_f \leq \gamma_k := \min_{T, S \subseteq G : |T|, |S \setminus T| \leq k} \frac{\sum_{v \in S \setminus T} f(v|T)}{f(S|T)}.$$

When both the numerator and denominator are equal to 0, the ratio is defined as 1.

It is worth noting that we consider the function f as an oracle. For (non-) submodular optimization, the number of calls to the oracle is widely used as the time complexity measurement.

3   Main Results

In this section, we propose two algorithms for solving (1), which generalize the Sieve-Streaming^{++} and Batch-Sieve-Streaming^{++}. The main observation of Sieve-Streaming^{++} is that in the process of guessing OPT, the previous algorithm (cf. Sieve-Streaming) uses Δ as a lower bound for OPT, where Δ is the maximum value of singleton elements observed so far. In fact, it has $OPT \geq LB = \max_{\tau} f(S_\tau)$ and as a result, there is no need to keep thresholds smaller than $LB/2k$. Also, for a threshold τ, there is at most LB/τ elements in set S_τ. These two important observations can reduce the memory complexity. Based on the idea of Sieve-Streaming^{++}, we present the following Algorithm for non-submodular function with diminishing-return ratio $\gamma \in [0, 1]$.

Now we analyze the approximation ratio of Algorithm 1.

Algorithm 1. Non-Submodular-Sieve-Streaming^{++}

Input: Non-submodular function f along with diminishing-return ratio $\gamma \in [0, 1]$, data stream V, cardinality constraint k, and error term ϵ.

1: $\tau_{\min} \leftarrow 0$, $\Delta \leftarrow 0$ and $LB \leftarrow 0$
2: **while** there is an incoming item e from V **do**
3:    $\Delta \leftarrow \max\{\Delta, f(\{e\})\}$
4:    $\tau_{\min} = \frac{\max\{LB, \Delta\}}{2^\gamma \cdot k}$
5:    Discard all sets S_τ with $\tau < \tau_{\min}$
6:    **for** $\tau \in T_\epsilon = \left\{ (1 + \epsilon)^i \mid \tau_{\min}/1 + \epsilon \leq (1 + \epsilon)^i \leq \Delta/\gamma \right\}$
7:       **if** τ is a new threshold **then** $S_\tau \leftarrow \emptyset$
8:       **if** $|S_\tau| < k$ and $f(\{e\}|S_\tau) \geq \gamma \cdot \tau$ **then**
9:          $S_\tau \leftarrow S_\tau \cup \{e\}$ and $LB \leftarrow \max\{LB, f(S_\tau)\}$
10: **return** argmax$_{S_\tau} f(S_\tau)$

Lemma 1. *Consider any iteration of Algorithm 1 with subset S_τ, we have*

$$f(S_\tau) \geq \gamma \cdot \tau \cdot |S_\tau|. \tag{2}$$

Let $\tau^* = \frac{OPT}{2^\gamma \cdot k}$, where OPT is the optimal value of (1). We have the following result.

Lemma 2. *For any given $\epsilon > 0$, there exists a value $\tau \in T_\epsilon$ such that*

$$(1 - \epsilon) \cdot \tau^* \leq \tau < \tau^*. \tag{3}$$

From the above lemmas, we can obtain the following theorem.

Theorem 1. *For any given $\epsilon \in (0,1)$, and non-submodular function f with diminishing-return ratio $\gamma \in [0,1]$. Algorithm 1 produces a set S such that $|S| \leq k$ and $f(S) \geq \min\left\{ \frac{(1-\epsilon)\gamma}{2^\gamma}, 1 - \frac{1}{2^\gamma} \right\} OPT$, where OPT is the optimal value of (1).*

As follows, we introduce the second algorithm for solving (1), which is the generalization of BATCH-SIEVE-STREAMING^{++}. The main difference of two algorithms is that BATCH-SIEVE-STREAMING^{++} is used to maximizing the submodular function, but Algorithm 2 proposed in this section is used to maximizing the nonsubmodular function with diminishing-return ratio $\gamma \in [0,1]$. BATCH-SIEVE-STREAMING^{++} has two important properties: (i) the number of adaptive rounds is near-optimal, and (ii) it has an optimal memory complexity (by adopting an idea similar to SIEVE-STREAMING^{++}). In algorithm 2, we persist the two merits.

Algorithm 2. NON-SUBMODULAR-BATCH-SIEVE-STREAMING^{++}

Input: Stream of data V, non-submodular set function f along with diminishing-return ratio $\gamma \in [0,1]$, cardinality constraint k, buffer $\mathcal{B}$ with a memory B, *Threshold*, error term ϵ.

1: $\Delta \leftarrow 0$, $\tau_{\min} \leftarrow 0$, $LB \leftarrow 0$ and $\mathcal{B} \leftarrow \emptyset$
2: **while** there is an incoming element e from V **do**
3:     Add e to $\mathcal{B}$
4:     **if** the buffer $\mathcal{B}$ is *Threshold* percent full **then**
5:         $\Delta \leftarrow \max\{\Delta, \max_{e \in \mathcal{B}} f(e)\}$, $\tau_{\min} = \frac{\max\{LB, \Delta\}}{2^\gamma k (1+\epsilon)}$
6:         Discard all sets S_τ with $\tau < \tau_{\min}$
7:         **for** $\tau \in T_\epsilon = \left\{ (1+\epsilon)^i \mid \tau_{\min} \leq (1+\epsilon)^i \leq \Delta/\gamma \right\}$
8:             If τ is a new threshold then assign a new set S_τ to it , i.e., $S_\tau \leftarrow \emptyset$
9:             **if** $|S_\tau| < k$ **then**
10:                 $T \leftarrow$ NON-SUBMODULAR-THRESHOLD-SAMPLING $(f_{S_\tau}, \mathcal{B}, k - |S_\tau|, \tau, \epsilon, \gamma)$
11:                 $S_\tau \leftarrow S_\tau \cup T$
12:         $LB = \max_{S_\tau} f(S_\tau)$ and $\mathcal{B} \leftarrow \emptyset$
13: **return** $\text{argmax}_{S_\tau} f(S_\tau)$

In the process of implementing Algorithm 2, we should call Algorithm 3 several times. Now we give the detailed description of Algorithm 3.

In the following, we put forward the analysis on the approximation ratio of Algorithm 2.

Algorithm 3. NON-SUBMODULAR-THRESHOLD-SAMPLING

Input: Non-submodular set function f along with diminishing-return ratio $\gamma \in [0,1]$, set of buffer items $\mathcal{B}$, cardinality constraint k, threshold τ, error term ϵ.

1: $S \leftarrow \emptyset$
2: **while** $|\mathcal{B}| > 0$ and $|S| < k$ **do**
3:    update $\mathcal{B} \leftarrow \{x \in \mathcal{B} : f(\{x\}|S) \geq \gamma \cdot \tau\}$ and filter out the rest.
4:    **for** $i = 1$ to $\lceil \frac{1}{\epsilon} \rceil$ **do**
5:      Sample x uniformly at random from $\mathcal{B} \setminus S$
6:      **if** $f(\{x\} \mid S) \leq (1 - \epsilon) \cdot \gamma \cdot \tau$ **then**
7:        **break** and go to Line 2
8:      **else**
9:        $S \leftarrow S \cup \{x\}$
10:       **if** $|S| = k$ **then return** S
11:    **for** $i = \lceil \log_{1+\epsilon}(1/\epsilon) \rceil$ to $\lceil \log_{1+\epsilon} k \rceil - 1$ **do**
12:      $t \leftarrow \min\{\lfloor (1 + \epsilon)^{i+1} - (1 + \epsilon)^i \rfloor, |\mathcal{B} \setminus S|, k - |S|\}$
13:      Sample a random set T of size t from $\mathcal{B} \setminus S$
14:      **if** $|S \cup T| = k$ **then**
15:        **return** $S \cup T$
16:      **if** $\frac{f(T|S)}{|T|} \leq (1 - \epsilon) \cdot \gamma \cdot \tau$ **then**
17:        $S \leftarrow S \cup T$ and **break**
18:      **else**
19:        $S \leftarrow S \cup T$
20: **return** S

Theorem 2. *For any given* $\epsilon \in (0, \frac{1}{3})$, *and non-submodular function* f *with diminishing-return ratio* $\gamma \in [0,1]$, *Algorithm 2 produces a set* S *such that* $|S| \leq k$ *and* $f(S) \geq \min\left\{\frac{(1-3\epsilon)\gamma}{2\gamma}, 1 - \frac{1}{2\gamma}\right\} OPT$, *where* OPT *is the optimal value of (1).*

Similar to the SIEVE-STREAMING^{++} and BATCH-SIEVE-STREAMING^{++}, Algorithm 1 can obtain $O(k/\epsilon)$ memory complexity and $O(\log(k/\gamma)/\epsilon)$ update time. Algorithm 2 can obtain $O(B + k/\epsilon)$ memory complexity and $O(N \log B \log (k/\gamma)/(B\epsilon))$ adaptive complexity. Table 1 gives a comparison of Algorithms 1-2 in this paper with the Algorithms 3-4 in [35].

Table 1. Comparison of Algorithms 1-2 in this paper with Algorithms 3-4 in [35]

Indexes	Algorithm 1	Algorithm 2	Algorithms 3-4
Appro. ratio	$\min\left\{\frac{(1-\epsilon)\gamma}{2\gamma}, 1 - \frac{1}{2\gamma}\right\}$	$\min\left\{\frac{(1-3\epsilon)\gamma}{2\gamma}, 1 - \frac{1}{2\gamma}\right\}$	$\min\left\{\frac{\alpha\gamma}{2\gamma}, 1 - \frac{1}{2\gamma}\right\}$
Memory	$O(\frac{k}{\epsilon})$	$O(B + \frac{k}{\epsilon})$	$O(\frac{k\log(k/\gamma)}{\epsilon})$
Update time	$O(\frac{\log(k/\gamma)}{\epsilon})$	----	$O(\frac{\log(k/\gamma)}{\epsilon})$

4 Numerical Examples

In this section, we apply Algorithms 1-2 to support selections for sparse linear regression. And we compare the calculating results with the results in [35] and [17]. The whole codes are written in Matlab R2018b. All the numerical results are carried out on a desktop computer with Intel(R) Core(TM) i7-8700, CPU 3.20 GHz and RAM 8.00 GB. The example is similar to that in [35]. For the convenience of readers, we give a detailed description as follows.

Given the feature matrix $X \in R^{n \times d}$ with n samples and d features, and the response vector $y \in R^n$, we need to select k features to maximize the R^2 statistic. Let S be the set of selected features, X_S be the columns of X indexed by S, and $P_S := X_S(X_S^T X_S)^{-1}X_S^T$. Then the problem can be formulated as the non-submodular maximization (1) with $f(S) := \|P_S y\|_2^2$.

Consider the distributed setting for sparse linear regression. The feature matrix X is generated by the following process:

$$X_{i,t+1} = \sqrt{1 - \alpha^2}X_{i,t} + \alpha \varepsilon_{it},$$

where $\alpha = 0.5$ and ε_{it} is i.i.d. Gaussian variable with mean 0 and variance α^2. Set the response vector y to

$$y = Xb + z,$$

where b is a k-sparse vector (there are k non-zero elements in a k-sparse vector) and z_i is i.i.d. Gaussian variable with mean 0 and variance $\|Xb\|_2^2/100$. The entries of non-zero elements of the vector b are random, and the value of each non-zero element is set to

$$b_s = (-1)^{\text{Bern}(0.5)} \cdot (5\sqrt{\frac{\log d}{n}} + \delta),$$

where δ is a standard Gaussian variable, and $\text{Bern}(p)$ is a random number which equals 1 with the probability p or -1 with the probability $1 - p$. We normalize y and each column of X, and reset the order of features randomly.

Similar to [35], we enumerate the diminishing-return ratio γ: $\gamma = i/20$ for all $i = 1, \ldots, 20$. And we set $n = d = 500$. Consider two different values of k, which are 50 and 100 respectively. For each value of k, we test 10 instances and compute average running times and average total number of querying oracle. Furthermore, we test the relationship between memory complexity and ϵ or k for Algorithms 1-2. Especially, Algorithm 1 and Algorithm 2 in this paper are named N-SMSS++ and N-SMBSS++, respectively. And the Streak algorithm in [17] is named Streak. The numerical results are shown in Fig. 1, 2 and 3. It is easy to see that the Streak does not depend on γ, so we use a horizontal line to show its running time and total number of queries in the corresponding figures.

From Fig. 1(a) and (b), it can be observed that Algorithm 1 in this paper and Algorithms 3-4 in [35] become faster as the value of γ increases, and Algorithm 1 is faster than Algorithms 3-4 with the same value of γ. This result is

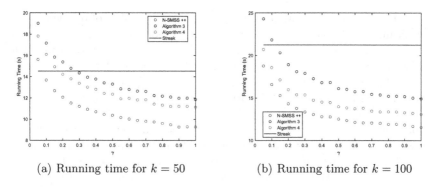

(a) Running time for $k = 50$ (b) Running time for $k = 100$

Fig. 1. Performance comparison of running time over all values of γ

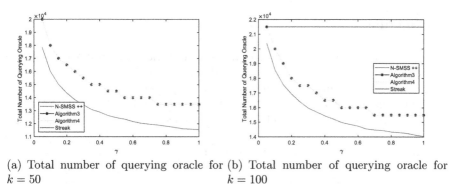

(a) Total number of querying oracle for   (b) Total number of querying oracle for
$k = 50$ $k = 100$

Fig. 2. Performance comparison of total number of querying oracle over all values of γ

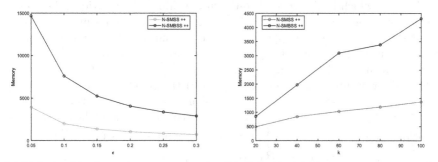

(a) Relationship between memory and ϵ (b) Relationship between memory and k
for $k = 100$ for $\epsilon = 0.2$

Fig. 3. The relationship between memory and ϵ or k

consistent with the theoretical result. Moreover, both Algorithm 1 in this paper
and Algorithms 3-4 in [35] are faster than the Streak algorithm as long as γ is
large enough.

From Fig. 2 (a) and (b), we can see that the total number of querying oracle of Algorithm 1 in this paper and Algorithms 3-4 in [35] become less as the value of γ increases, and Algorithm 1 is less than Algorithms 3-4 with the same value of γ. The main reason of this result is that the selection of $\tau_{\min}$ is related to the value of $f(S_\tau)$. Moreover, Algorithms 3-4 are exactly overlapping. And both Algorithms 1 in this paper and Algorithms 3-4 in [35] are less than the Streak algorithm as long as γ is large enough.

From Fig. 3 (a), we can observe that the memory decreases as the value of ϵ increases. And for the same ϵ, the memory of Algorithm 1 is less than Algorithm 2. From Fig. 3 (b), we can obtain that the memory increases as the value of k increases. And for the same k, the memory of Algorithm 1 is also less than Algorithm 2.

5   Conclusions

In this paper, we study the problem of maximizing a normalized monotone non-submodular set function subject to a cardinality constraint in the streaming model. Utilizing the diminishing-return ratio, we present two algorithms which are called NON-SUBMODULAR-SIEVE-STREAMING^{++} and NON-SUBMODULAR-BATCH-SIEVE-STREAMING^{++}. The two algorithms require only a single pass through the data.

Algorithm 1 is the generalization of SIEVE-STREAMING^{++} to non-submodular set function. We obtain the approximation ratio is min $\left\{ \frac{(1-\epsilon)\gamma}{2\gamma}, 1 - \frac{1}{2\gamma} \right\}$, which is related to the diminishing-return ratio of the objective function. The memory complexity is $O(\frac{k}{\epsilon})$ and the update time per element is $O(\frac{\log(k/\gamma)}{\epsilon})$, which is also related to the diminishing-return ratio of the objective function. Comparing Algorithm 1 with the Algorithms 3-4 in [35], we can see that they have the same approximation ratio and update time per element. But the memory complexity of Algorithm 1 is improved significantly.

The performance of Algorithm 2 is first to buffer a fraction of the data stream and then filtering procedure through a parallel threshold. It is the generalization of BATCH-SIEVE-STREAMING^{++} to non-submodular set function. Through utilizing the diminishing-return ratio of the objective function, the approximation ratio is min $\left\{ \frac{(1-3\epsilon)\gamma}{2\gamma}, 1 - \frac{1}{2\gamma} \right\}$, the memory complexity is $O(B + \frac{k}{\epsilon})$ and the adaptive complexity is $O(N \log B \log(k/\gamma)/(B\epsilon))$.

To illustrate the numerical effect of these algorithms, we give a numerical example in Sect. 4. The running time and the total number of querying oracle of Algorithm 1 in this paper are superior to Algorithms 3-4 in [35]. At the same time, they are all superior to the Streak algorithm in [17] when γ is large enough. Furthermore, the memory of Algorithm 1 is less than Algorithm 2 for the same ϵ or k.

There are many interesting problems about the streaming model. Further, we will study the streaming non-submodular maximization over sliding windows utilizing the diminishing-return ratio. We will also study the non-submodular maximization using submodularity ratio and curvature for streaming model.

Acknowledgements. The first author is supported by Natural Science Foundation of China (Nos. 11401438, 11571120). The third author is supported by Natural Science Foundation of Shandong Province (Nos. ZR2017LA002, ZR2019MA022) of China.

References

1. Ajtai, M., Jayram, T.S., Kumar, R., Sivakumar, D.: Approximate counting of inversions in a data stream. In: Proceedings of STOC, pp. 370–379 (2002)
2. Badanidiyuru, A., Mirzasoleiman, B., Karbasi, A., Krause, A.: Streaming submodular maximization: massive data summarization on the fly. In: Proceedings of SIGKDD, pp. 671–680 (2014)
3. Badanidiyuru, A., Vondrk, J.: Fast algorithms for maximizing submodular functions. In: Proceedings of SODA, pp. 1497–1514 (2014)
4. Balkanski, E., Rubinstein, A., Singer, Y.: An exponential speedup in parallel running time for submodular maximization without loss in approximation. In: Proceedings of SODA, pp. 283–302 (2019)
5. Breuer, A., Balkanski, E., Singer, Y.: The FAST algorithm for submodular maximization. arXiv: 1907.06173 (2019)
6. Buchbinder, N., Feldman, M., Schwartz, R.: Online submodular maximization with preemption. In: Proceedings of SODA, pp. 1202–1216 (2015)
7. Buchbinder, N., Feldman, M., Garg, M.: Deterministic $(\frac{1}{2}) + \epsilon$-approximation for submodular maximization over a matroid. In: Proceedings of SODA, pp. 241–254 (2019)
8. Chakrabarti, A., Kale, S.: Submodular maximization meets streaming: matchings, matroids, and more. Math. Program. **154**, 225–247 (2015). https://doi.org/10.1007/s10107-015-0900-7
9. Chan, T., Huang, Z., Jiang, S., Kang, N., Tang, Z.: Online submodular maximization with free disposal: Randomization beats 0.25 for partition matroids. (2016)
10. Chekuri, C., Gupta, S., Quanrud, K.: Streaming algorithms for submodular function maximization. In: Halldórsson, M.M., Iwama, K., Kobayashi, N., Speckmann, B. (eds.) ICALP 2015. LNCS, vol. 9134, pp. 318–330. Springer, Heidelberg (2015). https://doi.org/10.1007/978-3-662-47672-7_26
11. Das, A., Kempe, D.: Submodular meets spectral: greedy algorithms for subset selection, sparse approximation and dictionary selection. In: Proceedings of ICML, pp. 1057–1064 (2011)
12. Du, D., Li, Y., Xiu, N., Xu, D.: Simultaneous approximation of multi-criteria submodular function maximization. J. Oper. Res. Soc. China **2**, 271–290 (2014). https://doi.org/10.1007/s40305-014-0053-z
13. Dueck, D., Frey, B.J.: Non-metricaffinity propagation for unsupervised image categorization. In: Proceedings of ICCV, pp. 1–8 (2007)
14. El-Arini, K., Guestrin, C.: Beyond keyword search: discovering relevant scientific literature. In: Proceedings of SIGKDD, pp. 439–447 (2011)
15. Ene, A., Nguyen, H.: A nearly-linear time algorithm for submodular maximization with a knapsack constraint. In: Proceedings of ICALP, pp. 53:1–53:12 (2019)
16. El-Arini, K., Veda, G., Shahaf, D., Guestrin, C.: Turning down the noise in the blogosphere. In: Proceedings of SIGKDD, pp. 289–298 (2009)

17. Elenberg, E., Dimakis, A.G., Feldman, M., Karbasi, A.: Streaming weak submodularity: interpreting neural networks on the fly. In: Proceedings of NIPS, pp. 4044–4054 (2017)
18. Feige, U.: A threshold of $\ln n$ for approximating set cover. J. ACM **45**, 634–652 (1998)
19. Feldman, M., Karbasi, A., Kazemi, E.: Do less, get more: streaming submodular maximization with subsampling. In: Proceedings of ANIPS, pp. 730–740 (2018)
20. Filmus, Y., Ward, J.: Monotone submodular maximization over a matroid via non-oblivious local search. SIAM J. Comput. **43**, 514–542 (2014)
21. Goldengorin, B., Ghosh, D.: A multilevel search algorithm for the maximization of submodular functions applied to the quadratic cost partition problem. J. Glob. Optim. **32**, 65–82 (2005). https://doi.org/10.1007/s10898-004-5909-z
22. Gomes, R., Krause, A.: Budgeted nonparametric learning from data streams. In: Proceedings of ICML, pp. 391–398 (2010)
23. Guha, S., Mishra, N., Motwani, R., Ocallaghan, L.: Clustering data streams. In: Proceedings of FOCS, pp. 359–366 (2000)
24. Kazemi, E., Mitrovic, M., Zadimoghaddam, M., Lattanzi, S., Karbasi A.: Submodular streaming in all its glory: tight approximation, minimum memory and low adaptive complexity. In: Proceedings of ICML, pp. 3311–3320 (2019)
25. Krause, A., Golovin, D.: Submodular function maximization. In: Tractability: Practical Approaches to Hard Problems, pp. 71–104. Cambridge University Press, Cambridge (2014)
26. Krause, A., Singh, A., Guestrin, C.: Near-optimal sensor placements in Gaussian processes: theory, efficient algorithms and empirical studies. J. Mach. Learn. Res. **9**, 235–284 (2008)
27. Kulik, A., Shachnai, H., Tamir, T.: Approximations for monotone and nonmonotone submodular maximization with knapsack constraints. Math. Oper. Res. **38**, 729–739 (2013)
28. Lawrence, N., Seeger, M., Herbrich, R.: Fast sparse Gaussian process methods: the informative vector machine. In: Proceedings of NIPS, pp. 625–632 (2003)
29. Mirzasoleiman, B., Jegelka, S., Krause, A.: Streaming non-monotone sub-modular maximization: personalized video summarization on the fly. arXiv: 1706.03583 (2018)
30. Muthukrishnan, S.: Data streams: algorithms and applications. Found. Trends Theor. Comput. Sci. **1**, 117–236 (2005)
31. Nemhauser, G.L., Wolsey, L.A., Fisher, M.L.: Ananalysis of approximations for maximizing submodular set functions-I. Math. Program. **14**, 265–294 (1978)
32. Norouzi-Fard, A., Tarnawski J., Mitrovic, S., Zandieh, A., Mousavifar, A., Svensson, O.: Beyond 1/2-approximation for submodular maximization on massive data streams. In: Proceedings of ICML, pp. 3826–3835 (2018)
33. Sviridenko, M.: A note on maximizing a submodular set function subject to a knapsack constraint. Oper. Res. Lett. **32**, 41–43 (2004)
34. Vondrk, J.: Optimal approximation for the submodular welfare problem in the value oracle model. In: Proceedings of STOC, pp. 67–74 (2008)
35. Wang, Y., Xu, D., Wang, Y., Zhang, D.: Non-submodular maximization on massive data streams. J. Glob. Optim. **76**(4), 729–743 (2019). https://doi.org/10.1007/s10898-019-00840-8
36. Zoubin, G.: Scaling the Indian buffet process via submodular maximization. In: Proceedings of ICML, pp. 1013–1021 (2013)

On Fixed-Order Book Thickness Parameterized by the Pathwidth of the Vertex Ordering

Yunlong Liu[1]📵, Jie Chen[1]📵, Jingui Huang[1(✉)]📵, and Jianxin Wang[2]📵

[1] Hunan Provincial Key Laboratory of Intelligent Computing and Language
Information Processing, Hunan Normal University,
Changsha 410081, People's Republic of China
{ylliu,jie,hjg}@hunnu.edu.cn
[2] School of Computer Science and Engineering, Central South University,
Changsha 410083, People's Republic of China
jxwang@mail.csu.edu.cn

Abstract. Given a graph $G = (V, E)$ and a fixed linear order $\prec$ of V, the problem FIXED-ORDER BOOK THICKNESS asks whether there is a page assignment σ such that $\langle \prec, \sigma \rangle$ is a k-page book embedding of G. Recently, Bhore et al.(GD2019) presented an algorithm parameterized by the pathwidth of the vertex ordering (denoted by κ). In this paper, we first re-analyze the running time for Bhore et al.'s algorithm, and prove a bound of $2^{O(\kappa^2)} \cdot |V|$ improving on Bhore et al.'s bound of $\kappa^{O(\kappa^2)} \cdot |V|$. We further show that this parameterized problem does not admit a polynomial kernel unless NP $\subseteq$ coNP/poly. Finally, we show that the general FIXED-ORDER BOOK THICKNESS problem, in which a budget of at most c crossings over all pages was given, admits an algorithm running in time $(c + 2)^{O(\kappa^2)} \cdot |V|$.

1 Introduction

A *book embedding* of a graph G consists of placing the vertices of G in a line (called spine) in an order specified by a linear ordering $\prec$ of $V(G)$ and assigning edges of the graph to pages so that the edges assigned to the same page do not intersect. The minimum number of pages in which the graph G can be embedded is called the *book thickness* of G, denoted by bt(G) [1]. When the ordering of vertices in $V(G)$ along the spine is predetermined and fixed, the *book thickness* of G is specially called the *fixed-order book thickness* of G, denoted by fo-bt($G, \prec$) [2].

The problem FIXED-ORDER BOOK THICKNESS asks, given a graph $G = (V, E)$ and a positive integer k, whether fo-bt(G) $\leq k$. It is NP-complete in general [3].

This research was supported in part by the National Natural Science Foundation of China under Grant (No. 61572190, 61972423), 111 Project (No. B18059) and Hunan Provincial Science and Technology Program (No. 2018TP1018, 2018WK4001).

© Springer Nature Switzerland AG 2020
Z. Zhang et al. (Eds.): AAIM 2020, LNCS 12290, pp. 225–237, 2020.
https://doi.org/10.1007/978-3-030-57602-8_21

FIXED-ORDER BOOK THICKNESS is equivalent to determining whether a given circle graph can be properly vertex-colored by at most k colors [1], which was intensively studied in graph coloring [4–6]. Moreover, FIXED-ORDER BOOK THICKNESS originally arises in the context of sorting with parallel stacks, which has close relations and applications to VLSI design [7].

Pathwidth, as an important structural parameter of graphs, has been used to study parameterized complexity for many difficult problems [8–10]. For the problem FIXED-ORDER BOOK THICKNESS, Bhore et al. [2] considered the pathwidth of the vertex ordering, denoted by κ, as the parameter, and presented an algorithm running in time $\kappa^{O(\kappa^2)} \cdot |V|$. Bhore et al. [2] also posed a general version of FIXED-ORDER BOOK THICKNESS, which concerns the setting where edges are allowed to cross on the same page and the number of crossings over all pages is at most c. They mentioned the techniques in their algorithm can be extended to solving this general problem, but they did not elaborate further.

Our Results. Following the work recently done for FIXED-ORDER BOOK THICKNESS with respect to the vertex-cover number [11], our aim is to establish improved bound on algorithm also for its sister problem FIXED-ORDER BOOK THICKNESS parameterized by the pathwidth of the vertex ordering (abbreviated by BTPW). Our specific work includes three parts as follows.

(1). We re-analyze Bhore et al.'s algorithm in [2] and obtain an improved upper bound on its running time. By constructing an auxiliary graph, we re-estimate the size of the record set in that algorithm and prove that it can be reduced from $(\kappa + 2)^{\kappa^2}$ to 2^{κ^2}. This means that Bhore et al.'s algorithm can be done in time $2^{O(\kappa^2)} \cdot |V|$. Although the basic strategy used is similar to that of Liu et al. [11], the implementation tactics are quite distinct from those in [11].

(2). We show that FIXED-ORDER BOOK THICKNESS parameterized by the pathwidth of the vertex ordering does not admit a polynomial kernel unless NP $\subseteq$ coNP/poly. This kernel lower bound is derived from the framework called AND-cross-composition defined by Bodlaender et al. [12].

(3). We also investigate the general version of FIXED-ORDER BOOK THICKNESS (abbreviated by BTPW-CROSS). By expanding our analysis approach used in BTPW, we show that the problem BTPW-CROSS admits a parameterized algorithm running in time $(c + 2)^{O(\kappa^2)} \cdot |V|$.

2   Terminology and Notations

We consider only undirected graphs. Given a graph $G = (V, E)$, we use $V(G)$ to denote its vertex set and let $n = |V|$. For two vertices u, v in V, let uv denote the edge between u and v. For $r \in \mathbb{N}$, we use $[1, r]$ to denote the set $\{1, \ldots, r\}$.

Given a graph $G = (V, E)$ with a linear order $\prec$ of V such that $v_1 \prec v_2 \prec \ldots \prec v_n$, the *pathwidth* of $(G, \prec)$ is the minimum number κ such that for each vertex v_i ($i \in [1, n]$), there are at most κ vertices left of v_i that are adjacent to v_i or a vertex right of v_i. Formally, for each v_i we call the set $P_i = \{v_j \mid j < i, \exists q \geq i$

such that $v_j v_q \in E$} the *guard set* for v_i, and the pathwidth of $(G, \prec)$ is simply $\max_{i \in [1,n]} |P_i|$. The elements of the guard set are called the *guards* for v_i.

Let v_0 be a vertex with degree 0 and let v_0 be placed to the left of v_1 in $\prec$. For a vertex v_i, let $P^*_{v_i} = \{g^i_1, g^i_2, \ldots, g^i_m\}$ where for each $j \in [1, m-1]$, g^i_j is the j-th guard of v_i in reverse order of $\prec$, and $g^i_m = v_0$. For a vertex v_i, let $E_i = \{v_a v_b \mid v_a v_b \in E, b > i\}$ be the set of all edges with at least one endpoint to the right of v_i and let $S_i = \{g^i_j v_b \mid g^i_j \in P^*_{v_i}, g^i_j v_b \in E_i\}$ be the restriction of E_i to edges between a vertex to the right of v_i and a guard in $P^*_{v_i}$.

For ease of presentation, we define a special planer graph. A graph G is a *restricted plane graph with spine L and head h* if G satisfies the following properties: (1) all vertices lie in a horizontal line L with a fixed-order $\prec$; (2) all edges lie in the half-plane above L and are incident to the rightmost vertex (denoted as h) in $\prec$.

Some proofs are omitted due to space constrains; they will be given in the complete version of the paper.

3   Improved Upper-Bound on Bhore et al.'s Algorithm

In this section, we re-analyze Bhore et al.'s [2] algorithm for BTPW, and obtain an improved upper-bound on its running time.

We first restate some notations introduced in [2]. Let $i \in [1, n]$. A page assignment $\alpha : E_i \rightarrow [1, k]$ is called a *valid partial page assignment* if α maps the edges in E_i to pages in a non-crossing fashion. Let α be a valid partial page assignment of E_i, v_a be a vertex with $a \leq i$. A vertex v_x ($x < a$) is α-*visible* to v_a on a page p if it is possible to draw the edge $v_a v_x$ in page p without crossing any other edge mapped to p by α.

Let $a \leq i \leq n$, $p \in [1, k]$, and α be a valid partial page assignment of E_i. The edge $v_c v_d \in S_i$ (if exists) is a (α, i, p)-*important edge* of v_a if it satisfies the following properties: (1) $\alpha(v_c v_d) = p$; (2) $c < a$; and (3) $|a - c|$ is minimum among all such edges in S_i. Correspondingly, the vertex v_c is called the (α, i, p)-*important guard* of v_a. Bhore et al.'s [2] algorithm traverses vertices in a right-to-left order along $\prec$, and for each vertex, the algorithm stores a set of records containing some representative visibility vectors. More precisely, the visibility vector $U_i(v_a, \alpha)$ is defined as follows: the p-th entry is the (α, i, p)-important guard of v_a, and $\diamond$ if v_a has no (α, i, p)-important guard. The record set $\mathcal{Q}_i = \{(U_i(v_i, \alpha), U_i(g^i_1, \alpha), U_i(g^i_2, \alpha), \ldots, U_i(g^i_{m-1}, \alpha)) \mid \exists \text{ valid partial page assignment } \alpha : E_i \rightarrow [1, k] \}$. A mapping Λ_i from $\mathcal{Q}_i$ to a valid partial page assignments of E_i maps each tuple $\omega \in \mathcal{Q}_i$ to some α such that $\omega = (U_i(v_i, \alpha), U_i(g^i_1, \alpha), U_i(g^i_2, \alpha), \ldots, U_i(g^i_{m-1}, \alpha))$. Since there are at most $(\kappa + 2)^\kappa$ visibility vectors for each guard of v_i, $|\mathcal{Q}_i| \leq (\kappa + 2)^{\kappa^2}$ [2].

In Bhore et al.'s algorithm, the size of $\mathcal{Q}_i$ dominantly determines its running time. In the following, we focus on re-estimating $|\mathcal{Q}_i|$ by another approach.

Observe that the p-th components $U_i(g, \alpha)[p]$ for $g \in \{v_i\} \cup P^*_{v_i} \setminus \{g^i_m\}$ are not "independent", instead, they are uniformly determined by the guards of the vertex v_i. Observe further that the edges whose endpoints all lie on the right of

v_i have nothing to do with computing $U_i(x, \alpha)[p]$. Hence, we globally consider the combinations of edges that incident to the guards of v_i, and deduce a new function of κ bounding on $|\mathcal{Q}_i|$. For this target, we define a procedure called **Edge-Pruning-1** to adjust the edges on each page in α.

Let $i \in [1, n]$, $\alpha : E_i \to [1, k]$ be a valid partial page assignment, and $p \in [1, k]$. Let $g \in P_{v_i}^*$, and let $E(p, g)$ be the set of edges in E_i that are incident to g and assigned on page p by α. The steps in Edge-Pruning-1 are described as follows (Fig. 1).

Procedure Edge-Pruning-1(α)

1. **for** $p = 1$ to k **do**
2.   **for each** $g \in P_{v_i}^* \setminus \{g_m^i\}$ **do**
2.1   **if** $E(p, g) \neq \emptyset$ **then**
2.2   { **if** $gv_{i+1} \notin E(p, g)$ **then** add edge gv_{i+1} such that $\alpha(gv_{i+1})=p$;
2.3   **if** $E(p, g) \setminus \{gv_{i+1}\} \neq \emptyset$ **then** delete edges in $E(p, g) \setminus \{gv_{i+1}\}$; }
3.   **if** $v_i v_{i+1} \in E_i$ and $\alpha(v_i v_{i+1}) = p$ **then** delete the edge $v_i v_{i+1}$;
4. **delete** all vertex v_x for $x \geq i + 2$;
5. **output**(α').

Fig. 1. The main steps in Edge-Pruning-1 procedure

After executing the procedure Edge-Pruning-1 on a given assignment α, we obtain a simplified assignment α' such that there are at most κ edges with one common endpoint v_{i+1} on each page. Figure 2 shows an example of Edge-Pruning-1.

Fig. 2. An example on Edge-Pruning-1 from an original 2-page assignment of E_4 (a) to a simplified 2-page assignment (b).

Next, we show that each visibility vector for α' is equal to that for α, respectively.

Lemma 1. *For each $g \in \{v_i\} \cup P_{v_i}^* \setminus \{g_m^i\}$, $U_i(g, \alpha) = U_i(g, \alpha')$.*

Proof. According to the definition of visibility vector, it is sufficient to consider the visibility from g to v_x for $v_x \prec g$. Without loss of generality, we argue that $U_i(g, \alpha)[p] = U_i(g, \alpha')[p]$ on page p (for $p \in [1, k]$).

($\Rightarrow$) Assume that $U_i(g, \alpha)[p] = v_h$ ($v_h \in P_{v_i}^* \cup \{\diamond\}$). We distinguish two cases based on whether $v_h = \diamond$ or not. Case (1): $v_h = \diamond$. Then every vertex v_x with $v_x \prec g$ is α-visible to g. In other words, for any vertex v_x with $v_x \prec g$, the set $E(p, v_x) = \emptyset$. During the procedure Edge-Pruning-1, steps 2.1–2.3 will not be executed for the vertex v_x. Hence, every vertex v_x with $v_x \prec g$ is still α'-visible to g, that is, $U_i(g, \alpha')[p] = \diamond$. Case (2): $v_h \neq \diamond$. Without loss of generality, assume that v_h is incident to t ($t > 0$) edges in E_i, denoted as $v_h w_1, v_h w_2, \ldots, v_h w_t$. Since the edges incident to the vertex v_h are all in E_i, the vertex v_h must be one guard of v_i. During the procedure Edge-Pruning-1, steps 2.1–2.3 will be executed and the edges in $E(p, v_h)$ will be adjusted. After executing Edge-Pruning-1, the edge $v_h w_1$ either remains unchanged (in the case $w_1 = v_{i+1}$) or is replaced by $v_h v_{i+1}$ (in the case $w_1 \neq v_{i+1}$). Hence, the edge $v_h v_{i+1}$ must exist. It holds that $U_i(g, \alpha')[p] = v_h$.

($\Leftarrow$) Assume that $U_i(g, \alpha')[p] = v_h$ ($v_h \in P_{v_i}^* \cup \{\diamond\}$). We also distinguish two cases based on whether $v_h = \diamond$ or not. Case (1): $v_h = \diamond$. Then every vertex v_x with $v_x \prec g$ is α'-visible to g. We can infer that every vertex v_x with $v_x \prec g$ is α-visible to g. Assume towards a contradiction that there exists one vertex $v_z \prec g$ such that v_z is not α-visible to g. Then there must be a vertex v_a between v_z and g that is incident to at least one edge in E_i separating v_z and g. Note that the vertex v_a is also one guard of v_i. During the procedure Edge-Pruning-1, steps 2.1–2.3 will be executed and the edges in $E(p, v_a)$ will be adjusted. In the simplified assignment α', the edge $v_a v_{i+1}$ separates v_z and g on page p, contradicting the assumption that $U_i(g, \alpha')[p] = \diamond$. Case (2): $v_h \neq \diamond$. Then for any vertex v_x ($v_x \prec v_h$), the edge $v_h v_{i+1}$ on page p in α' separates v_x from g. Note that v_h is also one guard of v_i since $v_i \prec v_{i+1}$. By the description of Edge-Pruning-1, the existence of edge $v_h v_{i+1}$ in α' is due to the fact that $E(p, v_h) \neq \emptyset$ in α. So, in the original assignment α, v_x is separated from g by at least one edge in $E(p, v_h)$. It holds that $U_i(g, \alpha)[p] = v_h$. □

Based on Lemma 1, we re-estimate $|\mathcal{Q}_i|$ with an improved upper bound.

Lemma 2. *The size of $\mathcal{Q}_i$ can be bounded by 2^{κ^2}.*

Proof. Let L be a straight line joining $\kappa + 1$ vertices with a fixed order $\prec$, in which the rightmost vertex is denoted by h. Let $\mathcal{A} = \{B \,|\, B$ is a restricted plane graph with spine L and head $h\}$. We estimate the size of $\mathcal{A}$ according to the number of different combinations of its edges. For each vertex u lying on L except h, there are only two possible cases to be considered: (1) it is adjacent to the head h by one edge; (2) it is not adjacent to the head h. Hence, $|\mathcal{A}| = 2^\kappa$. Let $\mathcal{D} = \mathcal{A}_1 \times \mathcal{A}_2 \times \cdots \times \mathcal{A}_k$, where $\mathcal{A}_r = \mathcal{A}$ for $r \in [1, k]$. It follows that $|\mathcal{D}| = 2^{k\kappa}$.

Let $\mathcal{P}_i = \{\alpha \,|\, \alpha = \Lambda_i(\omega)$ and $\omega \in \mathcal{Q}_i\}$. Let α_1 and α_2 be two distinct assignments in $\mathcal{P}_i$. By the definition of $\mathcal{Q}_i$, there exists at least one vertex $g \in \{v_i\} \cup P_{v_i}^* \setminus \{g_m^i\}$ such that $U_i(g, \alpha_1) \neq U_i(g, \alpha_2)$. In the following, we show that there exists an injective function f from $\mathcal{P}_i$ to $\mathcal{D}$. (1) Let $\alpha \in \mathcal{P}_i$. The assignment α is a k-page book embedding which includes k half-planes. After executing the procedure Edge-Pruning-1 on α, the vertex v_i is an isolated vertex and each half-plane is translated into a restricted plane graph with spine

L and head h. Hence, there exists a unique tuple $(B_1, B_2, \ldots, B_k)$ in $\mathcal{D}$ such that $f(\alpha) = (B_1, B_2, \ldots, B_k)$. (2) For any two distinct assignments α_1 and α_2 in $\mathcal{P}_i$, it holds that $f(\alpha_1) \neq f(\alpha_2)$. Otherwise, for each $g \in \{v_i\} \cup P^*_{v_i} \setminus \{g^i_m\})$, $U_i(g, f(\alpha_1)) = U_i(g, f(\alpha_2))$. By Lemma 1, it follows that $U_i(g, \alpha_1) = U_i(g, \alpha_2)$ for $g \in \{v_i\} \cup P^*_{v_i} \setminus \{g^i_m\})$, contradicting the fact that α_1 and α_2 are two distinct assignments in $\mathcal{P}_i$.

As a consequence, the size of $\mathcal{Q}_i$ is no larger than that of $\mathcal{D}$. Note that $k < \kappa$ (In the case of $k \geq \kappa$, the problem BTPW is trivial [2]). Therefore, the size of $\mathcal{Q}_i$ can be bounded by 2^{κ^2}. □

Based on Lemma 2 and the fact that $\kappa^\kappa < 2^{\kappa^2}$, we arrive at our first result.

Theorem 1. *There is an algorithm which takes as input a graph $G = (V, E)$ with a vertex order $\prec$ and computes a page assignment σ of E such that $(\prec, \sigma)$ is a (fo-bt$(G, \prec)$)-page book embedding of G. The algorithm runs in $2^{O(\kappa^2)} \cdot |V|$ where κ is the pathwidth of $(G, \prec)$.*

4   On Kernel Lower Bound

In this section, we show that no polynomial kernel is possible for the problem BTPW unless NP $\subseteq$ coNP/poly by employing the framework named AND-cross-composition.

Definition 1. *(AND-cross-composition)([12]) Let $L \subseteq \Sigma^*$ be a set and let $Q \subseteq \Sigma^* \times \mathbb{N}$ be a parameterized problem. We say that L AND-cross-composes into Q if there is a polynomial equivalence relation R and an algorithm which, given t strings $x_1, x_2, \ldots, x_t$ belonging to the same equivalence class of R, computes an instance $(x^*, k^*) \in \Sigma^* \times \mathbb{N}$ in time polynomial in $\Sigma^t_{i=1}|x_i|$ such that:*

(1). $(x^*, k^*) \in Q$ if and only if for all i, $1 \leq i \leq t$, $x_i \in L$;
(2). k^* is bounded by a polynomial in $max^t_{i=1}|x_i| + \log t$.

Lemma 3. *([13]) Assume that an NP-hard language L AND-cross-composes into a parameterized language Q. Then Q does not admit a polynomial compression, unless NP $\subseteq$ coNP/poly.*

Theorem 2. *Book Thickness parameterized by the pathwidth of the vertex ordering does not admit a polynomial kernel unless NP $\subseteq$ coNP/poly.*

Proof. We prove this theorem by showing FIXED-ORDER BOOK THICKNESS AND-cross-composes into Book Thickness parameterized by the pathwidth of the vertex ordering; by Lemma 3 this is sufficient to establish this theorem. As we known, the problem FIXED-ORDER BOOK THICKNESS is NP-hard in general [3]. In the following, we show that there exists one AND-cross-composition algorithm from the former to the latter.

First of all, we define a polynomial equivalence relation R. An instance of FIXED-ORDER BOOK THICKNESS is a tuple $((G, \prec), k)$ and asks whether G admits a book embedding with k pages. The relation R is defined as follows: all pairs $((G_i, \prec_i), k_i)$, $((G_j, \prec_j), k_j)$ go to the same equivalence class if $|V(G_i)| = |V(G_j)|$ and $k_i = k_j$. The instance is malformed if either it is not the right format or if $|V(G)|^2 < 2k$. All the malformed instances were put into one equivalence class. We show that this relation meets the conditions in the polynomial equivalence relation. (1) Given two well-formed instances $((G_i, \prec_i), k_i)$, $((G_j, \prec_j), k_j)$, we can check in polynomial time if $|V(G_i)| = |V(G_j)|$ and $k_i = k_j$. (2) the number of equivalence classes is at most $n^3/2 + 1$ since $2k_i \leq |V(G_i)|^2$ in a well formed instance, where $n = \max_{i=1}^t |V(G_i)|$.

Next, we give an AND-cross-composition algorithm for instances belonging to the same equivalence class. Given a bunch of malformed instance, we output a trivial NO-instance. So, we assume that $((G_1, \prec_1), k)$, $((G_2, \prec_2), k)$, ..., $((G_t, \prec_t), k)$ are the instances in the same equivalence class and $|V(G_i)| = n$ for all $i \in [1,t]$. A composition algorithm is as follows. After being input with $((G_1, \prec_1), k)$, $((G_2, \prec_2), k)$, ..., $((G_t, \prec_t), k)$, it outputs an instance $((G, \prec), \kappa, k')$. In the following, we give some annotation for elements in the output instance. (1). G is a disjoint union of $G_1, G_2, \ldots, G_t$. To ease of presentation, we assume that $G_1, G_2, \ldots, G_t$ are arranged in a row from left to right according to their subscript indices. (2). The linear order $\prec$ of $V(G)$ is the union of $\prec_1, \prec_2, \ldots, \prec_t$ in a left-to-right fashion. Since the order $\prec_i$ of $V(G_i)$ for $i \in [1,t]$ is fixed, the order $\prec$ of $V(G)$ is fixed. (3). The pathwidth of $((G, \prec))$, denoted as κ, can be computed in time $O(tn^3)$. More precisely, for each vertex $v_i \in V(G)$ $(i \in [1, tn])$, we compute the number of vertices left of v_i that are adjacent to v_i or a vertex right of v_i in time $O(n^2)$ and denote it by n_{v_i}. It holds that $\kappa = \max_{i \in [1,tn]} n_{v_i}$. Let κ_i be the pathwidth of $((G_i, \prec_i))$ for $i \in [1,t]$. By the structure of $(G, \prec)$, it follows that $\kappa = \max_{i \in [1,t]} \kappa_i$. Since $\kappa_i \leq |V(G_i)|$ for $i \in [1,t]$, κ is bounded by a polynomial in $\max_{i=1}^t |V(G_i)| + \log t$. (4). We set $k' = k$.

Finally, we show that $((G, \prec), \kappa, k)$ is a YES instance of BTPW, if and only if $((G_i, \prec_i), k)$ was a YES instance of FIXED-ORDER BOOK THICKNESS for all $i \in [1,t]$.

($\Rightarrow$) Let $((G, \prec), \kappa, k)$ be a YES instance of BTPW. Then there exists a k-page book embedding $\langle \prec, \sigma \rangle$ for G. Since G is a disjoint union of $G_1, G_2, \ldots, G_t$, the k-page book embedding $\langle \prec, \sigma \rangle$ can be decomposed into t disjoint parts $\langle \prec_1, \sigma_1 \rangle$, $\langle \prec_2, \sigma_2 \rangle$, ..., $\langle \prec_t, \sigma_t \rangle$ such that $\langle \prec_i, \sigma_i \rangle$ is a k-page book embedding of G_i (for $i \in [1,t]$). Hence, $((G_i, \prec_i), k)$ is a YES instance of FIXED-ORDER BOOK THICKNESS for all $i \in [1,t]$.

($\Leftarrow$) Let $((G_i, \prec_i), k)$ be a YES instance of FIXED-ORDER BOOK THICKNESS for all $i \in [1,t]$. Then for each graph G_i, there exists a k-page book embedding $\langle \prec_i, \sigma_i \rangle$. Since G is a disjoint union of $G_1, G_2, \ldots, G_t$, any two k-page book embedding $\langle \prec_i, \sigma_i \rangle$ and $\langle \prec_j, \sigma_j \rangle$ are disjoint. Let p_h^i denote the h-th page in $\langle \prec_i, \sigma_i \rangle$ for $i \in [1,t]$ and $h \in [1,k]$. We can construct a k-page book embedding $\langle \prec, \sigma \rangle$ of G by setting a disjoint union of $p_h^1, p_h^2, \ldots, p_h^t$ as its the h-th page p_h for $h \in [1,k]$. Hence, $((G, \prec), \kappa, k)$ is a YES instance of BTPW. $\square$

5   A Parameterized Algorithm for the General Problem

A general version of FIXED-ORDER BOOK THICKNESS parameterized by both the pathwidth of the vertex order κ and the number c of crossings over all pages (i.e., BTPW-CROSS) is formally defined as follows.

Input: A tuple $(G, \prec)$, a non-negative integer c;

Parameters: κ, c;

Question: Can we find a k-page book drawing $(\prec, \sigma)$ of G such that the number of crossings over all pages is no more than c ?

   We first briefly present a specific algorithm for BTPW-CROSS by extending the techniques in [2], whose feasibility was mentioned by Bhore et al. [2]. Then, we pay more attention to employing our approach to analyze its running time.

5.1   Design of the Algorithm

We begin with expanding the notion of valid page assignment. Let $i \in [1, n]$. An assignment $\alpha : E_i \rightarrow [1, k]$ is a *valid page assignment* if α assigns the edges in E_i to pages such that the number of crossings over all pages is at most c. For ease of presentation, we introduce the notion of potential edge. Given a valid partial page assignment α of edges in E_i and two vertices v_x, v_a with $x < a \leq i$, we draw an edge between v_x and v_a on page p. The added edge $v_x v_a$ is called a *potential edge* with respect to the edges in E_i assigned on page p by α.

   To capture the information on the number of crossings generated by the potential edges, we also introduce the notion of *crossing number matrix*, which is originated from the notion of visibility vector in [2]. Given a valid partial page assignment α and a vertex v_a with $a \leq i$, we define a crossing number matrix $M_i(v_a, \alpha)$ with k rows and $c + 1$ columns. The entry (p, q) in $M_i(v_a, \alpha)$ is set by the following rule.

   If there exists a guard $v_z \in P_{v_i}^*$ ($z < a$) such that the potential edge $v_z v_a$ on page p exactly crosses q edges in E_i assigned to page p by α, **then** the entry $(p, q) = v_z$. Once there are at least two guards that satisfy this condition, we only choose the utmost guard to the left of v_a. **Otherwise**, $(p, q) = $ "null".

   Note that for a given tuple (i, v_a, α), it is straightforward to compute $M_i(v_a, \alpha)$ in polynomial time. Figure 3 shows a crossing number matrix for a 2-page assignment of E_5, where $c = 3$.

$$M_5(v_5, \alpha) = \begin{pmatrix} 0 & 1 & 2 & 3 \\ v_4 & v_2 & \text{null} & v_1 \\ v_4 & \text{null} & v_2 & v_0 \end{pmatrix}$$

Fig. 3. A partial 2-page assignment of the edges in E_5 (left) and the corresponding crossing number matrix (right).

We also introduce *a crossing number vector* $N_i(\alpha)$ as follows: the p-th component of $N_i(\alpha)$ is the number of crossings on the p-th page mapped by α (for $p \in [1, k]$).

For some consecutive vertices in $V(G)$, the corresponding crossing number matrices are actually the same one. Thus, we can obtain the following statement, which expanding Lemma 3 in [2].

Lemma 4. *Let α be a valid partial assignment of E_i, $v_a \prec v_i$, and assume $v_a \notin P_{v_i}^*$. Let $v_b \in P_{v_i}^* \cup \{v_i\}$ such that $b > a$ and $b - a$ is minimized. Then $M_i(v_a, \alpha) = M_i(v_b, \alpha)$.*

By Lemma 4, we can use $\kappa + 1$ crossing number matrices to capture the complete information about all of crossings on each page. Moreover, the information stored in the crossing vector is an important factor in distinguishing different page assignments. Hence, we define one *expanded record set* as follows: $\mathcal{Q}_i' = \{(N_i(\alpha), M_i(v_i, \alpha), M_i(g_1^i, \alpha), M_i(g_2^i, \alpha), \ldots, M_i(g_{m-1}^i, \alpha)) \mid \exists \text{ valid partial }$ page assignment $\alpha : E_i \to [1, k]\}$. For ease of presentation, the tuple $(N_i(\alpha), M_i(v_i, \alpha), M_i(g_1^i, \alpha), M_i(g_2^i, \alpha), \ldots, M_i(g_{m-1}^i, \alpha))$ is also called a *matrix queue* for α, and is denoted by $\omega_i(\alpha)$ in the rest of this paper. Along with $\mathcal{Q}_i'$, we also store a mapping Λ_i' from $\mathcal{Q}_i'$ to valid partial page assignments of E_i which maps each record $\omega_i(\alpha) \in \mathcal{Q}_i'$ to some α such that $\omega_i(\alpha) = (N_i(\alpha), M_i(v_i, \alpha), M_i(g_1^i, \alpha), M_i(g_2^i, \alpha), \ldots, M_i(g_{m-1}^i, \alpha))$.

Adapting the framework on dynamic programming for solving BTPW in [2], we can obtain an algorithm for solving BTPW-CROSS, denoted by ALPW. Specifically, the main steps in ALPW can be described as follows.

The basic strategy is to dynamically generate some k-book drawing containing at most c crossings along the order $\prec$ in a right-to-left fashion. Let $F_{i-1} = E_{i-1} \setminus E_i$. Assume that the record set $\mathcal{Q}_i'$ has been computed. Each page assignment β of edges in F_{i-1} and each record $\omega \in \mathcal{Q}_i'$ are branched. For each such β and $\alpha = \Lambda_i'(\omega)$, the combined assignment $\alpha \cup \beta$ is tested under two conditions: (1) the number of crossings on each page is no more than c; (2) the number of crossings over all page is no more than c. If $\alpha \cup \beta$ forms a valid partial page assignment, the corresponding record is computed and stored. The mapping Λ_{i-1}' is set to map this record to $\alpha \cup \beta$. Otherwise, the pair (α, β) is discarded.

Let α_1 and α_2 be two valid page assignments of E_i, β be a page assignment of edges in F_{i-1}, and let $\alpha_1 \cup \beta$ and $\alpha_2 \cup \beta$ be the intermediate assignments generated during the algorithm ALPW.

Lemma 5. *If $\omega_i(\alpha_1) = \omega_i(\alpha_2)$, then $\omega_{i-1}(\alpha_1 \cup \beta) = \omega_{i-1}(\alpha_2 \cup \beta)$.*

Based on Lemma 5, we can obtain the following conclusion.

Theorem 3. *If $((G, \prec), \kappa, c)$ contains at least one valid assignment, then the algorithm ALPW$((G, \prec), \kappa, c)$ returns a valid page assignment.*

5.2   Analysis on the Running Time

We adapt the approach used for BTPW to analyze its running time. First of all, we expand the notion of edge pruning and re-define a procedure **Edge-Pruning-2**. Let $i \in [1, n]$, $\alpha : E_i \rightarrow [1, k]$ be a valid partial page assignment, and $p \in [1, k]$. Let $g \in P_{v_i}^*$ and let $E(p, g)$ be the set of edges in E_i that are incident to g and assigned on page p by α. Assume that $E(p, g) \backslash \{gv_{i+1}\} = \{gw_1, gw_2, \ldots, gw_t\}$ and $g \prec w_1 \prec w_2 \prec \ldots \prec w_t$ in $\prec$. The description of Edge-Pruning-2 is identical to that for Edge-Pruning-1 in Sect. 3, except that step 2.3 becomes:

for $r = 1$ to t **do**: (1) delete the edge gw_r; (2) **if** the number of edges between g and v_{i+1} is less than $c + 1$, **then** add one multiple edge gv_{i+1}.

Given a valid partial assignment α, we execute the procedure Edge-Pruning-2 on it. After pruning some edges, we obtain a simplified assignment α' such that the edges on each page are incident to the common vertex v_{i+1}. Figure 4 gives an example of Edge-Pruning-2.

(a)   (b)

Fig. 4. An example of Edge-Pruning-2 from an original 2-page assignment of E_5 (a) to the simplified 2-page assignment, in which $c = 3$ (b).

Let $M_i(v_x, \alpha)$ be a crossing number matrix for an original assignment α and let $M_i(v_x, \alpha')$ be that for the simplified assignment α' $(v_x \in \{v_i\} \cup P_{v_i}^* \backslash \{g_m^i\})$.

Lemma 6. *For each $v_x \in \{v_i\} \cup P_{v_i}^* \backslash \{g_m^i\}$, $M_i(v_x, \alpha) = M_i(v_x, \alpha')$.*

Proof. By the definition of crossing number matrix, we only need to argue that any entry in $M_i(v_x, \alpha)$ is equal to the corresponding entry in $M_i(v_x, \alpha')$.

($\Rightarrow$) Assume that an entry (p, q) in $M_i(v_x, \alpha)$ is equal to z. Our aim is to show that (p, q) in $M_i(v_x, \alpha')$ is also equal to z. We distinguish two cases based on whether $z = $ "null" or not. Case (1): $z \neq$ "null" (let $z = v_b$). Then, the potential edge $v_b v_x$ exactly generates q crossings on page p. Without loss of generality, assume that there are r $(r \geq 1)$ guards of v_x lying between v_b and v_x, denoted as $v_b \prec v_1^x \prec v_2^x \prec \ldots \prec v_r^x \prec v_x$. Assume further that the vertex v_h^x (for $h \in [1, r]$) is incident to n_h^x edges in E_i that enclose v_x on page p. By the assumption that $(p, q) = v_b$, it follows that $\Sigma_{h=1}^r n_h^x = q$. Since $q \in [0, c]$, it holds that $n_h^x \leq c$. Note that the vertex v_h^x (for $h \in [1, r]$) is also a guard of v_i. During executing the procedure Edge-Pruning-2, the edges incident to v_h^x will be adjusted. After pruning edges, there are n_h^x multiple edges between v_h^x and v_{i+1}. It still holds that $\Sigma_{h=1}^r n_h^x = q$. Thus, the potential edge $v_b v_x$ still exactly crosses q edges on page p, which means that the entry (p, q) in $M_i(x, \alpha')$ is also equal to v_b. Case

(2): $z =$ "null". We first locate the entry (p, j) in $M_i(v_x, \alpha)$ such that (p, j) has the following properties: ① $(p, j) \neq$ "null", ② $j < q$, and ③ $q - j$ is minimized. Assume that $(p, j) = v_w$, and that v_w is incident to n_w edges in E_i on page p. By the assumption that $(p, q) =$ "null", it holds that $n_w \neq 1$. After pruning edges, the number of edges between v_w and v_{i+1} is either n_w (in the case $n_w \leq c + 1$) or $c + 1$ (in the case $n_w > c + 1$). Hence, in the simplified assignment α', it still holds that $n_w \neq 1$, which means that the entry $(p, q) =$ "null" in $M_i(v_x, \alpha')$.

($\Leftarrow$) Assume that an entry (p, q) in $M_i(v_x, \alpha')$ is equal to z. Our aim is to show that (p, q) in $M_i(v_x, \alpha)$ is also equal to z. We also distinguish two cases based on whether $z =$ "null" or not. Case (1): $z \neq$ "null" (let $z = v_b$). Without loss of generality, assume that there are r $(r \geq 1)$ guards of v_x between v_b and v_x, denoted as $v_b \prec v_1^x \prec v_2^x \prec \ldots \prec v_r^x \prec v_x$. Assume further that, for each guard v_h^x of v_x $(h \in [1, r])$, there are n_h^x multiple edges between v_h^x and v_{i+1}. By step 2.3 in Edge-Pruning-2, we can infer that the vertex v_h^x is incident to n_h^x edges in E_i that enclose the vertex v_x on page p in the assignment α. Hence, the potential edge $v_b v_x$ exactly generates q crossings on page p, which means (p, q) in $M_i(v_x, \alpha)$ is also equal to v_b. Case (2): $z =$ "null". Similarly, we first locate the entry (p, j) in $M_i(v_x, \alpha')$ such that (p, j) has the following properties: ① $(p, j) \neq$ "null", ② $j < q$, and ③ $q - j$ is minimized. Assume that $(p, j) = v_w$, and that the number of multiple edges between v_w and v_{i+1} is n_w. By the assumption that $(p, q) =$ "null" in $M_i(v_x, \alpha')$, it holds that $n_w \neq 1$. By step 2.3 in Edge-Pruning-2, we also can infer that the number of edges in E_i incident to v_w is either n_w or at least $c + 2$ on page p in the assignment α. Hence, the entry $(p, q) =$ "null" in $M_i(v_x, \alpha)$.   □

Based on Lemma 6, we obtain a bound for the size of $\mathcal{Q}_i'$.

Lemma 7. *The size of $\mathcal{Q}_i'$ can be bounded by $(c + 1)^k (c + 2)^{k\kappa}$.*

Proof. Let L be a straight line joining $\kappa + 1$ vertices with a fixed order, in which the rightmost vertex is denoted by h, and let $\mathcal{H} = \{B \mid B$ is a restricted plane graph with spine L and head $h\}$. We first estimate the size of $\mathcal{H}$ according to the number of different combinations of its edges. For each vertex u lying on L except h, there are $(c + 2)$ possible cases to be considered: (1) u is not adjacent to the head h; (2) there is one edge between u and h; (3) there are t (for $t \in [2, c + 1]$) multiple edges between u and h. Hence, $|\mathcal{H}| = (c + 2)^\kappa$.

Let N be a $k \times 1$ matrix, and for each $i \in [1, k]$, assume that $N[i] \in [0, c]$. Let $\mathcal{D}' = N \times \mathcal{H}_1 \times \mathcal{H}_2 \times \cdots \times \mathcal{H}_k$ $(\mathcal{H}_r = \mathcal{H}$ for $r \in [1, k])$. Then, $|\mathcal{D}'| = (c+1)^k (c+2)^{k\kappa}$.

Let $\mathcal{P}_i' = \{\alpha \mid \alpha = \Lambda_i'(\omega)$ and $\omega \in \mathcal{Q}_i'\}$. Based on Lemma 6, we can show that there exists an injective function f from $\mathcal{P}_i'$ to $\mathcal{D}'$ along the same lines in the proof of Lemma 2.

As a consequence, the size of $\mathcal{Q}_i'$ is no larger than that of $\mathcal{D}'$. Therefore, the size of $\mathcal{Q}_i'$ can be bounded by $(c + 1)^k (c + 2)^{k\kappa}$.   □

Based on Lemma 7, we obtain the flowing conclusion.

Theorem 4. *The algorithm ALPW for the problem BTPW-CROSS runs in time $(c + 1)^k (c + 2)^{k\kappa} \kappa^\kappa \cdot |V|$.*

Let fo-bt$(G, \prec, c)$ be the minimum k such that $(G, \prec, c)$ is a YES instance of the problem BTPW-CROSS. Since fo-bt$(G, \prec, c) \leq$ fo-bt$(G, \prec)$ and fo-bt$(G, \prec) < \kappa$, it follows that fo-bt$(G, \prec, c) < \kappa$. Now, we arrive at our third result.

Theorem 5. *There is an algorithm which takes as input a graph* $G = (V, E)$ *with a vertex order* $\prec$, *and an integer c, runs in time* $(c + 2)^{O(\kappa^2)} \cdot |V|$, *and computes a page assignment* σ *such that* $(\prec, \sigma)$ *is a (fo-bt$(G, \prec)$)-page book drawing of* G.

6 Conclusions

We further study parameterized algorithms for the problem FIXED-ORDER BOOK THICKNESS with respect to the pathwidth of the vertex ordering. We prove an improved running time bound for the algorithm given by Bhore et al. in [2], derive that this parameterized problem does not admit a polynomial kernel unless NP$\subseteq$ coNP/poly, and show that the general FIXED-ORDER BOOK THICKNESS problem admits a parameterized algorithm running in time $(c + 2)^{O((\kappa^2)} \cdot |V|$.

Acknowledgements. The authors thank the anonymous referees, whose comments improved the presentations of this paper.

References

1. Dujmović, V., Wood, D.R.: On linear layouts of graphs. Discret. Math. Theoret. Comput. Sci. **6**, 339–358 (2004)
2. Bhore, S., Ganian, R., Montecchiani, F., Nöllenburg, M.: Parameterized algorithms for book embedding problems. In: Archambault, D., Tóth, C.D. (eds.) GD 2019. LNCS, vol. 11904, pp. 365–378. Springer, Cham (2019). https://doi.org/10.1007/978-3-030-35802-0_28
3. Unger, W.: The complexity of colouring circle graphs. In: Finkel, A., Jantzen, M. (eds.) STACS 1992. LNCS, vol. 577, pp. 389–400. Springer, Heidelberg (1992). https://doi.org/10.1007/3-540-55210-3_199
4. Gyárfás, A.: On the chromatic number of multiple interval graphs and overlap graphs. Discret. Math. **55**(2), 161–166 (1985)
5. Ageev, A.A.: A triangle-free circle graph with chromatic number 5. Discret. Math. **152**(1–3), 295–298 (1996)
6. Kostochka, A., Kratochvíl, J.: Covering and coloring polygon-circle graphs. Discret. Math. **163**(1–3), 299–305 (1997)
7. Chung, F., Leighton, F., Rosenberg, A.: Embedding graphs in book: a layout problem with applications to VLSI design. SIAM J. Alg. Discr. Meth. **8**(1), 33–58 (1987)
8. Gutin, G., Jones, M., Wahlström, M.: The mixed Chinese postman problem parameterized by pathwidth and treedepth. SIAM J. Discret. Math. **30**(4), 2177–2205 (2016)
9. Cygan, M., Kratsch, S., Nederlof, J.: Fast Hamiltonicity checking via bases of perfect matchings. J. ACM **65**(3), 12:1–12:46 (2018)
10. Belmonte, R., Lampis, M., Mitsou, V.: Parameterized (approximate) defective coloring. In: Niedermeier, R., Vallée, B. (eds.) STACS 2018, LIPICS, pp. 11:1–11:24 (2018)

11. Liu, Y., Chen, J., Huang, J.: Fixed-order book thickness with respect to vertex-cover number: new observations and further analysis. In: Chen, J., Feng, Q., Xu, J. (eds.) TAMC 2020. LNCS. Springer, Heidelberg (2020, to appear)
12. Bodlaender, H.L., Jansen, B.M.P., Kratsch, S.: Kernelization lower bounds by cross-composition. SIAM J. Discret. Math. **28**(1), 277–305 (2014)
13. Drucker, A.: New limits to classical and quantum instance compression. SIAM J. Comput. **44**(5), 1443–1479 (2015)

Selfish Bin Packing with Parameterized Punishment

Weiwei Zhang[1], Alin Gao[1], and Ling Gai[2($\boxtimes$)]

[1] School of Management, Shanghai University,
Shanghai 201444, People's Republic of China
zthomas@shu.edu.cn, alin.gao@foxmail.com
[2] Glorious Sun School of Business and Management, Donghua University,
Shanghai 200051, People's Republic of China
lgai@dhu.edu.cn

Abstract. In this paper we consider the problem of selfish bin packing with parameterized punishment. Different from the classical bin packing problem, each item to be packed belongs to a selfish agent, who wants to maximize his utility by selecting an appropriate bin. The utility of the agent is defined as the total size of the items sharing the same bin with its item. If an item moves unilaterally to another bin, it may have to pay the punishment. A parameter is defined such that the items are classified whether or not they are fit for the punishment. We study three versions of punishment-full, expansile and partial punishment, and prove the corresponding bounds of PoA^1 (Price of Anarchy).

Keywords: Selfish bin packing · Price of Anarchy · Nash equlibrium · Punishment

1 Introduction

The selfish bin packing problem was first introduced by Bilò [2] in 2006. He presented a noncooperative version of the classical Minimum Bin Packing problem, each item is charged a cost according to the percentage of the bin used. So the selfish items are interested in being packed in one of the bins such that their cost are minimized. They proved that such a game always converges to a pure Nash equilibrium starting from any initial packing of the items. They also studied the bounds of PoA (Price of Anarchy), which is the ratio between the objective value of the worst Nash Equilibrium and that of the classical optimization problem.

Motivated by the One Belt One Road, a large economic union could benefit its every member, and the members together contribute to the whole union's utility. Any unilateral deviation or betrayal would hurt the other members, and more deviations maybe incurred because of it. A Nash Equilibrium is a state of profile that no member would like to deviate his economic union unilaterally,

---

Research supported by NSFC (11201333).

Z. Zhang et al. (Eds.): AAIM 2020, LNCS 12290, pp. 238–247, 2020.
https://doi.org/10.1007/978-3-030-57602-8_22

so the union numbers are fixed and stable. Some economic unions tried to prevent, or reduce the happening of deviations, by setting the rules of high tariff or other punishment methods. Such as Van Rompuy's speaking in the city of London in 2nd March 2013, reported by Reuters, pointed out that from a legal point of view, it is "not impossible for Britain to leave the EU.", but he also warned the Cameron administration that "it would be impossible to leave the EU without paying". Whether or not the punishment can help to get a better Nash Equilibrium has not been proved by formal methodology yet.

In this paper, we try to study the effect of punishment for the performance of Nash Equilibrium. Three kinds of punishment are to be considered, the first one is called *full punishment*, which means that the deviation member has to pay the sum of all his previous contribution to the union (other members) if he moves unilaterally; the other two kinds of punishment are directly related to the item size, we use a constant α to time with size s_i, to represent the punishment item i has to pay. The second version with $\alpha \geq 1$ is called *expansile punishment*; The third version is with $\alpha < 1$, we call it as *partial punishment*, the amount that the "betrayer" has to pay is a part of his contribution (affection) to the economy. We define the maximum utility version of selfish bin packing with different punishment rules, study the existence of Nash Equilibria, and consider the price of anarchy.

Formally, there are m items to be packed, s_i is the size of the item i, $i = 1, \ldots, m$. As in the classical bin packing problem, item sizes $0 \leq s_i \leq 1$, and the bin size is 1. For a bin B_h, we use $|B_h|$ to denote the number of items packed in bin B_h and $S(B_h)$ to denote the load of bin. Define the item's utility to be the bin load that it is packed in, $u_i = S_{i \in B_h}(B_h)$. If an item i moves to another bin unilaterally, it has to pay the punishment of p_i. We are interested to know if there is a stable state that no item wants to move unilaterally, and how about the performance (number of bins used) of the stable states. Related definitions are interpreted below:

Nash Equilibrium. In game theory, the Nash equilibrium, named after the mathematician John Forbes Nash Jr., is a proposed solution of a non-cooperative game involving two or more players in which each player is assumed to know the equilibrium strategies of the other players, and no player has anything to gain by changing only their own strategy [1].

Price of Anarchy. The Price of Anarchy (PoA) [8] is a concept in economics and game theory that measures how the efficiency of a system degrades due to selfish behavior of its agents. The Price of Anarchy measures the ratio between the worst equilibrium and the optimal centralized solution.

Our definition for the item utility u_i represents a situation that each member can enjoy the whole group benefit. The idea is very natural. For example, members do not have to share the tariff concession brought by their economic union; all students from a well-known university can benefit the high prestige of their university and previous graduates, they do not have to "share" the reputation, either.

Before proceeding our study on the selfish bin packing with punishment, we have to look at the following instance:

Instance 1. Suppose there are $2n$ items to be packed, each of them is with size of $1/n$. We can see for the profile $\{(1/n, 1/n), \ldots, (1/n, 1/n)\}$, no item has the motivation to change his bin, since the utility remains $2/n$ unchanged. So this profile is a Nash Equilibrium with $\frac{n}{2}$ bins used, while an optimal packing just use 2 bins. The Price of Anarchy is unlimited when $n \to \infty$.

Note that this instance is also applied for the punishment with $\alpha \geq 1$. Too much punishment makes the world worse. Motivated by this, in the following we introduce a parameter k, to separate the items deserving or not deserving the punishment. And the problem studied becomes *selfish bin packing with parameterized punishment.*

There are lots of results published since the selfish bin packing problem was introduced in 2006 [2]. Epstein [9] gave a survey on the previous results. In [4] each item has a positive weight, and costs are based on cost sharing proportional to the weights of items that share a bin. The PoA is equal to 1.7 in the case of general weights. Cristina et al. [3] studied the selfish 2-dimensional bin packing game, where the items to be packed are rectangles, and the bins are unit squares. The cost of an item is defined as the ratio between its area and the total occupied area of the respective bin. They showed that this game always converges to a Nash equilibrium, and for the selfish square packing case, the price of anarchy is at least 2.3634 and at most 2.6875. In [11] Zhang et al. found a more efficient mechanism for selfish bin packing, narrowed the performance gap between the optimization problem and a game model. They proposed a simple mechanism with $PoA = 1.5$, then showed that for a large class of mechanisms for the selfish bin packing problem, 1.5 is a lower bound of PoA. And they proposed a new mechanism with $PoA \leq 1.467$. Wang et al. [10] studied the bin packing game with an interest matrix, where a_{ij} stands for how much item i likes item j. The payoff of item i is the sum of a_{ij} over all items j in the same bin with item i, and each item wants to stay in a bin where it can fit and its payoff is maximized. They showed that if the matrix is symmetric, a Nash Equilibrium (NE) always exists. However the Price of Anarchy (PoA) may be very large, they gave some bounds for PoA in several special cases. [5] studied the case with proportional cost sharing selfish bin packing and present a new lower bound of PoA. [6] studied the strong equilibria of selfish bin packing and show the SPoA is between 1.69103 and 1.611824. [7] considered the case that each item is with a nonnegative weight and they gave a general lower bound of PoA holds for all possible weights. [12] studied the cost-sharing mechanisms for selfish bin packing in decentralized environments and proposed a new mechanism with $PoA \leq 22/15$.

The rest of the paper is organized as follows. The selfish bin packing with parameterized full punishment is studied in Section 2. In Sect. 3 we study the selfish bin packing with parameterized expansile punishment, then in Sect. 4 the version of partial punishment is studied. Conclusion is presented in Sect. 5.

2   The Selfish Bin Packing with Parameterized Full Punishment

In the case of selfish bin packing with parameterized punishment, we let the punishment p_i only be applicable when the item is big enough. That is, for a given parameter k, items with size bigger or equal to $1/k$ should pay p_i for deviating to other bins unilaterally; the small items are free to move. Specifically, if item i in bin B_h wants to move to other bins, then the punishment p_i for its unilaterally moving is defined as follows:

$$
p_i = \begin{cases} (|B_h| - 1)\, s_i & s_i \geq \frac{1}{k} \\ \\ 0 & s_i < \frac{1}{k} \end{cases}
$$

2.1   The Existence of Nash Equilibrium

Define the potential function as $P = \sum_{j \in n} [S(B_j)]^2 \leq n$, where n is the number of bins used. It means that the potential function is upper-bounded and its value strictly increases with the item's deviation for higher utility. Specifically, if the item i deviates to bin t from bin h, no matter it is a big item or small item, we have:

$$
\begin{aligned}
\Delta &= P' - P \\
&= [S(B_t) + s_i]^2 + [S(B_h) - s_i]^2 - S^2(B_h) - S^2(B_t) \\
&= [S(B_t)]^2 + s_i^2 + 2s_i S(B_t) + [S(B_h)]^2 + s_i^2 - 2s_i S(B_h) - [S(B_t)]^2 - [S(B_h)]^2 \\
&= 2s_i[s_i + S(B_t) - S(B_h)] > 0
\end{aligned}
$$

The existence of an Equilibrium is proved. For the number of convergence steps, let $R > 0$ be the minimal integer such that R_{s_j} is integer for all s_j, $j = 1, \ldots, m$. Then $R[s_i + s(B_t) - s(B_h)] \gg 1$, thus the potential function will increase at least $2s_i/R$ after any deviation step. Consequently, after at most $R_n/2s_i$ steps a configuration where no item has incentive to deviate anymore could be achieved, which means a Nash Equilibrium is reached.

2.2   The Upper Bound of the Price of Anarchy

Let B_i^* denote the largest bin that item s_i can move to from its current bin, for $i = 1, \ldots, m$. Let s_{B_h} denote the size of the smallest item in bin B_h.

From the definition of the full punishment, we know that a profile π is a Nash Equilibrium as long as for any item i

$$
\begin{cases} S(B_i^*) \leq S(B_h) + (|B_h| - 2)s_i, & s_i \geq \frac{1}{k} \\ \\ S(B_i^*) \leq S(B_h) - s_i, & s_i < \frac{1}{k} \end{cases}
$$

In order to fulfill the proof, we define a bin B_h as big if $S(B_h) \geq 1/2$, otherwise it is called small. Suppose n bins are used in some given Nash Equilibrium π.

Sort the bins in non-increasing order, such that $S(B_1) \geq S(B_2) \geq \cdots \geq S(B_n)$. Let n_l be the number of big bins and n_s be the number of small bins. Denote Opt as the optimum solution of the bin packing problem.

Lemma 1. *For any bin B_h containing some small items, it has $S(B_h) = S(B_1)$ or $S(B_j)_{j<h} > \frac{k-1}{k}$.*

Proof. For a small item with $s_i < \frac{1}{k}$, if it has the motivation to deviate from its current bin B_h, then $S(B_h^*) > S(B_h) - s_i$. So in any NE packing, if bin B_h contains item i, then either B_h is one of the largest bin or, item i cannot fit other bins B_j with $j < h$.

Lemma 2. *In any NE packing, if $n_s \geq 2$, there are at least two items in each small bin.*

Proof. For any small bin B_h, suppose $|B_h| = 1$. If $s_{B_h} \geq \frac{1}{k}$, then $S(B_h^*) \leq 0$, which means there could not be any other bin can pack this item, so B_h is the only small bin. Similarly, if $s_{B_h} < \frac{1}{k}$, we still have $S(B_h^*) \leq 0$, a contradiction. So we know that each small bin contains at least two items.

Theorem 1. *The Price of Anarchy is smaller or equal to $max\{4, k/2\}$.*

Proof. Case 1. $n_l \geq n_s$. We know $Opt \geq \frac{n_l}{2}$, then $n = n_l + n_s \leq 2n_l \leq 4Opt$;
Case 2. $n_l < n_s$.

If there are small items in the small bin, they can only be packed in the biggest small bin. So the size of any big bins are at least $\frac{k-1}{k}$. According to Lemma 2, the size of each small bin is larger than $\frac{2}{k}$, so we have

$$Opt(\pi) \geq \frac{k-l}{k} n_l + \frac{2}{k} n_s$$

then

$$n = n_l + n_s \leq \frac{n_l + n_s}{\frac{k-1}{k} n_l + \frac{2}{k} n_s} Opt = \left(\frac{\frac{k^2-3k}{k-1}}{(k-1)\frac{n_l}{n_s} + 2} + \frac{k}{k-1} \right) Opt \leq \frac{k}{2} Opt$$

If there is no small item in the small bins,

$$Opt \geq \frac{1}{2} n_l + \frac{2}{k} n_s$$

then

$$n = n_l + n_s \leq \frac{n_l + n_s}{\frac{1}{2} n_l + \frac{2}{k} n_s} Opt = \left(\frac{2k-8}{k \frac{n_l}{n_s} + 4} + 2 \right) Opt \leq \frac{k}{2} Opt$$

Above all, the upper bound of PoA is $max\{4, k/2\}$. If let $k = 8$, we can get a constant ratio of $PoA \leq 4$.

$\square$

2.3   The Lower Bound of Price of Anarchy

Instance 2. Suppose there are n bins, each of them contains two items. The size of each item is $\frac{1}{k}$. It is easy to prove that this is a NE packing and the Price of Anarchy is $\frac{k}{2}$.

3   The Selfish Bin Packing with Parameterized Expansile Punishment

Here for item i in bin B_h, the punishment p_i for its unilaterally moving is defined as

$$
p_i = \begin{cases} \alpha s_i & s_i \geq \frac{1}{k} \\ \\ 0 & s_i < \frac{1}{k} \end{cases}
$$

3.1   The Lower Bound of the Price of Anarchy

Instance 3. Suppose there are n bins each of which contains only one item, the size of each item equals to $\frac{1}{k}$. Then this is a NE packing with k bins used. We can see that an optimal packing just use one bin for these items, so the lower bound of PoA is k.

3.2   The Upper Bound of the Price of Anarchy

From the definition of punishment, we know that a profile π is a Nash Equilibrium as long as for any item i

$$
\begin{cases} S(B_i^*) \leq S(B_h) + (\alpha - 1)s_i & s_i \geq \frac{1}{k} \\ \\ S(B_i^*) \leq S(B_h) - s_i & s_i < \frac{1}{k} \end{cases}
$$

for the case of $s_h < \frac{1}{k}$, it is the same as the former case, Lemma 2 is still valid here.

Theorem 2. *The Price of Anarchy is smaller or equal to* $max\{4, k\}$.

Proof. If $n_l \geq n_s$, we know that $Opt \geq \frac{n_l}{2}$, then

$$
n = n_l + n_s \leq 2n_l \leq 4Opt
$$

If $n_l < n_s$, similarly, the small items can only be packed in the largest small bins. So we know that the size of all the small bins is at least $\frac{1}{k}$ and that of the big bins is larger than $\frac{k-1}{k}$.

Since

$$
Opt \geq \frac{k-1}{k}n_l + \frac{1}{k}n_s
$$

then

$$n = n_l + n_s \leq \frac{n_l + n_s}{\frac{k-1}{k}n_l + \frac{1}{k}n_s} Opt = \left(\frac{\frac{k^2-2k}{k-1}}{(k-1)\frac{n_l}{n_s} + 2} + \frac{k}{k-1} \right) Opt \leq kOpt.$$

If there is no small item in small bins, then the size of small bins should be larger than $\frac{1}{k}$ and the size of big bins should be larger than $\frac{1}{2}$. We can know that

$$Opt \geq \frac{1}{2}n_l + \frac{1}{k}n_s$$

then

$$n = n_l + n_s \leq \frac{n_l + n_s}{\frac{1}{2}n_l + \frac{1}{k}n_s} Opt = \left(\frac{2k-4}{k\frac{n_l}{n_s} + 2} + 2 \right) Opt \leq kOpt.$$

To sum up, the upper bound of PoA is $max\{4, k\}$. Let k equal to 4, $PoA = 4$.
□

4   The Selfish Bin Packing with Partial Punishment

For the case with $\alpha < 1$, each item has the same motivation to move or not no matter its size is larger or smaller than $\frac{1}{k}$. So in this section we just define the punishment of item i for its unilaterally moving as

$$p_i = \alpha s_i (\alpha < 1).$$

4.1   The Lower Bound of the Price of Anarchy

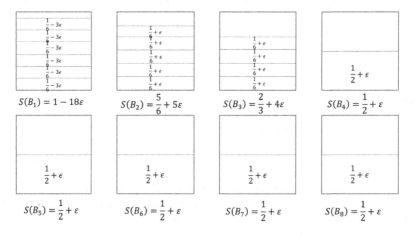

Fig. 1. The packing in a nash equilibrium

Instance 4. Figure 1 shows a NE packing with 8 bins, including 3 big bins and 5 small bins. First, we suppose that $\varepsilon < \frac{1}{102}$, then $\frac{1}{6} + \varepsilon + 1 - 18\varepsilon > 1$. For the size of item equals to $\frac{1}{6} - 3\varepsilon$, we get $1 - 18\varepsilon > \frac{2}{3} + 4\varepsilon + \frac{1}{6} - 3\varepsilon$, $\frac{1}{6} - 3\varepsilon + \frac{5}{6} + 5\varepsilon > 1$ and $1 - 18\varepsilon > \frac{2}{3} + 4\varepsilon + \frac{1}{6} - 3\varepsilon$. Therefore, for the items with size $\frac{1}{6} - 3\varepsilon$, they have no motivation to move. Similarly, we can get same conclusion for other items. So the packing is a Nash Equlibrium.

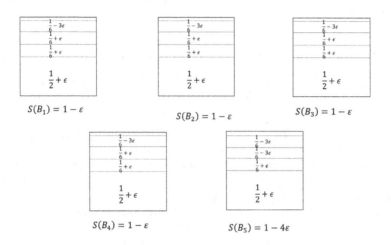

Fig. 2. The optimal packing

Figure 2 shows the optimal packing for these items. Thus, we get the lower bound of PoA as $\frac{8}{5}$.

4.2   Weight Function for Special Condition

Define a bin as mixed bin, if it contains at least 2 items. It is a single bin, if the bin contains one single item. It is clearly that the single bin always effects the efficiency of the packing. Let NE to denote a Nash equlibrium packing, and also the number of bins used in this equilibrium. OPT as the classical optimal packing, and also denotes the bins number.

Theorem 3. *The upper bound of the PoA is $(1 - b + a)b$, if except t special bins (single and mixed), the load of other bins are all larger than b.*

Assume that there are some single bins with loads between a and b $(a < b)$, and some bins with load at least b. Then let t be the number of rest bins, the single bins with load less than a and the mixed bins with load less than b.

The weight function is defined as

$$w(s_i) = \begin{cases} s_i + \delta(1 - s_i) & B_h \ is \ a \ single \ bin \ and \ s_i \in (a, b) \\ s_i & others \end{cases}$$

To make each bin's weight larger than b, $s_i + \delta(1 - s_i) \geq b$,then $\delta \geq \frac{b - s_i}{1 - s_i}$. Since $s_i \in (a, b)$, then $\delta \geq \frac{b - a}{1 - a}$. It is obvious that when $\delta \in (0, 1)$, the weight of each bin is less than 1. So $OPT > W(OPT)$, $W(NE) > bNE - t$. The change of the weight is caused by some single bins, therefore the weight of OPT plus all single bins' additional weights are larger than the weight of NE. Moreover, if the number of single bin is greater than bNE, then $OPT > bNE$, which implies that $W(OPT) + (1 - a)\delta bNE > W(NE)$. Above all, $OPT > (1 - (1 - a)\delta)bNE - t > (1 - b + a)bNE - t$. □

4.3   The Upper Bound of the Price of Anarchy

The weight function is designed to prove PoA under the condition that most bins are with load larger than b. To satisfy this condition, the parameters a and b should be identified.

Lemma 3. *There is at most one bin whose size is less than $\frac{1}{2}$ in a NE packing.*

Proof. Suppose there are two bins B_h and B_j whose sizes are less than $\frac{1}{2}$ in a NE packing and $S(B_h) \geq S(B_j)$. Consider an item s_i in the bin B_j, it fits bin B_h but it chooses not move. We can know that $S(B_j) + \alpha s_i \geq S(B_h) + s_i$. Since $\alpha < 1$, So $S(B_h) < S(B_j)$, a contradiction. The lemma is proved.

Lemma 3 implies that there is at most one single bin with the load less than $\frac{1}{2}$. Therefore, parameter a is equal to $\frac{1}{2}$.

Lemma 4. *There is at most one mixed bin whose size is less than $\frac{2}{3}$.*

Proof. Suppose there are two mixed bins whose sizes are less than $\frac{2}{3}$. Apparently such mixed bin contains at less one item i whose size is less than $\frac{1}{3}$. Suppose the bin $S(B_h) \geq S(B_j)$, for the item i according the definition of the Nash Equilibrium, here is $S(B_j) + \alpha s_i > S(B_h) + s_i$, which is a confliction. □

Lemma 4 indicates that there is at most one mixed bin with load less than $\frac{2}{3}$, which shows that the parameter t is at most 2, composed of one single bin with load less than $\frac{1}{2}$ and one mixed bin with load less than $\frac{2}{3}$. Hence parameter b equals to $\frac{2}{3}$.

Theorem 4. *The upper bound of the selfish bin packing with partial punishment is $\frac{9}{5}$.*

Proof. According to Theorem 3, Lemma 3 and Lemma 4, $a = 0.5$, $b = \frac{2}{3}$, then $OPT > \frac{9}{5}ALG - 2$.

5   Conclusion

In this paper, we study a new version of selfish bin packing problem. We define three versions of punishment for item unilaterally moving. A parameter is given to classify the items that deserve or not punishment. The existence of Nash Equlibrium is proved and the bounds of PoA for three kinds of punishment mechanism are studied. For the full punishment version, the PoA is proved to be $k/2$, where k is a given parameter to separate the items to be punished or not. For the expansile punishment version, PoA is showed to be k. As for the partial punishment version, we do not separate the items and let them all be punished if they move unilaterally, the lower bound of PoA is proved to be $\frac{8}{5}$ and the upper bound is $\frac{9}{5}$. We may realize that punishment does not always work, comparing with the selfish bin packing without punishment.

References

1. Dixit, A., Susan, S., David, R.: Games of Strategy, 3rd edn. W.W. Norton Company (2009)
2. Bilò, V.: On the packing of selfish items. In: Proceedings of the 20th International Parallel and Distributed Processing Symposium (IPDPS2006), pp. 45–54. IEEE, Rhodes (2006)
3. Fernandes, C.G., Ferreira, C.E., Miyazawa, F.K., Wakabayashi, Y.: Prices of anarchy of selfish 2D bin packing games. Int. J. Found. Comput. Sci. **30**(3), 355–374 (2019)
4. Dósa, G., Epstein, L.: Generalized selfish bin packing (2012). ArXiv:1202.4080 [cs.GT]
5. Dósa, G., Epstein, L.: A new lower bound on the price of anarchy of selfish bin packing. Inf. Process. Lett. **150**, 6–12 (2019)
6. Dósa, G., Epstein, L.: Quality of strong equilibria for selfish bin packing with uniform cost sharing. J. Sched. **22**(4), 473–485 (2018). https://doi.org/10.1007/s10951-018-0587-8
7. Dósa, G., Kellerer, H., Tuza, Z.: Using weight decision for decreasing the price of anarchy in selfish bin packing games. Eur. J. Oper. Res. **278**(1), 160–169 (2019)
8. Koutsoupias, E., Papadimitriou, C.: Worst-case equilibria. Comput. Sci. Rev. **3**(2), 65–69 (2009)
9. Epstein, L.: Selfish bin packing problems. In: Kao, M.Y. (eds) Encyclopedia of Algorithms. LNCS, pp. 1927–1930, Springer, Boston (2016). https://doi.org/10.1007/978-1-4939-2864-4
10. Wang, Z., Han, X., Dósa, G., Tuza, Z.: General bin packing game: interest taken into account. Algorithmica **80**, 1534–1555 (2018)
11. Zhang, C., Zhang, G.: Cost-sharing mechanisms for selfish bin packing. In: Gao, X., Du, H., Han, M. (eds.) COCOA 2017. LNCS, vol. 10627, pp. 355–368. Springer, Cham (2017). https://doi.org/10.1007/978-3-319-71150-8_30
12. Zhang, C., Zhang, G.: From packing rules to cost-sharing mechanisms. J. Comb. Optim. 1–16 (2020)

Multiple Facility Location Games
with Envy Ratio

Wenjing Liu[ID], Yuan Ding, Xin Chen, Qizhi Fang[(✉)], and Qingqin Nong[ID]

Ocean University of China, Qingdao 266100, China
liuwj_123@126.com, cxin0307@163.com,
{dingyuan,qfang,qqnong}@ouc.edu.cn

Abstract. We study deterministic mechanism design without money for k-facility location games with envy ratio on a real line segment, where a set of strategic agents report their locations and a social planner locates k facilities for minimizing the envy ratio. The objective of envy ratio, which is defined as the maximum over the ratios between any two agents' utilities, is derived from fair division to measure the fairness with respect to a certain facility location profile.

The problem is studied in two settings. In the homogeneous k-facility location game where k facilities serve the same purpose, we propose a $\frac{2k}{2k-1}$-approximate deterministic group strategyproof mechanism which is also the best deterministic strategyproof mechanism. In the heterogeneous k-facility location game where each facility serves a different purpose, when k is even, we devise the optimal and group strategyproof mechanism; when k is odd, we provide a $\frac{k+1}{k-1}$-approximate deterministic group strategyproof mechanism.

Keywords: Mechanism design · Facility location · Strategyproof · Fairness · Envy ratio

1 Introduction

In this paper, we study k-facility location games with envy ratio on a real line segment. The objective of envy ratio, which is derived from fair division [6,9,13, 21], is defined as the maximum over the ratios between any two agents' utilities and can be used to measure the fairness with respect to a certain facility location. In k-facility location games with envy ratio, a social planner is going to build k facilities based on the reported locations from a set of agents and aims to minimize the envy ratio. However, each agent who has her location as private information, is strategic and may misreport her location to minimize her own cost. Thus, the social planner seeks to design mechanisms that can minimize the

This research was supported in part by the National Natural Science Foundation of China (11971447, 11871442), the Natural Science Foundation of Shandong Province of China (ZR2019MA052) and the Fundamental Research Funds for the Central Universities (201964006).

© Springer Nature Switzerland AG 2020
Z. Zhang et al. (Eds.): AAIM 2020, LNCS 12290, pp. 248–259, 2020.
https://doi.org/10.1007/978-3-030-57602-8_23

envy ratio while guaranteeing truthful report from agents (i.e., strategyproof or group strategyproof).

The k-facility location game with envy ratio models well the real life scenario where a social planner taking fairness into account needs to locate multiple facilities to serve agents. For example, three shopping malls are to be built in a district based on the ideal locations reported by local residents. It is natural to assume that each resident only concerns her own distance to the malls and may lie if necessary. Meanwhile, for the consideration of fairness, the local government hopes that the three malls will be located while keeping the distance differences among residents as small as possible. To deal with issues similar to the above, mechanism design for k-facility location games with envy ratio is studied.

We discuss the problem in two settings. The fist one is the homogeneous k-facility location, where the k facilities serve the same purpose and the cost of each agent is her Euclidean distance to the nearest facility. For fitting this setting, consider the scenario that the local government plans to build k public parking lots in a street. All the agents in the street prefer living close to the parking lot and can park their cars at the nearest one. The second one is the heterogeneous k-facility location, where each facility serves a different purpose and the cost of each agent is the sum of her Euclidean distances to the k facilities. For fitting this setting, consider the scenario that the local government plans to build several public facilities in a street, such as a library, a park, a bus stop and so on. Then the cost of each agent in the street is the sum of her distances to the k facilities since she needs to read in the library, to walk in the park, to wait for a bus at the bus stop, etc.

For both settings, we are interested in designing deterministic strategyproof or group strategyproof mechanisms that can perform well in minimizing the envy ratio.

1.1 Our Results

This paper studies deterministic mechanism design without money for k-facility location games on the real line segment $[0, 1]$ with the objective of minimizing the envy ratio. The problem is considered in two settings which are described above.

Our key innovations and results are summarized as follows.

In Sect. 2, we formulate the k-facility location games with the objective of minimizing the envy ratio. To the best of our knowledge, it is the first time that the envy ratio is considered in the multiple facility location games for strategyproof mechanism design.

In Sect. 3, we concentrate the homogeneous k-facility location game with envy ratio, where the cost of each agent is her distance to the nearest facility. We show a lower bound of $\frac{2k}{2k-1}$ for the approximation ratio of any deterministic strategyproof mechanism and propose a $\frac{2k}{2k-1}$-approximate deterministic group strategyproof mechanism, which implies that the best deterministic strategyproof mechanism has been obtained.

In Sect. 4, we consider the heterogeneous k-facility location game with envy ratio, where the cost of each agent is the sum of her distances to the k facilities. When k is even, we devise an optimal and deterministic group strategyproof mechanism. When k is odd, we devise a $\frac{k+1}{k-1}$-approximate deterministic group strategyproof mechanism.

1.2   Related Work

Approximate mechanism design without money for facility location games was introduced by Procaccia & Tennenholtz [18]. They studied the facility location games on the real line with the social cost objective and the maximum cost objective in three settings: 1-facility, 2-facility and multiple locations per agent. Before them, Moulin [17] and Schummer & Vohra [19] provided a complete characterization of strategyproof mechanisms for the facility location game with the single peaked preference on line, tree and cycle networks. So far, mechanism design without money for facility location games has been well studied.

For the one-facility location game, Alon et al. [1] extended the facility location game to other networks for the maximum cost objective. Many variants of the problem have also been studied to adapt more realistic scenarios. Cheng et al. [5] introduced an obnoxious facility game on networks where every agent wants to stay far away from the facility. Zhang & Li [23] extended the facility location and the obnoxious facility location to games with weighted agents on a line. Feigenbaum & Sethuraman [8] and Zou & Li [24] studied the dual or hybrid preference game where some agents want to stay close to the facility while the others want to stay away from the facility. Mei et al. [16] introduced a happiness factor to measure the agent's satisfaction degree with respect to the facility location and Li et al. [12] studied the facility game with externalities where every agent's utility is affected by others.

For the multiple facility location game, Lu et al. [14,15] improved the results of [18] for the 2-facility location game. Fotakis & Tzamos [11] provided a complete characterization of deterministic strategyproof mechanisms for the 2-facility location game on a real line with the social cost objective. From then on, heterogeneous multiple facility location games where each facility serves a different purpose have been studied. Serafino & Ventre [20] and Yuan et al. [22] considered heterogeneous 2-facility location games with public location information but private optional preference. Zou & Li [24] and Chen et al. [4] studied 2-opposite-facility location game with limited distance. Fong et al. [10] proposed a fractional preference model for the facility location game with two facilities serving the same purpose on a line segment. Anastasiadis & Deligkas [2] studied strategyproof mechanism design for heterogeneous k-facility location games with the minimum utility objective. Duan et al. [7] introduced the minimum distance requirement to heterogeneous 2-facility location games with the social cost objective.

The work mentioned above mainly focused on the objective of minimizing the social cost or the maximum cost, where the former represents the utilitarianism and the latter represents the egalitarianism. Recently, motivated by fair division, several new objectives have been adopted to measure the fairness of the facility location, such as maximum envy and envy ratio. Cai et al. [3] studied one-facility location game with the objective of minimizing the maximum envy. Ding et al. [6] introduced the envy ratio to one-facility location games, devised the best deterministic strategyproof mechanism and gave a lower bound for randomized strategyproof mechanisms. This paper can be considered as an expanding direction of the work of [6], in which deterministic mechanism design for k-facility location games with envy ratio is studied.

2 Preliminaries

In this section, we introduce some definitions and notations used in the k-facility location games with envy ratio on the real line segment $I = [0, 1]$.

Let $N = \{1, 2, \ldots, n\}$ be a set of agents. Each agent $i \in N$ has a location $x_i \in I$, which is i's private information. The collection $\mathbf{x} = (x_1, x_2, \ldots, x_n)$ is referred to as a location profile or an instance. For $i \in N$, denote $\mathbf{x}_{-i} = (x_1, \ldots, x_{i-1}, x_{i+1}, \ldots, x_n)$, then $\mathbf{x} = (x_i, \mathbf{x}_{-i})$. For a nonempty set $S \subseteq N$, denote $\mathbf{x}_S = (x_i)_{i \in S}$ and $\mathbf{x}_{-S} = (x_i)_{i \notin S}$, then $\mathbf{x} = (\mathbf{x}_S, \mathbf{x}_{-S})$. For an instance $\mathbf{x}$, if there are q different locations $x_1, \ldots, x_q$ and N can be partitioned into q nonempty coalitions such that all agents in N_i occupy a same location x_i, $\mathbf{x}$ is called a q-location instance and is denoted as $(x_1 : N_1, \cdots, x_q : N_q)$.

Suppose the k facilities are located at $y_1, \ldots, y_k$ respectively, then the location profile of k facilities is denoted by $\mathbf{y} = (y_1, \ldots, y_k) \in I^k$. For any agent $i \in N$ at location x_i, let $cost(\mathbf{y}, x_i)$ denote the cost of agent i. In the homogeneous k-facility location, $cost(\mathbf{y}, x_i) = \min_{1 \leq j \leq k} |y_j - x_i|$. In the heterogeneous k-facility location, $cost(\mathbf{y}, x_i) = \sum_{j=1}^{k} |y_j - x_i|$.

Definition 1. *A (deterministic) **mechanism** f is a function that maps a location profile of n agents to that of k facilities, i. e., $f : I^n \to I^k$.*

Given a mechanism f and a location profile $\mathbf{x} \in I^n$, the cost of agent $i \in N$ at location x_i is denoted as $cost(f(\mathbf{x}), x_i)$.

In the k-facility location game, the social planner announces a mechanism, then asks each agent to report her location and output the k facility locations. Agents may misreport their locations to decrease their own costs. Therefore, it is important to ensure strategyproofness or group strategyproofness of the mechanism, which are defined as follows.

Definition 2. *A mechanism f is **strategyproof** if no agent can benefit from misreporting her location, regardless of the other agents' strategies.*

Formally, for every location profile $\mathbf{x} \in I^n$, every agent $i \in N$, and every $x_i' \in I$, $cost(f(x_i', \mathbf{x}_{-i}), x_i) \geq cost(f(\mathbf{x}), x_i)$.

Definition 3. *A mechanism f is **partial group strategyproof** if for any coalition of agents that occupy the same location, none of them can benefit from misreporting their locations simultaneously.*

Formally, for every location profile $\mathbf{x}$, every coalition of agents S occupying a same location x, and every $\mathbf{x}_S' \in I^{|S|}$, $cost(f(\mathbf{x}_S', \mathbf{x}_{-S}), x) \geq cost(f(\mathbf{x}), x)$.

Definition 4. *A mechanism f is* **group strategyproof** *if for any coalition of agents misreporting their locations, at least one of them can not benefit.*

Formally, for every location profile $\mathbf{x}$, *every coalition of agents* $S \subseteq N$, *and every* $\mathbf{x}'_S \in I^{|S|}$, *there exists some agent* $i \in S$ *such that* $cost(f(\mathbf{x}'_S, \mathbf{x}_S), x_i) \geq cost(f(\mathbf{x}), x_i)$.

Remark. By the definitions, any *group strategyproof* mechanism is also *partial group strategyproof*, and any *partial group strategyproof* mechanism is also *strategyproof*. Furthermore, in the k-facility location game, any *strategyproof* mechanim is also *partial group strategyproof* [15]. Therefore, *strategyproofness* will be regarded as equivalent to *partial group strategyproofness* in the following analysis.

In this paper, we are interested in pursuing locations of the k facilities that take fairness among agents into account. To capture the notion of fairness, we employ the concept of *envy ratio*, which is derived from fair division. In the context of fair division, agent i is said to *envy* agent j if she prefer the bundle allocated to j to her own and the *envy ratio* is defined by the utility of one agent for another agent's bundle over her utility for her own bundle [13]. Here, we define the utility of each agent as a constant minus her cost and use utilities to define the envy ratio [6][1].

Definition 5. *For a location profile* $\mathbf{x} \in I^n$, *the* **envy ratio** *of the location profile of k facilities* $\mathbf{y} \in I^k$ *is defined as*

$$ER(\mathbf{y}, \mathbf{x}) = \max_{1 \leq i \neq j \leq n} \frac{u(\mathbf{y}, x_i)}{u(\mathbf{y}, x_j)}, \tag{1}$$

where $u(\mathbf{y}, x_i)$ *is the utility of agent i. Specifically,* $u(\mathbf{y}, x_i) = 1 - cost(\mathbf{y}, x_i)$ *in the homogeneous k-facility location and* $u(\mathbf{y}, x_i) = k - cost(\mathbf{y}, x_i)$ *in the heterogeneous k-facility location.*

For a location profile $\mathbf{x} \in I^n$, let $OPT(\mathbf{x})$ be the optimal solution to the minimization problem $\min_{\mathbf{y} \in I^k} ER(\mathbf{y}, \mathbf{x})$ and $ER(OPT, \mathbf{x})$ be the optimal envy ratio. The envy ratio of mechanism f is denoted as $ER(f(\mathbf{x}), \mathbf{x})$. Without confusion, we denote $ER(f(\mathbf{x}), \mathbf{x})$ as $ER(f, \mathbf{x})$ for simplicity. Note that $ER(\mathbf{y}, \mathbf{x}) \geq 1$ for any $\mathbf{y} \in I^k$.

Now we present the approximation ratio which was introduced by [18] to measure the performance of a mechanism.

---

[1] Analogically, in the facility location setting, we can say agent i envies agent j if her cost is greater than j's, or equivalently her utility is less than j's. It seems that defining the envy ratio as the maximum over ratios between any two agents' cost can also represent fairness. We simply follow the way of [6] and give the utility version of the envy ratio. Besides, we conjecture there might not exist any positive results for the cost version of the envy ratio, although without verification.

Definition 6. *A mechanism f is said to have an approximation ratio of γ ($\gamma \geq$ 1), if it satisfies*

$$\gamma = \sup_{\mathbf{x} \in I^n} \frac{ER(f, \mathbf{x})}{ER(OPT, \mathbf{x})}. \tag{2}$$

For the k-facility location game with envy ratio, we are interested in strategyproof or group strategyproof mechanisms that also perform well in minimizing the envy ratio, i. e., with a small approximation ratio.

Notations. For a location profile $\mathbf{x} \in I^n$, denote $lm(\mathbf{x}) = \min_{i \in N} x_i$ which is the leftmost point of $\mathbf{x}$, $rm(\mathbf{x}) = \max_{i \in N} x_i$ which is the rightmost point of $\mathbf{x}$, and $L(\mathbf{x}) = rm(\mathbf{x}) - lm(\mathbf{x})$ which is the length of $\mathbf{x}$.

3   Homogeneous Facility Location Game

In this section, we consider the homogeneous k-facility location game ($k \geq 2$), where $cost(\mathbf{y}, x_i) = \min_{1 \leq j \leq k} |y_j - x_i|$ and $u(\mathbf{y}, x_i) = 1 - cost(\mathbf{y}, x_i)$.

We first show that any deterministic strategyproof mechanism has an approximation ratio of at least $2k/(2k-1)$, then provide a group strategyproof mechanism with approximation ratio of $2k/(2k-1)$. This implies that the best possible deterministic strategyproof mechanism has been obtained.

We start with an optimal solution of minimizing the envy ratio for any $(k+1)$-location instance, which will be used in the following analysis.

Lemma 1. *For any $(k + 1)$-location instance $\mathbf{x} = (x_1 : N_1, \ldots, x_{k+1} : N_{k+1})$, $ER(OPT, \mathbf{x}) = 1$.*

Proof. Without loss of generality, assume that $x_1 < \cdots < x_{k+1}$. Choose an $l \in \arg\min_{1 \leq i \leq k}(x_{i+1} - x_i)$ and denote $\delta = (x_{l+1} - x_l)/2$.

Let $\mathbf{y}^\star = (y_1^\star, \ldots, y_k^\star)$, where

$$y_j^\star = \begin{cases} x_j + \delta, & j < l \\ (x_j + x_{j+1})/2, & j = l \\ x_j - \delta, & j > l + 1 \end{cases} \tag{3}$$

It is straightforward that for any $i \in N$, $cost(\mathbf{y}^\star, x_i) = \delta$. Thus, $ER(\mathbf{y}^\star, \mathbf{x}) = 1$ and $\mathbf{y}^\star$ is the optimal solution.   □

Theorem 1. *For k-facility location game with $n \geq k + 1$ agents, any deterministic strategyproof mechanism has an approximation ratio of at least $2k/(2k - 1)$ for minimizing the envy ratio.*

Proof. Let f be any deterministic strategyproof mechanism for k-facility location with $n \geq k + 1$ agents. It suffices to show that f has an approximation ratio of at least $2k/(2k - 1)$.

Consider the $(k + 1)$-location instance $\mathbf{x} = (0 : N_1, 1/k : N_2, \cdots, (k - 1)/k : N_k, 1 : N_{k+1})$. Denote $f(\mathbf{x}) = (y_1, y_2, \cdots, y_k)$ and without loss of generality assume that $y_1 \leq y_2 \leq \cdots \leq y_k$. We analyze the approximation ratio of f through the following four cases.

Case 1: $y_k \in [0, \frac{k-1}{k}]$. In this case, we have

$$\min\{cost(f(\mathbf{x}), 0), cost(f(\mathbf{x}), 1/k), \cdots, cost(f(\mathbf{x}), (k-1)/k)\} \leq 1/k, \qquad (4)$$

then

$$\max\{u(f(\mathbf{x}), 0), u(f(\mathbf{x}), 1/k), \cdots, u(f(\mathbf{x}), (k-1)/k)\} \geq 1 - 1/k. \qquad (5)$$

In addition, $u(f(\mathbf{x}), 1) = 1 - cost(f(\mathbf{x})) \leq 1 - 1/k$. Thus, we have

$$ER(f, \mathbf{x}) \geq \frac{1 - 1/(2k)}{1 - 1/k} = \frac{2k - 1}{2k - 2} > \frac{2k}{2k - 1}. \qquad (6)$$

Case 2: $y_k \in (\frac{k-1}{k}, \frac{2k-1}{2k}]$.
 Consider $\mathbf{x}' = (y_k : N_k, \mathbf{x}_{-N_k})$. Denote $f(\mathbf{x}') = (y'_1, \cdots, y'_k)$ with $y_1 \leq \cdots \leq y'_k$. Then $y_k \in \{y'_1, \cdots, y'_k\}$; otherwise, all the agents of N_k at location y_k in $\mathbf{x}'$ can benefit from misreporting location $(k-1)/k$ simultaneously, which contradicts f's strategyproofness. Thus, $u(f(\mathbf{x}'), y_k) = 1$.

Case 2.1: $y'_k = y_k$. $u(f(\mathbf{x}'), 1) = y_k \leq (2k - 1)/(2k)$. Thus,

$$ER(f, \mathbf{x}') \geq \frac{u(f(\mathbf{x}'), y_k)}{u(f(\mathbf{x}'), 1)} \geq \frac{2k}{2k - 1}. \qquad (7)$$

Case 2.2: For some $j \in \{1, \cdots, k-1\}$, $y'_j = y_k$.
 In this case, either agents at location $(k-2)/k$ occupy the facility location $y'_j = y_k$ or agents at $k-1$ different locations $0, 1/k, \cdots, (k-2)/k$ jointly occupy no more than $j - 1(\leq k - 2)$ facility locations. In either case, it holds that

$$\max\{cost(f(\mathbf{x}'), 0), \cdots, cost(f(\mathbf{x}'), \frac{k-2}{k}\} \geq \frac{1}{2k}. \qquad (8)$$

Then,

$$\min\{u(f(\mathbf{x}'), 0), \cdots, u(f(\mathbf{x}'), \frac{k-2}{k}\} \leq 1 - \frac{1}{2k}. \qquad (9)$$

Thus, we have

$$ER(f, \mathbf{x}') \geq \frac{1}{1 - 1/(2k)} = \frac{2k}{2k - 1}. \qquad (10)$$

Case 3: $y_k \in (\frac{2k-1}{2k}, 1)$.
 Consider $\mathbf{x}'' = (y_k : N_{k+1}, \mathbf{x}_{-N_{k+1}})$. Denote $f(\mathbf{x}'') = (y''_1, \cdots, y''_k)$ with $y''_1 \leq \cdots \leq y''_k$. By f's strategyproofness, $y_k \in \{y''_1, \cdots, y''_k\}$. Thus, $u(f(\mathbf{x}''), y_k) = 1$. Either agents at location $(k-1)/k$ occupy the facility location y_k, or agents

at k different locations $0, 1/k, \cdots, (k-1)/k$ jointly occupy less than k facility locations. In either case, it holds that

$$\max\{cost(f(\mathbf{x}''), 0), \cdots, cost(f(\mathbf{x}''), \frac{k-1}{k})\} \geq \frac{1}{2k}. \tag{11}$$

Then,

$$\min\{u(f(\mathbf{x}''), 0), \cdots, u(f(\mathbf{x}''), \frac{k-1}{k})\} \leq 1 - \frac{1}{2k}. \tag{12}$$

Thus, we have

$$ER(f, \mathbf{x}'') \geq \frac{1}{1 - 1/(2k)} = \frac{2k}{2k-1}. \tag{13}$$

Case 4: $y_k = 1$. In this case, $u(f(\mathbf{x}), 1) = 1$ and

$$\max\{cost(f(\mathbf{x}), 0), \cdots, cost(f(\mathbf{x}), \frac{k-1}{k})\} \geq \frac{1}{2k}. \tag{14}$$

Then,

$$\min\{u(f(\mathbf{x}), 0), \cdots, u(f(\mathbf{x}), \frac{k-1}{k})\} \leq 1 - \frac{1}{2k}. \tag{15}$$

Thus, we have

$$ER(f, \mathbf{x}) \geq \frac{1}{1 - 1/(2k)} = \frac{2k}{2k-1}. \tag{16}$$

By Lemma 1, the optimal envy ratio for any $(k+1)$-location instance is 1. Therefore, f has an approximation ratio of $2k/(2k-1)$.   □

Mechanism 1. Given a location profile $\mathbf{x} \in I^n$, output

$$f(\mathbf{x}) = (\frac{1}{2k}, \frac{3}{2k}, \cdots, \frac{2k-1}{2k}). \tag{17}$$

Theorem 2. *For k-facility location game with $n \geq 2$ agents, Mechanism 1 is group strategyproof with approximation ratio of $2k/(2k-1)$ for minimizing the envy ratio.*

Proof. $f(\mathbf{x}) = (\frac{1}{2k}, \frac{3}{2k}, \cdots, \frac{2k-1}{2k})$ is group strategyproof since it does not depend on any information from the agents.

For every $\mathbf{x} \in I^n$ and every $i \in N$, it holds that $1 - 1/(2k) \leq u(f(\mathbf{x}), x_i) \leq 1$. Thus,

$$ER(f, \mathbf{x}) \leq \frac{1}{1 - 1/(2k)} = \frac{2k}{2k-1}. \tag{18}$$

Note that $ER(OPT, \mathbf{x}) \geq 1$, and we obtain

$$\sup_{\mathbf{x} \in I^n} \frac{ER(f, \mathbf{x})}{ER(OPT, \mathbf{x})} \leq \frac{2k}{2k-1}. \tag{19}$$

Furthermore, consider a 2-location instance $\mathbf{x}' = (0 : N_1, 1/(2k) : N_2)$. It is obvious that $ER(f, \mathbf{x}') = 2k/(2k-1)$ and $ER(OPT, \mathbf{x}') = 1$.

Therefore, f has an approximation ratio of $2k/(2k-1)$.   □

Combining Theorem 1 with Theorem 2, we observe that *Mechanism 3* is the best deterministic strategyproof mechanism for homogeneous k-facility location game with envy ratio. Specifically, we obtain the following result for the 2-facility location setting.

Corollary 1. *For the homogeneous 2-facility location game with $n \geq 3$ agents, any deterministic strategyproof mechanism has an approximation ratio of at least $4/3$ for minimizing the envy ratio and $f(\mathbf{x}) = (1/4, 3/4)$ for all $\mathbf{x} \in I^n$ is the best deterministic strategyproof mechanism, with approximation ratio of $4/3$.*

Mechanism 2. Given a location profile $\mathbf{x} \in I^n$, output $f(\mathbf{x}) = (lm(\mathbf{x}), rm(\mathbf{x}))$.

Remark 1. Recall that *Mechanism 2* is a well known group strategyproof mechanism and has been shown the best possible deterministic strategyproof mechanism for the homogeneous 2-facility location game with the objective of minimizing the social cost or the maximum cost [11,18]. But in our setting, this mechanism performs worse than simply locating at $1/4$ and $3/4$, with approximation ratio of 2. However, *Mechanism 2* can be extended to the heterogeneous k-facility location setting.

4 Heterogeneous Facility Location Game

In this section, we consider the heterogeneous k-facility location game, where $cost(\mathbf{y}, x_i) = \sum_{j=1}^{k} |y_j - x_i|$ and $u(\mathbf{y}, x_i) = k - cost(\mathbf{y}, x_i)$.

Mechanism 3. Given a location profile $\mathbf{x} \in I^n$, if k is even, the first $k/2$ facilities are located at $lm(\mathbf{x})$ and the last $k/2$ facilities are located at $rm(\mathbf{x})$; if k is odd, the first $(k+1)/2$ facilities are located at $lm(\mathbf{x})$ and the last $(k-1)/2$ facilities are located at $rm(\mathbf{x})$.

Theorem 3. *For the heterogeneous k-facility location game with envy ratio, Mechanism 3 is group strategyproof. Furthermore, if k is even, Mechanism 3 is optimal; if k is odd, Mechanism 3 has an approximation ratio of $\frac{k+1}{k-1}$.*

Proof. Denote *Mechanism 3* as f. We first concentrate on f's group strategyproofness. Let $S \subseteq N$ be any nonempty set and $\mathbf{x} = (\mathbf{x}_S, \mathbf{x}_{-S})$. We shall show that for any $\mathbf{x}'_S \in I^{|S|}$, there exists agent $i \in S$, such that

$$cost(f(\mathbf{x}'_S, \mathbf{x}_{-S}), x_i) \geq cost(f(\mathbf{x}_S, \mathbf{x}_{-S}), x_i). \tag{20}$$

Denote $\mathbf{x}' = (\mathbf{x}'_S, \mathbf{x}_{-S})$, $\Delta_1 = lm(\mathbf{x}) - lm(\mathbf{x}')$ and $\Delta_2 = rm(\mathbf{x}') - rm(\mathbf{x})$. The group strategyproofness of f will be verified through the following four cases.

Case 1: $\Delta_1 \geq 0, \Delta_2 \geq 0$.

If k is even, for every $i \in S$, it holds that

$$cost(f(\mathbf{x}'), x_i) = \frac{k}{2}(rm(\mathbf{x}') - lm(\mathbf{x}')) \tag{21}$$

$$= cost(f(\mathbf{x}), x_i) + \frac{k}{2}(\Delta_1 + \Delta_2) \tag{22}$$

$$\geq cost(f(\mathbf{x}), x_i). \tag{23}$$

If k is odd, for every $i \in S$, it holds that

$$cost(f(\mathbf{x}'), x_i) = \frac{k}{2}(rm(\mathbf{x}') - lm(\mathbf{x}')) + (x_i - lm(\mathbf{x}')) \tag{24}$$

$$= cost(f(\mathbf{x}), x_i) + \frac{k}{2}(\Delta_1 + \Delta_2) + \Delta_1 \tag{25}$$

$$\geq cost(f(\mathbf{x}), x_i). \tag{26}$$

Case 2: $\Delta_1 \geq 0, \Delta_2 < 0$. The rightmost agent must be a member of S. It is obvious that whether k is even or odd, this agent cannot benefit from the deviation, since the rightmost location and possibly the leftmost location move away from her.

Case 3: $\Delta_1 < 0, \Delta_2 \geq 0$. This case is symmetric to *Case 2*.

Case 4: $\Delta_1 < 0, \Delta_2 < 0$. Both the leftmost agent and the rightmost must be members of S.

Whether k is even or odd, it holds that

$$cost(f(\mathbf{x}'), lm(\mathbf{x})) + cost(f(\mathbf{x}'), rm(\mathbf{x})) \tag{27}$$
$$= k(rm(\mathbf{x}) - lm(\mathbf{x})) \tag{28}$$
$$= cost(f(\mathbf{x}), lm(\mathbf{x})) + cost(f(\mathbf{x}), rm(\mathbf{x})). \tag{29}$$

Thus, either the leftmost agent or the rightmost agent cannot benefit from the deviation, whether k is even or odd.

Now let us turn to proving the approximation ratio of f.

If k is even, for every $\mathbf{x} \in I^n$ and every $i \in N$, it holds that

$$cost(f(\mathbf{x}), x_i) = \frac{k}{2}(rm(\mathbf{x}) - lm(\mathbf{x})) = \frac{k}{2}L(\mathbf{x}), \tag{30}$$

which implies that $ER(f, \mathbf{x}) = 1$. Thus, f is optimal.

If k is odd, for every $\mathbf{x} \in I^n$ and every $i \in N$, it holds that

$$\frac{k-1}{2}L(\mathbf{x}) \leq cost(f(\mathbf{x}), x_i) = \frac{k-1}{2}L(\mathbf{x}) + (x_i - lm(\mathbf{x})) \leq \frac{k+1}{2}L(\mathbf{x}). \tag{31}$$

Thus,

$$ER(f, \mathbf{x}) \leq \frac{k - \frac{k-1}{2}L(\mathbf{x})}{k - \frac{k+1}{2}L(\mathbf{x})} \tag{32}$$

$$\leq \frac{k - (k-1)/2}{k - (k+1)/2} \tag{33}$$

$$= \frac{k+1}{k-1}, \tag{34}$$

where the second inequality holds because $g(t) = \frac{k - \frac{k-1}{2}t}{k - \frac{k+1}{2}t}$ monotonically increases on $[0, 1]$.

Note that $ER(OPT, \mathbf{x}) \geq 1$ and we have

$$\sup_{\mathbf{x} \in I^n} \frac{ER(f, \mathbf{x})}{ER(OPT, \mathbf{x})} \leq \frac{k+1}{k-1}. \tag{35}$$

Consider $\mathbf{x}' = (0 : N_1, 1 : N_2)$. It is obvious that $ER(OPT, \mathbf{x}') = 1$ and $ER(f, \mathbf{x}') = \frac{k+1}{k-1}$. Thus, f has an approximation ratio of $\frac{k+1}{k-1}$.   □

5   Conclusions and Future Work

In this paper, we studied deterministic mechanism design without money for k-facility location games with envy ratio on the line interval, where a set of strategic agents report their locations and a social planner locates k facilities to minimize the envy ratio. We discuss the problem in two settings: the homogeneous k-facility location where k facilities serve the same purpose and the heterogeneous k-facility location where each facility serves a different purpose. In the homogeneous location game, we devised a $\frac{2k}{2k-1}$-approximate group strategyproof mechanism which is also the best deterministic strategyproof mechanism. In the heterogeneous location game, when k is even, we proposed the optimal and group strategyproof mechanism; when k is odd, we obtained a $\frac{k+1}{k-1}$-approximate group strategyproof mechanism.

There are at least three directions for future research. The first is a deterministic lower bound for heterogeneous k-facility location game when k is odd. The second is randomized mechanism design for multiple facility location games with envy ratio. The third is the formulation of heterogeneous k-facility location game with envy ratio which can adapt more realistic scenarios, such as locations combined with fractional preferences [10].

References

1. Alon, N., Feldman, M., Procaccia, A.D., Tennenholtz, M.: Strategyproof approximation of the minimax on networks. Math. Oper. Res. **35**(3), 513–526 (2010)
2. Anastasiadis, E., Deligkas, A.: Heterogeneous facility location games. In: Proceedings of the 17th International Conference on Autonomous Agents and MultiAgent Systems, pp. 623–631 (2018)
3. Cai, Q., Filos-Ratsikas, A., Filos, A., Tang, P.: Facility location with minimax envy. In: Proceedings of the 25th International Joint Conference on Artificial Intelligence, pp. 137–143 (2016)
4. Chen, X., Hu, X., Jia, X., Li, M., Tang, Z., Wang, C.: Mechanism design for two-opposite-facility location games with penalties on distance. In: Deng, X. (ed.) SAGT 2018. LNCS, vol. 11059, pp. 256–260. Springer, Cham (2018). https://doi.org/10.1007/978-3-319-99660-8_24
5. Cheng, Y., Wu, W., Zhang, G.: Strategyproof approximation mechanisms for an obnoxious facility game on networks. Theor. Comput. Sci. **497**, 154–163 (2013)

6. Ding, Y., Liu, W., Chen, X., Fang Q., Nong, Q.: Facility location game with envy ratio. Working paper, Ocean University of China (2020)
7. Duan, L., Li, B., Li, M., Xu, X.: Heterogenious two-facility location games with minimum distance requirement. In: Proceedings of the 18th International Conference on Autonomous Agents and MultiAgent Systems, pp. 1461–1469 (2019)
8. Feigenbaum, I., Sethuraman, J.: Strategyproof mechanisms for one-dimensional hybrid and obnoxious facility location models. In: Workshop on Incentive and Trust in E-Communities at the 29th AAAI Conference on Artificial Intelligence, pp. 8–13 (2015)
9. Foley, D.: Resource allocation and the public sector. Yale Econ. Essays 7, 45–98 (1967)
10. Fong, K., Li, M., Lu, P., Todo, T., Yokoo, T.: Facility location games with fractional preferences. In: Proceedings of the 32nd AAAI Conference on Artificial Intelligence, pp. 1039–1046 (2018)
11. Fotakis, D., Tzamos, C.: On the power of deterministic mechanisms for facility location games. ACM Trans. Econ. Comput. 2(4), Article 15 (2014)
12. Li, M., Mei, L., Xu, Y., Zhang, G., Zhao, Y.: Facility location games with externalities. In: Proceedings of the 18th International Conference on Autonomous Agents and Multiagent Systems, pp. 1443–1451 (2019)
13. Lipton, R., Markakis, E., Mossel, E., Saberi, A.: On approximately fair allocations of indivisible goods. In: Proceedings of the 5th ACM Conference on Electronic Commerce, pp. 125–131 (2004)
14. Lu, P., Wang, Y., Zhou, Y.: Tighter bounds for facility games. In: Proceedings of the 5th International Workshop on Internet and Network Economics, pp. 137–148 (2009)
15. Lu, P., Sun, X., Wang, Y., Zhu, Z.: Asymptotically optimal strategy-proof mechanisms for two-facility games. In: Proceedings of the 11th ACM Conference on Electronic Commerce, pp. 315–324 (2010)
16. Mei, L., Li, M., Ye, D., Zhang, G.: Facility location games with distinct desires. Discrete Appl. Math. 264, 148–160 (2019)
17. Moulin, H.: On strategy-proofness and single peakedness. Pub. Choice 35(4), 437–455 (1980)
18. Procaccia, A.D., Tennenholtz, M.: Approximate mechanism design without money. In: Proceedings of the 10th ACM Conference on Electronic Commerce, pp. 177–186 (2009)
19. Schummer, J., Vohra, R.: Strategy-proof location on a network. J. Econ. Theory 104(2), 405–428 (2002)
20. Serafino, P., Ventre, C.: Heterogeneous facility location without money. Theor. Comput. Sci. 636, 27–46 (2016)
21. Varian, H.: Equity, envy and efficiency. J. Econ. Theory 9, 63–91 (1974)
22. Yuan, H., Wang, K., Fong, K., Zhang, Y., Li, M.: Facility location games with optional preference. In: Proceedings of the Twenty-Second European Conference on Artificial Intelligence, pp. 1520–1527 (2016)
23. Zhang, Q., Li, M.: Strategyproof mechanism design for facility location games with weighted agents on a line. J. Comb. Optim. 28(4), 756–773 (2013). https://doi.org/10.1007/s10878-013-9598-8
24. Zou, S., Li, M.: Facility location games with dual preference. In: Proceedings of the 14th International Conference on Autonomous Agents and Multiagent Systems, pp. 615–623 (2015)

Strategyproof Mechanisms for *2*-Facility Location Games with Minimax Envy

Xin Chen, Qizhi Fang, Wenjing Liu$^{(\boxtimes)}$ (iD), and Yuan Ding

Ocean University of China,
Qingdao 266100, Shandong Province, People's Republic of China
liuwj_123@126.com

Abstract. We study a fairness-based model for *2*-facility location games on the real line where the social objective is to minimize the maximum envy over all agents. All the agents seek to minimize their personal costs, and the envy between any two of them is the difference in their personal costs. We consider two cases of personal costs, called min-dist and sum-dist cost. We are interested in investigating strategyproof mechanisms for *2*-facility location games in both cases.

In the case of min-dist personal cost, we prove that a lower bound of the additive approximation for any deterministic strategyproof mechanism is $1/4$; then we propose a $1/2$-additive approximate deterministic group strategyproof mechanism and a $1/4$-additive approximate randomized strategyproof mechanism. In the case of sum-dist personal cost, we design an optimal and group strategyproof deterministic mechanism.

Keywords: Facility location game · Mechanism design · Minimax envy

1 Introduction

Facility location games [13,14,17], motivated by the social choice, has been extensively studied in algorithmic mechanism design [16]. In these settings, the problem input is a set of strategic agents with private location information and the agents might lie about their private locations. At the beginning, the social planner (or mechanism designer) announces an algorithm, which outputs facility locations by collecting all the agents' reported location information. All the agents proceed to strategically reporting their locations, based on the algorithm (or mechanism), to benefit themselves. Naturally, the mechanism designer wishes to encourage each agent's truthful report while optimizing a certain social objective. Thus, how to design efficient and truthful (or strategyproof) mechanisms has been the central topic in facility location games.

This research was supported partially by the National Natural Science Foundation of China (11871442, 11971447) and the Fundamental Research Funds for the Central Universities (201964006, 201861001).

Z. Zhang et al. (Eds.): AAIM 2020, LNCS 12290, pp. 260–272, 2020.
https://doi.org/10.1007/978-3-030-57602-8_24

Previous Work. Procaccia and Tennenholtz [17] initiated approximate mechanism design without money for facility location games. They studied approximate strategyproof mechanisms for *1*-facility location games under two social objective functions: the social cost and the maximum cost, and extended the results to two settings: *2*-facility location and multiple locations per agent. Subsequently, Lu et al. [10, 11] improved the lower and upper bounds of approximation under the social cost objective. Fotakis and Tzamos [7] studied *k*-facility location games and provided an elegant characterization of deterministic strategyproof mechanisms for *2*-facility location games. Zou and Li [20] explored the property of dual preference in facility location games and proposed two extended games, called the dual character facility location game and the two opposite-facility location game with limited distance. Mei et al. [12] introduced the happiness factor of each agent, which considers the agent's degree of satisfaction for the facility location. Duan et al. [5] studied the requirement of the minimum distance between two facilities and established several strategyproof mechanisms. Li et al. [9] proposed the optional preference model for the facility location games with two heterogeneous facilities on a line. Other extended work of classic facility location games can be referred to [2, 4, 6, 18, 19].

Motivations. Our work is motivated by the social objective of minimax-envy for facility location games. Prior to this work, Cai et al. [3] introduced a fairness criterion, called minimax-envy [8, 15] to *1*-facility location games and proposed desirable strategyproof mechanisms.

We study a fairness-based model for *2*-facility location games on the real line where the social objective is to minimize the maximum envy over all the agents. All the agents seek to minimize their personal costs, and the envy between any two of them is the difference in their personal costs. We consider two cases of personal costs, called min-dist cost and sum-dist cost. Intuitively, suppose that the government plans to build two homogeneous facilities (e.g. two supermarkets) on the central avenue and customers prefer to go to the nearest market from their home, which leads to the min-dist cost where each agent's personal cost is the minimization of her distances to both facilities. On the other hand, suppose that our government plans to build two heterogeneous facilities (e.g. one hospital and one pharmacy), customers are concerned about the sum of their distances to both facilities, which leads to sum-dist cost where agent's personal cost is the sum of her distances to both facilities.

Contributions. In this paper, we are interested in investigating strategyproof mechanisms for *2*-facility location games with minimax-envy objective. In Sect. 2, we formulate the *2*-facility game with minimax-envy objective. For any two agents, we define the envy as the difference of their personal costs, resulting in the maximum-envy objective. In Sect. 3, we focus on studying the case of min-dist personal cost. On the side of deterministic strategyproof mechanisms, we firstly show a lower bound of 1/4-additive approximation, then propose a 1/2-additive approximate group strategyproof mechanism. On the side of randomized strategyproof mechanisms, we propose a 1/4-additive approximate strategyproof mechanism. In Sect. 4, we study the case of sum-dist cost and we propose an optimal and group strategyproof mechanism. The last section concludes the results of this paper and discusses several open problems.

2 Preliminaries

2.1 The Basic Setting

INSTANCES. Let $N = \{1, 2, \ldots, n\}$ be a collection of agents where each agent $i \in N$ has a location $x_i \in \mathbb{R}$ as her private information. The vector $\mathbf{x} = (x_1, x_2, \ldots, x_n) \in \mathbb{R}^n$ is referred to as an *instance* (or a *location profile*). Specially, an instance $\mathbf{x} = (x_1 : N_1, \ldots, x_k : N_k)$ with $\cup_{i=1}^{k} N_i = N$ is referred to as a *k-location instance* if all the n locations (which could be the same) distributed at exactly k different locations.

For any instance $\mathbf{x} \in \mathbb{R}^n$, let $lm(\mathbf{x})$ (or $rm(\mathbf{x})$) be the *leftmost* (or the *rightmost*) location of $\mathbf{x}$, and let $L(\mathbf{x}) = rm(\mathbf{x}) - lm(\mathbf{x})$ be the *length* of $\mathbf{x}$.

MECHANISMS. A *deterministic mechanism* for 2-facility location game is a function $f : \mathbb{R}^n \to \mathbb{R}^2$, which maps an instance to the locations of two facilities. A *randomized mechanism* is a function $f : \mathbb{R}^n \to \Delta(\mathbb{R}^2)$ where $\Delta(\mathbb{R}^2)$ is the set of all probability distributions over $\mathbb{R}^2$, that is, the output of a randomized mechanism $f(\mathbf{x})$ is a probability distribution $\mathcal{P}$ over location profiles of two facilities.

PERSONAL OBJECTIVE. Given an instance $\mathbf{x} \in \mathbb{R}^n$, if f is deterministic, denote $f(\mathbf{x}) = (f_1(\mathbf{x}), f_2(\mathbf{x}))$. Agent i aims to minimize her personal cost which is defined as

(i) MIN-DIST COST: agent i's distance to the nearest facility, i.e.,

$$cost(f(\mathbf{x}), x_i) = \min\{|f_1(\mathbf{x}) - x_i|, |f_2(\mathbf{x}) - x_i|\}.$$

(ii) SUM-DIST COST: the sum of agent i's distances to the two facilities, i.e.,

$$cost(f(\mathbf{x}), x_i) = |f_1(\mathbf{x}) - x_i| + |f_2(\mathbf{x}) - x_i|.$$

Remark. If f is randomized, agent i's min-dist cost is her expected distance to the nearest facility; agent i's sum-dist cost is the expected sum of her distances to the two facilities.

STRATEGYPROOFNESS. The social planner hopes to design mechanisms that have no incentive for agents to strategically misreport locations. Here, we introduce three types of strategyproofness.

(i) A mechanism f is *strategyproof* if no agent can achieve less cost by misreporting her truthful location, regardless of the other agents' reports.

(ii) A mechanism f is *group strategyproof* if no coalition (which controls at least one location) of agents can simultaneously misreport such that every agent in the coalition achieves strictly less cost, regardless of the other agents' reports.

(iii) A mechanism f is *partial group strategyproof* if no coalition (which controls exactly one location) of agents can simultaneously misreport such that every agent in the coalition achieves strictly less cost, regardless of the other agents' reports.

Remark. By definitions and Lu et al. [10], we have the following:

- In a k-facility game, a group strategyproof mechanism is also partial group strategyproof.
- In a k-facility game, a mechanism is partial group strategyproof if and only if it is strategyproof.

MINIMAX-ENVY OBJECTIVE. Given an instance $\mathbf{x} \in \mathbb{R}^n$, let $\mathbf{y} = (y_1, y_2)$ be the locations of two facilities. The *envy* of agent i w.r.t. agent j is defined as $cost(\mathbf{y}, x_i) - cost(\mathbf{y}, x_j)$ and the maximum envy of $\mathbf{y}$ w.r.t. $\mathbf{x}$ is defined as $me(\mathbf{y}, \mathbf{x}) = \max_{i \neq j}\{cost(\mathbf{y}, x_i) - cost(\mathbf{y}, x_j)\}/L(\mathbf{x})$ if $L(\mathbf{x}) > 0$, $me(\mathbf{y}, \mathbf{x}) = 0$ if $L(\mathbf{x}) = 0$. We will concentrate on the case of $L(\mathbf{x}) > 0$, unless specified. Taking account of the fairness between agents, the social planner aims to minimize the *maximum envy* w.r.t $\mathbf{x}$ over all possible $\mathbf{y} \in \mathbb{R}^2$. In addition, let $opt(\mathbf{x}) = \min_{\mathbf{y} \in \mathbb{R}^2}\{me(\mathbf{y}, \mathbf{x})\}$ be the optimal value and $\mathbf{y}^*$ be an optimal solution w.r.t. $\mathbf{x}$.

On the other hand, from the perspective of mechanism design, let f be any mechanism. If f is deterministic, the maximum envy w.r.t. $\mathbf{x}$ is defined as

$$me(f, \mathbf{x}) = \frac{\max_{i \neq j}\{cost(f(\mathbf{x}), x_i) - cost(f(\mathbf{x}), x_j)\}}{L(\mathbf{x})}.$$

If f is randomized, the expected maximum envy w.r.t. $\mathbf{x}$ is defined as

$$me(f, \mathbf{x}) = \frac{\mathbb{E}_{\mathbf{y} \sim f(\mathbf{x})}[\max_{i \neq j}\{cost(\mathbf{y}, x_i) - cost(\mathbf{y}, x_j)\}]}{L(\mathbf{x})}.$$

Our objective is to seek strategyproof mechanisms that minimize the maximum envy for 2-facility location games.

2.2 The Performance of Strategyproof Mechanisms

In the previous literatures [6,11,17], the *approximation ratio* is typically used to quantify the performance of a mechanism in facility location games. However, in homogeneous 2-facility location games with min-dist cost, the approximation ratio is no longer suitable to measure the performance of mechanisms. In fact, it can be verified that no deterministic strategyproof mechanism can achieve a finite approximation ratio.

To show this negative result, we start with the optimal solution for minimizing the maximum envy on *3*-location instances.

Lemma 2.1. *Given any 3-location instance* $\mathbf{x} = (x_1 : N_1, x_2 : N_2, x_3 : N_3)$, *the optimal value* $opt(\mathbf{x}) = 0$ *holds for minimizing maximum envy w.r.t. either min-dist cost or sum-dist cost.*

Lemma 2.2. *In homogeneous 2-facility location games with min-dist cost, no deterministic strategyproof mechanism has a finite approximation ratio for minimizing the maximum envy.*

Proof. Let f be any deterministic strategyproof mechanism. Recall that the min-dist cost of each agent i is $cost(f(\mathbf{x}), x_i) = \min\{|f_1 - x_i|, |f_2 - x_i|\}$. Consider a *3*-location instance $\mathbf{x} = (x_1 : N_1, x_2 : N_2, x_3 : N_3)$, where $x_1 = 0$, $x_2 = 1/2$, $x_3 = 1$. Notice that $L(\mathbf{x}) = 1$. Based on Lemma 2.1, the optimal value $opt(\mathbf{x}) = 0$ (with $\mathbf{y}^* = (1/4, 3/4)$).
Let $f(\mathbf{x}) = (f_1, f_2)$, $f_1 \leq f_2$.
    (i) If $f_2 = x_i$, for some $i \in \{1, 2, 3\}$, it is obvious that

$$cost(f(\mathbf{x}), x_i) = 0; \max_{j \neq i}\{cost(f(\mathbf{x}), x_j)\} \geq 1/4.$$

Thus, the maximum envy $me(f, \mathbf{x}) \geq 1/4$. Combined with $opt(\mathbf{x}) = 0$, the approximation ratio $\gamma_0 \geq me(f, \mathbf{x})/opt(\mathbf{x}) = +\infty$.
    (ii) If $f_2 \neq x_i$, $i = 1, 2, 3$, we consider another *3*-location instance $\mathbf{x}' = (x_1' : N_1, x_2' : N_2, x_3' : N_3)$ where $x_1' = x_1 = 0$, $x_2' = f_2$, and $x_3' = x_3 = 1$. By Lemma 2.1, the optimal value is $opt(\mathbf{x}') = 0$. Let $f(\mathbf{x}') = (f_1', f_2')$. By f's strategyproofness, we have $f_2 \in \{f_1', f_2'\}$; otherwise, agents in N_2 with location f_2 can benefit by deviating from x_2' to x_2. Thus, $cost(f(\mathbf{x}'), f_2) = 0$, and it implies that $\max\{cost(f(\mathbf{x}'), 0), cost(f(\mathbf{x}'), 1)\} > 0$. Hence, the maximum envy $me(f, \mathbf{x}') > 0$, and the approximation ratio $\gamma_0 \geq me(f, \mathbf{x}')/opt(\mathbf{x}') = +\infty$. □

In this paper, we introduce the *additive approximation* [1] to measure the performance of mechanisms, which is defined as follows.

ADDITIVE APPROXIMATION. A mechanism f is said to have an additive approximation of ρ ($\rho \geq 0$) if $\sup\{me(f, \mathbf{x}) - opt(\mathbf{x}) \mid \mathbf{x} \in \mathbb{R}^n\} = \rho$, where $opt(\mathbf{x})$ is the optimal value for minimizing the maximum envy.

3   Min-Dist Personal Cost

In this section, we study homogeneous *2*-facility location games with min-dist cost (i.e., $cost(f(\mathbf{x}), x_i) = \min\{|f_1 - x_i|, |f_2 - x_i|\}$, $\forall i \in N$), and design deterministic and randomized strategyproof mechanisms.

3.1   Deterministic Strategyproof Mechanisms

In this subsection, we firstly show a lower bound of additive approximation of $1/4$ for any deterministic strategyproof mechanisms. Then, we design a deterministic group strategyproof mechanism with additive approximation of $1/2$.

Theorem 3.1. *In homogeneous 2-facility location games, any deterministic strategyproof mechanism has an additive approximation of at least 1/4 for minimizing the maximum envy.*

Proof. Consider an *3*-location instance $\mathbf{x} = (0 : N_1, 1/2 : N_2, 1 : N_3)$ with $x_1 = 0$, $x_2 = 1/2$, $x_3 = 1$. Notice that $L(\mathbf{x}) = 1$. Based on Lemma 2.1, the optimal value $opt(\mathbf{x}) = 0$. Let f be any deterministic strategyproof mechanism. Denote $f(\mathbf{x}) = (f_1, f_2)$ with $f_1 \leq f_2$.

Case 1. $f_1 \geq 1/2$.
In this case, we observe that the envy of agent 1 w.r.t. agent 2 in $\mathbf{x}$ is

$$cost(f(\mathbf{x}), x_1) - cost(f(\mathbf{x}), x_2) = |f_1 - 0| - |f_1 - 1/2| = 1/2,$$

which implies that the maximum envy $me(f, \mathbf{x}) \geq (1/2)/1 > 1/4$.

Case 2. $f_2 \leq 1/2$. This case is symmetric to Case 1.

Case 3. $f_1 < 1/2$ and $f_2 > 1/2$.

Case 3.1. $f_1 \leq 0$ and $f_2 > 1/2$.

Case 3.1.1. $cost(f(\mathbf{x}), 1/2) = |f_1 - 1/2|$. Observe that $cost(f(\mathbf{x}), 0) = |f_1 - 0|$. Thus, the maximum envy

$$me(f, \mathbf{x}) \geq (|f_1 - 1/2| - |f_1 - 0|)/1 = 1/2 > 1/4.$$

Case 3.1.2 $cost(f(\mathbf{x}), 1/2) = |f_2 - 1/2|$.
(i) If $f_2 \geq 1$, we get that $cost(f(\mathbf{x}), 1) = |f_2 - 1|$, $cost(f(\mathbf{x}), 1/2) = |f_2 - 1/2|$. Thus, the maximum envy

$$me(f, \mathbf{x}) \geq (|f_2 - 1| - |f_2 - 1/2|)/1 = 1/2 > 1/4.$$

(ii) If $1/2 < f_2 < 3/4$, we assume that $f_2 = 3/4 - \varepsilon$, $\varepsilon \in (0, 1/4)$.
Consider another *3*-location instance $\mathbf{x}' = (x_1' : N_1, x_2' : N_2, x_3' : N_3)$ with $x_1' = x_1 = 0$, $x_2' = 3/4 - \varepsilon$ and $x_3' = x_3 = 1$. Let $f(\mathbf{x}') = (f_1', f_2')$, $f_1' \leq f_2'$. It can be verified that $cost(f(\mathbf{x}'), x_2') = 0$. Otherwise if $cost(f(\mathbf{x}'), x_2') > 0$, agents in N_2 under $\mathbf{x}'$ can benefit by misreporting $1/2$ (which leads to $cost(f(\mathbf{x}), x_2') = 0$), in contradiction to f's strategyproofness.
When $f_1' = 3/4 - \varepsilon$, we have $cost(f(\mathbf{x}'), x_1') = |f_1' - 0| = 3/4 - \varepsilon$. Thus, the maximum envy

$$me(f, \mathbf{x}') \geq cost(f(\mathbf{x}'), x_1') - cost(f(\mathbf{x}'), x_2') = 3/4 - \varepsilon > 1/4.$$

When $f_2' = 3/4 - \varepsilon$, we have $cost(f(\mathbf{x}'), x_3') = |f_2' - 1| = 1/4 + \varepsilon$. Thus, the maximum envy

$$me(f, \mathbf{x}') \geq cost(f(\mathbf{x}'), x_3') - cost(f(\mathbf{x}'), x_2') = 1/4 + \varepsilon \geq 1/4.$$

(iii) If $3/4 \le f_2 < 1$, we assume that $f_2 = 3/4 + \varepsilon$, $\varepsilon \in [0, 1/4)$.

Consider another instance $\mathbf{x}'' = (x_1'' : N_1, x_2'' : N_2, x_3'' : N_3)$ with $x_1'' = x_1 = 0$, $x_2'' = x_2 = 1/2$ and $x_3'' = 3/4 + \varepsilon$. Notice that $L(\mathbf{x}'') = 3/4 + \varepsilon$. Let $f(\mathbf{x}'') = (f_1'', f_2'')$, $f_1'' \le f_2''$. It can be verified that $cost(f(\mathbf{x}''), x_3'') = 0$ by f's strategyproofness.

When $f_1'' = 3/4 + \varepsilon$, we have $cost(f(\mathbf{x}''), x_1'') = |f_1'' - 0| = 3/4 + \varepsilon$. Thus,

$$me(f, \mathbf{x}'') = (3/4 + \varepsilon - 0)/(3/4 + \varepsilon) = 1 > 1/4.$$

When $f_2'' = 3/4 + \varepsilon$. we have $\max_{i=1,2,3}\{cost(f(\mathbf{x}''), x_i'')\} \ge 1/4$. Thus,

$$me(f, \mathbf{x}'') \ge (1/4 - 0)/(3/4 + \varepsilon) > 1/4.$$

Case 3.2. $0 < f_1 < 1/2$ and $f_2 > 1/2$.

(i) If $f_1 < 1/4$, we assume that $f_1 = 1/4 - \varepsilon$, $\varepsilon \in [0, 1/4)$.

Consider another instance $\mathbf{x}' = (x_1' : N_1, x_2' : N_2, x_3' : N_3)$ with $x_1' = 1/4 - \varepsilon$, $x_2' = x_2 = 1/2$, and $x_3' = x_3 = 1$. Notice that $L(\mathbf{x}') = 3/4 + \varepsilon$. Let $f(\mathbf{x}') = (f_1', f_2')$, $f_1' \le f_2'$. By the strategyproofness of f, we can verify that $cost(f(\mathbf{x}'), x_1') = 0$.

When $f_2' = 1/4 - \varepsilon$, we have that $cost(f(\mathbf{x}'), x_3') = |f_2' - 1| = 3/4 + \varepsilon$. Thus,

$$me(f, \mathbf{x}') \ge (3/4 + \varepsilon - 0)/(3/4 + \varepsilon) = 1.$$

When $f_1' = 1/4 - \varepsilon$, we get that $\max_{i=1,2,3}\{cost(f(\mathbf{x}'), x_i')\} \ge 1/4$. Thus,

$$me(f, \mathbf{x}') \ge (1/4 - 0)/(3/4 + \varepsilon) > 1/4.$$

(ii) If $1/4 \le f_1 < 1/2$, we assume that $f_1 = 1/4 + \varepsilon$, $\varepsilon \in (0, 1/4)$.

Consider another instance $\mathbf{x}'' = (x_1'' : N_1, x_2'' : N_2, x_3'' : N_3)$ with $x_1'' = x_1 = 0$, $x_2'' = 1/4 + \varepsilon$, and $x_3'' = x_3 = 1$. Notice that $L(\mathbf{x}'') = 1$. Let $f(\mathbf{x}'') = (f_1'', f_2'')$, $f_1'' \le f_2''$. We can verify that $cost(f(\mathbf{x}''), x_2'') = 0$ by strategyproofness.

When $f_2'' = 1/4 + \varepsilon$, we have $cost(f(\mathbf{x}''), x_3'') = |f_2 - 1| = 3/4 - \varepsilon$. Thus,

$$me(f, \mathbf{x}'') = (3/4 - \varepsilon - 0)/1 = 3/4 - \varepsilon \ge 1/2 > 1/4.$$

When $f_1'' = 1/4 + \varepsilon$, we have $cost(f(\mathbf{x}''), x_1'') = |f_1 - 0| = 1/4 + \varepsilon$. Thus,

$$me(f, \mathbf{x}'') \ge (1/4 + \varepsilon)/1 \ge 1/4.$$

Considering the fact that $opt(\mathbf{x}) = 0$ for any *3*-location instance $\mathbf{x}$, f has an additive approximation of at least $1/4$.

□

We now concentrate on the upper bound of deterministic strategyproof mechanisms for the minimax envy objective. Specifically, we design a simple deterministic group strategyproof mechanism with additive approximation of $1/2$.

MECHANISM 1. *For any instance* $\mathbf{x} \in \mathbb{R}^n$, *select the leftmost location* $lm(\mathbf{x})$ *and the rightmost location* $rm(\mathbf{x})$.

Theorem 3.2. *In homogeneous 2-facility location games with min-dist cost, Mechanism 1 is a deterministic group strategyproof mechanism with additive approximation of* $1/2$.

3.2   Randomized Strategyproof Mechanisms

In this section, we turn to designing randomized strategyproof mechanisms for minimizing the maximum envy in homogeneous 2-facility location games.

MECHANISM 2. *For any instance* $\mathbf{x} \in \mathbb{R}^n$*, place two facilities at:*

(i) $lm(\mathbf{x})$ *and* $rm(\mathbf{x})$ *with probability 1/6;*

(ii) $lm(\mathbf{x}) + 1/4 \cdot L(\mathbf{x})$ *and* $rm(\mathbf{x}) - 1/4 \cdot L(\mathbf{x})$ *with probability 2/3;*

(iii) $lm(\mathbf{x}) - 3 \cdot L(\mathbf{x})$ *and* $rm(\mathbf{x}) + 3 \cdot L(\mathbf{x})$ *with probability 1/6.*

Theorem 3.3. *In homogeneous 2-facility location games with min-dist cost, Mechanism 2 is a randomized strategyproof mechanism with additive approximation of at most $1/4$ for minimizing the maximum envy.*

Proof. Let us first tackle the additive approximation of Mechanism 2. Let f denote Mechanism 2. Given an instance $\mathbf{x} \in \mathbb{R}^n$, let $cen(\mathbf{x}) = 1/2 \cdot (lm(\mathbf{x}) + rm(\mathbf{x}))$ be the center of $\mathbf{x}$. Consider any agent $i \in N$ and assume w.l.g. that $x_i \in [lm(\mathbf{x}), cen(\mathbf{x})]$. By Mechanism 2, agent i's cost in $\mathbf{x}$ is

$$cost(f(\mathbf{x}), x_i) = 1/6 \cdot (x_i - lm(\mathbf{x})) + 2/3 \cdot |x_i - (lm(\mathbf{x}) + 1/4 \cdot L(\mathbf{x}))| \quad (3\text{-}2)$$
$$+ 1/6 \cdot (x_i - (lm(\mathbf{x}) - 3 \cdot L(\mathbf{x}))).$$

To be detailed, if $x_i \in [lm(\mathbf{x}), lm(\mathbf{x}) + 1/4 \cdot L(\mathbf{x})]$, we have

$$cost(f(\mathbf{x}), x_i) = -1/3 \cdot x_i + 1/3 \cdot lm(\mathbf{x}) + 2/3 \cdot L(\mathbf{x}),$$

which implies that $cost(f(\mathbf{x}), x_i) \in [7/12 \cdot L(\mathbf{x}), 8/12 \cdot L(\mathbf{x})]$. Otherwise if $x_i \in (lm(\mathbf{x}) + 1/4 \cdot L(\mathbf{x}), cen(\mathbf{x})]$, we have

$$cost(f(\mathbf{x}), x_i) = x_i - lm(\mathbf{x}) + 1/3 \cdot L(\mathbf{x}),$$

which implies that $cost(f(\mathbf{x}), x_i) \in (7/12 \cdot L(\mathbf{x}), 10/12 \cdot L(\mathbf{x})]$.

To be concluded, for any agent $i \in N$, $cost(f(\mathbf{x}), x_i) \in [7/12 \cdot L(\mathbf{x}), 10/12 \cdot L(\mathbf{x})]$, and the maximum envy $me(f, \mathbf{x}) \leq 1/4$. Thus, the additive approximation $\rho = \sup\{me(f, \mathbf{x}) - opt(\mathbf{x}) \mid \mathbf{x} \in \mathbb{R}^n\} \leq 1/4$.

Let us now turn to proving the strategyproofness. Consider any agent $i \in N$ and assume w.l.g that $x_i \in [lm(\mathbf{x}), cen(\mathbf{x})]$. For any $x_i' \in R$, consider another instance $\mathbf{x}' = (x_{-i}', \mathbf{x}_{-i})$. We wish to prove that agent i can not benefit by deviation (from x_i to x_i'), i.e.,

$$cost(f(\mathbf{x}'), x_i) \geq cost(f(\mathbf{x}), x_i). \quad (3\text{-}3)$$

In fact, Eq. (3-3) can be proved by a case analysis.

Case 1. $x_i' \in (-\infty, lm(\mathbf{x}))$.

In this case, $lm(\mathbf{x}') < lm(\mathbf{x})$, $rm(\mathbf{x}') = rm(\mathbf{x})$ and $L(\mathbf{x}') > L(\mathbf{x})$. Denote $\Delta = lm(\mathbf{x}) - lm(\mathbf{x}')$, we have $L(\mathbf{x}') = L(\mathbf{x}) + \Delta$.

Case 1.1. $x_i \leq cen(\mathbf{x}')$.

$$
\begin{aligned}
cost(f(\mathbf{x}'), x_i) &= 1/6 \cdot (x_i - lm(\mathbf{x}')) + 2/3 \cdot |lm(\mathbf{x}') + 1/4 \cdot L(\mathbf{x}') - x_i| \\
&\quad + 1/6 \cdot (x_i - lm(\mathbf{x}') + 3 \cdot L(\mathbf{x}')) \\
&\geq cost(f(\mathbf{x}), x_i) + 1/6 \cdot \Delta + 2/3 \cdot (-\Delta - 1/4 \cdot \Delta) + 1/6 \cdot (\Delta + 3\Delta) \\
&\geq cost(f(\mathbf{x}), x_i),
\end{aligned}
$$

where the first inequality holds from Eq. (3-2).

Case 1.2 $x_i > cen(\mathbf{x}')$.

$$
\begin{aligned}
cost(f(\mathbf{x}'), x_i) &= 1/6 \cdot (rm(\mathbf{x}') - x_i) + 2/3 \cdot |rm(\mathbf{x}') - 1/4 \cdot L(\mathbf{x}') - x_i| \quad (3\text{-}4) \\
&\quad + 1/6 \cdot (rm(\mathbf{x}') + 3 \cdot L(\mathbf{x}') - x_i).
\end{aligned}
$$

By the initial assumption that $x_i \leq cen(\mathbf{x})$, we get that

$$
\begin{aligned}
|rm(\mathbf{x}') - 1/4 \cdot L(\mathbf{x}') - x_i| &\geq |rm(\mathbf{x}) - 1/4 \cdot L(\mathbf{x}) - x_i| - 1/4 \cdot \Delta \quad (3\text{-}5) \\
&\geq |x_i - lm(\mathbf{x}) - 1/4 \cdot L(\mathbf{x})| - 1/4 \cdot \Delta
\end{aligned}
$$

Combining Eq. (3-2) with Eq. (3-4), we have

$$
\begin{aligned}
&cost(f(\mathbf{x}'), x_i) - cost(f(\mathbf{x}), x_i) \\
&= 1/6 \cdot [(rm(\mathbf{x}') - x_i) - (x_i - lm(\mathbf{x}))] \\
&\quad + 2/3 \cdot [|rm(\mathbf{x}') - 1/4 \cdot L(\mathbf{x}') - x_i| - |x_i - lm(\mathbf{x}) - 1/4 \cdot L(\mathbf{x})|] \\
&\quad + 1/6 \cdot [(rm(\mathbf{x}') + 3 \cdot L(\mathbf{x}) - x_i) - (x_i - lm(\mathbf{x}) + 3 \cdot L(\mathbf{x}))] \\
&\geq 2/3 \cdot (-1/4 \cdot \Delta) + 1/6 \cdot 0 + 1/6 \cdot 3\Delta \\
&= 1/3 \cdot \Delta \\
&> 0,
\end{aligned}
$$

where the first inequality holds from Eq. (3-5) and $x_i \leq cen(\mathbf{x})$, the second inequality holds from $\Delta > 0$.

Case 2. $x_i' \in [lm(\mathbf{x}), rm(\mathbf{x})]$.

Case 2.1. $x_i = lm(\mathbf{x})$.

In this subcase, $lm(\mathbf{x}') \geq lm(\mathbf{x})$, $rm(\mathbf{x}') = rm(\mathbf{x})$ and $L(\mathbf{x}') \leq L(\mathbf{x})$. Denote $\Delta = lm(\mathbf{x}') - lm(\mathbf{x}) \geq 0$, we have $L(\mathbf{x}') = L(\mathbf{x}) - \Delta$. Notice that $x_i = lm(\mathbf{x}) \leq lm(\mathbf{x}') \leq cen(\mathbf{x}')$. By Mechanism 2,

$$
\begin{aligned}
cost(f(\mathbf{x}'), x_i) &= 1/6 \cdot (lm(\mathbf{x}') - x_i) + 2/3 \cdot (lm(\mathbf{x}') + 1/4 \cdot L(\mathbf{x}') - x_i) \\
&\quad + 1/6 \cdot (x_i - lm(\mathbf{x}') + 3 \cdot L(\mathbf{x}')) \\
&= cost(f(\mathbf{x}), x_i) + 1/6 \cdot \Delta + 2/3 \cdot (\Delta - 1/4 \cdot \Delta) + 1/6 \cdot (-\Delta - 3 \cdot \Delta) \\
&= cost(f(\mathbf{x}), x_i),
\end{aligned}
$$

where the second equality holds from Eq. (3-2).

Case 2.2. $x_i \in (lm(\mathbf{x}), cen(\mathbf{x})]$.

In this subcase, $lm(\mathbf{x}') = lm(\mathbf{x})$, $rm(\mathbf{x}') = rm(\mathbf{x})$ and $L(\mathbf{x}') = L(\mathbf{x})$. It follows that $f(\mathbf{x}') = f(\mathbf{x})$. Thus, $cost(f(\mathbf{x}'), x_i) = cost(f(\mathbf{x}), x_i)$.

Case 3. $x_i' \in (rm(\mathbf{x}), +\infty)$.

Case 3.1 $x_i = lm(\mathbf{x})$.

In this subcase, $lm(\mathbf{x}') \geq lm(\mathbf{x})$, $rm(\mathbf{x}') > rm(\mathbf{x})$. Denote $\Delta_1 = lm(\mathbf{x}') - lm(\mathbf{x})$, $\Delta_2 = rm(\mathbf{x}') - rm(\mathbf{x})$, we have $\Delta_1 \geq 0$, $\Delta_2 > 0$, $L(\mathbf{x}') = L(\mathbf{x}) - \Delta_1 + \Delta_2$.

$$cost(f(\mathbf{x}'), x_i) = 1/6 \cdot (lm(\mathbf{x}') - x_i) + 2/3 \cdot (lm(\mathbf{x}') + 1/4 \cdot L(\mathbf{x}') - x_i) \quad (3\text{-}6)$$
$$+ 1/6 \cdot |x_i - lm(\mathbf{x}') + 3 \cdot L(\mathbf{x}')|.$$

(i) If $x_i \geq lm(\mathbf{x}') - 3 \cdot L(\mathbf{x}')$, Eq. (3-6) can be rewritten as

$$\begin{aligned}
cost(f(\mathbf{x}'), x_i) &= 1/6 \cdot (lm(\mathbf{x}') - x_i) + 2/3 \cdot (lm(\mathbf{x}') + 1/4 \cdot L(\mathbf{x}') - x_i) \\
&\quad + 1/6 \cdot (x_i - lm(\mathbf{x}') + 3 \cdot L(\mathbf{x}')) \\
&= cost(f(\mathbf{x}), x_i) + 1/6 \cdot \Delta_1 + 2/3 \cdot (\Delta_1 + 1/4 \cdot (-\Delta_1 + \Delta_2)) \\
&\quad + 1/6 \cdot (-\Delta_1 + 3 \cdot (-\Delta_1 + \Delta_2)) \\
&= cost(f(\mathbf{x}), x_i) + 2/3 \cdot \Delta_2 \\
&> cost(f(\mathbf{x}), x_i),
\end{aligned}$$

where the last inequality holds since $\Delta_2 > 0$.

(ii) If $x_i < lm(\mathbf{x}') - 3 \cdot L(\mathbf{x}')$, we observe that $\Delta_1 \geq 3/4 \cdot (L(\mathbf{x}) + \Delta_2)$. Otherwise if $\Delta_1 < 3/4 \cdot (L(\mathbf{x}) + \Delta_2)$, then $lm(\mathbf{x}') - 3 \cdot L(\mathbf{x}') = lm(\mathbf{x}) + 4 \cdot \Delta_1 - 3 \cdot (L(\mathbf{x}) + \Delta_2) < lm(\mathbf{x}) = x_i$. which contradicts the assumption of this case. Thus,

$$\begin{aligned}
cost(f(\mathbf{x}'), x_i) &= 1/6 \cdot (lm(\mathbf{x}') - x_i) + 2/3 \cdot (lm(\mathbf{x}') + 1/4 \cdot L(\mathbf{x}') - x_i) \\
&\quad + 1/6 \cdot (-x_i + lm(\mathbf{x}') - 3 \cdot L(\mathbf{x}')) \\
&= cost(f(\mathbf{x}), x_i) + 1/6 \cdot \Delta_1 + 2/3 \cdot (\Delta_1 + 1/4 \cdot (-\Delta_1 + \Delta_2)) \\
&\quad + 1/6 \cdot (\Delta_1 - 3 \cdot (-\Delta_1 + \Delta_2)) - L(\mathbf{x}) \\
&= cost(f(\mathbf{x}), x_i) + 4/3 \cdot \Delta_1 - 1/3 \cdot \Delta_2 - L(\mathbf{x}) \\
&\geq cost(f(\mathbf{x}), x_i) + 2/3 \cdot \Delta_2 \\
&> cost(f(\mathbf{x}), x_i)
\end{aligned}$$

where the first inequality holds since $\Delta_1 \geq 3/4 \cdot (L(\mathbf{x}) + \Delta_2)$ and the second inequality holds since $\Delta_2 > 0$.

Case 3.2. $x_i \in (lm(\mathbf{x}), cen(\mathbf{x})]$.

In this subcase, $lm(\mathbf{x}') = lm(\mathbf{x})$, $rm(\mathbf{x}') \geq rm(\mathbf{x})$. Let $rm(\mathbf{x}') = rm(\mathbf{x}) + \Delta$, $\Delta > 0$. Notice that $L(\mathbf{x}') = L(\mathbf{x}) + \Delta$. By Mechanism 2,

$$
\begin{aligned}
cost(f(\mathbf{x}'), x_i) &= 1/6 \cdot (x_i - lm(\mathbf{x}')) + 2/3 \cdot |lm(\mathbf{x}') + 1/4 \cdot L(\mathbf{x}') - x_i| \\
&\quad + 1/6 \cdot (x_i - lm(\mathbf{x}') + 3 \cdot L(\mathbf{x}')) \\
&\geq cost(f(\mathbf{x}), x_i) + 2/3 \cdot (-1/4 \cdot \Delta) + 1/6 \cdot 3 \cdot \Delta \\
&= cost(f(\mathbf{x}), x_i) + 1/6 \cdot \Delta \\
&\geq cost(f(\mathbf{x}), x_i),
\end{aligned}
$$

where the first inequality holds from Eq. (3-2), and the second inequality holds since $\Delta \geq 0$. □

4   Sum-Dist Personal Cost

In this section, we focus on designing strategyproof mechanisms for heterogeneous 2-facility location games with sum-dist cost.

Given an instance $\mathbf{x} \in \mathbb{R}^n$, a deterministic mechanism f and each agent i's sum-dist cost is the sum of its distances to both facilities, i.e.,

$$
cost(f(\mathbf{x}), x_i) = |f_1(\mathbf{x}) - x_i| + |f_2(\mathbf{x}) - x_i|. \tag{4-1}
$$

Let us review Mechanism 1, i.e., select the leftmost location $lm(\mathbf{x})$ and the rightmost location $rm(\mathbf{x})$ for any instance $\mathbf{x} \in \mathbb{R}^n$. We can verify that Mechanism 1 is an optimal strategyproof mechanism for the heterogeneous case.

Theorem 4.1. *In heterogeneous games with sum-dist cost, Mechanism 1 is optimal and group strategyproof for minimizing maximum envy.*

5   Conclusions

In this paper, we studied 2-facility location games with minimax-envy on the real line. As for personal cost, we discussed two natural cases, called min-dist cost and sum-dist cost. For both cases, we proposed several desirable strategyproof mechanisms. In the case of min-dist cost, we propose a deterministic group strategyproof mechanism with 1/2-additive approximation and we show that any deterministic strategyproof mechanism has an additive approximation of at least 1/4; then we propose a randomized strategyproof mechanism with 1/4-additive approximation. In the case of sum-dist cost, we find an optimal and group strategyproof mechanism.

However, there still exist some open problems for further research. Firstly, on the case of min-dist, there still exists a gap between the upper bound of 1/2 and the lower bound of 1/4, over all deterministic approximate strategyproof mechanisms. Secondly, can we find a desirable lower bound for randomized strategyproof mechanisms in the min-dist case? At last, it would be interesting to investigate facility location games with other fairness criteria.

References

1. Alon, N., Shapira, A., Sudakov, B.: Additive approximation for edge-deletion problems. Ann. Math. **170**(1), 371–411 (2009)
2. Ben-Porat, O., Tennenholtz, M.: Multiunit facility location games. Math. Oper. Res. **44**(3), 865–889 (2019)
3. Cai, Q., Filos-Ratsikas A., Tang, P.: Facility location with minimax envy. In: Proceedings of the 25th International Joint Conference on Artificial Intelligence, pp. 137–143 (2016)
4. Cheng, Y., Yu, W., Zhang, G.: Mechanisms for obnoxious facility game on a path. In: Wang, W., Zhu, X., Du, D.-Z. (eds.) COCOA 2011. LNCS, vol. 6831, pp. 262–271. Springer, Heidelberg (2011). https://doi.org/10.1007/978-3-642-22616-8_21
5. Duan, L., Li, B., Li, M., Xu, X.: Heterogeneous two-facility location games with minimum distance requirement. In: Proceedings of the 18th International Conference on Autonomous Agents and MultiAgent Systems, pp. 1461–1469 (2019)
6. Fong, K., Li, M., Lu, P., Todo, T., Yokoo, T.: Facility location games with fractional preferences. In: Proceedings of the 32nd AAAI Conference on Artificial Intelligence, pp. 1039–1046 (2018)
7. Fotakis, D., Tzamos, C.: On the power of deterministic mechanisms for facility location games. ACM Trans. Econ. Comput. **2**(4), 1–37 (2014). Article 15
8. Lipton, R., Markakis, E., Mossel, E., Saberi, A.: On approximately fair allocations of indivisible goods. In: Proceedings of the 5th ACM Conference on Electronic Commerce, pp. 125–131 (2004)
9. Li, M., Lu, P., Yao, Y., Zhang, J.: Strategyproof mechanism for two heterogeneous facilities with constant approximation ratio. In: Proceedings of the International Joint Conference on Artificial Intelligence (2020)
10. Lu, P., Sun, X., Wang, Y., Zhu, Z.: Asymptotically optimal strategy-proof mechanisms for two-facility games. In: Proceedings of the 11th ACM Conference on Electronic Commerce, pp. 315–324 (2010)
11. Lu, P., Wang, Y., Zhou, Y.: Tighter bounds for facility games. In: Leonardi, S. (ed.) WINE 2009. Lecture Notes in Computer Science, vol. 5929, pp. 137–148. Springer, Heidelberg (2009). https://doi.org/10.1007/978-3-642-10841-9_14
12. Mei, L., Li, M., Ye, D., Zhang, G.: Facility location games with distinct desires. Discrete Appl. Math. **264**, 148–160 (2019)
13. Miyagawa, E.: Locating libraries on a street. Soc. Choice Welfare **18**, 527–541 (2001). https://doi.org/10.1007/s003550000074
14. Moulin, H.: On strategy-proofness and single peakedness. Public Choice **35**(4), 437–455 (1980). https://doi.org/10.1007/BF00128122
15. Nguyen, T.T., Rothe, J.: How to decrease the degree of envy in allocations of indivisible goods. In: Perny, P., Pirlot, M., Tsoukiàs, A. (eds.) ADT 2013. LNCS (LNAI), vol. 8176, pp. 271–284. Springer, Heidelberg (2013). https://doi.org/10.1007/978-3-642-41575-3_21
16. Nisan, N., Ronen, A.: Algorithmic mechanism design. Games Econ. Behav. **35**, 166–196 (2001)
17. Procaccia, A.D., Tennenholtz, M.: Approximate mechanism design without money. In: Proceedings of the 10th ACM Conference on Electronic Commerce, pp. 177–186 (2009)
18. Wada, Y., Ono, T., Todo, T., Yokoo, M.: Facility location with variable and dynamic populations. In: Proceedings of the 10th International Conference on Autonomous Agents and Multiagent Systems, pp. 336–344 (2018)

19. Yuan, H., Wang, K., Fong, K., Zhang, Y., Li, M.: Facility location games with optional preference. In: Proceedings of the Twenty-Second European Conference on Artificial Intelligence, pp. 1520–1527 (2016)
20. Zou, S., Li, M.: Facility location games with dual preference. In: Proceedings of the 14th International Conference on Autonomous Agents and Multiagent Systems, pp. 615–623 (2015)

Robustness and Approximation
for the Linear Contract Design

Guichen Gao[1,3], Xinxin Han[1,3], Li Ning[1], Hing-Fung Ting[2],
and Yong Zhang[1(✉)]

[1] Shenzhen Institutes of Advanced Technology, Chinese Academy of Sciences,
Shenzhen, People's Republic of China
{gc.gao,xx.han,li.ning,zhangyong}@siat.ac.cn
[2] Department of Computer Science, The University of Hong Kong,
Hong Kong, People's Republic of China
hfting@cs.hku.hk
[3] Shenzhen College of Advanced Technology,
University of Chinese Academy of Sciences,
Beijing, People's Republic of China

Abstract. We consider the contract design problem where a principal designs a contract to incentivize an agent to undertake n independent tasks. In the contract, the principal decides the payment $\boldsymbol{w} = \{w_0, w_1, ..., w_n\}$. The principal is associated with a non-decreasing revenue function $f(k)$, where k is the number of successful tasks. The objective of the contract design problem is to maximize the principal's expected profit, i.e., $\max_{\boldsymbol{s},\boldsymbol{w}} \{F(\boldsymbol{s}) - W(\boldsymbol{s},\boldsymbol{w})\}$, where $\boldsymbol{s}$ is the agent's strategy set, $F(\boldsymbol{s})$ is the expected revenue of the principal and $W(\boldsymbol{s},\boldsymbol{w})$ is the expected payment to the agent. Given the payment $\boldsymbol{w}$, the agent will choose her strategy from $\boldsymbol{s}$ to maximize her expected utility. For each task, the agent may work hard or shirk and her strategy is to decide the number of tasks to work hard on. If the principal knows the strategy set $\boldsymbol{s}$, the optimum contract can be found by linear programming. In [7], a more complicated model where the strategy set $\boldsymbol{s}$ is unknown to the principal is considered, they showed that the linear contract is robust and presented an $1/N$-approximation algorithm given $N \leq n$ dominating strategies. In this paper, we further analyze this model and prove the approximation factor can be upper bounded by $(1 - \alpha_N)/(1 - \alpha_N^N)$, where $\alpha_N \in [0, 1)$ is a given constant.

1 Introduction

The classic principal-agent model in contract theory is one of the most fundamental problems in economics and it gradually becomes a hot topic in the field of optimization. We consider the principal-agent problem that the principal wants an agent to undertake his n independent tasks so as to maximize the expected profit. To achieve this target, the principal will design a contract to incentivize the agent, who will also select the best strategy in response to the contract.

© Springer Nature Switzerland AG 2020
Z. Zhang et al. (Eds.): AAIM 2020, LNCS 12290, pp. 273–285, 2020.
https://doi.org/10.1007/978-3-030-57602-8_25

Such principal-agent problem has been well studied for many years. Contract design for explicit computational methods has appeared in [1,9]. Babaioff et al. [1] discussed a combinatorial variant of the classical principal-agent problem from economic theory. Unlike the classical principal-agent problem, the principal must incentivize a group of strategic agents, whose actions are hidden from the principal and each agent takes some cost with respect to her action. The main challenge of this new model is to determine the optimal amount of effort required for each agent and their focus is how the complex mix of agents' efforts affect the outcome. Ho et al. [9] studied the dynamic adjustment of task quality compensation for the principal. They considered a multi-round version of the principal-agent model, in which the agent makes strategic choices on the work levels that the principal cannot directly observe. They proposed a new algorithm with sublinear regret in the time horizon.

Following the above principal-agent problems, there are a lot of works on the variants and extensions of the principal-agent model. Schosser [12] analyzed how to allocate the risk so that the principal and the agent have the consistant incentive and he also pointed out this can be measured by performance indicators. In addition, Grant et al. [8] considered various dispute settlement mechanisms and analyzed how the best dispute settlement mechanism depends on the ambiguous attitudes of the parties. Under the condition of risk and ambiguity, various types of optimal contracts had been discussed. Under the incentive scheme, Cvitanić et al. [5] studied the general form of the principal-agent problem of lump-sum payment and provided a systematic method to solve this kind of problem. They reduced a nonzero sum stochastic differential game to a stochastic control problem that could be solved by a standard tool of the control theory. Corgent et al. [4] discussed the setting of the optimal contract under the condition of both monetary and non-monetary incentives. The model shows how a labor contract with a combination of weak monetary incentives and targets unrelated to wages can be optimal. When the demand distribution is continuous, Singham [10] and Cai [11] introduced the method of sample average approximation by solving a discrete distribution problem. The contract contains one or more options, under each of which the quantity of products transferred to the quantity received by the agent who may have hidden preferences that affect their demand for products. Different from the above demand distribution continuity, when the actions satisfy the continuity condition, Dai et al. [6] studied a model to describe the two-way principal-agent problem with asymmetric information. Bichler et al. [2] introduced a principal-agent model of enterprises participating in multi-unit auction. In each company, the agent bids on behalf of the principal. The agent hides her valuation of the goods and wants to maximize the utility, while the unwitting principal aims to maximize the profit by assigning goods to the agent with the budget constraints.

Recently, there are some interesting models introduced by Carroll [3] and Dütting et al. [7]. Carroll [3] considered a simple moral hazard problem in the case of risk-neutral and limited liability, in which the principal is uncertain about the techniques used by the agent. The principal is aware of some actions that the agent may take, however, there may exist some actions unknown to the principal. The principal evaluates the performance of the contract under the worst-case analysis. In general cases, they proved that the unique optimal contract must be linear. This model provides a new interpretation of the linear contract that is widely used in practice, and also provides a flexible and easily handled method for the moral hazard under non-quantified uncertainty. Dütting et al. [7] proved the robustness of the linear contract, roughly speaking, the linear contracts is no worse than the worst-case contract in different settings. In addition, under conditions of limited liability, individual-rational agent and incentive compatible, they analyzed the ratio between the linear contract and the optimal contract with respect to different parameters.

If the principal wants to maximize his expected revenue $F(s)$, intuitively, he might incentivize the agent to work hard on more tasks. However, more revenue may not lead to more profit. The objective is to maximize the principal's expected profit. In this paper, we further analyze the linear contract in algorithmic point of view. We firstly prove the robustness of the linear contract, i.e., the linear contract is no worse than the worst-case contract in achieving the principal's expected profit by a simple analysis. Furthermore, we prove the approximation factor can be upper bounded by $(1 - \alpha_N)/(1 - \alpha_N^N)$, which improves Dütting et al.'s result [7], where $\alpha_N \in [0, 1)$ is a given constant.

2   Problem Statement

There is one principal who wants to find an agent to undertake his n tasks. These tasks are independent and each task only has two states: success and failure. When the agent works hard, the success probability for each task is π, i.e., $p_w = \pi$ and when the agent work shirk, the success probability for each task is ε, i.e., $p_l = \varepsilon$ which satisfies $0 \leq \varepsilon < \pi \leq 1$. For the agent, she has different strategies to finish these tasks. We use p_s^n to denote the strategy set of the agent which indicates that the agent will work hard on arbitrary s tasks in the total n tasks. Sometimes, we abbreviate the strategy set p_s^n to s. The strategy set of the agent is hidden from principal. With the strategy set, the agent finally returns expected revenue $F(p_s^n) = \sum_{k=0}^n f(k) \cdot A_{n,k}(p_s^n)$ to the principal, where $f(k)$ is the principal's revenue function and $A_{n,k}(p_s^n)$ represents the probability that in the strategy set p_s^n the agent could achieve success exactly k tasks. For the probability expression $A_{n,k}(p_s^n)$, it satisfies that if $k > n$ or $k < 0$ then $A_{n,k}(p_s^n) = 0$. In addition, in the course of performing the task, the agent will incur corresponding cost $e(p_s^n)$, which is hidden from the principal and related to the strategy set p_s^n. For the principal, his revenue function $f(k)$ is nondecreasing, where k represents the number of successful tasks and the revenue function $f(k)$ is common information. When the agent returns the expected revenue $F(p_s^n)$ to the principal, the principal pays the agent expected payment $W(p_s^n, w) = \sum_{k=0}^n w_k \cdot A_{n,k}(p_s^n)$ where w_k represents the agent's wage in k successful tasks.

Due to the relationship between the principal and the agent, we consider the system is limited liability (LL), i.e. $w_k \geq 0$ for all $k = 0, 1, \ldots, n$. For the agent, her expected utility can be represented as $E[U_w] = W(\boldsymbol{p}_s^n, \boldsymbol{w}) - e(\boldsymbol{p}_s^n)$ and she is individual-rational (IR) which means the agent's expected utility satisfies $E[U_w] \geq 0$. In addition, the agent is incentive compatible (IC) which means the agent always choose the strategy to maximize her expected utility according to the principal's payment $\boldsymbol{w}$. For the principal, his expected profit can be represented as $E[R_{\boldsymbol{p}_s^n, w}] = F(\boldsymbol{p}_s^n) - W(\boldsymbol{p}_s^n, \boldsymbol{w})$ and he pays the agent wage only depending on the expected revenue $F(\boldsymbol{p}_s^n)$. In addition, $F(\boldsymbol{p}_s^n)$, $W(\boldsymbol{p}_s^n, \boldsymbol{w})$ and the cost $e(\boldsymbol{p}_s^n)$ increase as the size of the strategy set $\boldsymbol{p}_s^n$ increasing. And $A_{n,k}(\boldsymbol{p}_s^n)$ is the total probability of k, i.e., $\sum_{k=0}^n A_{n,k}(\boldsymbol{p}_s^n) = 1$.

Based on the above, the goal of this problem is to maximize the principal's expected profit without knowing the agent's strategy set. We give the model as following:

$$\max_{s,w} \quad F(\boldsymbol{p}_s^n) - W(\boldsymbol{p}_s^n, \boldsymbol{w})$$

$$s.t. \quad \boldsymbol{p}_s^n \in \arg\max_{s'} W(\boldsymbol{p}_{s'}^n, \boldsymbol{w}) - e(\boldsymbol{p}_{s'}^n) \quad (IC)$$

$$W(\boldsymbol{p}_s^n, \boldsymbol{w}) - e(\boldsymbol{p}_s^n) \geq 0 \quad (IR)$$

$$w_k \geq 0 \ \forall \ k = 0, 1, \ldots n \quad (LL)$$

Before analyzing the objective of the problem, we give some helpful definitions as following.

Definition 1. *We say that the probability expression* $A_{n,k}(\boldsymbol{p}_s^n)$ *is convertible between different strategy set, if*

$$A_{n,k}(\boldsymbol{p}_s^n) = p_w \cdot A_{n-1,k-1}(\boldsymbol{p}_{s-1}^{n-1}) + (1 - p_w) \cdot A_{n-1,k}(\boldsymbol{p}_{s-1}^{n-1})$$

$$A_{n,k}(\boldsymbol{p}_{s-1}^n) = p_l \cdot A_{n-1,k-1}(\boldsymbol{p}_{s-1}^{n-1}) + (1 - p_l) \cdot A_{n-1,k}(\boldsymbol{p}_{s-1}^{n-1}).$$

Definition 2. *We say* $F(\boldsymbol{p}_s^n) - e(\boldsymbol{p}_s^n)$ *is the First-Best (FB) solution, if*

$$F(\boldsymbol{p}_s^n) - e(\boldsymbol{p}_s^n) = \max_{s'} \left\{ F(\boldsymbol{p}_{s'}^n) - e(\boldsymbol{p}_{s'}^n) \right\}.$$

Definition 3. *We say the strategy set is distribution-ambiguous if the principal only gets the expected revenue* $F(\boldsymbol{p}_s^n)$ *without knowing the agent's clear successful probability* $\{\varepsilon, \pi\}^n \in [0,1]^n$ *in the case that IC, IR and LL satisfied.*

Definition 4. *We say the contract is worst-case contract if the principal's expected profit is minimized with the distribution-ambiguous strategy, subjected to the constraints IC, IR and LL.*

Definition 5. *We say the linear contract is optimal if the linear contract* $w_k = \alpha \cdot f(k)$, $\alpha \in [0,1]$ *maximizes the principal's expected profit from all the incentive linear contract, i.e.:*

$$\alpha = \arg\max_{\alpha'}\{F(\boldsymbol{p}_s^n) - W(\boldsymbol{p}_s^n, \boldsymbol{w})\} = \arg\max_{\alpha'}\{(1 - \alpha') \cdot F(\boldsymbol{p}_s^n)\}$$

Definition 6. *We say incentive strategy set A_N includes some strategies which maximize the agent's expected utility within some range of the α, $\alpha \in [0, 1]$, where N represents the number of incentive strategies of the A_N and $N \leq n$.*

In addition, we give one important assumption for this paper.

Assumption 1. $\forall \ \boldsymbol{p}_s^n, \boldsymbol{p}_{s'}^n$, *if* $F(\boldsymbol{p}_s^n) \leq F(\boldsymbol{p}_{s'}^n)$, *there must exist* $e(\boldsymbol{p}_s^n) \leq e(\boldsymbol{p}_{s'}^n)$.

We give the Assumption 1 to show that the agent's cost increases as the expected revenue increases.

Based on the above definitions and assumption, we discuss the robustness of the linear contract and the performance ratio of the optimal linear contract.

3   Robustness of the Linear Contract

In this section, we discuss the linear contract's robustness. Before analyzing robustness of the linear contract, there are some helpful information about the principal's expected revenue and profit.

Lemma 1. *Principal's expected revenue $F(\boldsymbol{p}_s^n)$ increases with the strategy $\boldsymbol{p}_s^n$ increasing.*

Proof.

$$F(\boldsymbol{p}_s^n) - F(\boldsymbol{p}_{s-1}^n) = \sum_{k=0}^{n} f(k) \cdot A_{n,k}(\boldsymbol{p}_s^n) - \sum_{k=0}^{n} f(k) \cdot A_{n,k}(\boldsymbol{p}_{s-1}^n) \tag{1}$$

$$= \sum_{k=0}^{n} f(k) \cdot [A_{n,k}(\boldsymbol{p}_s^n) - A_{n,k}(\boldsymbol{p}_{s-1}^n)] \tag{2}$$

$$= (\pi - \varepsilon) \cdot \sum_{k=0}^{n} f(k) \cdot [A_{n-1,k-1}(\boldsymbol{p}_{s-1}^{n-1}) - A_{n-1,k}(\boldsymbol{p}_{s-1}^{n-1})] \tag{3}$$

$$= (\pi - \varepsilon) \cdot \sum_{k=0}^{n-1} [f(k+1) - f(k)] \cdot A_{n-1,k}(\boldsymbol{p}_{s-1}^{n-1}) \tag{4}$$

$$\geq 0 \tag{5}$$

The Eq. (1) comes from the definition of the expected revenue and Eq. (3) is from the probability expression which is convertible between different strategy set. Recall that if $k > n$ *or* $k < 0$, then $A_{n,k}(\boldsymbol{p}_s^n) = 0$, Eq. (4) comes from rearranging the Eq. (3). In addition, $f(k+1) \geq f(k)$ and $\pi > \varepsilon$, then $F(\boldsymbol{p}_s^n) \geq F(\boldsymbol{p}_{s-1}^n)$, i.e., if the agent works in more tasks, the principal's expected revenue is higher.   □

According to the Lemma 1, we give a theoretical analysis of that the principal has to incentivize the agent to work on more tasks in order to obtain higher expected revenue. However, our goal is to maximize the principal's expected profit rather than his expected revenue. Then, the important and difficult one is how to pay the agent's wage. Based on the First-Best solution, we give one observation as following.

Observation 1. *When the First-Best solution is decided, we can solve the following linear programming to maximize the principal's expected profit.*

$$\min_{w} \quad W(\boldsymbol{p}_s^n, \boldsymbol{w}) - e(\boldsymbol{p}_s^n)$$

$$s.t. \quad W(\boldsymbol{p}_s^n, \boldsymbol{w}) \geq W(\boldsymbol{p}_{s'}^n, \boldsymbol{w}) - e(\boldsymbol{p}_{s'}^n), \ \forall \ \boldsymbol{p}_{s'}^n \neq \boldsymbol{p}_s^n \quad (IC)$$

$$W(\boldsymbol{p}_s^n, \boldsymbol{w}) - e(\boldsymbol{p}_s^n) \geq 0 \quad\quad\quad\quad\quad (IR)$$

$$w_k \geq 0, \quad \forall k = 0, 1, \ldots, n \quad\quad\quad\quad (LL)$$

In general, the result of this linear programming may be tight, i.e., $E[U_w] = W(\boldsymbol{p}_s^n, \boldsymbol{w}) - e(\boldsymbol{p}_s^n) = 0$. Thus, the agent's expected payment just equals to her cost and the principal undertakes the total cost. However, in order to solve this linear programming, the principal needs to know all the strategies of the agent which is the important and difficult point for this problem.

In order to solve the problem of the information asymmetry, based on the previous work, we introduce the affine contract and prove that the principal's expected profit achieved by the affine contract is no worse than the worst-case contract.

Lemma 2. *The principal gives the agent wage subjecting to the constraints of IC, IR and LL. For arbitrary distribution-ambiguous strategy, there always exists an affine contract $\alpha_1, \alpha_0 \geq 0, w_k = \alpha_1 \cdot f(k) + \alpha_0$, which can achieve the principal's expected profit, and then it is no worse than the worst-case contract $\boldsymbol{w}$.*

Proof. There are two case for this problem.

- *Case 1: $w_0 > w_n$:* Without loss of generality, we assume that $A_{n,i}(\boldsymbol{p}_s^n) = 0, \forall i \in [n]/\{0, n\}$ i.e., $A_{n,0}(\boldsymbol{p}_s^n), A_{n,n}(\boldsymbol{p}_s^n) \neq 0$. Suppose that two strategies $\boldsymbol{p}_s^n, \boldsymbol{p}_{s'}^n$ satisfy $F(\boldsymbol{p}_s^n) < F(\boldsymbol{p}_{s'}^n), e(\boldsymbol{p}_s^n) < e(\boldsymbol{p}_{s'}^n)$. Since $F(\boldsymbol{p}_s^n) < F(\boldsymbol{p}_{s'}^n)$, we can get the following

$$F(\boldsymbol{p}_s^n) - F(\boldsymbol{p}_{s'}^n) = \sum_{k=\{0,n\}} f(k) \cdot A_{n,k}(\boldsymbol{p}_s^n) - \sum_{k=\{0,n\}} f(k) \cdot A_{n,k}(\boldsymbol{p}_{s'}^n)$$

$$= \sum_{k=\{0,n\}} f(k) \cdot [A_{n,k}(\boldsymbol{p}_s^n) - A_{n,k}(\boldsymbol{p}_{s'}^n)] \quad\quad (6)$$

$$= [f(n) - f(0)] \cdot [A_{n,n}(\boldsymbol{p}_s^n) - A_{n,n}(\boldsymbol{p}_{s'}^n)] \quad\quad (7)$$

$$< 0 \quad\quad\quad\quad\quad\quad\quad\quad\quad\quad\quad\quad\quad\quad\quad (8)$$

Recall $A_{n,k}(\boldsymbol{p}_s^n)$ is the total probability of k, then $A_{n,0}(\boldsymbol{p}_s^n) + A_{n,n}(\boldsymbol{p}_s^n) = 1$. Equation (7) comes from $A_{n,0}(\boldsymbol{p}_s^n) - A_{n,0}(\boldsymbol{p}_{s'}^n) = A_{n,n}(\boldsymbol{p}_{s'}^n) - A_{n,n}(\boldsymbol{p}_s^n)$. Clearly, Eq. (8) is from $F(\boldsymbol{p}_s^n) < F(\boldsymbol{p}_{s'}^n)$. Because the revenue function satisfies $f(k+1) > f(k)$, we have $A_{n,n}(\boldsymbol{p}_s^n) < A_{n,n}(\boldsymbol{p}_{s'}^n)$ and $A_{n,0}(\boldsymbol{p}_s^n) > A_{n,0}(\boldsymbol{p}_{s'}^n)$.

Based on the above, we have,

$$W(\boldsymbol{p}_s^n, \boldsymbol{w}) - W(\boldsymbol{p}_{s'}^n, \boldsymbol{w}) = \sum_{k=\{0,n\}} w_k \cdot A_{n,k}(\boldsymbol{p}_s^n) - \sum_{k=\{0,n\}} w_k \cdot A_{n,k}(\boldsymbol{p}_{s'}^n)$$

$$= \sum_{k=\{0,n\}} w_k \cdot [A_{n,k}(\boldsymbol{p}_s^n) - A_{n,k}(\boldsymbol{p}_{s'}^n)] \tag{9}$$

$$= (w_0 - w_n) \cdot [A_{n,n}(\boldsymbol{p}_{s'}^n) - A_{n,n}(\boldsymbol{p}_s^n)] \tag{10}$$

$$> 0 \tag{11}$$

Equation (10) is the same as Eq. (7) which comes from $A_{n,0}(\boldsymbol{p}_s^n) - A_{n,0}(\boldsymbol{p}_{s'}^n) = A_{n,n}(\boldsymbol{p}_{s'}^n) - A_{n,n}(\boldsymbol{p}_s^n)$. Due to $w_0 > w_n$ and $A_{n,n}(\boldsymbol{p}_{s'}^n) > A_{n,n}(\boldsymbol{p}_s^n)$, Eq. (11) holds. Then, we have $W(\boldsymbol{p}_s^n, \boldsymbol{w}) > W(\boldsymbol{p}_{s'}^n, \boldsymbol{w})$. Recall assumption 1, $e(\boldsymbol{p}_s^n) \leq e(\boldsymbol{p}_{s'}^n)$ when $F(\boldsymbol{p}_s^n) < F(\boldsymbol{p}_{s'}^n)$. Therefore,

$$W(\boldsymbol{p}_s^n, \boldsymbol{w}) - e(\boldsymbol{p}_s^n) > W(\boldsymbol{p}_{s'}^n, \boldsymbol{w}) - e(\boldsymbol{p}_{s'}^n)$$

That means the strategy $\boldsymbol{p}_s^n$ is incentived. The agent's expected utility and the principal's expected profit can be presented as $E[U_{\boldsymbol{w}}] = W(\boldsymbol{p}_s^n, \boldsymbol{w}) - e(\boldsymbol{p}_s^n)$, $E[R_{\boldsymbol{p}_s^n, \boldsymbol{w}}] = F(\boldsymbol{p}_s^n) - W(\boldsymbol{p}_s^n, \boldsymbol{w})$ respectively. Based on the case 1, we analyze the special affine contract, i.e., linear contract $w_k = \alpha_1 \cdot f(k)$ is no worse than the worst-case contract $\boldsymbol{w}$.

(1) $F(\boldsymbol{p}_s^n) - e(\boldsymbol{p}_s^n) \geq F(\boldsymbol{p}_{s'}^n) - e(\boldsymbol{p}_{s'}^n)$: Because the linear contract $w_k = \alpha_1 \cdot f(k)$, the agent's expected utility can be written as

$$E[U_{\boldsymbol{w}}] = \sum_{k=0}^n w_k \cdot A_{n,k}(\boldsymbol{p}_s^n) - e(\boldsymbol{p}_s^n)$$

$$= \alpha_1 \sum_{k=0}^n f(k) \cdot A_{n,k}(\boldsymbol{p}_s^n) - e(\boldsymbol{p}_s^n)$$

$$= \alpha_1 \cdot F(\boldsymbol{p}_s^n) - e(\boldsymbol{p}_s^n) \tag{12}$$

And the principal's expected profit is represented as

$$E[R_{\boldsymbol{p}_s^n, \boldsymbol{w}}] = \sum_{k=0}^n f(k) \cdot A_{n,k}(\boldsymbol{p}_s^n) - \sum_{k=0}^n w_k \cdot A_{n,k}(\boldsymbol{p}_s^n)$$

$$= \sum_{k=0}^n f(k) \cdot A_{n,k}(\boldsymbol{p}_s^n) - \alpha_1 \sum_{k=0}^n f(k) \cdot A_{n,k}(\boldsymbol{p}_s^n)$$

$$= (1 - \alpha_1) \cdot F(\boldsymbol{p}_s^n) \tag{13}$$

According to the Eq. (12), when $\alpha_1 = 1$, we have the agent's expected utility $E[U_{\boldsymbol{w}}] = F(\boldsymbol{p}_s^n) - e(\boldsymbol{p}_s^n)$. And when $\alpha_1 = 0$, the agent's expected utility is $E[U_{\boldsymbol{w}}] = -e(\boldsymbol{p}_s^n)$. The change trend of the agent's expected utility with linear contract α_1 can be seen in Fig. 1 (a). The strategy $\boldsymbol{p}_s^n$

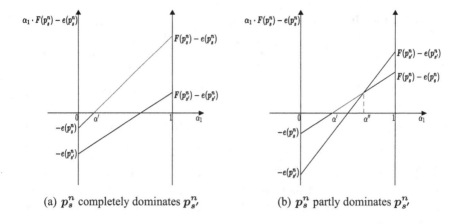

(a) $\boldsymbol{p}_s^n$ completely dominates $\boldsymbol{p}_{s'}^n$ (b) $\boldsymbol{p}_s^n$ partly dominates $\boldsymbol{p}_{s'}^n$

Fig. 1. The change trend of the agent's expected utility with linear contract α_1.

completely dominates the strategy $\boldsymbol{p}_{s'}^n$, i.e. the strategy $\boldsymbol{p}_s^n$ is incentived. Recall the payment $\boldsymbol{w}$, the strategy $\boldsymbol{p}_s^n$ is as well as incentived and it achieves the principal's expected profit $E[R_{\boldsymbol{p}_s^n,\boldsymbol{w}}] = F(\boldsymbol{p}_s^n) - W(\boldsymbol{p}_s^n,\boldsymbol{w})$. For the linear contract, there exists $\alpha' \in (0,1)$ which achieves the principal's expected profit $E[R_{\boldsymbol{p}_s^n,\boldsymbol{w}} \sim \alpha'] = F(\boldsymbol{p}_s^n) - e(\boldsymbol{p}_s^n)$. According to the constraint IR, i.e. $W(\boldsymbol{p}_s^n,\boldsymbol{w}) \geq e(\boldsymbol{p}_s^n)$, thus, the linear contract achieves the principal's expected profit no worse than the payment $\boldsymbol{w}$, i.e. $E[R_{\boldsymbol{p}_s^n,\boldsymbol{w}} \sim \alpha'] \geq E[R_{\boldsymbol{p}_s^n,\boldsymbol{w}}]$.

(2) $F(\boldsymbol{p}_s^n) - e(\boldsymbol{p}_s^n) < F(\boldsymbol{p}_{s'}^n) - e(\boldsymbol{p}_{s'}^n)$: We also show the change trend of the agent's expected utility in Fig. 1 (b).

Recall the payment $\boldsymbol{w}$ incentives the strategy $\boldsymbol{p}_s^n$ which corresponds to the red line, i.e., $\alpha_1 \in [\alpha', \alpha'']$ in the Fig. 1 (b). In the linear contract $\alpha_1 \in [\alpha', \alpha'']$, there exists one α' satisfying $E[R_{\boldsymbol{p}_s^n,\boldsymbol{w}} \sim \alpha'] = (1 - \alpha') \cdot F(\boldsymbol{p}_s^n) = F(\boldsymbol{p}_s^n) - e(\boldsymbol{p}_s^n)$. By the constraint IR, we have $F(\boldsymbol{p}_s^n) - e(\boldsymbol{p}_s^n) \geq F(\boldsymbol{p}_s^n) - W(\boldsymbol{p}_s^n,\boldsymbol{w})$ i.e., $E[R_{\boldsymbol{p}_s^n,\boldsymbol{w}} \sim \alpha'] \geq E[R_{\boldsymbol{p}_s^n,\boldsymbol{w}}]$.

Based on the above, when $w_0 > w_n$, the linear contract achieves the principal's expected profit which is no worse than the payment $\boldsymbol{w}$.

– *Case 2:* $w_0 < w_n$ it : In this part, we prove this problem from three aspects.

(1) When the payment $\boldsymbol{w}$ and the revenue function satisfy $w_k = \alpha_1 \cdot f(k) + \alpha_0$ and $\alpha_1, \alpha_0 \geq 0$, the Lemma 2 naturally formed.

(2) Without loss of generality, as shown in the Fig. 2 (a), there is a payment w_k strictly above the affine function of w_0 and w_n.

We assume the k-th payment of the linear contract is $w_{k'}$. Since $W(\boldsymbol{p}_s^n,\boldsymbol{w}) = \sum_{k=0}^{n} w_k \cdot A_{n,k}(\boldsymbol{p}_s^n)$ and $w_k > w_{k'}$, the agent's expected payment between the worst-case contract $\boldsymbol{w}$ and the linear contract $\boldsymbol{w}'$ satisfies $W(\boldsymbol{p}_s^n,\boldsymbol{w}) \geq W(\boldsymbol{p}_s^n,\boldsymbol{w}')$. When the strategy $\boldsymbol{p}_s^n$ is incentive, the principal's expected profit between the worst-case contract $\boldsymbol{w}$

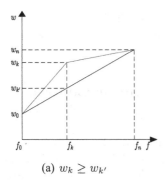

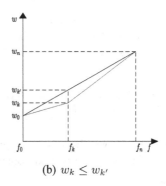

(a) $w_k \geq w_{k'}$ (b) $w_k \leq w_{k'}$

Fig. 2. The payment w_k is strictly above (a) and below (b) the affine function of w_0 and w_n

and the linear contract $\boldsymbol{w}'$ satisfies $E[R_{\boldsymbol{p}_s^n, \boldsymbol{w}'}] = F(\boldsymbol{p}_s^n) - W(\boldsymbol{p}_s^n, \boldsymbol{w}') \geq F(\boldsymbol{p}_s^n) - W(\boldsymbol{p}_s^n, \boldsymbol{w}) = E[R_{\boldsymbol{p}_s^n, \boldsymbol{w}}]$.

Thus, the principal's expected profit achieved by the affine function is no worse than the worst-case contract $\boldsymbol{w}$.

(3) Without loss of generality, there is a payment w_k strictly below the affine function of w_0 and w_n shown in the Fig. 2 (b).

We also assume the k-th payment of the linear contract is $w_{k'}$. Since $W(\boldsymbol{p}_s^n, \boldsymbol{w}) = \sum_{k=0}^n w_k \cdot A_{n,k}(\boldsymbol{p}_s^n)$ and $w_k < w_{k'}$, we have $W(\boldsymbol{p}_s^n, \boldsymbol{w}) \leq W(\boldsymbol{p}_S^n, \boldsymbol{w}')$. Recall the worst-case contract, i.e., $E[R_{\boldsymbol{p}_s^n, \boldsymbol{w}}] = \min_{\boldsymbol{w}''} E[R_{\boldsymbol{p}_s^n, \boldsymbol{w}''}]$. Then, there exists one strategy $\boldsymbol{p}_{s''}^n$ which satisfies $F(\boldsymbol{p}_{s''}^n) - W(\boldsymbol{p}_{s''}^n, \boldsymbol{w}'') \geq F(\boldsymbol{p}_s^n) - W(\boldsymbol{p}_s^n, \boldsymbol{w})$. According to the constraint IR, we have $W(\boldsymbol{p}_{s''}^n, \boldsymbol{w}'') \geq e(\boldsymbol{p}_{s''}^n)$. Therefore $F(\boldsymbol{p}_{s''}^n) - W(\boldsymbol{p}_{s''}^n, \boldsymbol{w}'') \leq F(\boldsymbol{p}_{s''}^n) - e(\boldsymbol{p}_{s''}^n)$. When the strategy $\boldsymbol{p}_{s''}^n$ is incentive, we can find the linear contract α' just like the Fig. 1 (a) or 1 (b) which achieving the principal's expected profit equals to $F(\boldsymbol{p}_{s''}^n) - e(\boldsymbol{p}_{s''}^n)$. It implies that the principal's expected profit achieved by the affine contract is no worse than the worst-case contract.

In conclusion, there is always an affine contract $\alpha_1, \alpha_0 \geq 0$, which makes the principal's expected profit no worse than the worst-case contract $\boldsymbol{w}$. ☐

Theorem 1. *In the distribution-ambiguous strategy, there exists an optimal linear contract to maximize the worst-case contract's expected profit with all the IC, IR and LL constraints.*

Proof. According to the Lemma 2, there always exists an affine contract $\alpha_1, \alpha_0 \geq 0$ to make the principal's expected profit no worse than the worst-case contract. We assume that the linear contract is $w_{k'} = \alpha_1 \cdot f(k)$. Because of $\alpha_1, \alpha_0 \geq 0$, we have $w_k = \alpha_1 \cdot f(k) + \alpha_0 \geq \alpha_1 \cdot f(k) = w_{k'}$. Besides, $W(\boldsymbol{p}_s^n, \boldsymbol{w}) \geq W(\boldsymbol{p}_s^n, \boldsymbol{w}')$. And in the same strategy, $F(\boldsymbol{p}_s^n)$ is consistent. Then, the principal's expected profit satisfies $F(\boldsymbol{p}_s^n) - W(\boldsymbol{p}_s^n, \boldsymbol{w}) \leq F(\boldsymbol{p}_s^n) - W(\boldsymbol{p}_s^n, \boldsymbol{w}')$.

Therefore, the linear contract is better than the affine contract. In addition, according to the Definition 5, the optimal linear contract maximizes the worst-case contract's expected profit. □

4   The Ratio Between the Optimal Linear Contract and the Optimal Contract

In this section, we discuss the approximate ratio of the optimal linear contract. We present the optimal contract solution as OPT which satisfies $OPT = \max_{s',w}\{F(\boldsymbol{p}_{s'}^n) - W(\boldsymbol{p}_{s'}^n, \boldsymbol{w})\}$. Recall the definition of the First-Best solution and the constraint IR, we have

$$FB = \max_{s'}\{F(\boldsymbol{p}_{s'}^n) - e(\boldsymbol{p}_s^n)\}$$
$$\geq \max_{s',w}\{F(\boldsymbol{p}_{s'}^n) - W(\boldsymbol{p}_{s'}^n, \boldsymbol{w})\}$$
$$= OPT$$

Then, we have that the OPT is no more than the First-Best solution (FB).

Recall the definition of the incentive strategy set A_N which includes some strategies maximizing the agent's expected utility within some range of the $\alpha \in [0,1]$. Because only the agent's strategy is incentive, the principal can obtain his expected profit. Then, the incentive strategy set A_N must include the optimal linear contract which means if the strategy $\boldsymbol{p}_s^n$ maximizes the principal's expected profit, it is incentive i.e. $\boldsymbol{p}_s^n \in A_N$.

Lemma 3. $\forall \, \boldsymbol{p}_s^n, \boldsymbol{p}_{s'}^n \in A_N$, if they satisfy $F(\boldsymbol{p}_s^n) \leq F(\boldsymbol{p}_{s'}^n)$ and $e(\boldsymbol{p}_s^n) \leq e(\boldsymbol{p}_{s'}^n)$. There must have $F(\boldsymbol{p}_s^n) - e(\boldsymbol{p}_s^n) \leq F(\boldsymbol{p}_{s'}^n) - e(\boldsymbol{p}_{s'}^n)$.

Proof. Let's prove it by contradiction. $\forall \, \boldsymbol{p}_s^n, \boldsymbol{p}_{s'}^n \in A_N$, we suppose that they satisfy $F(\boldsymbol{p}_s^n) \leq F(\boldsymbol{p}_{s'}^n)$ and $e(\boldsymbol{p}_s^n) \leq e(\boldsymbol{p}_{s'}^n)$. If $F(\boldsymbol{p}_s^n) - e(\boldsymbol{p}_s^n) \geq F(\boldsymbol{p}_{s'}^n) - e(\boldsymbol{p}_{s'}^n)$, the strategy $\boldsymbol{p}_s^n$ will completely dominant the strategy $\boldsymbol{p}_{s'}^n$ in the linear contract, just like the Fig. 1 (a). Then, the strategy $\boldsymbol{p}_{s'}^n$ will not be in the incentive strategy set A_N. Contradiction! □

Without loss of generality, $\forall \, \boldsymbol{p}_s^n \in A_N$ we redefine $F(\boldsymbol{p}_s^n)$ as $F(\boldsymbol{p}_{s_i}^n)$ and sort $F(\boldsymbol{p}_{s_i}^n)$ by increment, i.e. $\forall \, i \in \{1, 2, \ldots, N\}$, $F(\boldsymbol{p}_{s_1}^n) \leq F(\boldsymbol{p}_{s_2}^n) \leq \cdots \leq F(\boldsymbol{p}_{s_N}^n)$. And according to the assumation 1, we have $e(\boldsymbol{p}_{s_1}^n) \leq e(\boldsymbol{p}_{s_2}^n) \leq \cdots \leq e(\boldsymbol{p}_{s_N}^n)$. Then,

$$F(\boldsymbol{p}_{s_1}^n) - e(\boldsymbol{p}_{s_1}^n) \leq F(\boldsymbol{p}_{s_2}^n) - e(\boldsymbol{p}_{s_2}^n) \leq \cdots \leq F(\boldsymbol{p}_{s_N}^n) - e(\boldsymbol{p}_{s_N}^n)$$

Theorem 2. *For arbitrary distribution-ambiguous strategy, the approximation ratio of the optimal linear contract is at least $\frac{1-\alpha_N}{1-\alpha_N^N}$, $\alpha_N \in [0,1)$.*

Proof. For the linear contract $w_k = \alpha_1 \cdot f(k)$, $\alpha_1 \in [0,1)$, the principal's expected profit and the agent's expected utility are respectively $E[R_{\boldsymbol{p}_s^n, w}] = (1 - \alpha_1) \cdot$

$F(\boldsymbol{p}_s^n)$ and $E[U_w] = \alpha_1 \cdot F(\boldsymbol{p}_s^n) - e(\boldsymbol{p}_s^n)$. For any two adjacent strategies $\boldsymbol{p}_{s_{i-1}^n}, \boldsymbol{p}_{s_i}^n$ in the incentive strategy set A_N, there always exists one linear contract $\alpha_{i-1,i} \in [0,1)$, $\forall\, i \in \{1, 2, \ldots, N\}$ to incentivize both strategies, i.e.,

$$\alpha_{i-1,i} \cdot F(\boldsymbol{p}_{s_{i-1}}^n) - e(\boldsymbol{p}_{s_{i-1}}^n) = \alpha_{i-1,i} \cdot F(\boldsymbol{p}_{s_i}^n) - e(\boldsymbol{p}_{s_i}^n)$$

We propose that when the agent's expected utility is the same in different strategies, she will choose the strategy to maximize the principal's expected profit. Because of $\boldsymbol{p}_{s_{i-1}}^n, \boldsymbol{p}_{s_i}^n \in A_N$, we have $F(\boldsymbol{p}_{s_{i-1}}^n) \leq F(\boldsymbol{p}_{s_i}^n)$, $e(\boldsymbol{p}_{s_{i-1}}^n) \leq \boldsymbol{p}_{s_i}^n$ and $F(\boldsymbol{p}_{s_{i-1}}^n) - e(\boldsymbol{p}_{s_{i-1}}^n) \leq F(\boldsymbol{p}_{s_i}^n) - e(\boldsymbol{p}_{s_i}^n)$. Therefore, we abbreviate $\alpha_{i-1,i}$ to α_i which incentivizing the agent to adopt the strategy $\boldsymbol{p}_{s_i}^n$ to maximize the principal's expected profit. In addition, let OPT and ALG be the principal's expected profit of the optimal contract and the optimal linear contract respectively, i.e., $OPT \leq \max_{\boldsymbol{p}_s^n}\{F(\boldsymbol{p}_s^n) - e(\boldsymbol{p}_s^n)\}$ and $ALG = \max_{\boldsymbol{p}_{s_i}^n}\{(1-\alpha_i)F(\boldsymbol{p}_{s_i}^n)\}$. Recall the definition of the incentive strategy set, it includes the strategies which maximize the agent's expected utility. Then, only when the agent selects the strategy, the principal can obtain the corresponding expected profit. Therefore, we can only consider the strategies in A_N and then $OPT \leq \max_{\boldsymbol{p}_{s_i}^n}\{F(\boldsymbol{p}_{s_i}^n) - e(\boldsymbol{p}_{s_i}^n)\}$.

For $\forall\, \boldsymbol{p}_{s_i}^n \in A_N$, because of $\alpha_i \cdot F(\boldsymbol{p}_{s_i}^n) - e(\boldsymbol{p}_{s_i}^n) = \alpha_i \cdot F(\boldsymbol{p}_{s_{i-1}}^n) - e(\boldsymbol{p}_{s_{i-1}}^n)$, there is a recursive expression:

$$F(\boldsymbol{p}_{s_i}^n) - e(\boldsymbol{p}_{s_i}^n) = (1 - \alpha_i) \cdot F(\boldsymbol{p}_{s_i}^n) + [\alpha_i \cdot F(\boldsymbol{p}_{s_{i-1}}^n) - e(\boldsymbol{p}_{s_{i-1}}^n)] \quad (14)$$

Therefore, we have

$$FB = F(\boldsymbol{p}_{s_N}^n) - e(\boldsymbol{p}_{s_N}^n) \quad (15)$$
$$= (1 - \alpha_N)F(\boldsymbol{p}_{s_N}^n) + [\alpha_N F(\boldsymbol{p}_{s_{N-1}}^n) - e(\boldsymbol{p}_{s_{N-1}}^n)] \quad (16)$$
$$\leq (1 - \alpha_N)F(\boldsymbol{p}_{s_N}^n) + \alpha_N[F(\boldsymbol{p}_{s_{N-1}}^n) - e(\boldsymbol{p}_{s_{N-1}}^n)] \quad (17)$$
$$\leq (1 - \alpha_N)F(\boldsymbol{p}_{s_N}^n) + \alpha_N(1 - \alpha_{N-1})F(\boldsymbol{p}_{s_{N-1}}^n) + \cdots$$
$$+ \alpha_N \cdots \alpha_2(1 - \alpha_1)F(\boldsymbol{p}_{s_1}^n) \quad (18)$$
$$\leq (1 + \alpha_N + \alpha_N^2 + \cdots + \alpha_N^{N-1})\max_{\boldsymbol{p}_{s_i}^n}\{(1 - \alpha_i)F(\boldsymbol{p}_{s_i}^n)\} \quad (19)$$

$$\leq \frac{1 - \alpha_N^N}{1 - \alpha_N}ALG \quad (20)$$

We have $F(\boldsymbol{p}_{s_N}^n) - e(\boldsymbol{p}_{s_N}^n) = \max_{\boldsymbol{p}_{s_i}^n}\{F(\boldsymbol{p}_{s_i}^n) - e(\boldsymbol{p}_{s_i}^n)\}$ in the incentive set. Recall the definition of the First-Best solution and the agent's expected utility $E[U_w] = \alpha_1 F(\boldsymbol{p}_s^n) - e(\boldsymbol{p}_s^n)$ in the linear contract, the strategy of the First-Best solution must be in the incentive strategy set. Therefore, we have $FB = F(\boldsymbol{p}_{s_N}^n) - e(\boldsymbol{p}_{s_N}^n)$ and the Eq. (15) holds. Equation (16) comes from the recursive expression (14). Because of $\alpha_i \in [0,1)$, we relax the Eq. (16) to the inequation (17). Inequation (18) comes from the recursive expression (14) and the inequation

(17). In addition, in the linear contract just like the Fig. 3, there is $\alpha_{i-1} < \alpha_i$. Then, we have $\alpha_N = \max_i \alpha_i$ and we relax the inequation (18) to the inequation (19). As the sum of equal ratio sequence, the inequation (20) holds.

Thus $OPT \le [(1 - \alpha_N^N)/(1 - \alpha_N)]ALG$ and the approximation ratio of the optimal linear contract is at least $(1 - \alpha_N)/(1 - \alpha_N^N)$, $\alpha_N \in [0, 1)$. □

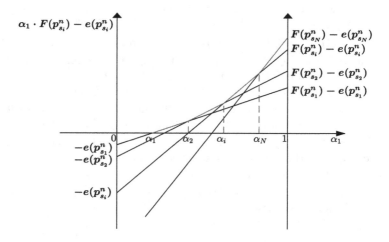

Fig. 3. The tendency of the agent's expected utility in the incentive strategy set.

5   Conclusion

In this paper, we consider the classic principal-agent problem of the contract design theory, and the objective is to maximize the principal's expected profit. Compared with the various contract design models in previous work, we discuss the robustness of the linear contract with a simple method. Moreover, we prove the linear contract is no worse than the worst-case contract. Our greatest contribution is that we prove the approximation factor can be upper bounded by $(1 - \alpha_N)/(1 - \alpha_N^N)$, which improves Dütting et al.'s result [7], where $\alpha_N \in [0, 1)$ is a given constant. A further study may start with the application of the optimal linear contract which includes the mechanism design of the truthfulness.

References

1. Babaioff, M., Feldman, M., Nisan, N.: Combinatorial agency. In: Proceedings of EC 2006, pp. 18–28 (2006)
2. Bichler, M., Paulsen, P.: A principal-agent model of bidding firms in multi-unit auctions. Games Econ. Behav. **111**, 20–40 (2018)
3. Carroll, G.: Robustness and linear contracts. Am. Econ. Rev. **105**(2), 536–563 (2015)

 4. Corgnet, B., Gómez-Miñambres, J., Hernán González, R.: Goal setting in the principal-agent model: weak incentives for strong performance. Games Econ. Behav. **109**, 311–326 (2018)
 5. Cvitanić, J., Possamaï, D., Touzi, N.: Dynamic programming approach to principal-agent problems. Finance Stoch. **22**(1), 1–37 (2018). https://doi.org/10.1007/s00780-017-0344-4
 6. Dai, S., Nie, G., Xiao, N.: The study of the two-way principal-agent model based on asymmetric information. Wirel. Pers. Commun. **102**(2), 629–639 (2018)
 7. Dütting, P., Roughgarden, T., Talgam-Cohen, I.: Simple versus optimal contracts. In: Proceedings of EC 2019, pp. 369–387 (2019)
 8. Grant, S., Kline, J.J., Quiggin, J.: Contracting under uncertainty: a principal-agent model with ambiguity averse parties. Games Econ. Behav. **109**, 582–597 (2018)
 9. Ho, C.-J., Slivkins, A., Vaughan, J.W.: Adaptive contract design for crowdsourcing markets: bandit algorithms for repeated principal-agent problems. J. Artif. Intell. Res. **55**, 317–359 (2016)
10. Singham, D.I.: Sample average approximation for the continuous type principal-agent problem. Eur. J. Oper. Res. **275**(3), 1050–1057 (2019)
11. Singham, D.I., Cai, W.: Sample average approximations for the continuous type principal-agent problem: an example. In: Proceedings of WSC 2017, pp. 2010–2020 (2017)
12. Schosser, J.: Consistency between principal and agent with differing time horizons: computing incentives under risk. Eur. J. Oper. Res. **277**(3), 1113–1123 (2019)

Scheduling Many Types of Calibrations

Hua Chen[1], Vincent Chau[2(✉)], Lin Chen[3], and Guochuan Zhang[1]

[1] Zhejiang University, Hangzhou, China
{chenhua_by,zgc}@zju.edu.cn
[2] Shenzhen Institutes of Advanced Technology, Chinese Academy of Sciences,
Shenzhen, China
vincentchau@siat.ac.cn
[3] Texas Tech University, Lubbock, TX, USA
chenlin198662@gmail.com

Abstract. Machines usually require maintenance after a fixed period. We need to perform a calibration before using the machine again. Such an operation requires a non-negligible cost. Thus finding a schedule minimizing the total cost of calibrations is of great importance.

This paper studies the following scheduling problem. We have a single machine, n jobs where each job j is characterized by its release time r_j, deadline d_j, and processing time p_j. Moreover, there are K types of calibrations, i.e., when the machine performs a calibration of type $k \in \{1, \ldots, K\}$ instantaneously, it can maintain calibrated for a fixed length T_k with a corresponding cost f_k. Jobs can only be processed when the machine is in the calibrated state. Our goal is to find a feasible schedule that minimizes the total cost of calibrations.

We consider two classes of models: the costs of the calibrations are arbitrary, and the costs of the calibrations are equal to their length. For the first model, we propose a pseudo-polynomial time algorithm and a $(2 + \epsilon)$-approximation algorithm when jobs have agreeable deadlines (later release time implies a later deadline). For the second model, we give a 2-approximation algorithm.

Keywords: Scheduling · Calibration · Approximation algorithms

1 Introduction

Scheduling is one of the most classical and important problems in combinatorial optimization. Recently a class of scheduling problems related to calibrations has been brought up by Bender et al. in their seminal paper [3]. The motivation of the problem comes from the Integrated Stockpile Evaluation (ISE) problem [4]. ISE is a program to test nuclear weapons so that they can function normally. Operating these tests needs precision, or safety mistakes can produce a significant

Hua Chen and Guochuan Zhang are supported by NSFC (No. 11531014). Vincent Chau is supported by the CAS President's International Fellowship Initiative n° 2020FYT0002, 2018PT0004.

Z. Zhang et al. (Eds.): AAIM 2020, LNCS 12290, pp. 286–297, 2020.
https://doi.org/10.1007/978-3-030-57602-8_26

loss. Meanwhile, there are testing machines for testing weapons. The testing machines need to be calibrated after running a fixed period to ensure that the testing tasks are processed smoothly.

Similarly, the calibration scheduling problem can be seen as a multi-agents game. For example, during a game, the agents may have abrasion resulting in the inaccurate shooting and need to be calibrated after using for a while. Generally, every agent can decide to charge after some time, and every charging has its corresponding cost and working time.

We formally define the ISE problem as follows: we are given a set J of n jobs (weapons) and m identical machines (testing machines). Each job $j \in J$ is defined by its release time r_j, its deadline d_j and its processing time p_j. We calibrate a machine instantaneously, and the machine can stay valid for $T \geq 2$ time units. The scheduling of all jobs must be feasible, i.e., (1) each job must be scheduled on one of the m identical machines and must be scheduled during p_j calibrated slots, (2) each job must be entirely scheduled between its release time r_j and its deadline d_j and (3) one machine can only process one job at the same time. The goal is to find a feasible schedule using the minimum number of calibrations, where a feasible schedule requires that the scheduling of all jobs should be feasible, and the calibrations used are non-overlapping.

Using the 3-field notation developed in [10], the problem can be denoted as $P|r_j, d_j, p_j, T|\#(calibrations)$.

Related Work

Bender et al. [3] studied the problem in which jobs have unit processing time. They gave a polynomial time algorithm to compute the optimal solution, while a 2-approximation algorithm is given for the multiple machine case. They pointed out that the complexity of the problem remained unknown. Recently, Chen et al. [8] proved that when the number of machines m is constant, the problem can be solved polynomially with dynamic programming. On the other hand, when m is part of the input, they gave a PTAS (polynomial-time approximation scheme).

Later, Fineman and Sheridan [9] considered the case in which jobs have arbitrary processing time, and the preemption of jobs is not allowed[1]. Note that it is NP-hard to decide whether a feasible schedule exists since it can be reduced from the decision version of the bin packing problem. They considered a resource-augmentation version of the problem, and they related it to the classical machine minimization problem [13]. When preemption of jobs is allowed, Angel et al. [1] generalized the algorithm from [3] and showed that it could be solved in polynomial time.

Chau et al. [5] considered the flow time problem with calibrations. They focused on the online version whose objective is to minimize the total flow time, the elapsed time between the release time of a job until its completion, as well as the calibration cost. They aimed to find a tradeoff between the flow time and the cost of the calibrations, and they gave several constant competitive online algorithms for different settings. Wang [14] studied the time-slot cost

---

[1] A job is not allowed to be interrupted once it has been started.

variant of the scheduling problem with calibrations. The cost of scheduling a job depends on the starting time. The goal is to compute a schedule of minimum cost with at most B calibrations. Wang [14] proposed dynamic programmings for different scenarios of this variant. Chau et al. [6] investigated the throughput variant of this scheduling problem: the goal is to maximize the total profit of scheduled jobs. They showed that the problem admits a constant approximation algorithm for arbitrary processing time jobs. Finally, Chau et al. [7] considered that calibrations could only occur simultaneously. They showed that the problem could be solved in polynomial time by giving a dynamic programming algorithm. They also proposed some fast approximation algorithms depending on the cost function of a batch of calibrations.

All the above problems considered one type of calibration. When there are K types of calibrations with respective length T_k and respective cost f_k for $k \in \{1, \ldots, K\}$, Angel et al. [1] proved that when jobs have unit processing time, the problem can be solved in polynomial time by providing a dynamic programming algorithm. However, when jobs have arbitrary processing time, the problem becomes NP-hard. They showed for the particular case in which all the jobs have the same release time and the same deadline. This particular case is similar to the Knapsack Cover Problem for which there exists a $(1 + \varepsilon)$-approximation algorithm [11].

Scheduling with calibrations has similarities with some other well-known scheduling problems, such as minimizing idle periods [2], and scheduling on cloud-based machines which must be rented to perform work [12].

Our Contributions

In this paper, we study the scheduling problem with K types of calibrations on a single machine. We have n jobs where each job j is characterized by its release time r_j, deadline d_j, and processing time p_j. Moreover, there are K types of calibrations, i.e., when the machine performs a calibration of type $k \in \{1, \ldots, K\}$ instantaneously, it can maintain calibrated for a fixed length T_k with a corresponding cost f_k. Jobs can only be processed when the machine is in the calibrated state. Our goal is to find a feasible schedule that minimizes the total cost of calibrations.

The problem is NP-hard even if all the jobs have common release time and common deadline. We investigate the following two generalized cases:

- *arbitrary calibration cost*: the cost of the calibrations does not depend on its length. In this work, we assume that every single job can entirely be scheduled into a single calibration, i.e., $\max_j p_j \leq \min_k T_k$, and jobs have *agreeable deadline*, i.e., for every pair of jobs i, j, we have $r_i \leq r_j$, if and only if $d_i \leq d_j$. In particular, we establish:
  - a pseudo-polynomial time algorithm whose running time is $O(n^5 K^2 P^2 f_{\min}^2)$ where $P = \sum_{j=1}^{n} p_j$ and $f_{\min} = \min_k f_k$ in Sect. 2.1.
  - a $(2 + \varepsilon)$-approximation algorithm in Sect. 2.2.
- *uniform calibration cost*: the cost of the calibrations is equal to its length. For this case, we give a 2-approximation algorithm in Sect. 3.

In the sequel, we suppose without loss of generality that jobs are sorted in non-decreasing order of their deadlines, i.e., $d_1 \leq d_2 \leq \ldots \leq d_n$. Similarly, we sort the calibration types in the non-decreasing order of their length, i.e., $T_1 < T_2 < \ldots < T_K$. Without loss of generality, we also have $f_1 < f_2 < \ldots < f_K$.

2   Arbitrary Calibration Cost

In this section, we investigate the problem with arbitrary calibration cost by proposing dynamic programming algorithms. We first give a pseudo-polynomial time algorithm, then we show how to adapt it into a polynomial running time by losing a constant factor on the objective function.

2.1   A Pseudo-Polynomial Time Algorithm

We first define several time points that are pertinent in any schedule. In [1], they showed some properties for the unit processing time jobs case. We obtain the following by dividing the jobs into unit processing time jobs, i.e., for each job j, we replace by p_j jobs with unit processing time.
    Let $\Phi := \{d_j - h | j = 1, \ldots, n; h = 1, \ldots, P\}$.

Proposition 1 (Proposition 1 [1]). *There exists an optimal schedule in which calibrations start at a time in Φ.*

In the sequel, we only consider schedules satisfying Proposition 1. Moreover, since jobs have agreeable deadlines, we have the following proposition.

Proposition 2 (Lemma 1 [14]). *There exists an optimal solution in which jobs are scheduled in the non-decreasing order of their deadline.*

    Let $\mathcal{F} := \{f_{\min}, f_{\min} + 1, \ldots, n f_{\min}\}$ be the set of cost of any schedule where $f_{\min} = \min_k f_k$.
    Because each job can fit into a single calibration, we know that the cost of the optimal solution OPT is at least $f_{\min}$ and at most $n f_{\min}$, so $OPT \in \mathcal{F}$. We are now ready to describe our dynamic programming.

Dynamic Programming. Let $c(j, f, t, k)$ be the minimum completion time of job j in a feasible schedule whose cost is at most f, such that:

– the first j jobs ($\{1, \ldots, j\}$) are scheduled into the opened calibrations;
– the starting time of the last calibration is t;
– the type of the last calibration is k.

    The idea of our dynamic programming is to compute the number of available time slots in the last calibration. Because jobs have agreeable deadline, we know that the job j will be scheduled after the job $j - 1$. The idea is to schedule the job j as early as possible in order to get the minimum completion time. We distinguish three cases (See Fig. 1 for an illustration of the different cases):

1. Job j is scheduled in the same calibration as the completion time of job $j-1$.
   (a) The job j starts immediately after the completion of the job $j-1$. The job j will be scheduled from $c(j-1,f,t,k)$ to $c(j-1,f,t,k)+p_j$.
   (b) There are some idle time slots between the completion time of the job $j-1$ and the release time of the job j. Then, the job j is scheduled from r_j to r_j+p_j.
2. Job j is scheduled in two calibrations: it starts in the same calibration as the completion of job $j-1$, and ends in another calibration.
   (a) The job j starts immediately after the completion of the job $j-1$. The job j will be executed from $c(j-1,f',t',k')$ to $t'+T_{k'}$, then from t to $t+p_j-(t'+T_{k'}-c(j-1,f',t',k'))$.
   (b) There are some idle time slots between the completion time of the job $j-1$ and the release time of the job j. Then, the job j starts from r_j to the end of the calibration at $t'+T_{k'}$, then from t to $t+p_j-(t'+T_{k'}-r_j)$.
3. Job j is scheduled in a different calibration as job $j-1$.
   (a) The job j starts in a different calibration containing the job $j-1$ and it starts at time t. The job j is executed from t to $t+p_j$.
   (b) The job j is executed in a different calibration containing the job $j-1$ and starts at its release time $r_j>t$. Thus, the job j is executed from r_j to r_j+p_j.

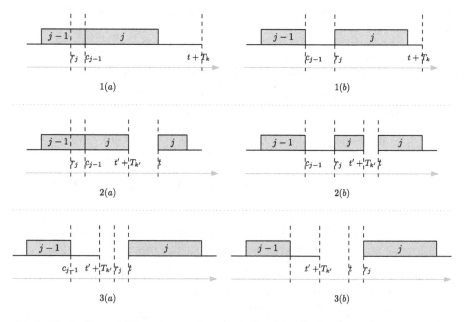

Fig. 1. Illustration of different cases for scheduling job j in the dynamic programming. c_{j-1} denotes the completion time of job $j-1$. In 1(a) and 1(b), $c_{j-1} := c(j-1,f',t,k)$. In the remaining cases, we have $c_{j-1} := c(j-1,f',t',k')$.

Hence, we have the following recursive function.

Proposition 3. *By convention, if the schedule is not feasible, the completion time of such a schedule is $+\infty$. We have $c(j, f, t, k)$*

$$
= \min
\begin{cases}
\min\{c_j \mid c_j = p_j + \max\{c(j-1, f, t, k), r_j\}, c_j \leq \min\{d_j, t + T_k\}\} \\[2mm]
\min\left\{c_j \;\middle|\; \begin{array}{l} c_j = p_j - t' - T_{k'} + \max\{c(j-1, f', t', k'), r_j\} + t, c_j \leq d_j, \\ r_j < t' + T_{k'}, f' + f_k \leq f, f' \in \mathcal{F}, t' \in \Phi, k' \in \{1, \dots, K\} \end{array} \right\} \\[2mm]
\min\{c_j \mid c_j = \max\{t, r_j\} + p_j, c_j \leq d_j, r_j \geq t' + T_{k'}, f' + f_k \leq f\} \\[2mm]
+\infty
\end{cases}
$$

We initialize the table as follows:

$$
c(1, f, t, k) = \min \begin{cases} \max\{t, r_1\} + p_1, & \text{if } f \geq f_k \text{ and } \max\{t, r_1\} + p_1 \leq d_1, \\ +\infty, & \text{otherwise.} \end{cases}
$$

The objective is to find the minimum cost f^* such that $c(n, f^*, t, k) \leq d_n$ for $f^* \in \mathcal{F}, t \in \Phi, k \in \{1, \dots, K\}$.

Theorem 1. *The dynamic programming algorithm in Proposition 3 computes an optimal solution for the arbitrary calibration cost scheduling problem.*

Proof. If the jobs $\{1, \dots, j\}$ cannot be scheduled into the opened calibrations whose total cost is at most f, then the schedule is not feasible, and we have $c(j, f, t, k) = +\infty$. In particular, if there are not enough time slots for the job j, i.e., $c(j, f, t, k) > d_j$, then the of the schedule is $+\infty$. It corresponds to the last line of the dynamic program.

We prove the claim by showing that in the dynamic program, we have tried every possibility of scheduling the job j, as well as the starting time of the calibrations. As described previously, we have six forms in mathematics (We assume $c(j, f, t, k) \leq d_j$ in the following):

1. Job j is scheduled in the same calibration as the completion of job $j - 1$.
   (a) If $r_j < c(j-1, f, t, k)$, then $c(j, f, t, k) = c(j-1, f', t, k) + p_j$.
   (b) If $r_j \geq c(j-1, f, t, k)$, then $c(j, f, t, k) = r_j + p_j$.
2. Job j is scheduled into two calibrations. Since we need to open a new calibration, we need to ensure that the schedule of the first $j - 1$ jobs is of cost at most $f' = f - f_k$.
   (a) If $r_j < t' + T_{k'}$ and $c(j-1, f', t', k') > r_j$, it means that the job j is scheduled right after the completion time of job $j - 1$ until the end of the current calibration, and then the remaining part of the job j, which is equal to $p_j - (t' + T_{k'} - c(j-1, f', t', k'))$, is scheduled in the last calibration that starts at time t. Hence, $c(j, f, t, k) = \min_{f', t', k'} \{p_j - (t' + T_{k'} - c(j-1, f', t', k')) + t\}$.

(b) If $r_j < t' + T_{k'}$ and $c(j-1, f', t', k') \leq r_j$, which means that the remaining part of the job, which is equal to $p_j - (t' + T_{k'} - r_j)$, is scheduled in a new calibration of type k starting at time t. Hence, $c(j, f, t, k) = \min\limits_{f', t', k'} \{p_j - (t' + T_{k'} - r_j) + t\}$.

3. Job j starts in a different calibration as job $j - 1$. As the previous case, we need to open a new calibration, and we need to ensure that the schedule of the first $j - 1$ jobs is of cost at most $f' = f - f_k$.

(a) If $r_j \geq t' + T_{k'}$ and $t > r_j$, it means that we have to schedule the job j no earlier than t, and thus $c(j, f, t, k) = t + p_j$.

(b) If $r_j \geq t' + T_{k'}$ and $t \leq r_j$, which means that the job j starts no earlier than its release time r_j, thus $c(j, f, t, k) = r_j + p_j$.

The optimal value is $\min\left\{ f \middle| \{c(n, f, t, k) : f \in \mathcal{F}, t \in \Phi, k \in \{1, \ldots, K\}\}\right\}$.
According to the dynamic programming algorithm above, its running time is $O(nK^2|\Phi|^2|\mathcal{F}|^2) = O(n^5K^2P^2f_{\min}^2)$, due to $|\Phi| = O(nP)$ and $|\mathcal{F}| = O(nf_{\min})$.
Hence, the running time is pseudo-polynomial. $\square$

2.2   A Constant Approximation Algorithm

To achieve a polynomial time algorithm, we aim to avoid going through all different parameter values in the dynamic program. So, we focus on the sets $\mathcal{F}$ and Φ whose sizes are pseudo-polynomial. We aim to reduce the size of such sets. We define the set of different objective values of the schedules as $\mathcal{F}'$.

Let $\mathcal{F}' := \{f_{\min} \cdot (1+\varepsilon)^q | q = 0, \ldots, \lceil \log_{1+\varepsilon} n \rceil\}$. We have $|\mathcal{F}'| = O(\log_{1+\varepsilon} n)$. We now show that considering the values in $\mathcal{F}'$ can lead to a solution whose cost is no more than $(1 + \varepsilon)$ times of optimal cost OPT.

Lemma 1. *If we restrict the cost f to $\mathcal{F}'$ and assume f^* attains the minimum value in all schedules of $\{c(n, f, t, k) | f \in \mathcal{F}', t \in \Phi, k \in \{1, \ldots, K\}\}$ after using the dynamic programming for $f \in \mathcal{F}', t \in \Phi, k \in \{1, \ldots, K\}$, then we have $f^* \leq (1 + \varepsilon)OPT$.*

Proof. We know that $f_{\min} \leq OPT \leq nf_{\min}$. Then there exists a q_0 such that $f_{\min} \cdot (1+\varepsilon)^{q_0-1} \leq OPT \leq f_{\min} \cdot (1+\varepsilon)^{q_0}$. Thus, $OPT \leq f_{\min} \cdot (1+\varepsilon)^{q_0} \leq OPT \cdot (1+\varepsilon)$. Since $f^* \leq f_{\min} \cdot (1+\varepsilon)^{q_0}$, we obtain $f^* \leq (1+\varepsilon)OPT$. $\square$

Similarly, we define the new set of the starting times of the calibrations as $\Phi' := \{d_j - aT_1 | j = 1, \ldots, n; a = 0, \ldots, n\}$.

Note that we have $|\Phi'| = O(n^2)$. Next, we show that if we restrict the starting times of calibrations to Φ' and do not restrict the costs, then a solution with no more than twice the optimal cost exists. We initially allow to have overlapping calibrations (a time slot can be covered by more than one calibration), which will be handled later without increasing the solution's cost.

Lemma 2. *For the scheduling problem with arbitrary calibration cost, there exists a 2-approximate solution such that the calibrations start at a time in Φ'.*

Proof. We denote OPT to be the optimal value of the problem. Let O be an optimal solution verifying Proposition 1, and we denote the sequence of calibrations in O as $\{C^O_{i_1}, C^O_{i_2}, \dots\}$, where $C^O_{i_u}$ represents the *u-th* calibration in O and its type is i_u where $i_u \in \{1, \dots, K\}$.

We show that when we restrict the starting times of the calibrations in S to Φ', we can get our desired conclusion. Let $C^O_{i_u}$ be a calibration in O such that it does not start at a time in Φ'. We replace it by two calibrations of the same type such that the first one starts at a time in Φ' and such that they cover (at least) the same interval as initially.

Since the schedule O verifies Proposition 1, it means that the starting time of the calibration $C^O_{i_u}$ is at most at a distance of P from a deadline d_h. We assume the distance between d_h and the starting time of $C^O_{i_u}$ is ℓ. See Fig. 2.

Moreover, we have $\min_j p_j \leq T_1$, so we have $\ell \leq P \leq nT_1$, so there is an integer point $t = d_h - aT_1 \in [d_h - \ell - T_{i_u}, d_h - \ell] \cap \Phi'$, where $a \in \{0, \dots, n\}$. Thus, we can replace such calibration by two calibrations starting respectively at t and $t + T_{i_u}$. See Fig. 3.

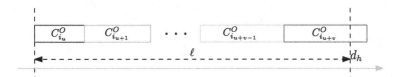

Fig. 2. Illustration of the calibration $C^O_{i_u}$ in O.

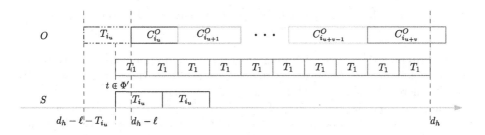

Fig. 3. Illustration of relation of O and the constructed S.

We repeat such modification as long as there is a calibration that does not start at a time in Φ' in the schedule (except the newly added calibrations).

For every calibration in O, there are two consecutive identical calibrations whose types are the same, and the starting time of the first one is in Φ'. Then, all the jobs stay at the same time as in O. Hence, the cost of such schedule is $2OPT$. □

Note that the schedule is feasible for jobs but not for the calibrations since there may exist calibrations that overlap each other. To make all calibrations used be non-overlapping, we need the following observation.

Observation 1. *We can transform a schedule with overlapping calibrations into a schedule without overlapping calibrations in polynomial time without increasing the cost of the solution.*

Indeed, when two calibrations overlap, we can change the starting time of the one that starts later to start when the first calibration ends. Meanwhile, all the jobs stay scheduled at their initial time. We modify at most $2n$ calibrations. Figure 4 illustrates an example to handle two overlapping calibrations.

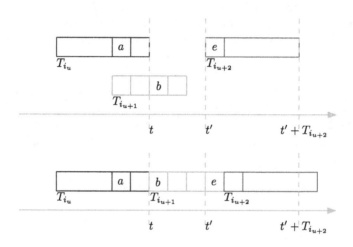

Fig. 4. An example of handling the overlapping calibrations.

The running time of the dynamic program is pseudo-polynomial in the number of choices of cost. Because of Lemma 2, we need to redefine the range of $\mathcal{F}'$ to $\mathcal{F}''$, where $\mathcal{F}'' := \{2f_{\min} \cdot (1+\varepsilon)^q | q = 0, \ldots, \lceil \log_{1+\varepsilon} n \rceil\}$.

First, we discretize the choices of the cost to $\mathcal{F}''$ instead of $\mathcal{F}'$. Then, we force the starting times of calibrations to be in a set Φ'. Finally, we use two calibrations every time to ensure consistency with the constructed S in Lemma 2.

Now we will show that when we restrict the cost to $\mathcal{F}''$ and the starting times of calibrations to Φ', there exists a feasible solution whose cost is no more than $(2+\varepsilon)OPT$.

Modified Dynamic Programming (MDP). We modify the dynamic programming proposed in Proposition 3 as follows:

- we restrict the choices of the cost to $\mathcal{F}''$;
- we restrict the choices of the starting times of calibrations to Φ';

– when a new job comes, there are two possibilities:
- there is not any new calibration to increase;
- increase two new consecutive calibrations whose types are identical.

The objective is to find the smallest f^* such that $c(n, f, t, k) \leq d_n$ for $f \in \mathcal{F}'', t \in \Phi', k \in \{1, \ldots, K\}$.

Theorem 2. *For the problem with arbitrary calibration cost, MDP computes a feasible schedule whose cost is no more than $(2 + \varepsilon)$ times of the optimal value, for $\varepsilon > 0$ arbitrarily small. The running time of MDP is polynomial in n and in $1/\varepsilon$.*

Proof. In order to prove the theorem, we combine Lemma 1 with Lemma 2. Note that the schedule returned by MDP allows the calibrations to overlap. So the cost f^* is a lower bound of the cost of such a schedule. Moreover, we have $f^* \leq 2OPT \cdot (1 + \varepsilon)$ by Lemma 1. So we have $f^* \leq 2(1 + \varepsilon)OPT = (2 + \varepsilon')OPT$, where $\varepsilon' = 2\varepsilon$.

Finally, we perform the same operations according to Observation 1 to make all calibrations non-overlapping. Thus, we get a feasible schedule of cost no more than f^*, which completes the proof of the approximation ratio $(2 + \varepsilon)$ for $\varepsilon > 0$ arbitrarily small.

Running Time. Since $|\mathcal{F}''| = O(\log_{1+\varepsilon} n)$, $|\Phi'| = O(n^2)$, and $|\{1, \ldots, K\}| = K$, the size of the table of MDP is $n \cdot |\mathcal{F}''| \cdot |\Phi'| \cdot |\{1, \ldots, K\}| = O(n^3 K \log_{1+\varepsilon} n)$. When the values of the table are fixed, the minimization is over the values f', t' and k', so the running time is $O(n^2 K \log_{1+\varepsilon} n)$. Hence, the overall time complexity is $O(n^5 K^2 \log^2_{1+\varepsilon} n)$. □

3   Uniform Calibration Cost

In this section, we consider the case in which the calibration cost is equal to its length. We show that there exists a 2-approximation algorithm. In particular, we use the Preemptive Lazy Binning (PLB) algorithm [1] with the shortest calibration.

Theorem 3. *For the problem $1|r_j, d_j, pmtn, \{T_1, \cdots, T_K\}|cost(calibrations)$ with $f_k = T_k$ for all $k \in \{1, \ldots, K\}$, PLB algorithm with the shortest calibration is a 2-approximate.*

Proof. Suppose we have the optimal sequence of calibrations type $O = \{T_{i_1}, T_{i_2}, \ldots\}$. Then we construct a feasible solution in which we only use the type 1.

We replace all the calibrations in O with type 1 such that they cover the initial calibrations $\{T_{i_1}, T_{i_2}, \ldots\}$, i.e., if a calibration has length T_k, then we replace it with $\lceil T_k/T_1 \rceil$ calibrations of type 1. See S' in Fig. 5. Jobs stay scheduled at the same time. Similarly, if some calibrations overlap, then we perform the operations as in Observation 1. A non-overlapping schedule S shown in Fig. 5 is obtained.

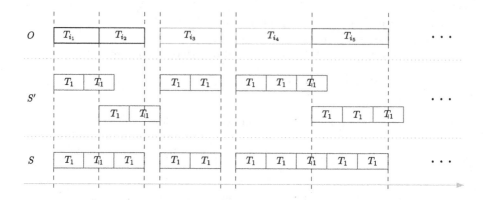

Fig. 5. Illustration of analysis.

Note that for any T_{i_u} in O, the new constructed schedule S' above will cover it by at most $T_{i_u} + T_1 < 2T_k$ since $T_1 < T_k$. So the cost of the non-overlapping schedule S is no more than the cost of S' whose cost is less than twice of the optimal value.

We recall that the PLB algorithm proposed in [1] returns a schedule with the minimum number of calibrations in polynomial time when there is only one type of calibration. If we are only allowed to use the shortest calibration of length T_1, S is a feasible schedule, while PLB algorithm returns a schedule with at most as many calibrations as in S.

Hence, by using the PLB algorithm with the shortest calibration, the cost of the returned schedule is no more than the cost of S and less than twice of the optimal value.   □

4   Conclusion

In this paper, we studied the scheduling problem with multiple types of calibration. When jobs have agreeable deadlines, we showed that the problem could be solved in pseudo-polynomial time, and we gave a $(2 + \varepsilon)$-approximation algorithm. We further studied the case in which the calibration cost is equal to its length, and we gave a simple 2-approximation algorithm. A natural question is to improve the approximation ratio. It would be mostly interesting if one can show a PTAS or exclude its existence.

Moreover, no approximation algorithm is known for the general case of jobs, a constant approximation algorithm would be of great interest.

References

1. Angel, E., Bampis, E., Chau, V., Zissimopoulos, V.: On the complexity of minimizing the total calibration cost. In: Xiao, M., Rosamond, F. (eds.) FAW 2017. LNCS, vol. 10336, pp. 1–12. Springer, Cham (2017). https://doi.org/10.1007/978-3-319-59605-1_1

2. Baptiste, P.: Scheduling unit tasks to minimize the number of idle periods: a polynomial time algorithm for offline dynamic power management. In: Proceedings of the 17th Annual ACM-SIAM Symposium on Discrete Algorithm (SODA), pp. 364–367. Society for Industrial and Applied Mathematics (2006)

3. Bender, M.A., Bunde, D.P., Leung, V.J., McCauley, S., Phillips, C.A.: Efficient scheduling to minimize calibrations. In: Proceedings of the 25th ACM Symposium on Parallelism in Algorithms and Architectures (SPAA), pp. 280–287 (2013)

4. Burroughs, C.: New integrated stockpile evaluation program to better ensure weapons stockpile safety, security, reliability (2006). http://www.sandia.gov/LabNews/060331.html

5. Chau, V., Li, M., McCauley, S., Wang, K.: Minimizing total weighted flow time with calibrations. In: Proceedings of the 29th ACM Symposium on Parallelism in Algorithms and Architectures (SPAA), pp. 67–76. ACM (2017)

6. Chau, V., Feng, S., Li, M., Wang, Y., Zhang, G., Zhang, Y.: Weighted throughput maximization with calibrations. In: Friggstad, Z., Sack, J.-R., Salavatipour, M.R. (eds.) WADS 2019. LNCS, vol. 11646, pp. 311–324. Springer, Cham (2019). https://doi.org/10.1007/978-3-030-24766-9_23

7. Chau, V., Li, M., Wang, E.Y., Zhang, R., Zhao, Y.: Minimizing the cost of batch calibrations. Theor. Comput. Sci. **828–829**, 55–64 (2020)

8. Chen, L., Li, M., Lin, G., Wang, K.: Brief announcement: approximation of scheduling with calibrations on multiple machines. In: Proceedings of the 31st ACM Symposium on Parallelism in Algorithms and Architectures (SPAA), pp. 237–239. ACM (2019)

9. Fineman, J.T., Sheridan, B.: Scheduling non-unit jobs to minimize calibrations. In: Proceedings of the 27th ACM Symposium on Parallelism in Algorithms and Architectures (SPAA), pp. 161–170 (2015)

10. Graham, R.L., Lawler, E.L., Lenstra, J.K., Kan, A.R.: Optimization and approximation in deterministic sequencing and scheduling: a survey. In: Annals of Discrete Mathematics, vol. 5, pp. 287–326. Elsevier (1979)

11. Ibarra, O.H., Kim, C.E.: Fast approximation algorithms for the Knapsack and sum of subset problems. J. ACM **22**(4), 463–468 (1975)

12. Mäcker, A., Malatyali, M., Meyer auf der Heide, F., Riechers, S.: Cost-efficient scheduling on machines from the cloud. J. Comb. Optimiz. **36**(4), 1168–1194 (2017). https://doi.org/10.1007/s10878-017-0198-x

13. Phillips, C.A., Stein, C., Torng, E., Wein, J.: Optimal time-critical scheduling via resource augmentation. In: Proceedings of the 29th Annual ACM Symposium on Theory of Computing (STOC), pp. 140–149 (1997)

14. Wang, K.: Calibration scheduling with time slot cost. Theor. Comput. Sci. **821**, 1–14 (2020)

Efficient Mobile Charger Scheduling in Large-Scale Sensor Networks

Xingjian Ding[1] , Wenping Chen[1], Yongcai Wang[1], Deying Li[1(✉)],
and Yi Hong[2]

[1] School of Information, Renmin University of China, Beijing 100872, China
{dxj,chenwenping,ycw,deyingli}@ruc.edu.cn
[2] School of Information, Beijing Forestry University, Beijing 100083, China
hongyi@bjfu.edu.cn

Abstract. As a promising way to replenish energy to sensor nodes, scheduling the mobile charger to travel through the network area to charge sensor nodes has attracted great attention recently. Most existing works study the mobile charger scheduling problem under the scenario that only the depot can recharge or replace the battery for the mobile charger. However, for large-scale sensor networks, this may be energy inefficient, as the mobile charger will travel for a long distance to charge each sensor node. In this paper, we consider the scenario that there are some service stations in the network area which can be used to replace the battery for the mobile charger, and we study the problem of Minimizing the number of used Batteries for a mobile chArger to charge a wireless sensor network (MBA). We first prove that the MBA problem is NP-hard, and then design an approximation algorithm to address it. We also give the theoretical analysis for the algorithm. We conduct extensive simulations to evaluate the performance of our algorithm, the simulation results show that our proposed algorithm is effective and promising.

Keywords: Mobile charger · Wireless sensor network · Wireless power transfer

1 Introduction

The recent breakthrough in wireless power transfer technology brings a novel method to replenish energy to sensor nodes [4]. As a promising way to prolong the lifetime of WSNs, wireless charging guarantees the continuous power supply for sensor nodes and is insensitive to surroundings. With the novel technology, some researchers study the problem of replenishing energy to sensor nodes in WSNs with a *mobile charger* (MC) [8,12] so that sensors can achieve continuous operation. Generally, there is a depot for maintaining the MC, and the MC will be periodically dispatched to traverse each sensor node and stay near each sensor

This work is supported by National Natural Science Foundation of China (Grant NO. 11671400, 61972404, 61672524).

Z. Zhang et al. (Eds.): AAIM 2020, LNCS 12290, pp. 298–310, 2020.
https://doi.org/10.1007/978-3-030-57602-8_27

node for a short time to charge it. For large-scale wireless sensor networks, using a single mobile charger may be very inefficient. The reason is that the energy capacity of the MC is limited in practice, the MC needs to return to the depot to charge itself after it charges a part of the sensor nodes, and thus it may take multiple rounds to charge all the sensor nodes. To address this problem, some researchers investigate the problem of charging a wireless sensor network with multiple mobile chargers [3,9], in which they schedule multiple wireless chargers from the depot to charge sensor nodes.

However, for extremely large-scale wireless sensor networks, it is energy inefficient if we just use the depot to maintain the MCs. It is because all the mobile chargers are dispatched from the depot, the mobile chargers will consume a lot of energy on traveling to reach these sensor nodes that are very far from the depot (or may not even be able to reach these sensors due to the limit energy capacity of each MC). In this paper, we consider the scenario that there are some service stations in the network area, which can be used to replace the battery of the MC with a fully charged battery under the help of some mechanical equipment. Under such a scenario, the depot dispatch a MC to traverse and charge each sensor node one by one, when the MC is about to run out of it energy, it will move to a nearby service station to replace a new battery and then go to traverse and charge the remaining sensor nodes. These service stations keep the replaced batteries and fully recharge them to maintain them for the next charging period. The number of used batteries of the MC during a charging cycle will significantly affect the total cost of the charging tour. How to design the charging tour for the MC to fully charge all the sensor nodes with the minimum number of used batteries, therefore, is a realistic and crucial problem in this scenario.

In this paper, we study the charging tour design problem for achieving the minimum number of used batteries during a charging cycle. The main contributions of our work are as follows.

- We consider the scenario that a network area has a set of service stations and define the MBA problem. We also prove the NP-hardness of the MBA problem.
- We design an approximation algorithm for the MBA problem, and give theoretical analysis for the proposed algorithm.
- We conduct extensive simulations to evaluate our proposed algorithm. The simulation results demonstrate the effectiveness of our algorithm.

The remainder of this paper is structured as follows. In Sect. 2, we introduce the related works of this paper. In Sect. 3, we formally define the MBA problem to be addressed, and prove the NP-hardness of problem MBA. In Sect. 4, we address the MBA problem and design an approximation algorithm for it. In Sect. 5, we conduct extensive simulations to evaluate our algorithm. And finally, we conclude this paper in Sect. 6.

2   Related Works

With the advance of the efficient wireless power transfer technology, using mobile chargers to replenish energy to wireless sensors has been widely studied in various contexts. Ma et al. [10] consider the scenario that multiple sensors can be charged simultaneously by a single charger, they aim to design a charging tour for the mobile charger to maximize the charging utility, while the consumed energy of the mobile charger is limited by its energy capacity. Wu et al. [11] study how to improve the charging utility by jointly considering the oriented sensor placement and the mobile charger scheduling problem. Some researchers investigate multiple mobile chargers scheduling problem. Lin et al. [6] study the temporal and spatial collaborative charging problem with multiple chargers, in which they focus on maximizing energy usage efficiency and survival rate simultaneously. Xu et al. [13] consider the problem of scheduling multiple mobile chargers to collaboratively charge sensors, their objective is to minimize the sum of traveling distance of these mobile chargers. For improving the overall energy usage efficiency of mobile chargers, Lin et al. [7] use the game theory model and propose a game theoretical collaborative charging scheduling method for on-demand charging architecture.

3   Model and Problem Formulation

3.1   Models and Assumptions

We consider a wireless rechargeable sensor network that contains m rechargeable sensor nodes $\mathcal{S} = \{s_1, s_2, \ldots, s_m\}$ and a stationary *base station* which is used to collect data and manage the entire network. All the sensor nodes are deployed in a 2-D bounded region, and their positions are fixed and can be known in advance. We assume that all sensor nodes are equipped with identical rechargeable batteries with energy capacity B. Each sensor node $s_i \in \mathcal{S}$ will periodically report its residual energy RE_i to the base station, thus the base station knows the battery status of the whole network.

To maintain a long-time operation of the network, the base station will dispatch a MC to charge the sensor nodes at certain time points. We assume there is a *depot* denoted by r next to the base station, and there is only one MC in the depot. The energy capacity of the MC is U which is used for both moving and charging sensors. The energy consumption rate of the MC for moving per unit distance is η. Generally, the energy capacity of the MC is limited and it can not fully charge all sensor nodes by one battery, thus the MC should be charged or replaced its battery before its energy is exhausted. We assume that there are a set of service stations $\mathcal{V} = \{v_1, v_2, \ldots, v_n\}$ in the network area, at which the MC can replace its battery with a fully charged battery under the help of some mechanical equipment. We also assume that the depot also can be used as a service station. Figure 1 shows an example of such a rechargeable sensor network.

In this paper, we consider the scenario that the MC moves to the site of the sensor nodes one by one and then begins to charge them wirelessly. The MC will move to the service station to replace its battery if its battery is going to run out of energy, and the MC must return to the depot after fully charging all the sensor nodes. We term such a charging process as a *charging tour*. Note that a sensor node can be charged multiple times during the charging tour. We ignore the sensor node energy consumption during a charging tour, as sensor nodes are low-power devices, the amount of energy consumed by each sensor node during a charging tour is several orders of magnitude less than its energy capacity.

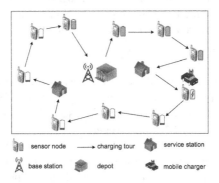

Fig. 1. Illustration of the rechargeable sensor network.

3.2   Problem Formulation

In this paper, we focus on scheduling a mobile charger equipped with an energy-limited battery to charge a wireless sensor network, such that all sensor nodes are fully charged after a charging tour. As mentioned before, the MC can not fully charge all sensor nodes by one battery, thus it needs to move to the service station to replace its battery during the charging tour. We aim to design a charging tour for the MC such that the total number of used batteries of the MC is minimized. We define the problem to be addressed in this paper as follows.

Problem 1. Minimizing the number of used Batteries for a mobile charger to chArge a wireless rechargeable sensor network (MBA). Given a wireless rechargeable sensor network that involves a set S of rechargeable sensor nodes with energy capacity B and residual energy $\{RE_1, RE_2, \ldots, RE_m\}$, a MC with battery capacity U located at depot r, and a set V of service stations, we aim to find a charging tour for the MC and determine when and where to replace its battery to fully charge all the sensor nodes, such that the number of used batteries is minimized.

Notice that the battery carried by the MC when it leaves the depot is also counted.

In the following, we will prove the NP-hardness of the MBA problem. We first introduce some problems before proving the NP-hardness of the MBA problem.

The Traveling Salesman Problem (TSP): Given a set $\mathcal{S}'$ of vertices, and the distance of each two vertices is known, the TSP problem is to find a closed tour C' that visits all the vertices in $\mathcal{S}'$, and the length of the tour C is minimized.

The Decision Version of the TSP Problem: Given a positive number l, does there exist a closed tour C' where the length of C' is at most l that visits all the vertices in $\mathcal{S}'$?

The Decision Version of the MBA Problem: Given an integer k, does there exist a charging tour C for the MC that fully charge all the sensor nodes, and the number of the total used batteries is not larger than k?

Theorem 1. *The MBA problem is NP-Hard.*

Proof. We prove the theorem by reducing the TSP problem to the MBA problem.

Consider such an instance of decision version of the TSP problem: we are given a positive number l and a vertex set $\mathcal{S}' = \{s'_1, s'_2, \ldots, s'_n\}$, and the distance between any two vertices in $\mathcal{S}'$ is an integer.

we construct an instance of the decision version of the MBA problem as follows. For each element $s'_i \in \mathcal{S}'$ we generate a sensor node s_i, the distance between each two sensor nodes s_i and s_j is equal to the distance between s'_i and s'_j. We set the energy capacity of each sensor node to an integer B and set the residual energy of sensor s_i to an integer RE_i, where $0 \le RE_i \le B$, then the total energy required to fully charge all the sensor nodes can be expressed by $E_c = \sum_{i=1}^{m}(B - RE_i)$, where E_c is also an integer. We set the energy capacity U of the MC to 1, and the energy consumption rate η of the MC for moving per unit distance is 1. Besides, we let $k = \frac{\lfloor l \rfloor + E_c}{U} = \lfloor l \rfloor + E_c$. We assume that there are enough service stations that the MC can replace its battery anywhere once its energy is exhausted.

If the instance of the decision version of the TSP problem has a "Yes" answer, i.e., there exists a closed tour C' that visits all the vertices in $\mathcal{S}'$, and the length of C' is no more than l. As the distance between any two vertices in $\mathcal{S}'$ is an integer, the length of C' must be an integer, and it is no more than $\lfloor l \rfloor$. We use C' as the charging tour C for the MC to charge all the sensor nodes, the MC replace its battery once its energy is exhausted on the tour C. The total energy consumed by the MC during the charging tour is no more than $\lfloor l \rfloor + E_c$, and thus the total number of used batteries by the MC is no more than $\frac{\lfloor l \rfloor + E_c}{U} = k$, that is, the instance of the decision version of the MBA problem has a "Yes" answer.

If the instance of the decision version of the MBA problem has a "Yes" answer, i.e., there is a charging tour C for the MC to fully charge all the sensor nodes, and the number of used batteries by the MC is at most k. All the used batteries provide at most $k * U$ energy for the MC, and the energy used for charging sensors is E_c, and thus the length of the charging tour C is at most $(k * U - E_c)/\eta = k * U - E_c = \lfloor l \rfloor$. We use C to be the tour C' for the decision version of the TSP problem, combining $\lfloor l \rfloor \leq l$, we know that the instance of the decision version of the TSP problem also has a "Yes" answer.

It is clear that the reduction can be finished in polynomial time. Since the TSP problem is a typical NP-Hard problem, we can conclude that the MBA problem is also NP-Hard.

4   Algorithm for the MBA Problem

In this section we address the MBA problem, we assume that the distance between any two service stations is no more than U/η. The general case that without any assumption will be our future work.

4.1   Algorithm Description

In the MBA problem, there is a set of service stations in the network area, we use $v(s_i)$ to denote the service station which is the nearest to the sensor node s_i, and use $L(s_i)$ to denote the distance between s_i and $v(s_i)$, we use L_{max} to denote the maximum $L(s_i)$, i.e., $L_{max} = \max_{s_i \in \mathcal{S}} L(s_i)$. We assumed that $L_{max} < \frac{U}{2*\eta}$ in this work, otherwise, there is no feasible solution.

We design an approximation algorithm, named FCMB, for the MBA problem. Our proposed algorithm mainly has three steps. The first step is to find a closed tour C on the set $\mathcal{S} \cup \{r\}$, we let the MC visits and charge each sensor node along the anticlockwise direction of the tour, then the tour C can be considered as a charging path. The second step is to transform the charging path C to a path C' by replacing each sensor node with an edge. The third step is to split the path C' into several sub-paths such that the MC will not exhaust its energy during each sub-path. In the following, we will describe the algorithm in details.

In the first step, we find a closed tour C by invoking the TSP algorithm [1] on the set $\mathcal{S} \cup \{r\}$. For the convenience of expression, we renumber the sensor nodes in the anticlockwise direction of the closed tour C beginning at r. Then the closed tour C can be represent as a path, i.e., $C = \{r, s_1, s_2, \ldots, s_m, r\}$, as shown in Fig. 2(a). We let the MC charge the sensor nodes one by one along the path C.

In the second step, we transform the path C to C' by using an edge (s_i', s_i'') to replace each sensor node s_i, and set the length of the edge (s_i', s_i'') to $E(s_i)/\eta$, where $E(s_i) = B - RE_i$, as shown in Fig. 2(b). Notice that (s_i', s_i'') is a virtual edge that represents the sensor node s_i, thus for any point p on edge (s_i', s_i''), its closest service station $v(p)$ is same to $v(s_i)$, correspondingly, $L(p) = L(s_i)$. For the ease of expression, we term these virtual edges as "charging edges", and other edges in C' as "traveling edges".

In the third step, we split the path C' into several sub-paths such that the MC can move to the closest service station before exhausting its energy during each sub-path. The split method is as follows. We let the MC move along the path C', then we try to find the farthest point p_1 from the starting point of C' on the "charging edges", where the MC has enough energy move to the closest service station $v(p_1)$, i.e., $L(r, p_1) + L(p_1) \leq \frac{U}{\eta}$, where $L(r, p_1)$ represents the distance from r to p_1 along the path C'. We term such a point as "leaving point" as the MC leaves the tour C' at the point. There are two cases for the "leaving point" p_1.

Case I: if we can find such a point p_1 on the "charging edge" (s'_i, s''_i), then we can split the first sub-path as $(r, \ldots, s'_i, p_1, v(p_1))$, and update the path C' to $(v(p_1), p_1, s''_i, \ldots, r)$.

Case II: we can not find such a point p_1 on the "charging edges", this case happens when some "traveling edges" are too long that it costs too much energy for the MC to travel it. Then we let the MC move to the service station $v(s_1)$ (the closest service station of the first sensor of C') directly, and we set $p_1 = s'_1$. Then the first sub-path is $(r, v(s_1))$, and we update the path C' as $(v(s_1), p_1, s''_1, \ldots, r)$.

It's worth mentioning that the case II happens only when the last "leaving point" is on the end of a "charging edge" or when we try to find the first "leaving point", otherwise, we can always find a "leaving point" on "charging edges" as we assumed $L_{max} < \frac{U}{2*\eta}$ in this work.

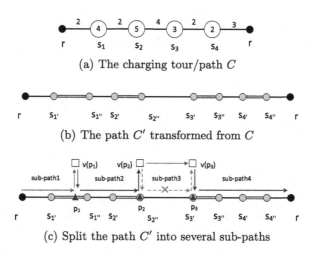

(a) The charging tour/path C

(b) The path C' transformed from C

(c) Split the path C' into several sub-paths

Fig. 2. Illustration of the FCMB algorithm.

We iteratively split the path C' in this way until the length of C' is no more than U/η. An example is shown in Fig. 2(c), we first find a "leaving point" p_1, and get the "sub-path1", then we update the C' as $(v(p_1), p_1, s''_1, \ldots, r)$; then we find the second "leaving point" p_2 and get the "sub-path2", the path C' is updated to $(v(p_2), p_2, s'_3, \ldots, r)$; however, we can't find the third "leaving point" on C' as the edge (s''_2, s'_3) is too large, i.e., $L(p_2) + L(s''_2, s'_3) + L(s'_3) > \frac{U}{\eta}$, then we set the third "leaving point" p_3 as s'_3 and let the MC move to $v(p_3)$ directly, we get the "sub-path3" as $(v(p_2), v(p_3))$, and the path C' is updated to $(v(p_3), p_3, s''_3, \ldots, r)$; as the length of the path C' is less than U/η now, so the split process is terminated, and we get "sub-path4".

After these three steps, we get a set of sub-paths of C', We can easily calculate the charging energy for sensor nodes in each sub-path, for example, in Fig. 2(c), the charging energy for sensor s_1 in "sub-path1" is $\eta * L(s'_1, p_1)$. We can also easily get a closed tour T beginning and ending at r for the MC by combining these sub-paths together, as shown in the Fig. 2(c), the closed tour T of the MC is $(r \to s_1 \to v(s_1) \to s_1 \to s_2 \to v(s_2) \to v(s_3) \to s_3 \to s_4 \to r)$.

Algorithm FCMB is described in Algorithm 1.

Algorithm 1. Find a closed Charging tour that used Minimum Batteries for the mobile charger(FCMB)

Input: $\mathcal{S}, \mathcal{V}, B, U, \eta$, service station r, and the residual energy RE_i for any $s_i \in \mathcal{S}$.

Output: A closed charging tour T including r and some service stations for the MC.

1: for each sensor node $s_i \in \mathcal{S}$, find its closest service station from $\mathcal{V} \cup \{r\}$ and denoted it by $v(s_i)$.

2: find a closed tour C for $\mathcal{S} \cup \{r\}$ by invoking the TSP algorithm proposed by Christofides [1];

3: renumber the sensor nodes in the anticlockwise direction of the closed tour C beginning at r, i.e., $C = \{r, s_1, s_2, \dots, s_m, r\}$;

4: compute $E(s_i)$ for each $s_i \in \mathcal{S}$, $E(s_i) = B - RE_i$;

5: transform C to the path C' by using an edge (s'_i, s''_i) to replace each sensor node s_i, and the weight of the edge equals $E(s_i)/\eta$;

6: split the path C' into several sub-paths as Fig. 2 shows;

7: calculate the charging energy for each sensor node in each sub-path;

8: combining these sub-paths together to get a closed tour T for the MC beginning and ending at r .

9: **return** the closed tour T;

4.2   Performance Analysis

Here we analyze the approximation ratio of the FCMB algorithm. In the MBA problem, we have assumed that $L_{max} < \frac{U}{2*\eta}$, otherwise there's no feasible solution, we assume that $L_{max} = \alpha * \frac{U}{2*\eta}$, where $0 \le \alpha < 1$. We have the following theorem.

Theorem 2. *The* FCMB *algorithm achieves a* $\lceil \frac{3}{1-\alpha} \rceil$-*approximation ratio for the MBA problem.*

Proof. We first analyze the lower bound of the optimal solution. We use L_{opt} to denote the distance of the optimal solution for the TSP problem on set $\mathcal{S} \cup \{r\}$, and use E_{opt} to denote the total consumed energy of the MC for both traveling and charging, clearly, we have

$$E_{opt} \ge \eta * L_{opt} + \sum_{i=0}^{m+1} E(s_i). \tag{1}$$

Assume the optimal solution uses N_{opt} batteries in total, where N_{opt} is a positive integer. Note that each battery provides at most U energy for the MC, and thus we have the following inequality.

$$N_{opt} \geq \left\lceil \frac{E_{opt}}{U} \right\rceil \geq \left\lceil \frac{\eta * L_{opt} + \sum_{i=1}^{m} E(s_i)}{U} \right\rceil . \tag{2}$$

In the following, we analyze the property of the solution obtained by FCMB. We use L_C to denote the length of the closed tour C we found in the first step of FCMB, as we get C by invoking the $\frac{3}{2}$-approximation algorithm, thus we have $L_C \leq \frac{3}{2} * L_{opt}$. The length of the path C' we get in the second step of FCMB is $L_{C'} = L_C + \sum_{i=1}^{m}(E(s_i)/\eta)$.

Without loss of generality, we assume that we totally find t "leaving points" in the third step of algorithm FCMB, where t is a positive integer. For the ease of expression, we label the beginning and ending point of the path C' as p_0 and p_{t+1} respectively, i.e., $r = p_0 = p_{t+1}$, and set that $L(p_0) = L(p_{t+1}) = 0$. Notice that $L(p_i) + L(p_i, p_{i+2}) + L(p_{i+2}) > U/\eta$ for any $0 \leq i \leq t - 1$, otherwise the "leaving point" p_{i+1} can be dropped. Thus we have the following inequality,

$$L(p_i, p_{i+2}) > \frac{U}{\eta} - L(p_i) - L(p_{i+2}) \geq \frac{U}{\eta} - 2 * L_{max}, \tag{3}$$

where $L(p_i, p_{i+2})$ is the distance between the two "leaving points" p_i and p_{i+2} along the path C'. Accumulating all the inequalities for any $0 \leq i \leq t - 1$, we have,

$$\sum_{i=0}^{t-1} L(p_i, p_{i+2}) > t * \left(\frac{U}{\eta} - 2 * L_{max} \right). \tag{4}$$

Notice that all the "leaving points" are selected on the path C', thus we have $\sum_{i=0}^{t-1} L(p_i, p_{i+2}) \leq 2 * L_{C'}$, combining inequality (4), we have,

$$
\begin{aligned}
t &< \frac{2 * L_{C'}}{\frac{U}{\eta} - 2 * L_{max}} = \frac{2 * (L_C + \sum_{i=1}^{m}(E(s_i)/\eta))}{\frac{U}{\eta} - 2 * L_{max}} \\
&\leq \frac{2 * \left(\frac{3}{2}\eta * L_{opt} + \sum_{i=1}^{m} E(s_i)\right)}{U - 2\eta * L_{max}} < \frac{3 * (\eta * L_{opt} + \sum_{i=1}^{m} E(s_i))}{(1 - \alpha)U}.
\end{aligned}
\tag{5}
$$

As t is an positive integer, inequality (5) can be rewritten as

$$t \leq \left\lceil \frac{3 * (\eta * L_{opt} + \sum_{i=1}^{m} E(s_i))}{(1 - \alpha)U} \right\rceil - 1. \tag{6}$$

We use N_A to denote the number of used batteries by the MC in the solution got by FCMB, as the battery carried by the MC when it leaves the depot is also counted, thus $N_A = t + 1$, then we can rewritten inequality (6) as follows.

$$N_A \leq \left\lceil \frac{3 * (\eta * L_{opt} + \sum_{i=1}^{m} E(s_i))}{(1 - \alpha)U} \right\rceil$$

$$\leq \left\lceil \frac{3}{1 - \alpha} \right\rceil * \left\lceil \frac{\eta * L_{opt} + \sum_{i=1}^{m} E(s_i)}{U} \right\rceil \tag{7}$$

$$\leq \left\lceil \frac{3}{1 - \alpha} \right\rceil * N_{opt}.$$

Thus the theorem holds.

5   Simulation Results

In this section, we conduct extensive simulations to evaluate the performance of our proposed algorithm. We compare our proposed algorithm with a lower bound of the optimal solution, which is used as a benchmark. All the data points plotted in this section are the average of 100 runs.

5.1   Experimental Settings

In our experiments, the basic parameters for the model are set as follows. The battery capacity B of theses sensor nodes is set to be 10.8 kJ [5], the residual energy of each sensor node is generated randomly within $(0, 10.8]$ kJ. The energy capacity of the mobile charger U is set to be 400 kJ. According to [2], an electric bike with 750 W power has a maximum speed of 32 km/h, we can calculate that the energy consumption rate of the electric bike is 93.75 J/m. In our simulations, we set the energy consumption rate η of the MC to be 100 J/m. We assume that there are 200 rechargeable sensor nodes randomly distributed over a 2000 × 2000 m² square area, and the depot r located at the left bottom of the network area.

Note that, the energy capacity U of the MC is 400 kJ, and the energy consumption rate η of the MC is 100 J/m. Then in a 2000 × 2000 m² square area, the distance of any two service stations is less than U/η, which means that our experimental settings are suitable for the MBA problem.

5.2   Performance Comparison

To evaluate the performance of our proposed algorithm, we use a lower bound of the optimal solution as the benchmark. According to the analysis in Theorem 2, we know that $\left\lceil \frac{\eta * L_{mst} + \sum_{i=1}^{m} E(s_i)}{U} \right\rceil$ is a lower bound of the optimal solution, where L_{mst} is the distance of the minimum spanning tree of set $\mathcal{S} \cup \{r\}$, we denote it as LB_OPT. Next, we compare our proposed algorithm with the benchmark under various settings.

1) Impact of the number of sensors (m): As shown in Fig. 3(a), with the increase of the number of sensors, both the FCMB and the LB_OPT will use more batteries, and the gap between our algorithm and the lower bound LB_OPT is stable,

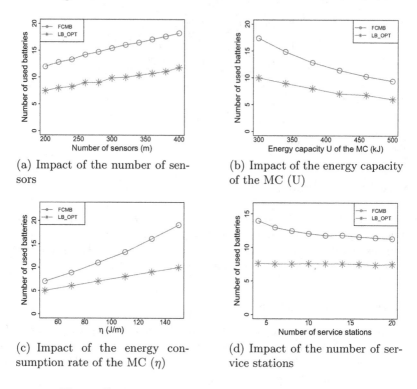

(a) Impact of the number of sensors

(b) Impact of the energy capacity of the MC (U)

(c) Impact of the energy consumption rate of the MC (η)

(d) Impact of the number of service stations

Fig. 3. Comparison between FCMB and the benchmark.

the number of used batteries obtained by our algorithm is about 60% more than that of the lower bound of the optimal solution.

2) Impact of the energy capacity of the MC (U): Figure 3(b) depicts the impact of the energy capacity U of the MC. We can see that the larger the U is, the less the number of batteries will be used. It also can be seen that with the increase of the energy capacity of the MC, the gap between our algorithm and the lower bound of the optimal solution becomes smaller and smaller, which implies that our algorithm achieves better performance when the energy capacity of the MC is large.

3) Impact of the energy consumption rate of the MC (η): As depicted in Fig. 3(c), when the energy consumption rate η increases from 50 J/m to 150 J/m, both of our algorithm and the *LB_OPT* will use more batteries, and the gap between our algorithm and the low bound grows up. This is because the MC will exhaust more energy for traveling. This observation also responds to the analysis of our algorithm FCMB in Theorem 2. The performance ratio of our algorithm is a function of α, where $\alpha = \frac{2\eta L_{max}}{U}$, a larger η will cause a larger α, and then it will decrease the performance of our algorithm.

4) Impact of the number of service stations: Figure 3(d) indicates that with the increase of the number of service stations, the number of used batteries delivered by our algorithm decreases, and the gap between our algorithm and *LB_OPT* also become smaller. The reason is that with the increase of the number of service stations, the MC is more easier to find a closer service station to replace it battery, and thus it will save more energy for charging sensors.

6 Conclusions

In this paper, we study the problem of minimizing the number of used batteries for a mobile charger to charge a wireless rechargeable sensor network (MBA). We prove the NP-hardness of the problem, and propose an approximation algorithm named FCMB for the MBA problem. We also give the theoretical analysis of our proposed algorithm. To evaluate the performance of our proposed algorithm, we conduct extensive simulations. Simulation results demonstrate that our proposed algorithm is effective and promising.

References

1. Christofides, N.: Worst-case analysis of a new heuristic for the travelling salesman problem. Technical report, Carnegie-Mellon University Pittsburgh Pa Management Sciences Research Group (1976)
2. Hicks, E.: Is My E-Bike Legal? USA Ebike Law, 23 April 2013. https://www.electricbike.com/electric-bike-law/. Accessed 02 Jan 2020
3. Jiang, G., Lam, S.K., Sun, Y., Tu, L., Wu, J.: Joint charging tour planning and depot positioning for wireless sensor networks using mobile chargers. IEEE/ACM Trans. Netw. **25**(4), 2250–2266 (2017)
4. Kurs, A., Karalis, A., Moffatt, R., Joannopoulos, J.D., Fisher, P., Soljačić, M.: Wireless power transfer via strongly coupled magnetic resonances. Science **317**(5834), 83–86 (2007)
5. Liang, W., et al.: Approximation algorithms for charging reward maximization in rechargeable sensor networks via a mobile charger. IEEE/ACM Trans. Netw. (TON) **25**(5), 3161–3174 (2017)
6. Lin, C., Wang, Z., Deng, J., Wang, L., Ren, J., Wu, G.: mTs: temporal-and spatial-collaborative charging for wireless rechargeable sensor networks with multiple vehicles. In: IEEE INFOCOM 2018-IEEE Conference on Computer Communications, pp. 99–107. IEEE (2018)
7. Lin, C., et al.: GTCCS: a game theoretical collaborative charging scheduling for on-demand charging architecture. IEEE Trans. Veh. Technol. **67**(12), 12124–12136 (2018)
8. Lin, C., Zhou, J., Guo, C., Song, H., Wu, G., Obaidat, M.S.: TSCA: a temporal-spatial real-time charging scheduling algorithm for on-demand architecture in wireless rechargeable sensor networks. IEEE Trans. Mob. Comput. **17**(1), 211–224 (2017)
9. Lin, C., Zhou, Y., Ma, F., Deng, J., Wang, L., Wu, G.: Minimizing charging delay for directional charging in wireless rechargeable sensor networks. In: IEEE INFOCOM 2019-IEEE Conference on Computer Communications, pp. 1819–1827. IEEE (2019)
10. Ma, Y., Liang, W., Xu, W.: Charging utility maximization in wireless rechargeable sensor networks by charging multiple sensors simultaneously. IEEE/ACM Trans. Netw. **26**(4), 1591–1604 (2018)

11. Wu, T., Yang, P., Dai, H., Xu, W., Xu, M.: Charging oriented sensor placement and flexible scheduling in rechargeable WSNs. In: IEEE INFOCOM 2019-IEEE Conference on Computer Communications, pp. 73–81. IEEE (2019)
12. Xu, W., Liang, W., Jia, X., Xu, Z., Li, Z., Liu, Y.: Maximizing sensor lifetime with the minimal service cost of a mobile charger in wireless sensor networks. IEEE Trans. Mob. Comput. 17(11), 2564–2577 (2018)
13. Xu, W., Liang, W., Lin, X., Mao, G.: Efficient scheduling of multiple mobile chargers for wireless sensor networks. IEEE Trans. Veh. Technol. 65(9), 7670–7683 (2016)

Revisit of the Scheduling Problems with Integrated Production and Delivery on Parallel Batching Machines

Long Wan$^{(\boxtimes)}$ and Jiajie Mei

School of Information Management, Jiangxi University of Finance and Economics,
Nanchang 310013, Jiangxi, People's Republic of China
cocu3328@163.com

Abstract. In the paper, we revisit the scheduling problems with integrated production and delivery on parallel batching machines. There are n jobs and m identical and parallel batching machines. The machines have identical capacities and the jobs have identical processing time. When a job is processed and delivered to customers in time, the company earns profit; otherwise, it earns nothing. A third party logistic (3PL) provider will be used to deliver the jobs. It provides certain vehicles with identical capacities at some certain time points. In the paper [Kai Li, Zhao-hong Jia, Joseph Y.-T. Leung (2015) Integrated production and delivery on parallel batching machines, European Journal of Operational Research, 247(3), 755-763.], the authors considered the scheduling problems and designed the algorithms to deal with them. But unfortunately, there are some wrong conclusions. Specifically, we construct counterexamples to show that both of Theorem 4 and Theorem 5 are invalid. Furthermore, we provide two faster algorithms than Algorithm Sch-Id-Size.

Keywords: Production and delivery · Total profit · Third party logistic · Scheduling · Batching machines

1 Introduction

Batching machine scheduling problems, as one of the important research areas in scheduling problems, have received more and more attention from researchers. Generally, one job has two variable parameters: size and processing time. For convenience, most researchers fix one of the parameters to be identical and make the other parameter arbitrary. Brucker et al. [2] and Potts et al. [13] have reviewed batching machine scheduling problems for a survey.

Ikura et al. [8] firstly studied these problems with jobs of identical processing time and sizes and presented an $O(n^2)$ algorithm to find a feasible schedule on

Supported by NSFC (11601198), Science and Technology Project of Jiangxi Provincial Department of Education (GJJ190250), NSFC (71761015) and NSFC (11701236).

a single machine. During the last 20 years, Wang et al. [17] developed a genetic algorithm with a random keys encoding scheme. Computational experiments showed that the algorithm has great average performance within reasonable computation time. Malve et al. [10] studied a similar problem as Wang et al. [17]. In their batch model, the jobs in the same family have the same processing time. Meanwhile, they presented a genetic algorithm and an improved heuristic. Computational experiments showed that the result is better than the iterative heuristic algorithm. Dupont et al. [6] proposed a branch-and-bound procedure for minimizing the makespan on a single-batch processing machine.

Recently, scheduling problems on identical and parallel machines have aroused increasing interest. Ma et al. [11] presented a $4(1+\varepsilon)$-competitive online algorithm to minimize total weighted completion time on uniform parallel machines. Experimental results showed that the algorithm for identical machines is efficient. Tian et al. [14] also considered online scheduling on m parallel-batch machines to minimize the makespan. They provided a new lower bound on the competitive ratio for dense-algorithms. Wan et al. [15] studied the problem of minimizing the maximum total completion time per machine on m parallel and identical machines. If m is a part of the input, they proved that the problem is strongly NP-hard; if m is fixed, they proposed a pseudo-polynomial time dynamic programming. Xu et al. [19] proposed a genetic algorithm to schedule a set of jobs on a set of machines with different capacities.

Production and delivery operations are two crucial steps in a supply chain. Very recently, scheduling problems with production and job delivery also received a lot of attention, but these researchers only dealt with job scheduling for classical models. Wang et al. [18] studied the issue of coordinating mail processing and distribution plans at mail processing centers. They showed that the problem was unary NP-hard, and designed some scheduling rules and heuristics. Wan et al. [16] considered coordinated scheduling on parallel identical machines with batch delivery to minimize the sum of job arrival time. They proved the problem is NP-hard and proposed the first approximation algorithm with the worst case ratio no more than 2.

Therefore, it is important to integrate the two operations for making profits. For many companies, there exists cost for holding inventories. The companies hope that the goods once produced can be distributed to the customers immediately. This situation is extensively applicable to the industries with the Make-To-Order (MTO) business model. In the MTO business model, products must be delivered to costumers with no delay. This calls for a closer link between production operation and delivery operation. Chen and Vairaktarakis [3] first formally introduced the integration of production and delivery to scheduling research. They combined customer service with delivery cost to produce a combinatorial objective which must be minimized. Since then, more and more researchers pay attention to the integration of production and delivery, see Chen [4] for a survey. However, these studies are only involved in classical scheduling models, where only a job can be processed by one machine.

Li et al. [9] first considered production-delivery scheduling on parallel batching machines. The model is stated as follows. There are m ($m \geq 1$) identical and parallel batching machines, we use $\mathcal{M} = \{M_1, M_2, \ldots, M_m\}$ to denote the machine set. Each machine has an identical capacity K ($K \geq 1$). There are n jobs which must be processed on the machines. We use $\mathcal{J} = \{J_1, J_2, \ldots, J_n\}$ to denote the job set. Each job J_j has a processing time p_j, a batch size $s_j^{(1)}$ and a delivery size $s_j^{(2)}$, a due date d_j and a profit R_j. Several jobs can be packed into a batch and processed by a single machine simultaneously with no violation of the rule that the total batch size of all the jobs in the batch doesn't exceed the capacity of the machine. If a job J_j can be processed and delivered to customers by its due date d_j, then a profit R_j will be made; otherwise, there will be no profit. As to delivery, a third party logistic (3PL) company will provide the delivery service to customers. The 3PL company will send certain vehicles at fixed time. At time T_k, the 3PL company provides v_k vehicles of capacity C ($C \geq 1$), $k = 1, 2, \ldots, z$. And each vehicle can deliver jobs with the total deliver size no more than the capacity of the vehicle. We assume that each job can't be split into pieces to process and deliver. The goal is to find a production-delivery schedule so that the profit is maximized. In other words, you must select a job subset $\mathcal{J}' \subseteq \mathcal{J}$ and decide how to batch, processing and delivering the jobs of $\mathcal{J}'$ to maximize the total profit $\sum_{J_j \in \mathcal{J}'} R_j$.

We introduce the five-field notation formulated by [4] and [5] to represent the scheduling problem under consideration: $\gamma_1 | \gamma_2 | \gamma_3 | \gamma_4 | \gamma_5$. γ_1 represents machine model and γ_2 represents production information and γ_3 represents vehicle configuration. γ_3 is usually in a form of (η_1, η_2), η_1 denotes the information of number and delivery time of vehicles and η_2 denotes the capacity of a vehicle. γ_4 represents the number of customers and γ_5 represents the objective function. Under the description above, the three problems we investigate can be formulated as below.

P1: $P|\text{p-batch}, p_j = p, s_j^{(1)}, s_j^{(2)} = 1, d_j, R_j = 1|(v_1, v_2, \ldots, v_z), (T_1, T_2, \ldots, T_z), C|1| \sum R_j$. Where $\{P\}$ means parallel machines, $\{\text{p-batch}\}$ means that jobs can be batched together, $\{p_j = p\}$ means that all the jobs have the identical processing time p, $\{s_j^{(1)}\}$ means that the jobs may have the different batch sizes, $\{s_j^{(2)} = 1\}$ means that the jobs have the identical deliver sizes equal to 1; $\{d_j\}$ means that the jobs may have the different due dates; $\{R_j = 1\}$ means that the jobs have the identical profits 1, $\{(v_1, v_2, \ldots, v_z), (T_1, T_2, \ldots, T_z)\}$ means that the 3PL company will provide $v_1, v_2, \ldots, v_z$ vehicles at time $T_1, T_2, \ldots, T_z$ respectively, $\{C\}$ means that each vehicle has a identical capacity C, $\{1\}$ of the fourth field means that there is only one customer and $\{\sum R_j\}$ of the fifth field means that the objective is to maximize the profit function.

P2: $P|\text{p-batch}, p_j = p, s_j^{(1)}, s_j^{(2)} = 1, d_j, R_j|(v_1, v_2, \ldots, v_z), (T_1, T_2, \ldots, T_z), C|1| \sum R_j$. Where $\{P\}$ means parallel machines, $\{\text{p-batch}\}$ means that jobs can be batched together, $\{p_j = p\}$ means that all the jobs have the identical processing time p, $\{s_j^{(1)}\}$ means that the jobs may have the different batch sizes, $\{s_j^{(2)} = 1\}$ means that the jobs have the identical deliver sizes equal to 1, $\{d_j\}$

means that the jobs may have the different due dates, $\{R_j\}$ means that the jobs may have the different profits, $\{(v_1, v_2, \ldots, v_z), (T_1, T_2, \ldots, T_z)\}$ means that the 3PL company will provide $v_1, v_2, \ldots, v_z$ vehicles at time $T_1, T_2, \ldots, T_z$ respectively, $\{C\}$ means that each vehicle has a identical capacity C, $\{1\}$ of the fourth field means that there is only one customer and $\{\sum R_j\}$ of the fifth field means that the objective is to maximize the profit function.

P3: $P|\text{p-batch}, p_j = p, s_j^{(1)} = s_j^{(2)} = 1, d_j, R_j|(v_1, v_2, \ldots, v_z), (T_1, T_2, \ldots, T_z), C|1|\sum R_j$. Where $\{P\}$ means parallel machines, $\{\text{p-batch}\}$ means that jobs can be batched together, $\{s_j^{(1)} = s_j^{(2)} = 1\}$ means that all the jobs have the identical batch sizes 1 and the identical deliver sizes 1, $\{p_j = p\}$ means that all the jobs have the identical processing time p, $\{d_j\}$ means that the jobs may have the different due dates, $\{R_j\}$ means that the jobs may have the different profits, $\{(v_1, v_2, \ldots, v_z), (T_1, T_2, \ldots, T_z)\}$ means that the 3PL company will provide $v_1, v_2, \ldots, v_z$ vehicles at time $T_1, T_2, \ldots, T_z$ respectively, $\{C\}$ means that each vehicle has a identical capacity C, $\{1\}$ of the fourth field means that there is only one customer and $\{\sum R_j\}$ of the fifth field means that the objective is to maximize the profit function.

By definition, we know that both of Problems P1 and P3 are two particular subproblems of Problem P2.

By a simple reduction from the well-known bin-packing problem which is stated as NP-hard in [7], we can easily show P1 is NP-hard, which means P2 is also NP-hard. Li et al. [9] considered the three problems and made some conclusions. In detail, they showed that there exists a $\frac{4}{3}$-approximation algorithm for P1 (Theorem 4 of [9]), 3-approximation algorithm for P2 (Theorem 5 of [9]) and an optimal algorithm within the running time of $O(n(z^2 + \log n))$ for P3. But regrettably, Theorems 4 and 5 of [9] are wrong. And for P3, we will design two faster algorithms with the running time of $O(nz + n \log n)$ and $O(n \log n)$ respectively.

This paper is organized as follows. In Sect. 2, we present some preliminaries. Section 3 points out the errors of Theorems 4 and 5 of [9] by counterexample. We design two faster algorithms for P1 in Sect. 4.

2   Preliminaries

Given an arbitrary instance I of problem P2, there are m machines and n jobs. $\mathcal{M} = \{M_1, M_2, \ldots, M_m\}$ and $\mathcal{J} = \{J_1, J_2, \ldots, J_n\}$. Each machine has a identical capacity K ($K \geq 1$). Job J_j has a processing time p_j ($p_j = p$), a batch size $s_j^{(1)}$ and a delivery size $s_j^{(2)} = 1$, a due date d_j and a profit R_j, $j = 1, 2, \ldots, n$. There are $v_1, v_2, \ldots, v_z$ vehicles of capacity C delivering processed jobs to customers at time $T_1, T_2, \ldots, T_z$ respectively, $T_1 < T_2 < \cdots < T_z$. Generally, we let $T_0 = 0$ and $T_{z+1} = +\infty$. For any non-dominated feasible schedule of Instance I, since all the jobs have the identical processing time p, the completion time of each job on machines must be an integral multiple of p. When vehicles arrive at T and deliver jobs immediately, these vehicles can only deliver jobs completed at time $p, 2p, \ldots, \lfloor \frac{T}{p} \rfloor p$. Hence, we can set the arriving time T as $\lfloor \frac{T}{p} \rfloor p$ without changing the problem. Without loss of generality, we can assume that T_i and d_j are integral multiples of p, $i = 1, 2, \ldots, z$ and $j = 1, 2, \ldots, n$. Otherwise, we can modify T_i to $\lfloor \frac{T_i}{p} \rfloor p$ and d_j to T_k if $T_k \leq d_j < T_{k+1}$ without change of the optimal solution, $i = 1, 2, \ldots, z$ and $j = 1, 2, \ldots, n$. If a job can be processed and delivered before its due date, then we call the job as *early job* (E-job for short). Otherwise, we call the job as *late job* (L-job for short). The objective is to find a subset $\mathcal{J}' \subseteq \mathcal{J}$ of E-jobs so that $\sum_{J_j \in \mathcal{J}'} R_j$ is maximal.

Now, we can characterize an optimal schedule of Problem P2. Suppose $\mathcal{J}' \subseteq \mathcal{J}$ is the set of E-jobs in an optimal solution of Problem P2. By simple pairwise interchange, we can easily have the following lemma.

Lemma 1. *For Problem P2, there exists an optimal schedule such that the jobs in $\mathcal{J}'$ fulfilling the following properties:*

- *If E-job J_j of $\mathcal{J}'$ is processed earlier than E-job J_k of $\mathcal{J}'$, then E-job J_j is delivered earlier than E-job J_k.*
- *All the jobs of $\mathcal{J}'$ are processed and delivered according to the nondecreasing order of the due dates (EDD order for short).*

3 Pointing Out of Errors

In the paper [9], the errors exist in section 5. Specifically, Theorem 4 for Problem P1 and Theorem 5 for Problem P2 are both wrong.

In the following, we construct instances to show that the two theorems are wrong.

For Theorem 4 in [9], counterexamples are constructed as follow. Without loss of generality, we assume t is an integral multiple of C. There are only two delivery time $T_1 = p$ and $T_2 = (t + 1)p$. And at each time we are provided $\frac{t}{C}$ vehicles, where C is the capacity of each vehicle, i.e., each vehicle can deliver at most C jobs. Moreover, we have $m = 2$ machines and $n = 2t$ jobs with due date T_2, where t jobs have size ϵ (ϵ is sufficiently small) and the other t jobs have size 1. The capacity of each batch is $K = 2$. Obviously, by algorithm Heu1, at time interval $[T_1, T_2]$, we first process and deliver the t small jobs, and at time interval $[0, T_1] = [0, p]$, since each batch can process 2 jobs and $m = 2$, there are 4 jobs to be processed and delivered. So we have $PF(h1) = t + 4$. But obviously $PF(opt) = 2t$, since we can process and deliver the t large jobs in time interval $[T_1, T_2]$ and the t small jobs in time interval $[0, T_1] = [0, p]$. Then $\frac{PF(opt)}{PF(h1)} = \frac{2n}{t+4} \to 2$ when $t \to \infty$.

For Theorem 5 in [9], counterexamples are constructed as follow. There are only one delivery time $T_1 = p$. And at this time we are only provided one vehicle of capacity $C = 2$. Moreover, we have $m = 2$ machines and 4 jobs with due date T_1, 2 jobs have size ϵ and profit 2ϵ, the other 2 jobs have size 1 and profit 1. Obviously, the ratio of profit to size of small jobs is 2 and the ratio of the 2 large jobs is 1. By algorithm Heu1, we process and deliver the 2 small jobs. So we have $PF(h1) = 4\epsilon$. But obviously $PF(opt) = 2$, since we can process and deliver the 2 large jobs. Then $\frac{PF(opt)}{PF(h1)} = \frac{2}{4\epsilon} \to +\infty$ when $\epsilon \to 0$.

4   Two Faster Algorithms

Given an arbitrary instance I of problem P3, there are m machines and n jobs. $\mathcal{M} = \{M_1, M_2, \ldots, M_m\}$ and $\mathcal{J} = \{J_1, J_2, \ldots, J_n\}$. Each machine has a identical capacity K ($K \geq 1$). Job J_j has a processing time $p_j = p$, a batch size $s_j^{(1)} = 1$ and a delivery size $s_j^{(2)} = 1$, a due date d_j and a profit R_j, $j = 1, 2, \ldots, n$. There are $v_1, v_2, \ldots, v_z$ vehicles of capacity C delivering processed jobs to customers at time $T_1, T_2, \ldots, T_z$ respectively, $T_1 < T_2 < \cdots < T_z$. Generally, we let $T_0 = 0$ and $T_{z+1} = +\infty$. By the discussion of Sect. 2, and P3 is a subproblem of P2, we can assume that T_i is integral multiple of p and d_j is some delivery time T_k, $i = 1, 2, \ldots, z$ and $j = 1, 2, \ldots, n$.

In the paper [9], Algorithm Sch-Id-Size is designed to solve Problem P3 and the running time is $O(n(z^2 + \log n))$. The algorithm is constructed backwards. This idea is feasible, but the design of the algorithm procedure is a bit clumsy. In fact, we can improve the realization of the idea by using better data structure.

In order to demonstrate the two algorithms clearly, we introduce some proper notations. For each $i = 1, 2, \ldots, z$, let $\mathcal{J}^{(i)}$ denote the job set of all the jobs with due date T_i and $t_i = |\mathcal{J}^{(i)}|$. We rearrange the jobs of $\mathcal{J}^{(i)}$ as $J_1^{(i)}, J_2^{(i)}, \ldots, J_{t_i}^{(i)}$ according to the non-increasing order of their profits. To break the tie, we select the job of the smallest job index. Let $R_j^{(i)}$ denote the profit of $J_j^{(i)}$, $i = 1, 2, \ldots, z$ and $j = 1, 2, \ldots, t_i$. Let $N_i = mK\frac{T_i - T_{i-1}}{p}$ and $H_i = v_i C$, $i = 1, 2, \ldots, z$. Let $J_{t_i+1}^{(i)}$ be a null job, $i = 1, 2, \ldots, z$.

Algorithm Backwards: For problem P3

Input: An instance $\mathcal{J} = \{J_1, J_2, \ldots, J_n\}$, $\mathcal{M} = \{M_1, M_2, \ldots, M_m\}$ with machine capacity K and the delivery data $\{(v_1, v_2, \ldots, v_z), (T_1, T_2, \ldots, T_z)\}$ with vehicle capacity C of P3.

Step 1: Modify delivery time $T_1, T_2, \ldots, T_z$ and due dates $d_1, d_2, \ldots, d_n$ as stated in Sect. 2. And set $T_0 := 0$, $T_{z+1} := +\infty$, $N_i = mK\frac{T_i - T_{i-1}}{p}$ and $H_i = v_i C$, $i = 1, 2, \ldots, z$.

Step 2: According to the modified due dates, we partition $\mathcal{J}$ into $\mathcal{J}^{(1)}, \mathcal{J}^{(2)}, \ldots, \mathcal{J}^{(z)}$. And rearrange the jobs of $\mathcal{J}^{(i)}$ as $J_1^{(i)}, J_2^{(i)}, \ldots, J_{t_i}^{(i)}$ according to the non-increasing order of their profits, $i = 1, 2 \ldots, z$.

Step 3: Initialize $r_j := 1$, $j = 1, 2, \ldots, z$, $s := z$ and $d := z$, $N := N_s$ and $H := H_d$, $\mathcal{J}' := \emptyset$ and $R := 0$.

Step 4: If $s \leq 0$ or $d \leq 0$, then Stop.

Step 5:

– If $N > 0$ and $H > 0$, we consider two cases.

    • Case 1: $\{J_{r_s}^{(s)}, J_{r_{s+1}}^{(s+1)}, \ldots, J_{r_z}^{(z)}\}$ has a non-null job. Then we select the job $J_{r_k}^{(k)}$ of the maximal profit out of $\{J_{r_s}^{(s)}, J_{r_{s+1}}^{(s+1)}, \ldots, J_{r_z}^{(z)}\}$, to break the tie, we select the job of the smallest job index. Schedule job $J_{r_k}^{(k)}$ to process in $[T_{s-1}, T_s]$ and deliver in time T_d. $\mathcal{J}' := \mathcal{J}' \cup \{J_{r_k}^{(k)}\}$, $R := R + R_{r_k}^{(k)}$, $r_k := r_k + 1$, $N := N - 1$ and $H := H - 1$. Go to Step 4.

- Case 2: $\{J_{r_s}^{(s)}, J_{r_{s+1}}^{(s+1)}, \ldots, J_{r_z}^{(z)}\}$ has no non-null job. Then $s := s - 1$, $N := N_s$.
- If $N > 0$ and $H = 0$ and $s < d$, then $d := d - 1$, $H := H_d$. Go to Step 4.
- If $N > 0$ and $H = 0$ and $s = d$, then $s := s - 1$ and $d := d - 1$, $N := N_s$ and $H := H_d$. Go to Step 4.
- If $N = 0$ and $H > 0$, then $s := s - 1$, $N := N_s$, Go to Step 4.
- If $N = 0$ and $H = 0$, then $s := s - 1$ and $d := d - 1$, $N := N_s$ and $H := H_d$. Go to Step 4.

Output: $\mathcal{J}'$ and R.

Explanation of Algorithm Backwards: during the implementation of Algorithm Backwards, s represents the interval time $[T_{s-1}, T_s]$ that the current job will be processed and d represents the delivery time T_d that the current job will be delivered. N_s represents the remain processable capacity of interval time $[T_{s-1}, T_s]$ and H_d represents the remain deliverable capacity of delivery time T_d. r_j represents the job of the maximal profit in the unconsidered jobs of $\mathcal{J}^{(j)}$, $j = 1, 2, \ldots, z$. $\mathcal{J}'$ represents the current job set of scheduled jobs and R represents the current total profit. $s \leq d$ holds thoroughly.

Algorithm Backwards is actually a variant of Algorithm Sch-Id-Size in [9]. For Algorithm Backwards, it takes $O(n)$ time to complete Step 1. It takes $O(n + \sum_{i=1}^{z} t_i \log t_i)$ time to complete step 2. Note that $\sum_{i=1}^{z} \log t_i = n$, we takes at most $O(n \log n)$ time to fulfil Step 2. Step 3 can be implemented in $O(n)$ time. In Step 5, since $\{J_{r_s}^{(s)}, J_{r_{s+1}}^{(s+1)}, \ldots, J_{r_z}^{(z)}\}$ has at most z jobs, we can select the job $J_{r_k}^{(k)}$ of the maximal profit out of $\{J_{r_s}^{(s)}, J_{r_{s+1}}^{(s+1)}, \ldots, J_{r_z}^{(z)}\}$ in $O(z)$. Therefore, it takes $O(nz)$ time to implement Step 4 and Step 5. In conclusion, the running time of Algorithm Backwards is $O(nz + n \log n)$ which is less than the running time of Algorithm Sch-Id-Size.

Theorem 1. *Algorithm Backwards solves Problem P3 in the running time of* $O(nz + n \log n)$.

In the following, let us start to design the other faster algorithm. The algorithm is very different to Algorithm Backwards. In Algorithm Backwards, the schedule is constructed backwards. On the contrary, we build the schedule forwards in the following new algorithm.

Given an arbitrary instance I of problem P3, by Lemma 1, we sort the jobs of $\mathcal{J}$ such that $d_1 \leq d_2 \leq \cdots \leq d_n$.

Algorithm Forwards: For problem P3

Input: An instance $\mathcal{J} = \{J_1, J_2, \ldots, J_n\}$, $\mathcal{M} = \{M_1, M_2, \ldots, M_m\}$ with machine capacity K and the delivery data $\{(v_1, v_2, \ldots, v_z), (T_1, T_2, \ldots, T_z)\}$ with vehicle capacity C of P3.

Step 1: Modify delivery time $T_1, T_2, \ldots, T_z$ and due dates $d_1, d_2, \ldots, d_n$ as stated in Sect. 2. And set $T_0 := 0$, $T_{z+1} := +\infty$, $N_i := mK \frac{T_i - T_{i-1}}{p}$ and $H_i := v_i C$, $i = 1, 2, \ldots, z$.

Step 2: Sort the jobs of $\mathcal{J}$ such that $d_1 \leq d_2 \leq \cdots \leq d_n$.

Step 3: Initialize $i := 1$, $s := 1$ and $d := 1$, $N := N_s$ and $H := H_d$, $\mathcal{J}' := \emptyset$ and $R := 0$.

Step 4: If $s > z$ or $d > z$ or $i > n$, then Stop.

Step 5:

- If $N > 0$ and $H > 0$, we consider two cases.
  - Case 1: $d_i \geq T_d$. Then we schedule job J_i to process in $[T_{s-1}, T_s]$ and deliver in time T_d. $\mathcal{J}' := \mathcal{J}' \cup \{J_i\}$, $R := R + R_i$, $i := i + 1$, $N := N - 1$ and $H := H - 1$. Go to Step 4.
  - Case 2: $d_i < T_d$. Then we find the job J_h of the minimal profit in $\mathcal{J}'$ and substitute J_h by J_i in the process-delivery schedule. $\mathcal{J}' = \{\mathcal{J}' \setminus \{J_h\}\} \cup \{J_i\}$, $R := R + R_i - R_h$, $i := i + 1$. Go to Step 4.
- If $N > 0$ and $H = 0$, then $d := d + 1$, $H := H_d$. Go to Step 4.
- If $N = 0$, $H > 0$ and $s < d$, then $s := s + 1$, $N := N_s$, Go to Step 4.
- If $N = 0$, $H > 0$ and $s = d$, then $s := s + 1$ and $d := d + 1$, $N := N_s$ and $H := H_d$. Go to Step 4.
- If $N = 0$ and $H = 0$, then $s := s + 1$ and $d := d + 1$, $N := N_s$ and $H := H_d$. Go to Step 4.

Output: $\mathcal{J}'$ and R.

Explanation of Algorithm Forwards: during the implementation of Algorithm Forwards, s represents the interval time $[T_{s-1}, T_s]$ that the current job will be processed and d represents the delivery time T_d that the current job will be delivered. N_s represents the remain processable capacity of interval time $[T_{s-1}, T_s]$ and H_d represents the remain deliverable capacity of delivery time T_d. i represents the current job J_i considered. $\mathcal{J}'$ represents the current job set of scheduled jobs and R represents the current total profit. $s \leq d$ holds thoroughly.

We try to schedule jobs to process and deliver. Once the current job can not be processed and delivered by its due date, we substitute the scheduled job of the minimal profit by the current job. This is feasible, since we proceed to schedule the jobs according to the non-decreasing order of their due dates. This idea is similar to the idea of the algorithm presented to cope with $1 || \sum U_j$ in [12]. You can also see the algorithm in Chapter 4 of [1].

If all the jobs of $\mathcal{W} \subseteq \mathcal{J}$ are E-jobs in some feasible schedule, then we call $\mathcal{W}$ *feasible*. That is, by the Lemma 1, the jobs of $\mathcal{W}$ can be processed and delivered according to the EDD order.

Lemma 2. *Let $\mathcal{J}'$ be the feasible job set returned by Algorithm Forwards, then $\mathcal{J}'$ is the feasible set of maximal cardinality. That is, $|\mathcal{J}'| \geq |\mathcal{W}|$ for any feasible set $\mathcal{W} \subseteq \mathcal{J}$.*

Proof. We complete the proof by contradiction. Assume that $\mathcal{J}$ is the job set of minimal cardinality violating the lemma and $\mathcal{J}'$ is the feasible job set returned by Algorithm Forwards. Let $\mathcal{U} \subseteq \mathcal{J}$ denote a feasible set of maximal cardinality, $\mathcal{L} = \mathcal{J} \setminus \{J_n\}$ and $\mathcal{L}'$ be the feasible job set returned by Algorithm Forwards on the instance with job set $\mathcal{L}$. Then $|\mathcal{U}| > |\mathcal{J}'|$. We discuss two cases.

Case 1: $J_n \notin \mathcal{U}$. Obviously, $\mathcal{U}$ is a feasible subset of $\mathcal{L}$ of maximal cardinality and $|\mathcal{U}| > |\mathcal{J}'| \geq |\mathcal{L}'|$, which means that the instance with job set $\mathcal{L}$ is also an instance violating the lemma. This breaks the minimal cardinality of set $\mathcal{J}$.

Case 2: $J_n \in \mathcal{U}$. Let $\mathcal{H} = \mathcal{U} \setminus \{J_n\}$. Since $\mathcal{J}$ has the minimal cardinality violating the lemma, $|\mathcal{H}| = \mathcal{U} - 1 \geq |\mathcal{J}'| \geq |\mathcal{L}'|$. Combined with Lemma 1, we know that $J_n \in \mathcal{J}'$ and $\mathcal{J}' = \mathcal{L}' \cup \{J_n\}$. Therefore, $|\mathcal{H}| > |\mathcal{L}'|$. Note that $\mathcal{H}$ is a feasible set of $\mathcal{L}$, we can infer that the instance with $\mathcal{L}$ is also an instance violating the lemma. This breaks the minimal cardinality of set $\mathcal{J}$. □

Let $\mathcal{J}'$ be the feasible job set returned by Algorithm Forwards on job set $\mathcal{J}$ and h be the maximal cardinality of feasible sets of $\mathcal{J}$. By Lemma 2, $|\mathcal{J}'| = h$.

For each $k = 1, 2, \ldots, h$, we construct a new Problem P3(k) which is to find a feasible job set of maximum profit out of the job set of the cardinality at most k. That is, we must find a feasible set T of the cardinality at most k in $\mathcal{J}$ such that the objective $\sum_{J_j \in T, |T| \leq k} R_j$ is maximized.

In order to cope with Problem P3(k), we add a counter of the cardinality of $\mathcal{J}'$ during the implementation of Algorithm Forwards.

Algorithm Forwards (k): For problem P3(k)

Input: An instance $\mathcal{J} = \{J_1, J_2, \ldots, J_n\}$, $\mathcal{M} = \{M_1, M_2, \ldots, M_m\}$ with machine capacity K and the delivery data $\{(v_1, v_2, \ldots, v_z), (T_1, T_2, \ldots, T_z)\}$ with vehicle capacity C of P3(k).

Step 1: Modify delivery time $T_1, T_2, \ldots, T_z$ and due dates $d_1, d_2, \ldots, d_n$ as stated in Sect. 2. And set $T_0 := 0$, $T_{z+1} := +\infty$, $N_i := mK\frac{T_i - T_{i-1}}{p}$ and $H_i := v_i C$, $i = 1, 2, \ldots, z$.

Step 2: Sort the jobs of $\mathcal{J}$ such that $d_1 \leq d_2 \leq \cdots \leq d_n$.

Step 3: Initialize $t := 0$, $i := 1$, $s := 1$ and $d := 1$, $N := N_s$ and $H := H_d$, $\mathcal{J}' := \emptyset$ and $R := 0$.

Step 4: If $s > z$ or $d > z$ or $i > n$, $t > k$, then Stop.

Step 5:

- If $N > 0$ and $H > 0$, we consider two cases.
  - Case 1: $d_i \geq T_d$. Then we schedule job J_i to process in $[T_{s-1}, T_s]$ and deliver in time T_d. $\mathcal{J}' := \mathcal{J}' \cup \{J_i\}$, $R := R + R_i$, $t := t + 1$, $i := i + 1$, $N := N - 1$ and $H := H - 1$. Go to Step 4.
  - Case 2: $d_i < T_d$. Then we find the job J_h of the minimal profit in $\mathcal{J}'$ and substitute J_h by J_i in the process-delivery schedule. $\mathcal{J}' = \{\mathcal{J}' \setminus \{J_h\}\} \cup \{J_i\}$, $R := R + R_i - R_h$, $i := i + 1$. Go to Step 4.
- If $N > 0$ and $H = 0$, then $d := d + 1$, $H := H_d$. Go to Step 4.
- If $N = 0$, $H > 0$ and $s < d$, then $s := s + 1$, $N := N_s$, Go to Step 4.
- If $N = 0$, $H > 0$ and $s = d$, then $s := s + 1$ and $d := d + 1$, $N := N_s$ and $H := H_d$. Go to Step 4.
- If $N = 0$ and $H = 0$, then $s := s + 1$ and $d := d + 1$, $N := N_s$ and $H := H_d$. Go to Step 4.

Output: $\mathcal{J}'$ and R.

Explanation of Algorithm Forwards (k): during the implementation of Algorithm Forwards (k), $1 \leq k \leq h$. s represents the interval time $[T_{s-1}, T_s]$ that the current job will be processed and d represents the delivery time T_d that the current job will be delivered. N_s represents the remain processable capacity of interval time $[T_{s-1}, T_s]$ and H_d represents the remain deliverable capacity of delivery time T_d. i represents the current job J_i considered. $\mathcal{J}'$ represents the current job set of scheduled jobs and R represents the current total profit, t represents the current cardinality of $\mathcal{J}'$ and the terminal number of t is k. $t \leq k$ and $s \leq d$ hold thoroughly.

For each $i = 1, 2 \ldots, h$, we define $\mathcal{J}_{i,k}$ as the current job set $\mathcal{J}'$ of scheduled jobs after job J_k is considered during the implementation of Algorithm Forwards (i) and let $e(i, k)$ is the minimum job index so that $J_{e(i,k)}$ is a job of minimal profit in $\mathcal{J}_{i,k}$. The following observation is crucial.

Lemma 3. *For each $i = 2, 3, \ldots, h$ and each $k = 1, 2, \ldots, n$, we have*

$$\mathcal{J}_{i,k} = \begin{cases} \mathcal{J}_{i-1,k}, & |\mathcal{J}_{i,k}| < i, \\ \mathcal{J}_{i-1,k} + \{J_{e(i,k)}\}, & |\mathcal{J}_{i,k}| = i. \end{cases}$$

Proof. If $|\mathcal{J}_{i,k}| < i$, neither before or after job J_k is considered, the implementation of Algorithm Forwards (i) is exact the the implementation of Algorithm Forwards $(i-1)$. Therefore $\mathcal{J}_{i,k} = \mathcal{J}_{i-1,k}$.

If $|\mathcal{J}_{i,k}| = i$. Let r is the maximum index so that $\mathcal{J}_{i,r} = \mathcal{J}_{i-1,r}$ before job k is considered. We know that $|\mathcal{J}_{i,r}| = |\mathcal{J}_{i-1,r}| = i - 1$, $e(i,r) = e(i-1,r)$, $|\mathcal{J}_{i,r+1}| = i$ and $|\mathcal{J}_{i-1,r+1}| = i - 1$, $\mathcal{J}_{i,r+1} = \mathcal{J}_{i,r} \cup \{J_{r+1}\}$. If $R_{r+1} \geq R_{e(i-1,r)}$, then $\mathcal{J}_{i-1,r+1} = \{\mathcal{J}_{i-1,r} \setminus \{J_{e(i-1,r)}\}\} \cup \{J_{r+1}\}$, $e(i-1,r) = e(i,r) = e(i, r+1)$ and $\mathcal{J}_{i,r+1} = \mathcal{J}_{i-1,r+1} \cup \{J_{e(i,r+1)}\}$. If $R_{r+1} < R_{e(i-1,r)}$, then $\mathcal{J}_{i-1,r+1} = \mathcal{J}_{i-1,r}$, $e(i, r+1) = r+1$ and $\mathcal{J}_{i,r+1} = \mathcal{J}_{i-1,r+1} \cup \{J_{e(i,r+1)}\}$. We complete the proof by induction on $t = r+1, r+2, \ldots, k$. When $t = r+1$, the theorem apparently holds. Assume that the theorem holds for $t = u$ with $r+1 \leq u < k$, which means that $\mathcal{J}_{i,u} = \mathcal{J}_{i-1,u} + \{J_{e(i,u)}\}$. If $R_{u+1} < R_{e(i,u)}$, then $\mathcal{J}_{i,u+1} = \mathcal{J}_{i,u}$, $\mathcal{J}_{i-1,u+1} = \mathcal{J}_{i-1,u}$ and $e(i, u+1) = e(i, u)$. Hence, $\mathcal{J}_{i,u+1} = \mathcal{J}_{i-1,u+1} + \{J_{e(i,u+1)}\}$, the theorem holds. If $R_{e(i,u)} \leq R_{u+1} < R_{e(i-1,u)}$, then $\mathcal{J}_{i,u+1} = \{\mathcal{J}_{i,u} \setminus \{J_{e(i,u)}\}\} \cup \{J_{u+1}\}$, $\mathcal{J}_{i-1,u+1} = \mathcal{J}_{i-1,u}$ and $e(i, u+1) = u+1$. Hence, $\mathcal{J}_{i,u+1} = \mathcal{J}_{i-1,u+1} + \{J_{e(i,u+1)}\}$, the theorem holds. If $R_{u+1} \geq R_{e(i-1,u)}$, then $\mathcal{J}_{i,u+1} = \{\mathcal{J}_{i,u} \setminus \{J_{e(i,u)}\}\} \cup \{J_{u+1}\}$, $\mathcal{J}_{i-1,u+1} = \{\mathcal{J}_{i-1,u} \setminus \{J_{e(i-1,u)}\}\} \cup \{J_{u+1}\}$ and $e(i, u+1) = e(i-1, u)$. Hence, $\mathcal{J}_{i,u+1} = \mathcal{J}_{i-1,u+1} + \{J_{e(i,u+1)}\}$, the theorem holds. □

For any job set $\mathcal{K}$, we define $f(\mathcal{K}) = \sum_{J_j \in \mathcal{K}} R_j$.

Theorem 2. *Algorithm Forwards (k) solves Problem P3(k), $k = 1, 2, \ldots, h$.*

Proof. When $k = 1$ and the cardinality of the job set under consideration is arbitrary, the theorem obviously holds.

In the following, let w denote the cardinality of the job set under consideration. We complete the proof of the theorem by induction on $k + w = 2, 3, \ldots, n+h$.

When $k + w = 2$, that is, $k = 1$ and $w = 1$ which means that the job set under consideration has only one job. The theorem obviously holds.

Assume that the theorem holds when $k + w \leq l - 1$. We consider Algorithm Forwards (i) and the instance of t jobs with $i + t = l$. Let $\mathcal{J} = \{J_1, J_2, \ldots, J_t\}$ with $d_1 \leq d_2 \leq \cdots \leq d_t$ be the job set under consideration. Let $\mathcal{L} = \mathcal{J} \setminus \{J_t\}$. We denote the returned job sets of Algorithm Forwards (i) on $\mathcal{J}$ and $\mathcal{L}$ by $\mathcal{J}'$ and $\mathcal{L}'$ respectively. Let $\mathcal{U}$ be an optimal job set of the instance with job set $\mathcal{J}$ of P3(i) and $\mathcal{W}$ be an optimal job set of the instance with job set $\mathcal{L}$ of P3(i). We discuss two cases.

Case 1: $J_t \notin \mathcal{U}$. Then $\mathcal{U}$ is also an optimal job set of the instance with job set $\mathcal{L}$ of P3(i), $f(\mathcal{U}) = f(\mathcal{W})$. Consider Algorithm Forwards (i) and the instance of job set $\mathcal{L}$, note that $i + |\mathcal{L}| = l - 1$, we know that $f(\mathcal{L}') = f(\mathcal{W})$. Obviously, $f(\mathcal{L}') \leq f(\mathcal{J}')$ and $f(\mathcal{J}') \leq f(\mathcal{U})$. So we have $f(\mathcal{J}') = f(\mathcal{U})$, the theorem holds.

Case 2: $J_t \in \mathcal{U}$. $|\mathcal{U}| \leq i$. If $|\mathcal{U}| \leq i - 1$, Consider Algorithm Forwards $(i - 1)$ and the instance of job set $\mathcal{J}$ note that $i - 1 + |\mathcal{J}| = l - 1$, let $\mathcal{Q}$ be the returned job set of Algorithm Forwards $(i - 1)$ on $\mathcal{J}$, $f(\mathcal{Q}) = f(\mathcal{U})$. Obviously, $f(\mathcal{Q}) \leq f(\mathcal{J}')$ and $f(\mathcal{J}') \leq f(\mathcal{U})$. So we have $f(\mathcal{J}') = f(\mathcal{U})$, the theorem holds. If $|\mathcal{U}| = i$, let $\mathcal{H} = \mathcal{U} \setminus \{J_t\}$. We know that $f(\mathcal{U}) = f(\mathcal{H}) + R_t$ and $\mathcal{H}$ is a feasible job set of the instance $\mathcal{L}$ of P3$(i - 1)$. Let $\mathcal{H}'$ be the returned job sets of Algorithm Forwards $(i - 1)$ on $\mathcal{L}$. Since $i - 1 + |\mathcal{L}| = l - 2 < l - 1$, by induction, we know that $f(\mathcal{H}) \leq f(\mathcal{H}')$. If $|\mathcal{L}'| < i$, then $\mathcal{J}' = \mathcal{L}' \cup \{J_t\}$. And by Lemma 3, we have $\mathcal{L}' = \mathcal{H}'$. Therefore, $f(\mathcal{J}') = f(\mathcal{L}') + R_t \geq f(\mathcal{H}') + R_t \geq f(\mathcal{H}) + R_t = f(\mathcal{U})$. So we have $f(\mathcal{J}') = f(\mathcal{U})$, the theorem holds. If $|\mathcal{L}'| = i$, let q is the minimum job index so that J_q is a job of minimal profit in $\mathcal{L}'$. By Lemma 3, $\mathcal{L}' = \mathcal{H}' \cup \{J_q\}$. By the Algorithm Forwards (i) on $\mathcal{J}$, we can infer that $f(\mathcal{J}') = \max\{f(\mathcal{L}'), f(\mathcal{H}') + R_t\} \geq f(\mathcal{H}') + R_t \geq f(\mathcal{H}) + R_t = f(\mathcal{U})$. So we have $f(\mathcal{J}') = f(\mathcal{U})$, the theorem holds. □

Apparently, Problem P3(h) is equivalent to Problem P3 and the implementation of Algorithm Forwards (h) is the exact implementation of Algorithm Forwards. And for the running time of Algorithm Forwards, it takes $O(n \log n)$ time to fulfil Step 2 and $O(n)$ time to fulfil Step 1, Step 3, Step 4 and Step 5. So the running time of Algorithm Forwards is $O(n \log n)$ which is less than the running time of Algorithm Sch-Id-Size. The following theorem follows.

Theorem 3. *Algorithm Forwards solves Problem P3 in $O(n \log n)$ time.*

References

1. Brucker, P.: Scheduling Algorithms. Springer, Heidelberg (2007). https://doi.org/10.1007/978-3-540-69516-5
2. Brucker, P., et al.: Scheduling a batching machine. J. Sched. **1**, 31–54 (1998)
3. Chen, Z.L., Vairaktarakis, G.L.: Integrated scheduling of production and distribution operations. Manag. Sci. **51**(4), 614–628 (2005)
4. Chen, Z.L.: Integrated production and outbound distribution scheduling: review and extensions. Oper. Res. **58**(1), 130–148 (2010)

5. Cheng, B.Y., Leung, J.Y.T., Li, K.: Integrated scheduling on a batch machine to minimize production, inventory and distribution costs. Eur. J. Oper. Res. **258**(1), 104–112 (2017)

6. Dupont, L., Dhaenens-Flipo, C.: Minimizing the makespan on a batch machine with non-identical job sizes. Comput. Oper. Res. **29**(7), 807–819 (2002)

7. Garey, M.R., Johnson, D.S.: Computers and Intractability: A Guide to the Theory of NP-Completeness. Freeman, San Francisco (1979)

8. Ikura, Y., Gimple, M.: Efficient scheduling algorithms for a single batch processing machine. Oper. Res. Lett. **5**, 61–65 (1986)

9. Li, K., Jia, Z.H., Leung, J.Y.T.: Integrated production and delivery on parallel batching machines. Eur. J. Oper. Res. **247**(3), 755–763 (2015)

10. Malve, S., Uzsoy, R.: A genetic algorithm for minimizing maximum lateness on parallel identical batch processing machines with dynamic job arrivals and incompatible job families. Comput. Oper. Res. **34**, 3016–3028 (2007)

11. Ma, R., Wan, L., Wei, L.J., Yuan, J.J.: Online bounded-batch scheduling to minimize total weighted completion time on parallel machines. Int. J. Prod. Econ. **156**, 31–38 (2014)

12. Moore, J.M.: An n job, one machine sequencing algorithm for minimizing the number of late jobs. Manag. Sci. **15**(1), 102–109 (1968)

13. Potts, C.N., Kovalyov, M.Y.: Scheduling with batching: a review. Eur. J. Oper. Res. **120**, 228–249 (2000)

14. Tian, J., Cheng, T.C.E., Ng, C.T., Yuan, J.J.: Online scheduling on unbounded parallel-batch machines to minimize the makespan. Inf. Process. Lett. **109**, 1211–1215 (2009)

15. Wan, L., Ding, Z.H., Li, Y.P., Chen, Q.Q., Tan, Z.Y.: Scheduling to minimize the maximum total completion time per machine. Eur. J. Oper. Res. **242**, 45–50 (2015)

16. Wan, L., Zhang, A.: Coordinated scheduling on parallel machines with batch delivery. Int. J. Prod. Econ. **150**, 199–203 (2014)

17. Wang, C.-S., Uzsoy, R.: A genetic algorithm to minimize maximum lateness on a batch processing machine. Comput. Oper. Res. **29**, 1621–1640 (2002)

18. Wang, Q., Batta, R., Szczerba, R.J.: Sequencing the processing of incoming mail to match an outbound truck delivery schedule. Comput. Oper. Res. **32**, 1777–1791 (2005)

19. Xu, S., Bean, J.C.: A genetic algorithm for scheduling parallel non-identical batch processing machines. In: Proceedings of the 2007 IEEE Symposium on Computational Intelligence in Scheduling, pp. 143–150 (2007)

Range Partitioning Within Sublinear Time in the External Memory Model

Baoling Ning[(⊠)] [iD], Jianzhong Li, and Shouxu Jiang

Harbin Institute of Technology, Harbin 150000, China
ningbaoling2009@163.com, {lijzh,jiangsx}@hit.edu.cn

Abstract. Range partitioning is a popular method for processing massive data, whose task is to divide the input N data items into k ranges of the same size. To avoid accessing the whole input, in the RAM model, sampling based (ϵ, δ)-approximation algorithms with $O(\frac{k \log(N/\delta)}{\epsilon^2})$ time cost have been well studied. However, massive data may be too large to be maintained in the main memory. Usually, they are stored in the external memory devices and need to design I/O efficient algorithms in the external memory model. Then, a natural question is whether or not there are efficient range partitioning algorithms with $O(\frac{k \log(N/\delta)}{B \epsilon^2})$ I/O cost. To answer the above question, this paper studies the range partitioning problem in the external memory model. Two lower bounds of the sampling cost required by the external sublinear range partitioning algorithms are proved, which show that it needs to make a full scan of the input in the worst case. Motivated by the hard instances utilized in the proof of lower bounds, a model for describing the inputs of the range partitioning problem in practical applications is proposed. Finally, for the special case that input data are generated by the proposed model, a nearly optimal algorithm with $O(\frac{k \log(N/\delta)}{w B \epsilon^2})$ I/O cost is introduced.

1 Introduction

Horizontal data partitioning is a wildly utilized strategy when processing massive data. It provides the ability of dividing the data into several physical parts in massive data applications, which can lead to many benefits such as improving the performance of complexity analytical queries, enhancing the storage ability by distributing data, tolerating the system failures and so on. The most commonly used data partitioning methods include: *range partitioning, hash partitioning, round-robin partitioning* and so on [14]. Compared with the other two approaches, range partitioning is semantics related and commonly used when the data need to be partitioned under the premise of clustering properties.

This paper focuses on the problem of range partitioning in the external memory model. Different from the classical RAM model, data is organized into blocks which are the unit of external memory access. The external memory model is more proper for massive data computing, since the most common case is that data is too large to be maintained in main memory.

© Springer Nature Switzerland AG 2020
Z. Zhang et al. (Eds.): AAIM 2020, LNCS 12290, pp. 323–335, 2020.
https://doi.org/10.1007/978-3-030-57602-8_29

Given N data items admitting a total order and an integer k, the range partitioning problem is to partition data into a set of continuous ranges such that each range is of size N/k. The exact version of the range partitioning problem can be solved efficiently based on an $O(\frac{N}{B} \log_{M/B} \frac{N}{B})$ sorting algorithm in external memory model, where B is the block size and M is the main memory size. Although such an algorithm is nearly optimal, it will still cause a huge I/O cost which is unacceptable for massive data.

Therefore, the approximation version of the range partitioning problem is studied by previous works [6,18]. It is shown that, in the RAM model, randomly selecting a sample set of size larger than $O(\frac{k}{\epsilon^2} \log \frac{N}{\delta})$ is enough to guarantee an (ϵ, δ)-approximation.

However, as shown by [6], simply applying the classical sampling methods in external memory model will make an I/O cost as large as $O(\frac{k}{\epsilon^2} \log \frac{N}{\delta})$. Generally speaking, if there is an $O(f(n))$ algorithm in the RAM model, it is often expected that an external algorithm with $O(f(n)/B)$ I/O cost exists. Thus, a natural question is whether or not it is possible to design a more efficient external approximate range partitioning algorithm.

In this paper, to answer the above challenging question, based on the external memory model, the lower bounds of the sampling size required are studied first. It is shown that in general case the lower bound can be $\Omega(N/B)$, which means that we need to scan all input data items in worst case. Motivated by the lower bounds, the partial dependent model describing the data distributions in practical applications are proposed, and an $O(\frac{k \log(N/\delta)}{wB\epsilon^2})$ external range partitioning algorithm is designed. Since the current best sublinear range partitioning algorithm in the RAM model is bounded by $O(\frac{k \log(N/\delta)}{\epsilon^2})$, the propose external algorithm is nearly optimal.

2 Preliminaries

2.1 The Range Partitioning Problem

The input of the range partitioning (RP for short) problem is supposed to be a multiset D of N data items admitting a total order. In the followings, $w.l.o.g.$, it is assumed that all data items come from the integer set $\mathbb{Z}$.

Definition 1 (k-Partition). *Given the input D, a k-partition, denoted by P^D, is a set $\{P_1, \ldots, P_k\}$ of disjoint subsets of D, such that the union of all subsets in P^D is equal to D, that is $\bigcup \{P_1, \ldots, P_k\} = D$.* □

Definition 2 (k-Range-Partition). *A k-range-partition $P^D = \{P_1, \ldots, P_k\}$ for the input D is a k-partition defined by a splitter set $\{s_1, \ldots, s_{k-1}\}$ satisfying $s_1 < s_2 < \cdots < s_{k-1}$. Here, the partitions in P^D can be defined as follows.*

$$P_i = \begin{cases} \{x | x < s_1, x \in D\}, & when\ i = 1 \\ \{x | s_{i-1} \le x < s_i, x \in D\}, & when\ 2 \le i \le k-1 \\ \{x | x \ge s_{k-1}, x \in D\}, & when\ i = k \end{cases}$$

Usually, it is also assumed that the input D is essentially a list which is proper for many real applications since D is often stored in a linear storage device such as disks. In this case, the input is also denoted by L, and a partition can be called a *sublist* or *range* r, and r can be represented by the pair (l_r, u_r) of *lower* and *upper* bounds of the partition. Then, the corresponding partition L_r can be also denoted by $L_{[l_r, u_r)}$ or $[l_r, u_r)$. Moreover, given a splitter s_i, $index(s_i)$ is the smallest index in L whose item is s_i.

Example 1. Suppose the input list L is $\{1,2,2,4,5,5,6,6,7\}$. Let k be 3, that is we need to partition L into three parts. A splitter set is $\{s_1 = 4, s_2 = 6\}$, and L is partitioned into 3 sublists: $\{1, 2, 2\}$, $\{4, 5, 5\}$ and $\{6, 6, 7\}$. Moreover, $index(s_1) = 4$ and $index(s_2) = 7$, which are the corresponding position index of s_1 and s_2. □

Given an input list L of size N, without loss of generality, it is assumed that N can be divided by k, then we can define the range partitioning problem studied by this paper.

Definition 3 (The k-Range Partitioning Problem, k-RP for short).
Given an input list L and an integer k, the k-range partitioning problem is to find $k - 1$ splitters $\{s_1, \ldots, s_{k-1}\}$ such that all partitions have the same size N/k. □

Obviously, if L contains no duplicate items, there is always a feasible solution for the k-RP problem.

2.2 Approximation Notions

Obviously, a sort based algorithm can be utilized to solve the k-RP problem trivially, which will produce an algorithm with $O(N \log N)$ time cost in the RAM model and an algorithm with $O(\frac{N}{B} \log_{\frac{M}{B}} \frac{N}{B})$ I/O cost in the external memory model [17]. To exactly solve the k-RP problem, the sort based algorithms have nearly matched the *lower bound*, since at least one full scan is needed.

This paper focuses on solving the *approximation* version of the k-RP problem by sampling based methods. The followings will introduced the related approximation notions, where some of them are adopted from previous works like [2].

In general, a computing problem can also be modelled as a function. For example, the k-RP problem essentially defines a function $\mathbb{N}^N \to \mathbb{N}^k$. Given a general function $f : \mathcal{X}^n \to \mathcal{Y}$, an approximation notion of f is a family of subsets $\{A_{f,\epsilon}(\mathbf{x})\}_{\mathbf{x} \in \mathcal{X}^n}$ of $\mathcal{Y}$. Here, $A_{f,\epsilon}(\mathbf{x})$ is the ϵ-*approximation set* of $\mathbf{x}$, which includes both $f(\mathbf{x})$ and the ϵ-*approximations* of $f(\mathbf{x})$. For more, the meaning of ϵ-approximation can adopt any reasonable definitions of approximation, and it should be satisfied that $A_{f,\epsilon}(\mathbf{x}) \subseteq A_{f,\epsilon'}(\mathbf{x})$ for any $0 \leq \epsilon \leq \epsilon'$.

Typically, there are two kinds of definition of approximation algorithms for the range partitioning problem.

Definition 4 (ϵ-Approximation of Range Length). *Given an input list L and an integer k, the ϵ-approximation of range length of the k-RP problem is $A_{kRP,\epsilon}(L) = \{S_i\}$, where each S_i is a set of $k-1$ splitters $\{s_{i,1}, \ldots, s_{i,k-1}\}$ and we have $(1-\epsilon)N/k \leq |L_r| \leq (1+\epsilon)N/k$ for each range r obtained by S_i.* □

Definition 5 (ϵ-Approximation of Splitter Position). *Given an input list L and an integer k, the ϵ-approximation of splitter position of the k-RP problem is $A_{kRP,\epsilon}(L) = \{I_i\}$, where each I_i is a set of $k-1$ splitter positions $\{i_{i,1}, \ldots, i_{i,k-1}\}$ and we have $-\epsilon N/k \leq i_{i,j} - index(s_{i,j}) \leq \epsilon N/k$ for $j \in [1, k-1]$.* □

Obviously, if there are no duplicate items, the above two definitions of the range partitioning problem are the same, otherwise, they are different.

3   The Hardness of Designing External Sublinear Partitioning Algorithms

3.1   External Memory Model and External Sampling Algorithms

To design efficient algorithms for massive data, it is highly required to study algorithms in external memory model, since massive data usually are not stored in main memory, but on the external memory devices (*e.g.* hard disks).

In the external memory model (or the I/O model) [17], the data is stored in *blocks* of fixed size B, which are further organized in a sequential way. The main difference between the classic RAM model and the I/O model is that each external memory access will return *one block* of data but not one word or data item. For a problem admitting a RAM algorithm with cost T, it is often expected to design an external memory algorithm with T/B I/O costs. Most of current sublinear algorithms, especially for sampling based methods, do not consider the external memory model, and it is usually challenging to design external sublinear algorithms [1]. For the k-RP problem, sampling based approximation algorithms in the RAM model have been studied well. In [6], the problem is studied in the name of *equi-height histogram construction*, and a sufficient large sample size $Sample_{RAM} = \frac{4k \ln(2N/\delta)}{\epsilon^2}$ is given. As a consequence, it is highly expected that an external partitioning algorithm with $O(Sample_{RAM}/B) = O(\frac{k \log(N/\delta)}{\epsilon^2 B})$ can be found.

However, trivially executing the RAM sampling based algorithm on the external memory model, will cause an I/O cost as large as $\frac{4k \ln(2N/\delta)}{\epsilon^2}$, whose performance is extremely terrible in the aspect of external algorithms.

Does there exist external memory partitioning algorithms smart enough to reduce the high I/O cost? The following analysis will give negative answers by searching the lower bounds of the sample size for the k-RP problem in external memory model.

3.2   A Sampling Lower Bound for Large Block Cases

In this part, considering the case that only relatively large blocks are allowed, a lower bound of the sampling size is given for the situation that there are no duplicate items in the list, which will induce a lower bound for the general case in consequence.

Intuitively, when the block size B is large enough to cover at least $2\epsilon N/k$ items, to make sure that only ϵ-approximations are outputted, enough external memory accesses are needed to gurantee special blocks are sampled, which will help us to build the lower bound.

To show the lower bound, the techniques introduced by [2] are utilized here. First, we introduce the related concepts.

Definition 6 (Disjoint Inputs). *Let $f : \mathcal{X}^n \to \mathcal{Y}$ be a function with approximation notion $A_{f,\epsilon}$. Two inputs $\mathbf{x}$ and $\mathbf{x}'$ are called* disjoint *or ϵ-disjoint, if $A_{f,\epsilon}(\mathbf{x}) \cap A_{f,\epsilon}(\mathbf{x}') = \emptyset$.* □

To illustrate how many items of the input we change are needed to affect the output significantly, the concept of ϵ-sensitive is needed.

Definition 7 (ϵ-sensitive). *A function $f : \mathcal{X}^n \to \mathcal{Y}$ is ϵ-sensitive to a subset of variables $I \subseteq [n]$ on input x, if there is a value combination $Q \in \mathcal{X}^{|I|}$ and x' is obtained by replacing the data items in I of x with Q which satisfy that x and x' are ϵ-disjoint.* □

Then, we introduce the concept of *block sensitivity*, which can be used to describe the characteristics of special functions.

Definition 8 (Block Sensitivity). *The ϵ-block sensitivity of a function f on input x, denoted by $bs_\epsilon(f,x)$, is the maximum number t of pairwise disjoint subsets $I_1, \ldots, I_t \subseteq [n]$, such that f is ϵ-sensitive to each of them on x. Then, the ϵ-block sensitivity of f, denoted by $bs_\epsilon(f)$, is $\max_{x \in \mathcal{X}^n} bs_\epsilon(f,x)$.* □

Then, by analyzing the block sensitivity of functions, the following theorem is introduced by [2] to obtain a lower bound of the sampling size.

Theorem 1 ([2]). *For every $\epsilon \geq 0$, $0 \leq \delta \leq 1/2$, and $f : \mathcal{X}^n \to \mathcal{Y}$, the expected sampling size needed by an (ϵ, δ)-approximation algorithm for f is at least $(1 - 2\delta)bs_\epsilon(f)$.* □

Finally, utilizing Theorem 1, we can obtain the following result about the lower bound of the expected sampling size. It should be note that this result is much stronger than a lower bound for the worst case.

Theorem 2. *When the block size B satisfies $B \geq 2\epsilon N/k$, the expected sampling size needed by an external memory (ϵ, δ)-approximation algorithm for the k-RP problem is at least $(1 - \delta)N/B$. If δ is considered to be a constant enough small, the lower bound of sampling size is $\Omega(N/B)$.* □

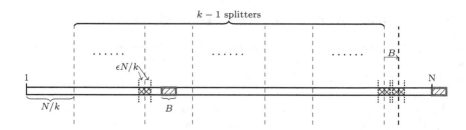

Fig. 1. An Example of the Proof of Theorem 2

Proof. Without loss of generality, it can be assumed that the length N can be divided by the block size B and the number of partitions k, and the length $(1 - \epsilon)N/k$ can be divided by B also.

First, we will construct the worst case input x, and define the sensitive blocks and the corresponding values Q.

Since there is a further restriction that no duplicate items appear in L, let x be the list $1, 2, \ldots, N$. As shown in Fig. 1, since N/k is an integer, let $s_1, s_2, \ldots, s_{k-1}$ be the $k - 1$ splitters which can perfectly partition L into ranges of size N/k.

For some ϵ-approximation solution S' of the k-RP problem on L defined by the range length, let $s'_1, \ldots, s'_{k-1}$ be the splitters in S'. By observing the relations between S and S', we can find out the set of ϵ-sensitive blocks by two parts I^1 and I^2 as follows. For each integer $i \in [2, \frac{N}{B}]$, the ith ϵ-sensitive block is defined to be $I_i = ((i - 1)B, iB]$, and let $I_1 = [1, B]$. For the integers within $[1, u = \frac{N - (1 - \epsilon)N/k}{B}]$, where u is obviously an integer according to the assumptions, the corresponding blocks are called I^1, while other blocks belong to I^2.

Obviously, for any different i, j, we have $I_i \cap I_j = \emptyset$, that is $I^1 \cup I^2$ is a set of pairwise disjoint blocks. Next, we will show that all blocks built are ϵ-sensitive.

- The k-RP problem can be represented by a function $f : \mathbb{Z}^N \to \mathbb{Z}^{k-1}$, where we can use the item value to represent the splitter since each item has an unique value. Let $\mathbb{Z}_B$ be the set of value combinations of B items, which can be defined by $\mathbb{Z}_B = \underbrace{\mathbb{Z} \times \mathbb{Z} \times \cdots \times \mathbb{Z}}_{B \text{ times in total}}$. The function can be reformed to be a function $g : \mathbb{Z}_B^{N/B} \to \mathbb{Z}^{k-1}$. Naturally, as a consequence, each input x of the k-RP problem can be transformed to an input x_g of g by separating the input using blocks of size B.
- Each block in $I^1 \cup I^2$ built above can be treated to be a block of size 1 for the function g.
- For each block $I_i \in I^1$, let the exchanging value Q be $(N + 1, \ldots, N + B)$, and the input x'_g is obtained by replacing the values in I_i of x_g with Q. Now, let us consider the output of ϵ-approximation of g on x_g and x'_g. As shown in Fig. 1, the distance between last splitters of x_g and x'_g must be B, since all Q items must be behind all current items of x. According to the definition of ϵ-approximation, for x_g the output $A_{g,\epsilon}(x_g) \subseteq [s_{k-1} - \epsilon N/B, s_{k-1} + \epsilon N/B]$,

while for x'_g the output $A_{g,\epsilon}(x'_g) \subseteq [s'_{k-1} - \epsilon N/B, s'_{k-1} + \epsilon N/B]$. Obviously, $A_{g,\epsilon}(x_g) \cap A_{g,\epsilon}(x'_g) = \emptyset$. That is, I_i is a ϵ-sensitive block of g on x_g.
- For each block $I_i \in I^2$, let the exchanging value Q be $(-B, \ldots, -1)$, and the input x'_g is obtained by replacing the values in I_i of x_g with Q. Similarly, by checking the first splitter, we can obtain that I_i is also a ϵ-sensitive block of g on x_g.

Then, it can be known that the ϵ-block sensitivity of g $bs_\epsilon(g)$ is at least $|I^1 \cup I^2|$, that is N/B.

Finally, by Theorem 1, the expected sampling size needed by an external memory (ϵ, δ)-approximation algorithm for k-RP is at least $(1 - \delta)N/B$. □

3.3 Bounding the Sampling Size for Small Block Cases

Since for the case of large block sizes the sampling lower bound of partitioning algorithms has been as large as $\Omega(N/B)$, a natural question is whether we can achieve a better case by considering only the small block sizes.

In the following of this part, it is assumed that $\epsilon N/k = cB$ where $c \geq 1$ is a constant. Also, we need several important concepts from [2]. A function f is ϵ-symmetric if for each input x and a permutation π of x, $A_{f,\epsilon}(x) = A_{f,\epsilon}(\pi(x))$, that is the ϵ-approximation of function f is invariant under permutations of the input items.

To introduce the lower bound, we still need a probabilistic view of the input. Given an input list $L \in \mathbb{Z}^n$, it can induces a distribution P_L on $\mathbb{Z}$ defined by picking the elements of L uniformly. Since we only consider the non-duplicate cases in this part, the distribution can be obtained by setting $P_L(e) = \frac{1}{|L|}$ for $e \in L$ and $P_L(e) = 0$ otherwise. The *Hellinger distance* between two distributions P and Q can be defined by $h(P, Q) = \left(\frac{1}{2} \sum_{e \in \Phi} (\sqrt{P(e)} - \sqrt{Q(e)})^2\right)^{\frac{1}{2}}$.

Definition 9 (Minimum Hellinger Distance). *The ϵ-minimum Hellinger distance of a function f is $h_\epsilon(f) = \min\{h(P_x, P_y) \mid x, y \text{ are } \epsilon\text{-disjoint}\}$.* □

Theorem 3 ([2]). *For all $\epsilon \geq 0$, $0 \leq \delta < 1/4$, and every ϵ-symmetric function $f : \mathcal{X}^n \to \mathcal{Y}$ with worst case sampling size $S_{\epsilon,\delta}(f) \leq n/4$ and $h_\epsilon(f) \leq 1/2$, we have $S_{\epsilon,\delta}(f) \geq \frac{1}{4h_\epsilon^2(f)} \ln \frac{1}{4\delta + O(1/n)}$.* □

Theorem 4. *When the block size B satisfies $\epsilon N/k = cB$, where c is a constant larger than 1, the worst case sampling size needed by an external memory (ϵ, δ)-approximation algorithm for the k-RP problem is at least $(1 - \delta)N/B$.* □

Proof. First, we will show that the RP problem defines a *symmetric function* essentially, which will let the lower bounds obtained by Theorem 3 can be generalized to the general sampling based algorithms. Intuitively, a *symmetric* function will keep invariant under permutations of the inputs.

Again, let the function corresponding the k-RP problem be $f^k_{RP} : \mathbb{Z}^N \to \mathbb{Z}^{k-1}$. It can be observed that the output of the k-RP problem can always be computed by first sorting the N input items and then selecting the proper $k - 1$

splitters. Obviously, the above computing procedure is independent from the input order, and the function f_{RP}^k is *symmetric*.

Then, to use Theorem 3, we will build a pair of disjoint inputs $\mathbf{x}$ and $\mathbf{y}$.

- Let $\mathbf{x}$ be the list of $(1, 2, \ldots, N)$.
- Let $\mathbf{y}$ be the list obtained by removing $(1, 2, \ldots, 2\epsilon N/k)$ and adding $(N + 1, N + 2, \ldots, N + 2\epsilon N/k)$.

Similar with what is done in Theorem 2, $\mathbf{x}$ and $\mathbf{y}$ can be proved to be ϵ-disjoint.

Again, we transform the function f into g, where the function g is defined by $g : \mathbb{Z}_B^{N/B} \rightarrow \mathbb{Z}^{k-1}$, and the inputs $\mathbf{x}$ and $\mathbf{y}$ can be transformed to $\mathbf{x}_g$ and $\mathbf{y}_g$ by using block size B. Obviously, under such a transformation, the obtained function g is still *symmetric*.

Considering the worst case sampling size $S_{\epsilon,\delta}(g)$, we can use the trivial sampling algorithm of f using at most $\frac{4k \ln(2N/\delta)}{\epsilon^2}$ I/O costs. Obviously, taking enough large N value, the cost will be smaller than $\frac{N}{4B}$.

Then, we can calculate the minimum Hellinger distance $h_\epsilon(g)$ as follows.

$$
\begin{aligned}
h_\epsilon^2(g) &\geq h(P_\mathbf{x}, P_\mathbf{y})^2 \\
&= \frac{1}{2} \sum_{e \in [1, N+2\epsilon N/k]^B} (\sqrt{P_\mathbf{x}(e)} - \sqrt{P_\mathbf{y}(e)})^2 \\
&= \frac{1}{2} \cdot \frac{4\epsilon N}{kB} (\sqrt{\frac{1}{N/B}} - \sqrt{0})^2 \\
&= \frac{2\epsilon}{k}
\end{aligned}
$$

Utilizing the Theorem 3, we can obtain a lower bound $\frac{k}{16\epsilon} \ln \frac{1}{4\delta + O(B/N)}$ of $S_{\epsilon,\delta}(g)$. Combining with $\epsilon N/k = cB$, we have $S_{\epsilon,\delta}(g) \geq \frac{N}{16cB} \ln \frac{1}{4\delta + O(B/N)}$. $\quad\square$

4 External Sublinear Range Partitioning Algorithm

Motivated by the results of sampling lower bounds, a special model for describing the input distributions is introduced first, then an efficient external sublinear range partitioning algorithm is designed.

4.1 The Model of Generating Hard Inputs

There are two extreme cases when considering the correlations between items in the list. As shown in [6], if all items are totally independent, an optimal algorithm with $O(Sample_{RAM}/B)$ I/O cost can be obtained trivially, and if all items are dependent a natural extension of RAM based algorithm may cause $O(Sample_{RAM})$ I/O cost in the worst case. Moreover, according to Theorem 2, it is impossible to design efficient external sublinear partitioning algorithms for the general setting.

However, if more common and practical correlations between items are considered, there are still some typical cases which may admit I/O efficient solutions. Intuitively, all hard results obtained above are based on exploiting the correlations between items in the list, that is the items in the input list depends on each other. There are two kinds of correlations, which are the one among blocks and within blocks respectively. The correlation between blocks can be resolved by randomly selecting blocks to access usually, and it is highly implied that the mainly hardness is oriented from the correlations within blocks.

Therefore, this part considers a special kind of inputs of the k-RP problem, which is also referred by [6]. They can be described by the *partial dependent model* which explains what the distribution of data in blocks is like and works as follows. Imagine that we have an ordered list $\hat{L}$ large enough, under the partial dependent model, to build the input list L of blocks, the data items in each block B_i are obtained by the followings.

(1) Assume a global parameter $0 \leq w \leq 1$ is known, which is a ratio explaining how to mix two kinds of data to build B_i.
(2) Randomly select a continuous part in $\hat{L}$ of size $(1 - w)B$, and add them to the block B_i.
(3) Randomly select wB data items from $\hat{L}$ in an independent way, and add them to B_i.

According to the construction steps introduced above, essentially, under the partial dependent model, within each block, there are wB data items highly correlated and the other data items are independent.

Algorithm PartitioningPDM (Partitioning for Partial Dependent Model)

Input: The input list L of blocks $B_1, \ldots, B_{N/B}$ and an integer k.
Output: A splitter set $S = \{s_1, \ldots, s_{k-1}\}$.

1. Let (ϵ, δ) be the approximation quality parameters;
2. Compute the total sample size $r = \frac{4k \ln(2N/\delta)}{\epsilon^2}$;
3. $R = \emptyset$;
4. **while** $r > 0$ **do**
5.   Pick two random blocks B' and B'' without replacement;
6.   Let $T = B' \cup B''$;
7.   Sort T and obtain a list L_T;
8.   **for** each *continuous* sublist ΔL in L_T s.t. $|\Delta L| > \alpha$ **do**
9.   Remove ΔL;
10.   Insert all data items in L_T to R;
11. $r = r - |L_T|$;
12. Sort R;
13. Select $k - 1$ splitters of R into S, s.t. each partition size is at most $|R|/k$;
14. **return** S;

Fig. 2. The PartitioningPDM Algorithm.

4.2   The PartitioningPDM Algorithm

Assuming the inputs are generated by the partial dependent model, the PartitioningPDM algorithm is designed. As shown in Fig. 2, the main idea is to select two random blocks and eliminate the highly dependent parts by checking the relations between the two blocks. Firstly, the approximation parameters are collected and the general sampling size for approximating the range partition problem is calculated (line 1–2). Then, a sample set of enough large size is built by iteratively selecting pairs of blocks and eliminating the useless parts (line 4–10). Let α be a constant smaller than $(1-w)B$. After each pair of blocks are selected, the related items are sorted, and the *continuous* sublists of size at least α in L_T are removed (line 7–9). Here, a continuous sublist is composed of data items all from B' or B''. Finally, when enough samples are collected, they are sorted and $k-1$ splitters for balance partitioning them are extracted and returned. (line 12–14).

Theorem 5. *If $\alpha < (1-w)B$ and $w = \Omega(\sqrt{\frac{1}{B}\log\frac{1}{B}})$, PartitioningPDM is an (ϵ, δ)-approximation algorithm of the k-RP problem with high probability, and the I/O cost can be bounded by $O(\frac{k\log(N/\delta)}{wB\epsilon^2})$ with high probability.* $\square$

Proof. Intuitively, all we need to do is to show that the continuous part built by the step (2) in the partial dependent model will be removed and all other parts will be kept in a high probability.

First, we will show that the probability that the continuous sublist of size larger than α is built by the random part of the partial dependent model in very a low probability. Let L' be the list obtained by sorting all items in B', suppose $a_1, \ldots, a_\alpha$ is a sublist in L' built by the random selections. Then, consider the whole input list L and let $p_{a1}, \ldots, p_{a\alpha}$ be the corresponding positions of all a_i in L. The process of selecting wB random samples is approximately equivalent to the process of selecting each item independently with probability $\frac{wB}{N}$. We will show the result by proving that for a length $l = c\frac{\alpha N}{wB}$ $(c < 1)$ the probability that it is produced by $a_1, \ldots, a_\alpha$ is very low. Let X_i $(1 \leq i \leq l)$ be the random variable identifying whether the ith item behind p_{a1} is selected, and let $X = \sum X_i$. Obviously, we have

$$\mathbf{E}[X] = l \cdot \mathbf{E}[X_i] = l \cdot \frac{wB}{N} = c\alpha.$$

Then, the event we focus on is equivalent to the fact that $X \geq \alpha$.

$$\mathbf{Pr}[X \geq \alpha] = \mathbf{Pr}[X \geq \mathbf{E}[X]\frac{\alpha}{c\alpha}] \leq e^{-\frac{(1/c-1)^2}{3}\mathbf{E}[X]} \leq e^{-\frac{c(1/c-1)^2\alpha}{3}}$$

Assume a small probability like $p = \frac{c'}{B}$, if we require $\mathbf{Pr}[X \geq \alpha] \leq p$, we only need to choose $\alpha \geq \frac{3\ln(c'/B)}{c(1/c-1)^2}$. Since there are at most B different continuous sublist in each block, the probability that there is a continuous sublist of size larger than α is built by the random part of the partial dependent model is at most c'.

Obviously, since $\alpha < (1-w)B$, we only need to show that the parts generated by random selection will be kept in high probability. Considering the process of merging B' and B'', it is equivalent to randomly insert every random generated item of B'' into B'. Then, for a special continuous sublist ΔL, it is removed is equivalent with the fact that there are no items in B'' are inserted into ΔL.

$$\mathbf{Pr}[\Delta L \text{ is removed.}] \leq (1 - \frac{l}{N})^{wB} \leq (\frac{1}{e})^{wB\frac{N}{l}} = e^{-w^2 B \frac{B}{c\alpha}} \leq e^{-w^2 B}$$

Requiring $\mathbf{Pr}[\Delta L \text{ is removed.}] \leq q$, where $q = \frac{c''}{B}$, we only need $w > \sqrt{\frac{\ln(c''/B)}{B}}$.

Based on the above two facts, all samples collected by the PartitioningPDM algorithm can be guaranteed to be selected randomly, the obtained algorithm is an (ϵ, δ)-approximation obviously.   □

5   Related Works

Range partitioning is one of the most commonly used partitioning methods in massive data computing [14], and has increased lots of research interests which focus on how to optimize the practical application goal by using range partitioning [4,8,11,13,15]. The most relevant work is studied in the name of *approximation histogram* [6], which essentially studies both the record level and page level sampling methods, and [5] exploits the idea of adaptive block sampling further. However, neither of them considers the problem focusing by us. Usually, the duplicate items in the input will let the partitioning problem much harder, therefore, [16] and [18] further study the more general range partitioning problem, and give a similar sampling size bound $O(\frac{1}{(\epsilon-\phi)^2} \cdot \log \frac{Nk}{\delta})$ where ϕ is the upper bound of frequency of a data item. Another series of related research works is sampling based sort algorithm, such as [3,7,9] and so on. Actually, the approximation requirements are usually more relax than the one used by the range partitioning problem. A typical sampling size is determined to be larger than $k \ln(Nk)$, which will guarantee that each range obtained is of size $O(N/k)$. The range partitioning problem is also related to the problem of computing quantiles. Most of these works focus on the data stream model [10,12], which needs to make at least a full scan. Comparing with them, our work focuses on solving the problem by accessing data as little as possible. Our work is also motivated by the work of external sampling [1], which studies the possibility of efficient external sampling. The essential idea of this paper can be treated as investigating whether the sampling based range partitioning algorithms can be adapted to the external memory model.

6   Conclusion

In this paper, the range partitioning problem is studied in the external memory model, and sampling based sublinear partitioning algorithms are our interests. For the general case, two lower bounds of the sampling sizes are proved, which shows that in the worst case we need to scan the whole input at least once. For the case that data is partially dependent within each block, a nearly optimal external range partitioning algorithm is designed.

Acknowledgement. This work was supported in part by the Key Program of the National Natural Science Foundation of China under grant No. 61832003, the Major Program of the National Natural Science Foundation of China under grant No. U1811461, the General Program of the National Natural Science Foundation of China under grant No. 61772157, and the China Postdoctoral Science Foundation under grant 2016M590284.

References

1. Andoni, A., Indyk, P., Onak, K., Rubinfeld, R.: External sampling. In: ICALP 2009, pp. 83–94 (2009)
2. Bar-Yossef, Z., Kumar, R., Sivakumar, D.: Sampling algorithms: lower bounds and applications. In: STOC 2001, pp. 266–275 (2001)
3. Blelloch, G.E., Leiserson, C.E., Maggs, B.M., Plaxton, C.G., Smith, S.J., Zagha, M.: A comparison of sorting algorithms for the connection machine CM-2. In: SPAA 1991, pp. 3–16 (1991)
4. Cai, Z., Miao, D., Li, Y.: Deletion propagation for multiple key preserving conjunctive queries: approximations and complexity. In: ICDE 2019, pp. 506–517 (2019)
5. Chaudhuri, S., Das, G., Srivastava, U.: Effective use of block-level sampling in statistics estimation. In: SIGMOD 2004, pp. 287–298 (2004)
6. Chaudhuri, S., Motwani, R., Narasayya, V.R.: Random sampling for histogram construction: how much is enough? In: SIGMOD 1998, pp. 436–447 (1998)
7. DeWitt, D.J., Naughton, J.F., Schneider, D.A.: Parallel sorting on a shared-nothing architecture using probabilistic splitting. In: PDIS 1991, pp. 280–291 (1991)
8. Du, J., Miller, R.J., Glavic, B., Tan, W.: Deepsea: progressive workload-aware partitioning of materialized views in scalable data analytics. In: EDBT 2017, pp. 198–209 (2017)
9. Frazer, W.D., McKellar, A.C.: Samplesort: a sampling approach to minimal storage tree sorting. J. ACM **17**(3), 496–507 (1970)
10. Greenwald, M., Khanna, S.: Space-efficient online computation of quantile summaries. In: SIGMOD 2001, pp. 58–66 (2001)
11. Liu, X., Cai, Z., Miao, D., Li, J.: Tree size reduction with keeping distinguishability. Theor. Comput. Sci. **749**, 26–35 (2018)
12. Manku, G.S., Rajagopalan, S., Lindsay, B.G.: Approximate medians and other quantiles in one pass and with limited memory. In: SIGMOD 1998, pp. 426–435 (1998)
13. Miao, D., Cai, Z., Li, J.: On the complexity of bounded view propagation for conjunctive queries. IEEE Trans. Knowl. Data Eng. **30**(1), 115–127 (2018)
14. Özsu, M.T., Valduriez, P.: Principles of Distributed Database Systems, 3rd edn. Springer, Heidelberg (2011). https://doi.org/10.1007/978-3-030-26253-2
15. Pavlo, A., Curino, C., Zdonik, S.B.: Skew-aware automatic database partitioning in shared-nothing, parallel OLTP systems. In: SIGMOD 2012, pp. 61–72 (2012)
16. Vasudevan, D., Vojnovic, M.: Random sampling for data intensive computations. Technical Report MSR-TR-2009-08, Microsoft, April 2010

17. Vitter, J.S.: External memory algorithms and data structures. ACM Comput. Surv. **33**(2), 209–271 (2001)
18. Vojnovic, M., Xu, F., Zhou, J.: Sampling based range partition methods for big data analytics. Technical report, Microsoft (2012)

New Results on the Complexity
of Deletion Propagation

Dongjing Miao[1], Jianzhong Li[1], and Zhipeng Cai[2]([✉])

[1] Faculty of Computing, Harbin Institute of Technology, Harbin, China
{miaodongjing,lijzh}@hit.edu.cn
[2] College of Arts and Sciences, Georgia State University, Atlanta, USA
zcai@gsu.edu

Abstract. The problem of deletion propagation in relational database has been studied in database community for decades, where tuples are deleted from the source database in order to realize a desired removal of tuples from the result of a certain query. The deletion may result in unexpected view and source side effects. To minimize the side effects, we study two problems: MaxDP which is to seek a deletion of source tuples that maximizes query result remaining after deleting some view tuples, and MinSD which is to seek a minimum set of source tuples that should be deleted. Two problems have been proved that they are polynomially tractable for conjunctive queries without existential variables (∀-CQs). However, for ∀-CQs, the complexity of MaxDP is still unknown for deletion forbidden restriction, and so does MinSD in the presence of inclusion dependencies. In this paper, new complexity results are obtained on both problems for ∀-CQs. MaxDP is turned out to be not only NP-complete, but also NP-hard to approximate within $O(n^{1/5-\epsilon})$ for any constant $\epsilon > 0$ when the deletion of some tuples is forbidden. We then show that even for linear queries, MinSD is no longer polynomially tractable in the presence of inclusion dependencies. The results shows that the complexity of deletion propagation is very sensitive to the presence of some simple constraints.

Keywords: Deletion propagation · Database · Complexity

1 Introduction

Deletion propagation is a class of view update problem raised in database area. Unfettered access to the database is sometimes not allowed in practice. In this case, database accesses are typically performed through the views [1,2]: a deletion on the view should be translated into a deletion on the source database. As a result, deletion propagation became the most essential problem that how to do the translation *properly*.

The reason why deletion propagation becomes non-trivial is the ambiguity caused by underspecification. A deletion on the view can be realized by possibly

© Springer Nature Switzerland AG 2020
Z. Zhang et al. (Eds.): AAIM 2020, LNCS 12290, pp. 336–345, 2020.
https://doi.org/10.1007/978-3-030-57602-8_30

very different deletions on the source database. Therefore, a great deal of research has been devoted to disambiguate the translation for updates through views [1, 2,5]. However, in some scenarios, disambiguate the translation is not necessary. Instead, the goal is to translate the deletion with as little as possible *side effect* [4, 7,8,10–12,15,16]. As we mentioned, a deletion on source database may result in the *side effect* which is the deletion of additional tuples from the source database or the view, besides the intentionally deleted tuples. Intuitively, an appropriate translation (deletion on source data) should be those with minimum the side effect on the source data or the view. Therefore, two problems are formally defined and widely studied for deletion propagation: MINSD (minimizing source deletion) and MAXDP (maximizing deletion propagation).

Firstly, the problem MINSD is to minimize the number of tuples deleted from the source data, which can be formulated as follows,

Minimizing Source Deletion MINSD($\mathcal{S}, Q$)	
Fixed	Database schema $\mathcal{S}$ and query Q.
Input	A source database I, the view $Q(I)$, an integer $k > 0$.
Output	A subset $J \subseteq I$ such that $Q(J) = \emptyset$ and $|I \setminus J| \leq k$.

Secondly, the problem MAXDP is to maximize the number of tuples that remain in the view after removing intentionally deleted tuples, which can be stated as follows,

Maximizing Deletion Propagation MAXDP($\mathcal{S}, Q$)	
Fixed	Database schema $\mathcal{S}$ and query Q.
Input	A source database I, the view $Q(I)$, view deletion $X \subseteq Q(I)$, an integer $k > 0$.
Output	A subset $J \subseteq I$ such that $X \subseteq Q(I) \setminus Q(J)$ and $|Q(J)| \geq k$.

MINSD is shown to be NP-hard for a large fragment of CQs, only polynomial solvable for those very restricted classes of CQs [4,6–8]. Especially, Freire *et al.* [8] identified the triad structure of CQs, and stated that MINSD is polynomial solvable for *linear conjunctive queries* which are CQs excluding triads. Moreover, even in the presence of database constraints, such as functional dependencies, *linear conjunctive queries* are still not able to make MINSD polynomial intractability.

Similarly, existing results on MAXDP also imply that it is polynomial solvable only for a very restricted class of CQs [12]. Kimelfeld *et al.* [12] introduced the *level-k head-domination* (hd) structure, and show that MAXDP is polynomial tractable if inputting *level-1 hd*-CQs, but NP-hard for other $k > 1$.

However, the complexities of MINSD and MAXDP are still open when incorporating integrity constraints and forbidding certain source deletions. Such two questions were asked in [11] and have not been answered yet. They are motivated by some requirements of modern database applications, like data cleaning [17] and provenance [3,5]. View deletions are always specified prior to source data deletions and even to guide how to clean the source data. In such scenarios, some tables in the input source database are correct, hence, any deletion on them are not allowed. On the other hand, the updated database should satisfy a set of given integrity constraints. Therefore, we revisit the two problems in this paper.

We try to answer the two questions in terms of inclusion dependencies and tuple-level deletion forbidden. In practice, inclusion dependencies are widely-used. Consider in the case of inclusion dependencies, like foreign keys, deletion of tuples may incur the consequent deletion of other tuples in the related table, we need to investigate if it will increase the hardness of deletion propagation. Moreover, some source deletions are not permitted, that is, not every tuple in the given source database is a candidate of source tuple deletion. The deletion of certain facts is not allowed, like data from the English or country-name dictionary.

Observe that, in the presence of integrity constraints and forbidden deletions, MINSD and MAXDP will definitely turn to be NP-hard when inputting linear CQs or level-k hd-CQs. Therefore, to help delineating the boundary between intractable and tractable cases, we focus on $\forall$-CQs without existential variables which a subclass of both linear CQs and level-k hd-CQs. In next section, we formally define $\forall$-CQs and study the complexity of MINSD and MAXDP for $\forall$-CQs when the database is restricted with inclusion dependencies and forbidden deletions.

2   Formal Settings

We introduce notations used in this section. Throughout the paper, we use the datalog-style notations for simplicity.

Database Schemas, Instances. We fix an infinite set Const of constants like a, b and c. A database schema $\mathcal{S}$ is a finite sequence $(R_1, \ldots, R_d)$ of distinct relation symbols, where each R_i has an arity $r_i > 0$. A database instance I (over $\mathcal{S}$) is a sequence $(R_1^I, \ldots, R_d^I)$, such that each R_i^I is a relation of arity r_i over Const. For simplicity, we use R_i to denote both the relation symbol and the relation R_i^I that interprets it. $R_i(\mathbf{a})$ is said to be a fact if $\mathbf{a} \in \mathsf{Const}^{r_i}$. An instance is a set of its facts, thus $R(\mathbf{a}) \in I$ for every fact $R(\mathbf{a})$ of I. We say that J is a sub-instance of I over $\mathcal{S}$, denoted $J \subseteq I$, if $R_i^J \subseteq R_i^I$ for all $i \in [d]$.

CQs, sjf-$\forall$-CQs, Linear Queries. Let Var be a set of variables which ar disjoint with Const, and variables in it are also denoted as lower-case letters like x, y and z. A conjunctive query (CQ) over a schema $\mathcal{S}$ is written as the form of $Q(\mathbf{y}) :\!- \varphi(\mathbf{x}, \mathbf{y}, \mathbf{a})$ where $\mathbf{x}$ and $\mathbf{y}$ are disjoint tuples of variables from Var, $\mathbf{a}$ is a tuple of constants from Const, and $\varphi(\mathbf{x}, \mathbf{y}, \mathbf{a})$ is a conjunction of atoms $\varphi_i(\mathbf{x}, \mathbf{y}, \mathbf{a})$, e.g., $Q(y) :\!- R_1(x, y, a) \wedge R_2(y, z)$. In the datalog style, we use ',' instead of '$\wedge$', so that it can be rewritten as $Q(y) :\!- R_1(x, y, a), R_2(y, z)$. Here, the LHS of a query is called *head*, and the RHS of a query is called *body*. For example, $Q(y)$ is the head of the query shown above, and $R_1(x, y, a), R_2(y, z)$ is the body.

A conjunctive query Q is said to be *self join free* if any relation symbol occurs at most once. Any variable not contained in the head of a query is called an existential variable of the query. Since all the existential variables are in the tuple $\mathbf{x}$, any conjunctive query Q without existential variables can be written in the form of

$$Q(\mathbf{y}) :- R_1(\mathbf{y}_1, \mathbf{a}_1), \ldots, R_m(\mathbf{y}_m, \mathbf{a}_m).$$

If the tuple $\mathbf{y}$ is exactly $(\mathbf{y}_1, \ldots, \mathbf{y}_m)$, then Q is called $\forall$-CQ, that is, it can be written in the form of $Q(\mathbf{y}) :- \varphi(\mathbf{y}, \mathbf{a})$. For example,

$$Q^*(y_1, y_1, y_2, y_3) :- R_1(y_1, a), R_2(y_1, y_2, b), R_3(y_3, c)$$

is an sjf-$\forall$-CQ, since all the variables in the body are contained in the head. For convenience, we ignore the variables in the head of any sjf-$\forall$-CQ. For example, Q^* could be written as

$$Q^* :- R_1(y_1, a), R_2(y_1, y_2, b), R_3(y_3, c).$$

An sjf-$\forall$-CQ Q is *linear* if its atoms may be arranged in a *linear* order such that each variable occurs in a contiguous sequence of atoms. Intuitively, let each variable be vertex, and each atom be a hyperedge, then a query can be drawn as a hypergraph which is called the *dual hypergraph* of the query. Then, geometrically, a query is linear if all the vertices of its *dual hypergraph* can be drawn along a straight line, and all the hyperedges can be drawn as convex regions. For example, the following query is linear: $Q :- R(w, x), S(x, y), T(y, z)$, since its dual hypergraph can be drawn along a line $w, x \rightarrow x, y \rightarrow y, z$.

Views. To formally define views, we begin with an assignment of a (conjunctive) query. An assignment μ for a query Q is a mapping from $\mathsf{Var}(Q)$ to Const, such that $\mu(\mathbf{y})$ is a tuple created by substituting each variable y in $\mathbf{y}$ with the constant $\mu(y)$. An assignment μ for an atom R of Q, $\mu(R_i)$ is a *fact* obtained by substituting every variable y with the constant $\mu(y)$.

Given a database instance I over schema $\mathcal{S}$, a match for Q in I is an assignment μ for Q where $\mu(R)$ is a fact from I for each atoms R of Q. If μ is a match for Q of I, then $\mu(y)$ is called an answer for Q. The set of all the answers for Q in I is called query result or view $Q(I)$.

Inclusion Dependencies. An inclusion dependency (IND) over a schema $\mathcal{S}$ is an expression in the form of $\phi : R.\mathbf{x} \subseteq S.\mathbf{y}$ where R and S are relations in $\mathcal{S}$, while $\mathbf{x}$ consists of some variables in R and $\mathbf{y}$ consists of some variables in S respectively. A database instance I is said to satisfy the IND ϕ if $Q_1(I) \subseteq Q_2(I)$ such that $Q_1(\mathbf{x}) :- R(\mathbf{x})$ and $Q_2(\mathbf{y}) :- S(\mathbf{y})$.

Deletion Forbidden. We follow the way expressing deletion forbidden used in modelling query answer causality [14]. A relation R is said to be *exogenous* if the deletion of any fact in R is not allowed. We use the footnote '$\times$' to label any exogenous relation, e.g., $R^\times$. Otherwise, the relation R is said to be *endogenous*.

Next, we investigate MINSD and MAXDP for linear queries in the presence of inclusion dependencies and deletion forbidden.

3   Complexity Results

We found that MINSD become NP-hard for linear queries if incorporating only one simple inclusion dependence, foreign key constraint in fact. Concretely, in the presence of inclusion dependencies, any feasible solution of MINSD should satisfy all the given INDs. Our next theorem shows this requirement increases the complexity of MINSD even for linear queries.

Theorem 1 (inclusion dependency makes linear query hard). *Let ϕ be an inclusion dependency defined on I, MINSD is NP-complete even if query Q is a linear query with respect to ϕ.*

Proof. We build a polynomial reduction from the hitting-set problem. Let $\mathcal{H} = \{S_1, S_2, \ldots, S_n\}$ be the given sets and $U = \bigcup_{1 \leq i \leq n} S_i = \{e_1, e_2, \ldots, e_m\}$ be the union of all the given sets. Given a number $k > 0$, the decision problem of the hitting set problem states whether there exists a set $U' \subseteq U$ with k elements such that at least one element of each set S_i is contained in U'. We denote an instance of hitting-set problem as $\langle \mathcal{H}, U \rangle$. We build the corresponding instance $\langle \mathcal{S}, I, Q, \phi, k' \rangle$ of MINSD problem where Q is linear query from $\langle \mathcal{H}, U \rangle$ of hitting set problem.

Schema and Database Instance. The schema $\mathcal{S}$ can be fixed with three binary relations in it: A is exogenous that is deletion forbidden while B and C are endogenous. Then we build the database instance I in the following way. For each relation, we add facts as follows,

- For each set S_i, add a tuple $(\mathcal{H}, S_i)$ into A,
- For each set S_i and every $e_j \in S_i$, add a tuple (S_i, e_j) into B,
- For each element $e_j \in U$, add nm tuples (e_j, j_l) into C where $l = 1, \ldots, nm$.

It is clear that I can be made in polynomial time of $O(nm^2)$ since there are exactly n facts in A, at most nm facts in B, and exactly nm^2 facts in C.

Inclusion Dependency. An inclusion dependency $\phi :\text{-} B.x \subseteq A.y$ where $B.x$ is the first attribute of B and $A.y$ is the second attribute of A.

Linear Query. Let Q is a linear query defined as the following form

$$Q :\text{-} A^{\times}(w, x), B(x, y), C(y, z)$$

It is easy to verify that the result of boolean query $Q(I)$ is true, *i.e.*, $I \models Q$.

Size of Source Side Effect. Let the size of $|X|$ be at most

$$k' = knm + r$$

where $r = \sum_{1 \leq i \leq n}(|s_i| - 1)$.

One can simply verify the correctness of proof. $\square$

Due to the inclusion dependency, a feasible deletion from the source data is restricted such that, after deleting some source tuples, the active domain of the endogenous variable is required to be a subset of the exogenous variable which it depends on. This motivates a concern on the relationship of variable domination, we formalize it as follows.

Active Domination $\leq_a$. Given two atoms $A^\times$ and B, if $A^\times.x \subseteq B.y$, we say that x actively dominates y, *i.e.*, $x \leq_a y$.

Theorem 1 shows the hardness is caused by the existence of active domination, that is, there exists an endogenous atom actively dominated by an exogenous atom.

Let A and B be two endogenous relations such that $A.x \leq_a B.y$ and $B.y \leq_a A.x$, then they are said to be an inclusion dependency cycle. We next show that, even without active domination, the problem MINSD is still NP-hard if there exists inclusion dependency cycle in linear queries.

Theorem 2. *Inclusion dependency cycles make the problem* MINSD *hard even for linear queries.*

Proof. Consider a linear query Q :- $V(x), E(x,y), U(y)$ and two inclusion dependencies $V.x \subseteq U.y$ and $U.y \subseteq V.x$. We build a reduction from vertex cover problem in which input a graph $G(V, E)$, it is to decide if there is a vertex set C such that $|C| \leq k$ and for every $(u, v) \in E$, either u or v is in C. Given a graph $G(V, E)$, we do the following two steps:

(1) for each vertex v_i, insert two unary tuples $V(i)$ and $U(i)$,
(2) for each edge (v_i, v_j), insert two tuples $E(i, j)$ and $E(j, i)$.

The reduction can be done in polynomial time. And we claim that there is a vertex cover of size at most k if and only if there is a solution of size at most $2k$ for MINSD problem.

$\Rightarrow$ Let C be a vertex cover of at most size k, for each vertex $v_i \in C$, we delete $V(i)$ and $U(i)$, then query on the data left will be evaluated as *false*, note that there are at most $2k$ deletions.

$\Leftarrow$ Let S be a solution of at most size $2k$ The inclusion dependency cycle requires that, if there is a tuple $V(i)$ in S, then there should be a tuple $U(i)$ in it. Therefore, if $V(i)$ in S, then add v_i to C. For tuple $E(i, j) \in S$, we delete $V(i)$ and $U(i)$, then query on the data left will be evaluated as *false*, note that there are at most $2k$ tuple deletions.   □

Therefore, the polynomial tractable condition of the problem MINSD should be very restricted when incorporating database constraints no matter the existence of deletion forbidden.

Our second major result is for the lower bound of the problem MAXDP. The problem MAXDP is extremely hard in terms of combined complexity. In fact, the hardness is mainly caused by the unbounded size of queries. To be more practical, we focus on the data complexity of the problem MAXDP instead of the combined complexity. The analysis to the complexity of MAXDP gives a way to study the problem where the size of the query is bounded by a certain constant. We next investigate the data complexity of this problem. We show that this problem does not admit constant approximation ratio even for a very simple setting.

Theorem 3. MAXDP *cannot be approximated within* $O(n^{1/5-\epsilon})$ *even for* $\forall$-*CQs and any constant* $\epsilon > 0$, *unless* $P = ZPP$.

We here show the gap-reserving reduction, and then finish the proof of this theorem. Given a graph $G(V, E)$ having n vertices and m edges, w.l.o.g., suppose there is no isolated vertex in G, the maximum independent set problem is to find a set of vertices in G of the largest possible size such that no two of which are adjacent.

Schema $\mathcal{S}$ and Instance I. Six relations R_1, S_1, T, S_1, R_2 are fixed, and the corresponding instance I is constructed as follows.

(a) For each edge $(i, j) \in E$, add tuple (i, j) into R_1 and R_2;
(b) For each vertex $i \in V$, add tuples $(i, e_1), \ldots, (i, e_l)$ into R_1 and R_2, where l is an integer depending on the size of G;
(c) For each vertex $i \in V$, add tuples $(i, =)$ and $(i, \neq)$ into S_1 and S_2;
(e) For each pair of vertices $i, j \in V$, add tuple $(i, =, i)$ into T if $i = j$, otherwise, add $(i, \neq, j)$ into T;

Query. Define the fixed $\forall$-CQ as the following query

$$Q :- R_1(x, z_1), S_1(x, w), T(x, w, y), S_2(y, w), R_2(y, z_2)$$

Views. Let the view set $\mathcal{V}$ be $\{V_1, V_2\}$, such that $V_1 = Q_1(I)$ and $V_2 = Q_2(I)$. The sizes of V_1 and V_2 are both polynomial of $|I|$.

View Deletion . Let the view deletion X consists of only deletions on V_1, which is defined as

$$X = \{(i, \neq, j, \neq), \forall (i, j) \in E\}$$

The following example shows the result built for an instant $G(V, E)$ of independent set problem, The inapproximability could be derived by the following claims, then the theorem follows. Observe that, any deletion not on $S_1(x, \neq)$ and $S_2(x, \neq)$ is unnecessary due to the monotonicity of conjunctive queries, thus implying the following claim.

Claim. Any minimal solution should be a subset of $\{S_1(x, \neq)\} \cup \{S_2(x, \neq)\}$.

In the following content, we only consider minimal solutions, which are subsets of $S_1(x, \neq) \cup S_2(x, \neq)$. In any minimal solution ΔI, let S_1' and S_2' be the resultants after a deletion ΔI applied to S_1 and S_2, then the intersection $S_1' \cap S_2'$ contains tuples of the forms $(i, =)$ and $(i, \neq)$. Let $S = \{(i, =), \forall i \in V\}$, assume P is the set of tuples of the form $(i, \neq)$, i.e.,

$$P = (S_1' \cap S_2') \setminus S,$$

then the following lemmas follows.

Lemma 1. $|P| \geq k$ *if and only if G has an independent set α with size k.*

Proof. Suppose G has an independent set α with size k. In I, we remove the tuples $(i, \neq)$ from S_1 and S_2 where $i \notin \alpha$, then we claim all results in X are removed. Otherwise, if some tuple (i, j) in R_1 is still joined with tuple (j, i) in R_2, R_1 and R_2 must be joined by $(i, \neq)$ and $(j, \neq)$. According to the construction of $I \setminus \Delta I$, i and j are incident to each other. This is a contradiction.

If $|P| \geq k$, the claim stated above implies the incident witnesses of the corresponding k vertices do not intersect in G, which implies these vertices are independent in G, otherwise there are still some results of X left, *i.e.*, $Q_1(I \setminus \Delta I) \cap X \neq \emptyset$.

Lemma 2. *If G must has an independent set α with size k, there is a solution for the database instance I with size $\Omega(k^2 l^2)$.*

Proof. By Lemma 1, we have $|P| \geq k$. Therefore, there are at least k vertices such that any two of them, say $i, j \in V$, are not incident with each other, *i.e.*, $(i, j) \notin E$, thus S_1' contains tuples $(i, \neq)$ and $(j, \neq)$, and so does S_2'. Then, for every such pair $(i, \neq) \in S_1'$ and $(j, \neq) \in S_2'$, there are at least $l + 1$ tuples in R_1 need to joined with it, and with at least another $l + 1$ tuples in R_2. The number of results survived is at least $\binom{k(l+1)}{2}$, which is $\Omega(k^2 l^2)$.

Lemma 3. *If there is a solution ΔI for database instance I with a size of $\Omega((n^2 + k \cdot n^{\frac{3}{2}})^2)$, G must has an independent set α with size at least k.*

Proof. In the resultant instance $I \setminus \Delta I$ after applying the deletion, R_1 (also, for R_2) can be partitioned into group as follows,

P_0 the set of tuples in the form of (i, z) such that $i \in V$ and $(i, \neq)$ is preserved in S_1'

P_i the set of tuples in the form of (i, z) such that $i \in V$ and $(i, \neq)$ is removed from S_1'

then let c_i denote the size of P_i, we have

$$c_i \leq \begin{cases} m + kl, & i = 0; \\ n - 1 + l, & i = 1, \dots, N. \end{cases}$$

since there is no loss in R_1, then

$$\sum_{i=0}^{n} c_i \leq 2m + nl$$

let $\mathcal{N}$ denote the size of resultant view $Q(I \setminus \Delta I)$, since each P_i contributes c_i^2 to $\mathcal{N}$, then we have

$$\mathcal{N} \leq \sum_{i=0}^{n} c_i^2$$
$$\leq \sum_{i=0}^{n} c_i \cdot (n - 1 + l) + c_0^2 - c_0 \cdot (n - 1 + l)$$
$$\leq (2m + nl)(n - 1 + l) + c_0^2 - c_0 \cdot (n - 1 + l)$$
$$\leq (2m + nl)(n - 1 + l) + (m + kl)^2 - (m + kl) \cdot (n - 1 + l)$$

Since $m \sim O(n^2)$ and $k \sim O(n)$, let $l = n^{\frac{3}{2}}$, then we could know that

$$\mathcal{N} \sim O\left(n^{\frac{5}{2}} \cdot n^{\frac{3}{2}} + (n^2 + k \cdot n^{\frac{3}{2}})^2 - (n^2 + k \cdot n^{\frac{3}{2}}) \cdot n^{\frac{3}{2}}\right)$$

i.e.,

$$\mathcal{N} \sim O\left((n^2 + k \cdot n^{\frac{3}{2}})^2\right)$$

So that c_0 dominates the value of query result preserved $Q(I \setminus \Delta I)$ for the database instance I. Therefore, if there is a solution preserving $\Omega((n^2 + k \cdot n^{\frac{3}{2}})^2)$ results, there should be an independent set of a size at least k in G.

Now we finish the proof of Theorem 3.

Proof. Let $\mathcal{N}^{opt}$ the size of result preserved by optimal data deletion, and $\mathcal{M}^{opt}$ the size of maximum independent set. Since there is no $n^{1-2\epsilon}$-approximation for the maximum independent set problem for any constant $\epsilon < 0.5$, unless $P=ZPP$, it can not be distinguished whether $\mathcal{M}^{opt} \geq n^{1-\epsilon}$ or $\mathcal{M}^{opt} \leq n^\epsilon$.

Assume $\mathcal{M}^{opt} = k^*$, then, (1) if $k^* \geq n^{1-\epsilon}$, by Lemma 2, $\mathcal{N}^{opt}$ is $\Omega(k^{*2}l^2)$, which is $\Omega(n^{5-2\epsilon})$ under setting of $l = n^{\frac{3}{2}}$; (2) if $k^* \leq n^\epsilon$, by Lemma 3, $\mathcal{N}^{opt}$ is $O((n^2 + k^* \cdot n^{\frac{3}{2}})^2)$ which is $O(n^4)$.

Therefore, it also can not be distinguished whether $\mathcal{N}^{opt}$ is $\Omega(n^{5-2\epsilon})$ or $O(n^4)$.

Suppose there is an $O(n^{\frac{1}{5}-\epsilon})$-approximation for our problem. Since the size of instance is $O(n^5)$, we have the following two claims,

(1) it is able to obtain a solution preserving $\Omega(n^{4+3\epsilon})$ results when $\mathcal{N}^{opt}$ is $\Omega(n^{5-2\epsilon})$;
(2) a trivial solution preserves $O(n^4)$ results when $\mathcal{N}^{opt}$ is $\Omega(n^4)$;

This means it is distinguished whether $\mathcal{N}^{opt}$ is $\Omega(n^{5-2\epsilon})$ or $O(n^4)$.
It contradicts, so that Theorem 3 follows immediately. □

4   Conclusions

New complexity results are obtained on problems MinSD and MaxDP for ∀-CQs in this paper. For linear queries, MinSD is no longer polynomially tractable in the presence of inclusion dependencies. MaxDP is NP-hard to be approximated within $O(n^{1/5-\epsilon})$ for any constant $\epsilon > 0$ in the presence of deletion forbidden. Our results shows that the complexity of deletion propagation is very sensitive to the presence of some simple integrity constraints or operation requirements.

Acknowledgement. This work is supported by the National Natural Science Foundation of China (NSFC) Grant NOs. 61972110, 61832003, U1811461, and the National Key R&D Program of China Grant NO. 2019YFB2101900.

References

1. Bancilhon, F., Spyratos, N.: Update semantics of relational views. ACM Trans. Database Syst. **6**, 557–575 (1981)
2. Bohannon, A., Pierce, B.C., Vaughan, J.A.: Relational lenses: a language for updatable views. In: Proceedings of 25th ACM Symposium Principles of Database Systems, pp. 338–347 (2006)
3. Buneman, P., Chapman, A., Cheney, J.: Provenance management in curated databases. In: Proceedings of the ACM International Conference Management of Data, pp. 539–550 (2006)
4. Buneman, P., Khanna, S., Tan, W.-C.: On propagation of deletions and annotations through views. In: Proceedings of 21th ACM Symposium Principles of Database Systems, pp. 150–158 (2002)
5. Cheney, J., Chiticariu, L., Tan, W.-C.: Provenance in databases: why, how, and where. Found. Trends databases **1**, 379–474 (2009)
6. Cong, G., Fan, W., Geerts, F.: Annotation propagation revisited for key preserving views. In: Proceedings of 15th International Conference Information and Knowledge Management, pp. 632–641 (2006)
7. Cong, G., Fan, W., Geerts, F., Li, J., Luo, J.: On the complexity of view update analysis and its application to annotation propagation. IEEE Trans. Knowl. Data Eng. **23**, 506–519 (2012)
8. Freire, C., Gatterbauer, W., Immerman, N., Meliou, A.: The complexity of resilience and responsibility for self-join-free conjunctive queries. Proc. VLDB Endow. **9**, 180–191 (2015)
9. Hastad, J.: Some optimal inapproximability results. J. ACM **48**, 798–859 (2001)
10. Kimelfeld, B.: A dichotomy in the complexity of deletion propagation with functional dependencies. In: Proceedings of 31st ACM Symposium Principles of Database Systems, pp. 191–202 (2012)
11. Kimelfeld, B., Vondrák, J., Williams, R.: Maximizing conjunctive views in deletion propagation. ACM Trans. Database Syst. **37**, 1–37 (2012)
12. Kimelfeld, B., Vondrák, J., Woodruff, D.P.: Multi-tuple deletion propagation: approximations and complexity. Proc. VLDB Endow. **6**, 1558–1569 (2013)
13. Meliou, A., Gatterbauer, W., Moore, K.F., Suciu, D.: The complexity of causality and responsibility for query answers and non-answers. Proc. VLDB Endow. **4**, 34–45 (2010)
14. Meliou, A., Roy, S., Suciu, D.: Causality and explanations in databases. Proc. VLDB Endow. **7**, 1715–1716 (2014)
15. Miao, D., Cai, Z., Li, J.: On the complexity of bounded view propagation for conjunctive queries. IEEE Trans. Knowl. Data Eng. **30**, 115–127 (2017)
16. Miao, D., Liu, X., Li, J.: On the complexity of sampling query feedback restricted database repair of functional dependency violations. Theor. Comput. Sci. **609**, 594–605 (2016)
17. Yakout, M., Elmagarmid, A.K., Nevillem, J., Ouzzani, M., Ilyas, I.F.: Guided data repair. Proc. VLDB Endow. **4**, 279–289 (2011)

Search Complexity: A Way for the Quantitative Analysis of the Search Space

Li Ning$^{(\boxtimes)}$ and Yong Zhang

Center for High Performance Computing,
Shenzhen Institutes of Advanced Technology, CAS, Shenzhen, China
{li.ning,zhangyong}@siat.ac.cn

Abstract. We considered the situations when search is performed in the structured spaces. In order to isolate the effort of reducing the uncertainty about the information of the space structure from the implementation cost of processing the acquired knowledge, we introduced the concept of *search complexity*, which describes the quantity order of the information that is necessary to discover the optimum. Furthermore, the proposed concept has been extended to the cases when an approximation to the optimum is acceptable, and we have investigated a situation, in which a solution with the multiplicative error $0 \leq \delta \leq 1$ and the additive error $\epsilon \geq 0$ can be found in a relatively low search complexity.

Keywords: Search complexity · Search space · Approximation

1 Introduction

Optimization is everywhere in engineering applications [7]. In some cases, the objective to optimize is given in the analytical expressions, and the attributes can be changed freely or under some constraints of specific forms, which makes it possible to derive the optimal objective value (and usually the corresponding configuration of attributes as well) without exhaustively exploring all the possibilities. However, there are still a lot of cases in which we only know some properties of the objective, not including the exact expressions. In such cases, the optimum can not be derived by directly solving the equations, and one have to make the sufficient use of the known properties, in order to avoid the unnecessary explorations.

It is common that the knowledge provides some clues, and before the optimum is correctly claimed, more than one configurations (of the attributes) should be checked and the uncertainty is reduced due to the checking results. Generally, such kind of searching processes alternates between checking if an optimum can be claimed based on the current knowledge and exploring to further reduce the uncertainty. While some searching strategies are guided with the solid theory about the space structures [1], the others are just heuristics [4], in which situation it is often difficult or impractical (if not impossible) to acquire enough

© Springer Nature Switzerland AG 2020
Z. Zhang et al. (Eds.): AAIM 2020, LNCS 12290, pp. 346–357, 2020.
https://doi.org/10.1007/978-3-030-57602-8_31

information to significantly reduce the amount of the exploring effort, and the optimal objective value can not be determined before checking every or almost every configurations.

In this work, to measure the computing cost that is necessary to discover the optimum, different from the classical complexity analysis (e.g. the time complexity and the space complexity) that considers all the operations during the searching process, we would like to focus on the structural properties of the search space, and isolate the effort of reducing the information uncertainty from the implementation cost. Inspired by the communication complexity [5] that quantitates the amount of necessary information to reveal some particular feature of a system, we introduce the concept of *search complexity*, in order to measure the quantity of the information (about the space structure) that is necessary to claim the optimal element. With this concept, it only focus on whether a piece of information is required, but not including the calculation cost to discover or to make use of such information. Thus the search complexity provides a lower bound for the time complexity of any search algorithm running with the same space. This fact will be claimed formally at the end of Sect. 3.

Although there are techniques (e.g. branch and bound) to avoid the unnecessary explorations in the exhaustive searching processes, e.g. tree-based depth-first search and breadth-first search [2], the time complexity cannot be reduced in general. Therefore for large search spaces, a compromise on the result quality is sometimes necessary to finish the searching process in the tolerable time. If the requirement to find the exact optimum is not strict, which means an error is acceptable to some degree, then the knowledge of the objective's properties could be used to search for a configuration close to the optimal one, which is known as an approximation [8].

To cover such cases, we extend the search complexity for deciding the exact optimum to the cases when an approximation is acceptable, and introduce the idea of (δ, ϵ)-*approximate search complexity*. In Theorem 7, we have shown a situation when a solution with the multiplicative error $0 \leq \delta \leq 1$ and the additive error $\epsilon \geq 0$ can be found in a (relatively) low search complexity.

Main Contributions. Considering the ubiquity of optimization problems, developing more efficient solving techniques is theoretically and practically meaningful. The proposed concept of *search complexity* provides another perspective to measure the effort devoting to exploration of the search space, and it gives some clues about what kind of property is useful to save the unnecessary searching effort, which is especially important when the calculation burden is heavy.

2   Preliminaries

In the arguments through this paper, we represent the set of numbers in blackboard-bold uppercase letters. In particular, $\mathbb{Z}$ denotes the set of all the integers and $\mathbb{R}$ denotes the set of all the reals. Attaching the superscripts $*$ and $+$ exclude the negative values and the non-positive values respectively. For example, $\mathbb{Z}^*$ is the set of all the non-negative integers, and $\mathbb{Z}^+$ is the set of all the positive integers.

Note that the notation $|\cdot|$ has two meanings in our arguments. For a real number, we use $|\cdot|$ to take the absolute value. That is, $|r| = r$ if $r \geq 0$ and $|r| = -r$ if $r < 0$. For a set S, we use $|\cdot|$ to take the size of S, i.e. the number of elements in S.

Recall that given a set S, the powerset of S is the set of all S's subsets, and it is denoted by 2^S.

3   Search Complexity

Definition 1 (Utility Oracle). *A utility oracle $\mathcal{U}$ is defined by a pair $\langle f_{\mathcal{U}}, \mathcal{C}_{\mathcal{U}} \rangle$, where*

- *the utility function $f_{\mathcal{U}}$ maps the elements from a universal set $\mathcal{N}$ to the nonnegative reals, i.e. $f_{\mathcal{U}} : \mathcal{N} \mapsto \mathbb{R}^*$;*
- *the utility context $\mathcal{C}_{\mathcal{U}}$ is a set of the finite subsets of $\mathcal{N}$, i.e. $\mathcal{C}_{\mathcal{U}} \subseteq 2^{\mathcal{N}}$ and $|S| < \infty$ for all $S \in \mathcal{C}_{\mathcal{U}}$.*

Definition 2 (Search Complexity). *Consider a utility oracle $\mathcal{U} = \langle f_{\mathcal{U}}, \mathcal{C}_{\mathcal{U}} \rangle$. Assume that for any set $S \in \mathcal{C}_{\mathcal{U}}$, calling $f_{\mathcal{U}}$ for $\mathcal{T}(|S|)$ times is sufficient and necessary to claim (correctly and deterministically) the element $s^* \in S$ of the maximum value, i.e. $f_{\mathcal{U}}(s^*) = \max_{s \in S} f_{\mathcal{U}}(s)$. Then $\mathcal{T}(\cdot)$ is called the search complexity of $\mathcal{U}$. As the exact form of $\mathcal{T}(\cdot)$ contains too much details in many cases, it is usually represented in the big-O notation.*

When we say "calling $f_{\mathcal{U}}$ for $\mathcal{T}(|S|)$" times, it is assumed that query to the oracle is the only way to get the utility value $f_{\mathcal{U}}(s)$ of any element $s \in S$, and for each element s, calling $f_{\mathcal{U}}$ for once is enough. However, in many cases, the expression of $f_{\mathcal{U}}$ is analytical or the utility value of some element s can be inducted based on the value (or values) of another element (or elements). In such cases, "calling $f_{\mathcal{U}}$ for element s" should be understood as "deriving the value $f_{\mathcal{U}}(s)$". Furthermore, for any element $s \in S$, there is no need to call $f_{\mathcal{U}}(s)$ for more than once during an independent single run of the searching process, as the value can be stored and accessed again after the first call of $f_{\mathcal{U}}(s)$.

Consider a simple example of utility oracle $\mathcal{U} = \langle f_{\mathcal{U}}, \mathcal{C}_{\mathcal{U}} \rangle$, in which $f_{\mathcal{U}}$ is defined on $\mathcal{N} = \mathbb{Z}$, and the context $\mathcal{C}_{\mathcal{U}}$ is defined as the set of the finite subsets of $\mathbb{R}$ such that the following conditions hold for S

- S always contains elements $\underline{s} \in \mathbb{R}$ and $\overline{s} \in \mathbb{R}$ with $\underline{s} < \overline{s}$;
- for every element $s \in S$, it holds that $\underline{s} < s < \overline{s}$ and $f_{\mathcal{U}}(s) < \max\{f_{\mathcal{U}}(\underline{s}), f_{\mathcal{U}}(\overline{s})\}$.

It is easy to check that for any set $S \in \mathcal{C}_{\mathcal{U}}$ of size k, the element of maximum value in S can be determined with the knowledge of the values $f_{\mathcal{U}}(\underline{s})$ and $f_{\mathcal{U}}(\underline{s})$. Thus the search complexity of $\mathcal{U}$ is $\mathcal{T}(k) = 2 = O(1)$. See Fig. 1 for a set $S \in \mathcal{C}_{\mathcal{U}}$ of size 10.

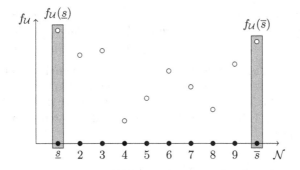

Fig. 1. A set $S \in \mathcal{C}_{\mathcal{U}}$ of size 10.

Although it is easy to give a search algorithm of time complexity $O(1)$ for this simple example, there are examples for which the search complexity is lower than the time complexity of any search algorithm. Consider a utility oracle $\mathcal{U} = \langle f_{\mathcal{U}}, \mathcal{C}_{\mathcal{U}} \rangle$, in which $f_{\mathcal{U}}$ is defined on a universal set of elements $\mathcal{N}$, and the context $\mathcal{C}_{\mathcal{U}}$ is defined as the set of the finite subsets of $\mathcal{N}$, such that in a set $S \in \mathcal{C}_{\mathcal{U}}$ of size k, there are at most $O(\log k)$ elements of positive values. Consequently, the search complexity of this utility oracle is $O(\log k)$. However, as there is no additional information about how $f_{\mathcal{U}}$ is defined, any search algorithm should call $f_{\mathcal{U}}(s)$ for every $s \in S$ to identify all the elements of positive values. Hence, the time complexity of the search algorithm is $\Omega(k)$. Note that the gap between the search complexity and the time complexity is not rare. Consider the longest increasing subsequence (LIS) problem for another example. In an instance of LIS problem, a sequence of unique integers $A = \{a_1, a_2, \ldots, a_n\}$ is given. An increasing subsequence is a sequence $A' = \{a_{i_1}, \ldots, a_{i_m}\}$ with $a_{i_p} < a_{i_q}$ and $i_p < i_q$ for any $p < q$. It is well known that the longest increasing subsequence can be determined in $O(n^2)$ time by a dynamic programming algorithm [2], in which we can define $L[i]$ as the longest increasing subsequence of $A_i = \{a_1, a_2, \ldots, a_i\}$, and it is sufficient to determine the longest one over A, with the knowledge of the length of n subsequences (the subsequences of length $L[i]$ for $i \in \{1, \ldots, n\}$). As there are $k = 2^n$ subsequences in total, the search complexity is $O(n) = O(\log k)$.

Formally, the following theorems (Theorems 1 and 2) conclude the general connection between the search complexity and the time complexity of the search algorithms.

Theorem 1. *Given a utility oracle $\mathcal{U} = \langle f_{\mathcal{U}}, \mathcal{C}_{\mathcal{U}} \rangle$, if there is a search algorithm that can return the element $s \in S$ of the maximum value in $\mathcal{T}(k)$ time, for any set $S \in \mathcal{C}_{\mathcal{U}}$ of size k, then the search complexity of utility oracle $\mathcal{U}$ is upper bounded by $\mathcal{T}(k)$.*

Proof. The proof is direct. Since for any set $S \in \mathcal{C}_{\mathcal{U}}$ of size $|S| = k$, there is a search algorithm that can find the element of the maximum value with time cost $\mathcal{T}(k)$. Therefore, during the running of the algorithm, $f_{\mathcal{U}}(\cdot)$ is called for at most $\mathcal{T}(k)$ times, which gives a upper bound for the search complexity. $\square$

Theorem 2. *Given a utility oracle $\mathcal{U} = \langle f_{\mathcal{U}}, \mathcal{C}_{\mathcal{U}} \rangle$, if the search complexity of utility oracle $\mathcal{U}$ is $\mathcal{T}(k)$, then for any search algorithm that can return the element $s \in S$ of the maximum value, the running time is at least $\mathcal{T}(k)$, where $k = |S|$.*

Proof. Assume that the conclusion is not true. Then there is a search algorithm that can return the element $s \in S$ of the maximum value for any set $S \in \mathcal{C}_{\mathcal{U}}$ of size k with time cost $\mathcal{T}'(k) < \mathcal{T}(k)$, then it follows that the search complexity is upper bounded by $\mathcal{T}'(k)$ due to the conclusion of Theorem 1, and hence there is a contradiction. □

4   Approximate Search Complexity

Definition 3 ((δ, ϵ)-Approximate Search Complexity). *Consider a utility oracle $\mathcal{U} = \langle f_{\mathcal{U}}, \mathcal{C}_{\mathcal{U}} \rangle$. Assume that there exist $0 \leq \delta \leq 1$ and $\epsilon \geq 0$, such that for any set $S \in \mathcal{C}_{\mathcal{U}}$, calling $f_{\mathcal{U}}$ for $\mathcal{T}(|S|)$ times is sufficient and necessary to claim (correctly and deterministically) an element $\hat{s} \in S$ with $f_{\mathcal{U}}(\hat{s}) \geq \delta \cdot \max_{s \in S} f_{\mathcal{U}}(s) - \epsilon$. Then $\mathcal{T}(\cdot)$ is called the (δ, ϵ)-approximate search complexity of $\mathcal{U}$, where δ is called the approximation ratio and ϵ is called the approximation error.*

To investigate the situations where an approximation can be made in relatively low search complexity, we at first introduce an idea to characterize how the utility value changes from element to element, when the distance is well defined for all the pairs.

Definition 4 (Bound of The Difference Quotient). *Given a utility function f defined on the universal set $\mathcal{N}$ and a distance function ℓ defined between every pair $p, q \in \mathcal{N}$, the difference quotient of f is bounded by $\rho > 0$ if it holds that*

$$|f(p) - f(q)| \leq \rho \cdot \ell(p, q),$$

for all $p, q \in \mathcal{N}$.

Definition 5 (Bounding Grid for A Set). *Consider the points in $\mathbb{R}^d$ with $d \in \mathbb{Z}^+$. A grid with anchor $a = (a_0, a_1, \ldots, a_{d-1}) \in \mathbb{R}^d$ and side length $l > 0$ is defined as the area g_a^l with*

$$g_a^l := \{(x_0, x_1, \ldots, x_{d-1}) \in \mathbb{R}^d \mid a_i \leq x_i < a_i + l, \ \forall i \in \{0, 1, \ldots, d-1\}\}.$$

Given a set $S \subset \mathbb{R}^d$, a bounding grid g_S for S is a grid that covers all the points in S, i.e. $S \subseteq g_S$.

Consider a utility oracle $\mathcal{U} = \langle f_{\mathcal{U}}, \mathcal{C}_{\mathcal{U}} \rangle$, in which $f_{\mathcal{U}}$ is defined on $\mathbb{R}^d$ with $d \in \mathbb{Z}^+$, and the context $\mathcal{C}_{\mathcal{U}}$ is defined as the set of the finite subsets of $\mathbb{R}^d$, such that for any set $S \in \mathcal{C}_{\mathcal{U}}$, there is a bounding grid g_S of side length $O(\log |S|)$.

See Fig. 2 for an example of $S \in \mathbb{R}^2$. Define the distance between points $p = (p_x, p_y), q = (q_x, q_y) \in \mathbb{R}^2$ by

$$\ell(p, q) = \|p - q\|_2 = \sqrt{(p_x - q_x)^2 + (p_y - q_y)^2}.$$

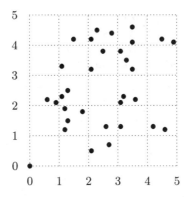

Fig. 2. A set of 2^5 points placed in a square of side length 5.

Assume that the difference quotient of $f_{\mathcal{U}}$ is bounded by $\rho > 0$, then it holds that

$$|f_{\mathcal{U}}(p) - f_{\mathcal{U}}(q)| \le O(\rho \cdot \log k),$$

for any points $p, q \in S$, as $\|p - q\|_2 \le \sqrt{2} \cdot O(\log k) = O(\log k)$. Thus for any element $s \in S$, it satisfies

$$f_{\mathcal{U}}(s) \ge \max_{s \in S} f_{\mathcal{U}}(s) - O(\rho \cdot \log k),$$

which implies the $(1, O(\rho \cdot \log k))$-approximate search complexity is $O(1)$. Furthermore, a better approximation can be achieved if more effort has been devoted.

Note that for any points p, q with $|p_x - q_x| \le 1$ and $|p_y - q_y| \le 1$, it holds that $\|p - q\|_2 \le \sqrt{2}$. Since the difference quotient of $f_{\mathcal{U}}$ is bounded by $\rho > 0$, it implies that

$$|f_{\mathcal{U}}(p) - f_{\mathcal{U}}(q)| \le \sqrt{2}\rho,$$

for any points p, q with $|p_x - q_x| \le 1$ and $|p_y - q_y| \le 1$. Thus after dividing the bounding grid of S into disjoint small grids of side length 1, we can select a representative for each small grid from its intersection with S, as long as the intersection is not empty. Let S' be the set of all the representatives. Then for any element $s \in S$, there exists a representative $s' \in S'$ with $\|s - s'\|_2 \le \sqrt{2}$. Consequently, we have

$$|f_{\mathcal{U}}(s) - f_{\mathcal{U}}(s')| \le \sqrt{2}\rho,$$

and hence

$$\max_{s' \in S'} f_{\mathcal{U}}(s') \ge \max_{s \in S} f_{\mathcal{U}}(s) - \sqrt{2}\rho.$$

Note that $|S'| \le O(\log k)^2$, and it follows that the $(1, \sqrt{2}\rho)$-approximate search complexity is $O(\log k)^2$.

It is direct to extend the argument above to the general form, as shown in Theorem 3.

Theorem 3. *Consider a utility oracle $\mathcal{U} = \langle f_{\mathcal{U}}, \mathcal{C}_{\mathcal{U}} \rangle$, in which $f_{\mathcal{U}}$ is defined on $\mathbb{R}^d$ with $d \in \mathbb{Z}^+$, and the context $\mathcal{C}_{\mathcal{U}}$ is defined as a set of finite subsets of $\mathbb{R}^d$, such that for any set $S \in \mathcal{C}_{\mathcal{U}}$, there is a bounding grid g_S of side length $O(|S|)$. If the difference quotient of $\mathcal{U}$ is bounded by $\rho > 0$, then with any $\epsilon > 0$, the $(1, \epsilon)$-approximate search complexity of $\mathcal{U}$ is $O\left(\frac{\rho \cdot \sqrt{d} \cdot \log k}{\epsilon}\right)^d$, where the distance between any points $p = (p_0, p_1, \ldots, p_{d-1}), q = (q_0, q_1, \ldots, q_{d-1}) \in \mathbb{R}^d$ is defined by*

$$\ell(p, q) = \|p - q\|_2 = \left(\sum_{i \in \{0, 1, \ldots, d-1\}} (p_i - q_i)^2 \right)^{\frac{1}{2}}.$$

Proof. Consider a set $S \in \mathcal{C}_{\mathcal{U}}$ of size k. Let g_S be the bounding grid of S with side length $\Delta = O(\log k)$, and denote its anchor by $\alpha = (\alpha_0, \ldots, \alpha_{d-1})$. Divide g_S into the disjoint small grids of side length η, where

$$\eta = \max\left\{ l \mid \frac{\Delta}{l} \in \mathbb{Z}^+, l \leq \frac{\epsilon}{\rho \cdot \sqrt{d}} \right\}.$$

For $i_0, \ldots, i_{d-1} \in \{0, 1, \ldots, \frac{\Delta}{\eta} - 1\}$, let $g_{i_0, \ldots, i_{d-1}}$ denote the grid that is anchored at $a_{i_0, \ldots, i_{d-1}} = (x_{i_0}, \ldots, x_{i_{d-1}})$ with $x_{i_j} = \alpha_j + j \cdot \eta$, for all $j \in \{0, 1, \ldots, d-1\}$. For each grid $g_{i_0, \ldots, i_{d-1}}$ that satisfies $g_{i_0, \ldots, i_{d-1}} \cap S \neq \emptyset$, select a representative $s_{i_0, \ldots, i_{d-1}} \in g_{i_0, \ldots, i_{d-1}} \cap S$. Let S' be the set of all the representatives. Then it follows that

$$|S'| \leq \left(\frac{\Delta}{\eta}\right)^d.$$

By the definition of η, it holds that

$$\frac{\Delta}{\eta} < \frac{\Delta \cdot \rho \cdot \sqrt{d}}{\epsilon} + 1.$$

Consequently, we have

$$|S'| = O\left(\frac{\rho \cdot \sqrt{d} \cdot \log k}{\epsilon}\right)^d,$$

which gives the $(1, \epsilon)$-approximate search complexity of $\mathcal{U}$, since for any element $s \in S$, there exists a representative $s' \in S'$ such that

$$\|s - s'\|_2 \leq \eta \cdot \sqrt{d} \leq \frac{\epsilon}{\rho},$$

which implies

$$|f_{\mathcal{U}}(s) - f_{\mathcal{U}}(s')| \leq \rho \cdot \|s - s'\|_2 \leq \epsilon,$$

and consequently

$$\max_{s' \in S'} f_{\mathcal{U}}(s') \geq \max_{s \in S} f_{\mathcal{U}}(s) - \epsilon.$$

$\square$

Furthermore, we can relax the assumptions in Theorem 3, and derive a similar conclusion (Theorem 4).

Theorem 4. *Consider a utility oracle $\mathcal{U} = \langle f_\mathcal{U}, \mathcal{C}_\mathcal{U} \rangle$, in which $f_\mathcal{U}$ is defined on $\mathbb{R}^d$ with $d \in \mathbb{Z}^+$, and the context $\mathcal{C}_\mathcal{U}$ is defined as a set of finite subsets of $\mathbb{R}^d$, such that for a set $S \in \mathcal{C}_\mathcal{U}$, there is a bounding grid of S with side length $O(|S|)$. Define the distance between any points $p = (p_0, p_1, \ldots, p_{d-1}), q = (q_0, q_1, \ldots, q_{d-1}) \in \mathbb{R}^d$ by*

$$\ell(p, q) = \max_{i \in \{0, 1, \ldots, d-1\}} |p_i - q_i|,$$

and assume that there exists $\eta > 0$ and $\rho > 0$, such that it holds

$$|f_\mathcal{U}(p) - f_\mathcal{U}(q)| \leq \rho,$$

for all $p, q \in \mathbb{R}^d$ with $\ell(p, q) \leq \eta$. Then the $(1, \rho)$-approximate search complexity of $\mathcal{U}$ is $O\left(\frac{\log k}{\eta}\right)^d$.

Proof. The argument is similar to the one for Theorem 3. Consider a set $S \in \mathcal{C}_\mathcal{U}$ of size k. Let g_S be the bounding grid of S of side length $\Delta = O(\log k)$. Divide g_S into the disjoint small grids of side length η, and select a representative for each grid if it has a non-empty intersection with S. Let S' be the set of all the representatives. Without loss of generality, we can assume that $\frac{\Delta}{\eta} \in \mathbb{Z}^+$, as for any $\eta' < \eta$, it holds that

$$|f_\mathcal{U}(p) - f_\mathcal{U}(q)| \leq \rho,$$

for all $p, q \in \mathbb{R}^d$ with $\ell(p, q) \leq \eta' < \eta$. Consequently, it follows that

$$|S'| \leq \left(\frac{\Delta}{\eta}\right)^d = O\left(\frac{\log k}{\eta}\right)^d,$$

which gives the $(1, \rho)$-approximate search complexity of $\mathcal{U}$, since for any element $s \in S$, there exists a representative $s' \in S'$ with

$$\ell(s, s') \leq \eta \text{ and } |f_\mathcal{U}(s) - f_\mathcal{U}(s')| \leq \rho,$$

which implies

$$\max_{s' \in S'} f_\mathcal{U}(s') \geq \max_{s \in S} f_\mathcal{U}(s) - \rho.$$

$\square$

Next, we consider the universal set $\mathcal{N}$ in which the elements are sets as well. Formally, let $\mathcal{I}$ be a set of items, and define $\mathcal{N}$ as the set of all the subsets of $\mathcal{I}$, i.e. $\mathcal{N} = 2^\mathcal{I}$. Then any function defined on $\mathcal{N}$ is a function of a set of items from $\mathcal{I}$.

Definition 6 (Monotone Function). *Given a function f defined on $\mathcal{N} = 2^\mathcal{I}$, f is monotone if and only if for any subset $A \subseteq \mathcal{I}$ and any item $a \in \mathcal{I}$, it holds that*

$$f(A \cup \{a\}) \geq f(A).$$

Definition 7 (Submodular Function). *Given a function f defined on $\mathcal{N} = 2^{\mathcal{I}}$, f is submodular if and only if for any subsets $A \subseteq A' \subseteq \mathcal{I}$ and any item $a \in \mathcal{I}$, it holds that*

$$f(A \cup \{a\}) - f(A) \geq f(A' \cup \{a\}) - f(A').$$

Theorem 5 ([3,6]). *Given a non-negative monotone and submodular function f defined on $\mathcal{N} = 2^{\mathcal{I}}$, construct a set A of size $K > 0$ by selecting items one by one, and let A_i be the intermedia set after the first i items have been selected and $A_0 := \emptyset$. For $i = \{0, 1, \ldots, K - 1\}$, the $(i + 1)$-th item a_{i+1} is selected by taking*

$$a_{i+1} = \arg \max_{a \in \mathcal{I} \backslash A_i} (f(A_i) \cup \{a\} - f(A_i)).$$

Then it holds that

$$f(A) \geq \left(1 - \frac{1}{e}\right) \cdot \max_{A' \subseteq \mathcal{I}, |A'| \leq K} f(A'),$$

where e is the base of the natural logarithm.

Theorem 6. *Consider a utility oracle $\mathcal{U} = \langle f_{\mathcal{U}}, \mathcal{C}_{\mathcal{U}} \rangle$, in which $f_{\mathcal{U}}$ is a non-negative monotone and submodular function defined on $\mathcal{N} = 2^{\mathcal{I}}$ where $\mathcal{I}$ is a set of $d \in \mathbb{Z}^+$ items, and the context $\mathcal{C}_{\mathcal{U}}$ is defined as the set of finite subsets of $2^{\mathcal{I}}$, such that for each $S \in \mathcal{C}_{\mathcal{U}}$, there exists $\mathcal{I}' \subseteq \mathcal{I}$ and $S := \{s \subseteq \mathcal{I}' \mid |s| \leq K\}$ for an integer constant $K \geq d$. Then the $(1 - \frac{1}{e}, 0)$-approximate search complexity of $\mathcal{U}$ is $O(d \cdot K)$.*

Proof. Consider a set $S \in \mathcal{C}_{\mathcal{U}}$ of size k. Let $\mathcal{I}'$ be the subset of $\mathcal{I}$ such that $S := \{s \subseteq \mathcal{I}' \mid |s| \leq K\}$. To find the element of the maximum value, it is required to find a subset $A^* \subseteq \mathcal{I}'$ with $|A^*| \leq K$ and

$$f_{\mathcal{U}}(A^*) = \max_{A \subseteq \mathcal{I}', |A| \leq k} f_{\mathcal{U}}(A).$$

Next, we construct an approximation to A^*, in an approach similar to the one described in Theorem 5. Initially, let $A_0 = \emptyset$. To construct A_{i+1} with $i \in \{0, 2, \ldots, K - 1\}$, define $A_{i+1} := A_i \cup \{a_{i+1}\}$, where

$$a_{i+1} := \arg \max_{a \in \mathcal{I}' \backslash A_i} (f_{\mathcal{U}}(A_i) \cup \{a\} - f_{\mathcal{U}}(A_i)).$$

Note that A_i and A_{i+1} are all elements in S, and it is equivalent to find A_{i+1} by

$$A_{i+1} := \arg \max_{s \in S, |s| = i+1, A_i \subset s} (f_{\mathcal{U}}(s) - f_{\mathcal{U}}(A_i)).$$

Then let $A := A_K$. Following the conclusion of Theorem 5, it holds that

$$f_{\mathcal{U}}(A) \geq \left(1 - \frac{1}{e}\right) \cdot \max_{A' \in S} f_{\mathcal{U}}(A').$$

It is direct to check that $f_{\mathcal{U}}$ is called for at most d times to find each A_i with $i = \{1, 2, \ldots, K\}$. Thus the conclusion follows.   □

In addition to the monotonicity and submodularity, we will discovery more into the relation between the utility function and the context. At first, consider encoding the subsets of items into the binary numbers. Given a set of d items, say $\mathcal{I} = \{a_0, a_1, \ldots, a_{d-1}\}$, we can encode the subsets of $\mathcal{I}$ in the binary numbers of d bits. In details, for a subset $s \subseteq \mathcal{I}$, it can be encoded into a binary number $b_s^{\mathcal{I}} = b_0 b_1 \cdots b_{d-1}$ of d bits, where $b_i = 1$ if and only if $a_i \in s$. Call $b_s^{\mathcal{I}}$ as the binary coding of set s under $\mathcal{I}$.

Definition 8 (η-Variance). *Given a function f defined on the subsets of a set $\mathcal{I}$ of d items, and a subset $S \subseteq 2^{\mathcal{I}}$. For any $\eta \in \mathbb{Z}^{+}$, the η-variance of f with respect to S is defined by*

$$V_f^{\eta}(S, \mathcal{I}) := \max_{s, s' \in S, |b_s^{\mathcal{I}} - b_{s'}^{\mathcal{I}}| < \eta} |f(s) - f(s')|.$$

Theorem 7. *Consider a utility oracle $\mathcal{U} = \langle f_{\mathcal{U}}, \mathcal{C}_{\mathcal{U}} \rangle$, in which $f_{\mathcal{U}}$ is a non-negative monotone and submodular function defined on $\mathcal{N} = 2^{\mathcal{I}}$ where $\mathcal{I}$ is a set of $d \in \mathbb{Z}^{+}$ items, and the context $\mathcal{C}_{\mathcal{U}}$ is defined as a set of finite subsets of $2^{\mathcal{I}}$, such that for each $S \in \mathcal{C}_{\mathcal{U}}$, there exists $\mathcal{I}' \subseteq \mathcal{I}$ and $S := \{s \subseteq \mathcal{I}' \mid |s| \leq K\}$ for an integer constant $K \geq d$. For $\epsilon > 0$, if there exists $\eta > 1$ satisfying*

$$\max_{S \in \mathcal{C}_{\mathcal{U}}} V_{f_{\mathcal{U}}}^{\eta}(S, \mathcal{I}) < \epsilon,$$

then the $\left(1 - \frac{1}{e}, \left(1 - \frac{1}{e}\right) \cdot \epsilon\right)$-approximate search complexity of $\mathcal{U}$ is $O((d - \log \eta) \cdot K)$.

Proof. The proof combines the ideas used in the arguments for Theorems 3 and 6. Consider a set $S \in \mathcal{C}_{\mathcal{U}}$ and $\mathcal{I}' = \{a_0', a_1', \ldots, a_{d'-1}'\}$ be the subset of $\mathcal{I}$ such that $S = \{s \subseteq \mathcal{I}' \mid |s| \leq K\}$. Let $k = |S| = \sum_{i \leq K} \binom{d'}{i}$. Define $L = \lfloor \log \eta \rfloor$ and $K' = d' - L$. Let $S' \subseteq S$ be the set of all the subsets not containing any element from $\{a_i' \in \mathcal{I}' \mid i \geq K'\}$ (Fig. 3).

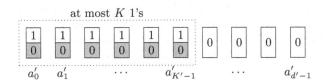

Fig. 3. S': subset of $\mathcal{I}'$ corresponding to binary codes with 1's only appearing before the K'-th bit.

Therefore, for any element $s' \in S'$, the last L bits of its binary coding $b_{s'}^{\mathcal{I}'}$ (under $\mathcal{I}'$) are all 0. Furthermore, for any element $s \in S$, there exists $s' \in S'$ with

$$|b_s^{\mathcal{I}'} - b_{s'}^{\mathcal{I}'}| < 2^L \leq \eta \text{ and } |f_{\mathcal{U}}(s) - f_{\mathcal{U}}(s')| < \epsilon.$$

Thus we know

$$\max_{s' \in S'} f_{\mathcal{U}}(s') \geq \max_{s \in S} f_{\mathcal{U}}(s) - \epsilon.$$

If $K' \leq K$, selecting all the first K' items of $\mathcal{I}'$ achieves $\max_{s' \in S'} f_{\mathcal{U}}(s')$, since $f_{\mathcal{U}}$ is non-negative monotone. Otherwise (i.e. $K' > K$), a set A of size K among the first K' items of $\mathcal{I}'$ can be constructed to satisfy

$$f_{\mathcal{U}}(A) \geq \left(1 - \frac{1}{e}\right) \cdot \max_{s' \in S'} f_{\mathcal{U}}(s') \geq \left(1 - \frac{1}{e}\right) \cdot \max_{s \in S} f_{\mathcal{U}}(s) - \left(1 - \frac{1}{e}\right) \cdot \epsilon.$$

Following an analysis similar to the one used in Theorem 6, we know that during the construction of A, it calls $f_{\mathcal{U}}$ for at most $K' \cdot K$ times. Then the conclusion follows since $K' = d' - L \leq d - \log \eta + 1$.   □

Following the conclusion of Theorem 7, we can pay less effort to approximate $\max_{s \in S} f_{\mathcal{U}}(s)$ with an additive error, if η is large enough.

5   Discussions

It is clear that the usefulness of Theorem 7 lays on the existence of η. With a large η, it means some items can be ignored when selecting a subset with the good enough evaluation. Thus the connection between the utility function and the binary coding of its input reveals the priorities among the items. Intuitively, with a collection of items, the marginal value is small if one or more items of low priorities are added. Thus it may results in a good enough result by considering only the items of high priorities, especially when the cost of exploration is an issue. In applications, to be inspired by the idea behind Theorem 7, it suggests to investigate if there are some items that always cause small variance of the set utilities.

On the other hand, as some people may noticed, it is also convenient to alternatively define the search complexity with respect to an algorithm, and then the problem difficulty can be revealed with the search complexity of the best known algorithms, as what have been done with the time complexity and the space complexity. This alternating definition is generally equivalent to the one proposed in this work, and we will make a further investigation in the future to see if there are situations in which one definition is preferred to the other.

References

1. Boyd, S., Boyd, S.P., Vandenberghe, L.: Convex Optimization. Cambridge University Press, Cambridge (2004)
2. Cormen, T.H., Leiserson, C.E., Rivest, R.L., Stein, C.: Introduction to Algorithms. MIT Press, Cambridge (2009)
3. Cornuejols, G., Fisher, M.L., Nemhauser, G.L.: Exceptional paper–location of bank accounts to optimize float: an analytic study of exact and approximate algorithms. Manag. Sci. **23**(8), 789–810 (1977)

4. Edelkamp, S., Schroedl, S.: Heuristic Search: Theory and Applications. Elsevier, Amsterdam (2011)
5. Kushilevitz, E.: Communication complexity. In: Advances in Computers, vol. 44, pp. 331–360. Elsevier (1997)
6. Nemhauser, G.L., Wolsey, L.A., Fisher, M.L.: An analysis of approximations for maximizing submodular set functions. Math. Program. **14**(1), 265–294 (1978). https://doi.org/10.1007/BF01588971
7. Rao, S.S.: Engineering Optimization: Theory and Practice. Wiley, Hoboken (2019)
8. Williamson, D.P., Shmoys, D.B.: The Design of Approximation Algorithms. Cambridge University Press, Cambridge (2011)

Multi-AGVs Pathfinding Based on Improved Jump Point Search in Logistic Center

Yunwang Zhang(iD) and Hejiao Huang(✉)

Harbin Institute of Technology, Shenzhen, China
huanghejiao@hit.edu.cn

Abstract. In recent years, automated guided vehicle (AGV) is becoming increasingly important for logistic center, which usually has tens of thousands of express packages to sort and transport every day. In order to achieve higher transportation and sorting throughput, we need to plan a feasible path for every AGV, so that sorting tasks can be finished with high performance. In this paper, we therefore study a lifelong version of the MAPF problem, called the multi-agent pickup and delivery (MAPD) problem. In the MAPD problem, one agent has to first move to a given pickup location and then to a given delivery location while avoiding collisions with other agents. We present an algorithm named Improved Jump Point Search which contains two stages. Offline phase computes all jump points and online phase plans conflict-free path for every AGV. Furthermore, we propose a strategy called Congestion Control which can guide AGVs in local congestion situations and reduce overlapping between paths. Compared with existing work like Cooperative A* and Jump Point Search, experimental results demonstrate that our algorithm has a higher throughput and fewer waiting steps.

Keywords: Multi-AGVs · Path planning · Jump point search

1 Introduction

With the rapid development of e-commerce, more and more commodities will be transported through express mails, bringing a new challenge for express companies to sort such large number of packages. Automated Guided Vehicle (AGV) is currently utilized in logistics systems, with the features of high flexibility and easy configurability. In order to apply AGVs to auto-sorting in logistics center, we need to plan collision-free paths for multiple AGVs and minimize the cost of the paths. This problem is known as the famous multi-agent path finding(MAPF) problem, which is hard to handle. Finding an optimal solution to such a problem is NP-hard and intractable [10].

This work is financially supported by National Key R&D Program of China under Grant No.2017YFB0803002 and No.2016YFB0800804, National Natural Science Foundation of China under Grant No. 61672195 and No. 61732022.

© Springer Nature Switzerland AG 2020
Z. Zhang et al. (Eds.): AAIM 2020, LNCS 12290, pp. 358–368, 2020.
https://doi.org/10.1007/978-3-030-57602-8_32

To solve this problem, existing work can be classified into centralized and decentralized approaches. The main advantage of centralized methods lies in that it can calculate the optimal schedule of AGV plan in theory and the bottleneck is that the central computer runs very slow with the scale of AGV grows. The centralized strategies include the following methods. In [11] the authors propose an approach called prioritized path planning, which assigns different priorities to the robots and plans the paths in sequence according to the allocated priority. Time window based dynamic routing is used in paper [9], which manages conflict-free routing by operations of AGVs. Paper [5] gives a method of generating the hierarchical roadmap which has a multi-layered structure and solves a path planning problem in a home environment. Paper [7] adopts a conflict-free method by using a new data structure called conflict tree. Every conflict can be solved well in branches of conflict tree but it costs too much with the large scale of AGVs. Paper [6,14] develops the algorithm based on pathfinding algorithms to find shortest paths and solve conflicts.

For the consideration of running speed in large-scale working scenario, decentralized algorithms are often adopted, in which each AGV determines their paths and avoids conflicts with others through information exchange and negotiate with nearby AGVs [1,12,13]. Therefore, there is no need for the central unit, which is replaced by several distributed controllers. Collective behaviors for groups of robots rely on distributed simultaneous estimation and control. An effective approach with cooperative path planning is discussed in [4]. These multi-AGVs coordinate their path effectively to accomplish complex and critical tasks for various applications.

The MAPF problem does not capture important characteristics of many real-world domains, such as automated warehouses, where agents are constantly engaged with new tasks [3]. In this paper, we therefore study a lifelong version of MAPF problem with pickup tasks and delivery tasks, called the multi-agent pickup and delivery (MAPD) problem. We mainly focus on finding feasible paths with higher throughput and fewer waiting times. Tasks can enter the system at any time and the assignment of tasks is controlled by the algorithm. Each AGV travels from starting port to pickup port, then delivers the cargo to delivery port. If one AGV finishes a delivery task, it can execute next pickup task immediately or move to a buffer port. For the consideration of running speed and throughput with multi-AGVs in large scenario, we propose an Improved Jump Point Search which includes two stages. Offline phase travels each node in each reachable direction and stores all the jump points into Jump Point Table(JPT). Online phase computes successors by JPT and generates feasible paths by waiting and replanning. After that, we use Congestion Control strategy to reduce times of local congestion. We evaluate the performance of our approach by transportation throughput and waiting times. Experiment results show that our method outperforms existing work.

The rest of this paper is organized as follows. In Sect. 2, we describe the MAPD problem. After that, we outline our approach in Sect. 3. In Sect. 4, we setup experiments and analyze the results. Finally, conclusions and future work are drawn in Sect. 5.

2   Problem Description

The problem consists of m agents $A = (a_1, a_2, ...a_m)$ and a grid map $G = (V, E)$ whose vertices V correspond to grids and whose edges E correspond to connections between grids that the agents can move along. Let $l_i(t) \in V$ denote the location of agent a_i in discrete timestep t. Agent a_i starts in its initial location $l_i(0)$. In each timestep t, the agent either stays in its current location $l_i(t)$ or moves to an adjacent location, that is $l_i(t+1) = l_i(t)$ or $(l_i(t), l_i(t+1)) \in E$. Agents need to avoid collisions with each other: two agents cannot be in the same location in the same timestep, that is for all agents a_i and a_j with $a_i \neq a_j$ and timestep t: $l_i(t) \neq l_j(t)$.

Consider a task Set Q that contains the set of unexecuted tasks. In each timestep, the system can add all new tasks to the task set. Each task $q_j \in Q$ is characterized by a pickup location $s_j \in S_{load}$ and a delivery location $g_j \in S_{unload}$. The agent with the task q_j has to move from its current location via the pickup location s_j to the delivery location g_j. Then the agent moves to a buffer point or executes the next task. The objective is to finish each task with higher throughput and fewer waiting times. Consequently, the effectiveness of a MAPD algorithm is evaluated by the throughput, the total number of waiting steps and the total number of moving steps.

3   Algorithm Improvement

As we discuss in Sect. 2, our goal is finding feasible paths with more throughput and fewer waiting times in the given logistic center. Standard Jump Point Search algorithm must recursively calculate jump points and overlapping paths are generated. In some scenarios, the recursion cost is greater than direct search and the overlapping paths may cause conflicts between AGVs. Therefore, we separate the calculation process of jump points from the process of path planning, and resolve conflicts during the path planning stage. Congestion Control strategy is proposed to reduce the overlapping paths. In this Section, we will describe the detail of the algorithm. The notations and definitions to be used are shown in Table 1.

3.1   Offline Phase

In pathfinding algorithms, it is a key operation to expand from one node to the surrounding to find the optimal successor node. Jump Point Search are proposed for speeding up optimal search by selectively expanding only certain nodes on a grid map which are termed jump points [2]. Standard Jump Point Search algorithm must recursively calculate jump points with start points and

Table 1. Notations and definitions

Notations	Definitions
A	The set of all AGVs
Q	The set of unexecuted tasks
q_j	The unexecuted task with index j
s_j	The pickup point of the unexecuted task with index j
g_j	The delivery point of the unexecuted task with index j
S_{load}	The set of pickup points
S_{unload}	The set of delivery points
$G = (V, E)$	The grid map of logistic center
$p(v)$	The parent node of node v
$f(v)$	The estimated cost from start point to the goal by node v
$h(v)$	The regular estimated cost from node v to the goal
$q(v)$	The overlapping estimated cost to represent density of paths through node v
$path_i$	Feasible path for AGV_i
$dirs_{node}$	Reachable directions in the node
W_{max}	The maximum value of estimated overlapping cost
JPT	Jump Point Table that stores all the jump points which can be used in online phase

goal points, so the recursive calculation of jump points and path planning are performed simultaneously. There is an essential property that jump points are only related to the positions of obstacles. It is worth noting that there is no start node and goal node of the task in the offline phase, that is, no path planning has been started. In the absence of a given start point and end point, we can use the property of jump points to calculate all the jump points which can be used for pathfinding in online phase.

In the given logistic center, the start node of delivery task is in S_{load} and the goal node is in S_{unload}. In order to reduce repeated calculation of pathfinding, we compute all the jump points and store them into Jump Point Table(JPT) in offline phase. For the constraints of AGVs, AGVs can not move diagonally and only straight moves are allowed. So the definitions about jump points can be simplified in our situations.

Definition 1. *A node $n \in neighbours(x)$ is a forced neighbour if n is not a obstacle, and the following condition is satisfied.*

$$len(< p(x), x, n >) < len(< p(x), ..., n > \setminus x) \tag{1}$$

where $< p(x), x, n >$ represents a path through $p(x)$ and x to node n which is a neighbour of x. $< p(x), ..., n > \backslash x$ represents a path that passes from $p(x)$ to node n through some nodes which do not include x.

Definition 2. *A node n is the jump point if n has at least one neighbour whose evaluation is a forced neighbour.*

Fig. 1. The jump example in a 4×4 map.

Figure 1 shows the jump example in a 4×4 map. The character F represents that this node is a forced neighbour and grey node means that this node is an obstacle. In the left of Fig. 1, we perform the jump process in the order of up, down, left, and right. The current node is v and the jump process is recursive, so it keeps recursing down until it finds a jump point as the red arrow describes. In our situation, moving from $p(x)$ to point marked F is not allowed, so the left node of x is a forced neighbour according to Definition 1. This means that passing point x to point F is the shortest path currently and there is no shorter path if we do not pass point x to point F. The neighbours of node marked x are jump points according to Definition 2, so we add $suc(v, DOWN) = x$ to JPT. It means that we can use this information to jump to x from v in the online phase and reduce the search for intermediate nodes. When the direction is left, the jump process reaches the border of the map and we also add $suc(v, LEFT) = a$. Although there are 2 obstacles in the figure on the right, the jump point situation is the same. Offline phase algorithm is described in Algorithm 1. In offline phase, the algorithm travels every reachable direction of each node to compute jump points(Line1~Line3). Then the set of jump points (v, dir) will be added to JPT. In online phase, given v and jump direction, we can get the successors immediately.

3.2 Online Phase

In online phase, we need to plan a feasible path for each AGV. When apply Jump Point Search, we always use the heuristic function $f(v) = g(v) + h(v)$, where $g(v)$ and $h(v)$ denote the real cost from the start point to the node v and the estimated cost from node v to the goal point. $h(v)$ represents the heuristic value that guides the AGV to the goal point in each step and influences the speed of pathfinding.

Algorithm 1: Offline Phase

Input: G(V,E), $dirs_j$
Output: Jump Point Table
1 **foreach** $node \subset G$ **do**
2   **foreach** $dir \subset dirs_{node}$ **do**
3          $points \leftarrow jump(node, dir)$;
4          $JPT \xleftarrow{add} points$;

5 **return** JPT;

In online phase, jump points generated by offline phase can be reused. If the cost of a jump point is small, it will be selected many times in tasks with similar paths and paths will have a lot of overlapping parts. So we introduce $q(v)$ to represent overlapping cost of paths through node v. The heuristic function is defined as:

$$f(v) = g(v) + h(v) + q(v) \qquad (2)$$

The details of the calculation of $q(v)$ will be discussed in Sect. 3.3.

In this paper, the online phase of IJPS is described in Algorithm 2. In this algorithm, we will assign AGV tasks and find each feasible path for each AGV. Then conflicts will be checked and different conflict solutions will be applied.

Firstly, a feasible path will be generated for each AGV by JPT(Line1 $\sim$ Line19). We search the successors of a node by jump points described as (v, dir) in JPT(Line6) and try move to the goal by jump points. It is difficult to guarantee global conflict-free situations and we only consider about local conflict-free situations. It means that we only solve the conflicts that will happen in the next time slot(Line20~Line27) and temporarily ignore conflicts that may occur at a later time. A feasible path described as $path_i$ determines the route of a_i and a_i tries to occupy the next node of $path_i$ in the next time slot. For all conflict situations, there are two cases: 1) AGV tries to occupy the node which is already occupied by the other AGV. 2) two different AGVs try to occupy the same node in this time slot. For case 1, conflicts are solved by replan (Line23) and conflicts are solved by waiting for a step for case 2(Line25).

3.3 Congestion Control

As described in Sect. 3, the overlapping paths may cause conflicts between AGVs. In this paper, we propose a Congestion Control strategy which changes the overlapping estimated cost of jump points based on the number of overlapping in jump points. When the AGV plans the path according to JPT, jump points with small cost will be selected repeatedly and paths will overlap in the area. It is necessary to add an overlapping cost in the calculation of the cost. This cost is related to jump points, so we also store it into JPT and dynamically update

---

Algorithm 2: Online Phase

Input: A

1 **for** $a_i \in A$ **do**
2 | **if** $path_i = \varnothing$ **then**
3 | | $a_i.task \leftarrow TaskPool.next$;
4 | | **while** $List_{open} \neq \varnothing$ **do**
5 | | | $C \leftarrow getMin(List_{open})$;
6 | | | $interP \xleftarrow{add} getSByJPT(C)$;
7 | | | **if** $C = a_i.end$ **then**
8 | | | | $path_i \leftarrow calcPath(end)$;
9 | | | **for** $node \in interP$ **do**
10 | | | | **if** $node \notin List_{open}$ **then**
11 | | | | | **if** $node \notin List_{close}$ **then**
12 | | | | | | $node.parent \leftarrow C$;
13 | | | | | | $node.F \leftarrow calcF(node)$;
14 | | | | | | $List_{open} \xleftarrow{add} node$;
15 | | | | **else**
16 | | | | | $tmpF \leftarrow calcF(node, target)$;
17 | | | | | **if** $tmpF < node.F$ **then**
18 | | | | | | $node.F \leftarrow tmpF$;
19 | | | | | | $node.parent \leftarrow C$;

20 **for** $a_i \in A$ **do**
21 | $n \leftarrow path_i.next$;
22 | **if** n *is blocked by* a_j **then**
23 | | Replan $path_i$ by IJPS;
24 | **else if** $n = path_j.next$ **then**
25 | | a_i wait for one step;
26 | **else**
27 | | $control(a_i)$;

---

it. The cost is defined by the following equation.

$$q(v) = \begin{cases} q(p(v)) + \frac{l(v)}{L} & \text{v is a jump point} \\ q(p(v)) & \text{v is not a jump point} \end{cases} \tag{3}$$

where $q(v)$ represents the overlapping estimated cost and $l(v)$ is the number of overlapping in node v between all paths. L is the length of $path_i$. Congestion Control is described in Algorithm 3 which computes the overlapping cost and stores the jump points with new cost into JPT. If the cost exceeds maximum, we change the destination to a buffer point, clean the former path and reset the cost(Line4~Line6). The next round of Online Phase will plan the path to the buffer point. A new cost of this jump point is updated in JPT(Line9) and this new cost will influence the next path planning of Online Phase.

Algorithm 3: Congestion Control

   Input: a_i
1 **foreach** $node \subset path_i$ **do**
2   **if** $node \in JPT$ **then**
3        $cost \leftarrow calcQ(node)$;
4   **if** $cost > W_{max}$ **then**
5   set the goal point of a_i to a buffer point;
6           $path_i \leftarrow \varnothing$;
7   updateJPT$(node,0)$;
8   **else**
9   updateJPT$(node,cost)$;

---

4 Simulations and Result Analysis

In this section, we compare our IJPS with existing work such as CA* [8] method and Jump Point Search [2] in different situations. CA* is a typical MAPF problem solver and the restrictions of pickup points and delivery points are added to solve MAPD problem. JPS algorithm is a pathfinding method like A* which can not solve conflicts for AGVs. In order to compare the performances of different algorithms in the same MAPD problem, we add a global conflict table to standard JPS.

4.1 Experiment Setup

The scale of the logistic center and the number of AGVs will influence the performance. In this paper, we set the scale of the test map is 30×50 and there are up to 60 AGVs in the warehouse. Our experiments were conducted on an Inter(R) Core i7-8550u which operates at 1.80 GHz and the computer has 8GB of RAM. All code for experiments was written in Java and situations of AGVs are visualized with Java Swing.

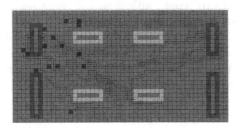

Fig. 2. The screenshot of the experiment.

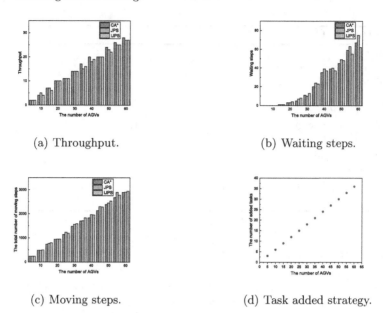

(a) Throughput.   (b) Waiting steps.

(c) Moving steps.   (d) Task added strategy.

Fig. 3. The experiment results in the first task added strategy.

4.2   Experiment Results

We compare IJPS with JPS and CA*. Figure 2 is the screenshot of the experiment. There are 5 AGVs which are running delivery tasks and the red lines show paths which are generated in Online Phase. The number of tasks influence the results of the experiment, so we use two tasks added strategies to simulate different situations.

In the first situation, the total number of tasks is fewer than the total number of AGVs as Fig. 3(d) describes. Some AGVs must be without tasks and these AGVs which stay at fixed positions are obstacles for other AGVs with tasks.

Figure 3(a) shows the throughput of CA*, JPS and IJPS in 100 time steps. We note that IJPS is without more throughput than others. It is because that there are enough spaces to avoid conflicts and it is not necessary to apply Congestion Control in this situation. For waiting times, it represents the number of stops in response to a conflict and Fig. 3(b) describes the waiting times of three algorithms. When the number of AGVs is up to 40, IJPS with our Congestion Control can greatly reduce the waiting times. It can be clearly seen that our method achieves fewer waiting steps than CA* and JPS. Figure 3(c) shows the relationship between AGV number and total moving steps. As we can see, the total number of moving steps is very close between CA* and JPS.

In the second situation, we change the tasks added strategy as Fig. 4(d) shows. In this situation, the number of tasks is saturated and each AGV will execute one task. Figure 4(a) shows the throughput of CA*, JPS and IJPS in 100 time steps. We note that IJPS is without particularly great advantages when

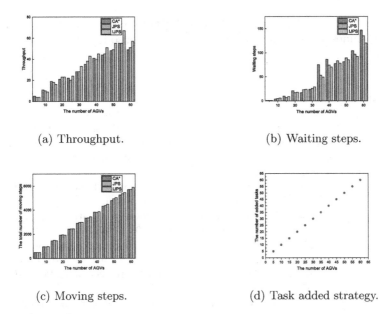

(a) Throughput.   (b) Waiting steps.

(c) Moving steps.   (d) Task added strategy.

Fig. 4. The experiment results in the second task added strategy.

the number of AGVs is less than 30. When the number of AGVs is up to 40, it can be clearly seen that our method achieves higher transportation throughput than CA* and JPS. Figure 4(b) describes the waiting times of three algorithms. There are tiny waiting time with less 10 AGV because there are lots of nodes that AGVs can occupy and AGVs can move without waiting. With the growing of AGV number, the total waiting times increase greatly but our method has relative fewer waiting times. It is because our method dynamically changes the weight of jump points and reduces the overlapping of paths. Figure 4(c) shows the relationship between AGV number and total moving steps. The total number of moving steps is slightly more than the steps of other algorithms. It is because that we introduce overlapping cost and the paths generated by IJPS are not necessarily the shortest paths. On the whole, our method finds feasible paths with more throughput and fewer waiting times.

5   Conclusion

In this paper, we studied the problem of multi-AGVs pathfinding in a logistic center. We propose the algorithm called Improved Jump Point Search which is a two-stage algorithm. In offline phase, jump points are computed and stored. In online phase, we focus on local conflicts and plan feasible path for each AGV. Congestion Control is added to reduce overlapping among paths and reduce local conflicts. Compared with existing work such as JPS and CA*, our method promotes the transportation throughput with fewer waiting times.

In the future, we will focus on more actual situations where every task of AGV not only includes a start point and a goal point, but also has the constraint with finished time.

References

1. Digani, V., Sabattini, L., Secchi, C., Fantuzzi, C.: Hierarchical traffic control for partially decentralized coordination of multi AGV systems in industrial environments. In: 2014 IEEE International Conference on Robotics and Automation (ICRA), pp. 6144–6149. IEEE (2014)
2. Harabor, D.D., Grastien, A.: Online graph pruning for pathfinding on grid maps. In: Twenty-Fifth AAAI Conference on Artificial Intelligence (2011)
3. Ma, H., Li, J., Kumar, T., Koenig, S.: Lifelong multi-agent path finding for online pickup and delivery tasks. arXiv preprint arXiv:1705.10868 (2017)
4. Miskovic, N., Bogdan, S., Petrovic, I., Vukic, Z.: Cooperative control of heterogeneous robotic systems. In: 2014 37th International Convention on Information and Communication Technology, Electronics and Microelectronics (MIPRO), pp. 982–986. IEEE (2014)
5. Park, B., Choi, J., Chung, W.K.: An efficient mobile robot path planning using hierarchical roadmap representation in indoor environment. In: 2012 IEEE International Conference on Robotics and Automation, pp. 180–186. IEEE (2012)
6. Qing, G., Zheng, Z., Yue, X.: Path-planning of automated guided vehicle based on improved Dijkstra algorithm. In: 2017 29th Chinese Control and Decision Conference (CCDC), pp. 7138–7143. IEEE (2017)
7. Sharon, G., Stern, R., Felner, A., Sturtevant, N.R.: Conflict-based search for optimal multi-agent pathfinding. Artif. Intell. **219**, 40–66 (2015)
8. Silver, D.: Cooperative pathfinding. AIIDE **1**, 117–122 (2005)
9. Smolic-Rocak, N., Bogdan, S., Kovacic, Z., Petrovic, T.: Time windows based dynamic routing in multi-AGV systems. IEEE Trans. Autom. Sci. Eng. **7**(1), 151–155 (2009)
10. Surynek, P.: An optimization variant of multi-robot path planning is intractable. In: AAAI, pp. 1–3 (2010)
11. Van Den Berg, J.P., Overmars, M.H.: Prioritized motion planning for multiple robots. In: 2005 IEEE/RSJ International Conference on Intelligent Robots and Systems, pp. 430–435. IEEE (2005)
12. Yang, P., Freeman, R.A., Lynch, K.M.: Multi-agent coordination by decentralized estimation and control. IEEE Trans. Autom. Control **53**(11), 2480–2496 (2008)
13. Zheng, K., Tang, D., Gu, W., Dai, M.: Distributed control of multi-AGV system based on regional control model. Prod. Eng. Res. Devel. **7**(4), 433–441 (2013). https://doi.org/10.1007/s11740-013-0456-4
14. Zheng, X., Tu, X., Yang, Q.: Improved JPS algorithm using new jump point for path planning of mobile robot. In: 2019 IEEE International Conference on Mechatronics and Automation (ICMA), pp. 2463–2468. IEEE (2019)

Distance-Based Adaptive Large Neighborhood Search Algorithm for Green-PDPTW

Jinying Lu[ID] and Hejiao Huang[✉]

Harbin Institute of Technology, Shenzhen, China
huanghejiao@hit.edu.cn

Abstract. Green Pickup-and-Delivery Problem with Time-Windows (Green-PDPTW) is a new sub-problem of the Capacitated Vehicle Routing Problem (CVRP). It aims to solve PDPTW in a way that emits the least amount of greenhouse gases. Adaptive Large Neighborhood Search (ALNS) is a commonly used algorithm to solve such problems, but usually, it focuses more on expanding the search range rather than giving a clear search direction. Therefore, we propose Distance-based ALNS (DALNS), using the distance between customers as an important factor when generating initial solution and destroy solutions to searching. We also add a heuristic on the number of orders to be removed in each iteration of DALNS. From simulation experiments, we draw the conclusion that DALNS has a significant effect on reducing greenhouse gas emissions and retaining higher economic benefits for the enterprise at the same time. In addition, we find that DALNS shows great performance on instances where customers are clustered and a load of vehicles is high.

Keywords: Green-PDPTW · DALNS · Greenhouse gas emission

1 Introduction

Vehicle routing is a significant part of logistics. Challenges to be solved for all capacitated vehicle routing problems (CVRP) include total distance traveled, total travel time, and the number of vehicles used. The most classic CVRP is proposed by Dantzig and Ramser [3] based on these challenges. It's an NP-hard problem [16]. Pickup-and-Delivery Problem with Time-Windows (PDPTW) is a sub-problem of CVRP. In PDPTW, vehicles can only start order service in each customer during a specific time window varies from all customers. In addition to the objective of the shortest total distance traveled, solving PDPTW needs to pay more attention to how to minimize the number of vehicles used.

Green-PDPTW is derived from PDPTW. It focuses on how to reduce greenhouse gas emissions for protecting the environment. To our knowledge, the research on Green-PDPTW is in its infancy. Kunnapadeelert and Kachitvichyanukul [7] define Green-PDPTW and using differential evolution algorithm (DE) to solve. A feasible solution for Green-PDPTW is also a feasible

© Springer Nature Switzerland AG 2020
Z. Zhang et al. (Eds.): AAIM 2020, LNCS 12290, pp. 369–380, 2020.
https://doi.org/10.1007/978-3-030-57602-8_33

solution to PDPTW, but the greenhouse gas emission is less in Green-PDPTW than in PDPTW.

Heuristic algorithms are commonly used to solve CVRPs (Montoya et al. [9]; Affi et al. [1]; Roberto et al. [12]), and Large Neighborhood Search (LNS) is one of the popular heuristic algorithms. It creates a large neighborhood search area by destroying and repairing a feasible solution. Based on LNS, Christiaens and Vanden [2] propose slack induction by string removals algorithm (SISR). Hemmelmayr et al. [5] and Prescott-Gagnon et al. [10] also use LNS to solve CVRP.

ALNS is an improvement of LNS, it comprehensively considers the performance of all destroy strategies and repair strategies in the previous iteration, and adds a heuristic to select the destroy strategy and repair strategy used in the next iteration [13]. However, the process of search is usually blind. No clear search guide to the search process.

In order to create a search direction, we propose Distance-based Adaptive Large Neighborhood Search (DALNS) to make better use of the distance between customers. We introduce the concept of the order pool when generating the initial solution. The elements of each pool are determined by the distance and time windows of all customers. The initial solution is easy to be improved greatly because of the use of distance. What's more, we add a heuristic in the number of orders to be removed in each iteration, which can help improve the search efficiency. Our experiment results show that DALNS effectively reduces the greenhouse gas emissions. At the same time, we find that DALNS has a better performance in solving the transportation problem of clustered customers with high load demands.

2   Problem Description for Green-PDPTW

Minimizing greenhouse gas emission is the objective of Green-PDPTW. As a mathematic model derived from logistics, the feasible solution of Green-PDPTW must simultaneously meet the requirements on vehicle capacity and order service time window. With a reasonable number of vehicles, all orders are distributed is the basic requirement of this problem. During the transportation, a pickup order and its corresponding delivery order are distributed to only one vehicle. No delay and no overload can be tolerated during transportation. Also, each vehicle can only visit their customers once.

Green-PDPTW is defined as a complete and undirected graph $G = (N^*, A)$, where N is the union of the pickup customer set $P^+ = \{1, 2, \ldots, L\}$ and the delivery customer set $P^- = \{L+1, L+2, \ldots, 2L\}$. N^* is the union of set N and the depot. A is a set of arcs, $A = \{(i, j) : i, j \in N^*, i \neq j\}$. Each arc is associated with a travel distance D_{ij} and emission of greenhouse gas C_{ij}. Speed of all vehicles is supposed to be a constant. Each vehicle $k \in M$ has the same capacity Q. Z_{ik} is the load of vehicle $k \in M$ when leaving from node $i \in N^*$. Each customer $i \in N$ has the earliest order service start time e_i, the latest order service start time l_i and the real order service start time t_i. Time spent in each customer is S_i. Load demand of each order is q_i. q_i means how much load vehicle need to carry, and it can be negative. X_{ijk} is the binary variable that defined as 1 if the vehicle $k \in M$ travels from node $i \in N^*$ to node $j \in N^*$ and defined as 0 otherwise. Another binary variable is y_{ik}, which is defined as 1 if the vehicle $k \in M$ passes by customer $i \in N^*$, otherwise defined as 0.

The mathematical model of Green-PDPTW is shown as follows:

$$Min \sum_{k \in M} \sum_{i \in N^*} \sum_{j \in N^*} C_{ij} X_{ijk}$$

$$\text{s.t.} \sum_{k \in M} y_{ik} = 1, \forall i \in N^* \tag{1}$$

$$\sum_{k \in M} \sum_{j \in N^*} X_{ijk} = 1, \forall i \in N^* \tag{2}$$

$$\sum_{j \in N^*} X_{ijk} = \sum_{j \in N^*} X_{jik}, \forall i \in N^*, \forall k \in M \tag{3}$$

$$\sum_{j \in N^*} X_{ijk} = \sum_{j \in N^*} X_{j,L+i,k}, \forall i \in P^+, \forall k \in M \tag{4}$$

$$\sum_{i \in P^+} X_{0ik} = 1, \forall k \in M \tag{5}$$

$$\sum_{i \in P^-} X_{i0k} = 1, \forall k \in M \tag{6}$$

$$e_i \leq t_i \leq l_i, \forall i \in N^* \tag{7}$$

$$e_i \leq t_j = \sum_{i=j} \sum_{k \in M} X_{jik}(t_i + t_{ij} + S_i) \leq l_j, \forall i \in N^* \tag{8}$$

$$t_i + t_{i,L+i} \leq t_{L+i}, \forall i \in P^+ \tag{9}$$

$$0 \leq Z_{ik} \leq Q, \forall i \in N^* \tag{10}$$

$$(Z_{ik} + q_j)X_{ijk} \leq Z_{jk}, \forall i,j \in N^*, \forall k \in M \tag{11}$$

Minimizing the greenhouse gas emission is the objective of Green-PDPTW. Constraint (1) ensures that each customer only be visited once. Constraint (2) ensures that each arc can only be passed by once. Constraint (3) enforces that each arc is undirected. Constraint (4) ensures that each order is distributed by only one vehicle. Constraint (5) enforces that all vehicles depart from the depot. Constraint (6) ensures that all vehicle eventually return to the depot. Constraint (7) guarantees arriving time constraint of each customer. Constraint (8) guarantees order service time constraint of each customer. Constraint (9) enforces the delivery order must be distributed after the pickup order has been distributed. Constraint (10) guarantees vehicle capacity feasibility. Constraint (11) ensures that the load of each vehicle can only be changed at customers.

$$G(Z_{ik}) = 9.4 \times Z_{ik}/Z + 29.6 \tag{12}$$

Equation (12) is the formula for the fuel consumed by a 10-ton capacity vehicle per 100 km. The vehicle k departing from customer i consumes $G(Z_{ik})$ liters of fuel for every 100 km traveled. The greenhouse gas emission of diesel is 2.6569 kg per liter. The greenhouse gas emission of Liquefied Petroleum Gas (LPG) is 1.5301 kg per liter. Table 1 is the estimation of greenhouse gas emission factor of diesel and LPG under different vehicle loads. Both Table 1 and Equation (12) refers from Kunnapapdeelert, S. et al. [7], whose table of greenhouse gas emission is obtained from Ubeda et al. [17] and Defra [4].

Table 1. Estimation of greenhouse gas emission for a 10 tons capacity trunk

Vehicle state	Weight laden (%)	Emission factor for diesel (kg CO_2)	Emission factor for LPG (kg CO_2)
Empty	0	0.786442	0.452910
Low loaded	25	0.850208	0.489632
Half loaded	50	0.913974	0.526354
High loaded	75	0.975082	0.561547
Full loaded	100	1.036191	0.596739

3   Distance-Based Adaptive Large Neighborhood Search

Based on Adaptive Large Neighborhood Search (ALNS), we propose Distance-based Adaptive Large Neighborhood Search (DALNS). DALNS similarly performs destroying and repairing to obtain a better solution, but the distance between customers will show more influence in each iteration, comparing with ALNS.

The overall algorithm flow of this paper is shown in Algorithm 1. The initial solution is formed by Poolseperate and Distance-based insert (line 1 and 2). Then the feasible solution is operated by DALNS repeatedly (line 7 to 17). We use four strategies to destroy (line 8 to 10), and adaptive regret-k to repair (line 11 to 12). q is the number of orders to be removed. We add the heuristic of q to improve the efficiency of the search (line 9). The whole algorithm uses simulate annealing (SA) to judge the newly generated solution for the next iteration (line 13 to 16). In simulate annealing, a worse solution may be accepted in a random probability, when $U(0,1)$ represents the continuous uniform distribution (line 13). The initial temperature T is set to T_0 (line 5) and updated by a cooling constant c (line 18), which is introduced by Kirkpatrick, Gelatt and Vecchi [6]. The procedure of DALNS stops when the temperature T is smaller than T_f, which relates to the number of cooling iteration f.

3.1   Initial Solution Generation

Considering distance is an important factor that can cause high emissions when constructing the initial solution, we try to find a way that can easily shorten the total distance traveled when searching in the solution's neighborhood. PD-pair means a pickup order and its corresponding delivery order. The method

Algorithm 1: Distance-based Adaptive Large Neighborhood Search

---

Input: Set P, a set of all orders
Output: Best Solution S_{best}

1 $P_{separate} \leftarrow$ PoolSeparate(P);
2 $S_{init} \leftarrow$ DistanceBasedInsert($P_{separate}$);
3 $S_{cur} \leftarrow S_{init}$;
4 $S_{best} \leftarrow S_{init}$;
5 $T \leftarrow T_0$;
6 initialize $removeHeuristic$, $qHeuristic$, $kHeuristic$;
7 **while** $T > T_f$ **do**
8   choose a remove way by $removeHeuristic$;
9   determine q by $qHeuristic$;
10     $S_{removed} \leftarrow$ Remove(S_{cur}, q);
11   determine k by $kHeuristic$;
12     $S_{new} \leftarrow$ Repair($S_{removed}$);
13   **if** $emission(S_{new}) < emission(S_{cur}) - T ln(U(0,1))$ **then**
14         $S_{cur} \leftarrow S_{new}$;
15   **if** $emission(S_{new}) < emission(S_{best})$ **then**
16         $S_{best} \leftarrow S_{new}$;
17   update $removeHeuristic$, $qHeuristic$, $kHeuristic$;
18 $T \leftarrow cT$;
19 **return** S_{best};

---

of constructing an initial solution is mainly divided into two steps: (1) pre-processing all orders and divide all PD-pairs; (2) forming solution by using order pools.

The division is made as follows: To group multiple orders into several order pools based on the distance of pickup customer and delivery customer, we sort PD-pairs by Euclidean Distance of pickup customer and delivery customer from the furthest to the nearest at first. An order pool has only one master PD-pair, but it can be empty if no PD-pair belongs to it. A master PD-pair cannot belong to any other order pool, but a PD-pair which is not a master PD-pair can belong to many different order pools at the same time. PD-pairs in the same order pool have two characteristics. One is that the time window of each order in the pool cannot over the earliest service time of the master pickup customer and the latest service time of the master delivery customer. Another is, any customer of the PD-pair in the pool can be reached by a vehicle starting from the master pickup customer and end at the master delivery customer, within the earliest service time of the master pickup customer and the latest service time of the master delivery customer. We define $TWdist$ as the longest distance that a vehicle can run within this time range.

Figure 1 is an example of forming an order pool. Node P is the master pickup customer and node D is the master delivery customer. The distance of P_3P and P_3D in Fig. 1b is shorter than the $TWdist$ of this order pool, and distance of D_3P

Fig. 1. An example of forming *pool*(*P*).

and D_3D also shorter than $TWdist$, so the PD-pair corresponding to customer P_3 and D_3 will be added to the pool(P). However, P_1D_1 and P_2D_2 in Fig. 1c and Fig. 1d do not meet the requirements of $TWdist$, so their PD-pair will not be added to this order pool.

After gathering geographically adjacent customers, the second step in forming the initial solution is to use the order pool to form the longest route in each order pool. A route forming operation performs when there is undistributed PD-pairs. Only one route is generated per order pool one time, and an order pool may generate several routes finally. When generating routes, to gather customers together, we propose distance-based insertion to achieve this purpose and show the pseudo-code in Algorithm 2.

Distance-based insertion performs when there is undistributed PD-pairs. Insertion will be repeated in every pool until all PD-pairs are distributed (line 1 to 4). For each undistributed PD-pair, if it can be successfully inserted into a route, then the new route will be compared with the best route with the largest number of inserted customers. If there are more customers in the new route, then the best route will be replaced by the new route (line 12 to 14). If the PD-pair cannot be inserted into the forming route (line 7), and there have been n insertion failures (line 8), then one of the PD-pairs in the current forming route will be randomly removed (line 9), and the insertion will restart from the first n PD-pairs that fail to insert (line 10). After the route has been formed, the newly formed route will try to merge with other routes (line 16 to 17). If it fails to merge, the newly formed route will be added to the initial solution as an independent route (line 19). After all PD-pairs have been distributed, an initial solution is formed.

3.2 Destroy Strategies

In this step, we set four strategies, including random removal, worst removal, Shaw removal [14], and distance-based removal. We propose distance-based removal to comprehensively consider the distance between a customer and its neighbor customers.

Algorithm 3 shows how distance-based removal works.

First of all, we need to build an adjacent list for every customer, because all adjustments are based on distance. Each adjacent list is used to store the position of m customers near the list owner customer (line 1). A seed customer is randomly selected to decide which region in the solution will be adjusted (line

---

Algorithm 2: Distance Based Insertion

Input: Set $P_{separate}$, a set of separated orders
Output: Initial solution S_{init}

1 Initialize S_{init};
2 **while** S_{init} *is infeasible* **do**
3   **for** *each order pool* $pool(p^+)$ **do**
4   **if** $pool(p^+)$ *has undistributed order* **then**
5   **for** *all I undistributed orders in* $pool(p^+)$ **do**
6                  $p'^+p'^- \leftarrow pool(p^+)[i]$;
7   **if** $p'^+p'^-$ *can't insert in* $Route_{pool(p+)}$ **then**
8   **if** *this is the* n^{th} *failure of insertion* **then**
9   remove a pair of customers randomly from $Route_{pool(p+)}$;
10                     $i \leftarrow i - n$;
11   **else**
12   insert $p'^+p'^-$ in $Route_{pool(p+)}$;
13   **if** $Route_{pool(p+)}.length > Routebest_{pool(p+)}.length$ **then**
14                     $Routebest_{pool(p+)} = Route_{pool(p+)}$;
15   **for** $Route_s \leftarrow S_{init}$ **do**
16   **if** $Routebest_{pool(p+)}$ *can merge with* $Route_s$ **then**
17                 $Route_s = \text{merge}(Routebest_{pool(p+)}, Route_s)$;
18   **else**
19   add $Routebest_{pool(p+)}$ in S_{init};

20 **return** S_{init};

---

2). After deciding the seed customer, the customer, and its pick up or delivery customer will be added in the remove list (line 3). The adjacent list of the seed customer will be traversed (line 4). Starting from the customer closest to the seed customer, two judgments are performed: (1) whether the route where the neighbor customer is located has been damaged (line 6); (2) whether there are other neighbor customers located on the same route (line 8). If more than one neighbor customer can be found on the same undamaged route, then the first two neighbor customers of the seed customer are used as the endpoints of a string, and the entire string of the route will be added into remove list (line 9), otherwise just remove N_l and its corresponding pickup or delivery customer (line 11). If the remove list stores more than q orders, then randomize the remove list and remove pairs of customers randomly until the number of orders reaches to q (line 14 to 15).

Usually, the number of orders needed to be removed is random. However, when the search range is converging, a large number of removed orders cause a long time to repair the solution, which decreases the efficiency of searching. Thus we add a heuristic on deciding how many orders will be removed in this step. We regard the value range of q as an evaluable object. Evaluation happens

---

Algorithm 3: Distance Based Removal

    Input: Solution S, solution need to be destroyed; Integer q, number of PD-pairs
        need to be removed

    Output: Solution $S_{destroyed}$, an infeasible solution

1 Create a list $L_{adjacent}$, a set of list of each customer's m neighbor customers;

2 N_{seed} = a random customer;

3 insert N_{seed} into L_{remove};

4 while $L_{adjacent}.length < q$ **do**

5   **for** *each customer N_l in $L_{adjacent}(N_{seed})$* **do**

6   **if** *Route(N_l) has not been destroyed* **then**

7   **for** *each customer N_r in Route(N_l)* **do**

8   **if** *N_r is in $L_{adjacent}(N_{seed})$* **then**

9   **return** string $N_l N_r$ and customers associated

10   **if** *no customer of $L_{adjacent}(N_{seed})$ in Route(N_l) besides N_l* **then**

11   **return** N_l and its pickup or delivery customer

12   add customers that need to be removed in L_{remove};

13   **if** $L_{removed}.length > q$ **then**

14 $Randomize(L_{remove})$;

15   keep q PD-pairs in L_{remove};

16 return $S_{destroyed}$;

---

every 100 iterations. We evaluate each range from weights and update weights for the next 100 iterations. The range of values is the percentage range of the total number of orders. In our experiment, the percentage ranges are 20%–45%, 35%–65%, 60%–75% and 80%–85%. After the percentage range is determined, the value of q will be randomly obtained within the value range.

3.3   Repair Strategies

The third step of DALNS is to repair the damaged solution, that is, to insert the customers removed from the last step back into the solution. The algorithm used at this step is an adaptive regret-k algorithm [11]. The algorithm is an improvement based on the optimal insertion. The value of regret-k means the gap between the customer's lowest inserting cost and the customer's k^{th} low inserting cost. In regret-k, the chosen customer to be inserted in each iteration is the customer who has the largest regret-k value, and the value of k is usually a constant. But regret-k heuristic will set k to a value that performs well in the former iterations. In our repair, the value of k is one of 1, 2, 3, 4, and random.

4   Results and Discussion

Our experiments are executed using the 100-customer instances of Li and Lim's benchmark [8], in which the vehicle capacity of each instance is fixed to 200, the maximum number of the vehicle used shall not exceed 25, and the vehicle speed is 1. Vary from the distribution of customers, Li and Lim's instance can be divided into three types: half-random and half-clustered distribution, clustered distribution, and random distribution. NV in each table represents the number of vehicles used and TD represents the total distance traveled. The experimental environment is Intel Core i7-8700 (3.20 GHz, 8 GB RAM). The code of the algorithm is written in C++, and the integrated development environment is Visual Studio 2017.

4.1 Results for pdp_100_lrc1 Instances

The best results of the DE algorithm is given in the study of Kunnapadeelert and Kachitvichyanukul [7]. It can be seen from Table 2 that three instances show a great reduction of CO_2 emission. And we can get a conclusion from Table 2 that DALNS is a better choice not only in terms of environmental friendliness but also in terms of economic benefits.

Table 2. Results for pdp_100_lrc1 instances in Green-PDPTW

Instance	Algorithm	NV	TD	CO_2 emissions from diesel (kg. CO_2)	CO_2 emissions From LPG (kg. CO_2)	Gap of emission
lrc101	DALNS	15	1703.21	1363.17	785.047	21.37%
	DE	16	1872.64	**1123.19**	**646.843**	
lrc102	DALNS	12	1558.07	1249.32	719.48	11.21%
	DE	17	2194.68	**1123.39**	**646.965**	
lrc103	DALNS	11	1258.74	1007.97	580.487	4.38%
	DE	13	1629.16	**965.699**	**556.142**	
lrc104	DALNS	10	1128.40	**905.077**	**521.231**	**−44.76%**
	DE	11	1325.74	1638.42	943.563	
lrc105	DALNS	13	1637.62	1309.42	754.089	21.93%
	DE	15	1859.26	**1073.94**	**618.476**	
lrc106	DALNS	11	1424.73	1143.41	658.484	4.65%
	DE	14	1880.24	**1092.57**	**629.212**	
lrc107	DALNS	11	1230.14	**990.092**	**570.191**	**−27.20%**
	DE	16	2233.55	1360.05	783.245	
lrc108	DALNS	10	1147.43	**925.184**	**532.81**	**−40.71%**
	DE	13	1648.92	1560.56	898.727	
average	DALNS	12	1386.04	**1111.71**	**640.23**	**−10.51%**
	DE	15	1830.52	1242.23	715.40	

4.2   Results for pdp_100_3large and pdp_100_4large Instances

In the original instance of Li and Lim's benchmark, the load demand of each order has little influence on a load of vehicles, while the greenhouse gas emission of the vehicle is not only affected by the distance, but also by a load of vehicles. Therefore, based on the original instances, we generate pdp_100_3large and pdp_100_4large instances. $Xlarge$ means the load demand of each order is X times of the original instance.

In the former experiments, DALNS gets all the optimal solutions of pdp_100 instances shown on the Sintef website [15], so we use DALNS to obtain the optimal solutions of pdp_100_3large and pdp_100_4large under PDPTW, which is the comparison result of DALNS under Green-PDPTW.

Table 3 shows the average results for the newly generated instances. It shows that the increase of the load demand indeed leads to the increase of greenhouse gas emissions, and more vehicles are used to meet a larger load demand, which results in a longer total distance traveled. However, it also shows that the performance of pdp_100_4large is not better than that of pdp_100_3large. In our analysis, the reason is: When searching for the optimal solution with the shortest total distance traveled, the algorithm performance in both two models is very similar. However, if the vehicle is too close to full load, it will limit the adjustable space of routes, which greatly affects the reduction of carbon emissions.

Table 3. Average results for newly generated instances

Instance	Model	NV	TD	Emissions from diesel (kg. CO_2)	Decrease from diesel (kg. CO_2)	Emissions from LPG (kg. CO_2)	Decrease from LPG (kg. CO_2)	Gap of emission
3large_lc1	PDPTW	11	1136.14	950.63	121.58	547.46	70.02	−11.79%
	GPDPTW	13	1013.22	829.05		477.44		
3large_lr1	PDPTW	13	1264.94	1054.71	4.22	607.41	2.43	−0.37%
	GPDPTW	13	1260.22	1050.49		604.98		
3large_lrc1	PDPTW	13	1448.40	1202.14	13.47	692.31	7.72	−1.13%
	GPDPTW	13	1436.59	1188.67		684.55		
4large_lc1	PDPTW	12	1219.30	1027.20	100.98	591.56	58.16	−8.79%
	GPDPTW	14	1121.69	926.22		533.41		
4large_lr1	PDPTW	13	1332.87	1128.89	11.36	650.12	6.54	−1.00%
	GPDPTW	14	1324.65	1117.52		643.58		
4large_lrc1	PDPTW	13	1531.15	1283.42	8.09	739.12	4.66	−0.58%
	GPDPTW	14	1523.67	1275.33		734.46		

Another meaningful discovery is that, in clustered-customer instances, though the number of vehicles used is increasing, the total distance traveled is shortened. This is more conducive to DALNS to have a good performance on optimization.

In conclusion, DALNS can achieve good performance in the instances of Li and Lim's benchmark. Besides, from the second experiment, we draw the conclusion that DALNS is more suitable for the problem with clustered customers and high vehicle load. It should be noted that Green-PDPTW does not perform better with the increase of vehicle load. On the contrary, a full load may make the reduction of greenhouse gas emissions more difficult due to the difficulty of vehicle load adjustment.

5   Conclusion and Future Work

Green-PDPTW is an extension of PDPTW and Green-CVRP. As a sub-problem of CVRP, it takes order distribution, time window, and greenhouse gas emissions into consideration at the same time. A feasible solution to this problem effectively guarantees the efficiency of the logistics transportation process and reduces the negative impact of transportation on the environment. Therefore, research of Green-PDPTW has a very positive impact on real industrial production.

In this paper, we study Green-PDPTW and use DALNS to obtain the optimal solution. We introduce the concept of the order pool and propose an initial solution generation algorithm based on distance. All orders are divided into order pools at first, and then routes are generated based on order pools. We propose distance-based insertion to generate routes. The objective of the distance-based insertion is to provide good modification conditions for subsequent improvement. When destroying the solution, the concept of the adjacent list is introduced to help search the neighbor customers of a randomly selected seed customer, which makes the process of shortening the total distance traveled efficiently. What's more, we add a heuristic on the number of orders to be removed to improve the search efficiency.

Another contribution of this paper is the comparative study of the application of DALNS. In the simulation experiment, we find that DALNS is more suitable for solving the problem with clustered customers. Besides, when shortening the total distance traveled, if the load of the vehicle is too close to full load, it would be hard to make a significant change.

Compared with other researches, we take more consideration about geographical factors in solving vehicle routing problems, which is beneficial to decrease the total distance traveled. What's more, we form the initial solution for being improved easily, which helps get the real best solution in searching.

Greenhouse gas emission is not only affected by the distance, but also by a load of all vehicles. DALNS is a distance-based optimization algorithm, so it is more sensitive to the change of distance. In the further study of Green-PDPTW, to reduce greenhouse gas emission by emphasizing the balance of a load of vehicles is a good entrance.

Acknowledgement. This work is financially supported by National key R&D program under Grant No. 2017YFB0803002 and No. 2016YFB0800804, National Natural Science Foundation of China under Grant No. 61672195 and No. 61732022.

References

1. Affi, M., Derbel, H., Jarboui, B.: Variable neighborhood search algorithm for the green vehicle routing problem. Int. J. Ind. Eng. Comput. **9**(2), 195–204 (2018)

2. Christiaens J., Vanden Berghe, G.: Slack induction by string removals for vehicle routing problems. Institute for Operations Research and the Management Sciences (02) (2019)

3. Dantzig, G.B., Ramser, J.H.: The truck dispatching problem. Manag. Sci. **6**(1), 80–91 (1959)

4. Defra, J.: Environmental reporting guidelines: including mandatory greenhouse gas emissions reporting guidance, report. Department for Environment Food & Rural Affairs. https://www.gov.uk/government/publications. Accessed 17 Apr 2020

5. Hemmelmayr, C.V., Cordeau, J.F., Cranic, T.: An adaptive large neighborhood search heuristic for two-echelon vehicle routing problems arising in city logistics. Comput. Oper. Res. **39**, 3215–3228 (2012)

6. Kirkpatrick, S., Gelatt, C., Vecchi, M.P.: Optimization by simulated annealing. Science **220**(4598), 671–680 (1983)

7. Kunnapapdeelert, S., Kachitvichyanukul, V.: New enhanced differential evolution algorithms for solving multi-depot vehicle routing problem with multiple pickup and delivery requests. Int. J. Serv. Oper. Manag. **31**(3), 370–395 (2018)

8. Li, H., Lim, A.: A metaheuristic for the pickup and delivery problem with time windows. Int. J. Artif. Intell. Tools **12**(02), 173–186 (2003)

9. Montoya, A., Gueret, C., Mendoza, J.E., Villegas, J.G.: A multi-space sampling heuristic for the green vehicle routing problem. Transp. Res. Part C: Emerg. Technol. **70**, 113–128 (2016)

10. Prescott-Gagnon, E., Desaulniers, G., Rousseau, L.-M.: Heuristics for an oil delivery vehicle routing problem. Flex. Serv. Manuf. J. **26**(4), 516–539 (2012). https://doi.org/10.1007/s10696-012-9169-9

11. Ribeiro, G.M., Laporte, G.: An adaptive large neighborhood search heuristic for the cumulative capacitated vehicle routing problem. Comput. Oper. Res. **39**, 728–735 (2012)

12. Roberto, D., Mauceri, S., Carroll, P.: A genetic algorithm for a green vehicle routing problem. Electron. Notes Discret. Math. **64**, 65–74 (2018)

13. Ropke S., Pisinger D.: An adaptive large neighborhood search heuristic for the pickup and delivery problem with time windows. Technical report, Department of Computer Science, University of Copenhagen (2004)

14. Shaw, P.: A new local search algorithm providing high quality solutions to vehicle routing problems. APES Group. Dept of Computer Science, University of Strathclyde, Scotland, UK (1997)

15. Li, S., Benchmark, L. https://www.sintef.no/projectweb/top/pdptw/li-lim-benchmark/100-customers. Accessed 17 Apr 2020

16. Toth, P., Vigo, D. (eds): The Vehicle Routing Problem. Monographs on Discrete Mathematics and Applications, p. 9. Society for Industrial and Applied Mathematics, Philadelphia (2002)

17. Ubeda, S., Faulin, J., Serrano, A., Arcelus, F.J.: Solving the green capacitated vehicle routing problem using a tabu search algorithm. Lect. Notes Manag. Sci. **6**(1), 141–149 (2014)

The Theories of a Novel Filled Function Method for Non-smooth Global Optimization

Wei-xiang Wang[1] , You-lin Shang[2]([⊠]) , and Shuo Li[2]

[1] Shanghai Polytechnic University, Shanghai 201209, China
zhesx@126.com
[2] Henan University of Science and Technology, Luoyang 471003, China
mathshang@sina.com, 3216748517@qq.com

Abstract. This paper proposes a novel filled function method for non-smooth box constrained global optimization. The constructed filled function contains two parameters, which could be easily adjusted during the process of terations. The theoretical and numerical properties of the filled function are studied, and a filled function algorithm is given. Finally, several numerical results, including the application of the filled function method in solving nonlinear equations, are reported.

Keywords: Non-smooth global optimization · Filled function method · Global minimizer · Nonlinear equations · NCP problem

1 Introduction

Lots of problems in science and engineering fields are increasingly dependent on the need to find out the global optimizers, and global optimization now becomes one of the most hot topics in optimization. There are many existing algorithms could be used to obtain those global optimizers, including stochastic methods and deterministic methods. One of the favourable deterministic methods is filled function method, which was firstly proposed by Ge [1] for continuous smooth global optimization problem $(P) : \min_{x \in X} f(x)$, where X is a box set. The filled function method uses an auxiliary function called filled function to escape from a given local minimizer x^*. It contains two phases. Phase 1 searches for one local minimizer of (P) by any local minimization procedure. When the phase 1 finished, filled function method switches to phase 2. Phase 2 constructs a filled function and then minimizes it to obtain an improved starting point for phase 1. These two phases repeated until no better minimizers could be found for the original problem. The filled function proposed in [1] contains an exponential term, which might give rise to failure of computation when the values of the

Supported by organization National Natural Science Foundation of China (No. 11471102), Basic research projects for key scientific research projects in Henan Province (No. 20ZX001).

Z. Zhang et al. (Eds.): AAIM 2020, LNCS 12290, pp. 381–390, 2020.
https://doi.org/10.1007/978-3-030-57602-8_34

exponential term increases rapidly. Later, the filled function was reconsidered in [2–6]. Note that the filled function defined in [2] is an improved version in [1]. One major issue associated with the filled function method in [2] is that the minimization of filled function is not performed in the problem domain, but on a line connecting the current optimizer and a point in some neighborhoods of the next better optimizer which is unknown. Finding out such a direction will cause a high degree of computational difficulty. The filled functions given in [3–6] were originally proposed for smooth global optimization. However, in practice, many problems are non-smooth global optimization problems. In this paper, we extend the filled function methods for smooth global optimization to include non-smooth case.

Generally speaking, there are two difficulties faced by the global optimization, one issue is how to escape from the current optimizer to locate a better optimizer, and the another is how to check the current optimizer is a global one. This paper focuses only on the former issue.

This paper is organized as follows: In Sect. 2, we make some assumptions on the problem (P), and construct a filled function and discuss its theoretical properties. In Sect. 3, we give an corresponding filled function algorithm. Finally, we provide several numerical results, including the application of the filled function method in solving nonlinear equations.

2   A Filled Function and Its Properties

In this section, we first make some assumptions on the objective function and then define a filled function for non-smooth global optimization.

Assumption 1. The function $f(x)$ Lipschitz continuous on X with a rank $L > 0$.

Assumption 2. The problem (P) has at least one global minimizer and has a finite number of different minimal function values.

The main tool used in the non-smooth filled function method is Clark generalized gradient. For more details about its properties, please refers to [9].

Denote $L(P)$ the set of the minimizers of the problem (P) and let $x^* \in L(P)$.

Definition 1. *A function $P(x, x^*)$ is called a filled function of $f(x)$ at $x^* \in L(P)$, if it has the following properties:*

1. *x^* is a strictly maximizer of $P(x, x^*)$;*
2. *Any point $x^* \neq x \in X$ satisfying $f(x) \geq f(x^*)$ cannot be a local minimizer or a saddle point of $P(x, x^*)$;*
3. *If x^* is not a global minimizer, then $P(x, x^*)$ has at least one minimizer in the set $S_2 = \{x \in X : f(x) < f(x^*)\}$.*

Now, we present an auxiliary function as follows:

$$P(x, x^*, q, r) = -\ln(1 + \|x - x^*\|^2) + q\max^2(0, f(x) - f(x^*)) \\ + r\arctan(\min^3(0, f(x) - f(x^*))). \tag{1}$$

where $q > 0$ and $r > 0$ are two parameters. Let $D = \max_{x_1, x_2 \in X} \|x_1 - x_2\|$.
In the following, we will prove that $P(x, x^*, q, r)$ is a filled function.

Theorem 1. *Let $x^* \in L(P)$. If $q \geq 0$ is chosen to be suitable small, then x^* is a strict local maximizer of $P(x, x^*, q, r)$.*

Proof. Since $x^* \in L(P)$, there exists a neighborhood $N(x^*, \sigma^*)$ of x^* such that $f(x) \geq f(x^*)$ for all $x \in N(x^*, \sigma^*) \cap X$, where $0 < \sigma^* < 1$ is a constant.

From the inequality that: $-\ln(1 + y) \leq \dfrac{x^2}{2} - y$, for every $y \geq 0$.

We have, for $x^* \neq x \in N(x^*, \sigma^*) \cap X$, if $q < \dfrac{1 - 0.5(\sigma^*)^2}{L^2}$, then

$$\begin{aligned} P(x, x^*, q, r) &= q(f(x) - f(x*))^2 - \ln(1 + \|x - x^*\|^2) \\ &\leq qL^2\|x - x^*\|^2 - \|x - x^*\|^2 + \frac{1}{2}\|x - x^*\|^4 \\ &\leq (qL^2 - 1 + \frac{1}{2}(\sigma^*)^2)\|x - x^*\|^2 \\ &< 0 = P(x^*, x^*, q, r). \end{aligned} \tag{2}$$

This shows that x^* is a strict local maximizer of $P(x, x^*, q, r)$.

Theorem 2. *Let $x^* \in L(P)$. If $0 \leq q < \dfrac{1}{L^2(1 + D^2)}$, then for any $x \neq x^*$ and $f(x) \geq f(x^*)$, we have $0 \notin \partial P(x, x^*q, r)$.*

Proof. By the conditions, it holds

$$\partial P(x, x^*, q, r) \subseteq 2q(f(x) - f(x^*))\partial f(x) - \dfrac{2(x - x^*)}{1 + \|x - x^*\|^2}. \tag{3}$$

Thus, we have

$$\begin{aligned} \langle x - x^*, \partial P(x, x^*, q, r) \rangle &\leq -\dfrac{2\|x - x^*\|^2}{1 + \|x - x^*\|^2} + 2qL^2\|x - x^*\|^2 \\ &\leq 2\|x - x^*\|^2(qL^2 - \dfrac{1}{1 + D^2}) \\ &< 0. \end{aligned} \tag{4}$$

Hence, we have $0 \notin \partial P(x, x^*q, r)$.

Theorem 3. *Let $x^* \in L(P)$, and suppose that x_1 and x_2 are two points such that $\|x_1 - x^*\| < \|x_2 - x^*\|$ and $f(x_2) > f(x_1) > f(x^*)$. If $0 \leq q < \dfrac{1}{LM(1 + D^2)}$, where*

$$M \geq L \frac{\|x_2 - x_1\|}{\|x_2 - x^*\| - \|x_1 - x^*\|}$$
$$\geq \frac{\|x_2 - x_1\|}{\|x_2 - x^*\| - \|x_1 - x^*\|} \sup_{0 \leq \lambda \leq 1} \max_{\varsigma \in \partial f(x_1 + \lambda(x_2 - x_1))} \left| \left\langle \varsigma, \frac{x_2 - x_1}{\|x_2 - x_1\|} \right\rangle \right|. \tag{5}$$

Then $P(x_2, x^*, q, r) < P(x_1, x^*, q, r) < P(x^*, x^*, q, r)$.

Proof. By the mean value theorem, there exists a constant $\theta \in (0,1)$, such that

$$\begin{aligned}
&\ln(1 + \|x_2 - x^*\|^2) - \ln(1 + \|x_1 - x^*\|^2) \\
&= \frac{1}{1 + \theta \|x_2 - x^*\|^2 + (1 - \theta)\|x_1 - x^*\|^2} (\|x_2 - x^*\|^2 - \|x_1 - x^*\|^2) \\
&\geq \frac{1}{1 + D^2} (\|x_2 - x^*\|^2 - \|x_1 - x^*\|^2).
\end{aligned} \tag{6}$$

Thus, we have

$$\begin{aligned}
&P(x_2, x^*, q, r) - P(x_1, x^*, q, r) \\
&\leq (\|x_2 - x^*\|^2 - \|x_1 - x^*\|^2)(-\frac{1}{1 + D^2} + q\frac{f(x_2) - f(x_1)(f(x_2) - f(x^*) + f(x_1) - f(x^*))}{\|x_2 - x^*\|^2 - \|x_1 - x^*\|^2}) \\
&\leq (\|x_2 - x^*\|^2 - \|x_1 - x^*\|^2)(-\frac{1}{1 + D^2} + \frac{qL^2 \|x_2 - x_1\|}{\|x_2 - x^*\| - \|x_1 - x^*\|}). \\
&\leq (\|x_2 - x^*\|^2 - \|x_1 - x^*\|^2)(-\frac{1}{1 + D^2} + qLM) < 0
\end{aligned}$$

Therefore, the theorem follows from the proof of theorem1.

Theorem 4. *Let* $x^* \in L(p)$, *If* $q \geq 0$, *Then*

$$P(x_2, x^*, q, r) < P(x_1, x^*, q, r) < P(x^*, x^*, q, r). \tag{7}$$

Proof. Since the following holds

$$\begin{aligned}
&P(x_2, x^*, q, r) - P(x_1, x^*, q, r) \\
&= -\ln\left(1 + \|x_2 - x^*\|^2\right) + \ln\left(1 + \|x_1 - x^*\|^2\right) \\
&\quad + q((f(x_2) - f(x^*))^2 - (f(x_1) - f(x^*))^2) \\
&< 0
\end{aligned} \tag{8}$$

thus, the theorem follows from the proof of theorem 1.

Theorem 5. *Let* $x^* \in L(p)$, *and* x_1 *be a point such that* $f(x_1) > f(x^*)$. *If* $q > 0$ *is chosen to be suitable small, then for any small* $\varepsilon > 0$, *there exists d such that*

$$0 < \|d\| \leq \varepsilon, \|x_1 - d - x^*\| < \|x_1 - x^*\| < \|x_1 + d - x^*\|, f(x_1 \pm d) \geq f(x_1)$$

and

$$P(x_1 + d, x^*, q, r) < P(x_1, x^*, q, r) < P(x_1 - d, x^*, q, r) < P(x^*, x^*, q, r). \tag{9}$$

Proof. For a given $\varepsilon > 0$, let $d = \varepsilon \dfrac{x_1 - x^*}{2\|x_1 - x^*\|}$. Then $0 < \|d\| \leq \varepsilon$, Furthermore, if $\varepsilon > 0$, is sufficiently small and the condition on M in the Theorem 3 is satisfied, then we have

$$\|x_1 + d - x^*\| = \|x_1 - x^*\|\left(1 + \tfrac{\varepsilon}{2}\|x_1 - x^*\|\right) > \|x_1 - x^*\|$$
$$\|x_1 - d - x^*\| = \|x_1 - x^*\|\left(1 - \tfrac{\varepsilon}{2}\|x_1 - x^*\|\right) < \|x_1 - x^*\| \tag{10}$$
$$f(x_1 \pm d) \geq f(x^*).$$

Thus, if $q \geq 0$ is chosen to be suitable small, then the following inequality $P(x_1 + d, x^*, q, r) < P(x_1, x^*, q, r) < P(x_1 - d, x^*, q, r) < P(x^*, x^*, q, r)$ follows directly from the Theorem 3 and Theorem 4.

Theorem 6. *Assume that $x^* \in L(p)$ is not a global minimizer, then there exists a point $x_0 \in S_2 = \{x \in X : f(x) < f(x^*)\}$ such that x_0 is a local minimizer of $P(x, x^*, q, r)$.*

Proof. Let $S_3 = \{x \in X : f(x) \leq f(x^*)\}$ and $\partial S_2 = \{x \in X : f(x) = f(x^*)\}$. Obviously, both ∂S_2 and S_3 are compact sets. For any $x_1 \in \partial S_2$,

$$P(x, x^*, q, r) = -\ln\left(1 + \|x - x^*\|^2\right). \tag{11}$$

Since ∂S_2 is a compact set, there exists a point $x_1 \in \partial S_2$, such that

$$\min_{x \in \partial S_2} -\ln\left(1 + \|x - x^*\|^2\right) = -\ln\left(1 + \|x_1 - x^*\|^2\right)$$
$$= P(x_1, x^*, q, r). \tag{12}$$

On the other hand, since $x^* \in L(P)$ is not a global minimizer, there exists a point $x_2 \in S_2$ such that

$$P(x_2, x^*, q, r) = -\ln\left(1 + \|x_2 - x^*\|^2\right) + r\arctan\left(f(x_2) - f(x^*)\right)^3. \tag{13}$$

Therefore, when

$$r > \frac{\ln\left(1 + \|x_1 - x^*\|^2\right) - \ln\left(1 + \|x_2 - x^*\|^2\right)}{\arctan\left(f(x^*) - f(x_2)\right)^3}, \tag{14}$$

We have

$$P(x_2, x^*, q, r) < P(x_1, x^*, q, r).$$

Fix r, since S_3 is a compact set, $P(x, x^*, q, r)$ must have a global minimizer x_0 on S_3. Note that

$$\min_{x \in S_3} P(x, x^*, q, r) = \min_{x \in S_2} P(x, x^*, q, r)$$
$$= P(x_0, x^*, q, r) \tag{15}$$
$$< P(x_1, x^*, q, r),$$

We have $x_0 \in S_2$, and x_0 is a local minimizer of $P(x, x^*, q, r)$.

3   Filled Function Algorithm

Based on the theoretical results of filled function obtained in Sect. 2, we give a filled function algorithm below.

Filled function algorithm

Initialization step:
Let q_l be the lower bound of parameter q, r_u the upper bound of parameter r, x_1 the initial point and $e_1, e_2, \cdots, e_{2n}$ the positive and negative coordinate directions. Set $k = 1$, and go to the main step.

Main step:

1. Starting from x_1, minimize (P) by any non-smooth local minimization procedure to find a local minimizer x_1^* and go to 2;
2. Set $q = 1$ and $r = 1$;
3. Construct a filled function $P(x, x_1^*, q, r)$ and go to 3;
4. If $k > 2n$, then go to 7; otherwise, set $x = x_1^* + 0.1e_k$, and take x as an initial point to find a local minimizer x_k of the following problem:

$$\min_{y \in X} P(y, x_1^*, q, r);$$

5. If $x_k \notin X$, then set $k = k + 1$, and go to 4; otherwise, go to 6;
6. If $f(x_k) < f(x_1^*)$, then, (a) set $x = x_k, k = 1$. (b) Use x as a new initial point and minimize (P) to find its another local minimizer x_2^* with $f(x_2^*) < f(x_1^*)$. (c) Set $x_1^* = x_2^*$ and go to 2; Else if $f(x_k) \geq f(x_1^*)$, then go to 7;
7. Reduce q by setting $q = 0.1q$. If $q \geq q_l$, then set $k = 1$, and go to 3; otherwise, go to 8;
8. Increase r by setting $r = 10r$. If $r \leq r_u$, then set $k = 1$, and go to 3; otherwise, take x_1^* as a global minimizer, and the algorithm stops.

Remarks:

1. The proposed filled function method can also be applied to smooth box constrained global optimization problem.
2. There are two phases in the filled function method: local minimization and filling. In phase 1, a local minimizer x^* is located by any non-smooth local minimization algorithms, such as Hybrid Hooke and Jeeves-Direct Method for Non-smooth Optimization [8], Mesh Adaptive Direct Search Algorithms for Constrained Optimization [7], Bundle methods, Powell's method, etc. In particular, the Hybrid Hooke and Jeeves-Direct Method is more preferable to others, since it is guaranteed to find a local minimum of a non-smooth function subject to simple bounds. In phase 2, the constructed filled function $P(x, x^*, q, r)$ is minimized. During the minimization, if a point x_k is found such that $f(x_k) < f(x^*)$, then phase 2 stops and the algorithm returns to phase 1 to find a better optimizer for $f(x)$. The aforementioned process repeats until the global minimizer is identified.

4   Numerical Experiment

The proposed filled function method has lots of applications. In this section, we perform a few numerical tests including two applications of the filled function method in solving nonlinear equations.

Problem 1:

$$\min f(x) = \left|\frac{x-1}{4}\right| + \left|\sin(\pi(1 + \frac{x-1}{4}))\right| + 7, |x| \le 10. \tag{16}$$

The algorithm successfully found a global solution: $x^* = 1$ with $f(x^*) = 7$. Table 1 records the numerical results of Problem 1.

Table 1. Computational results for Problem 1.

k	x_k^0	$f(x_k)$	x_k^*	$f(x_k^*)$
1	6.0000	8.9571	5.0000	8.0001
2	0.9678	7.0333	0.9998	7.0001

Problem 2

$$\min f(x) = \max\{5x_1 + x_2, -5x_1 + x_2, x_1^2 + x_2^2 + 4x_2\}, -4 \le x_1, x_2 \le 4 \tag{17}$$

The algorithm successfully found a global solution: $x^* = (0, -3)$ with $f(x^*) = -3$. Table 2 records the numerical results of Problem 2.

Table 2. Computational results for Problem 2.

k	x_k^0	$f(x_k)$	x_k^*	$f(x_k^*)$
1	(1, 1)	6.0000	(0.0000, 0.0000)	0.0000
2	(−0.0002, −0.9725)	−0.9715	(−0.0002, −0.9725)	−0.9715
3	(−0.0003, −2.5644)	−2.5487	(0.0000, −3.0000)	−3.0000

Problem 3

$$\min f(x) = \max_{j=1,\cdots,10} \sum_{i=1}^{10} \frac{(ix_i - 1)^2}{i + j + 1}$$
$$+ \min_{j=1,\cdots,10} \frac{(ix_i - 1)^2}{i + j + 1}, |x_i| \le 10. \tag{18}$$

The algorithm successfully found a global solution: $x^* = (1, 0.5, \cdots, 0.1)$, with $f(x^*) = 0$. Table 3 records the numerical results of Problem 3.

Table 3. Computational results for Problem 3.

k	x_k^0	$f(x_k)$	x_k^*	$f(x_k^*)$
1	$\begin{pmatrix} 5 \\ 5 \\ 5 \\ 5 \\ 5 \\ 5 \\ 5 \\ 5 \\ 5 \\ 5 \end{pmatrix}$	1823.8490	$\begin{pmatrix} 1.0000 \\ 0.5000 \\ 0.5000 \\ 0.5000 \\ 0.5000 \\ 0.5000 \\ 0.5000 \\ 0.5000 \\ 0.5000 \\ 0.5000 \end{pmatrix}$	9.3968
2	$\begin{pmatrix} 1.0000 \\ 0.4923 \\ 0.3714 \\ 0.3411 \\ 0.4909 \\ 0.1613 \\ 0.2356 \\ 0.1201 \\ 0.1981 \\ 0.1974 \end{pmatrix}$	0.9581	$\begin{pmatrix} 1.0000 \\ 0.5000 \\ 0.3333 \\ 0.2500 \\ 0.2000 \\ 0.1667 \\ 0.1429 \\ 0.1250 \\ 0.1111 \\ 0.1000 \end{pmatrix}$	0.0000

Problem 4:

$$\min f(x) = \left[\begin{array}{l} 10\sin^2(\pi y_1) + (y_n - 1)^2 \\ + \sum_{i=1}^{n-1} (y_i - 1)^2 \left(1 + 10\sin^2(\pi y_{i+1}) \right) \end{array} \right] \pi/n, \qquad (19)$$

The algorithm successfully found a global solution: $x^* = (1, 1, 1, 1, 1, 1, 1)$, with $f(x^*) = 0$ for $n = 7$. Table 4 records the numerical results of Problem 4.

The application of the filled function method in solving nonlinear equations.

Consider the following nonlinear equations $(NE) : G(x) = 0, x \in X$, where the mapping $G(x) = (f_1(x), f_2(x), \cdots, f_m(x))^T : R^n \to R^m$ is continuous, and $X \subset R^n$ is a box set.

Let $f(x) = \sum_{k=1}^m |f_k(x)|$, then the solution of problem (NE) may be obtained through solving the following reformulated global optimization problem $(P) : \min_{x \in X} f(x)$. In particular, suppose that the problem (NE) has at least one root, then each global minimizer of the problem (P) with zero function value corresponds to one root of the (NE).

Table 4. Computational results for Problem 4.

k	x_k^0	$f(x_k)$	x_k^*	$f(x_k^*)$
1	$\begin{pmatrix} 6 \\ 6 \\ 6 \\ 6 \\ 6 \\ 6 \\ 6 \end{pmatrix}$	78.5398	$\begin{pmatrix} 0.9961 \\ -2.0137 \\ -2.9958 \\ -2.9982 \\ -2.9961 \\ -3.0011 \\ -2.9953 \end{pmatrix}$	39.9687
2	$\begin{pmatrix} 1.0241 \\ 0.9999 \\ 0.9898 \\ -0.2180 \\ 0.2243 \\ -2.9832 \\ -1.9997 \end{pmatrix}$	14.9231	$\begin{pmatrix} 1.003 \\ 0.9999 \\ 1.0000 \\ 1.0032 \\ 0.2284 \\ -2.9343 \\ -1.9997 \end{pmatrix}$	11.3654
3	$\begin{pmatrix} 0.0099 \\ 0.9999 \\ 1.0029 \\ 1.0024 \\ 1.0026 \\ 1.0001 \\ 1.0078 \end{pmatrix}$	0.4443	$\begin{pmatrix} 1.0000 \\ 1.0000 \\ 1.0000 \\ 1.0000 \\ 1.0000 \\ 1.0000 \\ 1.0000 \end{pmatrix}$	0.0000

Problem 5:

$$10^4 x_1 x_2 = 1, \quad e^{-x_1} + e^{-x_2} = 1.001,$$
$$s.t \quad 5.49 \times 10^{-6} \le x_1 \le 4.553, 2.196 \times 10^{-3} \le x_2 \le 18.21. \tag{20}$$

The algorithm successfully found its solution $x^* = (1.450 \times 10^{-5}, 6.8933335)$. Table 5 records its numerical results.

Table 5. Computational results for Problem 5.

k	x_k^0	$f(x_k)$	x_k^*	$G(x_k^*)$
1	$\begin{pmatrix} 3.0000 \\ 3.0000 \end{pmatrix}$	$\begin{pmatrix} 0.00001457 \\ 6.87403875 \end{pmatrix}$	7.11982×10^{-6}	$\begin{pmatrix} -4.6287 \times 10^{-6} \\ -2.4911 \times 10^{-6} \end{pmatrix}$
2	$\begin{pmatrix} 1.4523 \times 10^{-5} \\ 6.89330451 \end{pmatrix}$	$\begin{pmatrix} 0.000014509 \\ 6.89330448 \end{pmatrix}$	0.00000003	$\begin{pmatrix} 0.00000000 \\ 0.00000003 \end{pmatrix}$

Acknowledgement. This work was supported by the NNSF of China (No. 11001248, 11001248, 51776116).

References

1. Ge, R.P., Qin, Y.F.: A class of filled functions for finding a global minimizer of a function of several variables. J. Optim. Theory Appl. **54**(2), 241–252 (1987)
2. Zhang, L.S., Ng, C.K., Li, D.: A new filled function method for global optimization. J. Glob. Optim. **28**, 17–43 (2004)
3. Xu, Z., Huang, H.X., Pardalos, P.M., Xu, C.X.: Filled functions for unconstrained global optimization. J. Glob. Optim. **20**, 49–65 (2001)
4. Wu, Z.Y., Zhang, L.S., Teo, K.L., Bai, F.S.: New modified function for global optimization. J. Optim. Theory Appl. **125**(1), 181–203 (2005)
5. Wu, Z.Y., Li, D., Zhang, L.S.: Global descent methods for unconstrained global optimization. J. Glob. Optim. **50**, 379–396 (2011)
6. Yang, Y.J., Shang, Y.L.: A new filled function method for unconstrained global optimization. Appl. Math. Comput. **173**, 501–512 (2006)
7. Audet, C., Dennis, Jr., J.E.: Mesh adaptive direct search algorithms for constrained optimization. SIAM J. Optim. **17**, 188–217 (2006)
8. Price, C.J., Robertson, B.L., Reale, M.: A hybrid Hooke and Jeeves - direct method for non-smooth optimization. Adv. Model. Optim. **11**(1), 43–61 (2009)
9. Clarke, F.H.: Optimization and Non-smooth Analysis. Wiley, New York (1983)

Complexity of Tree-Coloring Interval Graphs Equitably

Bei Niu(ID), Bi Li$^{(\boxtimes)}$(ID), and Xin Zhang(ID)

School of Mathematics and Statistics, Xidian University, Xi'an 710071, China
beiniu@stu.xidian.edu.cn, {libi,xzhang}@xidian.edu.cn

Abstract. An equitable tree-k-coloring of a graph is a vertex k-coloring such that each color class induces a forest and the size of any two color classes differ by at most one. In this work, we show that every interval graph G has an equitable tree-k-coloring for any integer $k \geq \lceil (\Delta(G)+1)/2 \rceil$, solving a conjecture of Wu, Zhang and Li (2013) for interval graphs, and furthermore, give a linear-time algorithm for determining whether a proper interval graph admits an equitable tree-k-coloring for a given integer k. For disjoint union of split graphs, or $K_{1,r}$-free interval graphs with $r \geq 4$, we prove that it is $W[1]$-hard to decide whether there is an equitable tree-k-coloring when parameterized by number of colors, or by treewidth, number of colors and maximum degree, respectively.

Keywords: $W[1]$-hardness · Linear-time algorithm · Equitable tree-coloring · Interval graph · Communication network

1 Introduction

A minimization model in graph theory so-called the equitable tree-coloring can be used to formulate a structure decomposition problem on the communication network with some security considerations [17]. Namely, an *equitable tree-k-coloring* of a (finite, simple and undirected) graph G is a mapping $c : V(G) \rightarrow \{1, 2, \cdots, k\}$ so that $c^{-1}(i)$ induces a forest for each $1 \leq i \leq k$, and $\left| |c^{-1}(i)| - |c^{-1}(j)| \right| \leq 1$ for each pair of $1 \leq i < j \leq k$. The notion of the equitable tree-k-coloring was introduced by Wu, Zhang and Li [14], who conjectured that every graph G has an equitable tree-k-coloring for any integer $k \geq \lceil (\Delta(G)+1)/2 \rceil$. This conjecture (equitable vertex arboricity conjecture, EVAC for short) is known to have an affirmative answer in some cases including:

- G is complete or bipartite [14];
- $\Delta(G) \geq (|G| - 1)/2$ [16,18];
- $\Delta(G) \leq 3$ [15];

Supported by the National Natural Science Foundation of China (11871055, 11701440) and the Youth Talent Support Plan of Xi'an Association for Science and Technology (2018-6).

X. Zhang—This author shares a co-first authorship.

Z. Zhang et al. (Eds.): AAIM 2020, LNCS 12290, pp. 391–398, 2020.
https://doi.org/10.1007/978-3-030-57602-8_35

- G is 5-degenerate [1];
- G is d-degenerate with $\Delta(G) \geq 10d$ [17];
- G is IC-planar with $\Delta(G) \geq 14$ or $g(G) \geq 6$ [12];
- G is a d-dimensional grid with $d \in \{2, 3, 4\}$ [3].

EVAC is still widely open.

Algorithmically, the following EQUITABLE TREE COLORING is NP-complete [7].

EQUITABLE TREE COLORING
Instance: A graph G and the number of colors k.
Question: Is there an equitable tree-k-coloring of G?

Recently in [10], the last two authors proved that EQUITABLE TREE COL-ORING problem is W[1]-hard when parameterized by treewidth, and that it is polynomial solvable in the class of graphs with bounded treewidth, and in the class of graphs of bounded vertex cover number.

This paper focuses on interval graphs. A graph G is an *interval graph* if there exist an *interval representation* of G, i.e., a family $\{T_v | v \in V(G)\}$ of intervals on the real line such that u and v are adjacent vertices in G if and only if $T_u \cap T_v \neq \emptyset$. For any vertex $v \in V(G)$, $L(v)$ and $R(v)$ denote the left point and the right point of its corresponding interval T_v, respectively. For two vertices $u, v \in V(G)$, if $L(u) < L(v)$, or $L(u)=L(v)$ and $R(u) \leq R(v)$, then we write $u < v$. For any three vertices $u, v, w \in V(G)$, it is clear that

$$\text{if } u < v < w \text{ and } uw \in E(G), \text{ then } uv \in E(G). \tag{1}$$

Olariu [13] shows that a graph is an interval graph if and only if it has a *linear order* $<$ on $V(G)$ satisfying (1); and that the order can be found in linear time. Using this fact, we give the following result as a quick start of this paper, confirming EVAC for interval graphs.

Theorem 1. *Every interval graph G has an equitable tree-k-coloring for any integer $k \geq \lceil \frac{\Delta(G)+1}{2} \rceil$, where the lower bound of k is sharp.*

Proof. Sort the vertices of G into $v_0 < v_1 < \cdots < v_{n-1}$ so that (1) holds, and for each $0 \leq i \leq n-1$, let $c(v_i) = i \pmod{k}$. It is clear that c is an equitable k-coloring of G. If there is a monochromatic cycle in color i, then there are three vertices $v_{i+\alpha k}$, $v_{i+\beta k}$ and $v_{i+\gamma k}$ with $0 \leq \alpha < \beta < \gamma$ such that $v_{i+\alpha k}v_{i+\beta k}, v_{i+\alpha k}v_{i+\gamma k} \in E(G)$. By (1), $v_{i+\alpha k}v_j \in E(G)$ for any $i+\alpha k < j \leq i+\gamma k$, which implies $d_G(v_{i+\alpha k}) \geq (\gamma-\alpha)k \geq 2k \geq \Delta(G)+1$, a contradiction. Hence, there is no monochromatic cycle under c. This implies that c is an equitable tree-k-coloring of G. Since the complete graph K_{2s} is an interval graph and it does not admit an equitable tree-k-coloring for any $k \leq s - 1$, the lower bound of k in this result is sharp.

On the other hand, if we are given an integer $k < \lceil \frac{\Delta(G)+1}{2} \rceil$, determining whether an interval graph admits an equitable tree-k-coloring is not easy. Precisely, the next two theorems will be proved in Sect. 2.

Theorem 2. EQUITABLE TREE COLORING *of the disjoint union of split graphs parameterized by number of colors is* $W[1]$-*hard.*

Theorem 3. EQUITABLE TREE COLORING *of* $K_{1,r}$-*free interval graph with* $r \geq 4$ *parameterized by treewidth, number of colors and maximum degree is* $W[1]$-*hard.*

Here, a *split graph* is a graph in which the vertices can be partitioned into a clique and an independent set, and a $K_{1,r}$-*free graph* is a graph that does not contain the star $K_{1,r}$ as an induced subgraph.

However, the situation is much better if we are working with an *proper interval graph*, that is an interval graph that has an interval representation in which no interval properly contains any other interval. Actually, we have the following theorem, which will be proved in Sect. 3.

Theorem 4. *There is a linear-time algorithm to determine whether a proper interval graph admits an equitable tree-k-coloring for a given integer* k.

To end this section, we collect some notations that will be used in the next sections. For any two graphs G and H, their *sum* $G \oplus H$ is the graph given by $V(G \oplus H) = V(G) \cup V(H)$ and $E(G \oplus H) = E(G) \cup E(H) \cup \{uv | u \in V(G), v \in V(H)\}$, and their *union* $G \cup H$ is the graph given by $V(G \cup H) = V(G) \cup V(H)$ and $E(G \cup H) = E(G) \cup E(H)$. By nG, we denote the n disjoint copies of G, and $[k]$ stands for $\{1, 2, \cdots, k\}$. Other undefined notations follow [2].

2   W[1]-Hardness: The Proofs of Theorems 2 and 3

All of our reductions involve the following BIN-PACKING problem, which is NP-hard in the strong sense [5], and is $W[1]$-hard when parameterized by the number of bins [8,9].

BIN-PACKING
Instance: A set of n items $A = \{a_1, a_2, ..., a_n\}$ and a bin capacity B.
Parameter: The number of bins k.
Question: Is there a k-partition φ of A such that, $\forall i \in [k], \sum_{a_j \in \varphi_i} a_j = B$?

Proof of Theorem 2. Given an instance of BIN-PACKING as above, Our strategy is to construct a disjoint union of split graph G such that the answer of the BIN-PACKING is YES if and only if G admits an equitable tree-k-coloring. Here,

$$G = \bigcup_{j \in [n]} H(a_j, k),$$

where $a_1, a_2, \cdots, a_n$ are arbitrarily given integers in the instance of BIN PACK-ING, and

$$H(a, k) = K_{2k-1} \oplus (a + 1) K_1$$

defines a split graph for integers a and k. For $j \in [n]$, let I_j be the independent set of size $a_j + 1$ in $H(a_j, k)$, and let c_j be a fixed vertex in the clique part K_{2k-1} of $H(a_j, k)$.

Suppose that there is a k-partition φ of $A = \{a_1, a_2, ..., a_n\}$ such that, $\forall$ $i \in [k]$, $\sum_{a_j \in \varphi_i} a_j = B$. For any $i \in [k]$ and for any j satisfying $a_j \in \varphi_i$, color c_j and all vertices in I_j with color i. For any $j \in [n]$ such that $a_j \in \varphi_i$, color the $2k - 2$ vertices in $V(H(a_j, k)) \setminus (I_j \cup \{c_j\})$ with $k - 1$ distinct colors in $[k] \setminus i$ so that each color is used exactly twice. At this moment, for any $j \in [n]$, $H(a_j, k)$ has been colored with k colors so that the set of all vertices with color i (here i is the integer such that $a_j \in \varphi_i$) induces a forest, which is actually a star with center c_j, and each of another $k - 1$ colors besides i is used for exactly two vertices. Hence this gives a tree-k-coloring of $H(a_j, k)$ for any $j \in [n]$, and thus finally gives a tree-k-coloring of G. To see that this coloring is an equitable tree-k-coloring of G, we denote the set of vertices with color i as V_i for any color $i \in [k]$. Clearly,

$$|V_i| = \sum_{a_j \in \varphi_i} (a_j + 2) + \sum_{a_j \notin \varphi_i} 2 = \sum_{a_j \in \varphi_i} a_j + 2n = B + 2n$$

for any $i \in [k]$, which implies that such a coloring is equitable.

On the other direction, if G admits an equitable tree-k-coloring ψ, then in the clique part K_{2k-1} of each $H(a_j, k)$ with $j \in [n]$, there is a color i appearing on exactly one vertex, and each of another $k - 1$ colors appears on exactly two vertices. It follows that all vertices in I_j of $H(a_j, k)$ are colored with i, since any other color classes contains two vertices in the clique part K_{2k-1}, which are adjacent to all vertices in I_j. Therefore, taking any one vertex $v_j \in I_j$ together with the clique part K_{2k-1} induces a clique K_{2k} containing exactly two vertices in each color class. Let ψ_i be the vertices of G colored with i under the coloring ψ. We show that $\sum_{I_j \subseteq \psi_i} a_j = B$, which indicates that the answer for the BIN-PACKING is YES.

Since there are

$$n(2k - 1) + \sum_{j \in [n]} (a_j + 1) = k(2n + B)$$

vertices in G (note that in BIN-PACKING we always assume that $\sum_{j \in [n]} a_j = kB$), each color class of ψ contains exactly $2n + B$ vertices, which consists of, for each $j \in [n]$, two vertices in the clique K_{2k} of $H(a_j, k)$ as chosen above and a_j vertices in $I_j \setminus \{v_j\}$ if $I_j \subseteq \psi_i$. So

$$2n + B = |\psi_i| = 2n + \sum_{I_j \subseteq \psi_i} a_j,$$

which gives $\sum_{I_j \subseteq \psi_i} a_j = B$.

Lemma 5. *[4, 11] A graph is an interval graph if and only if its maximal cliques can be ordered as $M_1, M_2, \cdots, M_k$ such that for any $v \in M_i \cap M_k$ with $i < k$, it holds that $v \in M_j$ for any $i \leq j \leq k$.*

Proof of Theorem 3. We prove the theorem for parameter number of colors; and the theorem for another two parameters can be proved in the same way. Given an instance of the BIN-PACKING, our strategy is to construct a $K_{1,r}$-free interval graph G with $r \geq 4$ such that the answer of the BIN-PACKING is YES if and only if G admits an equitable tree-k-coloring. Here,

$$G = \bigcup_{j \in [n]} J(a_j, k)$$

with $a_1, a_2, \cdots, a_n$ being arbitrarily given integers in the instance of BIN PACK-ING, and

$$J(a, k) = \left(\bigcup_{i \in [a]} (Q_i \oplus y_i) \right) \bigcup \left(\bigcup_{i \in [a]} (Q'_i \oplus y_i) \right) \bigcup \left(\bigcup_{i \in [a-1]} (Q_{i+1} \oplus y_i) \right),$$

where $S = \{Q_1, Q'_1, \cdots, Q_a, Q'_a\}$ is a set of cliques such that $Q_i \simeq Q'_i \simeq K_{2k-1}$ and $Y = \{y_1, \cdots, y_a\}$ is a set of vertices. Note that the vertices of G with largest degree are the ones contained in $Y \setminus \{y_a\}$, which have degrees equal to $3(2k-1)$, and the treewidth of G is $2k - 1$.

We claim first that G is an interval graph. Indeed, it is sufficient to show that $J(a, k)$ is an interval graph for any positive integer a. By the definition of $J(a, k)$, one can see that it has $3a - 1$ maximal cliques $M_1, M_2, \cdots, M_{3a-1}$ such that

$$M_i = \begin{cases} Q_i \oplus y_i & \text{if } i \equiv 1 \ (\mathrm{mod}\ 3) \\ Q'_i \oplus y_i & \text{if } i \equiv 2 \ (\mathrm{mod}\ 3) \\ Q_{i+1} \oplus y_i & \text{if } i \equiv 0 \ (\mathrm{mod}\ 3). \end{cases}$$

Since $M_i \cap M_j = \emptyset$ for any $i \equiv 1 \ (\mathrm{mod}\ 3)$ and $j \geq i+3$, $M_i \cap M_j = \emptyset$ for any $i \not\equiv 1 \ (\mathrm{mod}\ 3)$ and $j \geq i+2$, and $M_i \cap M_{i+1} = M_i \cap M_{i+2} = M_{i+1} \cap M_{i+2} = \{y_i\}$ for any $1 \leq i \leq 3a - 5$ with $i \equiv 1 \ (\mathrm{mod}\ 3)$, the ordering $M_1, M_2, \cdots, M_{3a-1}$ satisfies the property described by Lemma 5, and therefore, $J(a, k)$ is an interval graph.

Suppose that there is a k-partition φ of A such that, $\forall\ i \in [k]$, $\sum_{a_j \in \varphi_i} a_j = B$. For any $i \in [k]$ and for any j satisfying $a_j \in \varphi_i$, color all vertices of Y_j, corresponding to the vertex set Y in $J(a_j, k)$, with color i. For any $j \in [n]$ such that $a_j \in \varphi_i$, color each $(k-1)$-clique of S_j, corresponding to the set S of cliques in $J(a_j, k)$, so that the color i is used for exactly one vertex, and each of the remaining $k-1$ colors in $[k] \setminus i$ are used for exactly two vertices. Clearly, this gives a tree-k-coloring of $J(a_j, k)$, where $3a_j$ vertices consisting of Y_j and one vertex in each clique in S_j, are colored with color i, and each of another $k - 1$ colors is used for exactly two vertices in each clique of S_j. Hence a tree-k-coloring of G is given now. To see that this is an equitable tree-k-coloring of G, we denote the set of vertices with color i as V_i for any color $i \in [k]$. The fact that

$$|V_i| = \sum_{a_j \in \varphi_i} 3a_j + \sum_{a_j \notin \varphi_i} 4a_j = 3B + 4(kB - B) = (4k - 1)B$$

for any $i \in [k]$ implies the equability of this coloring.

On the other direction, if G admits an equitable tree-k-coloring ψ, then in each clique of S_j for $j \in [n]$, there is a color appearing on exactly one vertex, and each of another $k - 1$ colors appears on exactly two vertices. Suppose that the color i appears on exactly one vertex of the first clique $Q_1 \in S_j$ for some $j \in [n]$. It follows that y_1 should be colored with i because y_1 is adjacent to every vertices of Q_1 and each color in $[k]\backslash i$ already appears on two vertices of Q_1. Consequently, the color being used exactly once for the vertices of the second clique $Q_1' \in S_j$ and the third clique $Q_2 \in S_j$ is indeed i, which implies that y_2 shall be colored with i. Following this process, we can conclude that each vertex of Y_j is colored with i, and in each clique of S_j, the color i appears on exactly one vertex, and each of another $k - 1$ colors in $[k]\backslash i$ appears on exactly two vertices. Let ψ_i be the vertices of G colored with i under the coloring ψ. We show that $\sum_{Y_j \subseteq \psi_i} a_j = B$, which indicates that the answer for the BIN-PACKING is YES.

Since there are

$$\sum_{j \in [n]} \left(a_j + 2a_j(2k - 1) \right) = kB + (4k - 2)kB = k(4k - 1)B$$

vertices in G, for each $i \in [k]$,

$$|\psi_i| = (4k - 1)B,$$

and by the way of the coloring ψ as described above, we also see that

$$|\psi_i| = \sum_{Y_j \subseteq \psi_i} 3a_j + \sum_{Y_j \cap \psi_i = \emptyset} 4a_j = \sum_{j \in [n]} 4a_j - \sum_{Y_j \subseteq \psi_i} a_j = 4kB - \sum_{Y_j \subseteq \psi_i} a_j.$$

Combining the two expression gives $\sum_{Y_j \subseteq \psi_i} a_j = B$.

3   Linear-Time Algorithm: The Proof of Theorem 4

Lemma 6. *Let G be a proper interval graph with $V(G) = \{v_1, v_2, \cdots, v_n\}$, where $v_1 < v_2 < \cdots < v_n$. If $v_i v_j \in E(G)$, then $\{v_i, v_{i+1}, \cdots, v_{j-1}, v_j\}$ induces a clique of size $j - i + 1$.*

Proof. By the definition of the proper interval graph, for any $i \leq s < \ell \leq j$, $L(v_s) < L(v_\ell)$, and if $v_i v_j \in E(G)$, then $L(v_i) \leq L(v_s) < L(v_\ell) \leq L(v_j) \leq R(v_i) \leq R(v_s) < R(v_\ell) \leq R(v_j)$. This implies that the interval $[L(v_s), R(v_s)]$ intersects the interval $[L(v_\ell), R(v_\ell)]$ and thus $v_s v_\ell \in E(G)$. Hence any two vertices among $\{v_i, v_{i+1}, \cdots, v_{j-1}, v_j\}$ are adjacent and such a vertex set induces a clique.

Lemma 7. *A proper interval graph has an equitable tree-k-coloring if and only if its maximum clique has size at most $2k$.*

Proof. If c is an equitable tree-k-coloring of G, then there is no clique on at least $2k + 1$ vertices, because otherwise there is a color appearing at least three times on this clique, implying the existence of a monochromatic triangle, a contradiction. Hence the maximum clique of G has size at most $2k$.

On the other direction, if the maximum clique of G has size at most $2k$, then sort the vertices of G into $v_0, v_1, \cdots, v_{n-1}$ so that $v_i < v_j$ if $i < j$. Let $c(v_i) = i \pmod{k}$. It is clear that c is an equitable k-coloring of G. If there is a monochromatic cycle under c, then there are two adjacent vertices v_i and v_j with $j - i = \beta k$ with $\beta \geq 2$. By Lemma 6, G contains a clique of size $j - i + 1 = \beta k + 1 \geq 2k + 1$ as a subgraph, a contradiction. This implies that there is no monochromatic cycle under c, and thus c is an equitable tree-k-coloring of G.

Algorithm 1: Linear-time Algorithm for proper interval graphs

Input: A proper interval graph $G = (V, E)$ on n vertices; a set of k colors $\{0, 1, \cdots, k - 1\}$;

Output: *Answ*;

1 $Answ \leftarrow YES$;
2 Sort the vertices of G into $v_0 < v_1 < \cdots < v_{n-1}$, where $<$ is a linear order on $V(G)$;
3 **for** $i = 0$ *to* $n - 1$ **do**
4 $\quad\lfloor$ Color vertex v_i with the color $c(i) = i \pmod{k}$;
5 **if** *there is a monochromatic cycle in any color class* **then**
6 $\quad\lfloor$ Output *NO*;
7 Output *Answ*.

Theorem 8. *Given a proper interval graph G and an integer $k > 0$, Algorithm 1 outputs YES in linear time if and only if there exists an equitable tree-k-coloring of G; moreover, if YES, it gives an equitable tree-k-coloring of G.*

Proof. From Algorithm 1, one sees that the given coloring is an equitable tree-k-coloring of G if it outputs YES. If there is an equitable tree-k-coloring of G, then the size of the maximum clique of G is at most $2k$ by Lemma 7. In any iterative step of Algorithm 1, every color class induces disjoint unions of paths, because if not, there are two adjacent vertices $v_s v_\ell$ with $\ell - s \geq 2k$, which gives a clique of size $2k + 1$ by Lemma 6, a contradiction. So the algorithm outputs YES. The time complexity dominates by Line 3, which takes $O(|V| + |E|)$ time, see [13, Theorem 6].

Proof of Theorem 4. This is an immediate corollary from Theorem 8.

Remark. Lemma 7 implies that EQUITABLE TREE COLORING of proper interval graphs is equivalent to determine whether $2k$ is the upper bound of its clique number. We know that to calculate all maximal cliques of a triangulated graph $G = (V, E)$ (i.e, a graph without induced cycles on at least four vertices) can be done in $O(|V| + |E|)$ time [6, Theorem 4.17], and any proper interval graph is a triangulated graph by Lemma 6. This also proves Theorem 4, however, without giving an equitable tree-k-coloring of G if the algorithm outputs YES.

Acknowledgements. The last author would like to acknowledge the supports provided by China Scholarship Council (CSC) under the grant number 201906965003 and by Institute for Basic Science (IBS, South Korea) during a visit of him to Discrete Mathematics Group, IBS.

References

1. Chen, G., Gao, Y., Shan, S., Wang, G., Wu, J.: Equitable vertex arboricity of 5-degenerate graphs. J. Comb. Optim. **34**(2), 426–432 (2017)
2. Diestel, R.: Graph Theory, 5th edn. Springer, Heidelberg (2016). https://doi.org/10.1007/978-3-662-53622-3
3. Drgas-Burchardt, E., Dybizbański, J., Furmańczyk, H., Sidorowicz, E.: Equitable list vertex colourability and arboricity of grids. Filomat **32**(18), 6353–6374 (2018)
4. Fishburn, P.: Interval Orders and Interval Graphs: A Study of Partially Ordered Sets. Wiley-Interscience Series in Discrete Mathematics. Wiley, New York (1985)
5. Garey, M.R., Johnson, D.S.: Computers and Intractability: A Guide to the Theory of NP-Completeness. W.H. Freeman & Co., USA (1990)
6. Golumbic, M.C.: Triangulated graphs. In: Golumbic, M.C. (ed.) Algorithmic Graph Theory and Perfect Graphs, pp. 81–104. Academic Press (1980)
7. Gomes, G., Lima, C., Dos Santos, V.: Parameterized complexity of equitable coloring. Discret. Math. Theor. Comput. Sci. **21**(1), #8 (2019)
8. Jansen, K., Kratsch, S., Marx, D., Schlotter, I.: Bin packing with fixed number of bins revisited. In: Kaplan, H. (ed.) SWAT 2010. LNCS, vol. 6139, pp. 260–272. Springer, Heidelberg (2010). https://doi.org/10.1007/978-3-642-13731-0_25
9. Jansen, K., Kratsch, S., Marx, D., Schlotter, I.: Bin packing with fixed number of bins revisited. J. Comput. Syst. Sci. **79**(1), 39–49 (2013)
10. Li, B., Zhang, X.: Tree-coloring problems of bounded treewidth graphs. J. Comb. Optim. **39**(1), 156–169 (2020)
11. McMorris, F.: Interval orders and interval graphs – a study of partially ordered sets (Peter C. Fishburn). SIAM Rev. **29**(3), 484–486 (1987)
12. Niu, B., Zhang, X., Gao, Y.: Equitable partition of plane graphs with independent crossings into induced forests. Discret. Math. **343**(5), #111792 (2020)
13. Olariu, S.: An optimal greedy heuristic to color interval graphs. Inf. Process. Lett. **37**(1), 21–25 (1991)
14. Wu, J.-L., Zhang, X., Li, H.: Equitable vertex arboricity of graphs. Discret. Math. **313**(23), 2696–2701 (2013)
15. Zhang, X.: Equitable vertex arboricity of subcubic graphs. Discret. Math. **339**(6), 1724–1726 (2016)
16. Zhang, X., Niu, B.: Equitable partition of graphs into induced linear forests. J. Comb. Optim. **39**(2), 581–588 (2020)
17. Zhang, X., Niu, B., Li, Y., Li, B.: Equitable vertex arboricity of d-degenerate graphs. arXiv:1908.05066v2 [math.CO] (2019)
18. Zhang, X., Wu, J.-L.: A conjecture on equitable vertex arboricity of graphs. Filomat **28**(1), 217–219 (2014)

Approximation Algorithms for Balancing Signed Graphs

Zhuo Diao[1] and Zhongzheng Tang[2(✉)]

[1] School of Statistics and Mathematics,
Central University of Finance and Economics, Beijing 100081, China
diaozhuo@amss.ac.cn
[2] School of Sciences, Beijing University of Posts and Telecommunications,
Beijing 100876, China
tangzhongzheng@amss.ac.cn

Abstract. Structural balance theory is an important theory in signed graphs. We consider the optimization problems: given a signed graph, the maximum number of edges that needed to be kept to make it balanced is called $K(G)$. We firstly prove the computation of $K(G)$ is NP-hard. Next we design four approximation algorithms to compute $K(G)$.

Keywords: Balanced · Signed graph · NP-hard · Approximation algorithms

1 Introduction

In the area of graph theory in mathematics, a signed graph is a graph in which each edge has a positive or negative sign. The name "signed graph" and the notion of balance appeared first in a mathematical paper of Frank Harary in 1953 [7]. Denes Konig had already studied equivalent notions in 1936 under a different terminology but without recognizing the relevance of the sign group [11]. At the Center for Group Dynamics at the University of Michigan, Dorwin Cartwright and Frank Harary generalized Fritz Heider's psychological theory of balance in triangles of sentiments to a psychological theory of balance in signed graphs [2].

Signed graphs have been rediscovered many times because they come up naturally in many unrelated areas [18]. For instance, they enable one to describe and analyze the geometry of subsets of the classical root systems. They appear in topological graph theory and group theory. They are a natural context for questions about odd and even cycles in graphs. They appear in computing the ground state energy in the non-ferromagnetic Ising model; for this one needs to

This research is supported part by National Natural Science Foundation of China under Grant No.11901605, and by the disciplinary funding of Central University of Finance and Economics.

Z. Zhang et al. (Eds.): AAIM 2020, LNCS 12290, pp. 399–410, 2020.
https://doi.org/10.1007/978-3-030-57602-8_36

find a largest balanced edge set. They have been applied to data classification in correlation clustering.

Structural balance theory dated back to the work of Heider in the 1940s [9], and generalized and extended to the language of graphs beginning with the work of Cartwright and Harary in the 1950s [2,5,7]. James A. Davis gave the model some applications and explained the model in detail [3], and proposed a weaker form of structural balance. Tibor Antal, Paul Krapivsky and Sidney Redner proposed a dynamic structural balance model [1]. Jure Leskovec, Dan Huttenlocher and Jon Kleinberg applied structural balance model to big data analysis [13,14]. By far, there have been numerous articles on structural balance theory [6,12,15]. David Easley and Jon Kleinberg have introduced and summarized the structural balance theory systematically in chapter 5 of their monograph [4].

The essential idea of structural balance theory is as follows: Consider a group in which any two persons are familiar with each other and they are friends or enemies. Everyone is susceptible to the influence of his friends or enemies. Therefore, if A and B are friends and B and C are friends, A and C will easily become friends, that is, a friend's friend is a friend. If A and B are enemies and B and C are enemies, A and C will easily become friends, that is, an enemy of an enemy is a friend. For a stable and balanced structure, it needs to follow the following rules: **a friend of a friend is a friend, and an enemy of an enemy is a friend**. Based on these two basic rules, it is easy to further derive the following two rules: **An enemy of a friend is an enemy, and a friend of an enemy is an enemy**.

The above rules could be further expressed in graph theory. Let us consider the relationship among three persons. Using a plus sign for friendships and a minus sign for enemies, according to the number of friends or enemies, there are four basic situations as shown in Fig. 1 (a) is a stable structure in which three persons are friends with each other; (b) is an unstable structure because it violates the basic principle that a friend of a friend is a friend; (c) is a stable structure, two persons are each other's friends and they have a common enemy; (d) is an unstable structure because it violates the basic principle that an enemy of an enemy is a friend.

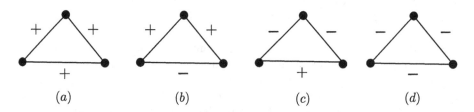

Fig. 1. Four basic structures in the classical structural balance model

Our Results: Given a signed graph, the minimum number of edges needed to be changed to make it balanced is called $C(G)$; the maximum number of edges that needed to be kept to make it balanced is called $K(G)$. Obviously, the minimum number of changed edges and the maximum number of kept edges form a partition of the edges set. Thus $C(G)+K(G) = m$ where m is the edge number of graph G. We firstly prove the computation of $C(G)$ or $K(G)$ is NP-hard. Next we design four approximation algorithms to compute $K(G)$. The first and second algorithms are deterministic algorithms with approximation ratio $1/2$, belonging to the greedy algorithm and local search algorithm. The third algorithm is a randomized algorithm with approximation ratio $1/2$ in expectation sense. The last algorithm is a semidefinite programming algorithm with approximation ratio 0.87856. All these approximation algorithms are derived from the approximation algorithms to compute Max-Cut problem.

2   Characterizing the Structure of Balanced Networks

2.1   Completed Graphs

Given a complete signed graph, Frank Harary has proved the following Balance Theorem in 1953 [2, 7].

Theorem 1 [2,7]. *Balance Theorem: If a labeled complete graph is balanced, then either all pairs of nodes are friends, or else the nodes can be divided into two groups, X and Y, such that every pair of nodes in X like each other, every pair of nodes in Y like each other, and everyone in X is the enemy of everyone in Y.*

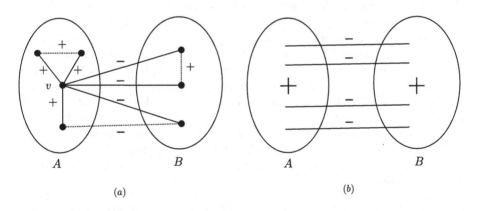

Fig. 2. Characterizing the structure of balanced networks

2.2   General Graphs

There are many ways to extend the structural balance theory from complete graphs to general graphs. Here we take the following approach in classical structural balance theory [2,7]: Treat the missing edges in incomplete graphs as implicit cases and still follow the structural balance rules of complete graphs: **friends of friends are friends, and enemies of enemies are friends.** If a general graph can be restored to a complete graph with balance structure conforming to the structural balance rules, then the general graph is structurally balanced; otherwise, it is non-structurally balanced. The following two theorems give the sufficient and necessary conditions for signed general networks to be structurally balanced.

Theorem 2 [2,7]. *The structure balance theorem in general graphs: a general signed network is structurally balanced if and only if all nodes can be divided into two sets, the inner edges of the same set are positive, and the edges between different sets are negative. (See Fig. 2 for an illustration.)*

Theorem 3 [2,7]. *The structure balance theorem in general graphs: a general signed network is structurally balanced if and only if it does not contain circles with odd negative edge labels.*

3   Optimization Problems

Given a general signed graph $G(V, E)$, consider the following optimization problem: how many edges should be changed at least to make the network balanced. This parameter, called $C(G)$, actually reflects the stability of the network. We will prove that calculating $C(G)$ in polynomial time is NP-hard problem. First, we notice another parameter: how many edges should be kept at most to make the network balanced. This parameter is called $K(G)$. Obviously, the minimum number of changed edges and the maximum number of kept edges form a partition of the edges set, i.e. $K(G) + C(G) = m$. Therefore, if we can prove that calculating $K(G)$ in polynomial time is NP-hard, then calculating $C(G)$ in polynomial time is also NP-hard. Max-Cut problem is one of Karp's 23 classical NPC problems and we will reduce Max-Cut problem to calculating $K(G)$.

Max-Cut problem: Given a simple graph $G(V, E)$, partition the vertex set V into two parts so that the number of edges joining vertices in different parts is as large as possible.

Theorem 4. *Given a general signed graph $G(V, E)$, calculating $C(G)$ and $K(G)$ is NP-hard.*

Proof. We only need to show that calculating $K(G)$ is NP-hard. The method is to reduce the maximum cut problem to calculating $K(G)$. Given any graph $G(V, E)$, we label all its edges with negative signs to form a signed network. We will prove the maximum cut of $G(V, E)$ is exactly $K(G)$ of the signed network. On one hand, given any partition V_1, V_2 of V, forming a cut set C. Keep negative sign of·the edges in the cut set C and change sign of the edges joining vertices in the same part from negative to positive. According to Theorem 2, the signed network after the change is balanced. On the other hand, according to Theorem 2, in a balanced structure, the vertices are divided into two sets. The inner edges of the same set are positive and the edges between different sets are negative. The two sets form a partition V_1, V_2 of V and a cut set C. Therefore, the number of edges in the cut set C is exactly the number of edges that keep the negative sign unchanged. The maximum number of edges in the cut set is exactly the maximum number of edges that keep the negative sign unchanged, that is, $K(G)$. Thus Max-Cut problem is reduced to calculating $K(G)$.

Remark 1. $G(V, E)$ is a signed graph. Give each vertex a value of $+1$ or -1; we call this a state of $G(V, E)$. An edge is called satisfied if it is positive and both endpoints have the same value, or it is negative and the endpoints have opposite values. An edge that is not satisfied is called frustrated. The smallest number of frustrated edges over all states is called the frustration index (or line index of balance) of $G(V, E)$. The largest number of satisfied edges over all states is called the satisfaction index of $G(V, E)$. The vertex set with value $+1$ and the vertex set with value -1 form a partition of all vertex set. The frustrated edges are exactly the edges which signs needed to be changed to form a balance structure and the satisfied edges are exactly the edges which signs needed to be kept to form a balance structure. Thus the frustration index is exactly $C(G)$ and the satisfaction index is exactly $K(G)$. This is a vertex version of the optimization problem.

Remark 2. A Hopfield network is a form of recurrent artificial neural network popularized by John Hopfield in 1982. Hopfield networks serve as content addressable memory systems with binary threshold nodes. They are guaranteed to converge to a local minimum. Hopfield networks also provide a model for understanding human memory. Stability and equilibrium are important properties of Hopfield neural networks, and their relations with structural balance networks can be discussed and studied.

4   Approximation Algorithms

Given a general signed graph $G(V, E)$, the maximum number of edges whose signs are kept unchanged to make it balanced is called $K(G)$. This parameter reflects the balanced degree of a signed graph. The previous section has shown that calculating $K(G)$ in polynomial time is NP-hard. This section presents four approximation algorithms for calculating $K(G)$:

- The first approximation algorithm is a deterministic algorithm with an approximation ratio of $1/2$, which belongs to the greedy algorithm. The basic idea is derived from the greedy algorithm of Max-Cut problem.
- The second approximation algorithm is a deterministic algorithm with an approximation ratio of $1/2$, which belongs to the local search algorithm. The basic idea is derived from the local search algorithm of Max-Cut problem.

Algorithm 1. Greedy Algorithm for $K(G)$

Input: An edge-signed graph $G(V, E)$.

Output: A balanced signed graph $\bar{G}$ and the edge set $C \subseteq E$ whose signs are kept.

1: Take a vertex order $v_1, \ldots, v_n$ arbitrarily.

2: Initialize $A = B = \emptyset$.

3: **for** $i = 1$ to n **do**

4:     Put v_i into A or B such that $\bar{G}[v_1, \ldots, v_i]$ is balanced and the edges

5:     incident to v_i are kept signs unchanged as many as possible.

6: Output $\bar{G} = \bar{G}[A, B]$ and edge set $C \subseteq E$ whose signs are kept.

- The third approximation algorithm is a randomized algorithm, and the approximation ratio is $1/2$ in the expectation. The basic idea is derived from the randomized algorithm of Max-Cut problem.
- The fourth approximation algorithm is a randomized algorithm with an approximation ratio of 0.87856, which belongs to the semidefinite programming algorithm. The basic idea is derived from the semidefinite programming algorithm of Max-Cut problem.

4.1   Greedy Algorithm

Algorithm 1: Given a general signed graph $G(V, E)$, take a vertex order $v_1, \ldots, v_n$ arbitrarily and initialize two vertex sets $A = B = \emptyset$. Place the vertices one by one to the set A or B in this order and the rules for placement are as follows: When a vertex v_i is put into one set A or B, choose the set which could keep the edges' signs unchanged as many as possible. Each time after the allocation of a vertex, the two sets A and B are maintained to form a balanced structure, which satisfies the edges between internal vertices in A or B are positive and the edges between A and B are negative. Finally, the two vertex sets A and B form a balanced structure and output a subset of edges $C \subseteq E$ which keep their signs unchanged. Let $\bar{G}[A, B]$ denote a balanced graph with partition (A, B). The greedy algorithm for $K(G)$ is as shown in Algorithm 1.

Theorem 5. *Algorithm 1 output a 1/2-approximation algorithm: For a signed graph $G(V, E)$, $|C| \geq K(G)/2$.*

Proof. We just need to explain $|C| \geq |E|/2 \geq K(G)/2$. In fact, each time after the allocation of a vertex, the two sets A and B are maintained to form a balanced structure, which satisfies the edges between internal vertices in A or B are positive and the edges between A and B are negative. When a vertex v_i is put into one set A or B, according to the placement rules: choose the set

---

Algorithm 2. Local Search Algorithm for $K(G)$

Input: An edge-signed graph $G(V, E)$.

Output: A balanced signed graph $\bar{G}$ and the edge set $C \subseteq E$ whose signs are kept.

1: Take a vertex order $v_1, \ldots, v_n$ arbitrarily.

2: Initialize $A = V$, $B = \emptyset$.

3: **while** there exists v_i that convert its set can increase unchanged edges **do**

4:    Convert v_i's set and update A and B.

5: Output $\bar{G} = \bar{G}[A, B]$ and edge set $C \subseteq E$ whose signs are kept.

---

which could keep the edges' signs unchanged as many as possible. So, at least half of the edges incident to v_i stay at the same signs. After n rounds, there are at least half of edges keeping their signs unchanged. The final set of edges $|C| \geq |E|/2 \geq K(G)/2$.

4.2   Local Search Algorithm

Algorithm 2: Given a general signed graph $G(V, E)$, take a vertex order $v_1, \ldots, v_n$ arbitrarily and initialize two vertex sets $A = V, B = \emptyset$. Check the vertices one by one in this order. If a vertex v_i could keep more edges' signs unchanged by converting its set, then v_i is put into another set. This completes a round of local search. Repeat the local search process several rounds until the two sets A and B satisfy any vertex v_i cannot increase the number of edges keeping their signs unchanged by converting its set. Finally, the two sets A and B form a balanced structure satisfying the edges between internal vertices in A or B are positive and the edges between A and B are negative. Output a subset of edges $C \subseteq E$ which keep their signs unchanged. Let $\bar{G}[A, B]$ be a balanced graph with partition (A, B). The local search algorithm for $K(G)$ is shown in Algorithm 2.

Theorem 6. *Algorithm 2 output a 1/2-approximation algorithm: For a signed graph $G(V, E)$, $|C| \geq K(G)/2$.*

Proof. First of all, we notice that after each round of conversion, the new two sets A and B form a balanced structure satisfying the number of edges keeping the signs unchanged increases strictly. The parameter $K(G) \leq m \leq n^2/2$. Therefore, the algorithm must terminate at most $n^2/2$ rounds. And then we just need to show $|C| \geq |E|/2 \geq K(G)/2$. The final two sets A and B satisfy any vertex v_i cannot increase the number of edges keeping their signs unchanged by converting its set. This means that for any vertex v_i, at least half of its incident edges keep their signs unchanged. For any edge e incident to v_i, there is exactly one set in A and B keeping its sign unchanged. Thus $K_{v_i} \geq C_{v_i}$. Here K_{v_i}, C_{v_i} are the number of edges keeping the signs unchanged and the number of edges changing

---

Algorithm 3. Randomized Algorithm for $K(G)$

Input: An edge-signed graph $G(V, E)$.

Output: A balanced signed graph $\bar{G}$ and the edge set $C \subseteq E$ whose signs are kept.

1: Take a vertex order $v_1, \ldots, v_n$ arbitrarily.

2: Initialize $A = B = \emptyset$.

3: **for** $i = 1$ to n **do**

4:    Put v_i into A or B with equal probability $\frac{1}{2}$.

5: Output $\bar{G} = \bar{G}[A, B]$ and edge set $C \subseteq E$ whose signs are kept.

---

the signs. They form a partition of the incident edges of vi. Go through all the vertices and add all n inequalities. Since each edge is computed twice, we have

$$2|C| = \sum_{i \in [n]} K_{v_i} \geq \sum_{i \in [n]} C_{v_i} = 2(|E| - |C|) \Rightarrow |C| \geq |E|/2 \geq K(G)/2.$$

4.3 Randomized Algorithm

Algorithm 3: Given a general signed graph $G(V, E)$, take a vertex order $v_1, \ldots, v_n$ arbitrarily and initialize two vertex sets $A = B = \emptyset$. Place the vertices one by one to the set A or B in this order and the rules for placement are as follows: every vertex v_i is put into the set A or B with probability $1/2$. Each time after the allocation of a vertex, the two sets A and B are maintained to form a balanced structure, which satisfies the edges between internal vertices in A or B are positive and the edges between A and B are negative. Finally, the two vertex sets A and B form a balanced structure and output a subset of edges $C \subseteq E$ which keep their signs unchanged. Let $\bar{G}[A, B]$ be a balanced graph with partition (A, B). The randomized algorithm for $K(G)$ is shown in Algorithm 3.

Theorem 7. *Algorithm 3 output a 1/2-approximation random algorithm: For a signed graph $G(V, E)$, $\mathbf{E}[|C|] \geq K(G)/2$, here $\mathbf{E}[|C|]$ is the expectation of $|C|$.*

Proof. We just need to explain $\mathbf{E}[|C|] = |E|/2 \geq K(G)/2$. In fact, according to the placement rule: every vertex v_i is put into the set A or B with one half probability. Different vertices are placed independently of each other. For any edge $e(u, v)$, define a random variable Xe. If the edge e keeps its sign unchanged, $X_e = 1$ and if the edge e changes its sign, $X_e = 0$. Thus we have $|C| = \sum_{e \in E} X_e$. Since the vertices u and v are put into the set A or B with one half probability independently of each other, therefore, regardless the sign of edge e, the probability of keeping the sign is $1/2$. Thus the expectation of X_e is $1/2$. $\mathbf{E}[X_e] = 1/2$. Therefore $\mathbf{E}[|C|] = \sum_{e \in E} \mathbf{E}[X_e] = |E|/2$.

4.4   Semidefinite Programming Algorithm

Let us give a strict quadratic program for this optimization problem. Let x_i be an indicator variable for vertex v_i which will be constrained to be either $+1$ or -1. The partition (A, B) is defined as follows: $A = \{v_i \mid x_i = 1\}$ and $B = \{v_i \mid x_i = -1\}$. If edge $v_i v_j$ has a plus sign, then the contribution to the unchanged edge number is $(1 + x_i \cdot x_j)/2$. If edge $v_i v_j$ has a minus sign, then the contribution to the unchanged edge number is $(1 - x_i \cdot x_j)/2$. For ease of presentation, define $E_+ = \{v_i v_j \mid v_i v_j$ has a plus sign $\}$ and $E_- = \{v_i v_j \mid v_i v_j$ has a minus sign $\}$. Hence, a strict quadratic program is established as follows:

$$
\begin{aligned}
\max \quad & \frac{1}{2} \sum_{v_i v_j \in E_+} (1 + x_i \cdot x_j) + \frac{1}{2} \sum_{v_i v_j \in E_-} (1 - x_i \cdot x_j) \\
\text{s.t.} \quad & x_i^2 = 1 \quad v_i \in V \\
& x_i \in \mathbb{Z} \quad v_i \in V
\end{aligned}
\tag{1}
$$

We will relax this program to a vector program. We use n vector variables in $\mathbb{R}^n$, say $v_1, \ldots, v_n$ to replace $x_i, \ldots, x_n$, and replace each product $x_i \cdot x_j$ with the corresponding inner product $v_i \cdot v_j$. Then, we obtain the following vector program for this problem.

$$
\begin{aligned}
\max \quad & \frac{1}{2} \sum_{v_i v_j \in E_+} (1 + v_i \cdot v_j) + \frac{1}{2} \sum_{v_i v_j \in E_-} (1 - v_i \cdot v_j) \\
\text{s.t.} \quad & v_i \cdot v_i = 1 \quad v_i \in V \\
& v_i \in \mathbb{R}^n \quad v_i \in V
\end{aligned}
\tag{2}
$$

This vector program is similar to the vector program for Max-Cut problem in [17], and the vector program (2) is equivalent to a semidefinite program. For any $\epsilon > 0$, semidefinite programs can be solved within an additive error of ϵ, in polynomial time in n and $\log(1/\epsilon)$, using the ellipsoid algorithm [17].

Algorithm 4: Given a general signed graph $G(V, E)$, solve the corresponding vector program (2). Let $a_1, \ldots, a_n$ be a solution with error ϵ. Then, pick r to be a uniformly distributed vector on the unit sphere S_{n-1}. Place the vertices one by one to the set A or B in this order and the rules for placement are as follows: $A = \{v_i \mid a_i \cdot r \geq 0\}$ and $B = \{v_i \mid a_i \cdot r < 0\}$. The two sets A and B are maintained to form a balanced structure, which satisfies the edges between internal vertices in A or B are positive and the edges between A and B are negative. Finally, the two vertex sets A and B form a balanced structure and output a subset of edges $C \subseteq E$ which keep their signs unchanged. Let $\bar{G}[A, B]$ denote a balanced graph with partition (A, B). The semidefinite programming algorithm for $K(G)$ is as shown in Algorithm 4.

Then we show that performance gurantee $\alpha > 0.87856$.

Let OPT and OPT_v denote the optimal value of program (1) and (2), respectively. It is easy to know that $OPT \geq |E|/2$. Let θ_{ij} be an angle between vectors a_i and a_j.

Algorithm 4. Semidefinite Programming Algorithm for $K(G)$

Input: An edge-signed graph $G(V, E)$.

Output: A balanced signed graph $\bar{G}$ and the edge set $C \subseteq E$ whose signs are kept.

1: Solve vector program (2). Let $a_1, \ldots, a_n$ be a solution with error ϵ.
2: Pick r to be a uniformly distributed vector on the unit sphere S_{n-1}.
3: Let $A = \{v_i \mid a_i \cdot r \geq 0\}$ and $B = \{v_i \mid a_i \cdot r < 0\}$.
4: Output $\bar{G} = \bar{G}[A, B]$ and edge set $C \subseteq E$ whose signs are kept.

Theorem 8. $\mathbf{E}[\|C\|] \geq 0.87856 \, OPT$.

Proof. First, pick $\epsilon = 4 \times 10^{-6} |E|$. Thus, we have give the expression of OPT_v:

$$(1 - 8 \times 10^{-6}) OPT_v \leq OPT_v - \epsilon \leq \frac{1}{2} \sum_{v_i v_j \in E_+} (1 + \cos \theta_{ij}) + \frac{1}{2} \sum_{v_i v_j \in E_-} (1 - \cos \theta_{ij})$$

Define v_i and v_j are separated if v_i and v_j are on opposite sides of the partition (A, B). According to the definition of θ_{ij}, we have:

$$\mathbf{Pr}[v_i \text{ and } v_j \text{ are separated}] = \frac{\theta_{ij}}{\pi}$$

Then, we get

$$\mathbf{E}[\|C\|] = \sum_{v_i v_j \in E_+} \mathbf{Pr}[v_i \text{ and } v_j \text{ are not separated}]$$

$$+ \sum_{v_i v_j \in E_-} \mathbf{Pr}[v_i \text{ and } v_j \text{ are separated}]$$

$$= \sum_{v_i v_j \in E_+} (1 - \frac{\theta_{ij}}{\pi}) + \sum_{v_i v_j \in E_-} \frac{\theta_{ij}}{\pi}.$$

Let $\alpha = \frac{2}{\pi} \min_{0 \leq \theta \leq \pi} \frac{\theta}{1 - \cos \theta}$. We also know that $\alpha = 2 \min_{0 \leq \theta \leq \pi} \frac{1 - \frac{\theta}{\pi}}{1 + \cos \theta}$. Use elementary calculus, that $\alpha > 0.8785672$.

Then we derive the performance ratio as follows:

$$\mathbf{E}[\|C\|] \geq \alpha \cdot \left[\frac{1}{2} \sum_{v_i v_j \in E_+} (1 + \cos \theta_{ij}) + \frac{1}{2} \sum_{v_i v_j \in E_-} (1 - \cos \theta_{ij}) \right]$$

$$\geq \alpha \cdot (1 - 8 \times 10^{-6}) OPT_v \geq 0.87856 \, OPT_v \geq 0.87856 \, OPT.$$

Remark 3. If the unique games conjecture is true, 0.87856 is almost the best possible approximation ratio for maximum cut [10]. Without such unproven assumptions, it has been proven to be NP-hard to approximate the max-cut value with an approximation ratio better than $16/17 = 0.941...$ [8,16]. Thus these inapproximability results are also applied to the computation of $K(G)$.

5 Conclusion and Future Work

Given a signed graph, the maximum number of edges that needed to be kept to make it balanced is called $K(G)$. We firstly prove the computation of $K(G)$ is NP-hard. Next we design four approximation algorithms to compute $K(G)$. All these algorithms are derived from the approximation algorithms of Max-Cut problem. The best approximation ratio is about 0.87856, which is the best possible if the unique games conjecture is true.

The balance structure model on the edge weights could be considered. The weights on the edges represent the degree of the relationship. A balanced structure is needed to be defined for the weighted case.

Acknowledge. The authors are indebted to Professor Xujin Chen, Professor Xiaodong Hu and three anonymous referees for their invaluable suggestions and comments.

References

1. Antal, T., Krapivsky, P.L., Redner, S.: Social balance on networks: the dynamics of friendship and enmity. Phys. D **224**(1–2), 130–136 (2006)
2. Cartwright, D., Harary, F.: Structural balance: a generalization of Heider's theory. Psychol. Rev. **63**(5), 277 (1956)
3. Davis, J.A.: Structural balance, mechanical solidarity, and interpersonal relations. Am. J. Sociol. **68**(4), 444–462 (1963)
4. Easley, D., Kleinberg, J., et al.: Networks, Crowds, and Markets, vol. 8. Cambridge University Press, Cambridge (2010)
5. Fritz, H., et al.: The Psychology of Interpersonal Relations. Wiley, New York (1958)
6. Guha, R., Kumar, R., Raghavan, P., Tomkins, A.: Propagation of trust and distrust, pp. 403–412 (2004)
7. Harary, F., et al.: On the notion of balance of a signed graph. Michigan Math. J. **2**(2), 143–146 (1953)
8. Håstad, J.: Some optimal inapproximability results. J. ACM (JACM) **48**(4), 798–859 (2001)
9. Heider, F.: Attitudes and cognitive organization. J. Psychol. **21**(1), 107–112 (1946)
10. Khot, S., Kindler, G., Mossel, E., O'Donnell, R.: Optimal inapproximability results for MAX-CUT and other 2-variable CSPs? SIAM J. Comput. **37**(1), 319–357 (2007)
11. König, D.: Akademische verlagsgesellschaft (1936)
12. Kunegis, J., Lommatzsch, A., Bauckhage, C.: The slashdot zoo: mining a social network with negative edges, pp. 741–750 (2009)

13. Leskovec, J., Huttenlocher, D., Kleinberg, J.: Signed networks in social media, pp. 1361–1370 (2010)
14. Leskovec, J., Lang, K.J., Dasgupta, A., Mahoney, M.W.: Statistical properties of community structure in large social and information networks, pp. 695–704 (2008)
15. Marvel, S.A., Strogatz, S.H., Kleinberg, J.M.: Energy landscape of social balance. Phys. Rev. Lett. 103(19), 198701 (2009)
16. Trevisan, L., Sorkin, G.B., Sudan, M., Williamson, D.P.: Gadgets, approximation, and linear programming. SIAM J. Comput. 29(6), 2074–2097 (2000)
17. Vazirani, V.V.: Approximation Algorithms. Springer, Heidelberg (2013)
18. Zaslavsky, T.: A mathematical bibliography of signed and gain graphs and allied areas. Electron. J. Comb. DS8-Dec (2012)

Computing the One-Visibility Copnumber of Trees

Boting Yang[✉][iD] and Tanzina Akter

Department of Computer Science, University of Regina, Regina, SK, Canada
Boting.Yang@uregina.ca, tanzina.akter10@cuet.ac.bd

Abstract. In this paper, we prove a lower bound for the one-visibility copnumber of trees. We give a linear-time algorithm for computing the one-visibility copnumber of trees. We also present relations between zero-visibility and one-visibility copnumbers on trees.

1 Introduction

Graph searching provides mathematical models of many real-world problems. The cops and robber game was introduced by Nowakowski and Winkler [7] and Quilliot [8] independently. As a major model in the area of graph searching, it has received much attention in recent years. In this game, the cops and robber occupy only vertices, and they move alternatively to their neighbours. Both opponents have full information about each other's location as well as the structure of the graph. The cops move through the graph attempting to capture the robber, while the robber moves to avoid the cops. A broad overview of many graph searching models are given in [2] and many aspects of the cops and robber game can be found in [1].

The zero-visibility cops and robber game was introduced by Tošić [9], which can be considered as a hybrid of the cops and robber game [7,8] and the edge searching model [6]. Like the cops and robber game, the cops and robber take turns alternatively and each individual moves from the current vertex to one of its neighbours. Like the edge searching model, the robber is invisible. Dereniowski et al. [4] established a relationship between the zero-visibility copnumber and the pathwidth of a graph. Dereniowski et al. [5] gave a linear-time algorithm for computing the zero-visibility copnumber of trees. Recent results on the zero-visibility cops and robber game can be found in [10,11].

In [3], Clarke et al. considered a variation of cops and robber game, called the ℓ-visibility cops and robber game. This game has the same setting as the cops and robber game except that the cops have the information about the location of the robber only when the distance between the cops and the robber is less than or equal to ℓ. There are two sub-tasks for cops: seeing and capturing. In the first phase, the cops move within the distance ℓ of the robber and in the second

B. Yang—Research supported in part by an NSERC Discovery Research Grant, Application No.: RGPIN-2018-06800.

Z. Zhang et al. (Eds.): AAIM 2020, LNCS 12290, pp. 411–423, 2020.
https://doi.org/10.1007/978-3-030-57602-8_37

phase, they capture the robber. They used classes of subtrees to characterize the trees for which k cops can capture the robber for all $\ell \geq 1$. Since each of those classes contains exponential number of trees, this characterization is not suitable for designing polynomial time algorithms that can find the minimum number of cops to capture the robber for any $\ell \geq 1$. The one-visibility cops and robber game was considered in [13].

In this paper, we investigate the one-visibility cops and robber game on trees. In Sect. 3, we prove an essential theorem on the lower bound of the copnumber for trees. In Sect. 4 we propose a linear-time algorithm for computing the one-visibility copnumber of trees. This bottom-up algorithm allows us to find the copnumber of all rooted subtrees of a rooted tree. In Sect. 5, we establish relations between the one-visibility copnumber and zero-visibility copnumber of trees.

2   Preliminaries

Let G be a graph. The vertex set of G is denoted by $V(G)$. We use $u_1 \cdots u_m$ to denote a path with end vertices u_1 and u_m. The *length of a path* is the number of edges on the path. The *distance* between u and v, denoted by $\mathrm{dist}_G(u, v)$, is the length of the shortest path between u and v in G. Let H be a subgraph of G. The distance between u and H is defined to be $\mathrm{dist}_G(u, H) := \min\{\mathrm{dist}_G(u, v) \mid v \in V(H)\}$. The *neighbourhood* of v is the set $N_G(v) := \{u \in V(G) \mid \mathrm{dist}_G(u, v) = 1\}$. The *closed neighbourhood* of v is the set $N_G[v] := \{u \in V(G) \mid \mathrm{dist}_G(u, v) \leq 1\}$. For $k \geq 0$, we generalize this concept to the *k-th closed neighbourhood* of v, which is the set

$$N_G^k[v] := \{u \in V(G) \mid \mathrm{dist}_G(u, v) \leq k\}.$$

The *closed neighbourhood* of $U \subseteq V(G)$ is defined as the set $N_G[U] = \{u \in V(G) \mid \mathrm{dist}_G(u, U) \leq 1\}$.

The *degree* of v is the number of edges incident on v, denoted $\deg_G(v)$. A *leaf* is a vertex that has degree one. For $U \subseteq V(G)$, we use $G[U]$ to denote the subgraph induced by U, which consists of all vertices of U and all of the edges that connect vertices of U in G. We use $G - U$ to denote the subgraph $G[V(G) - U]$. If U contains a single vertex u, then for simplicity, we use $G - u$ for $G - \{u\}$.

A rooted tree is a tree where a single vertex is marked as the root. Let $T^{[r]}$ denote a rooted tree T with root r. Every vertex $v \neq r$ of the tree is connected with root r by a unique path where the *parent* of v is the sole neighbour of v in the unique path. If u is the parent of v, then v is a *child* of u. For a vertex $v \in V(T^{[r]})$, if a vertex u is on the unique path from r to v, then we say that v is a *descendant* of u, and u is an *ancestor* of v. For a vertex $v \in V(T^{[r]})$, we will use $T^{[v]}$ to denote the subtree of $T^{[r]}$ induced by v and all its descendants, where v is the root of this subtree. We will extensively use the notation of $T^{[v]} - v$ in Sect. 4 to denote the forest induced by $V(T^{[v]}) - \{v\}$, where $T^{[v]}$ is a rooted subtree of $T^{[r]}$. Note that each component in the forest $T^{[v]} - v$ is rooted at the vertex that is a child of v in $T^{[v]}$.

The one-visibility cops and robber game is played on a graph by two players: cop player and robber player. The cop player controls a set of cops and the robber player controls a single robber. The robber has full information about the locations of all cops, but the cops have the information about the location of the robber only when there is a cop whose distance to the robber is at most one. The game is played over a sequence of rounds. Each *round* consists of a cops' turn followed by a robber's turn. At round 0, the cops are placed on a set of vertices and then the robber is placed on a vertex. At each of the following rounds, the cops move first and the robber move next. At round i, $i \geq 1$, each cop either moves from the current vertex to a neighbouring vertex or stays still, then the robber does the same. The cops *see* the robber if the closed neighbourhood of the cops contains the robber. The cops *capture* the robber if one of them occupies the same vertex as the robber. If this happens in a finite number of rounds, then the *cops win*; otherwise, the *robber wins*. The *one-visibility cop number* of a graph G, denoted by $c_1(G)$, is the minimum number of cops required to capture the robber on G.

If G is not connected, from the above definition, we know that $c_1(G)$ is the sum of the one-visibility cop number of each component of G. We define $c_1^*(G)$ to be the largest possible $c_1(G')$, where G' is a component in G.

The zero-visibility cops and robber game has the same setting as the one-visibility cops and robber game except that the cops have no information about the location of the robber at any time, i.e., the robber is invisible to the cops. However, if a cop occupies the same vertex as the robber at some moment, then the robber is captured. The *zero-visibility cop number* of a graph G, denoted by $c_0(G)$, is the minimum number of cops required to capture the robber on G.

We say that a cop *vibrates* between two adjacent vertices x and y for a consecutive sequence of rounds if in these rounds, the cop alternates two actions: "sliding from x to y" and "sliding from y to x". A subgraph known to not contain the robber is called *cleared*; otherwise, the subgraph is *dirty*.

3   Lower Bound on $c_1(T)$

The following result is implied by Corollary 2.2 in [3].

Proposition 1. *For a tree T and a subtree H of T, $c_1(H) \leq c_1(T)$.*

Note that this proposition does not hold for general graphs. Similar to Lemma 4.4 in [3], we can prove the following result for general trees.

Lemma 1. *Let T be a tree and $v \in V(T)$. Let H be a component in the forest $T - v$. If $c_1(H) = k$, then in any search strategy to clear T, there is a moment at which at least k cops are on $N_T[V(H)]$.*

Proof. Let u be the unique neighbour of v in H, and let H_v be the subtree obtained by adding the vertex v and edge uv to H. Note that $N_T[V(H)] = V(H_v)$. Assume, for the sake of contradiction, that there is a cops' strategy S_T

for clearing T such that at any moment there are at most $k-1$ cops on H_v. We construct a pseudo strategy S to clear H_v as follows: In S_T, whenever a cop is initially placed in H_v, in S we place the corresponding cop on the same vertex. In S_T, whenever a cop moves within H_v, in S the corresponding cop has the same action in H_v. In S_T, whenever a cop enters H_v, in S we place the corresponding cop on v. In S_T, whenever a cop leaves H_v, in S we remove the corresponding cop from v. We can easily modify the pseudo strategy S to a cops' strategy S' for H_v such that the number of cops used in S' is the same as that in S. From the assumption, at any moment in S there are at most $k-1$ cops on H_v. So H_v can be cleared by S' using at most $k-1$ cops. It follows from Proposition 1 that H is cleared by at most $k-1$ cops. Thus $c_1(H) \leq k-1$, which is a contradiction.

Let T be a tree and let S be a cops' strategy that clears T in m rounds. A vertex v is *dirty* at some moment if v is occupied by the robber or the cops do not know if the robber is on v at the moment. Let V_b^i, $1 \leq i \leq m$, be the set of dirty vertices of T before the cops' move in round i, let V_a^i, $1 \leq i \leq m$, be the set of dirty vertices of T just after the cops' move in round i. and let V_c^i, $0 \leq i \leq m$, be the set of vertices of T occupied by cops after the cops' move in round i. In the next lemma we show that the robber's territory is not far away from the set of vertices occupied by cops.

Lemma 2. *Let T be a tree and let S be a cops' strategy that clears T in m rounds. Then*

(i) *for every component H of $T[V_b^i]$, $1 \leq i \leq m$, there is a vertex $v \in V_c^{i-1}$ such that $\text{dist}_T(v, H) \leq 2$; and*

(ii) *for every component H of $T[V_a^i]$, $1 \leq i \leq m-1$, there is a vertex $v \in V_c^i$ such that $\text{dist}_T(v, H) \leq 3$.*

Proof. Let V_r^i, $0 \leq i \leq m-1$, be the set containing a single vertex that is occupied by the robber after the robber's move in round i. At round 0, cops are placed on vertices of V_c^0, and then the robber is placed on the vertex in V_r^0. So at round 1, $V_b^1 = (V(T) - N_T[V_c^0]) \cup V_r^0$, and thus, statement (i) holds. Since every cop can move to a neighbouring vertex or stays still, it is easy to see that statement (ii) also holds. Suppose both statements hold at round i. We will show that both of them are true at round $i+1$.

At round $i+1$, we have $V_b^{i+1} = (N_T[V_a^i] - N_T[V_c^i]) \cup V_r^i$. Let H be a component of $T[V_b^{i+1}]$. Since $V_a^i \subseteq V_b^{i+1}$, there exists a component H' in $T[V_a^i]$ such that H' is a subgraph of H. By the assumption, there is $v \in V_c^i$ such that $\text{dist}_T(v, H') \leq 3$. If $\text{dist}_T(v, H') \leq 2$, then $\text{dist}_T(v, H) \leq 2$. If $\text{dist}_T(v, H') = 3$, then there is a vertex $x \in N_T[V_a^i]$ such that $\text{dist}_T(x, v) = 2$, and thus $x \in V_b^{i+1}$. Hence $\text{dist}_T(v, H) \leq 2$ and so statement (i) is true. Just after the cops' move at round $i+1$, we have $V_a^{i+1} = (V_b^{i+1} - N_T[V_c^{i+1}]) \cup V_r^i$. Let H be a component in $T[V_a^{i+1}]$. Since $V_a^{i+1} \subseteq V_b^{i+1}$, there is a component H'' in $T[V_b^{i+1}]$ such that H is a subgraph of H''. By the above, there is $v \in V_c^i$ such that $\text{dist}_T(v, H'') \leq 2$. Then for any vertex $x \in N(v) \cap V_c^{i+1}$, $\text{dist}_T(x, v) \leq 3$. Thus statement (ii) is true.

From the above two lemmas, we can show an essential theorem of this section, which can be considered as an extension of Lemma 4.9 in [3] for general trees in the 1-visibility cops and robber game.

Theorem 1. *Let T be a tree and let k be a positive integer. If there is a vertex $v \in V(T)$ such that the forest $T - N_T^3[v]$ contains three components with copnumber at least k and the path in T connecting any pair of these three components contains v, then $c_1(T) \geq k + 1$.*

Proof. Let X_1, X_2, X_3 be three components in $T - N_T^3[v]$ with $c_1(X_i) \geq k$, $1 \leq i \leq 3$. For $1 \leq i \leq 3$, let $p_i \in V(X_i)$ with $\text{dist}_T(p_i, v) = 4$. Let T_1 be a subtree of T which is formed from the disjoint union of X_i, $1 \leq i \leq 3$, together with the paths of length four from p_i to v. For the sake of contradiction, assume T_1 can be cleared by k cops. Then there is a round t_1 such that just after the cops' move in this round, only one of the three components, say X_1, is cleared and it remains cleared through the following rounds. There is also a round t_2 such that just after the cops' move, one of the other two components, say X_2, is cleared. Since $c_1(X_2) \geq k$, from Lemma 1, there is a round t_3 between t_1 and t_2 such that at the moment t^* just after the cops' move in round t_3, all k cops are simultaneously present in $N_T[V(X_2)]$. As X_3 is contaminated at t^* and $\text{dist}_{T_1}(v, N_T[V(X_2)]) = 3$, by Lemma 2, v is contaminated at t^*. So the robber will recontaminate X_1, which derives a contradiction. Therefore $c_1(T_1) \geq k + 1$. It follows from Proposition 1 that $c_1(T) \geq k + 1$.

4   Algorithm for Computing $c_1(T)$

From Sect. 2, we know $c_1^*(G) = \max\{c_1(G') \mid G'$ is a component in $G\}$.

For simplicity, we will use $T^{[v]} - N^3[v]$ for $T^{[v]} - N_{T^{[v]}}^3[v]$, which is the forest obtained from the rooted tree $T^{[v]}$ by deleting the vertices of $N_{T^{[v]}}^3[v]$. Similarly, if there is no ambiguity we will simply use $\text{dist}(u, v)$, $N[v]$, $N^2[v]$ and $N^3[v]$ without subscripts.

Definition 1 (k-pre-branching, k-weakly-branching, k-branching). Let $T^{[v]}$ be a rooted tree with $c_1(T^{[v]}) = k \geq 1$. We call v a *k-pre-branching vertex* if $c_1^*(T^{[v]} - N^2[v]) = k$ and $c_1(T_{2v}^{[u]}) = k$, where $T_{2v}^{[u]}$ is a tree obtained from two copies of $T^{[v]}$ by connecting each root v to a new root u.

We call v a *k-weakly-branching vertex* if one of the three forests, $T^{[v]} - v$, or $T^{[v]} - N[v]$, or $T^{[v]} - N^2[v]$, has exactly two components whose root is a k-pre-branching vertex in the component.

We call v a *k-branching vertex* if $c_1^*(T^{[v]} - N^2[v]) = k$, the forest $T^{[v]} - N^2[v]$ has exactly one component whose root is a k-weakly-branching vertex in the component, and the forest $T^{[v]} - N^3[v]$ has no component whose root is a k-weakly-branching vertex.

Let u be a child of v in $T^{[v]}$. If u is a k-pre-branching vertex (resp. k-weakly-branching vertex, k-branching vertex) in $T^{[u]}$, then we say that u is a *k-pre-branching child* (resp. *k-weakly-branching child*, *k-branching child*) of v. Similarly, we can define the k-*pre-branching descendant*, k-*weakly-branching descendant*, and k-*branching descendant* of v.

Definition 2. Let $T^{[v]}$ be a rooted tree with $c_1(T^{[v]}) = k \geq 1$. The k-*pre-branching indicator* $I_{\mathrm{pb}}^k(v)$ and the k-*weakly-branching indicator* $I_{\mathrm{wb}}^k(v)$ are defined to be:

$$I_{\mathrm{pb}}^k(v) = \begin{cases} 1, & \text{if } v \text{ is a } k\text{-pre-branching vertex in } T^{[v]}; \\ 0, & \text{otherwise.} \end{cases}$$

$$I_{\mathrm{wb}}^k(v) = \begin{cases} 1, & \text{if } v \text{ is a } k\text{-weakly-branching vertex in } T^{[v]}; \\ 0, & \text{otherwise.} \end{cases}$$

Definition 3. Let $T^{[v]}$ be a rooted tree with $c_1(T^{[v]}) = k \geq 1$. The k-*initial-counter* $J^k(v)$ and the k-*weakly-counter* $J_{\mathrm{w}}^k(v)$ are defined as follows:

$$J^k(v) = \begin{cases} 0, & I_{\mathrm{pb}}^k(v) = 0 \text{ and } c_1^*(T^{[v]} - v) = k - 1; \\ 1, & I_{\mathrm{pb}}^k(v) = 0, c_1^*(T^{[v]} - N[v]) = k - 1 \text{ and } c_1^*(T^{[v]} - v) = k; \\ 2, & I_{\mathrm{pb}}^k(v) = 0, c_1^*(T^{[v]} - N^2[v]) = k - 1 \text{ and } c_1^*(T^{[v]} - N[v]) = k; \\ 0, & \text{otherwise.} \end{cases}$$

$$J_{\mathrm{w}}^k(v) = \begin{cases} 0, & \text{if } I_{\mathrm{wb}}^k(v) = 1 \text{ and } v \text{ has exactly two } k\text{-pre-branching children} \\ & \text{and no } k\text{-weakly-branching child;} \\ 1, & \text{if } I_{\mathrm{wb}}^k(v) = 1 \text{ and } v \text{ has exactly one } k\text{-weakly-branching child} \\ & \text{and this child } u \text{ has } J_{\mathrm{w}}^k(u) = 0; \\ 2, & \text{if } I_{\mathrm{wb}}^k(v) = 1 \text{ and } v \text{ has exactly one } k\text{-weakly-branching child} \\ & \text{and this child } u \text{ has } J_{\mathrm{w}}^k(u) = 1; \\ 0, & \text{otherwise.} \end{cases}$$

Definition 4. (label $L_{T^{[v]}}(v)$, value $|L_{T^{[v]}}(v)|$) Let $T^{[v]}$ be a rooted tree. The *label* of v in $T^{[v]}$, denoted by $L_{T^{[v]}}(v)$, is a sequence

$$(s_1, v_1; s_2, v_2; \ldots; s_m, v_m; I_{\mathrm{wb}}^{s_m}(v), J_{\mathrm{w}}^{s_m}(v); I_{\mathrm{pb}}^{s_m}(v), J^{s_m}(v)),$$

where s_i and v_i are defined in the following procedure:

1. If $T^{[v]}$ contains only one vertex, then $s_1 = 1$, $v_1 = \perp$, $I_{\mathrm{wb}}^{s_i}(v) = J_{\mathrm{w}}^{s_i}(v) = I_{\mathrm{pb}}^{s_i}(v) = J^{s_i}(v) = 0$, and return $L_{T^{[v]}}(v) = (1, \perp; 0, 0; 0, 0)$; otherwise, set $i \leftarrow 1$ and $T_1^{[v]} \leftarrow T^{[v]}$.
2. Set $s_i \leftarrow c_1(T_1^{[v]})$. Then we have one of the following cases:
   (a) If v is an s_i-branching vertex in $T_1^{[v]}$, then $I_{\mathrm{wb}}^{s_i}(v) = J_{\mathrm{w}}^{s_i}(v) = I_{\mathrm{pb}}^{s_i}(v) = J^{s_i}(v) = 0$, and return $L_{T^{[v]}}(v) = (s_1, v_1; \ldots; s_i, v; 0, 0; 0, 0)$.

(b) If v has an s_i-branching descendant in $T_1^{[v]}$, let v_i be this vertex. Set $T_1^{[v]} \leftarrow T_1^{[v]} - V(T_1^{[v_i]})$, $i \leftarrow i + 1$, and go back to Step 2.

(c) If v is an s_i-weakly-branching vertex in $T_1^{[v]}$, then $I_{pb}^{s_i}(v) = J^{s_i}(v) = 0$, $I_{wb}^{s_i}(v) = 1$, and $J_w^{s_i}(v)$ can be determined by Definition 3; return $L_{T^{[v]}}(v) = (s_1, v_1; \ldots; s_i, \bot; 1, J_w^{s_i}(v); 0, 0)$.

(d) If v is an s_i-pre-branching vertex in $T_1^{[v]}$, then $I_{wb}^{s_i}(v) = J_w^{s_i}(v) = J^{s_i}(v) = 0$, and $I_{pb}^{s_i}(v) = 1$; return $L_{T^{[v]}}(v) = (s_1, v_1; \ldots; s_i, \bot; 0, 0; 1, 0)$.

(e) $I_{wb}^{s_i}(v) = J_w^{s_i}(v) = I_{pb}^{s_i}(v) = 0$, and $J^{s_i}(v)$ can be determined by Definition 3; return $L_{T^{[v]}}(v) = (s_1, v_1; \ldots; s_i, \bot; 0, 0; 0, J^{s_i}(v))$.

The *value* of $L_{T^{[v]}}(v)$, denoted by $|L_{T^{[v]}}(v)|$, is equal to s_1.

Definition 5. Let $T^{[u]}$ be a tree with root u whose children are $v_1, \ldots, v_d$. Suppose $c_1^*(T^{[u]} - u) = k \geq 1$. The counters $\#_{pb}^k(T^{[u]} - u)$, $\#_{wb}^k(T^{[u]} - u)$, $\#_c^k(T^{[u]} - u)$, $h^k(T^{[u]} - u)$ and $h_w^k(T^{[u]} - u)$ are defined as follows:

$$\#_{pb}^k(T^{[u]} - u) = \sum_{j=1}^{d} I_{pb}^k(v_j),$$

$$\#_{wb}^k(T^{[u]} - u) = \sum_{j=1}^{d} I_{wb}^k(v_j),$$

$$\#_c^k(T^{[u]} - u) = \left| \left\{ j \mid c_1(T^{[v_j]}) = k \text{ for } j \in \{1, \ldots, d\} \right\} \right|,$$

$$h^k(T^{[u]} - u) = \max \left\{ J^k(v_j) \mid j \in \{1, \ldots, d\} \right\},$$

$$h_w^k(T^{[u]} - u) = \max \left\{ J_w^k(v_j) \mid j \in \{1, \ldots, d\} \right\}.$$

The labels in Definition 4 have the following properties.

Theorem 2. *Let $T^{[u]}$ be a tree with root u whose children are $v_1, \ldots, v_d$. Suppose that $c_1^*(T^{[u]} - u) = k \geq 1$ and for $1 \leq j \leq d$,*

$$L_{T^{[v_j]}}(v_j) = (t^{v_j}, \bot; I_{wb}^{t^{v_j}}(v_j), J_w^{t^{v_j}}(v_j); I_{pb}^{t^{v_j}}(v_j), J^{t^{v_j}}(v_j)).$$

Then the label $L_{T^{[u]}}(u)$ must be of the form $(t^u, x^u, I_{wb}^{t^u}(u), J_w^{t^u}(u); I_{pb}^{t^u}(u), J^{t^u}(u))$ which can be determined as follows:

(1) If $\#_{wb}^k(T^{[u]} - u) > 1$, then $L_{T^{[u]}}(u) = (k+1, \bot; 0, 0; 0, 0)$.

(2) If $\#_{wb}^k(T^{[u]} - u) = 1$ and $\#_c^k(T^{[u]} - u) \geq 2$, then

    (2.1) if $h_w^k(T^{[u]} - u) = 2$, then $L_{T^{[u]}}(u) = (k+1, \bot; 0, 0; 0, 0)$.

    (2.2) if $h_w^k(T^{[u]} - u) = 1$ and $h^k(T^{[u]} - u) \geq 1$, then $L_{T^{[u]}}(u) = (k+1, \bot; 0, 0; 0, 0)$.

    (2.3) if $h_w^k(T^{[u]} - u) = 1$ and $h^k(T^{[u]} - u) = 0$, then $L_{T^{[u]}}(u) = (k, \bot; 1, 2; 0, 0)$.

    (2.4) if $h_w^k(T^{[u]} - u) = 0$ and $h^k(T^{[u]} - u) = 2$, then $L_{T^{[u]}}(u) = (k+1, \bot; 0, 0; 0, 0)$.

(2.5) *if* $h_{\mathrm{w}}^k(T^{[u]} - u) = 0$ *and* $h^k(T^{[u]} - u) \le 1$, *then* $L_{T^{[u]}}(u) = (k, \perp; 1, 1; 0, 0)$.

(3) *If* $\#_{\mathrm{wb}}^k(T^{[u]} - u) = 1$ *and* $\#_{\mathrm{c}}^k(T^{[u]} - u) = 1$, *then*

(3.1) *if* $h_{\mathrm{w}}^k(T^{[u]} - u) = 2$, *then* $L_{T^{[u]}}(u) = (k, u; 0, 0; 0, 0)$.

(3.2) *if* $h_{\mathrm{w}}^k(T^{[u]} - u) = 1$, *then* $L_{T^{[u]}}(u) = (k, \perp; 1, 2; 0, 0)$.

(3.3) *if* $h_{\mathrm{w}}^k(T^{[u]} - u) = 0$, *then* $L_{T^{[u]}}(u) = (k, \perp; 1, 1; 0, 0)$.

(4) *If* $\#_{\mathrm{wb}}^k(T^{[u]} - u) = 0$, *then*

(4.1) *if* $\#_{\mathrm{pb}}^k(T^{[u]} - u) \ge 3$, *then* $L_{T^{[u]}}(u) = (k + 1, \perp; 0, 0; 0, 0)$.

(4.2) *if* $\#_{\mathrm{pb}}^k(T^{[u]} - u) = 2$, *then* $L_{T^{[u]}}(u) = (k, \perp; 1, 0; 0, 0)$.

(4.3) *if* $\#_{\mathrm{pb}}^k(T^{[u]} - u) = 1$, *then* $L_{T^{[u]}}(u) = (k, \perp; 0, 0; 1, 0)$.

(4.4) *if* $\#_{\mathrm{pb}}^k(T^{[u]} - u) = 0$, *then*

(4.4.1) *if* $h^k(T^{[u]} - u) = 2$, *then* $L_{T^{[u]}}(u) = (k, \perp; 0, 0; 1, 0)$.

(4.4.2) *if* $h^k(T^{[u]} - u) = 1$, *then* $L_{T^{[u]}}(u) = (k, \perp; 0, 0; 0, 2)$.

(4.4.3) *if* $h^k(T^{[u]} - u) = 0$, *then* $L_{T^{[u]}}(u) = (k, \perp; 0, 0; 0, 1)$.

In Algorithm 1, we will compute the copnumber of subtrees in the reverse order $s_m, s_{m-1}, \ldots, s_1$ of Definition 4. For convenience, in the rest of the paper we let $t_i = s_{m-i+1}$ and $x_i = v_{m-i+1}$; i.e.,

$$L_{T^{[v]}}(v) = (t_m^v, x_m^v; \ldots; t_1^v, x_1^v; I_{\mathrm{wb}}^{t_1^v}(v), J_{\mathrm{w}}^{t_1^v}(v); I_{\mathrm{pb}}^{t_1^v}(v), J^{t_1^v}(v)), \qquad (1)$$

where the superscript v in t_i^v and x_i^v are used to refer to the vertex v. So, $|L_{T^{[v]}}(v)| = t_m^v$. Note that only x_1^v can be a "$\perp$" sign, which means that neither v is an t_1^v-branching vertex in $T^{[v]} - \bigcup_{i=2}^{m} V(T^{[x_i^v]})$ nor it has an t_1^v-branching descendant in $T^{[v]} - \bigcup_{i=2}^{m} V(T^{[x_i^v]})$. We call the first pair (t_m^v, x_m^v) an *item* associated with $T^{[v]}$, and call each pair (t_i^v, x_i^v), $1 \le i < m$, an *item* associated with subtree $T^{[v]} - \bigcup_{j=i+1}^{m} V(T^{[x_j^v]})$, where t_i^v is called the *key* of the item and x_i^v is the *attribute*.

Algorithm 1 is a bottom-up approach for computing 1-visibility copnumber of a tree. In this algorithm, we first assign labels to each vertex that has no children. Then for each vertex whose children have been labeled, we compute the label of this vertex using the rules proved in Theorems 2, 3 and 4. Finally, the first component in the label of the root is the 1-visibility copnumber of the tree.

Theorem 3. *Let* $T^{[u]}$ *be a rooted tree and* $v_1, \ldots, v_d$ *be the children of the root* u. *Suppose that* $L_{T^{[v_j]}}(v_j) =$

$$
\begin{cases}
(t_1^{v_j}, \perp; I_{\mathrm{wb}}^{t_1^{v_j}}(v_j), J_{\mathrm{w}}^{t_1^{v_j}}(v_j); I_{\mathrm{pb}}^{t_1^{v_j}}(v_j), J^{t_1^{v_j}}(v_j)), & \text{if } 1 \le j \le d_1, \\
(t_{m_j}^{v_j}, x_{m_j}^{v_j}; \ldots; t_1^{v_j}, \perp; I_{\mathrm{wb}}^{t_1^{v_j}}(v_j), J_{\mathrm{w}}^{t_1^{v_j}}(v_j); I_{\mathrm{pb}}^{t_1^{v_j}}(v_j), J^{t_1^{v_j}}(v_j)), & \text{if } d_1 < j \le d_2, \\
(t_{m_j}^{v_j}, x_{m_j}^{v_j}; \ldots; t_1^{v_j}, x_1^{v_j}; 0, 0; 0, 0), & \text{if } d_2 < j \le d,
\end{cases}
$$

Algorithm 1. Computing the 1-visibility copnumber of a tree

Input: A tree T with $n \geq 3$ vertices.

Output: $c_1(T)$.

1: Pick a vertex of T as its root.
2: Sort the vertices of T to a list $u_1, \ldots, u_n$ such that every vertex is before its parent in the list. For each vertex that has no child, set its label as $(1, \perp; 0, 0; 0, 0)$.
3: If the root u_n has obtained a label $L_{T^{[u_n]}}(u_n)$, then return the first component in this label; otherwise, let u be the first unlabeled vertex in the list currently. Run Steps 4 to 10 to compute $L_{T^{[u]}}(u)$.
4: Let v_j, $1 \leq j \leq d$, be all children of u with labels $L_{T^{[v_j]}}(v_j)$ in the form of Eq. (1). Let $I_\perp$ be the subset of children whose label contains $\perp$ and let I_b be the subset of children whose label does not contain $\perp$.
5: Compute $L_{T_1^{[u]}}(u)$, where $T_1^{[u]} = T^{[u]} - \bigcup\limits_{y \in I_b} V(T^{[x_1^y]}) - \bigcup\limits_{y \in I_\perp} V(T^{[x_2^y]})$ and let $k = |L_{T_1^{[u]}}(u)|$.
6: For $1 \leq j \leq d$, if $v_j \in I_\perp$, let L_j be a list obtained from $L_{T^{[v_j]}}(v_j)$ by deleting the last six components and items whose key is less than k; if $v_j \in I_b$, let L_j be a list obtained from $L_{T^{[v_j]}}(v_j)$ by deleting the last four components and items whose key is less than k. Let L_{d+1} be a list containing only the first item of $L_{T_1^{[u]}}(u)$.
7: If no key in $L_1, \ldots, L_d, L_{d+1}$ is repeated, then $L_{T^{[u]}}(u) \leftarrow L_{T_1^{[u]}}(u)$ and insert the items of $L_1, \ldots, L_d$ into $L_{T^{[u]}}(u)$. Go to Step 3.
8: Find the largest repeated key k^* in the lists $L_1, \ldots, L_d, L_{d+1}$.
9: Let $K = (k_1, \ldots, k_\ell)$ be a list containing the distinct keys from $L_1, \ldots, L_{d+1}$ satisfying that the keys in K are decreasing and are greater than or equal to k^*.
10: Find the smallest index h in K, where $1 \leq h \leq \ell$, such that $k_h = k_{h+1} + 1 = \cdots = k_\ell + (\ell - h)$. Update $K \leftarrow (k_1, \ldots, k_{h-1}, k_h')$ where $k_h' = k_h + 1$. Create a list $X = (Q_1, \ldots, Q_{h-1}, Q_h)$, where $Q_i = (k_i, x_i)$, $1 \leq i \leq h - 1$, is an item with key k_i and attribute x_i (note that x_i, $1 \leq i \leq h - 1$, is a k_i-branching vertex in some subtree) and $Q_h = (k_h', \perp)$. Insert $(0, 0; 0, 0)$ at the end of X. Set $L_{T^{[u]}}(u) \leftarrow X$. Go to Step 3.

where $x_1^{v_j}$, $d_2 < j \leq d$, are $t_1^{v_j}$-branching vertices of $T^{[u]}$. Let $k = c_1(T^{[u]} - \bigcup\limits_{j=d_1+1}^{d_2} V(T^{[x_2^{v_j}]}) - \bigcup\limits_{j=d_2+1}^{d} V(T^{[x_1^{v_j}]}))$. If $t_{m_j}^{v_j} < k$ for each $j \in \{d_1 + 1, \ldots, d\}$, then $c_1(T^{[u]}) = k$.

The following theorem is an extension of Theorem 1.

Theorem 4. *Let $T^{[u]}$ be a rooted tree and $v_1, \ldots, v_d$ be the children of the root u. Suppose that for $1 \leq j \leq d$,*

$$L_{T^{[v_j]}}(v_j) = (t_{m_j}^{v_j}, x_{m_j}^{v_j}; \ldots; t_1^{v_j}, x_1^{v_j}; 0, 0; 0, 0),$$

where $t_{m_1}^{v_1} \geq t_{m_2}^{v_2} \geq \cdots \geq t_{m_d}^{v_d}$ and $x_1^{v_j} \neq \perp$. Let $k = c_1^*(T^{[u]} - u)$. If $c_1(T^{[v_1]}) = c_1(T^{[v_2]}) = k$, then $c_1(T^{[u]}) = k + 1$.

From Theorems 2, 3 and 4, we can prove the correctness of Algorithm 1.

Theorem 5. *For a tree T with at least three vertices, Algorithm 1 computes $c_1(T)$.*

In the remainder of this section, we analyze the running time of Algorithm 1.

Lemma 3. *Suppose that each $S_i = (t^i_{m_i}, t^i_{m_i-1}, \ldots, t^i_1)$, $1 \le i \le d$, is a list of strictly decreasing positive integers. Let s be the largest number in $S_1, \ldots, S_d$ which occurs in at least two lists if one exists; otherwise, let $s = 0$. Then s can be determined in $O(\max\{t^1_{m_1}, \ldots, t^d_{m_d}\} + d)$ time.*

Lemma 4. *Let $f(n)$ be a function defined on the positive integers by the recurrence equation*

$$f(n) = \begin{cases} c, & n = 1 \\ f(n_1) + c, & d = 1, n \ge 2, \\ \max_M\{\sum_{i=1}^d f(n_i) + c(\lceil \log n_1 \rceil + d)\}, & d \ge 2, n \ge 3, \end{cases}$$

where $M = \{(n_1, \ldots, n_d) \mid n_1 \ge \cdots \ge n_d \ge 1, \text{ and } \sum_{i=1}^d n_i = n - 1\}$, and $c \ge 1$ is a constant. Then $f(n)$ is $O(n)$.

Lemma 4 can be proved similarly to Lemma 3.13 in [12].

From Lemmas 3 and 4, we can show the running time of Algorithm 1.

Theorem 6. *Algorithm 1 can be implemented in linear time.*

Proof. Let T be a tree with $n \ge 3$ vertices. In Algorithm 1, Step 2 requires $O(n)$ time by the topological sort. In each iteration of the loop from Steps 4 to 10, we compute the label of u in the subtree $T^{[u]}$. Let $v_1, \ldots, v_d$ be the children of u whose labels have been calculated. When we calculate those labels, we use a linked list for each label. We also use a flag associated with every label to indicate whether the label contains $\perp$. So Step 4 requires $O(d)$ time to find $I_\perp$ and I_b.

In Step 5, we construct the subtree $T_1^{[u]}$ and compute the label of u in this subtree from the labels of $v_1, \ldots, v_d$ using Theorem 2. In order to save time for scanning the labels, we compress the label representation. For a sublist of items with consecutive keys in a label, we use an interval to represent them. So by Theorem 2, it takes $O(d)$ time to compute $L_{T_1^{[u]}}(u)$. In Step 6, it takes $O(d)$ time to modify the labels to obtain the lists $L_1, \ldots, L_{d+1}$. Let $t^i_{m_i}$, $1 \le i \le d+1$, be the largest key in L_i. In Steps 7 and 8, from Lemma 3, it takes $O(\max\{t^1_{m_1}, \ldots, t^{d+1}_{m_{d+1}}\} + d)$ time to determine if all keys in $L_1, \ldots, L_{d+1}$ are different or find the largest repeated key k^* in the lists. Using the compressed representation of labels, the runtime of Steps 9 and 10 is $O(\max\{t^1_{m_1}, \ldots, t^{d+1}_{m_{d+1}}\} + d)$.

From Lemma 4, the total runtime of the loop from Steps 4 to 10 is $O(n)$. Thus Algorithm 1 can be implemented in linear time.

5 Relations Between $c_1(T)$ and $c_0(T)$

We first give a relation between $c_1(T)$ and $c_0(T)$ for a general tree T.

Theorem 7. *For any tree T, $c_1(T) \leq c_0(T) \leq 2c_1(T)$.*

A tree is called a *caterpillar* if removing all degree-one vertices produces a path or an empty graph. The next lemma shows that the lower bound in Theorem 7 is tight.

Lemma 5. *If T is a caterpillar, then $c_0(T) = c_1(T) = 1$.*

Note that adding a pendant edge to a tree may cause the copnumber to increase by one. The next result gives a case where the copnumber will not increase after adding a pendant edge.

Lemma 6. *Let T be a tree which has a path $v_1 v_2 v_3 v_4 u$ with $\deg_T(v_1) = 1$ and $\deg_T(v_i) = 2$, $2 \leq i \leq 4$. Let H be a tree obtained from T by adding a pendant edge $v_0 v_1$ to the vertex v_1. Then $c_1(H) = c_1(T)$.*

From Lemma 6, we can show that the lower bound in Theorem 7 is tight if for any two non-degree-2 vertices the distance between them is at least 8.

Lemma 7. *Let T be a tree such that the distance between any pair of non-degree-2 vertices is at least 8. Then $c_0(T) = c_1(T)$.*

The following lemma shows that the upper bound in Theorem 7 is also tight.

Lemma 8. *Let T be a caterpillar with at least one vertex of degree 3. Let H be a tree obtained from T by replacing every edge by a path of length 3. Then $c_0(H) = 2c_1(H) = 2$.*

In a rooted tree, the *height of a vertex* is the distance from the root to this vertex. The *height of a rooted tree* is the largest distance from the root to a leaf.

A *perfect k-ary tree* is a rooted tree in which every internal vertex has k children and all leaves have the same height. We use T_h^k to denote a perfect k-ary tree with height h. Notice that the 1-visibility copnumber of a perfect k-ary tree can be computed by Algorithm 1. However, the next two theorems give formulas for perfect binary trees and k-ary trees.

Theorem 8. *Let T_h^2 be a perfect binary tree with $h \geq 0$. Then $c_1(T_h^2) = \lceil \frac{h+1}{5} \rceil$ and $c_0(T_h^2) = \lceil \frac{h+1}{3} \rceil$.*

Proof. We first show $c_1(T_h^2) = \lceil \frac{h+1}{5} \rceil$ using induction. If $0 \leq h \leq 4$, it is easy to see that one cop can clear the tree, and so the claim holds. Assume that the claim is true when $5m \leq h \leq 5m + 4$, $m \geq 0$. We only need to show that the claim is also true when $5m + 5 \leq h \leq 5m + 9$.

Consider the case where $h = 5m + 5$. Let v be a child of the root u in T_{5m+5}^2. In the forest $T_{5m+5}^2 - N^3[v]$, there are at least two components which are isomorphic to T_{5m}^2 and there is at least one component which is isomorphic to T_{5m+2}^2 such that the path between any pair of the three components contains v. From the assumption and Theorem 1, we have $c_1(T_{5m+5}^2) \geq m+2$. Note that we can clear T_{5m+5}^2 with $m+2$ cops. Thus $c_1(T_{5m+5}^2) = m+2$. Similarly, we can show that $c_1(T_{5m+i}^2) = m+2$ for $6 \leq i \leq 9$. Therefore, $c_1(T_h^2) = \lceil \frac{h+1}{5} \rceil$ for any $h \geq 0$.

Similarly to the above, we can show $c_0(T_h^2) = \lceil \frac{h+1}{3} \rceil$ by Theorem 3.6 in [5].

Theorem 9. *Let T_h^k be a perfect k-ary tree with $k \geq 3$ and $h \geq 0$. Then $c_1(T_h^k) = \lceil \frac{h+1}{4} \rceil$, and $c_0(T_h^k) = \lceil \frac{h+1}{2} \rceil$.*

Proof. We use induction to show $c_1(T_h^k) = \lceil \frac{h+1}{4} \rceil$. It is easy to see that the claim holds when $0 \leq h \leq 3$. Assume that the claim is true when $4m-4 \leq h \leq 4m-1$, $m \geq 1$. We will show that the claim is also true when $4m \leq h \leq 4m+3$. First consider the case where $h = 4m$. Let u be the root of T_{4m}^k. In the forest $T_{4m}^k - N^3[u]$, there are at least three components which are isomorphic to T_{4m-4}^k such that the path between any pair of the three components contains u. From the assumption and Theorem 1, we have $c_1(T_{4m}^k) \geq m+1$. Since we can clear T_{4m}^k with $m+1$ cops. Thus $c_1(T_{4m}^k) = m+1$. Similarly, we can show that $c_1(T_{4m+i}^k) = m+1$ for $1 \leq i \leq 3$. Hence $c_1(T_h^k) = \lceil \frac{h+1}{4} \rceil$ for any T_h^k with $k \geq 3$ and $h \geq 0$.

Similarly, from Theorem 3.6 in [5], we can show that $c_0(T_h^k) = \lceil \frac{h+1}{2} \rceil$.

References

1. Bonato, A., Nowakowski, R.J.: The Game of Cops and Robbers on Graphs. American Mathematical Society, Providence (2011)
2. Bonato, A., Yang, B.: Graph searching and related problems. In: Pardalos, P.M., Du, D.-Z., Graham, R.L. (eds.) Handbook of Combinatorial Optimization, pp. 1511–1558. Springer, New York (2013). https://doi.org/10.1007/978-1-4419-7997-1_76
3. Clarke, N.E., Cox, D., Duffy, C., Dyer, D., Fitzpatrick, S., Messinger, M.-E.: Limited visibility Cops and Robbers. Discrete Appl. Math. **282**, 53–64 (2020)
4. Dereniowski, D., Dyer, D., Tifenbach, R., Yang, B.: Zero-visibility Cops & Robber and the pathwidth of a graph. J. Comb. Optim. **29**, 541–564 (2015)
5. Dereniowski, D., Dyer, D., Tifenbach, R., Yang, B.: The complexity of zero-visibility Cops and Robber. Theor. Comput. Sci. **607**, 135–148 (2015)
6. Megiddo, N., Hakimi, S.L., Garey, M., Johnson, D., Papadimitriou, C.H.: The complexity of searching a graph. J. ACM **35**, 18–44 (1988)
7. Nowakowski, R.J., Winkler, P.: Vertex-to-vertex pursuit in a graph. Discrete Math. **43**, 235–239 (1983)
8. Quilliot, A.: Jeux et pointes fixes sur les graphes, Thèse de 3ème cycle, Université de Paris VI, pp. 131–145 (1978)
9. Tošić, R.: Vertex-to-vertex search in a graph. In: Proceedings of the Sixth Yugoslav Seminar on Graph Theory, pp. 233–237, University of Novi Sad (1985)
10. Xue, Y., Yang, B., Zilles, S.: New results on the zero-visibility Cops and Robber game. In: Du, D.-Z., Li, L., Sun, X., Zhang, J. (eds.) AAIM 2019. LNCS, vol. 11640, pp. 316–328. Springer, Cham (2019). https://doi.org/10.1007/978-3-030-27195-4_29

11. Xue, Y., Yang, B., Zhong, F., Zilles, S.: A partition approach to lower bounds for zero-visibility Cops and Robber. In: Colbourn, C.J., Grossi, R., Pisanti, N. (eds.) IWOCA 2019. LNCS, vol. 11638, pp. 442–454. Springer, Cham (2019). https://doi.org/10.1007/978-3-030-25005-8_36
12. Yang, B., Zhang, R., Cao, Y., Zhong, F.: Search numbers in networks with special topologies. J. Interconnection Netw. **19**, 1–34 (2019)
13. Yang, F.: 1-visibility Cops and Robber Problem (Honour's thesis), University of Prince Edward Island (2012)

Maximum Subgraphs in Ramsey Graphs

Yan Li[1] , Yusheng Li[1] , and Ye Wang[2](✉)

[1] Tongji University, Shanghai 200092, China
[2] Harbin Engineering University, Harbin 150001, China
wima.wy@gmail.com

Abstract. For graphs G, F and H, let $G \rightarrow (F, H)$ signify that any edge coloring of G by red and blue contains either a red F or a blue H. Thus the Ramsey number $R(F, H)$ is $\min\{r \mid K_r \rightarrow (F, H)\}$. In this note, we consider an optimization problem as follows. For an integer $k \geq 1$, let $\mathbb{G} = \{G_k, G_{k+1}, \dots\}$ be a class of graphs G_n with $\delta(G_n) \geq 1$. We define the critical Ramsey number $R_\mathbb{G}(F, H)$ as $\max\{n \mid K_r \setminus G_n \rightarrow (F, H), G_n \in \mathbb{G}\}$, where $r = R(F, H)$. For some pairs F and H, we shall determine $R_\mathbb{G}(F, H)$, where $\mathbb{G}$ consists of books, matchings and complete graphs, respectively.

Keywords: Maximum subgraph · Ramsey number · Critical Ramsey graph

1 Introduction

For graphs G, F and H, let $G \rightarrow (F, H)$ signify that any red-blue edge coloring of G contains either a red F or a blue H. Thus $G \nrightarrow (F, H)$ means that there exists a red-blue edge coloring of G that contains neither a red F nor a blue H. For any graphs F and H, there is a graph G of large order such that $G \rightarrow (F, H)$, and the Ramsey number $R(F, H)$ is the smallest order of such G. Let $v(G)$ denote the order of G.

Definition 1. *For graphs G, F and H, we call G a Ramsey graph for F and H if $G \rightarrow (F, H)$ and $v(G) = R(F, H)$.*

Let G be a subgraph of K_r, where $r = R(F, H)$. Denote by $K_r \setminus G$ the graph obtained from K_r by removing an edge set of G from K_r. A natural problem is to find the maximum G such that $K_r \setminus G \rightarrow (F, H)$. To specify the types of the "maximum" graphs G, let us focus on some families of graphs as follows. Let $B_n = K_2 + nK_1$ and $M_n = nK_2$. Denote

$$\mathbb{B} = \{B_1, B_2, \dots\}, \ \mathbb{M} = \{M_1, M_2, \dots\}, \ \mathbb{K} = \{K_2, K_3, \dots\}.$$

Supported in part by NSFC.

Z. Zhang et al. (Eds.): AAIM 2020, LNCS 12290, pp. 424–435, 2020.
https://doi.org/10.1007/978-3-030-57602-8_38

Definition 2 [16]. *Let $k \geq 1$ be an integer and $\mathbb{G} = \{G_k, G_{k+1}, \ldots\}$ be a class of graphs G_n, where each graph $G_n \in \mathbb{G}$ has minimum degree $\delta(G_n) \geq 1$. Define the critical Ramsey number $R_{\mathbb{G}}(F, H)$ of F and H with respect to $\mathbb{G}$ as*

$$R_{\mathbb{G}}(F, H) = \max\{n \mid K_r \setminus G_n \to (F, H), \ G_n \in \mathbb{G}\},$$

where $r = R(F, H)$.

We shall call $R_{\mathbb{B}}(F, H)$, $R_{\mathbb{M}}(F, H)$ and $R_{\mathbb{K}}(F, H)$ the book-critical Ramsey number, the matching-critical Ramsey number and the complete-critical Ramsey number, respectively. Note that if $\mathbb{G}$ is a class of stars, then $R_{\mathbb{G}}(F, H)$ becomes the star-critical Ramsey number, which was introduced by [12] in a different form, and it has attracted much attention, see [9–14, 18, 19].

For vertex disjoint graphs F and H, denote by $F + H$ the join of F and H that is obtained from F and H by adding edges connecting $V(F)$ and $V(H)$ completely, and $F \cup H$ the union of F and H whose edge set as $E(F) \cup E(H)$. Denote by $s(F)$ the chromatic surplus of F that is the size of the smallest color class in a $\chi(F)$-coloring of F. If H is a connected graph and $v(H) \geq s(F)$, color the edges of $K_{(\chi(F)-1)(v(H)-1)+s(F)-1}$ red and blue such that the blue graph is isomorphic to $(\chi(F)-1)K_{v(H)-1} \cup K_{s(F)-1}$, which implies $K_{(\chi(F)-1)(v(H)-1)+s(F)-1} \not\to (F, H)$, and thus $R(F, H) \geq (\chi(F) - 1)(v(H) - 1) + s(F)$. Call a connected graph H with $v(H) \geq s(F)$ as F-good if

$$R(F, H) = (\chi(F) - 1)(v(H) - 1) + s(F).$$

A path in a graph is called suspended if the degree of each internal vertex is two. Bondy and Erdős [1] proved that a long cycle C_n (hence a long path) is C_m-good and $K_r(t)$-good. Furthermore, Burr [2] showed that H is F-good for any fixed F if H contains a sufficiently long suspended path. For a graph H, let H_n be a graph of order n that contains a suspended path of order $n - v(H) + 2$ obtained from H by adding $n - v(H)$ extra vertices to an edge. In this note, we have the following results, where star $S_n = K_{1,n}$.

Theorem 1. *Let F be a graph with $\chi(F) \geq 2$ and H a connected graph. Then*

$$R_{\mathbb{B}}(F, H_n) = n + C_n$$

for sufficiently large n, where $C_n = C_n(F, H)$ such that $|C_n| \leq v^2(F) + v(H)$.

It is often to call a vertex with degree greater than 1 as internal vertex, and we shall call it as non-leaf vertex in the following result to avoid confusion with that in a suspended path that has degree 2.

Theorem 2. *For integers $m \geq 2$ and $n \geq 3$, it holds*

$$R_{\mathbb{B}}(T_n, K_m) = \begin{cases} n - 4 & \text{if any non-leaf vertex of } T_n \text{ is adjacent to at most one leaf in } T_n, \\ n - 3 & \text{otherwise.} \end{cases}$$

Theorem 3. *Let m and n be positive integers. Then*

$$R_{\mathbb{M}}(S_m, S_n) = \begin{cases} 0 & \text{if } m \text{ and } n \text{ are both even,} \\ \left\lfloor \frac{m+n-1}{2} \right\rfloor & \text{otherwise.} \end{cases}$$

Theorem 4. *Let m and n be integers with $n \geq m \geq 1$ and $n \geq 2$. Then*

$$R_{\mathbb{M}}(M_m, M_n) = \left\lfloor \frac{2n+m-1}{2} \right\rfloor.$$

In addition to $R_{\mathbb{K}}(S_1, S_1) = 0$, we shall determine all other $R_{\mathbb{K}}(S_m, S_n)$.

Theorem 5. *Let m and n be positive integers with $m + n \geq 3$. Then*

$$R_{\mathbb{K}}(S_m, S_n) = \begin{cases} 0 & \text{if } m \text{ and } n \text{ are both even,} \\ m + n - 1 & \text{otherwise.} \end{cases}$$

Theorem 6. *Let m and n be integers with $n \geq m \geq 1$ and $n \geq 2$. Then*

$$R_{\mathbb{K}}(M_m, M_n) = n.$$

Let us defer the proofs to the following sections.

2   Book-Critical Ramsey Numbers

Before proceeding to proofs, we need some notation. If $V(F) \subseteq V(K_n)$, let $K_n \backslash F$ be the graph obtained from K_n by deleting the edges of F from K_n. Slightly abusing notation, for a graph G and a vertex subset S of G, we write $G \setminus S$ for the subgraph of G induced by $V(G) \setminus S$.

If there is a red-blue edge coloring of G that contains neither a red F nor a blue H, we call such a coloring an (F, H)-free coloring and the graph G is called an (F, H)-free graph. For a red-blue edge colored G, the subgraphs of G induced by red edges and by blue edges are denoted by G^R and G^B, respectively. For a vertex x of G, we denote the set of all neighbors of x in G^R and G^B by $N_G^R(x)$ and $N_G^B(x)$, respectively. Let $d_G(x)$ denote the degree of vertex x in G, and $d_G^R(x)$, $d_G^B(x)$ denote the degree of vertex x in G^R and G^B, respectively. Thus $d_G^R(x) = |N_G^R(x)|$, $d_G^B(x) = |N_G^B(x)|$ and $d_G^R(x) + d_G^B(x) = d_G(x)$. If there is no confusion, we will write $d(x)$, $d^R(x)$ and $d^B(x)$ for $d_G(x)$, $d_G^R(x)$ and $d_G^B(x)$, respectively and simply. Let $G[S]$ be the subgraph of G induced by $S \subseteq V(G)$. Note that each subgraph of G admits a red-blue edge coloring preserved from that of G.

Let $U_1, U_2, \ldots, U_{\chi(F)}$ be the color classes of vertex-coloring of F by $\chi(F)$ colors such that $|U_{\chi(F)}| = s(F)$, where each U_i is an independent set. Denote by $\tau(F)$ the minimum degree of vertices of $U_{\chi(F)}$ among all such vertex-colorings of F.

Lemma 1. *Let F be a graph with $\chi(F) \geq 2$ and H a connected graph of order $v(H) \geq s(F)$. If H is F-good, then*

$$R_{\mathbb{B}}(F, H) \leq \max\{s(F) - 3, v(H) + s(F) - \delta(H) - \tau(F) - 2\}.$$

Proof. For convenience, let $\delta = \delta(H)$, $\chi = \chi(F)$, $\tau = \tau(F)$ and $s = s(F)$. First, we assume $s \geq 2$. Consider the graph $G = K_r \setminus B_n$, where $r = R(F, H) = (\chi - 1)(v(H) - 1) + s$. We shall prove that there is a red-blue edge coloring of G such that G contains neither a red F nor a blue H. Define $K_\chi(v(H) - 1, \ldots, v(H) - 1, s - 2)$ on vertex set $V = \bigcup_{1 \leq i \leq \chi} V_i$ where $|V_i| = v(H) - 1$ for $1 \leq i \leq \chi - 1$ and $|V_\chi| = s - 2$. Let G_1 be the subgraph K_{r-2} of G. Color the edges of G_1 such that

$$G_1^R = K_\chi(v(H) - 1, \ldots, v(H) - 1, s - 2), \ G_1^B = K_{r-2} \setminus G_1^R.$$

Take a subset $S \subseteq V_{\chi-1}$ of size $\delta - 1$ and a subset $T \subseteq V_{\chi-1} \setminus S$ of size $\min\{\tau - 1, v(H) - \delta\}$.

Now consider two vertices in $V(G) \setminus V(G_1)$, denoted by v_1 and v_2. Let G^B be obtained from G_1^B by adding all edges between $\{v_1, v_2\}$ and S. Let G^R be obtained from G_1^R by adding all edges between $\{v_1, v_2\}$ and $V_1 \cup \ldots \cup V_{\chi-2} \cup T$. Note that G^B contains no H, as any subgraph of G^B of order $v(H)$ has the minimum degree at most $\delta - 1$ and $v(H) \geq s$. Furthermore, G^R contains no F as $\tau(G^R) \leq \tau - 1$. Thus,

$$R_\mathbb{B}(F, H) \leq s - 2 + v(H) - 1 - (\delta - 1) - \min\{\tau - 1, v(H) - \delta\} - 1$$
$$= \max\{s - 3, v(H) + s - \delta - \tau - 2\}.$$

If $s = 1$, let $r = R(F, H) = (\chi - 1)(v(H) - 1) + 1$. Similarly, consider the graph $G = K_r \setminus B_n$. Define $K_{\chi-1}(v(H) - 1, \ldots, v(H) - 1, v(H) - 2)$ on vertex set $V = \bigcup_{1 \leq i \leq \chi-1} V_i$ where $|V_i| = v(H) - 1$ for $1 \leq i \leq \chi - 2$ and $|V_{\chi-1}| = v(H) - 2$. Let G_1 be the subgraph of K_{r-2} of G, and color the edges of G_1 such that

$$G_1^R = K_{\chi-1}(v(H) - 1, \ldots, v(H) - 1, v(H) - 2), \ G_1^B = K_{r-2} \setminus G_1^R.$$

Take a subset $S \subseteq V_{\chi-1}$ of size $\delta - 1$ and a subset $T \subseteq V_{\chi-1} \setminus S$ of size $\min\{\tau - 1, v(H) - \delta - 1\}$. Denote by v_1 and v_2 the two vertices in $V(G) \setminus V(G_1)$. Let G^B be obtained from G_1^B by adding all edges between $\{v_1, v_2\}$ and S. Let G^R be obtained from G_1^R by adding all edges between $\{v_1, v_2\}$ and $V_1 \cup \ldots \cup V_{\chi-2} \cup T$. Similarly, G contains neither a blue H nor a red F. Thus,

$$R_\mathbb{B}(F, H) \leq v(H) - 2 - (\delta - 1) - \min\{\tau - 1, v(H) - \delta - 1\} - 1$$
$$\leq v(H) - \delta - \tau - 1,$$

completing the proof. $\qquad\qquad\qquad\qquad\qquad\qquad\qquad\qquad\qquad\qquad\qquad\quad \square$

Recall that H_n is a graph of order n obtained from H that contains a suspended path of order $n - v(H) + 2$ defined previously.

Lemma 2 [2]. *For graph F and connected graph H_n, H_n is F-good when n is sufficiently large.*

For the proof of Theorem 1, we shall employ an algorithmic proof as follows. Set $m = v(F)$, $s = s(F)$, $\chi = \chi(F)$ and $n \geq v(H) + (m - 2)(m - t) + t$, where t is the number of vertices in a largest color class among a proper vertex coloring of F. Let $G = K_r \setminus B_{p_1}$, where $r = R(F, H_n)$ and $p_1 = n - v(H) - s(s + t - 2) - t - 3$.

Lemma 3. *Let F be any bipartite graph on vertex sets V_1 and V_2 with $s = |V_1|$, $t = |V_2|$ and $s \le t$, and H_n a connected graph of order n containing a suspended path of order $n - v(H) + 2$. If $n \ge v(H) + s(s + t - 3) + 4$, then*

$$R_\mathbb{B}(F, H_n) \ge n - v(H) - s(s + t - 2) - t - 3.$$

Proof. By Lemma 2, we have $r = R(F, H_n) = n + s - 1$. Let $G = K_r \setminus B_{p_1}$ with $p_1 = n - v(H) - s(s + t - 2) - t - 3$. We shall show that any red-blue edge coloring of G contains either a red F or a blue H_n.

Let $\{v_1, v_2\}$ be the vertices of K_2 in the deleted book $K_2 + p_1 K_1$ and H_{n-i} a graph from H_n with the suspended path shortening by i. As $R(F, H_{n-2}) = n + s - 3$, we are done unless there is a blue H_{n-2} in $G \setminus \{v_1, v_2\}$. Let $Y = V(G) \setminus V(H_{n-2})$. Note that $v_1, v_2 \in Y$, and denote by X_1 the vertex set of the suspended path of length $n - v(H) - 1$ in the H_{n-2} with

$$X_1 = \{x_1, x_2, \ldots, x_{n-v(H)}\}$$

in order. Write $X_2 = \{x_1, x_2, \ldots, x_{n-v(H)-1}\} \subseteq X_1$. For any $y \in Y \setminus \{v_1, v_2\}$, consider the $n - v(H) - 1$ edges between y and X_2.

If there are two consecutive blue edges yx_i and yx_{i+1}, then there is a blue H_{n-1}. Denote by X_1' the vertex set of the suspended path in H_{n-1} with $X_1' = \{x_1', x_2', \ldots, x_{n-v(H)+1}'\}$ in order. Let

$$Y' = V(G) \setminus V(H_{n-1}) \text{ and } X_2' = \{x_1', x_2', \ldots, x_{n-v(H)}'\} \subseteq X_1'.$$

For any $y' \in Y' \setminus \{v_1, v_2\}$, we assume that no two consecutive edges $y'x_i$ and $y'x_{i+1}$ are both blue, otherwise we have a blue H_n. Furthermore, suppose that $s + t - 1$ edges are blue, say $y'x_{i_1}', y'x_{i_2}', \ldots, y'x_{i_{s+t-1}}'$. Then for $j < k$, if $x_{i_j+1}' x_{i_k+1}'$ is blue, then G contains a blue H_n with new suspended path

$$x_1' \ldots x_{i_j}' y' x_{i_k}' x_{i_k-1}' \ldots x_{i_j+1}' x_{i_k+1}' \ldots x_{n-v(H)+1}'.$$

Thus assume that all edges $x_{i_j+1}' x_{i_k+1}'$ are red, then $x_{i_1+1}', x_{i_2+1}', \ldots, x_{i_{s+t-1}+1}'$ and y' will induce a red K_{s+t} hence a red F. Consequently, we may assume that any $y' \in Y' \setminus \{v_1, v_2\}$ is adjacent to X_2' by at most $s + t - 2$ blue edges. By the similar argument, v_i is adjacent to X_2' by at most $s + t - 1$ blue edges, for $i = 1, 2$. As v_i is adjacent to at least $n - v(H) - p_1$ vertices in X_2', at least

$$n - v(H) - p_1 - 2(s + t - 1) - (s - 2)(s + t - 2) \ge t$$

vertices in X_2' are adjacent to each vertex of Y' in red completely. As $|Y'| = s$, these vertices and Y' yield a red F.

So we assume that there are no two consecutive blue edges yx_i and yx_{i+1}. Suppose there is a vertex $y_0 \in Y \setminus \{v_1, v_2\}$ such that y_0 is adjacent to X_2 by at least $s + t - 1$ blue edges, say $y_0 x_{i_1}, y_0 x_{i_2}, \ldots, y_0 x_{i_{s+t-1}}$. Since there is no red F, we can find a blue edge $x_{i_j+1} x_{i_k+1}$ with $j < k$. Then G contains a blue H_{n-1} with new suspended path

$$x_1 x_{i_j} y_0 x_{i_k} x_{i_k-1} \ldots x_{i_j+1} x_{i_k+1} \ldots x_{n-v(H)}.$$

Similarly as above, we can obtain a red F. Thus we may assume that for any $y \in Y \setminus \{v_1, v_2\}$, y is adjacent to X_2 by at most $s + t - 2$ blue edges. If v_1 or v_2 has two consecutive blue neighbors in X_2, say v_1, then we get a blue H_{n-1}, and v_2 is adjacent to X_2 by at most $s + t$ blue edges. As v_2 is adjacent to at least $n - v(H) - 1 - p_1$ vertices in X_2 and $|Y \setminus \{v_1\}| = s$, at least

$$n - v(H) - 1 - p_1 - (s + t) - (s - 1)(s + t - 2) = t$$

vertices in X_2 are adjacent to each vertex of $Y \setminus \{v_1\}$ in red completely, yielding a red F. Thus both v_1 and v_2 have no consecutive blue neighbors in X_2. Furthermore, v_1 or v_2 is adjacent to X_2 by at most $s + t$ blue edges. Otherwise suppose both v_1 and v_2 are adjacent to X_2 by $s + t + 1$ blue edges, say $N_{X_2}^B(v_1) = \{x_{i_1}, x_{i_2}, \ldots, x_{i_{s+t+1}}\}$ and $N_{X_2}^B(v_2) = \{x_{i'_1}, x_{i'_2}, \ldots, x_{i'_{s+t+1}}\}$ in order. Since there is no red F, we can find a blue edge $x_{i_j+1} x_{i_k+1}$ for some $j < k$, and then $V(H_{n-2}) \cup \{v_1\}$ induces a blue H_{n-1}. Apart from the left neighbor of v_1 in the path of H_{n-1}, we can find a blue edge $x_{i'_{j'}+1} x_{i'_{k'}+1}$ for some $j' < k'$, then $V(H_{n-1}) \cup \{v_2\}$ shall induce a blue H_n. So we may assume that v_1 is adjacent to X_2 by at most $s + t$ blue edges. Then $V(H_{n-2}) \cup \{v_1\}$ induces a blue H_{n-1}. Similarly, $|Y \setminus \{v_2\}| = s$ and at least t vertices in X_2 are adjacent to each vertex of $Y \setminus \{v_2\}$ in red completely, yielding a red F, completing the proof. $\square$

Lemma 4. *Let F be any graph of order m with $\chi(F) \geq 3$ and H_n a connected graph of order n containing a suspended path of length $n - v(H) + 1$. Let $F_{\chi(F)-1}$ be a graph from F by deleting all the t vertices in a color class among a proper vertex coloring of F. If $n \geq v(H) + (m - 2)(m - t) + t$, then*

$$R_{\mathbb{B}}(F, H_n) \geq \min\{R_{\mathbb{B}}(F_{\chi(F)-1}, H_n), n - v(H) - (m - 2)(m - t) - t - 2\}.$$

Proof. Set $s = s(F)$ and $\chi = \chi(F)$. By Lemma 2, we have $r = R(F, H_n) = (\chi - 1)(n - 1) + s$. Let G be the graph $K_r \setminus B_p$, where $p = \min\{R_{\mathbb{B}}(F_{\chi-1}, H_n), n - v(H) - (m - 2)(m - t) - t - 2\}$. We shall show that any red-blue edge coloring of G contains either a red F or a blue H_n. Let v_1 and v_2 be the vertices of the K_2 in the deleted book $K_2 + pK_1$, and H_{n-i} the graph obtained from H_n with the suspended path shortening by i. As $R(F, H_{n-1}) \leq r - 2$, we are done unless there is a blue H_{n-1} in $G \setminus \{v_1, v_2\}$. Delete $n - 1$ vertices of this blue H_{n-1}, then there are at least $r(F_{\chi-1}, H_n)$ vertices left. As $p \leq R_{\mathbb{B}}(F_{\chi-1}, H_n)$, we may assume that there is a red $F_{\chi-1}$. Thus we obtain a blue H_{n-1} and a red $F_{\chi-1}$. Let $X = V(H_{n-1})$ and $Y = V(F_{\chi-1})$ with $|X| = n - 1$ and $|Y| = m - t$. Denote the vertices of the suspended path of length $n - v(H)$ in H_{n-1} by

$$X_1 = \{x_1, x_2, \ldots, x_{n-v(H)+1}\}$$

in order. Write $X_2 = \{x_1, x_2, \ldots, x_{n-v(H)}\} \subseteq X_1$. Note that $v_1, v_2 \notin X_1$. Now we consider two cases.

Case 1. If $v_1 \notin Y$ and $v_2 \notin Y$, by the similar argument as in the proof of Lemma 3, for any $y \in Y$, y is adjacent to X_2 by at most $m - 2$ blue edges, and thus at least

$$n - v(H) - (m - 2)(m - t) \geq t$$

vertices in X_2 are adjacent to each vertex of Y in red completely, which yields a red F.

Case 2. If $v_1 \in Y$ and $v_2 \notin Y$, similarly, for any $y \in Y \setminus \{v_1\}$, y is adjacent to X_2 by at most $m - 2$ blue edges, and v_1 is adjacent to X_2 by at most $m - 1$ blue edges. Then at least

$$n - v(H) - p - (m - 1) - (m - 2)(m - t - 1) = t + 1$$

vertices in X_2 are adjacent to each vertex of Y in red completely, which yields a red F.

Case 3. If $v_1 \in Y$ and $v_2 \in Y$, similarly, for any $y \in Y \setminus \{v_1, v_2\}$, y is adjacent to X_2 by at most $m - 2$ blue edges. Both v_1 and v_2 are adjacent to X_2 by at most $m - 1$ blue edges. Then at least

$$n - v(H) - p - 2(m - 1) - (m - 2)(m - t - 2) = t$$

vertices in X_2 are adjacent to each vertex of Y in red completely. These vertices and Y yield a red F, completing the proof.   □

Proof of Theorem 1. Set $\delta = \delta(H)$, $m = v(F)$, $s = s(F)$, $\chi = \chi(F)$ and $\tau = \tau(F)$. Let $V_1, V_2, \ldots, V_\chi$ be the color classes of F under a proper vertex coloring using χ colors and assume that $|V_1| \geq |V_2| \geq \ldots \geq |V_\chi|$. Choose a vertex coloring such that $|V_\chi|$ is as small as possible, then let $|V_\chi| = s$ and $|V_1| = t$. We shall show that for sufficiently large n, it holds

$$n - v(H) - (m-2)(m-t) - t - 3 \leq R_{\mathbb{B}}(F, H_n) \leq \max\{s(F) - 3, n + s - \delta - \tau - 2\}. \quad (1)$$

The upper bound in (1) follows from Lemma 1. For the lower bound in (1), as it is trivial when $\chi = 2$ by Lemma 3, we assume $\chi \geq 3$. Define $F_{\chi - i} = F \setminus (V_1 \cup V_2 \cup \ldots \cup V_i)$, for $i = 1, 2, \ldots, \chi - 2$. By Lemma 4, we have

$$R_{\mathbb{B}}(F, H_n) \geq \min\{R_{\mathbb{B}}(F_{\chi - 1}, H_n), n - v(H) - (m - 2)(m - t) - t - 2\}.$$

Let $|V_{\chi - 1}| = t_1$. As $m \geq s + t_1 + t$, applying Lemma 4 repeatedly and by Lemma 3, we have

$$R_{\mathbb{B}}(F, H_n) \geq \min\{R_{\mathbb{B}}(F_2, H_n), n - v(H) - (m - 2)(m - t) - t - 2\}$$
$$\geq n - v(H) - (m - 2)(m - t) - t - 3,$$

which yields the claimed lower bound.   □

Let $r = R(T_n, K_m)$. Denote by $\{v_1, v_2\}$ the vertex set of K_2 in the deleted book. As it is trivial for star, so we focus on the case that T_n is not a star. We achieve the red T_n in a red-blue edge colored $G = K_r \setminus B_p$ containing no blue K_m by the following steps. Let $p = n - 4$ if we want return a tree with non-leaf vertex being adjacent to at most one leaf in T_n, otherwise $p = n - 3$.

Step 1: Apply Lemma 8, find $m - 1$ red cliques $U_1, U_2, \ldots, U_{m-1}$ with $|U_i| = n - 1$ for $1 \leq i \leq m - 2$ and $|U_{m-1}| = n - 2$.

Step 2: If we can find two red common neighbors of v_1 and v_2 in U_{m-1}, return a red T_n with any non-leaf vertex being adjacent to at most one leaf in T_n.

Step 3: Otherwise, find one common red neighbor of v_1 and v_2 in U_{m-1}.

Step 4: Return a red T_n with at least one non-leaf vertex being adjacent to at least two leaves in T_n.

Lemma 5 [4]. *For any tree on n vertices, $R(T_n, K_m) = (n-1)(m-1)+1$.*

Lemma 6 [16]. *For integers $n, m \geq 2$, $R_{\mathbb{K}}(K_{1,n}, K_m) = n$.*

Lemma 7 [17]. *If T is a tree on n vertices and H is a simple graph with $\delta(H) \geq n-1$, then T is a subgraph of H.*

Definition 3. *For integers $n, m \geq 2$, let $r = R(T_n, K_m) = (n-1)(m-1)+1$. Define the graph G to be a red-blue edge colored K_{r-2} such that*

$$G^R = K_{m-1}(n-1, \ldots, n-1, n-2), \quad G^B = K_{r-2} \setminus G^R.$$

Lemma 8. *For integers $n, m \geq 2$, let $r = R(T_n, K_m) = (n-1)(m-1)+1$. If c is a (T_n, K_m)-free coloring of K_{r-2} where T_n is not a star, then the resulting graph must be graph G as in Definition 3.*

Proof. We shall proceed the proof by induction. It is trivial for $m = 2$, and we may assume $m \geq 3$. Let G be a red-blue edge colored K_{r-2}. For any vertex $v \in V(G)$, if $d^B(v) \leq (n-1)(m-2) - 2$, then $d^R(v) = r - 3 - d^B(v) \geq n-1$. By Lemma 7, G contains a red T_n. So assume that there is a vertex u such that $d_G^B(u) \geq (n-1)(m-2) - 1$.

By induction, as $G[N_G^B(u)]$ is (T_n, K_{m-1})-free, $G[N_G^B(u)]$ has the structure of the graph described in Definition 3. Denote $U = N_G^B(u)$ and $U = U_1 \cup U_2 \cup \ldots \cup U_{m-2}$ with $|U_1| = |U_2| = \ldots = |U_{m-3}| = n-1$ and $|U_{m-2}| = n-2$. If $V(G) \setminus U$ induces a red clique, as there is no red T_n, the edges between $V(G) \setminus U$ and U are all blue, which yields the structure as required. Then we assume that there is a blue edge $v_1 v_2$ with $v_1, v_2 \in V(G) \setminus U$. For any vertex $x \in U_{m-2}$, either $x v_1$ or $x v_2$ is red, otherwise we have a blue K_m. If v_1 is adjacent to U_{m-2} in red completely, $\{v_1\} \cup U_{m-2}$ induces a red clique of order $n-1$, then $V(G) \setminus (U \cup \{v_1\})$ induces a red clique of order $n-2$, yielding the structure as required. So there exists $x_1 \in U_{m-2}$ such that $v_1 x_1$ is blue, then $v_2 x_1$ is red. Similarly, there exists $x_2 \in U_{m-2}$ such that $v_2 x_2$ is blue, then $v_1 x_2$ is red. As T_n is not a star, we can get a red T_n with leaves v_1 and v_2. Then $V(G) \setminus U$ induces a red clique and any vertex in $V(G) \setminus U$ is adjacent to U by blue completely, completing the proof. $\square$

Proof of Theorem 2. Let $r = R(T_n, K_m) = (n-1)(m-1)+1$, and we consider two cases.

Case 1. Any non-leaf vertex is adjacent to at most one leaf in T_n. For the upper bound, consider the graph $G = K_r \setminus B_{n-3}$. Let G_1 be the K_{r-2} in G, and $V(G) \setminus V(G_1) = \{v_1, v_2\}$. Color the edges of G_1 in red and blue such that

$$G_1^R = K_{m-1}(n-1, \ldots, n-1, n-2), \quad G_1^B = K_{r-2} \setminus G_1^R.$$

Then denote $V(G_1) = U_1 \cup U_2 \cup \ldots \cup U_{m-1}$ with $|U_1| = \ldots = |U_{m-2}| = n-1$ and $|U_{m-1}| = n - 2$. Color all the edges between $\{v_1, v_2\}$ and $U_1 \cup U_2 \cup \ldots \cup U_{m-2}$ in blue, and color all the edges between $\{v_1, v_2\}$ and one vertex from U_{m-1} in red. As any non-leaf vertex is adjacent to at most one leaf in T_n, there is no red T_n in G, which implies $R_\mathbb{B}(T_n, K_m) \le n - 4$.

For the lower bound, consider the graph $G = K_r \setminus B_{n-4}$. Let G_1 be the K_{r-2} in G, and $V(G) \setminus V(G_1) = \{v_1, v_2\}$. By Lemma 8, G_1 has the structure in Definition 3, where $V(G_1) = U_1 \cup U_2 \cup \ldots \cup U_{m-1}$ with $|U_1| = \ldots = |U_{m-2}| = n-1$ and $|U_{m-1}| = n-2$. As v_1 and v_2 can only have blue neighbors in U_i where $1 \le i \le m - 2$, then v_1 and v_2 have at least two red common neighbors in U_{m-1}, yielding a red T_n.

Case 2. There is an non-leaf vertex which is adjacent to at least two leaves in T_n. By Lemma 1, we have $R_\mathbb{B}(T_n, K_m) \le n-3$. For the lower bound, if T_n is a star, by Lemma 6, it is easy to see that $R_\mathbb{B}(K_{1,n-1}, K_m) \ge R_\mathbb{K}(K_{1,n-1}, K_m) - 2 = n - 3$. If T_n is not a star, consider the graph $G = K_r \setminus B_{n-3}$. Let G_1 be the K_{r-2} in G, and $V(G) \setminus V(G_1) = \{v_1, v_2\}$. By Lemma 8, G_1 has the structure in Definition 3, where $V(G_1) = U_1 \cup U_2 \cup \ldots \cup U_{m-1}$ with $|U_1| = \ldots = |U_{m-2}| = n - 1$ and $|U_{m-1}| = n - 2$. Thus we have v_1 and v_2 has at least one red common neighbor in U_{m-1}, yielding a red T_n. □

3 Matching-Critical Ramsey Numbers

Lemma 9 [3,5]. *Let m and n be positive integers. Then*

$$R(S_m, S_n) = \begin{cases} m + n - 1 & \text{if } m \text{ and } n \text{ are both even,} \\ m + n & \text{otherwise.} \end{cases}$$

Proof of Theorem 3. If m and n are both even, as shown in [8] that $K_{m+n-1} \setminus K_2 \not\to (S_m, S_n)$, then we have $R_\mathbb{M}(S_m, S_n) = 0$. If m or n is odd, $r = R(S_m, S_n) = m + n$, then we consider two cases.

Case 1. If $m + n$ is odd, for graph $G = K_r \setminus M_{(m+n-1)/2}$, then G must contain a vertex, denoted by v, such that $d(v) = m + n - 1$. Then we have $d^R(v) \ge m$ or $d^B(v) \ge n$, producing a red S_m or a blue S_n. Thus $R_\mathbb{M}(S_m, S_n) = (m+n-1)/2$.
Case 2. If $m + n$ is even, for graph $G = K_r \setminus M_{(m+n-2)/2}$, then G must contain a vertex, denoted by v, such that $d(v) = m + n - 1$, producing a red S_m or a blue S_n. So we have $R_\mathbb{M}(S_m, S_n) \ge (m + n - 2)/2$. For the upper bound, set $Z_{m+n} = \{0, 1, \ldots, m + n - 1\}$,

$$A_1 = \left\{ \pm 1, \pm 2, \ldots, \pm \frac{m-1}{2} \right\}, \quad A_2 = \left\{ \pm \frac{m+1}{2}, \pm \frac{m+3}{2}, \ldots, \pm \frac{m+n-2}{2} \right\}.$$

For graph $H = K_r \setminus M_{(m+n)/2}$, define an (S_m, S_n)-free coloring of H on Z_{m+n}, in which the edge uv is red if $u - v \in A_1$ and the edge uv is blue if $u - v \in A_2$. Then H^R is $(m - 1)$-regular and H^B is $(n - 1)$-regular, yielding H is (S_m, S_n)-free. □

Lemma 10 [6,7,15]. *Let m and n be integers with $n \geq m \geq 1$. Then $R(M_m, M_n) = 2n + m - 1$.*

Claim. $K_5 \setminus M_2 \rightarrow (M_2, M_2)$.

Proof. Let $G = K_5 \setminus M_2$ with $V(G) = \{v_1, v_2, v_3, v_4, v_5\}$ and $v_1 v_2, v_3 v_4 \notin E(G)$. Suppose that $v_2 v_3$ is red. Then $v_1 v_4$ and $v_4 v_5$ are both blue, otherwise there is a red M_2. Similarly, we have $v_1 v_3$ and $v_2 v_4$ are both red, yielding a red M_2. $\square$

Proof of Theorem 4. We shall prove $R_\mathbb{M}(M_m, M_n) = \lfloor (2n + m - 1)/2 \rfloor$ for $n \geq 2$ and $n \geq m \geq 1$. By Lemma 10, $r = R(M_m, M_n) = 2n + m - 1$ for $n \geq m \geq 1$. We will prove the claimed equality as follows.

Case 1. $m = 1$ and $n \geq 2$. In this case, as $r = R(M_1, M_n) = 2n$, the claimed equality can be seen easily.

Case 2. $m = n = 2$. In this case, by the claim, any edge coloring of $K_5 \setminus M_2$ in red and blue contains either a red M_2 or a blue M_2.

Claim. If the claimed equality holds for (M_m, M_n), then it holds for (M_{m+1}, M_{n+1}).

Proof. Let graph $G = K_{r'} \setminus M_{\lfloor r'/2 \rfloor}$ with $r' = R(M_{m+1}, M_{n+1}) = 2n + m + 2$. Note that if any vertex in G has no blue neighbors or no red neighbors, there would be a red M_{m+1} or a blue M_{n+1} in G, respectively. Then we may assume that there is a vertex v in G with a red neighbor a and a blue neighbor b. Let $H = G \setminus \{v, a, b\}$. Since $K_r \setminus M_{\lfloor r/2 \rfloor} \subseteq H$ with $r = R(M_m, M_n) = 2n + m - 1$, H contains either a red M_m or a blue M_n. Along with edge va or vb, we get either a red M_{m+1} or a blue M_{n+1}. $\square$

Combining Case 1, Case 2 and the above claim, the claimed equality in (M_m, M_n) can be reduced to that in (M_1, M_{n-m+1}) or (M_2, M_2), completing the proof. $\square$

4   Complete-Critical Ramsey Numbers

Proof of Theorem 5. Lemma 9 tells us the Ramsey number $R(S_m, S_n)$. Assume m or n is odd. As graph $G = K_{m+n} \setminus K_{m+n-1} = S_{m+n-1}$, for the center vertex $v \in V(S_{m+n-1})$, we have $d^R(v) + d^B(v) = m + n - 1$. Then $d^R(v) \geq m$ or $d^B(v) \geq n$, producing a red S_m or a blue S_n. If m and n are both even, as shown in [8] that $K_{m+n-1} \setminus K_2 \nrightarrow (S_m, S_n)$, then we obtain $R_\mathbb{K}(S_m, S_n) = 0$. $\square$

Proof of Theorem 6. Lemma 10 yields $r = R(M_m, M_n) = 2n + m - 1$ for $n \geq m \geq 1$. For the upper bound, let graph $G = K_{2n+m-1} \setminus K_{n+1} = K_{n+m-2} + (n+1)K_1$ with $G^R = K_{m-1} + 2nK_1$ and $G^B = (m-1)K_1 \cup (K_{n-1} + (n+1)K_1)$. It is easy to see that G is (M_m, M_n)-free.

For the lower bound, we will prove the claimed equality as follows.

Case 1. $m = 1$ and $n \geq 2$. In this case, any red-blue edge coloring of $K_{2n} \setminus K_n$ contains either a red M_1 or a blue M_n.

Case 2. $m = n = 2$. In this case, by the claim, any red-blue edge coloring of $K_5 \setminus K_2$ contains either a red M_2 or a blue M_2.

Claim. If the claimed equality holds for (M_m, M_n), then it holds for (M_{m+1}, M_{n+1}).

Proof. Let graph $G = K_{2n+m+2} \setminus K_{n+1} = K_{n+m+1} + (n+1)K_1$. Denote K_{n+m+1} by G_1 and $(n+1)K_1$ by G_2. Then there must be a vertex v in G_1 with two edges in different colors adjacent to vertex a in G_1 and b in G_2, respectively. Otherwise there will be a red M_{m+1} or a blue M_{n+1}. Let $H = G \setminus \{v, a, b\}$. Since $H = K_r \setminus K_n$ with $r = R(M_m, M_n) = 2n + m - 1$, the graph H contains either a red M_m or a blue M_n. Along with edge va or vb, we get either a red M_{m+1} or a blue M_{n+1}. □

Combining Case 1, Case 2 and the above claim, the claimed equality in (M_m, M_n) can be reduced to that in (M_1, M_{n-m+1}) or (M_2, M_2), completing the proof. □

References

1. Bondy, J., Erdős, P.: Ramsey numbers for cycles in graphs. J. Combin. Theory Ser. B **14**(1), 46–54 (1973)
2. Burr, S.A.: Ramsey numbers involving graphs with long suspended paths. J. London Math. Soc. **24**(3), 405–413 (1981)
3. Burr, S.A., Roberts, J.: On Ramsey numbers for stars. Utilitas Math. **4**, 217–220 (1973)
4. Chvátal, V.: Tree-complete graph Ramsey numbers. J. Graph Theory **1**(1), 93 (1977)
5. Chvátal, V., Harary, F.: Generalized Ramsey theory for graphs II, small diagonal numbers. Proc. Am. Math. Soc. **32**, 389–394 (1972)
6. Cockayne, E.J., Lorimer, P.J.: On Ramsey graph numbers for stars and stripes. Can. Math. Bull. **18**(1), 31–34 (1975)
7. Cockayne, E.J., Lorimer, P.J.: The Ramsey number for stripes. J. Aust. Math. Soc. Ser. A **19**, 252–256 (1975)
8. Erdős, P., Faudree, R.: Size Ramsey functions. In: Halasz, G., et al. (eds.) Sets, Graphs and Numbers. North-Holland Publishing Co., Amsterdam, pp. 219–238 (1992). Colloq. Math. Soc. Janos Bolyai 60
9. Haghi, S., Maimani, H., Seify, A.: Star-critical Ramsey numbers of F_n versus K_4. Discrete Appl. Math. **217**(2), 203–209 (2017)
10. Hao, Y., Lin, Q.: Star-critical Ramsey numbers for large generalized fans and books. Discrete Math. **341**(12), 3385–3393 (2018)
11. Hook, J.: Critical graphs for $R(P_n, P_m)$ and the star-critical Ramsey number for paths. Discuss. Math. Graph Theory **35**(4), 689–701 (2015)
12. Hook, J., Isaak, G.: Star-critical Ramsey numbers. Discrete Appl. Math. **159**(5), 328–334 (2011)
13. Li, Y., Li, Y., Wang, Y.: Minimal Ramsey graphs on deleting stars for generalized fans and books. Appl. Math. Comput. **372**, 125006 (2020)
14. Li, Z., Li, Y.: Some star-critical Ramsey numbers. Discrete Appl. Math. **181**, 301–305 (2015)
15. Lorimer, P.J.: The Ramsey numbers for stripes and one complete graph. J. Graph Theory **8**(4), 177–184 (1984)

16. Wang, Y., Li, Y.: Deleting edges from Ramsey graphs. Discrete Math. **343**(3), 111743 (2020)
17. West, D.B.: Introduction to Graph Theory, 2nd edn, p. 387. Prentice Hall, Upper Saddle River (2000)
18. Wu, Y., Sun, Y., Radziszowski, S.: Wheel and star-critical Ramsey numbers for quadrilateral. Discrete Appl. Math. **186**, 260–271 (2015)
19. Zhang, Y., Broersma, H., Chen, Y.: On star-critical and upper size Ramsey numbers. Discrete Appl. Math. **202**, 174–180 (2016)

Independent Perfect Domination Sets in Semi-Cayley Graphs

Xiaomeng Wang⬤, Shou-Jun Xu$^{(\boxtimes)}$⬤, and Xianyue Li⬤

School of Mathematics and Statistics, Lanzhou University, Lanzhou, China
{wangxm2015,shjxu,lixianyue}@lzu.edu.cn

Abstract. An independent perfect domination set in a graph Γ is an independent set S of $V(\Gamma)$ such that every vertex of $V(\Gamma) \setminus S$ is adjacent to exactly one vertex in S. In this paper, we first give a necessary and sufficient condition for the existence of independent perfect domination sets in Semi-Cayley graph $SC(G; R, R, T)$ over finite group G. Further, we obtain a necessary and sufficient condition for Cayley graphs on two class non-abelian groups to have independent perfect domination sets.

Keywords: Cayley graphs · Semi-Cayley graphs · Independent perfect domination sets · Dihedral groups · Dicyclic groups

1 Introduction

The operation of a group will be written multiplicatively, unless specifically stated otherwise. For a graph Γ, denote by $V(\Gamma)$ and $E(\Gamma)$ its vertex set and edge set. For any set S, we define $|S|$ to be the cardinality of set S and $[n] := \{1, \dots, n\}$.

Let G be a finite group with identity element e and A, B, C be subsets of G with $A = A^{-1}$, $B = B^{-1}$ and $e \notin A \cup B$. The *Cayley graph* $Cay(G, A)$ is a graph with vertex set $V(Cay(G, A)) := G$ and edge set $E(Cay(G, A)) := \{g \sim h \mid h^{-1}g \in A \text{ and } g, h \in G\}$. A graph is said to be a *Semi-Cayley graph* of a group G if it admits G as a semiregular automorphism group with two orbits. Resmini and Jungnickel [12] gave the structure representation of Semi-Cayley graphs. The *Semi-Cayley graph* $SC(G; A, B, C)$ is graph with vertex set $G \times \{0, 1\}$, and with vertices (g, i), (h, j) adjacent if and only if one of the following three possibilities occurs:

(1) $i = j = 0$ and $h^{-1}g \in A$,
(2) $i = j = 1$ and $h^{-1}g \in B$,
(3) $i = 0, j = 1$ and $h^{-1}g \in C$.

It is well known the study of Semi-Cayley graphs is a part of a larger project which aims at obtaining a deeper understanding of various classes of symmetric graphs.

Supported by the National Numerical Windtunnel Project (No. NNW2019ZT5-B16), the National Natural Science Foundation of China (Grant No. 11571155).

Z. Zhang et al. (Eds.): AAIM 2020, LNCS 12290, pp. 436–447, 2020.
https://doi.org/10.1007/978-3-030-57602-8_39

Let Γ be a connected graph. For a vertex $v \in V(\Gamma)$, denote by $N(v)$ and $N[v]$ its *neighborhood* and *closed neighborhood*. A subset S of $V(\Gamma)$ is called *perfect domination set* of Γ if for each $v \in V(\Gamma)$ there exists a unique element $s \in S$ such that v and s are adjacent [7]. If a set S of $V(\Gamma)$ is both a *perfect domination set* and an *independent set*, then S is called an *independent perfect domination set* or an *efficient domination set*. In [1,2], Chelvam et al. obtained efficient domination sets in circulant graphs. Obradovič gave necessary and sufficient conditions for the existence of efficient domination sets in circulant graphs of degree 3 and 4 [11]. Deng got a necessary and sufficient condition for the existence of efficient domination sets in $Cay(Z_n, A)$ ($\frac{n}{|A|+1} = p$ with p is prime and $n = p^k q, p^2 q^2, pqr, p^2 qr, pqrs, p, q, s$ are distinct primes, k is positive integer) [3,4]. Kumar and MacGillivray characterized efficient domination sets in Cayley graph $Cay(Z_n, A)$ with $\frac{n}{|A|+1} = 2$ and 3 [8].

Let $\tilde{\Gamma}$, Γ be two graphs and let ϕ be a homomorphism from $\tilde{\Gamma}$ to Γ. A homomorphism ϕ from $\tilde{\Gamma}$ to Γ is a *local isomorphism* if for each vertex $v \in V(\Gamma)$, the induced mapping from the set of neighbours of a vertex in $\phi^{-1}(v)$ to the neighbours of v is an bijection. We call ϕ a *covering map* if it is a local isomorphism, in which case we say that $\tilde{\Gamma}$ *covers* Γ [6]. If the covering map $\phi : \tilde{\Gamma} \to \Gamma$ is m-to-one, then ϕ is called m-fold covering projection. A covering projection $\phi : \tilde{\Gamma} \to \Gamma$ is called S-covering if there is a subgroup $\mathcal{A}$ of the automorphism group of $\tilde{\Gamma}$ acting freely on $\tilde{\Gamma}$, and there exists an isomorphism h from the graph Γ to the quotient graph $\tilde{\Gamma}/\mathcal{A}$ such that the quotient map $\tilde{\Gamma} \to \tilde{\Gamma}/\mathcal{A}$ is the composition $h \circ \phi$ of ϕ and h. The *fiber* of an edge or a vertex is its preimage under ϕ [6,9]. Lee studied the existence of independent perfect domination sets in Cayley graphs using covering projections [10].

Domination sets play an important role in the design and computer networks. For any node, domination sets can be used to decide the placement of limited resources. Cayley graphs and Semi-Cayley graphs are two class important graphs in algebraic graph theory. The problem of domination sets of Cayley graph was investigated by several authors [1–4,8,10,11]. These motivate us to write this paper.

In this paper, we first consider the Semi-Cayley graph $SC(G; R, R, T)$ and obtain that Semi-Cayley graph $SC(G; R, R, T)$ has an independent perfect domination set if and only if it is a covering of a complete graph $K_{|R|+|T|+1}$ (Theorem 1). When T is a empty set, we can obtain the result given by Lee [10]. Thanks to the Lemmas 4 and 5 (i.e., Lemmas 4.2 and 4.4 in [5]), we obtain the result that independent perfect domination sets of Cayley graphs on two non-abelian groups (dihedral groups and dicyclic groups) (Theorems 2, 3).

2   Preliminaries

In this section, we shall list some notation and known results which will be used in this paper. Let us now get back to results obtained by Lee in [10].

Lemma 1 *[10, Lemma 1]*

1. Let S_1 and S_2 be two independent perfect domination sets of a graph Γ. Then $|S_1| = |S_2|$.
2. Let $S_1, \ldots, S_n$ be n independent perfect domination sets of a graph Γ which are pairwise mutually disjoint. Then the subgraph H induced by $S_1 \cup \cdots \cup S_n$ is an m-fold covering graph of the complete graph K_n, where $m = |S_i|$ for each $i = 1, 2, \ldots, n$.

Lemma 2 *[10, Lemma 2]*. Let $\phi : \tilde{\Gamma} \to \Gamma$ be a covering and S a perfect domination set of Γ. Then $\phi^{-1}(S)$ is a perfect domination set of $\tilde{\Gamma}$. Moreover, if S is independent, then $\phi^{-1}(S)$ is independent.

Lemma 3 *[10, Theorem 1]*. Let Γ be a graph and n a natural number. Then Γ is a covering of the complete graph K_n if and only if Γ has a vertex partition $\{S_1, \ldots, S_n\}$ such that S_i is an independent perfect domination set for each $i = 1, 2, \ldots, n$.

We now turn our attention to Cayley graphs on two non-abelian groups (dihedral groups and dicyclic groups). For Semi-Cayley graphs, we have the following results.

Lemma 4 *[5, Lemma 4.2]*. Let D_n be a dihedral group and $Cay(D_n, H)$ a Cayley graph over dihedral group D_n with $H'' = \{i \mid a^i \in H\}$.

(1) If $H \cap \{a^i x \mid 0 \le i \le n-1\} = \emptyset$, then $Cay(D_n, H) \cong SC(Z_n; H'', H'', \emptyset)$;
(2) If $H \cap \{a^i x \mid 0 \le i \le n-1\} \ne \emptyset$, let $a^{i_0} x \in H$, then $Cay(D_n, H) \cong SC(Z_n; H'', H'', T')$, where $T' = \{i \mid a^{i_0 - i} x \in H\}$.

Lemma 5 *[5, Lemma 4.4]*. Let DC_{2n} be a dicyclic group and $Cay(DC_{2n}, H)$ a Cayley graph over dicyclic group DC_{2n} with $H'' = \{i \mid a^i \in H\}$.

(1) If $H \cap \{a^i x \mid 0 \le i \le 2n-1\} = \emptyset$, then $Cay(DC_{2n}, H) \cong SC(Z_{2n}; H'', H'', \emptyset)$;
(2) If $H \cap \{a^i x \mid 0 \le i \le 2n-1\} \ne \emptyset$, let $a^{i_0} x \in H$, then $Cay(DC_{2n}, H) \cong SC(Z_{2n}; H'', H'', T')$, where $T' = \{i \mid a^{i_0 - i} x \in H\}$.

For the sake of convenience and familiarity, the following notations are needed.

For a subset S of a finite group G, we define $S^0 = S \cup \{e\}$, where e is the identity element in G. For any two subsets S, T of G, we define $ST = \{st \mid s \in S, t \in T\}$. For any set

$$S = \{(s_1, i_1), \ldots, (s_m, i_m)\} \in V(SC(G; R, R, T)), i_j = 0, 1, j = 1, \ldots, m,$$

we define

$$\bar{S} = \{(s_1, (i_1 + 1)(mod\ 2)), \ldots, (s_m, (i_m + 1)(mod\ 2))\}, i_j = 0, 1, j = 1, \ldots, m,$$

and

$$S' = \{s_1, s_2, \ldots, s_m\}.$$

Obviously, S' does not have to be a subset of G. For any $r \in R$,

$$rS = \{(rs_1, i_1), \ldots, (rs_m, i_m)\} \in V(SC(G; R, R, T)), i_j = 0, 1, j = 1, \ldots, m,$$

and

$$Sr = \{(s_1 r, i_1), \ldots, (s_m r, i_m)\} \in V(SC(G; R, R, T)), i_j = 0, 1, j = 1, \ldots, m.$$

For any $t \in T$, we have

$$tS = t\bar{S} \text{ and } St = \bar{S}t.$$

3 Independent Perfect Domination Sets in Semi-Cayley Graphs $SC(G; R, R, T)$

In this section, we consider Semi-Cayley graphs $SC(G; R, R, T)$ where $R = R^{-1} = \{r_1, \ldots, r_m\}$ and $T = T^{-1} = \{t_1, \ldots, t_n\}$ are subsets of the group G.

Proposition 1. *Let G be a finite group and let $S = \{(s_1, i), \ldots, (s_k, j)\}, i, j = 0, 1$, be an independent perfect domination set of $SC(G; R, R, T)$. Then so is $\bar{S}$.*

Proof. Let $S = \{(s_1, 0), (s_2, 0), \ldots, (s_l, 0), (s_{l+1}, 1), \ldots, (s_k, 1)\}$ be an independent perfect domination set. Now, we prove

$$\bar{S} = \{(s_1, 1), (s_2, 1), \ldots, (s_l, 1), (s_{l+1}, 0), \ldots, (s_k, 0)\}$$

is an independent perfect domination set. If $\bar{S}$ is not an independent set, then we have the following cases.

Case 1. $(s_i, 1) \sim (s_j, 1) \in \bar{S}$ or $(s_i, 0) \sim (s_j, 0) \in \bar{S}$. Without loss of generality, we consider $(s_i, 1) \sim (s_j, 1) \in \bar{S}$. Then $s_j = s_i r$ for some $r \in R$ and $(s_i, 0) \sim (s_j, 0) \in S$. This contradicts to the fact that S is an independent set.

Case 2. $(s_i, 0) \sim (s_j, 1) \in \bar{S}$. Then $s_j = s_i t^{-1}$ for some $t^{-1} \in T$ and $(s_i, 1) \sim (s_j, 0) \in S$ by $T = T^{-1}$. This contradicts to the fact that S is an independent set.

For any vertex $(g, i) \in V(SC(G; R, R, T)) \setminus \bar{S}, i = 0, 1$, there are the following cases.

Case 1. $(g, i) \in S$. Then there is $(g, 0) \in S, (g, 1) \notin S$ or $(g, 0) \notin S, (g, 1) \in S$. Without loss of generality, considered $(g, 0) \in S, (g, 1) \notin S$. Then $(g, 1) \in \bar{S}, (g, 0) \notin \bar{S}$. Thus $(g, 1) \in N((s_j, k))$ for $(s_j, k) \in S, k = 0, 1$ and $s_j = gr_k$ or $s_j = gt_l$. Hence, $(g, 0) \in N((s_j, (i+1)(mod\ 2)))$.

Case 2. $(g, 0) \in N((s_i, 0))$, for $(s_i, 0) \in S$.

Subcase 2.1. $(g,1) \in N((s_i,0))$, for $(s_i,0) \in S$. From $(g,0),(g,1) \in N((s_i,0))$, we have $s_i = gr_j$, $s_i = gt_j$ for some $r_j \in R, t_j \in T$ and $s_i^{-1}g = t_j^{-1}$ by $T = T^{-1}$. Thus, $(g,0) \in N((s_i,1))$, for $(s_i,1) \in \bar{S}$.

Subcase 2.2. $(g,1) \in N((s_j,1))$, for $(s_j,1) \in S$. Then $s_i = gr_k$, $s_i = gr_l$ for some $r_j, r_l \in R$. Thus, $(g,0) \in N((s_i,1))$, for $(s_i,1) \in \bar{S}$.

Subcase 2.3. $(g,1) \in N((s_j,0))$, for $(s_j,0) \in S$. Then $s_i = gr_k$, $s_j = gt_l$ for some $r_k \in R, t_l \in T$. Hence, $(g,0) \in N((s_j,1))$ by $T = T^{-1}$, for $(s_j,1) \in \bar{S}$.

Case 3. $(g,1) \in N((s_i,1))$, for $(s_i,0) \in S$.

Subcase 3.1. $(g,0) \in N((s_i,1))$, for $(s_i,0) \in S$. From $(g,0),(g,1) \in N((s_i,1))$, there is $s_i = gr_j$, $s_i = gt_j$ for some $r_j \in R, t_j \in T$ and $s_i^{-1}g = t_j^{-1}$ by $T = T^{-1}$. Thus, $(g,1) \in N((s_i,0))$, for $(s_i,1) \in \bar{S}$.

Subcase 3.2. $(g,0) \in N((s_j,1))$, for $(s_j,1) \in S$. Then $s_i = gr_k$, $s_j = gt_l$ for some $r_k \in R, t_l \in T$. Thus, $(g,1) \in N((s_j,0))$, for $(s_j,0) \in \bar{S}$.

To complete the proof, we need to show that $|\bar{S} \cap N[(g,i)]| = 1$, for any $(g,i) \in V(SC(G; R, R, T))$. We consider the following cases.

Case 1. $(g,0) \sim (s_i,1)$ and $(g,0) \sim (s_j,1)$, for $(s_i,1),(s_j,1) \in \bar{S}$. Then $s_i = gt_k, s_j = gt_l$ for some $t_k, t_l \in T$ by $T = T^{-1}$. Thus, $(g,1) \sim (s_i,0)$ and $(g,1) \sim (s_j,0)$, for $(s_i,0),(s_j,0) \in S$.

Case 2. $(g,0) \sim (s_i,0)$, for $(s_i,0) \in \bar{S}$.

Subcase 2.1. $(g,0) \sim (s_j,0)$, for $(s_j,0) \in \bar{S}$. Then $s_i = gr_k, s_j = gr_l$ for some $r_k, r_l \in R$. Thus, $(g,1) \sim (s_i,1)$ and $(g,1) \sim (s_j,1)$, for $(s_i,1),(s_j,1) \in S$.

Subcase 2.2. $(g,0) \sim (s_j,1)$, for $(s_j,1) \in \bar{S}$. Then $s_i = gr_k, s_j = gt_l$ for some $r_k \in R, t_l \in T$. Thus, $(g,1) \sim (s_i,1)$ and $(g,1) \sim (s_j,0)$, for $(s_i,1),(s_j,0) \in S$.

Case 3. $(g,1) \sim (s_i,1)$, for $(s_i,1) \in \bar{S}$.

Subcase 3.1. $(g,1) \sim (s_j,1)$, for $(s_j,1) \in \bar{S}$. Then $s_i = gr_k, s_j = gr_l$ for some $r_k, r_l \in R$. Thus, $(g,0) \sim (s_i,0)$ and $(g,0) \sim (s_j,0)$, for $(s_i,0),(s_j,0) \in S$.

Subcase 3.2. $(g,1) \sim (s_j,0)$, for $(s_j,0) \in \bar{S}$. Then $s_i = gr_k, s_j = gt_l$ for some $r_k \in R, t_l \in T$. Thus, $(g,0) \sim (s_i,0)$ and $(g,0) \sim (s_j,1)$, for $(s_i,0),(s_j,1) \in S$.

Case 4. $(g,1) \sim (s_i,0)$ and $(g,1) \sim (s_j,0)$, for $(s_i,0),(s_j,0) \in \bar{S}$. Then $s_i = gt_k, s_j = gt_l$ for some $t_k, t_l \in T$. Thus, $(g,0) \sim (s_i,1)$ and $(g,0) \sim (s_j,1)$, for $(s_i,1),(s_j,1) \in S$.

Overall, all cases are contradict to the fact that S is a perfect domination set. Thus $|\bar{S} \cap N[(g,i)]| = 1$, for any $(g,i) \in V(SC(G; R, R, T))$. The proof is finished.

Lemma 6. *Let G be a finite group and let S be an independent perfect domination set of $SC(G; R, R, T)$. Then so are $r_i S$ and $t_j S, i = 1, \ldots, m, j = 1, \ldots, n$.*

Proof. Let $S = \{(s_1,0), \ldots, (s_l,0), (s_{l+1},1), \ldots, (s_k,1)\}$ be an independent perfect domination set. Then

$$r_i S = \{(r_i s_1, 0), \ldots, (r_i s_l, 0), (r_i s_{l+1}, 1), \ldots, (r_i s_k, 1)\},$$

and

$$t_j S = \{(t_i s_1, 1), \ldots, (r_i s_l, 1), (r_i s_{l+1}, 0), \ldots, (r_i s_k, 0)\}.$$

We only prove that $r_i S$ is an independent perfect domination set. With the similar manner as in the proof of $r_i S$, we can prove that $t_j S$ is an independent perfect domination set. For this, we consider two cases.

Case 1. $(r_i s_f, 0) \sim (r_i s_h, 0)$ or $(r_i s_f, 1) \sim (r_i s_h, 1)$, for $(s_f, i), (s_h, i) \in S$, $i = 0, 1$. Without loss of generality, considered $(r_i s_f, 0) \sim (r_i s_h, 0)$. Then $r_i s_h = r_i s_f r_t$ for some $r_t \in R$. Thus $s_h = s_f r_t$ and $(s_f, 0) \sim (s_h, 0)$.

Case 2. $(r_i s_f, 0) \sim (r_i s_h, 1)$, for $(s_f, 0), (s_h, 1) \in S$. Then $r_i s_h = r_i s_f t_t$ for some $t_t \in T$ by $T = T^{-1}$. Thus $s_f = s_h t_t$ and $(s_f, 0) \sim (s_h, 1)$.

Cases 1 and 2 contradict to the fact that S is an independent set. For any $(g, l) \in V(SC(G; R, R, T)) \setminus r_i S, l = 0, 1$, we have $g = r_i r_i^{-1} g$. If $(r_i^{-1} g, 0) \in N((s_t, 0))$, then $r_i^{-1} g = s_t r_k$ and $g = r_i s_t r_k$. Thus, $(g, 0) \in N((r_i s_t, 0))$. With the similar manner as in the proof of $(g, 0) \in N((r_i s_t, 0))$, we can obtain that $r_i S$ is domination set.

Thus, for any vertex $(g, l) \in V(SC(G; R, R, T)), l = 0, 1$, we have $(g, l) \in N[r_i S]$ for $r_i \in R, l = 0, 1$. To complete the proof, we need to show that $|r_i S \cap N[(g, l)]| = 1$ for $l = 0, 1$. The following cases are needed.

Case 1. $(g, 0) \sim (r_i s_f, 0)$. Then $g = r_i s_f r_k$ and $r_i^{-1} g = s_f r_k$. Thus,
Subcase 1.1. $(g, 0) \sim (r_i s_h, 0)$. Then $g = r_i s_h r_l$ and $r_i^{-1} g = s_h r_l$. Thus, $(r_i^{-1} g, 0) \sim (s_f, 0)$ and $(r_i^{-1} g, 0) \sim (s_h, 0)$.
Subcase 1.2. $(g, 0) \sim (r_i s_h, 1)$. Then $g = r_i s_h t_l$ and $r_i^{-1} g = s_h t_l$. Thus, $(r_i^{-1} g, 0) \sim (s_f, 0)$ and $(r_i^{-1} g, 0) \sim (s_h, 1)$.
Case 2. $(g, 0) \sim (r_i s_f, 1)$ and $(g, 0) \sim (r_i s_h, 1)$. Then $g = r_i s_f t_k$, $g = r_i s_h t_l$ and $r_i^{-1} g = s_f t_k$, $r_i^{-1} g = s_h t_l$. Thus, $(r_i^{-1} g, 0) \sim (s_f, 1)$ and $(r_i^{-1} g, 0) \sim (s_h, 1)$.
Case 3. $(g, 1) \sim (r_i s_f, 1)$. Then $g = r_i s_f r_k$ and $r_i^{-1} g = s_f r_k$. Thus,
Subcase 3.1. $(g, 1) \sim (r_i s_h, 1)$. Then $g = r_i s_h r_l$ and $r_i^{-1} g = s_h r_l$. Thus, $(r_i^{-1} g, 1) \sim (s_f, 1)$ and $(r_i^{-1} g, 1) \sim (s_h, 1)$.
Subcase 3.2. $(g, 1) \sim (r_i s_h, 0)$. Then $g = r_i s_h t_l^{-1}$ and $r_i^{-1} g = s_h t_l^{-1}$. Thus, $(r_i^{-1} g, 1) \sim (s_f, 1)$ and $(r_i^{-1} g, 1) \sim (s_h, 0)$.
Case 4. $(g, 1) \sim (r_i s_f, 0)$ and $(g, 1) \sim (r_i s_h, 0)$. Then $g = r_i s_f t_k^{-1}$, $g = r_i s_h t_l^{-1}$ and $r_i^{-1} g = s_f t_k^{-1}$, $r_i^{-1} g = s_h t_l^{-1}$. Thus, $(r_i^{-1} g, 1) \sim (s_f, 0)$ and $(r_i^{-1} g, 1) \sim (s_h, 0)$.

All cases are contradict to the fact that S is a perfect domination set. Thus $r_i S$ is an independent perfect domination set. This completes the proof.

In [10], Lee gave the partition of Cayley graph. Similarly, we obtain a partition of Simi-Cayley graph $SC(G; R, R, T)$.

Lemma 7. *Let G be a finite group and let S be an independent perfect domination set of $SC(G; R, R, T)$. Then $\{S, Sr_1, \ldots, Sr_m, St_1, \ldots, St_n\}$ is a partition of $SC(G; R, R, T)$.*

Proof. If $(g, l) \in V(SC(G; R, R, T)) \setminus S, l = 0, 1$ and $(g, l) \in N[S]$, then $(g, l) = (sr_i, l)$ or $(g, l) = (st_i, (l + 1)(mod\ 2))$ for some $(s, l) \in S$ and $r_i \in R, t_i \in T$ by S is an independent perfect domination set. Thus

$$V(G; R, R, T) \subseteq S \cup Sr_1 \cup \ldots \cup Sr_m \cup St_1 \cup \ldots \cup St_n.$$

To complete the proof, we need to show that $S, Sr_1, \ldots, Sr_m, St_1, \ldots, St_n$ are pairwise disjoint. Since S is independent set, $S \cap Sr_i \neq \emptyset$ and $S \cap St_j \neq \emptyset$ for $i = 1, \ldots, m, j = 1, \ldots, n$. The following cases need to be considered.

Case 1. $Sr_i \cap Sr_j \neq \emptyset$, for $i, j \in 1, \ldots, n$. Then there exist $(s, 0), (s', 0) \in S$ such that $sr_i = s'r_j$. This implies that the vertex $(sr_i, 0) = (s'r_j, 0)$ is adjacent to $(s, 0)$ and $(s', 0)$. This is a contradiction.

Case 2. $Sr_i \cap St_j \neq \emptyset$, for $i \in 1, \ldots, m, j \in 1, \ldots, n$. Then there are $(s, 0), (s', 1) \in S$ such that $sr_i = s't_j$. This implies that the vertex $(sr_i, 0) = (s't_j, 0)$ is adjacent to $(s, 0)$ and $(s', 1)$. This is a contradiction.

Case 3. $St_i \cap St_j \neq \emptyset$, for $i, j \in 1, \ldots, n$. Then there exist $(s, 0), (s', 0) \in S$ such that $st_i = s't_j$. This implies that the vertex $(st_i, 1) = (s't_j, 1)$ is adjacent to $(s, 0)$ and $(s', 0)$. This is a contradiction.

We can obtain the following result by Lemmas 1, 3 and 7.

Lemma 8. *Let G be a finite group and let $S = \{(s_1, i), \ldots, (s_k, j)\}, i, j = 0, 1$, be an independent perfect domination set of $SC(G; R, R, T)$. If $r_i S = Sr_i$ and $t_j S = St_j$ for every $r_i \in R, t_j \in T$, then there exists a covering $\phi : SC(G; R, R, T) \rightarrow K_{|R|+|T|+1}$ such that $S, Sr_1, \ldots, Sr_m, St_1, \ldots, St_n$ are the vertex fibers of ϕ. Moreover, Let G be a finite abelian group and let $S = \{(s_1, i), \ldots, (s_k, i)\}, i = 0, 1$, be an independent perfect domination set of $SC(G; R, R, T)$. If S' is a subgroup of G, then the covering $\phi : SC(G; R, R, T) \rightarrow K_{|R|+|T|+1}$ is an S-covering.*

Proof. From Lemmas 1, 3 and 7, we can obtain the existence of covering ϕ. Next, we only need to prove the covering ϕ is an S-covering. If G is a finite abelian group and $S' = \{s_1, \ldots, s_k\}$ is a subgroup of G, then we have $r_i S = Sr_i$ and $t_j S = St_j$ for each $i = 1, \ldots, m, j = 1, \ldots, n$. For any $(s_i, j) \in S, i \in [k], j = 0, 1$, the vertex (s_i, j) is an automorphism by

$$(s_i, j) : V(SC(G; R, R, T)) \rightarrow V(SC(G; R, R, T)), \quad (s, l) \mapsto (s_i s, j + l), l = 0, 1,$$

and it is a subgroup of the automorphism group of $SC(G; R, R, T)$ by normal subgroup S'. For any vertex $v \in K_{m+n+1}$, without loss of generality, we suppose $\phi^{-1}(v) \in Sr_i, r_i \in R$ or $\phi^{-1}(v) \in St_j, t_j \in T$ and construct

$$h : V(K_{m+n+1}) \rightarrow V(SC(G; R, R, T)/S),$$

by the following

$$h(v) := \begin{cases} r_i S, & \phi(v) \in Sr_i, \\ t_j S, & \phi(v) \in St_j. \end{cases} \tag{1}$$

Thus, $\phi : SC(G; R, R, T) \rightarrow K_{m+n+1}$ is an S-covering.

Next, we give the main result in this section.

Theorem 1. *Let G be a finite group and let $S' = \{s_1, \ldots, s_k\}$ be a subset of G with $r_i S = Sr_i$ and $t_j S = St_j$ for $r_i \in R, t_j \in T$. Then the following are equivalent.*

(1) $S = \{(s_1, i), \ldots, (s_k, j)\}, i, j = 0, 1$, *is an independent perfect domination set of the Semi-Cayley graph* $SC(G; R, R, T)$.

(2) *There exists a covering* $\phi : SC(G; R, R, T) \to K_{|R|+|T|+1}$ *such that* $\phi^{-1}(v) = S$ *for any* $v \in V(K_{|R|+|T|+1})$.

(3) $| S | = \frac{2|G|}{|R|+|T|+1}$ *and*

$$S \cap \{S((R^0 R^0) \setminus \{e\})\} = \emptyset, S \cap \{S(R^0 T)\} = \emptyset, S \cap \{S((TT) \setminus \{e\})\} = \emptyset.$$

Proof. ((1)$\Rightarrow$(2)) It follows from Lemma 8.

((2)$\Rightarrow$(3)) Let $\phi : SC(G; R, R, T) \to K_{|R|+|T|+1}$ such that $\phi^{-1}(v) = S$. Then $| S | = \frac{2|G|}{|R|+|T|+1}$ by lemmas 1, 2, 6 and 7. To complete the proof, we consider the following cases.

Case 1. $S \cap \{S((R^0 R^0) \setminus \{e\})\} \neq \emptyset$. Then there exist $(s, i), (s', i) \in S, i = 0, 1$, such that $s = s'rr'$ for some $r, r' \in R^0$ with $r \neq r'^{-1}$. It implies that $(sr^{-1}, i) = (s'r, i), i = 0, 1$ is adjacent to both (s, i) and (s', i). This is a contradiction.

Case 2. $S \cap \{S(R^0 T)\} \neq \emptyset$. Then there exist $(s, i), (s', j) \in S, i \neq j, i, j = 0, 1$, such that $s = s'rt$ for some $r \in R^0, t \in T$. It implies that $(st^{-1}, j) = (s'r, j), j = 1, 2$, is adjacent to both (s, i) and (s', j). This is a contradiction.

Case 3. $S \cap \{S((TT) \setminus \{e\})\} \neq \emptyset$. Then there exist $(s, i), (s', i) \in S, i = 0, 1$, such that $s = s'tt'$ for some $t, t' \in T$ with $t \neq t'^{-1}$. It implies that $(st'^{-1}, j) = (s't, j), i \neq j$ and $i, j = 0, 1$, is adjacent to both (s, i) and (s', i). This is a contradiction.

((3)$\Rightarrow$(1)) Since $S \cap \{S((R^0 R^0) \setminus \{e\})\} = \emptyset, S \cap \{S(R^0 T)\} = \emptyset, S \cap \{S((TT) \setminus \{e\})\} = \emptyset$. Then S is an independent set by $S \cap (Sr) = \emptyset, S \cap (St) = \emptyset$ for $r \in R$, $t \in T$. For every elements $(g, i), i = 1, 2$, the vertex $(g, i), i = 1, 2$, can be adjacent to at most one element in S by $(Sr) \cap (Sr') = \emptyset, (Sr) \cap (St) = \emptyset, (St) \cap (St') = \emptyset$. To complete the proof, it remains to show that $|N[S]| = 2|G|$. We have

$$N[S] = \bigcup_{s \in S} N[s] = \left(\bigcup_{s \in S}(sR^0)\right) \bigcup \left(\bigcup_{s \in S}(sT)\right) = \left(\bigcup_{r_i \in R^0}(Sr_i)\right) \bigcup \left(\bigcup_{t_j \in T}(St_j)\right)$$

and

$$|N[S]| = \sum_{r_i \in R^0} |Sr_i| + \sum_{t_j \in T} |St_j|.$$

Due to $|S| = |Sr_i| = |St_j|$ for each $r_i \in R^0, t_j \in T$ and $|S|(|R| + |T| + 1) = 2|G|$, there is

$$|N[S]| = \sum_{r_i \in R^0} |Sr_i| + \sum_{t_j \in T} |St_j| = |S||R^0| + |S||T| = |S|(|R| + 1 + |S|) = 2|G|.$$

This completes the proof.

If we suppose $T = \emptyset$, then we can obtain the result [10, Theorem 2]. Moreover, we can obtain the following case.

Corollary 1. *Let G be a finite abelian group and let $S' = \{s_1, \ldots, s_k\}$ be a subgroup of G. Then the following are equivalent.*

(1) $S = \{(s_1, i), \ldots, (s_k, i)\}, i = 0, 1$, is an independent perfect domination set of the Semi-Cayley graph $SC(G; R, R, T)$.
(2) $SC(G; R, R, T)$ is an S-covering graph of the complete graph $K_{|R|+|T|+1}$.
(3) $|G/S'| = (|R| + |T| + 1)/2$ and

$$S' \cap (R^0 R^0) = \{e\}, S' \cap (R^0 T) = \emptyset, S' \cap (TT) = \{e\}.$$

Proof. ((1)⇔(2)⇔(3)) It follows from Lemma 8 and Theorem 1. We only need to show that

$$S \cap \{S((R^0 R^0) \setminus \{e\})\} = \emptyset, S \cap \{S(R^0 T)\} = \emptyset, S \cap \{S((TT) \setminus \{e\})\} = \emptyset,$$

if and only if

$$S' \cap (R^0 R^0) = \{e\}, S' \cap (R^0 T) = \emptyset, S' \cap (TT) = \{e\}.$$

For the necessity, we consider three cases,

Case 1. $S' \cap (R^0 R^0) \neq \{e\}$. Then there exist $s' \in S'$ and $s' \in R^0 R^0$. For any $s \in S'$, $ss' \in S'$ by S' is a normal subgroup. Thus there exists $(s_i, i) \in S$ such that $(s_i, i) = (ss', i)$. This contradicts to the fact that $S \cap \{S((R^0 R^0) \setminus \{e\})\} = \emptyset$.
Case 2. $S' \cap (R^0 T) \neq \emptyset$. Then there exist $s' \in S'$ and $s' \in R^0 T$. For any $s \in S'$, $ss' \in S'$ by S' is a normal subgroup. Thus there exists $(s_i, i) \in S$ such that $(s_i, i) = (ss', (i+1)(mod\ 2))$. This contradicts to the fact that $S \cap \{S(R^0 T)\} = \emptyset$.
Case 3. $S' \cap (TT) \neq \{e\}$. Then there exist $s' \in S'$ and $s' \in TT$. For any $s \in S'$, $ss' \in S'$ by S' is a normal subgroup. Thus there exists $(s_i, i) \in S$ such that $(s_i, i) = (ss', i)$. This contradicts to the fact that $S \cap \{S((TT) \setminus \{e\})\} = \emptyset$.

For the sufficiency, we consider the following cases,

Case 1. $S \cap \{S((R^0 R^0) \setminus \{0\})\} \neq \emptyset$. Then there exist $(s, i), (s', i) \in S, i = 1, 2$, such that $s = s'rr'$ for $r, r' \in R^0$. Since S' is a subgroup of G, $e \neq rr' \neq ss'^{-1} \in S'$, Hence $S' \cap \{S'((R^0 R^0) \setminus \{0\})\} = \emptyset$.
Case 2. $S \cap \{S((R^0 T) \setminus \{0\})\} \neq \emptyset$. Then there exist $(s, i), (s', j) \in S, i \neq j, i, j = 1, 2$, such that $s = s'rt$ for $r \in R^0$, $t \in T$. Since S' is a subgroup of G, $e \neq rt = ss'^{-1} \in S'$.
Case 3. $S \cap \{S((TT) \setminus \{e\})\} \neq \{e\}$. Then there exist $(s, i), (s', i) \in S, i = 1, 2$, such that $s = s'tt'$ for $t, t' \in T$. Since S' is a subgroup of G, $e \neq tt' = ss'^{-1} \in S'$.

4   Some Special Cases

In this section, we consider that the independent perfect domination set of Cayley graph over dihedral groups D_n and dicyclic groups DC_{2n}, respectively. Throughout this section, additive notation is used to deal with groups Z_n.

4.1   Independent Perfect Domination Dets of Cayley Graph over Dihedral Groups

Let $D_n = \langle a, x \mid a^n = x^2 = e, \; ax = xa^{-1} \rangle$ be the dihedral group and $H \subseteq D_n \setminus \{e\}$ with $H = H^{-1}$. For convenience, we define

$$H' = H \cap \{a^i \mid 0 \le i \le n - 1\}, H'' = \{i \mid a^i \in H\}, T = H \cap \{a^i x \mid 0 \le i \le n - 1\}.$$

Denoted by Γ_1 the subgraph induced by $\{a^i \mid 0 \le i \le n - 1\}$ in $Cay(D_n, H)$. Obviously, the subgraph $\Gamma_1 \cong Cay(Z_n, H'')$. In this subsection, we suppose

$$H = \{a^{i_1}, \ldots, a^{i_p}, a^{j_1}x, \ldots, a^{j_q}x\},$$

$$\{a^{j_1}x, a^{j_2}x, \ldots, a^{j_q}x\} = \{a^{j_1}x, a^{2j_1-j_2}x, \ldots, a^{2j_1-j_q}x\},$$

where $j_1 = \min\{j_1, \ldots, j_q\}$. Then $H' = \{a^{i_1}, \ldots, a^{i_p}\}$, $T = \{a^{j_1}x, \ldots, a^{j_q}x\}$.

In this subsection, we give the independent perfect domination set of Cayley graph over dihedral groups D_n.

Lemma 9. *Let $Cay(D_n, H)$ be a connected Cayley graph over dihedral group D_n and $H = H' \cup T$. Then the Cayley graph $Cay(D_n, H)$ has an independent perfect domination set $S = \{a^{s_1}, \ldots, a^{s_m}\}$ with $a^{s_i}, a^{-s_i} \in S$ if and only if $Cay(Z_n, H'')$ has an independent perfect domination set $S' = \{s_1, \ldots, s_m\}$ and there exists a subset $T' \subseteq Z_n$ such that $\mid T' \mid = \mid H' \mid + 1$ and $s_i \ne s_j + t_r + t_s$ for any $s_i, s_j \in S'$, $t_r, t_s \in T'$, $t_r \ne -t_s$.*

Proof. Suppose $Cay(D_n, H)$ has an independent perfect domination set $S = \{a^{s_1}, \ldots, a^{s_m}\}$ with $a^{s_i}, a^{-s_i} \in S$. Then $\{a^{s_1}, \ldots, a^{s_m}\}$ is an independent perfect domination set of subgraph Γ_1. Thus, $Cay(Z_n, H'')$ has an independent perfect domination set $S' = \{s_1, \ldots, s_m\}$ by $\Gamma_1 \cong Cay(Z_n, H'')$. We have $|H'| = \frac{n-m}{m} = \frac{n}{m} - 1$ and $|T'| = \frac{n}{m}$, thus $|T'| = |H'| + 1$. The Cayley graph $Cay(D_n, H) \cong SC(Z_n; H'', H'', T')$ by Lemma 4, then $S'' = \{(s_1, 0), \ldots, (s_m, 0)\}$ is an independent perfect domination set of $SC(Z_n; H'', H'', T')$ and S'' is a normal subset. Thus, $s_i \ne s_j + j_r - j_s$ for any $s_i, s_j \in S'$, $j_r, j_s \in T$ by Theorem 1.

($\Leftarrow$) Assume $Cay(Z_n, H'')$ has an independent perfect domination set $S' = \{s_1, \ldots, s_m\}$. Then the Semi-Cayley graph $SC(Z_n; H'', H'', T')$ has an independent perfect domination set $S'' = \{(s_1, 0), \ldots, (s_m, 0)\}$ by Theorem 1. Thus the Cayley graph $Cay(D_n, H)$ has an independent perfect domination set by $Cay(D_n, H) \cong SC(Z_n; H'', H'', T')$.

We give a necessary and sufficient condition for the existence of Cayley graph $Cay(D_n, H)$.

Theorem 2. *Let $Cay(D_n, H)$ be a connected Cayley graph over dihedral group D_n and $H = H' \cup T$. Then the Cayley graph $Cay(D_n, H)$ has an independent*

perfect domination set $S = \{a^{s_1}, \ldots, a^{s_l}, a^{s_{l+1}}x, \ldots, a^{s_m}x\}$ *with normal subset* S
if and only if $m = \frac{2n}{|H'|+|T|+1}$ *and*

$$S \cap \{S((H'^0 H'^0) \setminus \{e\})\} = \emptyset, \ S \cap \{S(H'^0 T)\} = \emptyset, \ S \cap \{S(TT) \setminus \{e\}\} = \emptyset.$$

Proof. The Cayley graph $\mathrm{Cay}(D_n, H) \cong SC(Z_n; H'', H'', T')$ by Lemma 4. Then
the Cayley graph $\mathrm{Cay}(D_n, H)$ has an independent perfect domination set

$$S = \{a^{s_1}, \ldots, a^{s_l}, a^{s_{l+1}}x, \ldots, a^{s_m}x\}$$

if and only if the Semi-Cayley graph $SC(Z_n; H'', H'', T')$ has an independent
perfect domination set $S' = \{(s_1, 0), \ldots, (s_l, 0), (j_1 - s_{l+1}, 1), \ldots, (j_1 - s_m, 1)\}$.
From Theorem 1, the Semi-Cayley graph $SC(Z_n; H'', H'', T')$ has an indepen-
dent perfect domination set if and only if $m = \frac{2n}{|H'|+|T'|+1}$ and

$$S' \cap \{S' + ((H''^0 + H''^0) \setminus \{e\})\} = \emptyset, \ S' \cap \{S' + (H''^0 + T')\} = \emptyset,$$

$$S' \cap \{S' + ((T' + T') \setminus \{e\})\} = \emptyset.$$

Thus,

$$S \cap \{S((H'^0 H'^0) \setminus \{e\})\} = \emptyset, \ S \cap \{S(H'^0 T)\} = \emptyset, \ S \cap \{S(TT) \setminus \{e\}\} = \emptyset.$$

4.2 Independent Perfect Domination Sets of Cayley Graph over Dicyclic Groups

In this subsection, Independent perfect domination sets of Cayley graph over
dicyclic groups will be considered. Let $DC_{2n} = \langle a, x \mid a^{2n} = e, x^2 = a^n, ax = xa^{-1} \rangle$ be the dicyclic group and $H \subseteq DC_{2n} \setminus \{e\}$ with $H = H^{-1}$. We define

$$H' = H \cap \{a^i \mid 0 \le i \le 2n - 1\}, \ H'' = \{i \mid a^i \in H\}, \ T = H \cap \{a^i x \mid 0 \le i \le 2n - 1\}.$$

Similarly, we suppose

$$H = \{a^{i_1}, \ldots, a^{i_p}, a^{j_1}x, \ldots, a^{j_q}x\},$$

$$\{a^{j_1}x, a^{j_2}, \ldots, a^{j_q}x\} = \{a^{j_1}x, a^{2j_1-j_2}, \ldots, a^{2j_1-j_q}x\},$$

where $j_1 = \min\{j_1, \ldots, j_q\}$. Then $H' = \{a^{i_1}, \ldots, a^{i_p}\}$, $T = \{a^{j_1}x, \ldots, a^{j_q}x\}$.

Now we a necessary and sufficient condition for the existence of independent
perfect domination sets of $Cay(DC_{2n}, H)$.

Theorem 3. *Let* $Cay(DC_{2n}, H)$ *be a connected Cayley graph over dicyclic group*
DC_{2n} *and* $H = H' \cup T$. *Then the Cayley graph* $Cay(DC_{2n}, H)$ *has an inde-*
pendent perfect domination set $S = \{a^{s_1}, \ldots, a^{s_l}, a^{s_{l+1}}x, \ldots, a^{s_m}x\}$ *with normal*
subset S *if and only if* $m = \frac{4n}{|H'|+|T|+1}$ *and*

$$S \cap \{S((H'^0 H'^0) \setminus \{e\})\} = \emptyset, \ S \cap \{S(H'^0 T)\} = \emptyset, \ S \cap \{S(TT) \setminus \{e\}\} = \emptyset.$$

Proof. By Lemma 5, the Cayley graph $Cay(DC_{2n}, H) \cong SC(Z_{2n}; H'', H'', T')$. Then the Cayley graph $Cay(DC_{2n}, H)$ has an independent perfect domination set $S = \{a^{s_1}, \ldots, a^{s_l}, a^{s_{l+1}}x, \ldots, a^{s_m}x\}$ if and only if the Semi-Cayley graph $SC(Z_{2n}; H'', H'', T')$ has an independent perfect domination set $\{(s_1, 0), \ldots, (s_l, 0), (j_1 - s_{l+1}, 1), \ldots, (j_1 - s_m, 1)\}$ and from Theorem 1, the Semi-Cayley graph $SC(Z_{2n}; H'', H'', T')$ has an independent perfect domination set if and only if $m = \frac{4n}{|H'|+|T'|+1}$ and

$$S' \cap \{S' + ((H''^0 + H''^0) \setminus \{e\})\} = \emptyset, \ S' \cap \{S' + (H''^0 + T')\} = \emptyset,$$

$$S' \cap \{S' + ((T' + T') \setminus \{e\})\} = \emptyset.$$

Thus,

$$S \cap \{S((H'^0 H'^0) \setminus \{e\})\} = \emptyset, \ S \cap \{S(H'^0 T)\} = \emptyset, \ S \cap \{S(TT) \setminus \{e\}\} = \emptyset.$$

References

1. Chelvam, T.T., Rani, I.: Dominating sets in Cayley graphs on Z_n. Tamkang. Math. **37**(4), 341–345 (2007)
2. Chelvam, T.T., Mutharasu, S.: Subgroups as efficient dominating sets in Cayley graphs. Discrete Appl. Math. **161**(9), 1187–1190 (2013)
3. Deng, Y.: Efficient dominating sets in circulant graphs with domination number prime. Inf. Process. Lett. **114**(12), 700–702 (2014)
4. Deng, Y., Sun, Y., Liu, Q., Wang, H.: Efficient domination sets in circulant graphs. Discrete Math. **340**(7), 1503–1507 (2017)
5. Gao, X., Luo, Y.: The spectrum of semi-Cayley graphs over abelian groups. Linear Algebra Appl. **432**(11), 2974–2983 (2010)
6. Godsil, C., Royle, G.: Algebraic Graph Theory, 1st edn. Springer, New York (2001). https://doi.org/10.1007/978-1-4613-0163-9
7. Haynes, T.W., Hedetniemi, S.T., Slater, P.J.: Fundamentals of Domination in Graphs. Marcel Dekker Inc., New York (1998)
8. Kumar, K.R., MacGillivray, G.: Efficient domination in circulant graphs. Discrete Math. **313**(6), 767–771 (2013)
9. Kwak, J.H., Chun, J.-H., Lee, J.: Enumeration of regular graph coverings having finite abelian covering transformation groups. SIAM J. Discrete Math. **11**(2), 273–285 (1998)
10. Lee, J.: Independent perfect domination sets in Cayley graphs. J. Graph Theory. **37**(4), 213–219 (2001). https://doi.org/10.1023/A:1022476321014
11. Obradović, N., Peters, J., Ružić, G.: Efficient domination in circulant graphs with chord lengths. Inf. Process. Lett. **102**(6), 253–258 (2007)
12. Resmi, M.J.D., Jungnickel, D.: Strongly regular Semi-Cayley graphs. J. Algebra Combi. **1**, 171–195 (1992). https://doi.org/10.1023/A:1022476321014

A Parallel Algorithm for Constructing Two Edge-Disjoint Hamiltonian Cycles in Crossed Cubes

Kung-jui Pai[✉] [ID]

Department of Industrial Engineering and Management, Ming Chi University of Technology,
New Taipei City, Taiwan
poter@mail.mcut.edu.tw

Abstract. The n-dimensional crossed cube CQ_n, a variation of the hypercube Q_n, has the same number of vertices and the same number of edges as Q_n, but it has only about half of the diameter of Q_n. In the interconnection network, some efficient communication algorithms can be designed based on edge-disjoint Hamiltonian cycles. In addition, two edge-disjoint Hamiltonian cycles also provide the edge-fault tolerant Hamiltonicity for the interconnection network. Hung [Discrete Applied Mathematics 181, 109–122, 2015] designed a recursive algorithm to construct two edge-disjoint Hamiltonian cycles on CQ_n in $O(n2^n)$ time. In this paper, we provide an $O(n)$ time algorithm for each vertex in CQ_n to determine which two edges were used in Hamiltonian cycles 1 and 2, respectively. With the information of each vertex, we can construct two edge-disjoint Hamiltonian cycles in CQ_n with $n \geq 4$.

Keywords: Edge-disjoint Hamiltonian cycles · Crossed cubes · Interconnection networks

1 Introduction

The design of an interconnection network is an important issue for the multicomputer system. The hypercube [17, 18] is one of the most popular interconnection networks because of its attractive properties, including regularity, node symmetric, link symmetric, small diameter, strong connectivity, recursive construction, partition capability, and small link complexity. The architecture of an interconnected network is usually modeled by a graph with vertices representing processing units and edges representing communication links. We will use graph and network interchangeably in this paper.

The n-dimensional crossed cube CQ_n, proposed first by Efe [4, 5], is a variant of an n-dimensional hypercube. One advantage of CQ_n is that the diameter is only about one half of the diameter of an n-dimensional hypercube. Hung et al. [11] showed that CQ_n contains a fault-free Hamiltonian cycle, even if there are up to $2n - 5$ edge faults. Wang studied the embedding of the Hamiltonian cycle in CQ_n [19]. For more properties of CQ_n, the reader can refer to [4–6, 12].

© Springer Nature Switzerland AG 2020
Z. Zhang et al. (Eds.): AAIM 2020, LNCS 12290, pp. 448–455, 2020.
https://doi.org/10.1007/978-3-030-57602-8_40

A Hamiltonian cycle is a graph cycle through a graph that visits each node exactly once. The ring structure is important for the multicomputer system, and its benefits can be found in [13]. Two Hamiltonian cycles in the graph are said to be edge-disjoint if they do not share any common edges. Edge-disjoint Hamilton cycles can provide advantages for algorithms using ring structures, and their application can be found in [16]. Further, edge-disjoint Hamiltonian cycles also provide the edge-fault tolerant hamiltonicity for the interconnection network. That is, when one edge in the Hamiltonian cycle fails, the other edge-disjoint Hamiltonian cycle can be adopted to replace it for transmission.

Previous related works are described below. Barth and Raspaud [3] showed that the butterfly networks contain two edge-disjoint Hamiltonian cycles. Bae et al. [1] studied edge-disjoint Hamiltonian cycles in k-ary n-cubes and hypercubes. Then, Barden et al. [2] constructed the maximum number of edge-disjoint spanning trees in a hypercube. Petrovic and Thomassen [15] characterized the number of edge-disjoint Hamiltonian cycles in hypertournaments. Hung presented how to construct two edge-disjoint Hamiltonian cycles in locally twisted cubes [7], augmented cubes [8], twisted cubes [10], transposition networks, and hypercube-like networks [9], respectively. In [9], Hung designed a recursive algorithm to construct two edge-disjoint Hamiltonian cycles in CQ_n. In this paper, we provide a parallel algorithm to construct two edge-disjoint Hamiltonian cycles in CQ_n with $n \geq 4$. Each vertex of CQ_n can simultaneously run this algorithm to know which two edges were used in Hamiltonian cycles 1 and 2, respectively. The recursive algorithm [9] can be adopted by one vertex to constructs two Hamiltonian cycles and this vertex must transfer this message to all other vertices. However, according to our algorithm, each vertex can calculate to get the message, which is more helpful for implementation.

The rest of the paper is organized as follows: In Sect. 2, the structure of crossed cubes is introduced and some notations are given. Section 3 presented two edge-disjoint Hamiltonian cycles in CQ_4. Based on this result, we further show a parallel algorithm to construct two edge-disjoint Hamiltonian cycles in CQ_n with $n \geq 4$. Finally, Sect. 4 is the concluding remarks of this paper.

2 Preliminaries

Interconnection networks are usually modeled as undirected simple graphs $G = (V, E)$, where the vertex set $V (=V(G))$ and the edge set $E (=E(G))$ represent the set of processing units and the set of communication links between nodes, respectively. The neighborhood of a vertex v in a graph G, denoted by $N(v)$, is the set of vertices adjacent to v in G. A cycle C_k of length k in G, denoted by $v_0 - v_1 - v_2 - \ldots - v_{k-2} - v_{k-1} - v_0$, is a sequence $(v_0, v_1, v_2, \ldots, v_{k-1}, v_0)$ of nodes such that $(v_{k-1}, v_0) \in E$ and $(v_i, v_{i+1}) \in E$ for $0 \leq i \leq k - 2$.

Now, we introduce crossed cubes. A vertex of the n-dimensional crossed cube CQ_n is represented by a binary string of length n. A binary string b of length n is denoted by $b_{n-1}b_{n-2} \cdots b_1 b_0$, where b_{n-1} is the most significant bit. Suppose that G is a labeled graph whose vertices are associated with distinct binary strings, and let G^x be the graph obtained from G by prefixing the binary string on every node with x. Two binary strings $x = x_1 x_0$ and $y = y_1 y_0$ are pair-related, denoted $x \sim y$, if and only if $(x, \in y\{(00, 00), (10,$

10), (01, 11), (11, 01)}. In [5], Efe introduced the notion of pair-related to obtain that the diameter of CQ_n is only about one half of the diameter of Q_n.

Definition 1 (Efe [5].) *The n-dimensional crossed cube CQ_n is the labeled graph with the following recursive fashion:*

1) *CQ_1 is the complete graph on two vertices with labels 0 and 1.*

2) *For $n \geq 2$, CQ_n is composed of two subcubes CQ_{n-1}^0 and CQ_{n-1}^1 such that two vertices $x = 0x_{n-2}\cdots\cdots x_1 x_0 \in V(CQ_{n-1}^0)$ and $y = 1y_{n-2}\cdots\cdots y_1 y_0 \in V(CQ_{n-1}^1)$ are joined by an edge if and only if*

(i)   *$x_{n-2} = y_{n-2}$ if n is even, and*
(ii) *$x_{2i+1}x_{2i} \sim y_{2i+1}y_{2i}$ for $0 \leq i < \lfloor (n-1)/2 \rfloor$,*

where x and y are called the $(n-1)$-neighbors to each other, and denote as $N_{n-1}(x) = y$ or $N_{n-1}(y) = x$.

For conciseness, an edge $(x, N_j(x))$ is denoted as $e_j(x)$, and an e_j-edge is an edge $(x, N_j(x))$ in G. Obviously, there are 8 e_3-edges, 8 e_2-edges, 8 e_1-edges and 8 e_0-edges in CQ_4. For example, Fig. 1 shows a 4-dimensional crossed cube CQ_4. A CQ_4 is composed of two subcubes CQ_3^0 (the left half in Fig. 1) *and* CQ_3^1 (the right half in Fig. 1). According to Definition 1 and the notion of pair-related, there exists an e_3-edge connecting vertices 0000 and 1000, and so on. In this paper, sometimes the labels of vertices are changed to their decimal.

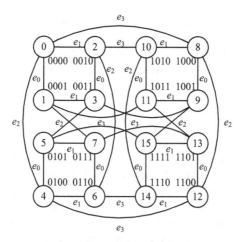

Fig. 1. A 4-dimensional crossed cube CQ_4.

3 Main Results

3.1 Two Edge-Disjoint Hamiltonian Cycles in CQ_4

Hung [9] provided two edge-disjoint Hamiltonian cycles in CQ_4. The first cycle C_{16} is equal to 0100–0000–1000–1001–1111–1101–0111–0101–0011–0001–1011–1010– 0010–0110–1110–1100–0100, and there are 6 e_3-edges, 4 e_2-edges, 4 e_1-edges and 2 e_0-edges in it. Since the cycle adopts a different number of edges in each dimension, the parallel construction algorithm that will be presented in the next section becomes more complicated. Fortunately, we found another set of two edge-disjoint Hamiltonian cycles in CQ_4, and each cycle has the same number of edges in each dimension. We described this result in Proposition 2, and its validness can check by Fig. 2.

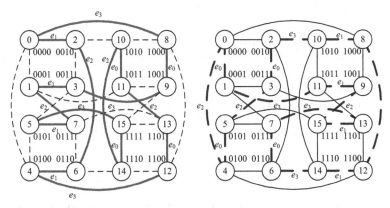

Fig. 2. (a) The first Hamiltonian cycle HC_1 in CQ_4, and (b) the second Hamiltonian cycle HC_2 in CQ_4, where the thick lines indicate the cycle.

Proposition 2. Let HC_1 = 0000–0010–0110–0100–1100–1101–1011–1010–1110– 1111–0101–0111–0001–0011–1001–1000–0000, and HC_2 = 0000–0001–1011–1001– 1111–1101–0111–0110–1110–1100–1000–1010–0010–0011–0101–0100–0000. HC_1 and HC_2 form two edge-disjoint Hamiltonian cycles in CQ_4.

3.2 Constructing Two Edge-Disjoint Hamiltonian Cycles in CQ_4

Based on the previous results, we now design an algorithm called Algorithm 2HCBase to construct two edge-disjoint Hamiltonian cycles in CQ_4. Each vertex (processing unit) in CQ_4 calls this algorithm and inputs its label to get which two edges are used in Hamiltonian cycles 1 and 2, respectively.

```
Algorithm 2HCBase
Input: b3 b2 b1 b0 in B //B : the label of this vertex
Output: H1 and H2 //Hi : edge set of the i-th Hamiltonian cycle
```

```
step 1. if b₃ = 0 then H₁ ← {e₁} else H₁ ← {e₀};
step 2. if b₃ = 0 then
step 3.     if b₂ = 0 then
step 4.         if b₁ xor b₀ = 0 then H₁ ← H₁∪{e₃} else H₁ ← H₁∪{e₂};
step 5.     else
step 6.         if b₁ = 0 then H₁ ← H₁∪{e₃} else H₁ ← H₁∪{e₂};
step 7.     end if
step 8. else
step 9.     if b₂ = 0 then
step 10.        if b₁ = 0 then H₁ ← H₁∪{e₃} else H₁ ← H₁∪{e₂};
step 11.    else
step 12.        if b₁ xor b₀ = 0 then H₁ ← H₁∪{e₃} else H₁ ← H₁∪{e₂};
step 13.    end if
step 14. end if
step 15. H₂ ← {e₃, e₂, e₁, e₀} \ H₁;
```

In Algorithm 2HCBase, each vertex determines which two edges are used in the first Hamiltonian cycles H_1. In step 1, either e_0-edge or e_1-edge will be selected into H_1. Then, it will add e_2-edge or e_3-edge into H_1 according to steps 2 to 14. Finally, the remaining two edges will be adopted in the second Hamiltonian cycle. Since there is no loop in Algorithm 2HCBase, we have the following lemma.

Lemma 3. The time complexity of Algorithm 2HCBase is $O(1)$.

Lemma 4. *By inputting the label of each vertex into Algorithm 2HCBase, we can obtain 2 cycles, which form two edge-disjoint Hamiltonian cycles in CQ_4.*

Proof. According to step 1 in the algorithm, each vertex selects either e_0-edge or e_1-edge into H_1 by b_0. Then, we provide the decision tree shown in Fig. 3 to illustrate steps 2 through 14 in the algorithm. The two edges selected for each vertex can be checked by Fig. 2. □

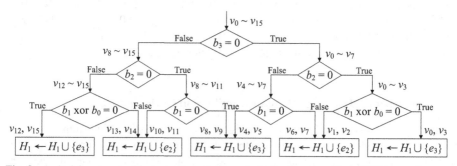

Fig. 3. A decision tree to illustrate steps 2 through 14 in the Algorithm 2HCBase, where v_i represents the vertex i in CQ_4.

3.3 Constructing Two Edge-Disjoint Hamiltonian Cycles in CQ_n for $n \geq 5$

We all know that CQ_{n+1} is composed of CQ_n^0 and CQ_n^1. As described in the previous subsection, there exist two edge-disjoint Hamiltonian cycles in CQ_4. The construction method of Hamiltonian cycle in CQ_{n+1} is as follows. First, we have the Hamiltonian cycle HC_A in CQ_n^0 and the Hamiltonian cycle HC_B in CQ_n^1. Second, remove an edge in HC_A (respectively, HC_B) to get the Hamiltonian path HP_A (respectively, HP_B). Third, connect one end vertex of HP_A and one end vertex of HP_B with an e_n-edge. Finally, connect the other end vertex of HP_A and the other end vertex of HP_B to obtain a Hamiltonian cycle in CQ_{n+1}. Figure 4 illustrates the construction of two such edge-disjoint Hamiltonian cycles in CQ_{n+1}. Base on this method, we design Algorithm 2HC as shown below.

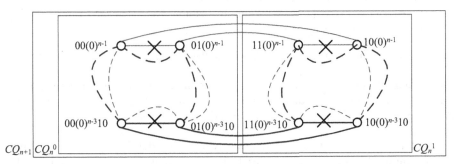

Fig. 4. The construction of two edge-disjoint Hamiltonian cycles in CQ_{n+1} while $n \geq 4$, where thin red lines (respectively, thick blue lines) indicate the first (respectively, the second) Hamiltonian cycle.

Algorithm 2HC
Input: $B (= b_{n-1}b_{n-2} \cdots \cdots b_1 b_0)$ and n //B : the label of this vertex, n : the dimension
Output: HC_1 and HC_2 //HC_i : edge set of the i-th Hamiltonian cycle
step 1. By calling **Algorithm 2HCBase**, $HC_1 \leftarrow H_1$ and $HC_2 \leftarrow H_2$;
step 2. **if** $b_2 = 0$ **and** $b_1 = 0$ **and** $b_0 = 0$ **then**
step 3.   flag $\leftarrow$ 1
step 4.   **for** $i = 3$ to $n - 3$ **do**
step 5.   **if** $b_i = 1$ **then**
step 6. $HC_1 \leftarrow HC_1 \setminus \{e_3\} \cup \{e_{i+1}\}$; flag $\leftarrow$ 0; **break for;**
step 7.   **end if**
step 8.   **end for**
step 9.   **if** flag $= 1$ **then** $HC_1 \leftarrow HC_1 \setminus \{e_3\} \cup \{e_{n-1}\}$;
step 10. **end if** //end if at step 2
step 11. **if** $b_2 = 0$ **and** $b_1 = 1$ **and** $b_0 = 0$ **then**
step 12.   flag $\leftarrow$ 1
step 13.   **for** $i = 3$ to $n - 3$ **do**
step 14.   **if** $b_i = 1$ **then**
step 15. $HC_2 \leftarrow HC_2 \setminus \{e_2\} \cup \{e_{i+1}\}$; flag $\leftarrow$ 0; **break for;**
step 16.   **end if**

```
step 17.      end for
step 18.      if flag = 1 then HC₂ ← HC₂ \ {e₂}∪{eₙ₋₁};
step 19. end if //end if at step 11
```

In Algorithm 2HC, each vertex will call Algorithm 2HCBase to obtain HC_1 and HC_2, where HC_i is the edge set of the i-th Hamiltonian cycle. In steps 2 to 10, if this vertex is one end vertex of the Hamiltonian path, it will modify its HC_1. For example, in CQ_6, vertex 000000 obtains $HC_1 = \{e_1, e_3\}$ by call Algorithm 2HCBase. Since vertex 000000 is one end vertex of the Hamiltonian path, $HC_1 = \{e_1, e_3\} \setminus \{e_3\} \cup \{e_5\}$. By the same way, in steps 11 to 19, if this vertex is one end vertex of Hamiltonian path, it will modify its HC_2. For example, in CQ_6, vertex 000010 obtains $HC_2 = \{e_1, e_2\}$ by call Algorithm 2HCBase. Since vertex 000010 is one end vertex of the Hamiltonian path, $HC_2 = \{e_1, e_2\} \setminus \{e_2\} \cup \{e_5\}$. According to Lemma 3 and steps 4, 13 in Algorithm 2HC, we have the following lemma.

Lemma 5. The time complexity of Algorithm 2HC is $O(n)$.

Theorem 6. *By inputting the label of each vertex into Algorithm 2HC, we can obtain 2 cycles, which form two edge-disjoint Hamiltonian cycles in CQ_n with $n \geq 4$ in $O(n2^n)$ time. In particular, it can be parallelized on CQ_n to run in $O(n)$ time.*

Proof. Each vertex of CQ_n can simultaneously run Algorithm 2HC to know which two edges were used in Hamiltonian cycles 1 and 2, respectively. By lemma 5, it can be parallelized on CQ_n to run in $O(n)$ time.

Without loss of generality, we consider vertex 0 to be the starting vertex in CQ_n. By Algorithm 2HC, vertex 0 obtains two edges used in the Hamiltonian cycle. With the dimension of the edge, we can obtain the next vertex of the Hamiltonian cycle. In the same way, the vertex sequence of the Hamiltonian cycle can be get. Since there are 2^n vertices in CQ_n, two edge-disjoint Hamiltonian cycles in CQ_n can be obtained in $O(n2^n)$ time.   □

For the convenience of checking the correctness of all results, we provide an interactive verification at the website [14]. It shows the usefulness and efficiency of our algorithms in practical settings.

4 Concluding Remarks

In this paper, we first present two edge-disjoint Hamiltonian cycles in CQ_4, and each cycle has the same number of edges in each dimension. Then, we provide an $O(n)$ time algorithm for each vertex in CQ_n to determine which two edges were used in Hamiltonian cycles 1 and 2, respectively. It is interesting to see if there are three edge-disjoint Hamiltonian cycles in CQ_n for $n \geq 6$. So far it is still an open problem.

Acknowledgments. This research was partially supported by MOST grants 107-2221-E-131-011 from the Ministry of Science and Technology, Taiwan.

References

1. Bae, M.M., Bose, B.: Edge disjoint Hamiltonian cycles in k-ary n-cubes and hypercubes. IEEE Trans. Comput. **52**, 1271–1284 (2003)
2. Barden, B., Libeskind-Hadas, R., Davis, J., Williams, W.: On edge-disjoint spanning trees in hypercubes. Inf. Process. Lett. **70**, 13–16 (1999)
3. Barth, D., Raspaud, A.: Two edge-disjoint Hamiltonian cycles in the butterfly graph. Inf. Process. Lett. **51**, 175–179 (1994)
4. Efe, K.: A variation on the hypercube with lower diameter. IEEE Trans. Comput. **40**, 1312–1316 (1991)
5. Efe, K.: The crossed cube architecture for parallel computing. IEEE Trans. Parallel Distrib. Syst. **3**, 513–524 (1992)
6. Efe, K., Blackwell, P.K., Slough, W., Shiau, T.: Topological properties of the crossed cube architecture. Parallel Comput. **20**, 1763–1775 (1994)
7. Hung, R.W.: Embedding two edge-disjoint Hamiltonian cycles into locally twisted cubes. Theor. Comput. Sci. **412**, 4747–4753 (2011)
8. Hung, R.W.: Constructing two edge-disjoint Hamiltonian cycles and two-equal path cover in augmented cubes. J. Comput. Sci. **39**, 42–49 (2012)
9. Hung, R.W.: The property of edge-disjoint Hamiltonian cycles in transposition networks and hypercube-like networks. Discrete Appl. Math. **181**, 109–122 (2015)
10. Hung, R.W., Chan, S.J., Liao, C.C.: Embedding two edge-disjoint Hamiltonian cycles and two equal node-disjoint cycles into twisted cubes. Lect. Notes Eng. Comput. Sci. **2195**, 362–367 (2012)
11. Hung, H.S., Fu, J.S., Chen, G.H.: Fault-free Hamiltonian cycles in crossed cubes with conditional link faults. Inf. Sci. **177**, 5664–5674 (2007)
12. Kulasinghe, P.D., Bettayeb, S.: Multiplytwisted hypercube with 5 or more dimensions is not vertex transitive. Inf. Process. Lett. **53**, 33–36 (1995)
13. Lin, T.J., Hsieh, S.Y., Juan, J.S.-T.: Embedding cycles and paths in product networks and their applications to multiprocessor systems. IEEE Trans. Parallel Distrib. Syst. **23**, 1081–1089 (2012)
14. Pai, K.J.: An interactive verification of constructing two edge-disjoint Hamiltonian cycles in crossed cubes (2020). http://poterp.iem.mcut.edu.tw/2HC_in_CQn/
15. Petrovic, V., Thomassen, C.: Edge-disjoint Hamiltonian cycles in hypertournaments. J. Graph Theory **51**, 49–52 (2006)
16. Rowley, R., Bose, B.: Edge-disjoint Hamiltonian cycles in de Bruijn networks. In: Proceedings of the 6th Distributed Memory Computing Conference, pp. 707–709 (1991)
17. Saad, Y., Schultz, M.H.: Topological properties of hypercubes. IEEE Trans. Comput. **37**, 867–872 (1988)
18. Wang, D.: A low-cost fault-tolerant structure for the hypercube. J. Supercomput. **20**, 203–216 (2001)
19. Wang, D.: On embedding Hamiltonian cycles in crossed cubes. IEEE Trans. Parallel Distrib. Syst. **19**, 334–346 (2008)

Antipodal Radio Labelling of Full Binary Trees

Satabrata Das[1] , Laxman Saha[1(✉)] , and Kalishankar Tiwary[2]

[1] Department of Mathematics, Balurghat College, Balurghat 733101, India
sdas1012@gmail.com, laxman.iitkgp@gmail.com
[2] Department of Mathematics, Raiganj University, Raiganj 733134, India
tiwarykalishankar@yahoo.com

Abstract. Let G be a graph with diameter d and $k \leq d$ be a positive integer. A radio k-labelling of G is a function f that assigns to each vertex with a non-negative integer such that the following holds for all vertices u, v: $|f(u) - f(v)| \geq k + 1 - d(u, v)$, where $d(u, v)$ is the distance between u and v. The span of f is the absolute difference of the largest and smallest values in $f(V)$. The radio number of G is the minimum span of a radio labelling admitted by G. In this article, we study radio $(d-1)$-labelling problem for full binary trees.

Keywords: Frequency assignment problem · Radio k-labelling · Radio antipodal number · Span · Tree

1 Introduction

Frequency Assignment Problem (FAP) consists into the assignments of frequencies to the transmitters in a network, ensuring that there are no interferences, namely close transmitters do not have close frequencies. Radio k-labelling of a simple connected graph is a variation of FAP. In 1980, Hale [5] has modeled FAP as a Graph labelling problem (in particular as a generalized graph labelling problem) and is an active area of research now. Given a simple connected graph $G = (V(G), E(G))$ and a positive integer k with $1 \leqslant k \leqslant \operatorname{diam}(G)$, a radio k-labelling of G is a mapping $f : V(G) \to \{0, 1, 2, \ldots\}$ such that $|f(u) - f(v)| \geqslant k + 1 - d(u, v)$ for each pair of distinct vertices u and v of G, where $\operatorname{diam}(G)$ is the diameter of G and $d(u, v)$ is the distance between u and v in G. The span of f, denoted by $\operatorname{span}_f(G)$, is the absolute difference of the largest and smallest values in $f(V)$. Without loss of generality, we may assume $\min f(V)$ is zero. The radio k-chromatic number of G is the minimum span among all radio k-labellings of G. Motivated by FM channel assignments problem, the radio k- labelling problem was introduced in [2,3] and studied further in [7,8]. The *radio antipodal number*, denoted by $an(G)$, is the minimum

Supported by National Board of Higher Mathematics (NBHM), India, with grants no. 2/48(22)/R & D II/4033, 2017.

Z. Zhang et al. (Eds.): AAIM 2020, LNCS 12290, pp. 456–468, 2020.
https://doi.org/10.1007/978-3-030-57602-8_41

span of a radio (diam(G) − 1)-labeling of G and the *radio number*, denoted by $rn(G)$, is the minimum span of a radio diam(G)-labeling of G.

Determining the radio k-chromatic number of a graph is an interesting yet difficult combinatorial problem with potential applications to FAP. So far it has been explored for a few basic families of graphs and values of k near to diameter. The radio number of any hypercube was determined in [10] by using generalized binary Gray codes. For two positive integers $m \geq 3$ and $n \geq 3$, the Toroidal grids $TG_{m,n}$ are the cartesian product of cycle C_m with cycle C_n. Saha et al. [12] have given exact value for radio number of $TG_{m,n}$ when $mn \equiv 0 \pmod 2$. For a cycle C_n, the radio number was determined by Liu and Zhu [7], and the antipodal number is known only for $n \equiv 1, 2, 3 \pmod 4$ (see [4,6]).

Surprisingly, even for paths finding the radio number was a challenging task. It is envisaged that in general determining the radio number would be difficult even for trees, despite a general lower bound for trees given in [8,16]. Till now, the radio number is known for very limited families of trees. For path P_n and full m-ary trees, the exact values of radio number were determined in [7,18]. The results for path were generalized [7] to spiders, leading to the exact value of the radio number in certain special cases. In [9], Reddy et al. gave an upper bound for the radio number of some special type of trees. For tree structured network T the antipodal number is known only when T is a path (see, [11]).

In this article, we determine the exact value of antipodal number for full binary trees T_h of any height h. Rest of this paper is organized as follows: In Sect. 2, we give some preliminary results on tree. In Sect. 3, we give a lower bound for antipodal number of full binary trees and finally, in Sect. 4 we show this lower bounds are exactly same as the antipodal number of T_h.

2   Preliminaries

A full binary tree T_h is a rooted tree such that each vertex of degree greater than one has exactly two children and all degree-one vertices are of equal distance (height) to the root.

Definition 1. Let T be any tree. The measure of *separability* of a vertex $v \in V(T)$, denoted by $\beta_T(v)$, is the size of maximum connected component of $T-\{v\}$. A vertex is called *centroid* if it has minimal separability over all vertices in T.

Let T be a tree with centroid S. The level of $u \in V(T)$ denoted by $L(u)$, is the distance of u from S (i.e. $L(u) = d(S,u)$). A vertex u of T is in level ℓ if $L(u) = \ell$. For distinct $u,v \in V(T)$, the length of the common part of the paths of T from S to u and v is denoted by $\phi(u,v)$.

Lemma 1. *Let T be a tree rooted at r. Then for distinct $u,v \in V(T)$ the following $(a) - (b)$ hold.*

(a) $d(u,v) = L(u) + L(v) - 2\phi(u,v)$
(b) $\phi(u,v) = 0$ if and only if $r \in \{u,v\}$ or u and v belong to the different branches.

Lemma 2. *For an n-vertex tree T, the following $(a) - (c)$ hold.*

(a) *If a vertex v is centroid, then $\beta_T(v) \leqslant \lfloor \frac{n}{2} \rfloor$*
(b) *A tree with odd number of vertices has exactly one centroid.*
(c) *A tree T with even number of vertices has two centroids S_1 and S_2 which are neighbors and $\sum\limits_{u \in V(T)} d(S_1, u) = \sum\limits_{u \in V(T)} d(S_2, u)$.*

Notation 1. For an n-vertex tree T and a centroid S of T, we call the number $\sum\limits_{u \in V(T)} d(S, u)$ the *weight of T* and denoted it by $w(T)$.

The *uv-radio labelling f_{uv}* of a tree T is a radio labelling f with minimum and maximum labels are at u and v, respectively. An *uv-radio number* of a tree T, denoted by $rn_{uv}(T)$, is the minimum span over all uv-radio labellings of T. Although Liu [8] have presented the following lemma and corollary using different symbols, here we give a complete proof for better understanding of our main results.

Lemma 3. *Let f be a radio labeling of an n-vertex tree T with first and last colored vertices u and v. Then*

$$\operatorname{span}_{uv}(T) \geq (n-1)(\operatorname{diam}(T) + 1) - 2w(T) + d(S, u) + d(S, v)$$

where $w(T)$ denotes the weight of T and S denotes a centroid of T.

Proof: Let d be the diameter of T. Since f is a radio labelling of T, f induces a linear order $u_0, u_1, u_2, \ldots, u_{n-1}$ of the vertices of T such that $f(u) = f(u_0) < f(u_1) < f(u_2) < \cdots < f(u_{n-1}) = f(v)$. Then uv-span of f is given by

$$\operatorname{span}_{uv}(T) = f(v) - f(u) = \sum_{i=0}^{n-2} [f(u_{i+1}) - f(u_i)]$$

$$\geq \sum_{i=0}^{n-2} [d + 1 - d(u_i, u_{i+1})]$$

$$\geq (n-1)(d+1) - \sum_{i=0}^{n-2} d(u_i, u_{i+1})$$

$$\geq (n-1)(d+1) - \sum_{i=0}^{n-2} [d(S, u_i) + d(S, u_{i+1})]$$

$$= (n-1)(d+1) - 2w(T) + d(S, u) + d(S, v).$$

Corollary 1. *Let T be an n-vertex tree with diameter d and centroid S. Then*

$$rn(T) \geqslant (n-1)(d+1) - 2w(T) + 1.$$

Moreover, the equality holds if and only if there exist a centroid S and a radio labelling f with $f(u_0) = 0 < f(u_1) < \ldots < f(u_{n-1})$, where all the following hold (for all $0 \leqslant i \leqslant n-2$):

(a) $\phi(u_i, u_{i+1}) = 0$, $S \in \{u_0, u_{n-1}\}$ and $d(u_0, u_{n-1}) = 1$.
(b) $f(u_{i+1}) = f(u_i) + d + 1 - L_S(u_i) - L_S(u_{i+1})$.

Proof: From Lemma 3, as $f(u_0) = 0$ and $d(S, u_0) + d(S, u_{n-1}) \geq 1$,

$$rn(T) = \mathrm{span}_{u_0 u_{n-1}}(T) \geqslant (n-1)(d+1) - 2w(T) + 1.$$

Remark 1. To compute the lower bound presented in above, first we search for a centroid S and then calculate the distances of other vertices from it. A linear time algorithm can be presented for finding a centroid and calculating the distances form this centroid.

Example 1. Even paths P_{2k} have radio numbers equal to the bound in Corollary 1, as one can find a radio labeling satisfying Corollary 1 (cf. [7]).

Example 2. Consequences of Corollary 1 include the radio number for full m-ary tree $T_{m,h}$, $m \geqslant 3$ (which was settled in [18] by a different approach).

3   Lower Bound for Antipodal Number of Full Binary Trees

Here we present a lower bound of antipodal number of full binary trees T_h of any height h and the sharpness of this lower bound has been presented in next section. We need the following definitions and results to present a lower bound for antipodal number of T_h,

Definition 2. A subgraph H of a graph G is said to be *maximal k-diameteral subgraph* if diameter of H is k and it contains maximum number of vertices of G.

Definition 3. Let $f : E \to F$ be a mapping from a set E to a set F. For a set $A \subset E$, we call the mapping $f|_A : A \to F$ as the *restriction of f on A*.

Lemma 4. *Let G be a graph with diameter d and H be a k-diameteral subgraph of G with $k < d$. If $rc_k(G)$ and $rn(H)$ be the radio k-chromatic number of G and the radio number of H, respectively, then $rc_k(G) \geq rn(H)$.*

Proof: Let f be a radio k-labelling of G. Here the diameter of H is k with $k < d$. Thus $V(H) \subset V(G)$. Let $g = f|_{V(H)}$ be the restriction of f on $V(H)$. Then $\mathrm{span}_f(G) \geqslant \mathrm{span}_g(H)$ and this is true for any radio k-labelling of G and so for restriction $g = f|_{V(H)}$ also. Since the diameter of H is k, we obtain the required result.

Lemma 5. *Let T_h be a full binary tree of height h. Then for $k = 2h - 1$, the following are hold*

(a) T_h has exactly two maximal k-diameteral subgraph (say T^u and T^v with centroid at u and v, respectively)

(b) Each maximal k-diameteral subgraph has $3 \cdot 2^{h-1} - 1$ vertices.

(c) The weight of each maximal k-diameteral subgraph is $3h \cdot 2^{h-1} - 5 \cdot 2^{h-1} + 3$.

Proof: Let r be the root vertex of T_h and u, v be two children of r. So T_h has two branches with respect to the root r, namely, the left branch $L(T_h)$ and the right branch $R(T_h)$.

(a) To reduce diameter of T_h by one, we have to delete at least all the pendant vertices of either branch, otherwise the diameter will remains unchanged. Removing all the pendant vertices of $R(T_h)$, we get a maximal k-diameteral subgraph T^u with a centroid at u. Similarly, removing all the pendant vertices of $L(T_h)$, we get another maximal k-diameter subgraph T^v with a centroid at v. Thus T_h has exactly two maximal k-diameteral subgraphs.

(b) and (c): Note that T^u and T^v are identical. The number of vertices of T^u and T^v are $|V(T^u)| = |V(T^v)| = |V(T_{h-1})| + |V(T_{h-2})| + 1 = 2^h + 2^{h-1} - 1$ and weights of T^u and T^v are $w(T^u) = w(T^v) = w(T_{h-1}) + w(T_{h-2}) + 2 \cdot |V(T_{h-2})| + 1 = 3h \cdot 2^{h-1} - 5 \cdot 2^{h-1} + 3$.

Theorem 1. For a full binary tree T_h of height h, $an(T_h) \geq 5 \cdot 2^h - 4h - 4$.

Proof: For $k = 2h-1$, let f be an arbitrary radio k-labelling of T_h. Let two maximal k-diameteral subgraphs of T_h are T^u and T^v with centroid at u and v, respectively. Let u_0 and u_{n-1} be the first and last colored vertex of T_h. Also let $f|_{V(T^u)}$ and $f|_{V(T^v)}$ are restrictions of f on $V(T^u)$ and $V(T^u)$, respectively. Since diameter of T^u and T^v is k, using Lemma 4 we have $rc_k(T_h) \geq \max\{rn(T^u), rn(T^v)\}$. Now applying Lemma 3 to both T^u and T^v and using the results of Lemma 5, we have the inequalities

$$rn(T^u) \geqslant 5 \cdot 2^h - 4h - 6 + f(x_u) + d(u, x_u) + d(u, y_u); \tag{1}$$

$$rn(T^v) \geqslant 5 \cdot 2^h - 4h - 6 + f(x_v) + d(v, x_v) + d(v, y_v); \tag{2}$$

where x_w and y_w denote the minimum and maximum colored vertices of T^w under the restriction $f|_{V(T^w)}$ for $w \in \{u, v\}$. If one of x_u and x_v is not equal to u_0 or one of y_u and y_v is not equal to u_{n-1}, then the result follows. For illustration, say $x_w \neq u_0$ for some $w \in \{u, v\}$. Then $f(x_w) \geq 1$ for some $w \in \{u, v\}$. Since $d(u, x_u) + d(u, y_u) \geqslant 1$ and $d(v, x_v) + d(v, y_v) \geqslant 1$, at least one of $f(x_u) + d(u, x_u) + d(u, y_u)$ and $f(x_v) + d(v, x_v) + d(v, y_v)$ must be greater than or equal to 2. Hence the theorem is proved in this case. Similar argument can be used of the case when $y_w \neq u_{n-1}$ for some $w \in \{u, v\}$. Thus we have consider the case when f and its restrictions $f|_{V(T^u)}, f|_{V(T^v)}$ attain the minimum color at same vertex as well as maximum color at same vertex i.e., $x_w = u_0$ and $y_w = u_{n-1}$ for $w \in \{u, v\}$. Then using $rc_k(T_h) \geq \max\{rn(T^u), rn(T^v)\}$ and inequalities (1) and (2), we have the following

$$rc_k(T_h) \geq 5 \cdot 2^h - 4h - 6 + f(u_0) + \max\{d(u, u_0) + d(u, u_{n-1}), d(v, u_0) + d(v, u_{n-1})\}. \tag{3}$$

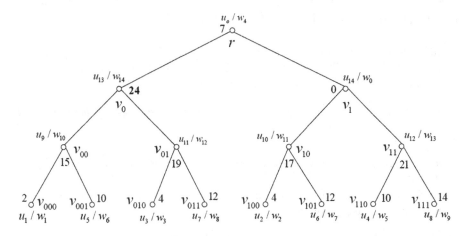

Fig. 1. Vertex-indices and antipodal labeling of T_3.

Now we determine the least value of

$$\max\{d(u, u_0) + d(u, u_{n-1}), d(v, u_0) + d(v, u_{n-1})\}$$

depending on the position of minimum colored vertex u_0 and maximum colored vertex u_{n-1}. For this we consider the following cases.

Case-I: $u_0 \in \{u, v\}$ and $u_{n-1} \notin \{u, v\}$. Let $w = \{u, v\} \setminus u_0$. In this case, $d(w, u_0) = 2$ and $d(w, u_{n-1}) \geq 1$ and consequently, $\max\{d(u, u_0) + d(u, u_{n-1}), d(v, u_0) + d(v, u_{n-1})\} \geq 3$.

Case-II: $u_0, u_{n-1} \in \{u, v\}$. First we take $u_0 = u, u_{n-1} = v$. In this case, $d(u, u_0) = 0, d(v, u_0) = 2, d(u, u_{n-1}) = 2$ and $d(v, u_{n-1}) = 0$ and hence

$$\max\{d(u, u_0) + d(u, u_{n-1}), d(v, u_0) + d(v, u_{n-1})\} = 2.$$

Similarly, $\max\{d(u, u_0) + d(u, u_{n-1}), d(v, u_0) + d(v, u_{n-1})\} = 2$ if $u_0 = v, u_{n-1} = u$.

Case-III: $u_0 = r$. In this case, $d(u, u_0) = 1, d(v, u_0) = 1$ and either $d(u, u_{n-1}) \geq 2$ or $d(v, u_{n-1}) \geq 2$. Thus $\max\{d(u, u_0) + d(u, u_{n-1}), d(v, u_0) + d(v, u_{n-1})\} \geq 3$.

Case-IV: $u_0 \notin \{r, u, v\}$. In this case, $\max\{d(u, u_0) + d(u, u_{n-1}), d(v, u_0) + d(v, u_{n-1})\} \geq 4$ because $d(v, u_0) \geq 3$ and $d(v, u_{n-1}) \geq 1$.

On account of the four cases above, the inequality in (3) gives

$$an(T_h) \geq 5 \cdot 2^h - 4h - 4.$$

4 Antipodal Number of Full Binary Trees

In this section, we present an algorithm for construction of antipodal labelling of full binary trees. The algorithm has been build up to produce a suitable arrangement of vertices for T_h that facilitate an antipodal labelling.

Theorem 2. *For a full binary tree T_h there exists an antipodal labelling f with* $span_f(T_h) = 5 \cdot 2^h - 4h - 4$.

Proof: In Algorithm 1 produces an antipodal labelling f with $span_f(T_h) = 5 \cdot 2^h - 4h - 4$.

For illustration, we give an optimal antipodal labelling f for full binary tree T_3 in Fig. 1 by using Algorithm 1.

Observation 1. From Step-II of Algorithm 1, we have the following

(a) u_0 is the root vertex of T_h
(b) $L_h = \{u_1, u_2, \ldots, u_{2^h}\}$ and $L(u_j) < h$ for $j > 2^h$.

Observation 2. For subsets $S_1 = \{w_i : 2^h + 2 \leq i \leq 2^{h+1} - 2\}$, $S_2 = \{w_i : 1 \leq i \leq 2^{h-1} - 1\}$ and $S_3 = \{w_i : 2^{h-1} + 1 \leq i \leq 2^h + 1\}$ following are true

(a) $S_1 \subset L_p$ for some $p \leq h - 1$;
(b) $S_2 \cup S_3 \subset L_h$;

where L_t denotes the set of vertices of T_h which are at level t.

Proof: (a) Let $w_j \in S_1$. Then $2^h + 2 \leq j \leq 2^{h+1} - 2$. Then from Step-III of Algorithm 1, $w_j = u_{j-1}$ as $2^h + 2 \leq j \leq 2^{h+1} - 2$. Observation 1 (b) implies that u_{j-1} is in level p for some positive $p < h$. Hence the result is true for w_j with $2^h + 2 \leq j \leq 2^{h+1} - 2$.

(b) Let $w_i \in S_2$ and $w_j \in S_3$. Then $1 \leq i \leq 2^{h-1} - 1$ and $2^{h-1} + 1 \leq j \leq 2^h + 1$, which implies that $w_i = u_i$ and $w_j = u_{j-1}$ from Step-III of Algorithm 1. Thus from Observation 1, both w_i and w_j are in level h.

Correctness of Algorithm 1:

From Theorem 1, $an(T_h) \geq 5 \cdot 2^h - 4h - 4$. Thus to prove Algorithm 1 produces an optimal antipodal labelling it is sufficient to show f is an antipodal labelling with span $5 \cdot 2^h - 4h - 4$.

Lemma 6. *The mapping f defined in Algorithm 1 has the span $5 \cdot 2^h - 4h - 4$.*

Algorithm 1: An optimal antipodal labelling of T_h.

---

begin

    Input : A full binary tree T_h of height h.

    Output: An antipodal labelling f of T_h with minimum span.

    Initialization : Start with the root vertex r of full binary tree T_h.

    Step-I : Indices the vertices of T_h by the following rule :

    (a) Denote the two children of r by v_0 and v_1.

    (b) Denote two children of v_0 by v_{00}, v_{01}; and two children of v_1 by v_{10}, v_{11}.

    (c) In general, for $1 \leq t \leq \ell < h$ and $i_t \in \{0,1\}$, two children of a level ℓ vertex $v_{i_1 i_2 i_3 \ldots i_\ell}$ of T_h are denoted by $v_{i_1 i_2 i_3 \ldots i_\ell i_{\ell+1}}$, where $i_{\ell+1} \in \{0,1\}$.

    Step-II : Create a new arrangement $u_0, u_1, u_2, \ldots, u_{n-1}$ of $V(T_h)$ by changing the suffices of the vertices that are produced in Step-I under the following rule :

    (a) rename r by u_0

    (b) rename a level ℓ $(1 \leq \ell \leq h)$ vertex $v_{i_1 i_2 i_3 \ldots i_\ell}$ by u_j, where

$$j = 1 + i_1 + i_2 \cdot 2 + i_3 \cdot 2^2 + \ldots + i_\ell \cdot 2^{\ell-1} + \sum_{\ell+1 \leqslant t \leqslant h} 2^t.$$

    Step-III : Give a modified ordering $\{w_0, w_1, w_2, \ldots, w_{n-1}\}$ of $V(T_h)$ by $w_i = u_{\sigma(i)}$, where σ is a permutation of $\{0, 1, \ldots, 2^{h+1} - 1\}$ defined as

$$\sigma(0) = 2^{h+1} - 2;$$
$$\sigma(2^{h-1}) = 0;$$
$$\sigma(i) = \begin{cases} i, & \text{if } 1 \leq i \leq 2^{h-1} - 1; \\ i - 1, & \text{if } 2^{h-1} + 1 \leq i \leq 2^{h+1} - 2. \end{cases}$$

    Step-IV : Define a mapping $f : \{w_i : 0 \leq i \leq 2^{h+1} - 2\} \to \{0, 1, 2, \ldots\}$ by

$$f(w_0) = 0; \ f(w_1) = h - 1,$$
$$f(w_{2i+1}) = f(w_{2i}) = h + 2i - 1, \quad 0 \leq i \leq 2^{h-2} - 1;$$
$$f(w_{2^{h-1}}) = 2^{h-1} + 2h - 3;$$
$$f(w_{a+2}) = f(w_{a+1}) = a + 3h - 3, \quad 0 \leq j \leq 2^{h-2} - 1;$$
$$f(w_{2^h+1}) = 2^h + 3h - 3;$$
$$f(w_{i+1}) = f(w_i) + 2h - \{L(w_i) + L(w_{i+1})\}, \quad 2^h + 1 \leq i \leq 2^{h+1} - 3,$$

    where $a = 2^{h-1} + 2j$.

Proof: From definition of f in Algorithm 1, $f(w_i) \leq f(w_j)$ for $i < j$ and so from Algorithm 1,

$$\text{span}_f(T_h) = f(w_{2^{h+1}-2})$$

$$= [f(w_{2^{h+1}-2}) - f(w_{2^h+1})] - f(w_{2^h+1})$$

$$= \sum_{j=2^h+1}^{2^{h+1}-3} [f(w_{j+1}) - f(w_j)] - f(w_{2^h+1})$$

$$= 2h(2^{h+1} - 2^h - 3) - \sum_{j=2^h+1}^{2^{h+1}-3} [L(w_j) + L(w_{j+1})] - (2^h + 3h - 3)$$

$$= 2h(2^{h+1} - 2^h - 3) - 2\sum_{j=2^h+2}^{2^{h+1}-4} L(w_j) - L(w_{2^h+1}) - L(w_{2^{h+1}-3}) - (2^h + 3h - 3)$$

$$= 2^h + 3h - 3 + 2h(2^{h+1} - 2^h - 3) - 2\{\omega(T_{h-1}) - 2\} - h - 1$$

$$= 2^h + 3h - 3 + 2h(2^{h+1} - 2^h - 3) - 2^{h+1}(h - 2) - h - 1$$

$$= 5 \cdot 2^h - 4h - 4.$$

Lemma 7. *The mapping f defined in Algorithm 1 is an antipodal labelling of T_h.*

Proof: To prove f is an antipodal labelling of T_h, we first partitioned the vertex set $V(T_h)$ into four partite sets S_1, S_2, S_3 and S_4, where S_i's are defined as follow:

$$S_1 = \{w_i : 2^h + 2 \le i \le 2^{h+1} - 2\}$$
$$S_2 = \{w_i : 1 \le i \le 2^{h-1} - 1\}$$
$$S_3 = \{w_i : 2^{h-1} + 1 \le i \le 2^h + 1\}$$
$$S_4 = \{w_0, w_{2^h-1}\}.$$

Then from the construction of f in Algorithm 1, we have the following

(a) $f(w_a) = f(w_{a-1}) + 2h - \{L(w_a) + L(w_{a-1})\}$ for all $w_a \in S_1$

(b) For any $w_a \in S_2$

$$f(w_a) = \begin{cases} h + a - 1, & \text{if } a \text{ is even}; \\ h + a - 2, & \text{if } a \text{ is odd}. \end{cases}$$

(c) For any $w_a \in S_3$

$$f(w_a) = \begin{cases} 3h + a - 4, & \text{if } a \text{ is odd}; \\ 3h + a - 5, & \text{if } a \text{ is even}. \end{cases}$$

(d) $f(w_0) = 0$ and $f(w_{2^h-1}) = 2^{h-1} + 2h - 3$. From the definition of f our labelling scheme is

$$w_0 \to S_2 \to w_{2^h-1} \to S_3 \to S_1.$$

Let w_a and w_b be arbitrary two distinct vertices of T_h. We consider the following cases depending on the position of w_a and w_b in S_i, $i = 1, 2, 3, 4$.

Case-1: $w_a \in S_4 = \{w_0, w_{2^h-1}\}$ $and w_b \in \cup_{i=1}^4 S_i$. First we consider $w_a = w_0$. Note that the position of w_0 is the root vertex of right branch of T_h. So it is clear that $|f(w_a) - f(w_b)| \geq 2h - d(w_0, w_a)$ for all $w_b \in S_1 \cup S_3 \cup S_4$ and $w_a = w_0$. If $w_b \in S_2$, then

$$d(w_0, w_b) = \begin{cases} h + 1, & \text{if } a \text{ is odd} \\ h - 1, & \text{if } a \text{ is even} \end{cases}$$

because odd or even index vertices of S_2 are the left or right branch pendant vertices of T_h, respectively. Therefore, $|f(w_b) - f(w_0)| \geq 2h - d(w_0, w_a)$ for all $w_b \in S_2$. Now we consider $w_a = w_{2^h-1}$, which is the root vertex of T_h. Since $f(w_{2^h-1}) = 2^{h-1} + 2h - 3$ and $d(w_b, w_{2^h-1}) = h$ for all $w_b \in S_2 \cup S_3$, w_{2^h-1} satisfies antipodal conditions over the set $S_2 \cup S_3$. Again from the definition of f, $f(w_b) - f(w_{2^h-1}) \geq 2h$ for all $w_b \in S_1$.

Case-2: $w_a, w_b \in S_2$. Let $\eta = \phi(w_a, w_b)$. If exactly one of a and b is even, then $\eta = 0$ and hence antipodal condition is satisfied. If both a and b are either even or odd then

$$
\begin{aligned}
f(w_b) - f(w_a) &= b - a \\
&\geq 2^\eta \\
&\geq 2\eta \\
&= 2h - d(w_a, w_b), \quad \text{since } d(w_a, w_b) = \ell_a + \ell_b - 2\eta = 2h - 2\eta.
\end{aligned}
$$

Case-3: $w_a, w_b \in S_3$. Then by the same argument as Case-2 we can show that vertices of S_3 satisfy antipodal condition.

Case-4: $w_a, w_b \in S_1$. Let $\phi(w_a, w_b) = \eta$. From definition of f, we have

$$
\begin{aligned}
f(w_b) - f(w_a) &= 2h(b - a) - \left\{ L(w_a) + L(w_b) + 2 \sum_{j=a+1}^{b-1} L(w_j) \right\} \\
&= 2h - d(w_a, w_b) + \delta(a, b);
\end{aligned}
$$

where

$$\delta(a, b) = 2h(b - a - 1) - 2 \sum_{j=a+1}^{b-1} L(w_j) - 2\eta. \tag{4}$$

Since $w_a, w_b \in S_1$ and S_1 contains consecutive w_i's, $w_j \in S$ for $a \leq j \leq b$. Then Observation 2 gives $L(w_j) \leq h - 1$ for each $j \in \{a, a+1, \ldots, b\}$ which reduces the inequality (4) as

$$
\begin{aligned}
\delta(a, b) &\geq 2h(b - a - 1) - 2(h - 1)(b - a - 1) - 2\eta \\
&= 2(b - a - 1) - 2\eta \\
&\geq 2(2^\eta - \eta - 1), \quad \text{as } b - a \geq 2^\eta \\
&\geq 0.
\end{aligned}
$$

Case-5: $w_a \in S_2$ and $w_b \in S_1$. Clearly, $b > a$. First we consider $b = 2^h + 2$. Then for even integer a, we have

$$
\begin{aligned}
f(w_b) - f(w_a) &= 2^h + 3h - 2 - (h + a - 1) \\
&= 2^h + 2h - a - 1 \\
&= 2h + b - a - 3 \\
&\geq (2h - 3) + 2^\eta \quad \text{as } b - a \geq 2^\eta \text{ due to Lemma} \\
&\geq 2\eta + 1 \\
&= 2h - d(w_a, w_b).
\end{aligned}
$$

Similarly, if a is odd, then $f(w_a) = h + a - 2$ and $f(w_b) - f(w_a) \geq 2h - d(w_a, w_b)$. Now we consider $b > 2^h + 2$ and a is even. Then the values of f at the points b and a are given

$$
f(w_b) = f(w_{2^h+2}) + (b - 2^h - 2)2h - \{L(w_b) + L(w_{2^h+2}) + 2 \sum_{j=2^h+3}^{b-1} L(w_j);
$$

$$
f(w_a) = h + a - 1.
$$

After simple calculations with the fact $L(w_a) = h$, $L(w_{2^h+2}) = h - 1$ and $\eta = \phi(w_a, w_b)$; the difference $f(w_b) - f(w_a)$ reduces to

$$
f(w_b) - f(w_a) = 2h - d(w_a, w_b) + \delta(a, b),
$$

where $d(w_a, w_b) = L(w_a) + L(w_b) - 2\eta = L(w_a) + L(w_b) - 2\phi(w_a, w_b)$ and $\delta(a, b) = f(w_{2^h+2}) + (b - 2^h - 2)2h - 2 \sum_{j=2^h+3}^{b-1} L(w_j) - h - a - 2\eta + 2$. As of our previous cases, here we also show that $\delta(a, b) > 0$. Since $w_{2^h+3}, w_b \in S_1$, $\{w_{2^h+3}, \ldots, w_b\} \subset S_1$ and Observation 2 gives $L(w_j) \leq h - 1$ for all $j \in \{2^h + 3, \ldots, b\}$. With these highest values of $L(w_j)$'s, we obtain an inequality for $\delta(a, b)$ as

$$
\begin{aligned}
\delta(a, b) &\geq 2^h + 3h - 2 + 2h(b - 2^h - 2) - 2(h - 1)(b - 2^h - 3) - h - a - 2\eta + 2 \\
&= 2^h + 2h + 2h(b - 2^h - 2 - b + 2^h + 6) + 2(b - 2^h - 3) - a - 2\eta \\
&= 2^h + 10h + 2b - 2 \cdot 2^h - 6 - a - 2\eta \\
&= 10h + (b - 2^h) + (b - a) - 2\eta - 6 \\
&> 0 \quad \text{since } b > 2^h \text{ and } b - a > 2^\eta \geq 2\eta.
\end{aligned}
$$

Similarly, we can show that $f(w_b) - f(w_a) \geq 2h - d(w_a, w_b)$ whenever $b > 2^h + 2$ and a is an odd integer.

Case-6: $w_a \in S_3$ and $w_b \in S_1$. Using similar arguments as used in Case-5, we can show that the antipodal condition is also satisfied for these two vertices w_a and w_b.

Case-7: $w_a \in S_2$ $and w_b \in S_3$. In this case $b - a \geq 2$. From the definition of f,

$$f(w_b) - f(w_a) \geq 3h + b - 5 - (h + a - 1)$$
$$\geq 2h + b - a - 4$$

From the above inequality, it is clear that antipodal conditions are satisfied whenever $b - a \geq 3$. So our remaining case is $b - a = 2$ and this is true only when $b = 2^{h-1} + 1$ and $a = 2^{h-1} - 1$. In this case

$$f(w_b) - f(w_a) \geq 3h + b - 4 - (h + a - 2)$$
$$\geq 2h + b - a - 2$$
$$= 2h$$

and hence antipodal conditions are also satisfied when $b - a = 2$.

References

1. Bantva, D., Vaidya, S., Zhou, S.: Radio number of trees. Discrete Appl. Math. **317**, 110–122 (2017)
2. Chartrand, G., Erwin, D., Harary, F., Zhang, P.: Radio labelings of graphs. Bull. Inst. Comb. Appl. **33**, 77–85 (2001)
3. Chartrand, G., Erwin, D., Zhang, P.: A graph labeling problem suggested by FM channel restrictions. Bull. Inst. Comb. Appl. **43**, 43–57 (2005)
4. Chartrand, G. Erwin, D., Zhang, P.: Radio antipodal colorings of cycles. In: 2000 Proceedings of the Thirty-First Southeastern International Conference on Combinatorics, Graph Theory and Computing, vol. 144, pp. 129–141, Boca Raton (2000)
5. Hale, W.: Frequency assignment theory and application. Proc. IEEE **68**(12), 1497–1514 (1980)
6. Juan, J.S.-T., Liu, D.D.-F.: Antipodal labelings for cycles. Ars Comb. **103**, 81–96 (2012)
7. Liu, D.D.-F., Zhu, X.: Multi-level distance labelings for paths and cycles. SIAM J. Discrete Math. **19**(3), 610–621 (2005)
8. Liu, D.D.-F.: Radio number for trees. Discrete Math. **308**(7), 1153–1164 (2008)
9. Reddy Palagiri, V.S., Iyer, K.V.: Upper bounds on the radio number of some trees. Int. J. Pure Appl. Math. **71**(2), 207–215 (2011)
10. Khennoufa, R., Togni, O.: The radio antipodal and radio numbers of the hypercube. Ars Comb. **102**, 447–461 (2011)
11. Khennoufa, R., Togni, O.: A note on radio antipodal colourings of paths. Math. Bohem. **130**(3), 277–282 (2005)
12. Saha, L., Panigrahi, P.: On the radio number of Toroidal grids. Aust. J. Combin. **55**, 273–288 (2013)
13. Saha, L., Panigrahi, P.: A lower bound for radio k-chromatic number. Discrete Appl. Math. **192**, 87–100 (2015)
14. Saha, L., Panigrahi, P.: A graph radio k-coloring algorithm. In: Arumugam, S., Smyth, W.F. (eds.) IWOCA 2012. LNCS, vol. 7643, pp. 125–129. Springer, Heidelberg (2012). https://doi.org/10.1007/978-3-642-35926-2_15
15. Sarkar, U., Adhikari, A.: On characterizing radio k-labelling problem by path covering problem. Discrete Math. **338**(4), 615–620 (2015)

16. Das, S., Ghosh, S.C., Nandi, S., Sen, S.: A lower bound technique for radio k-labelling. Discrete Math. **340**(5), 855–861 (2017)
17. Zhou, S.: A channel assignment problem for optical networks modelled by Cayley graphs. Theor. Comput. Sci. **310**, 501–511 (2004)
18. Li, X., Mak, V., Zhou, S.: Optimal radio labellings of complete m-ary trees. Discrete Appl. Math. **158**, 507–515 (2010)

Total Coloring of Outer-1-planar Graphs: The Cold Case

Weichan Liu[1,2,3] and Xin Zhang[1(✉)]

[1] School of Mathematics and Statistics, Xidian University, Xi'an 710071, China
liuweichan19@mails.ucas.ac.cn,xzhang@xidian.edu.cn
[2] Academy of Mathematics and System Sciences, Chinese Academy of Sciences, Beijing, China
[3] University of Chinese Academy of Sciences, Beijing, China

Abstract. A graph is outer-1-planar if it has a drawing in the plane so that its vertices are on the boundary face and each edge is crossed at most once. Zhang (2013) proved that the total chromatic number of every outer-1-planar graph with maximum degree $\Delta \geq 5$ is $\Delta + 1$, and showed that there are graphs with maximum degree 3 and total chromatic number 5. For outer-1-planar graphs with maximum degree 4, Zhang (2017) confirmed that its total chromatic number is at most 5 if it admits an outer-1-planar drawing in the plane so that any two pairs of crossing edges share at most one common end vertex. In this paper, we prove that the total chromatic number of every Anicop graph with maximum degree 4 is at most 5, where an Anicop graph is an outer-1-planar graph that admits a drawing in the plane so that if there are two pairs of crossing edges sharing two common end vertices, then any of those two pairs of crossing edges would not share any end vertex with some other pair of crossing edges. This result generalizes the one of Zhang (2017) and moves a step towards the complete solving of the cold case.

Keywords: Outer-1-planar graph · Total coloring · Maximum degree

1 Introduction

A *total k-coloring* of a graph G is an assignment of k colors to all vertices and edges of G so that no two adjacent or incident elements receive the same color. The *total chromatic number* $\chi''(G)$ of a graph G is the minimum integer k so that G has a total k-coloring. In any total coloring of a graph G with maximum degree Δ, it is easy to see that we shall use $\Delta + 1$ colors to color the vertex of degree Δ and its incident edges. This implies that $\chi''(G) \geq \Delta(G) + 1$ for every graph G. On the other hand, looking for a general upper bound in terms of $\Delta(G)$ for $\chi''(G)$ seems interesting and challenging. Actually, Behzad [3] and Vizing [10] independently conjectured at least fifty years ago that $\chi''(G) \leq \Delta(G) + 2$ for

Supported by NSFC (11871055) and the Youth Talent Support Plan of Xi'an Association for Science and Technology (2018-6).

Z. Zhang et al. (Eds.): AAIM 2020, LNCS 12290, pp. 469–480, 2020.
https://doi.org/10.1007/978-3-030-57602-8_42

every graph G. This conjecture was now confirmed for graphs with maximum degree at most 3 by Rosenfeld [7] and Vijayaditya [9], 4 and 5 by Kostochka [5,6], and verified for planar graphs with maximum degree 7 by Sanders and Zhao [8], 8 by Andersen [1], and at least 9 by Borodin [4]. However, the conjecture itself is still quite open, even for planar graphs with maximum degree 6.

In the literature, there are some well-established subclasses of planar graphs including

- *outerplanar graphs:* graphs that can be drawn in the plane so that all the vertices are on the outer face (equivalently, graphs that do not contain $K_{2,3}$ or K_4 as a minor);
- *series-parallel graphs:* graphs that do not contain K_4 as a minor;
- *outer-1-planar graphs:* graphs that can be drawn in the plane so that all the vertices are on the outer face and each edge is crossed at most once.

Outplanar graphs and series-parallel graphs are planar due to the well-known Wagner's theorem which says that a graph is planar if and only if it does not contain $K_{3,3}$ or K_5 as a minor. But the planarity of outer-1-planar graphs is not trivially proved—such a proof was given by Auer et al. [2], who also pointed out that the class of outer-1-planar graphs is not minor-closed. A graph is *quasi-Hamiltonian* if each of its block is Hamiltonian. Zhang, Liu, and Wu [19] showed that the intersection of the class of quasi-Hamiltonian outer-1-planar graphs and the class of series-parallel graphs is indeed the class of outerplanar graphs.

Zhang, Zhang, and Wang [20] showed in 1988 that the $\chi''(G) = \Delta(G) + 1$ for every outerplanar graph with maximum degree at least 3. The same result also holds for series-parallel graphs, which was proved in 2004 by Wu and Hu [12]. In 2011, Zhang and Liu [18] proved the total coloring conjecture for outer-1-planar graphs, and moreover, showed that $\chi''(G) = \Delta(G) + 1$ for every outer-1-planar graph with maximum degree at least 5, and this result was later generalized to its list version by Zhang [13] in 2013. In [13,18], the authors also pointed out that there are outer-1-planar graphs G with $\Delta(G) = 3$ and $\chi''(G) = 5$, and whether outer-1-planar graphs G with $\Delta(G) = 4$ satisfy $\chi''(G) = \Delta(G) + 1 = 5$ is unknown.

For this cold case, Zhang [15] considered the *Nicop graphs*, i.e., outer-1-plane graphs so that any two pairs of crossing edges share at most one common end vertex. Here, an *outer-1-plane graph* is a drawing of outer-1-planar graph in the plane so that its outer-1-planarity is preserved and the number of crossings is as small as possible. Zhang [15] proved the following

Theorem 1 [15]. *If G is a Nicop graph with $\Delta(G) = 4$, then $\chi''(G) = 5$.*

In this paper, we aim to generalize this result to a larger class of graphs $\mathcal{G}$. Here, a graph G belongs to $\mathcal{G}$ if and only if

- G is an outer-1-plane graph, and
- if there are two pairs of crossing edges sharing two common end vertices, then any of those two pairs of crossing edges would not share any end vertex with some other pair of crossing edges.

From now on, a graph $G \in \mathcal{G}$ is called an *outer-1-plane graph with almost-near-independent crossings*, or an *Anicop graph* for short. Our main result is stated as follows:

Theorem 2. *If G is an Anicop graph with $\Delta(G) = 4$, then $\chi''(G) = 5$.*

Since Nicop graphs are Anicop graphs, Theorem 2 implies Theorem 1. Actually, we believe that the same conclusion holds for every outer-1-planar graph with maximum degree 4, so we end this section with the following conjecture.

Conjecture 3. *If G is an outer-1-planar graph with $\Delta(G) = 4$, then $\chi''(G) = 5$.*

2 Reducibilities: The Proof of Theorem 2

From now on, when we mention an outer-1-planar graph G, we always refer to its *outer-1-planar diagram*, i.e, a drawing of G in the plane so that the outer-1-planarity of G is preserved and this drawing has the minimum number of crossings among all such outer-1-planar drawings.

To begin with, we define *base graphs* Π_i^1 and Π_i^2 with $1 \leq i \leq 3$ by Fig. 1. In each picture of this figure besides Π_1^1, all vertices are lying consecutively in an outer-1-planar diagram G as where they are drawn in that picture (i.e., the boundary edges incident with the black vertices in that picture form a sub-drawing of the outer-face of G). The two white vertices in each picture of Fig. 1 are called the *handles*.

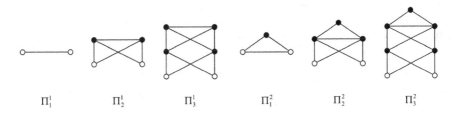

Π_1^1 Π_2^1 Π_3^1 Π_1^2 Π_2^2 Π_3^2

Fig. 1. Base graphs

Given two base graphs, say Π_i^j with handles u, v, and $\Pi_{i'}^{j'}$ with handles u', v', we have two operations:

$\Pi_i^j \circ \Pi_{i'}^{j'}$ Identifying v with v', see Fig. 2, and in the resulting graph let the degree of the vertex w corresponding v and v' be the number of edges incident with it in this partial drawing. The vertices u and u' in the resulting graph are called *linking handles*;

$\Pi_i^j \otimes \Pi_{i'}^{j'}$ Adding edges vv', uv' and $u'v$ so that uv' crosses $u'v$, see Fig. 2, and in the resulting graph let the degree of the vertex v or v' be the number of edges incident with it in this partial drawing. The vertices u and u' in the resulting graph are called *crossed-linking handles*.

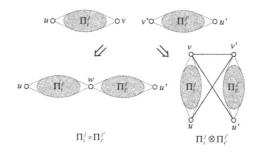

Fig. 2. Two operations generated by Π_i^j and $\Pi_{i'}^{j'}$

Note that $\Pi_i^j \circ \Pi_{i'}^{j'}$ or $\Pi_i^j \otimes \Pi_{i'}^{j'}$ is still an outer-1-planar diagram. We prove the following

Theorem 4. *Every 2-connected Anicop graph with maximum degree at most 4 contains one of the configurations among*

(C1) *a vertex u of degree 2 adjacent to a vertex v of degree at most 3;*
(C2) *a cycle of length 4 with two nonadjacent vertices of degree 2;*
(C3) *a triangle uvw with $d(v) = 2$ and u adjacent to a vertex x of degree 2;*
(C4) $\Pi_1^1 \otimes \Pi_1^2$;
(C5) $\Pi_1^2 \otimes \Pi_1^2$;
(C6) Π_3^1;
(C7) Π_2^1 *or* Π_2^2 *or* Π_3^2, *with a handle of degree at most 3;*
(C8) Π_2^1 *or* Π_2^2 *or* Π_3^2, *with a handle adjacent to a vertex of degree 2;*
(C9) Π_2^1 *or* Π_2^2 *or* Π_3^2, *with the two handles being adjacent;*
(C10) $\Pi_2^1 \circ \Pi_2^1$, *or* $\Pi_2^1 \circ \Pi_2^2$, *or* $\Pi_2^2 \circ \Pi_2^2$;
(C11) $\Pi_1^1 \otimes \Pi_2^1$, *or* $\Pi_1^1 \otimes \Pi_2^2$, *or* $\Pi_1^2 \otimes \Pi_2^1$, *or* $\Pi_1^2 \otimes \Pi_2^2$.

In this section, we apply Theorem 4 to prove the following theorem, which is slightly stronger than Theorem 2.

Theorem 5. *If G is an Anicop graph with maximum degree at most 4, then* $\chi''(G) \leq 5$.

Proof. (sketch). Let G be a counterexample with the minimum number of vertices. Clearly, G is 2-connected. It is sufficient to prove that G does not contain the configuration (Ci) for each $1 \leq i \leq 11$, contradicting Theorem 4. The proof of each item proceeds as follows. First, we construct a graph G' with $\Delta(G') \leq 4$ and $|G'| < |G|$ via removing some vertices appearing in (Ci) from G (we suppose, to the contrary, that (Ci) occurs), and after that, adding non-crossed edges inside the outer boundary (this operation applies sometimes, not always). Next, we prove that a total 5-coloring of G' can be extended to a total 5-coloring of G (sometimes the recoloring shall be involved). Note that if we remove vertices from an Anicop graph, or add non-crossed edges inside the outer boundary of an Anicop graph, the resulting graph is still an Anicop graph. So, by the minimality of G, G' is total-5-colorable, which implies $\chi''(G) \leq 5$, a contradiction.

3 Structures: The Proof of Theorem 4

3.1 Preliminaries

We first review some useful notations that were often used in many papers including [13–19].

Given a 2-connected Anicop graph G, by $v_1, v_2, \ldots, v_{|G|}$ we denote the vertices of G that lie in a clockwise sequence on the outer boundary. Let $\mathcal{V}[v_i, v_j] = \{v_i, v_{i+1}, \ldots, v_j\}$ and $\mathcal{V}(v_i, v_j) = \mathcal{V}[v_i, v_j]\backslash\{v_i, v_j\}$, where the subscripts are taken modulo $|G|$. Set $\mathcal{V}[v_i, v_i] = V(G)$ and $\mathcal{V}(v_i, v_i) = V(G) \setminus \{v_i\}$.

A vertex set $\mathcal{V}[v_i, v_j]$ is a *non-edge* if $j = i + 1 \pmod{|G|}$ and $v_i v_j \notin E(G)$, and is a *path* if $v_i v_{i+1} \cdots v_j$ (the subscripts are taken modulo $|G|$) forms a path. An edge $v_i v_j$ is a *chord* if $j = i + 1 \pmod{|G|}$. By $\mathcal{C}[v_i, v_j]$, we denote the set of chords xy with $x, y \in \mathcal{V}[v_i, v_j]$.

Let $v_i v_j$ and $v_k v_l$ be two chords in an Anicop graph G so that $v_i v_j$ crosses $v_k v_l$ and v_i, v_k, v_j and v_l lie in a clockwise sequence on the outer boundary of G. We say that $v_i v_j$ *co-crosses* $v_k v_l$, and $v_i v_j, v_k v_l$ are *co-crossed chords*, if $v_i v_k, v_k v_j, v_j v_l \in E(G)$, $l - j = k - i = 1 \pmod{|G|}$, and $j - k = 1$ and $d(v_k) = d(v_j) = 3$ (see the 1st picture of Fig. 3), or $j - k = 2$, $v_k v_{k+1}, v_{k+1} v_j \in E(G)$, $d(v_k) = d(v_j) = 4$ and $d(v_{k+1}) = 2$ (see the 2nd picture of Fig. 3).

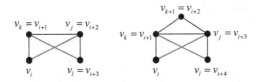

Fig. 3. $v_i v_j$ co-crosses $v_k v_l$

By the partial drawings of G as showed in Fig. 4, we define different types of *clusters* that will be frequently used in the following arguments. In any picture of this figure, vertices are all distinct, the edges drawn as crossed have to be crossed in G, and the curving edges are chords. Note that any graph in Fig. 4 contains a base graph as a subgraph.

We call H an *I-cluster* in G if H is either a left I^1-cluster, or a right I^1-cluster, or a left I^2-cluster, or a right I^2-cluster. The *II-cluster*, *III-cluster* and *IV-cluster* are defined similarly. The *width* of a cluster is the value of $|\mathcal{V}[v_L, v_R]|$, where L and R are the subscripts of the far left vertex and the far right vertex on the outer boundary (see in a clockwise direction from left to right). For convenience, we use $\{v_L, v_R\}_1$, $\{v_L, v_R\}_2$, $\{v_L, v_R\}_3$, and $\{v_L, v_R\}_4$ to represent a I-cluster, II-cluster, III-cluster, and IV-cluster, respectively. For example, the width of the left I^1-cluster $\{v_j, v_{i+3}\}_1$ is $(i + 3) - j + 1 = i - j + 4 \pmod{|G|}$, and the width of the right I^1-cluster $\{v_i, v_j\}_1$ is $j - i + 1 \pmod{|G|}$. Note that for a cluster, say a III-cluster for example, the left-type can be transferred to the right-type just by taking inversion.

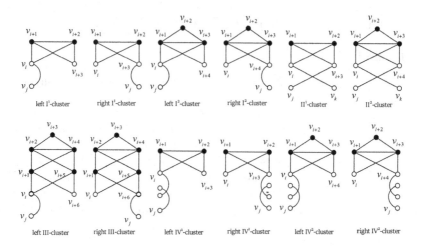

Fig. 4. The definitions of different types of clusters

The following three lemmas were originally proved for outer-1-plane graphs or Nicop graphs, and there is no double that their proofs also work for Anicop graphs.

Lemma 6 [19, Claim 1]. *Let v_i and v_j be vertices of a 2-connected outer-1-plane graph (or Anicop graph) G. If there is no crossed chord in $C[v_i, v_j]$ and no edge between $\mathcal{V}(v_i, v_j)$ and $\mathcal{V}(v_j, v_i)$, then $\mathcal{V}[v_i, v_j]$ is either a non-edge or a path.*

Lemma 7. *Let $v_i v_j$ and $v_k v_l$ with $i < k < j < l$ be two crossed chords in a 2-connected outer-1-plane graph (or Anicop graph) G with $\Delta(G) \leq 4$ so that $v_i v_j$ crosses $v_k v_l$ and there is no other pair of crossed chords contained in the drawing induced by $\mathcal{V}[v_i, v_l]$. We have*

(1) *at most one of $\mathcal{V}[v_i, v_k], \mathcal{V}[v_k, v_j]$ and $\mathcal{V}[v_j, v_l]$ is a non-edge [19, Claim 3];*
(2) *if one of $\mathcal{V}[v_i, v_k], \mathcal{V}[v_k, v_j]$ and $\mathcal{V}[v_j, v_l]$ is a non-edge, then G has a subgraph isomorphic to one of the configurations among (C1), (C2), and (C3) [19, Claims 2 and 4];*
(3) *if all of $\mathcal{V}[v_i, v_k], \mathcal{V}[v_k, v_j]$ and $\mathcal{V}[v_j, v_l]$ are paths, then either $v_i v_j$ co-crosses $v_k v_l$ in G, or G has a subgraph isomorphic to one of the configurations among (C1), (C2), (C3), (C4), and (C5) [19, Claims 2 and 5].*

Lemma 8 [15, Lemma 2.2]. *Let $\mathcal{V}[v_i, v_j]$ with $j - i \geq 3$ be a path in a 2-connected Nicop graph (or Anicop graph) G with $\Delta(G) \leq 4$. If there is no crossed chord in $C[v_i, v_j]$ and no edges between $\mathcal{V}(v_i, v_j)$ and $\mathcal{V}(v_j, v_i)$, then G contains (C1) or (C2).*

3.2 Proofs by Combinatorial Analyses

Let G be a 2-connected Anicop graph with $\Delta(G) \leq 4$. If G does not contain a crossing, then G is an outerplane graph, and the following is immediate.

Lemma 9 [11, Corollary 2.5]. *If G does not contain a crossing, then it contains (C1) or (C3).*

If G contains a crossing, then choose one pair of crossed chords $v_i v_j$ and $v_k v_l$ such that $v_i v_j$ crosses $v_k v_l$, and $\mathcal{C}[v_i, v_l]$ contains no other crossed chord besides $v_i v_j$ and $v_k v_l$. Applying Lemmas 6 and 7, one can conclude that $v_i v_j$ co-crosses $v_k v_l$ unless G contains one of the configurations among (C1), (C2), (C3), (C4), and (C5).

Hence in the following we assume that $v_i v_j$ co-crosses $v_k v_l$ with $1 = i < k < j < l$, and G does not contain any configurations among (C6)—(C11) (otherwise we win).

Since (C7) and (C7) are absent, $d(v_l) \geq 4$ and thus there is a chord $v_l v_s$ with $l < s \leq n$. In this case the drawing induced by $\mathcal{V}[v_i, v_l]$ and $v_l v_s$ is a I-cluster $\{v_i, v_s\}_1$. We make the following assumption, otherwise we can choose the shorter one I-cluster to replace $\{v_i, v_s\}_1$.

Assumption 1. $\{v_i, v_s\}_1$ *is the shortest I-cluster contained in the drawing induced by $\mathcal{V}[v_i, v_s]$.*

Lemma 10. *Suppose that v_a and v_b are two vertices with $l \leq a < b \leq s$. If there is no edge between $\mathcal{V}(v_a, v_b)$ and $\mathcal{V}(v_b, v_a)$, and there is a pair of chords $v_{i'} v_{j'}$ and $v_{k'} v_{l'}$ with $a \leq i' < k' < j' < l' \leq b$, then there is a II-cluster contained in the drawing induced by $\mathcal{V}[v_a, v_b]$ unless $\{i', l'\} = \{a, b\}$ and $v_{i'} v_{j'}$ co-crosses $v_{k'} v_{l'}$.*

Proof. Suppose that $v_{i'} v_{j'}$ does not co-cross $v_{k'} v_{l'}$. By Lemmas 6 and 7, there is another pair of crossed chords besides $v_{i'} v_{j'}$ and $v_{k'} v_{l'}$, say $v_{i''} v_{j''}$ and $v_{k''} v_{l''}$ with $i' \leq i'' < k'' < j'' < l' \leq l'$, in $\mathcal{C}[v_{i'}, v_{l'}]$. We choose $v_{i''} v_{j''}$ and $v_{k''} v_{l''}$ carefully so that there is no other pair of crossed chords in $\mathcal{C}[v_{i''}, v_{j''}]$ besides them. This implies that $v_{i''} v_{j''}$ co-crosses $v_{k''} v_{l''}$, because otherwise one of the configurations among (C1), (C2), (C3), (C4), and (C5) would appear by Lemmas 6 and 7. Since $\{i'', l''\} \neq \{i', l'\}$, $\{i'', l''\} \neq \{a, b\}$. By the absences of (C7) and (C9), and by Assumption 1, there are chords $v_{i''} v_{t''}$ and $v_{l''} v_{s''}$ with $l'' < t'' \leq b$ and $a \leq s'' < i''$. Therefore, a II-cluster $\{v_{s''}, v_{t''}\}_2$ is found in the drawing induced by $\mathcal{V}[v_a, v_b]$.

On the other hand, we assume that $v_{i'} v_{j'}$ co-crosses $v_{k'} v_{l'}$ but $\{i', l'\} \neq \{a, b\}$. Actually, one can see that $v_{i'} v_{j'}$ and $v_{k'} v_{l'}$ play the same role as $v_{i''} v_{j''}$ and $v_{k''} v_{l''}$ in the previous paragraph. Therefore, we can again find a II-cluster in the drawing induced by $\mathcal{V}[v_a, v_b]$.

In the following proofs, we distinguish two major cases.

The First Case: $v_l v_s$ is Non-crossed

Lemma 11. *There exists a II-cluster contained in the drawing induced by $\mathcal{V}[v_l, v_s]$.*

Proof. If there is no crossed chord in $\mathcal{C}[v_l, v_s]$, then $\mathcal{V}[v_l, v_s]$ is a path by Lemma 6. If $s - l = 2$, then $d(v_{l+1}) = 2$ and (C8) appears. If $s - l \geq 3$, then (C1) or (C2) appears by Lemma 8. Hence there is a pair of crossed chords $v_{i'}v_{j'}$ and $v_{k'}v_{l'}$ with $l \leq i' < k' < j' < l' \leq s$, and by Lemma 10, there is a II-cluster contained in the drawing induced by $\mathcal{V}[v_l, v_s]$ unless $\{i', l'\} = \{l, s\}$ and $v_{i'}v_{j'}$ co-crosses $v_{k'}v_{l'}$, which case would not occur because otherwise $d(v_l) \geq 5$.

By Lemma 11, there are chords $v_{i'}v_{j'}$ and $v_{k'}v_{l'}$ with $l < i' < k' < j' < l' < s$ so that $v_{i'}v_{j'}$ co-crosses $v_{k'}v_{l'}$, and moreover, there are chords $v_{i'}v_{t'}$ with $l' < t' \leq s$ and $v_{l'}v_{s'}$ with $l \leq s' < i'$. In other words, this structure is indeed a II-cluster $\{v_{s'}, v_{t'}\}_2$. Typically, the following assumption is natural.

Assumption 2. $\{v_{s'}, v_{t'}\}_2$ *is the shortest II-cluster contained in the drawing induced by* $\mathcal{V}[v_{s'}, v_{t'}]$.

Lemma 12. *The drawing induced by* $\mathcal{V}[v_{s'}, v_{t'}]$ *has a copy of* Π_3^2 *with handles* $v_{s'}$ *and* $v_{t'}$.

Proof. By Lemma 10 and by the fact that $\Delta(G) \leq 4$, there is no crossed chord in $\mathcal{C}[v_{l'}, v_{t'}]$, because otherwise we would find in the drawing induced by $\mathcal{V}[v_{s'}, v_{t'}]$ a shorter II-cluster than $\mathcal{V}[v_{s'}, v_{t'}]$, contradicting Assumption 2. By Lemma 6, $\mathcal{V}[v_{l'}, v_{t'}]$ is non-edge or path. If $\mathcal{V}[v_{l'}, v_{t'}]$ is a non-edge, then $d(v_{l'}) = 3$ and (C7) appears. Hence $\mathcal{V}[v_{l'}, v_{t'}]$ is a path. If $t' - l' \geq 3$, then by Lemma 8, G contains (C1) or (C2). If $t' - l' = 2$, then $d(v_{l'+1}) = 2$ and (C8) appears. Hence $t' - l' = 1$ and $v_{l'}v_{t'} \in E(G)$. By symmetry, $i' - s' = 1$ and $v_{s'}v_{i'} \in E(G)$. This implies that the drawing induced by $\mathcal{V}[v_{s'}, v_{t'}]$ contains either Π_3^1 or Π_3^2 with handles $v_{s'}$ and $v_{t'}$. However, Π_3^1 is forbidden in G, so it must be a copy of Π_3^2 with handles $v_{s'}$ and $v_{t'}$.

Since (C7) and (C9) are absent from G, there are chords $v_{t'}v_p$ and $v_{s'}v_q$ with $p \neq s', i'$ and $q \neq t', l'$. Since $s' \neq l$ and $v_{t'}v_p$ cannot cross $v_{s'}v_q$ by the definition of the Anicop graphs, either $t' < p \leq s$ or $l \leq r < s'$. We assume, without loss of the generality, the former, and in this case there is a III-cluster, say $\{v_{s'}, v_p\}_3$, contained in the drawing induced by $\mathcal{V}[v_{s'}, v_p]$. Again, we do the following natural assumption.

Assumption 3. $\{v_{s'}, v_p\}_3$ *is the shortest III-cluster contained in the drawing induced by* $\mathcal{V}[v_{s'}, v_p]$.

Lemma 13. *There is no crossed chord in* $\mathcal{C}[v_{t'}, v_p]$.

Proof. Suppose, to the contrary, that there is a pair of crossed chords there. By Lemma 11, there exists a II-cluster contained in the drawing induced by $\mathcal{V}[v_{t'}, v_p]$. Here one shall note that t' would not be incident with any crossed edge in the drawing induced by $\mathcal{V}[v_{t'}, v_p]$ by the definition of the Anicop graphs. Assume that $\{v_{s''}, v_{t''}\}$ with $t' < s'' < t''$ is the shortest II-cluster contained in the drawing induced by $\mathcal{V}[v_{t'}, v_p]$. By similar arguments as in the proof of Lemma 12, the drawing induced by $\mathcal{V}[v_{s''}, v_{t''}]$ contains a copy of Π_3^2 with handles

$v_{s''}$ and $v_{t''}$. Again, by the absences of (C7) and (C9) and by the definition of the Anicop graphs, there is a chord $v_{t''}v_{p'}$ with $t'' < p' \leq p$ or a chord $v_{s''}v_{q'}$ with $t' \leq q' < s''$. In each case we find in the drawing induced by $\mathcal{V}[v_{t'}, v_p] \subset \mathcal{V}[v_{s'}, v_p]$ a shorter III-cluster than $\{v_{s'}, v_p\}_3$, contradicting Assumption 3.

By Lemmas 6 and 13, $\mathcal{V}[v_{t'}, v_p]$ is a path. If $p - t' = 2$, then $d(v_{t'+1}) = 2$ and (C8) appears. If $p - t' \geq 3$, then (C1) or (C2) appears by Lemma 8. This is the end of the discussions for the first case.

The Second Case: $v_l v_s$ is Crossed. Suppose that $v_l v_s$ is crossed by a chord $v_r v_t$ with $l < r < s$, where $t = i$ is possible. Recall that when Assumption 1 is applied (in the proof of Lemma 10, for example), we actually only use the fact that there is no I-cluster in the drawing induced by $\mathcal{V}[v_l, v_s]$ with width at most $s - l$. Therefore, if we assume that there is no I-cluster in the drawing induced by $\mathcal{V}[v_s, v_t]$ with width at most $t - s$, then Lemma 10 still holds while l is replaced by s and s is replaced by t.

Lemma 14.

(1) *There is no crossed chord in $\mathcal{C}[v_l, v_r]$;*
(2) *There is no crossed chord in $\mathcal{C}[v_r, v_s]$;*
(3) *If there is no I-cluster in the drawing induced by $\mathcal{V}[v_s, v_t]$ with width at most $t - s$, then there is no crossed chord in $\mathcal{C}[v_s, v_t]$.*

Proof. The proof can be completed by similar arguments as we had presented in Sect. 3.2. We summary the idea for the readers.

Suppose that there is a pair of crossed chords $v_{i'}v_{j'}$ and $v_{k'}v_{l'}$ with $i' < k' < j' < l'$ in $\mathcal{C}[v_l, v_r]$ (or $\mathcal{C}[v_r, v_s]$, or $\mathcal{C}[v_s, v_t]$). If there is a II-cluster contained in the drawing induced by $\mathcal{V}[v_l, v_r]$ (or $\mathcal{V}[v_r, v_s]$, or $\mathcal{V}[v_s, v_t]$), then we choose one, say $\{v_{s'}, v_{t'}\}_2$, with the shortest width. Next, we prove that the drawing induced by $\mathcal{V}[v_{s'}, v_{t'}]$ has a copy of Π_3^2 with handles $v_{s'}$ and $v_{t'}$ (note that by the definition of the Anicop graphs, $s' \neq l, r, s$), based on which we can find a III-cluster in the drawing induced by $\mathcal{V}[v_l, v_r]$ (or $\mathcal{V}[v_r, v_s]$, or $\mathcal{V}[v_s, v_t]$). Again, choose the shortest III-cluster, say $\{v_{s'}, v_p\}_3$, and we can finally find some configuration that is forbidden in the graph induced by $\mathcal{V}[v_{s'}, v_p]$.

On the other hand, if no II-cluster is contained in the drawing induced by $\mathcal{V}[v_l, v_r]$ (or $\mathcal{V}[v_r, v_s]$, or $\mathcal{V}[v_s, v_t]$), then by Lemma 10, we conclude that $\{i', l'\} = \{l, r\}$ (or $\{i', l'\} = \{r, s\}$, or $\{i', l'\} = \{s, t\}$) and $v_{i'}v_{j'}$ co-crosses $v_{k'}v_{l'}$, which is impossible by the definition of the Anicop graphs.

Lemma 15. $r - l = 1$.

Proof. By Lemmas 6 and 14(1), $\mathcal{V}[v_l, v_r]$ is a non-edge or a path. If $\mathcal{V}[v_l, v_r]$ is a non-edge, then it is trivial that $r - l = 1$. If $\mathcal{V}[v_l, v_r]$ is a path, then by Lemma 8 and the absence of (C8), we also have $r - l = 1$.

Lemma 16. $\mathcal{V}[v_r, v_s]$ *is a path such that $s - r \leq 2$ and $v_r v_s \in E(G)$. Moreover, if $s - r = 2$, then $v_l v_r \in E(G)$.*

Proof. By Lemmas 6 and 14(2), $\mathcal{V}[v_r, v_s]$ is a non-edge or a path. If it is a non-edge, then $v_l v_r \in E(G)$ by the 2-connectedness of G. Hence $d(v_r) = 2$ by Lemma 15, and thus (C8) occurs. If $\mathcal{V}[v_r, v_s]$ is a path, then $s-r \leq 2$ by Lemma 8. If $s - r = 1$, then $v_r v_s \in E(G)$, because otherwise $v_l v_r \in E(G)$ and $d(v_r) = 2$ by the 2-connectedness of G and by Lemma 15, which implies that (C8) occurs. If $s - r = 2$, then $d(v_{r-1}) = 2$. Since (C1) is forbidden, $d(v_r) \geq 4$, which implies that $v_l v_r, v_r v_s \in E(G)$.

Lemma 17. $t \neq i = 1$.

Proof. Suppose, to the contrary, that $t = i$. If $s - r = 2$, then by Lemma 16, one can see that the drawing induced by $\mathcal{V}[v_i, v_s]$ contains a copy of $\Pi_1^2 \otimes \Pi_2^1$ or $\Pi_1^2 \otimes \Pi_2^2$ with crossed-linking handles v_i and v_s. If $s - r = 1$, then $v_l v_r \in E(G)$, because otherwise $d(v_r) = 2$. Since $v_i v_r \in E(G)$, (C8) appears. In this case, the drawing induced by $\mathcal{V}[v_i, v_s]$ has a copy of $\Pi_1^1 \otimes \Pi_2^1$ or $\Pi_1^1 \otimes \Pi_2^2$ with crossed-linking handles v_i and v_s. So we say that (C11) occurs.

Until now, we have actually proved the following result, which will be frequently used during the remaining arguments.

Lemma 18. *If $v_i v_j$ co-crosses $v_k v_j$ and $v_l v_s$ is a chord with $i < k < j < l < s$ such that $\{v_i, v_s\}_1$ is the shortest I-cluster contained in the drawing induced by $\mathcal{V}[v_i, v_s]$, then $v_l v_s$ is crossed by a chord $v_r v_t$ so that*

(1) $s < t \neq i$;
(2) $r - l = 1$;
(3) $\mathcal{V}[v_r, v_s]$ *is a path with $s - r \leq 2$ and $v_r v_s \in E(G)$, and if $s - r = 2$, then $v_l v_r \in E(G)$.*

Lemma 19. *There is a I-cluster in the drawing induced by $\mathcal{V}[v_s, v_t]$ with width at most $t - s$.*

Proof. If the opposite holds, then by Lemmas 6 and 14(3), there is no crossed chord in $\mathcal{C}[v_s, v_t]$, and thus $\mathcal{V}[v_s, v_t]$ is a non-edge or a path. If it is a non-edge, then $s - r = 1$ and $v_r v_s \in E(G)$, because otherwise $d(v_{s-1}) = 2$ and $d(v_s) = 3$, which implies that (C1) occurs. However, if $s - r = 1$ and $v_r v_s \in E(G)$, then $d(v_r) \leq 3$ and $d(v_s) = 2$ by Lemma 15, again implying the appearance of (C1). Hence $\mathcal{V}[v_s, v_t]$ is a path, and by Lemma 8, $t - s \leq 2$.

Suppose that $t - s = 2$. It follows that $d(v_{s+1}) = 2$. If $v_s v_t \in E(G)$, then $s - r = 1$ and $v_r v_s \in E(G)$, because otherwise $v_{s-1} v_s \in E(G)$ and $d(v_{s-1}) = 2$ by Lemma 16, which implies that (C3) appears. Similarly, $v_l v_r \in E(G)$, because otherwise $d(v_r) = 2$ and (C3) occurs again. In this case, the drawing induced by $\mathcal{V}[v_l, v_t]$ has a copy of $\Pi_1^1 \otimes \Pi_1^2$ with crossed-linking handles v_l and v_t, and thus (C11) occurs. On the other hand, if $v_s v_t \notin E(G)$, then $s - r = 2$ because otherwise $d(v_s) = 3$ and (C1) occurs. However, if $s - r = 2$, then $v_r v_s \in E(G)$ and $d(v_{s-1}) = 2$, which implies the appearance of (C3).

Hence $t - s = 1$ and $v_s v_t \in E(G)$. If $s - r = 2$, then by Lemma 16, the drawing induced by $\mathcal{V}[v_l, v_t]$ is a copy of Π_2^2 with handles v_l and v_t, and thus the drawing induced by $\mathcal{V}[v_i, v_t]$ has a copy of $\Pi_2^1 \circ \Pi_2^2$ or $\Pi_2^2 \circ \Pi_2^2$ with linking handles v_i and v_t. If $s - r = 1$, then $v_l v_r \in E(G)$ because otherwise $d(v_r) = 2$ and $d(v_s) = 3$, which implies that (C1) appears. In this case, the drawing induced by $\mathcal{V}[v_l, v_t]$ is a copy of Π_2^1 with handles v_l and v_t, and thus the drawing induced by $\mathcal{V}[v_i, v_t]$ has a copy of $\Pi_2^1 \circ \Pi_2^1$ or $\Pi_2^1 \circ \Pi_2^2$ with linking handles v_i and v_t. So we say that (C10) occurs.

Note that the drawing induced by $\mathcal{V}[v_i, v_l]$ and chords $v_l v_s, v_r v_t$ is a IV-cluster, say $\{v_i, v_t\}_4$, such that $t \neq i$, and the drawing induced by $\mathcal{V}[v_l, v_s]$ has the properties described by Lemmas 15 and 16. We call such a IV-cluster a determined IV-cluster. We do the following assumption.

Assumption 4. $\{v_i, v_t\}_4$ *is the shortest determined IV-cluster contained in the drawing induced by* $\mathcal{V}[v_i, v_t]$.

According to Lemma 19, we assume, without loss of generality, that $v_{i'} v_{j'}$ co-crosses $v_{k'} v_{l'}$ and $v_{l'} v_{s'}$ is a chord such that $s \leq i' < k' < j' < l' < s' \leq t$ (i.e., there is a I-cluster $\{v_{i'}, v_{s'}\}_1$ in the drawing induced by $\mathcal{V}[v_s, v_t]$). Clearly, we can carefully choose, in advance, i', k', j', l' and s' so that

Assumption 5. $\{v_{i'}, v_{s'}\}_1$ *is the shortest I-cluster contained in the drawing induced by* $\mathcal{V}[v_{i'}, v_{s'}]$.

By Lemma 18, $v_{l'} v_{s'}$ is crosses by a chord $v_{r'} v_{t'}$ with $l' < r' < s'$. If $s' < t' \leq t$, then there is a determined IV-cluster with width $t' - i' + 1 < t - i + 1$, say $\{v_{i'}, v_{t'}\}_4$, contained in the drawing induced by $\mathcal{V}[v_{i'}, v_{v_{t'}}]$, contradicting Assumption 4. Hence $s \leq t' < i'$.

By the absences of (C7) and (C9), there is a chord $v_{q'} v_{i'}$ with $t' \leq q' < i'$. If the I-cluster $\{v_{q'}, v_{l'}\}_1$ is the shortest one contained in the drawing induced by $\mathcal{V}[v_{q'}, v_{l'}]$, then by Lemma 18, $v_{q'} v_{i'}$ is crossed by a chord $v_{y'} v_{p'}$ with $t' \leq y' < q' < p' < i'$, and furthermore, there is a determined (left) IV-cluster with width $l' - y' + 1 < t - i + 1$, say $\{v_{y'}, v_{l'}\}_4$, contained in the drawing induced by $\mathcal{V}[v_{y'}, v_{v_{l'}}]$, contradicting Assumption 4. Hence there is a shorter I-cluster contained in the drawing induced by $\mathcal{V}[v_{q'}, v_{l'}]$. Among those I-clusters contained in the drawing induced by $\mathcal{V}[v_{q'}, v_{l'}]$, we choose the shortest one, say $\{v_{i''}, v_{s''}\}_1$ for example. Precisely, $v_{i''} v_{j''}$ co-crosses $v_{k''} v_{l''}$ and $v_{l''} v_{s''}$ is a chord with $q' \leq i'' < k'' < j'' < l'' < s'' \leq i'$. By Lemma 18, $v_{l''} v_{s''}$ is crossed by a chord $v_{r''} v_{t''}$ with $l'' < r'' < s''$. If $s'' < t'' \leq i'$, then there is a determined IV-cluster with width $t'' - i'' + 1 < t - i + 1$, say $\{v_{i''}, v_{t''}\}_4$, contained in the drawing induced by $\mathcal{V}[v_{i''}, v_{v_{t''}}]$, contradicting Assumption 4. Hence $q' \leq t'' < i''$. We reset $\{t', i', k', j', l', r', s'\} := \{t'', i'', k'', j'', l'', r'', s''\}$ and come back to the beginning of this paragraph. Since $s'' - t'' < s' - t'$ and the graph is finite, this iteration can stop somewhere.

References

1. Andersen, L.: Total colouring of simple graphs (in Danish). Master's thesis, University of Aalborg (1993)

2. Auer, C., et al.: Outer 1-planar graphs. Algorithmica **74**(4), 1293–1320 (2016)
3. Behzad, M.: Graphs and their chromatic numbers. Ph.D. thesis, MichiganState University (1965)
4. Borodin, O.: On the total coloring of planar graphs. Journal fur die Reine und Angewandte Mathematik **1989**(394), 180–185 (1989)
5. Kostochka, A.: The total coloring of a multigraph with maximal degree 4. Discrete Math. **17**(2), 161–163 (1977)
6. Kostochka, A.: The total chromatic number of any multigraph with maximum degree five is at most seven. Discrete Math. **162**(1–3), 199–214 (1996)
7. Rosenfeld, M.: On the total coloring of certain graphs. Israel J. Math. **9**(3), 396–402 (1971)
8. Sanders, D., Zhao, Y.: On total 9-coloring planar graphs of maximum degree seven. J. Graph Theory **31**(1), 67–73 (1999)
9. Vijayaditya, N.: On total chromatic number of a graph. J. Lond. Math. Soc. **s2-3**(3), 405–408 (1971)
10. Vizing, V.G.: Some unsolved problems in graph theory. Russ. Math. Surv. **23**(6), 125–141 (1968)
11. Wang, W., Zhang, K.: δ-matchings and edge-face chromatic numbers. Acta Mathematicae Applicatae Sinica **22**(2), 236–242 (1999). (in Chinese)
12. Wu, J.-L., Hu, D.: Total coloring of series-parallel graphs. Ars Combinatoria **73**, 215–217 (2004)
13. Zhang, X.: List total coloring of pseudo-outerplanar graphs. Discrete Math. **313**(20), 2297–2306 (2013)
14. Zhang, X.: The edge chromatic number of outer-1-planar graphs. Discrete Math. **339**(4), 1393–1399 (2016)
15. Zhang, X.: Total coloring of outer-1-planar graphs with near-independent crossings. J. Comb. Optim. **34**(3), 661–675 (2017)
16. Zhang, X., Lan, J., Li, B., Zhu, Q.: Light paths and edges in families of outer-1-planar graphs. Inf. Process. Lett. **136**, 83–89 (2018)
17. Zhang, X., Li, B.: Linear arboricity of outer-1-planar graphs. J. Oper. Res. Soc. China (2019). https://doi.org/10.1007/s40305-019-00243-2
18. Zhang, X., Liu, G.: Total coloring of pseudo-outerplanar graphs. arXiv 1108.5009 (2011)
19. Zhang, X., Liu, G., Wu, J.-L.: Edge covering pseudo-outerplanar graphs with forests. Discrete Math. **312**(18), 2788–2799 (2012)
20. Zhang, Z., Zhang, J., Wang, J.: The total chromatic number of some graphs. Sci. Sinica Ser. A **31**(12), 1434–1441 (1988)

Edge-Face List Coloring of Halin Graphs

Xin Jin[1], Min Chen[1(✉)] ⓘ, Xinhong Pang[1], and Jingjing Huo[2]

[1] Department of Mathematics, Zhejiang Normal University,
Jinhua 321004, China
chenmin@zjnu.cn
[2] Department of Mathematics, Hebei University of Engineering,
Handan 056038, China

Abstract. A plane graph G is k-edge-face colorable if the elements of $E(G) \cup F(G)$ can be colored with k colors such that any two adjacent or incident elements receive different colors. G is edge-face L-list colorable if for a given list assignment $L = \{L(x)|x \in E(G) \cup F(G)\}$, there exists a proper edge-face coloring π of G such that $\pi(x) \in L(x)$ for all $x \in E(G) \cup F(G)$. If G is edge-face L-list colorable for any list assignment with $|L(x)| = k$ for all $x \in E(G) \cup F(G)$, then G is edge-face k-choosable. The edge-face list chromatic number is defined to be the smallest integer k such that G admits an edge-face k-list coloring.

In this paper, we first use the famous Combinatorial Nullstellensatz to characterize the edge-face list chromatic number of wheel graphs by using Matlab. Then we show that every Halin graph G with $\Delta(G) \geq 6$ is edge-face $\Delta(G)$-choosable and this bound is sharp. Our proof demonstrates how edge-face choosability problems can numerically be approached by the use of computer algebra systems and the Combinatorial Nullstellensatz.

Keywords: Combinatorial Nullstellensatz · Halin graph · Wheel graph · Edge-face list coloring.

1 Introduction

All graphs considered in this paper are finite, loopless, and without multiple edges unless otherwise stated. A *plane* graph is a particular drawing in the Euclidean plane in such a way that its edges intersect only at their endpoints. For a plane graph G, we denote its vertex set, edge set, face set and maximum vertex degree by $V(G)$, $E(G)$, $F(G)$ and $\Delta(G)$, respectively.

The edge-face colorings of plane graphs were first studied by Jucovič [6] (1969) and Fiamčík [4] (1971). A plane graph G is *edge-face* k-*colorable* if the elements of $E(G) \cup F(G)$ can be colored with k colors such that any two adjacent or incident elements receive different colors. The *edge-face chromatic number* of

M. Chen—Supported by ZJNSFC (No. LY19A010015), NSFC (Nos. 11971437 and 11701136) and NSFHB (No. A2020402006).

© Springer Nature Switzerland AG 2020
Z. Zhang et al. (Eds.): AAIM 2020, LNCS 12290, pp. 481–491, 2020.
https://doi.org/10.1007/978-3-030-57602-8_43

G, denote by $\chi_{ef}(G)$, is defined to be the least integer k such that G is edge-face k-colorable.

A graph G is *edge-face L-list colorable* if for a given list assignment $L = \{L(x)|x \in E(G) \cup F(G)\}$, there is an edge-face coloring π such that $\pi(x) \in L(x)$, for any $x \in E(G) \cup F(G)$. If G is edge-face L-list colorable for any list assignment L with $|L(x)| = k$ for all $x \in E(G) \cup F(G)$, then G is *edge-face k-choosable*. The *edge-face list chromatic number* of G, denoted by $\chi^l_{ef}(G)$, is the smallest integer k such that G is edge-face k-choosable. Clearly, $\chi^l_{ef}(G) \geq \chi_{ef}(G)$ for any plane graph G. However, the equivalence dose not always hold. Wang and Lih [14] showed that there exists a plane graph G such that $\chi^l_{ef}(G) > \chi_{ef}(G)$. By considering colorings for $V(G)$ and $E(G)$, we can define analogous notions such as vertex k-choosability, and edge k-choosability.

In 1975, Mel'nikov [8] conjectured that every plane graph G is edge-face $(\Delta(G) + 3)$-colorable. Two similar, yet independent, proofs of the Mel'nikov's conjecture were given by Waller [12], Sanders and Zhao [9]. Both proofs made use of the Four-Color Theorem. Without employing the Four-Color Theorem, Wang and Lih [13] gave a new proof of this conjecture. In [14], Wang and Lih further extended this result to the edge-face list coloring situation by proving that every plane graph G is edge-face $(\Delta(G) + 3)$-choosable.

In 2001, Sanders and Zhao [11] proposed a strong conjecture which states that every plane graph G with $\Delta(G) \geq 3$ is edge-face $(\Delta(G) + 2)$-colorable. They confirmed it for the cases $\Delta(G) = 3$ and $\Delta(G) \geq 7$ in [10] and in [11], and left for the cases $\Delta \in \{4, 5, 6\}$ as a challenging open problem. Recently, Chen, Raspaud and Wang [3] settled the case $\Delta = 6$ for this conjecture. Now it remains open when $\Delta \in \{4, 5\}$.

A *Halin graph* is a plane graph G constructed as follows. Let T be a tree of order at least 4. All vertices of T are either of degree 1, called *leaves*, or of degree at least 3. Let C be a cycle connecting the leaves of T in such a way that C forms the boundary of the unbounded face. The tree T and the cycle C are called the *characteristic tree* and the *adjoint cycle* of G, respectively. We usually write $G = T \cup C$ to make the characteristic tree and the adjoint cycle of G explicit. For $n \geq 3$, the *wheel graph* W_n is a particular Halin graph whose characteristic tree is the complete bipartite graph $K_{1,n}$. As far as we know, there is no work concerning on the edge-face list chromatic number of wheel graphs and Halin graphs even with large maximum degree.

In this paper, we shall first characterize the edge-face list chromatic number of wheel graphs by using the powerful Combinatorial Nullstellensatz. Then we obtain an upper bound of the edge-face list choosability of Halin graph with given maximum degree.

2   Preliminaries

Now we collect the notation and basic definitions used in the subsequent sections. Let G be a plane graph. The unique unbounded face of G is called its *outer face*, denote by f^o, while other faces are *inner faces*. Edges of $E(f^o)$ are called *outer edges* of G and edges of $E(G) - E(f^o)$ are called *inner edges*. We may similarly define *outer vertices* and *inner vertices* of G. We say that an inner vertex is a *semi-leaf* if it is adjacent to an outer vertex. If a semi-leaf is adjacent to exactly one inner vertex, then it is called *good*. Two faces of G are said to be *adjacent* if they share at least one common boundary edge. We denote by f_e^o the inner face which is adjacent to f^o by a common outer edge e. An edge is said to be *incident* to a face if it lies on the boundary of the face. For $v \in V(G)$, let $d_G(v)$ denote the degree of a vertex v and we use $N_G(v)$ to denote the neighborhood of v in G: For $n \geq 3$, we use C_n to denote a cycle on n vertices. A vertex v is a *vertex of maximum degree* if $|N_G(v)| = \Delta(G)$. For $S \subseteq V(G)$, let $G[S]$ denote the subgraph of G induced by S. All notation not defined in this paper can be found in the book [2].

Let $P(x_1, x_2, \cdots, x_n)$ be a polynomial in n variables, where $n \geq 1$. Let $c_p(x_1^{k_1} x_2^{k_2} \ldots x_n^{k_n})$ denote the coefficient of the monomial $x_1^{k_1} x_2^{k_2} \ldots x_n^{k_n}$ in $P(x_1, x_2, \cdots, x_n)$, where k_i ($1 \leq i \leq n$) is a nonnegative integer. To derive our result, we need the following elegant formulation of the Combinatorial Nullstellensatz, which has wide application in list coloring, see [7].

Lemma 1 ([1], Combinatorial Nullstellensatz). *Let $\mathbb{F}$ be an arbitrary field, and let $P = P(x_1, x_2, \cdots, x_n)$ be a polynomial in $\mathbb{F}[x_1, x_2, \cdots, x_n]$. Suppose the degree $deg(P)$ of P equals $\sum_{i=1}^{n} k_i$, where each k_i is a nonnegative integer, and suppose $c_p(x_1^{k_1} x_2^{k_2} \ldots x_n^{k_n}) \neq 0$. If $S_1, S_2, \cdots, S_n$ are subsets of $\mathbb{F}$ with $|S_i| > k_i$, there are $s_1 \in S_1, s_2 \in S_2, \cdots, s_n \in S_n$ so that $P(s_1, s_2, \cdots, s_n) \neq 0$.*

3   Wheel Graphs

This section is devoted to the study of the edge-face list chromatic number of wheel graphs.

Theorem 1. *For the wheel W_n, we have*

$$\chi_{ef}^l(W_n) = \begin{cases} 5, & \Delta(W_n) \in \{3, 4, 5\}; \\ \Delta(W_n), & \Delta(W_n) \geq 6. \end{cases}$$

Proof. Write $W_n = K_{1,n} \cup C_n$. Let v denote the inner vertex of W_n, and we use $v_1, v_2, \cdots, v_n$ to denote all outer vertices of W_n in clockwise order. Clearly, $\Delta(W_n) = n$. For each $x \in E(W_n) \cup F(W_n)$, let $L(x)$ be a list assignment such that

$$|L(x)| = \begin{cases} 5, & n \in \{3, 4, 5\}; \\ n, & n \geq 6. \end{cases}$$

Clearly, $|L(x)| \geq n$ for each $x \in E(W_n) \cup F(W_n)$. We need to discuss two cases below in terms of the value of n.

Case 1. $n \geq 6$.

Since $\chi_{ef}^l(W_n) \geq \chi_{ef}(W_n) \geq \Delta(W_n) = n$, it suffices to prove that W_n has an edge-face L-coloring. First, we choose a possible color $a_i \in L(vv_i)$ for each inner edge vv_i such that $a_i \neq a_j$ for each pair $\{i, j\} \subset \{1, 2, \cdots, n\}$. Then, we assign a color $b \in L(f^o)$ to the outer face f^o. In what follows, let $L'(v_iv_{i+1}) = L(v_iv_{i+1}) \backslash \{a_i, a_{i+1}, b\}$, where $i \in \{1, 2, \cdots, n\}$ and the indices are taken modulo n. Obviously, $|L'(v_iv_{i+1})| \geq n - 3 \geq 3$ for each $i \in \{1, 2, \cdots, n\}$. So it enables us to properly color each outer edge v_iv_{i+1} with $c_i \in L'(v_iv_{i+1})$ such that $c_i \neq c_{i+1}$. Next, we write $f_i = [vv_iv_{i+1}]$, and let $L'(f_i) = L(f_i) \backslash \{a_i, a_{i+1}, c_i, b\}$. If $n \geq 7$, then $|L'(f_i)| \geq 3$. If $n = 6$, then $|L'(f_i)| \geq 2$. So in each case, we can always find a proper way to color each f_i with a color belonging to $L'(f_i)$ such that adjacent inner faces have distinct colors. Therefore, we obtain an edge-face L-coloring for W_n.

Case 2. $n \in \{3, 4, 5\}$.

In each of following case, we will construct a new graph G' so that the vertex set of G' is defined to be $V(G') = \{y | y \in E(W_n) \cup F(W_n)\}$ and two vertices, say y_i and y_j, are adjacent in G' if and only if their corresponding elements in W_n are adjacent or incident. We notice that an edge-face list 5-coloring of W_n is equivalent to a vertex list 5-coloring of G'.

Next, for each $n \in \{3, 4, 5\}$, we are firstly going to show that G' has a vertex list 5-coloring, which implies that $\chi_{ef}^l(W_n) \leq 5$.

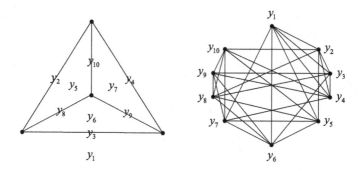

Fig. 1. W_3 and the new graph G'.

- $n = 3$. It is obvious that W_3 is isomorphic to the complete graph K_4. Moreover, G' is 6-regular which has 10 vertices, depicted in Fig. 1.

  Let $i \in \{1, 2, \cdots, 10\}$. By definition, $|L(y_i)| = 5$ for each $y_i \in V(G')$. Let S_i denote the available color set of y_i. Thus, $|S_i| = 5$. Associate with y_i a variable x_i. Based on the coloring condition, that is, adjacent vertices have different colors, we obtain the following polynomial Q_1:

  $$Q_1(x_1, x_2, \cdots, x_{10}) =$$

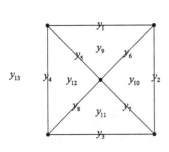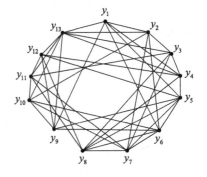

Fig. 2. W_4 and the new graph G'.

$(x_1-x_2)(x_1-x_3)(x_1-x_4)(x_1-x_5)(x_1-x_6)(x_1-x_7)(x_2-x_3)(x_2-x_4)(x_2-x_5)$
$(x_2-x_8)(x_2-x_{10})(x_3-x_4)(x_3-x_6)(x_3-x_8)(x_3-x_9)(x_4-x_7)(x_4-x_9)$
$(x_4-x_{10})(x_5-x_6)(x_5-x_7)(x_5-x_8)(x_5-x_{10})(x_6-x_7)(x_6-x_8)(x_6-x_9)$
$(x_7-x_9)(x_7-x_{10})(x_8-x_9)(x_8-x_{10})(x_9-x_{10})$. By Matlab, we cal-
culate that $c_{Q_1}(x_1^3x_2^4x_3^2x_4^3x_5^4x_6^3x_7^3x_8^1x_9^4x_{10}^3) = -1$. Since $k_i < |S_i|$ for each
$i \in \{1, 2, \cdots, 10\}$, by Lemma 1, we get a desired vertex list 5-coloring of
G'.

- $n = 4$. Then G' is a graph with 13 vertices, depicted in Fig. 2.
  Similarly, we use S_i to denote the available color set of each $y_i \in V(G')$.
  Again, $|S_i| = 5$ due to $|L(y_i)| = 5$. Associate with y_i a variable x_i. By the
  coloring condition, we have the following polynomial Q_2:
  $Q_2(x_1, x_2, \cdots, x_{13}) =$
  $(x_1-x_2)(x_1-x_4)(x_1-x_5)(x_1-x_6)(x_1-x_9)(x_1-x_{13})(x_2-x_3)(x_2-x_6)(x_2-x_7)$
  $(x_2-x_{10})(x_2-x_{13})(x_3-x_4)(x_3-x_7)(x_3-x_8)(x_3-x_{11})(x_3-x_{13})(x_4-x_5)$
  $(x_4-x_8)(x_4-x_{12})(x_4-x_{13})(x_5-x_6)(x_5-x_7)(x_5-x_8)(x_5-x_9)(x_5-x_{12})$
  $(x_6-x_7)(x_6-x_8)(x_6-x_9)(x_6-x_{10})(x_7-x_8)(x_7-x_{10})(x_7-x_{11})(x_8-x_{11})$
  $(x_8-x_{12})(x_9-x_{10})(x_9-x_{12})(x_9-x_{13})(x_{10}-x_{11})(x_{10}-x_{13})(x_{11}-x_{12})$
  $(x_{11}-x_{13})(x_{12}-x_{13})$
  By Matlab, we are easy to derive that $c_{Q_2}(x_1^4x_2^4x_3^4x_4^4x_5^4x_6^3x_7^2x_8^2x_9^4x_{10}^3x_{11}^2x_{12}^2x_{13}^4)$
  $= -2$. Therefore, by Lemma 1, we obtain a desired vertex list 5-coloring of
  G'.

- $n = 5$. Then G' has 16 vertices, depicted in Fig. 3. We may similarly define S_i
  and know that $|S_i| = 5$, where $i \in \{1, 2, \cdots, 16\}$. Associate with y_i a variable
  x_i. By the coloring condition, we deduce the following polynomial Q_3:
  $Q_3(x_1, x_2, \cdots, x_{16}) =$
  $(x_1-x_2)(x_1-x_5)(x_1-x_6)(x_1-x_{10})(x_1-x_{11})(x_1-x_{16})(x_2-x_3)(x_2-x_6)$
  $(x_2-x_7)(x_2-x_{12})(x_2-x_{16})(x_3-x_4)(x_3-x_7)(x_3-x_8)(x_3-x_{13})(x_3-x_{16})$
  $(x_4-x_5)(x_4-x_8)(x_4-x_9)(x_4-x_{14})(x_4-x_{16})(x_5-x_9)(x_5-x_{10})(x_5-x_{15})$
  $(x_5-x_{16})(x_6-x_7)(x_6-x_8)(x_6-x_9)(x_6-x_{10})(x_6-x_{11})(x_6-x_{12})(x_7-x_8)$
  $(x_7-x_9)(x_7-x_{10})(x_7-x_{12})(x_7-x_{13})(x_8-x_9)(x_8-x_{10})(x_8-x_{13})(x_8-x_{14})$
  $(x_9-x_{10})(x_9-x_{14})(x_9-x_{15})(x_{10}-x_{11})(x_{10}-x_{15})(x_{11}-x_{12})(x_{11}-x_{15})$
  $(x_{11}-x_{16})(x_{12}-x_{13})(x_{12}-x_{16})(x_{13}-x_{14})(x_{13}-x_{16})(x_{14}-x_{15})$

$(x_{14} - x_{16})(x_{15} - x_{16})$

By using Matlab, we have that $c_{Q_3}(x_1^3 x_2^4 x_3^4 x_4^4 x_5^4 x_6^3 x_7^4 x_8^3 x_9^2 x_{10}^1 x_{11}^4 x_{12}^4 x_{13}^4 x_{14}^4 x_{15}^4 x_{16}^4) = -15$. Hence, we obtain a vertex list 5-coloring of G' by applying Lemma 1.

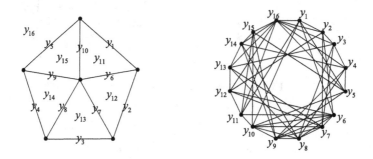

Fig. 3. W_5 and the new graph G'.

On the other hand, the following proposition tells us that for each $n \in \{3, 4\}$, W_n cannot be edge-face 4-colorable.

Proposition 1. *For $n \in \{3, 4\}$, $\chi_{ef}(W_n) \neq 4$.*

Proof. Noting that each W_n contains a 3-face as a subgraph, say $f = [v v_1 v_n]$, we see that these four elements $v v_1, v v_n, v_1 v_n$ and f need four distinct colors. Thus, $\chi_{ef}(W_n) \geq 4$. In what follows, we will make use of contradictions to demonstrate that W_n cannot be edge-face 4-colorable.

Suppose to the contrary that W_n admits an edge-face 4-coloring π. Let $C = \{1, 2, 3, 4\}$ denote its color set. For $x \in E(W_n) \cup F(W_n)$, if $n = 3$, then we take a look of the left graph in Fig. 1. W.l.o.g., assume that $\pi(y_1) = 1$, $\pi(y_5) = 2$, $\pi(y_6) = 3$ and $\pi(y_7) = 4$. Then $\pi(y_3) \in \{2, 4\}$. By symmetry, assume that $\pi(y_3) = 2$. It follows immediately that $\pi(y_9) = 1$ and $\pi(y_{10}) = 3$, implying that $\pi(y_4) = 2$, which is impossible.

Now suppose that $n = 4$. We look at the left graph in Fig. 2. W.l.o.g., assume that $\pi(y_1) = 1$, $\pi(y_5) = 2$, $\pi(y_6) = 3$ and $\pi(y_9) = 4$. So $\pi(y_{13}) \in \{2, 3\}$. By symmetry, assume that $\pi(y_{13}) = 2$. Then we obtain that $\pi(y_2) = 4$, $\pi(y_7) = 1$ and $\pi(y_8) = 4$. It follows that $\pi(y_3) = \pi(y_{11}) = 3$, a contradiction. $\square$

Since $\chi_{ef}^l(W_5) \geq \chi_{ef}(W_5) \geq 5$, together with Proposition 1, we derive that $\chi_{ef}^l(W_n) \geq 5$ for each case $n \in \{3, 4, 5\}$, Combining the previous discussion, we obtain that $\chi_{ef}^l(W_n) = 5$, and therefore we complete the proof of Theorem 1. $\square$

4   Halin Graphs

Gong and Wu [5] proved that every Halin graph G with $\Delta(G) \geq 6$ is edge-face $\Delta(G)$-colorable. Our purpose in this section is to extend this result to the list edge-face coloring situation. The main theorem is the following.

Theorem 2. *Every Halin graph G with $\Delta(G) \geq 6$ is edge-face $\Delta(G)$-choosable.*

Proof. Let $G = T \cup C$ be a Halin graph with $\Delta(G) \geq 6$. We prove the theorem by induction on $|V(G)|$. Obviously, $|V(G)| \geq 7$. Moreover, if $|V(G)| = 7$, then G is a wheel graph, and hence $\chi^l_{ef}(G) = \Delta(G)$ by Theorem 1. Next, we assume that $|G| \geq 8$, and Theorem 2 is true for the case that $8 \leq |V(G)| \leq k$. Now we consider the case that $|V(G)| = k + 1$. By Theorem 1, we may further suppose that G is not a wheel graph.

Proposition 2. *There are at least two good semi-leaves in G.*

Proof. Let $P = z_1 z_2 \cdots z_m$ denote the longest path in $G - V(C)$. If z_1 is adjacent to another inner vertex, say z', then $z' \notin V(P)$ and thus there exists a path $z' z_1 \cdots z_m$ which is longer than P, a contradiction. Hence, z_1 is a good semi-leaf. Similarly, we may prove that z_m is also a good semi-leaf.   □

By Proposition 2, we may choose a good semi-leaf v satisfying that $d_G(v) < \Delta(G)$ unless all good semi-leaves are of degree $\Delta(G)$. For convenience, denote its neighbors located on C by $v_1, v_2, \cdots, v_t$ in a cyclic order. By definition, we see that $d_G(v) = t + 1$.

In what follows, for each $x \in E(G) \cup F(G)$, let L be an assignment of G that satisfies $|L(x)| = \Delta(G)$. To complete the proof of Theorem 2, we need to handle two cases below depending on the value of t.

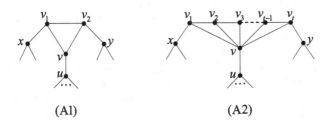

(A1)   (A2)

Fig. 4. Two configurations (A1) and (A2).

Case 1. G contains the configuration (A1).

Let $G^* = G - \{v_1, v_2\} + \{xv, yv\}$. Then G^* is also a Halin graph with $\Delta(G^*) = \Delta(G)$ and $|V(G^*)| = |V(G)| - 2 = k - 1 < k$. We use f^* and f^*_e to denote the outer face of G^* and the adjacent face to f^* by a common edge e, respectively. Now, we define an edge-face list assignment L^* of G^* as follows:

$L^*(f^*) = L(f^o);$

$L^*(f^*_{xv}) = L(f^o_{xv_1});$

$L^*(f^*_{yv}) = L(f^o_{yv_2});$

$L^*(xv) = L(xv_1);$

$L^*(yv) = L(yv_2);$

$L^*(s) = L(s)$, for $s \in E(G^*) \cup F(G^*) \setminus \{f^*, f^*_{xv}, f^*_{yv}, xv, yv\}$.

By inductive hypothesis, G^* has an edge-face L^*-coloring π^*. Then we will show how to extend π^* to G so that G has an edge-face L-coloring π. Let $\mathcal{A} = \{xv_1, f^o_{xv_1}, yv_2, f^o_{yv_2}, f^o\}$ and $\mathcal{S} = \{vv_1, v_1v_2, vv_2, f^o_{v_1v_2}\}$. First, we assign the colors of f^*, xv, f^*_{xv}, yv, f^*_{yv} in G^* to their corresponding elements f^o, xv_1, $f^o_{xv_1}$, yv_2, $f^o_{yv_2}$ in G. For any $s \in E(G) \cup F(G) \setminus (\mathcal{A} \cup \mathcal{S})$, let $\pi(s) = \pi^*(s)$. Next, we only need to show that the elements in $\mathcal{S}$ can also be given proper colors from their own color list L.

For the sake of discussion, let $s_1 = vv_1$, $s_2 = v_1v_2$, $s_3 = vv_2$ and $s_4 = f^o_{v_1v_2}$. Namely, $\mathcal{S} = \{s_1, s_2, s_3, s_4\}$. For each $i \in \{1, 2, 3, 4\}$, we denote the available color set of s_i in G by $A(s_i)$.

It is straightforward that $A(s_1) = L(s_1) \setminus \{\pi(xv_1), \pi(f^o_{xv_1}), \pi(uv)\}$, $A(s_2) = L(s_2) \setminus \{\pi(xv_1), \pi(yv_2), \pi(f^o)\}$, $A(s_3) = L(s_3) \setminus \{\pi(yv_2), \pi(f^o_{yv_2}), \pi(uv)\}$, and $A(s_4) = L(s_4) \setminus \{\pi(f^o), \pi(f^o_{xv_1}), \pi(f^o_{yv_2})\}$. Let $i \in \{1, 2, 3, 4\}$. Since $|L(s_i)| = \Delta(G) \geq 6$, one may deduce that $|A(s_i)| \geq 3$. Obviously, if $|A(s_i)| \geq 4$ for some fixed i, then we can easily color all the elements of $\mathcal{S}$ properly. In subsequent discussion, it suffices to deal with the case that $|A(s_i)| = 3$ for each $i \in \{1, 2, 3, 4\}$.

Case 1a. There exists a pair $\{i, j\} \subseteq \{1, 2, 3, 4\}$ such that $A(s_i) \neq A(s_j)$.

W.l.o.g., assume that $A(s_1) \neq A(s_2)$. Then we may color s_1 with $a \in A(s_1) \setminus A(s_2)$, s_3 with $b \in A(s_3) \setminus \{a\}$, s_4 with $c \in A(s_4) \setminus \{a, b\}$, and finally color s_2 with $d \in A(s_2) \setminus \{b, c\}$. It is easy to verify that the resulting coloring of G is an edge-face L-coloring.

Case 1b. $A(s_i) = A(s_j)$ for each pair $\{i, j\} \subseteq \{1, 2, 3, 4\}$.

For each $i \in \{1, 2, 3, 4\}$, let $A(s_i) = \{c_1, c_2, c_3\}$. Since xv, yv, uv are mutually adjacent in G^*, we may set $\pi(xv_1) = 1$, $\pi(yv_2) = 2$ and $\pi(uv) = 3$. Noting that f^*, f^*_{xv}, f^*_{yv} are also mutually adjacent in G^*, we may set $\pi(f^o) = a$, $\pi(f^o_{xv_1}) = b$ and $\pi(f^o_{yv_2}) = c$. It follows that $L(s_1) = \{c_1, c_2, c_3, 1, 3, b\}$, $L(s_2) = \{c_1, c_2, c_3, 1, 2, a\}$, $L(s_3) = \{c_1, c_2, c_3, 2, 3, c\}$ and $L(s_4) = \{c_1, c_2, c_3, a, b, c\}$. Note that both x and y are 3-vertices. We denote by x_1, x_2 the other two neighbors of x different from v_1 and y_1, y_2 the other two neighbors of y different from v_2. Firstly, we recolor xv_1 by $\alpha \in L(xv_1) \setminus \{1, a, b, \pi(xx_1), \pi(xx_2)\}$. Then, color s_1 with 1, s_2 with $\beta \in \{c_1, c_2, c_3\} \setminus \{\alpha\}$. Afterwards, we may select two possible colors belonging to $\{c_1, c_2, c_3\} \setminus \{\beta\}$ to color s_3, s_4 in succession.

Case 2. G contains the configuration (A2).

At this moment, $t \geq 3$. Let $G^* = G - \{v_2\} + \{v_1v_3\}$. It is clear that G^* is a Halin graph with $|V(G^*)| = |V(G)| - 1 = k$. By the choice of v, we are sure that $\Delta(G^*) = \Delta(G)$. Similarly, denote by f^* and f^*_e the outer face of G^* and the adjacent face of f^* sharing an edge e, respectively.

We define an edge-face list assignment L^* of G^* as follows:

$L^*(f^*) = L(f^o)$;
$L^*(f^*_{v_1v_3}) = L(f^o_{v_1v_2})$;
$L^*(v_1v_3) = L(v_1v_2)$;
$L^*(s) = L(s)$, for $s \in E(G^*) \cup F(G^*) \setminus \{f^*, f^*_{v_1v_3}, v_1v_3\}$.

By inductive hypothesis, G^* has an edge-face L^*-coloring π^*. Then we shall obtain an edge-face L-coloring π for G. First, we erase both colors of v_1v_3 and $f^*_{v_1v_3}$. Let $\mathcal{A} = \{v_1v_2, v_2v_3, vv_2, f^o_{v_1v_2}, f^o_{v_2v_3}\}$. Then, let $\pi(f^o) = \pi^*(f^*)$ and $\pi(s) = \pi^*(s)$ for each $s \in E(G) \cup F(G) \setminus \mathcal{A}$. In the following, we are going to show that all elements belonging to $\mathcal{A}$ can be given a proper color from their own color list L.

We first color vv_2 with $a \in L(vv_2) \setminus \{\pi(vv_1), \pi(vv_3), \cdots, \pi(vv_t), \pi(uv)\}$. Then color $f^o_{v_1v_2}$ with $b \in L(f^o_{v_1v_2}) \setminus \{a, \pi(vv_1), \pi(f^o), \pi(f^o_{xv_1})\}$. Let $A(v_1v_2) = L(v_1v_2) \setminus \{a, b, \pi(xv_1), \pi(vv_1), \pi(f^o)\}$, $A(v_2v_3) = L(v_2v_3) \setminus \{a, \pi(vv_3), \pi(v_3v_4), \pi(f^o)\}$ and $A(f^o_{v_2v_3}) = L(f^o_{v_2v_3}) \setminus \{a, b, \pi(vv_3), \pi(f^o), \pi(f^o_{v_3v_4})\}$. Note that $|A(v_1v_2)| \geq 1$, $|A(v_2v_3)| \geq 2$ and $|A(f^o_{v_2v_3})| \geq 1$.

- If $|A(f^o_{v_2v_3})| \geq 2$, then we can properly color v_1v_2, v_2v_3, $f^o_{v_2v_3}$ in succession.
- If $|A(v_1v_2)| \geq 2$, then we can properly color $f^o_{v_2v_3}$, v_2v_3, v_1v_2 in succession.
- If $|A(v_2v_3)| \geq 3$, then we can properly color v_1v_2, $f^o_{v_2v_3}$, v_2v_3 in succession.

So now, we may assume that $|A(v_1v_2)| = |A(f^o_{v_2v_3})| = 1$ and $|A(v_2v_3)| = 2$. Furthermore, it can be deduced that $A(v_1v_2) \neq A(f^o_{v_2v_3})$ since otherwise it is possible to properly color v_1v_2, $f^o_{v_2v_3}$, v_2v_3 in order. W.l.o.g., assume that $A(v_1v_2) = \{\alpha\}$ and $A(f^o_{v_2v_3}) = \{\beta\}$. At this moment, one may deduce that $A(v_2v_3) = \{\alpha, \beta\}$. Now $L(v_1v_2) = \{a, b, \alpha, \pi(xv_1), \pi(vv_1), \pi(f^o)\}$, $L(f^o_{v_2v_3}) = \{a, b, \beta, \pi(vv_3), \pi(f^o_{v_3v_4}), \pi(f^o)\}$, and $L(v_2v_3) = \{a, \alpha, \beta, \pi(vv_3), \pi(v_3v_4), \pi(f^o)\}$. We may color both v_1v_2 and $f^o_{v_2v_3}$ with b, and v_2v_3 with α, and then reassign a color belonging to $L(f^o_{v_1v_2}) \setminus \{a, b, \pi(f^o_{xv_1}), \pi(f^o), \pi(vv_1)\}$ to $f^o_{v_1v_2}$. It is easy to verify that the obtained coloring is an edge-face L-coloring of G.

This completes the proof of Theorem 2.   □

Appendix

% Input
syms x_1 x_2 x_3 x_4 x_5 x_6 x_7 x_8 x_9 x_{10}
% Theorem 1 ($n = 3$)
$Q_1 = (x_1 - x_2) * (x_1 - x_3) * (x_1 - x_4) * (x_1 - x_5) * (x_1 - x_6) * (x_1 - x_7) * (x_2 - x_3) *$
$\quad (x_2 - x_4) * (x_2 - x_5) * (x_2 - x_8) * (x_2 - x_{10}) * (x_3 - x_4) * (x_3 - x_6) * (x_3 - x_8) *$
$\quad (x_3 - x_9) * (x_4 - x_7) * (x_4 - x_9) * (x_4 - x_{10}) * (x_5 - x_6) * (x_5 - x_7) *$
$\quad (x_5 - x_8) * (x_5 - x_{10}) * (x_6 - x_7) * (x_6 - x_8) * (x_6 - x_9) * (x_7 - x_9) *$
$\quad (x_7 - x_{10}) * (x_8 - x_9) * (x_8 - x_{10}) * (x_9 - x_{10})$

$c_1 \quad = \mathrm{diff}(\mathrm{diff}(\mathrm{diff}(\mathrm{diff}(\mathrm{diff}(\mathrm{diff}(\mathrm{diff}(\mathrm{diff}(\mathrm{diff}(\mathrm{diff}(Q_1, x_1, 3), x_2, 4), x_3, 2), x_4, 3),$
$x_5, 4)$

$\quad , x_6, 3), x_7, 3), x_8, 1), x_9, 4), x_{10}, 3)$ /factorial(3)/factorial(4)/factorial(2)
$\quad$ /factorial(3)/factorial(4)/factorial(3)/factorial(3)/factorial(1)/factorial(4)
$\quad$ /factorial(3)

% Input
syms x_1 x_2 x_3 x_4 x_5 x_6 x_7 x_8 x_9 x_{10} x_{11} x_{12} x_{13}
% Theorem 1 ($n = 4$)

$$Q_2 = (x_1-x_2)*(x_1-x_4)*(x_1-x_5)*(x_1-x_6)*(x_1-x_9)*(x_1-x_{13})*(x_2-x_3)*$$
$$(x_2-x_6)*(x_2-x_7)*(x_2-x_{10})*(x_2-x_{13})*(x_3-x_4)*(x_3-x_7)*$$
$$(x_3-x_8)*(x_3-x_{11})*(x_3-x_{13})*(x_4-x_5)*(x_4-x_8)*(x_4-x_{12})*$$
$$(x_4-x_{13})*(x_5-x_6)*(x_5-x_7)*(x_5-x_8)*(x_5-x_9)*(x_5-x_{12})*$$
$$(x_6-x_7)*(x_6-x_8)*(x_6-x_9)*(x_6-x_{10})*(x_7-x_8)*(x_7-x_{10})*$$
$$(x_7-x_{11})*(x_8-x_{11})*(x_8-x_{12})*(x_9-x_{10})*(x_9-x_{12})*(x_9-x_{13})*$$
$$(x_{10}-x_{11})*(x_{10}-x_{13})*(x_{11}-x_{12})*(x_{11}-x_{13})*(x_{12}-x_{13})$$

c_2 $= \mathrm{diff}(\mathrm{diff}(\mathrm{diff}(\mathrm{diff}(\mathrm{diff}(\mathrm{diff}(\mathrm{diff}(\mathrm{diff}(\mathrm{diff}(\mathrm{diff}(\mathrm{diff}(\mathrm{diff}(\mathrm{diff}(Q_2,x_1,4),x_2,4),$
$x_3,4)$

$,x_4,4),x_5,4),x_6,3),x_7,2),x_8,2),x_9,4),x_{10},3),x_{11},2),x_{12},2),x_{13},4)$
$/\mathrm{factorial}(4)/\mathrm{factorial}(4)/\mathrm{factorial}(4)/\mathrm{factorial}(4)/\mathrm{factorial}(4)/\mathrm{factorial}(3)$
$/\mathrm{factorial}(2)/\mathrm{factorial}(2)/\mathrm{factorial}(4)/\mathrm{factorial}(3)/\mathrm{factorial}(2)/\mathrm{factorial}(2)$
$/\mathrm{factorial}(4)$

% Input
syms x_1 x_2 x_3 x_4 x_5 x_6 x_7 x_8 x_9 x_{10} x_{11} x_{12} x_{13} x_{14} x_{15} x_{16}
% Theorem 1 ($n = 5$)

$A = (x_1 - x_{16})*(x_2 - x_{16})*(x_3 - x_{16})*(x_4 - x_{16})*(x_5 - x_{16})*(x_{11} - x_{16})*$
$(x_{12}-x_{16})*(x_{13}-x_{16})*(x_{14}-x_{16})*(x_{15}-x_{16})*(x_5-x_{15})*(x_9-x_{15})*(x_{10}-$
$x_{15})*(x_{11}-x_{15})*(x_{14}-x_{15})*(x_4-x_{14})*(x_8-x_{14})*(x_9-x_{14})*(x_{13}-x_{14})*$
$(x_3-x_{13})*(x_7-x_{13})*(x_8-x_{13})*(x_{12}-x_{13})*(x_2-x_{12})*(x_6-x_{12})*(x_7-$
$x_{12})*(x_{11}-x_{12})*(x_1-x_{11})*(x_6-x_{11})*(x_{10}-x_{11});$
$\mathrm{g} = \mathrm{diff}(\mathrm{A}, x_{16}, 4);$
$\mathrm{f} = \mathrm{diff}(\mathrm{g}, x_{15}, 4);$
$\mathrm{m} = \mathrm{diff}(\mathrm{f}, x_{14}, 4);$
$\mathrm{l} = \mathrm{diff}(\mathrm{m}, x_{13}, 4);$
$\mathrm{q} = \mathrm{diff}(\mathrm{l}, x_{12}, 4);$
$\mathrm{s} = \mathrm{diff}(\mathrm{q}, x_{11}, 4);$
$G = (x_1-x_5)*(x_4-x_5)*(x_5-x_9)*(x_5-x_{10})*(x_3-x_4)*(x_4-x_8)*(x_4-x_9)*$
$(x_2-x_3)*(x_3-x_7)*(x_3-x_8)*(x_1-x_2)*(x_2-x_6)*(x_2-x_7)*(x_1-x_6)*(x_1-x_{10});$
$\mathrm{t} = \mathrm{diff}(\mathrm{s}*G, x_5, 4);$
$\mathrm{r} = \mathrm{diff}(\mathrm{t}, x_4, 4);$
$\mathrm{p} = \mathrm{diff}(\mathrm{r}, x_3 ,4);$
$\mathrm{n} = \mathrm{diff}(\mathrm{p}, x_2, 4);$
$\mathrm{h} = \mathrm{diff}(\mathrm{n}, x_1, 3);$
$N = (x_6 - x_7)*(x_6 - x_8)*(x_6 - x_9)*(x_6 - x_{10})*(x_7 - x_8)*(x_7 - x_9)*(x_7 -$
$x_{10})*(x_8 - x_9)*(x_8 - x_{10})*(x_9 - x_{10});$
$\mathrm{k} = \mathrm{diff}(\mathrm{h}*N, x_6, 3);$
$\mathrm{u} = \mathrm{diff}(\mathrm{k}, x_7, 4);$
$\mathrm{v} = \mathrm{diff}(\mathrm{u}, x_8, 2);$
$\mathrm{w} = \mathrm{diff}(\mathrm{v}, x_9 ,2);$
$\mathrm{y} = \mathrm{diff}(\mathrm{w}, x_{10}, 1);$
$c_3 = \mathrm{y}/\mathrm{factorial}(3)/\mathrm{factorial}(4)/\mathrm{factorial}(4)/\mathrm{factorial}(4)/\mathrm{factorial}(4)/\mathrm{factorial}(3)$

/factorial(4)/factorial(2)/factorial(2)/factorial(1)/factorial(4)/factorial(4)
/factorial(4)/factorial(4)/factorial(4)/factorial(4)

References

1. Alon, N.: Combinatorial Nullstellensatz. Comb. Probab. Comput. **8**, 7–29 (1999)
2. Bondy, J.A., Murty, U.S.R.: Graph Theory with Applications. MacMillan, New York (1976)
3. Chen, M., Raspaud, A., Wang, W.F.: Plane graphs with maximum degree 6 are edge-face 8-colorable. Graphs Comb. **30**, 861–874 (2014)
4. Fiamčík, J.: Simultaneous colouring of 4-valent maps. Mat. Časopis Sloven. Akad. Vied **21**, 9–13 (1971)
5. Gong, Q.Y., Wu, J.L.: Edge-face coloring of Halin graphs. Inform. Tech. **7**, 64–67 (2008)
6. Jucovič, E.: On a problem in map colouring. Mat. Časopis Sloven. Akad. Vied **19**, 225–227 (1969)
7. Kaul, H., Mudrock, J.A.: Combinatorial Nullstellensatz and DP-coloring of graphs. arXiv:2003.01112v1 (2020)
8. Mel'nikov, L.S.: Problem 9. In: Fiedler, M. (ed.) Proceedings of the Second Czechoslovak Symposium on Recent Advances in Graph Theory, Prague, June 1974, pp. 543. Academia Prague, Czechoslovak (1975)
9. Sander, D.P., Zhao, Y.: On simultaneous edge-face colorings of plane graphs. Combinatorica **17**, 441–445 (1997)
10. Sander, D.P., Zhao, Y.: A five-color theorem. Discrete Math. **220**, 279–281 (2000)
11. Sander, D.P., Zhao, Y.: On improving the edge-face coloring theorem. Graphs Comb. **17**, 329–341 (2001)
12. Waller, A.O.: Simultaneously colouring the edges and faces of plane graphs. J. Comb. Theory Ser. B **69**, 219–221 (1997)
13. Wang, W.F., Lih, K.: A new proof of Melnikov's conjecture on the edge-face coloring of plane graphs. Discrete Math. **253**, 87–95 (2002)
14. Wang, W.F., Lih, K.: The edge-face choosability of plane graphs. European J. Comb. **25**, 935–948 (2004)

Injective Coloring of Halin Graphs

Bu Yuehua[1,2(✉)] , Qi Chentao[1] , and Zhu Junlei[3]

[1] Department of Mathematics, Zhejiang Normal University, Jinhua 321004, China
yhbu@zjxu.edu.cn
[2] Zhejiang Normal University Xingzhi College, Jinhua 321004, China
[3] College of Mathematics, Physics and Information Engineering, Jiaxing University,
Jiaxing 321004, China
zhujl-001@163.com

Abstract. An injective k-coloring of a graph G is a mapping $f : V(G) \to \{1, 2, \ldots, k\}$ such that for any two vertices $v_1, v_2 \in V(G)$, if v_1 and v_2 have a common neighbor, then $f(v_1) \neq f(v_2)$. The injective chromatic number of a graph G, denoted by $\chi_i(G)$, is the smallest integer k such that G has an injective k-coloring. In this paper, we prove that for a Halin graph G, if $\Delta(G) \leq 5$, then $\chi_i(G) \leq \Delta(G) + 2$; if $\Delta(G) \geq 6$, then $\chi_i(G) \leq \Delta(G) + 1$.

Keywords: Injective coloring · Halin graph · Maximum degree

1 Introduction

An injective k-coloring of a graph G is a mapping $f : V(G) \to \{1, 2, \ldots, k\}$ such that for any two vertices $v_1, v_2 \in V(G)$, if v_1 and v_2 have a common neighbor, then $f(v_1) \neq f(v_2)$. The smallest integer k such that G has an injective k-coloring is called the injective chromatic number of G, denoted by $\chi_i(G)$. The injective coloring was introduced by Hahn, Kratochvíl, Širáň and Sotteau in 2002 [10]. Injective colorings have their origin in complexity theory, but related concepts had been studied earlier in [12].

For a graph G, it is trivial that $\Delta(G) \leq \chi_i(G) \leq |V(G)|$. Using a greedy algorithm, Hahn et al. [10] gave a trivial upper bound $\Delta(\Delta - 1) + 1$ on $\chi_i(G)$ in terms of the maximum degree Δ for general graphs. For planar graphs, the upper bound can be further strengthened to $\Delta^2 - \Delta$ if $\Delta \geq 3$ [11]. Specially, for a planar graph with maximum degree 3, $\chi_i(G) \leq 6$. Earlier results on planar graphs were given by Doyon, Hahn and Raspaud [9]. They obtained that for a planar graph G with maximum degree Δ, $\chi_i(G) \leq \Delta + 3$ if $g(G) \geq 7$, $\chi_i(G) \leq \Delta + 4$ if $g(G) = 6$ and $\chi_i(G) \leq \Delta + 8$ if $g(G) = 5$. Later, these upper bound were improved to $\Delta + 2$ by Lužar et al. [6,15], $\Delta + 3$ by Dong and Lin [7] and $\Delta + 6$ by Dong and Lin [8], respectively.

Supported by National Science Foundation of China under Grant Nos. 11901243, 11771403 and Zhejiang Provincial Natural Science Foundation of China under Grant No. LQ19A010005.

Z. Zhang et al. (Eds.): AAIM 2020, LNCS 12290, pp. 492–500, 2020.
https://doi.org/10.1007/978-3-030-57602-8_44

For planar graphs with maximum degree at least 4, Lužar [14] presented some planar graphs with maximum degree $5 \leq \Delta \leq 9$ and $\chi_i(G) = \Delta + 5$ and planar graphs with $diam(G) = 2$, maximum degree $\Delta \geq 8$ and $\chi_i(G) = \lfloor \frac{3\Delta}{2} \rfloor + 1$.

It is easy to see that $\chi_i(G) \leq \chi_2(G)$, where $\chi_2(G)$ is the smallest integer k such that G has a 2-distance k-coloring. A 2-distance k-coloring of a graph G is a mapping $f : V(G) \to \{1, 2, \ldots, k\}$ such that for any two vertices $v_1, v_2 \in V(G)$, $f(v_1) \neq f(v_2)$ if $1 \leq d(v_1, v_2) \leq 2$. Lužar [14] posed a variation for the injective chromatic number of planar graphs following Wegner's conjecture as follows.

Conjecture 1. Let G be a planar graph with maximum degree Δ. Then $\chi_i(G) \leq 5$ if $\Delta = 3$, $\chi_i(G) \leq \Delta + 5$ if $4 \leq \Delta \leq 7$ and $\chi_2(G) \leq \lfloor \frac{3\Delta}{2} \rfloor + 1$ if $\Delta \geq 8$.

Hahn et al. [10] showed that for any $k \geq 3$, it is a NP-complete problem to decide wether $\chi_i(G) \leq k$. For planar graphs, more results can be seen in [1–5,13]. For a K_4-minor-free graph G, Hahn, Raspaud and Wang [11] showed that $\chi_i(G) \leq \lceil \frac{3\Delta}{2} \rceil$.

In this paper, we investigate the injective chromatic number of Halin graphs. A Halin graph $G = T \cup C$ is a planar graph G constructed as follows. Let T be a tree of order at least 4. All vertices of T are either of degree 1 or of degree at least 3. Let C be a cycle connecting the leaves of T in such a way that C forms the boundary of the unbounded face. The tree T and the cycle C are called the characteristic tree and the adjoint cycle of G, respectively. We will prove the following two theorems.

Theorem 1. *If G is a Halin graph with maximum degree at most 5, then $\chi_i(G) \leq \Delta(G) + 2$.*

Theorem 2. *If G is a Halin graph with maximum degree at least 6, then $\chi_i(G) \leq \Delta(G) + 1$.*

2 Halin Graphs with Maximum Degree at Most 5

In a Halin graph $G = T \cup C$, v is called an inter vertex if $v \in T$ but $v \notin C$. Let $V_{in}(G)$ be the set of inter vertices of G. Let $|F(v)|$ be the number of colors that cannot be used for the vertex v. Let $|S(v)|$ be the number of colors that can be used for the vertex v. If there is a path $v_1 v_2 \cdots v_i$ ($i \leq k$) in G, then we call it k^--path. We shall prove Theorem 1.1 by three steps as follows.

Lemma 1. *If $G = T \cup C$ is a Halin graph with $\Delta(G) = 3$, then $\chi_i(G) \leq 5$.*

Proof. We prove the lemma by induction on the number of inter vertices of G. If $|V_{in}(G)| = 1$, then G is a wheel W_3 and thus the conclusion is obvious. Now let $|V_{in}(G)| \geq 2$ and let $P = v_1 v_2 \cdots v_l$ be a maximum path in T, $l \geq 4$. For convenience, we denote $u = v_2$, $v = v_3$, $w = v_4$ and $u_1 = v_1$. Let u_1, u_2 be two neighbors of u in $V(C)$. Since P is a maximum path in T, then there exists a 2^--path $P_1 = v v^1 \cdots v^k$ ($1 \leq k \leq 2$) between v and C, where $E(P) \cap E(P_1) = \emptyset$, $v^k \in V(C)$ is a neighbor of u_1 or u_2.

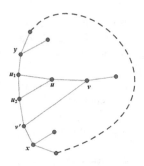

Fig. 1. G of case 1

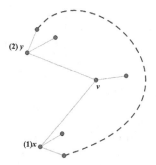

Fig. 2. Step 1 of case 1: there is a injective 5-coloring c' of G' ((k) is the color in $C = \{1, 2, 3, 4, 5\}$) (Color figure online)

Case 1: There exists a 1-path $P_1 = vv^1$ between v and C, where $E(P) \cap E(P_1) = \emptyset$.

W.l.o.g, we assume $u_2 v^1 \in C$, $x \neq u_2$ is a neighbor of v^1 and $y \neq u_2$ is neighbor of u_1, $x, y \in V(C)$. Let $G' = (G - \{u, u_1, u_2, v^1\}) \cup \{vx, vy\}$, then G' is a Halin graph with $\Delta(G') = 3$ and with fewer inter vertices than G. By induction, G' has an injective 5-coloring c'. Denote $C = \{1, 2, 3, 4, 5\}$. Now we shall color $V(G)$ using the colors in C. First, we erase the color on v and let $c(x') = c'(x')$ for any $x' \in V(G) - \{v, u, u_1, u_2, v^1\}$.

W.l.o.g, we assume $c(x) = 1, c(y) = 2$. Let $c'(u) = 1$, then we have that $|F(v^1)| \leq 4$, $|F(u_1)| \leq 3$, $|F(v)| \leq 3$, $|F(u_2)| \leq 2$ and thus we can color v^1, u_1, v, u_2 in order, see Figs. 1, 2, 3, 4, 5 and 6. Hence, G has an injective 5-coloring.

Case 2: There exists a 2-path $P_1 = vv^1v^2$ between v and C, where $E(P) \cap E(P_1) = \emptyset$.

W.l.o.g, we assume $u_2 v^2 \in C$. Since P is a maximum path in T, v^1 and v^2 has a common neighbor z in C. Let $x \neq v^2 \in V(C)$ be a neighbor of z and let $y \neq u_2$ be a neighbor of u_1, $x, y \in V(C)$. Let $G' = (G - \{u, u_1, u_2, v^1, v^2, z\}) \cup \{vx, vy\}$, then G' is a Halin graph with $\Delta(G') = 3$ and with fewer inter vertices than G. By induction, G' has an injective 5-coloring c'. Denote $C = \{1, 2, 3, 4, 5\}$. Now

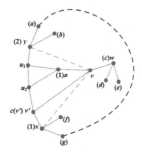

Fig. 3. Step 2 of case 1: use (1) to color u and use $c(v') \in C - (f) - (g) - (c) - (1)$ to color v' (Color figure online)

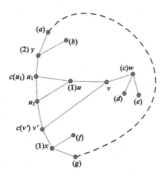

Fig. 4. Step 3 of case 1: use $c(u_1) \in C - (a) - (b) - c(v')$ to color u_1 (Color figure online)

we shall color $V(G)$ using the colors in C. First, we erase the color on v and let $c(x') = c'(x')$ for any $x' \in V(G) - \{v, u, u_1, u_2, v^1, v^2, z\}$.

Case 2.1: $c'(v) \notin \{c'(x), c'(y)\}$. W.l.o.g, we assume that $c(v) = 1, c(y) = 2, c(x) = 3$. Let $c'(u) = 3, c'(v^1) = 2, c'(u_1) = c'(z) = 1$, then we have $F(u_2) = F(v^2) = \{1, 2, 3\}$. Color u_2 and v^2 by color 4 and then we have $|F(v)| \leq 4$. Thus, we can get an injective 5-coloring of G.

Case 2.2: $c'(v) \in \{c'(x), c'(y)\}$. By symmetry, we assume that $c(v) = c(y) = 1, c(x) = 2$. Let $c'(u) = 2, c'(u_1) = c'(z) = 1$, then $F(u_2) = F(v^2) = \{1, 2\}$. Color u_2 and v^2 by color 3 and then we have $|F(v)| \leq 4, |F(v^1)| \leq 4$. Thus, we can get an injective 5-coloring of G.

Lemma 2. *If $G = T \cup C$ is a Halin graph with $\Delta(G) \leq 4$, then $\chi_i(G) \leq 6$.*

Proof. We prove it by induction on the number of inter vertices of G. If $|V_{in}(G)| = 1$, then G is a wheel W_3 or W_4. Thus, the conclusion holds. Now let $|V_{in}(G)| \geq 2$ and let $P = v_1 v_2 \cdots v_l$ be a maximum path in T, $l \geq 3$. For convenience, we denote $u = v_2, v = v_3, w = v_4$ and $u_1 = v_1$. Let $u_1, u_2, \ldots, u_k$ be neighbors of u in $V(C)$, where $k \in \{2, 3\}$. Since P is a maximum path in T,

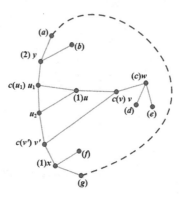

Fig. 5. Step 4 of case 1: use $c(v) \in C - (d) - (e) - c(u_1) - (1)$ to color v (Color figure online)

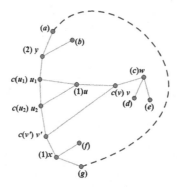

Fig. 6. Step 5 of case 1: use $c(u_2) \in C - (1) - (2) - c(u_1) - c(v)$ to color u_2 (Color figure online)

then there exists a 2^--path $P_1 = vv^1 \cdots v^k$ $(1 \le k \le 2)$ between v and C, where $E(P) \cap E(P_1) = \emptyset$, $v^k \in V(C)$ is a neighbor of u_1 or u_k.

Case 1.1: $d(u) = 3$ and there exists a 1-path $P_1 = vv^1$.

W.l.o.g, we assume $u_2v^1 \in C$, $x \ne u_2$ is a neighbor of u_1, $x \in V(C)$. Let $G' = (G - \{u_1, u_2\}) \cup \{ux, uv^1\}$, then G' is a Halin graph with $\Delta(G') \le 4$ and with fewer inter vertices than G. By induction, G' has an injective 6-coloring c'. Denote $C = \{1, 2, 3, 4, 5, 6\}$. Now we shall color $V(G)$ using the colors in C. First, we erase the color on u and let $c(x') = c'(x')$ for any $x' \in V(G) - \{u, u_1, u_2\}$.

Since u, v, v^1 are in the same 3-cycle, w.l.o.g, we assume $c(x) = 1, c(v^1) = 2, c(v) = 3, c(u) = k, k \in \{1, 4, 5\}$. Let $c'(u_1) = k$, then we have that $|F(u)| \le 5$, $|F(u_2)| \le 4$ and thus we can color u, u_2 in order. Hence, G has an injective 6-coloring.

Case 1.2: $d(u) = 3$ and there exists a 2-path $P_1 = vv^1v^2$ between v and C.

W.l.o.g, we assume $u_2v^2 \in C$, $x \ne u_2$ is a neighbor of u_1 and $y \ne u_2$ is a neighbor of v^2, $x, y \in V(C)$. Since P is a maximum path in T, $v^1y \in T$. Let

$G' = (G - \{u_1, u_2\}) \cup \{ux, uv^2\}$, then G' is a Halin graph with $\Delta(G') \leq 4$ and with fewer inter vertices than G. By induction, G' has an injective 6-coloring c'. Denote $C = \{1, 2, 3, 4, 5, 6\}$. Now we shall color $V(G)$ using the colors in C. First, we erase the color on u and let $c(x') = c'(x')$ for any $x' \in V(G) - \{u, u_1, u_2\}$.

W.l.o.g, we assume $c(x) = 1, c(v^2) = 2$. Since v^1, v^2, y are in the same 3-cycle, $c(v^1), c(v^2)$ and $c(y)$ are distinct. Let $c'(u_2) = 2$, then we have $|F(u)| \leq 5$, $|F(u_1)| \leq 4$ and thus we can color u, u_1 in order. Hence, G has an injective 6-coloring.

Case 2.1: $d(u) = 4$ and there exists a 1-path $P_1 = vv^1$ between v and C.

W.l.o.g, we assume $u_3v^2 \in C$, $x \neq u_2$ is a neighbor of u_1, $x \in V(C)$. Let $G' = (G - \{u_1, u_2, u_3\}) \cup \{ux, uv^1\}$, then G' is a Halin graph with $\Delta(G') \leq 4$ and with fewer inter vertices than G. By induction, G' has an injective 6-coloring c'. Denote $C = \{1, 2, 3, 4, 5, 6\}$. Now we shall color $V(G)$ using the colors in C. First, we erase the color on u and let $c(x') = c'(x')$ for any $x' \in V(G) - \{u, u_1, u_2, u_3\}$.

W.l.o.g, we assume $c(x) = 1, c(v^1) = 2$. Now we have $|F(u)| \leq 4, |F(u_1)| \leq 3$, $|F(u_2)| \leq 3, |F(u_3)| \leq 2$ and thus we can color u, u_1, u_2, u_3 in order. Hence, G has an injective 6-coloring.

Case 2.2: $d(u) = 4$ and there exists a 2-path $P_1 = vv^1v^2$ between v and C.

W.l.o.g, we assume $u_3v^2 \in C$, $x \neq u_2$ is a neighbor of u_1 and $y \neq u_3$ is a neighbor of v^2, $x, y \in V(C)$. Since P is a maximum path in T, $v^1y \in T$. Let $G' = (G - \{u_1, u_2, u_3\}) \cup \{ux, uv^2\}$, then G' is a Halin graph with $\Delta(G') \leq 4$ and with fewer inter vertices than G. By induction, G' has an injective 6-coloring c'. Denote $C = \{1, 2, 3, 4, 5, 6\}$. Now we shall color $V(G)$ using the colors in C. First, we erase the color on u and let $c(x') = c'(x')$ for any $x' \in V(G) - \{u, u_1, u_2, u_3\}$.

W.l.o.g, we assume $c(x) = 1, c(v^2) = 2$. Since v^1, v^2, y are in the same 3-cycle, $c(v^1), c(v^2), c(y)$ are distinct. Let $c'(u_3) = 2$, then we have $|F(u)| \leq 5$, $|F(u_1)| \leq 3, |F(u_2)| \leq 3$ and thus we can color u, u_1, u_2 in order. Hence, G has an injective 6-coloring.

Lemma 3. *If $G = T \cup C$ is a Halin graph with $\Delta(G) \leq 5$, then $\chi_i(G) \leq 7$.*

Proof. We prove it by induction on the number of inter vertices of G. If $|V_{in}(G)| = 1$, then G is a wheel and thus the conclusion holds. Now let $|V_{in}(G)| \geq 2$ and let $P = v_1v_2 \cdots v_l$ be a maximum path in T, $l \geq 3$. For convenience, we denote $u = v_2$, $v = v_3$, $w = v_4$ and $u_1 = v_1$. Let $u_1, u_2, \ldots, u_k$ be neighbors of u in $V(C)$, where $2 \leq k \leq 4$. Since P is a maximum path in T, then there exists a 2^--path $P_1 = vv^1 \cdots v^k$ $(1 \leq k \leq 2)$ between v and C, where $E(P) \cap E(P_1) = \emptyset$, $v^k \in V(C)$ is a neighbor of u_1 or u_k.

Case 1: $d(u) = 3$.

Let $G' = (G - \{u_1, u_2\}) \cup \{ux, uy\}$, then G' is a Halin graph with $\Delta(G') \leq \Delta(G) \leq 5$ and with fewer inter vertices than G. By induction, G' has an injective 7-coloring c'. Denote $C = \{1, 2, \ldots, 7\}$. Now we shall color $V(G)$ using the colors in C. First, we erase the color on u and let $c(x') = c'(x')$ for any $x' \in V(G) - \{u, u_1, u_2\}$. Then we have that $|S(u)| \geq 7 - 4 - 2 = 1, |S(u_1)| \geq 7 - 4 = 3$,

$|S(u_2)| \geq 7 - 4 = 3$ and thus we can color u, u_1, u_2 in order. Hence, G has an injective 7-coloring.

Case 2: $d(u) = 4$.

Let $G' = (G - \{u_1, u_2, u_3\}) \cup \{ux, uy\}$, then G' is a Halin graph with $\Delta(G') \leq \Delta(G) \leq 5$ and with fewer inter vertices than G. By induction, G' has an injective 7-coloring c'. Denote $C = \{1, 2, \ldots, 7\}$. Now we shall color $V(G)$ using the colors in C. First, we erase the color on u and let $c(x') = c'(x')$ for any $x' \in V(G) - \{u, u_1, u_2, u_3\}$. Then we have that $|S(u)| \geq 7 - 4 - 2 = 1$, $|S(u_1)| \geq 7 - 3 = 4$, $|S(u_2)| \geq 7 - 3 = 4$, $|S(u_3)| \geq 7 - 4 = 4$ and thus we can color u, u_1, u_3, u_2 in order. Hence, G has an injective 7-coloring.

Case 3: $d(u) = 5$.

Let $G' = (G - \{u_1, u_2, u_3, u_4\}) \cup \{ux, uy\}$, then G' is a Halin graph with $\Delta(G') \leq \Delta(G) \leq 5$ and with fewer inter vertices than G. By induction, G' has an injective 7-coloring c'. Denote $C = \{1, 2, \ldots, 7\}$. Now we shall color $V(G)$ using the colors in C. First, we erase the color on u and let $c(x') = c'(x')$ for any $x' \in V(G) - \{u, u_1, u_2, u_3, u_4\}$. Then we have that $|S(u)| \geq 7 - 4 - 2 = 1$, $|S(u_1)| \geq 7 - 3 = 4$, $|S(u_2)| \geq 7 - 2 = 5$, $|S(u_3)| \geq 7 - 2 = 5$, $|S(u_4)| \geq 7 - 3 = 4$ and thus we can color u, u_1, u_4, u_2, u_3 in order. Hence, G has an injective 7-coloring.

By the analysis above, we have proven Theorem 1.1.

3   Halin Graphs with Maximum Degree at Least 6

We prove Theorem 1.2 by induction on the number of inter vertices of G. If $|V_{in}(G)| = 1$, then G is a wheel, $\chi_i(G) = \Delta(G) + 1$. Thus, the conclusion holds. Now let $|V_{in}(G)| \geq 2$ and let $P = v_1 v_2 \cdots v_l$ be a maximum path in T, $l \geq 3$. For convenience, we denote $u = v_2$, $v = v_3$, $w = v_4$ and $u_1 = v_1$. Let $u_1, u_2, \ldots, u_k$ be neighbors of u in $V(C)$ in clockwise, where $2 \leq k \leq \Delta(G) - 1$. Since P is a maximum path in T, then there exists a 2^--path $P_1 = vv^1 \cdots v^t$ ($1 \leq t \leq 2$) between v and C, where $E(P) \cap E(P_1) = \emptyset$, $v^t \in V(C)$ is a neighbor of u_1 or u_k.

Case 1: $d(u) = k + 1, k \geq 2$ and there exists a 1-path $P_1 = vv^1$ between v and C.

W.l.o.g, we assume $u_k v^1 \in C$, $x \neq u_2$ is a neighbor of u_1, $x \in V(C)$. Let $G' = (G - \{u_1, u_2, \ldots, u_k\}) \cup \{ux, uv^1\}$, then G' is a Halin graph with $\Delta(G') \leq \Delta(G)$ and with fewer inter vertices than G. By induction, G' has an injective $(\Delta(G)+1)$-coloring c'. Denote $C = \{1, 2, \ldots, \Delta(G)+1\}$. Now we shall color $V(G)$ using the colors in C. First, we erase the color on u and let $c(x') = c'(x')$ for any $x' \in V(G) - \{u, u_1, u_2, \ldots, u_k\}$. Then $|F(u)| \leq \Delta(G)$ and thus we can color u. Now u is colored. If $d(u) = 3$, then $|S(u_1)| \geq \Delta(G)+1-5 \geq 2$, $|S(u_2)| \geq \Delta(G) + 1 - 4 \geq 3$. Thus we can color u_1 and u_2 in order. If $d(u) = 4$, then $|S(u_1)| \geq \Delta(G) + 1 - 4 \geq 3$, $|S(u_2)| \geq \Delta(G) + 1 - 4 \geq 3$, $|S(u_3)| \geq \Delta(G) + 1 - 3 \geq 4$. Thus we can color u_1, u_2, u_3 in order. If $d(u) \geq 5$, then $|S(u_1)| \geq \Delta(G) + 1 - 4 \geq 3$, $|S(u_2)| \geq \Delta(G) + 1 - 3 \geq 4$, $|S(u_t)| \geq \Delta(G) + 1 - 2 \geq \Delta(G) - 1, 3 \leq t \leq k-2$,

$|S(u_{k-1})| \geq \Delta(G) + 1 - 3 \geq 4$, $|S(u_k)| \geq \Delta(G) + 1 - 3 \geq 4$. Thus we can color u_1, u_2, u_{k-1}, u_k in order. Now $k - 4 \leq \Delta(G) - 5$ vertices are uncolored and each vertex has $\Delta(G) - 1 - 4 = \Delta(G) - 5$ colors available. Hence, G has an injective $(\Delta(G) + 1)$-coloring.

Case 2: $d(u) = k + 1, k \geq 2$ and there exists a 2-path $P_1 = vv^1v^2$ between v and C.

W.l.o.g, we assume $u_k v^2 \in C$, $x \neq u_2$ is a neighbor of u_1, $x \in V(C)$. Let $d(v^1) = m + 1$, $m \geq 2$. Since P is a maximum path in T, then all the neighbor of v^1 other that v are 1-vertices. For convenience, we denote $v^2 = w_1$. Let $w_1, w_2, \ldots, w_m \in V(G)$ be neighbors of v^1 and $w_1, w_2, \ldots, w_m$ are in clockwise. Let $y \neq w_{m-1}$ be a neighbor of w_m, $y \in C$.

Let $G' = (G - \{u, v^1, u_1, u_2, \ldots, u_k, w_1, w_2, \ldots, w_m\}) \cup \{vx, vy\}$, then G' is a Halin graph graph with $\Delta(G') \leq \Delta(G)$ and with fewer inter vertices than G. By induction, G' has an injective $(\Delta(G) + 1)$-coloring c'. Denote $C = \{1, 2, \ldots, \Delta(G) + 1\}$. Now we shall color $V(G)$ using the colors in C. Let $c(x') = c'(x')$ for any $x' \in V(G) - \{u, v^1, u_1, u_2, \ldots, u_k, w_1, w_2, \ldots, w_m\}$. Let $c(u) = c(y)$, $c(v^1) = c(x)$, then $|S(u_1)| \geq \Delta(G) + 1 - 4 \geq 3$, $|S(u_2)| \geq \Delta(G) + 1 - 3 \geq 4$, $|S(u_t)| \geq \Delta(G) + 1 - 2 \geq \Delta(G) - 1, 3 \leq t \leq k - 1$, $|S(u_k)| \geq \Delta(G) + 1 - 4 \geq 3$. Thus we can color u_1, u_2, u_k in order. Now we have $k - 3 \leq \Delta(G) - 4$ vertices uncolored and each vertex has $\Delta(G) - 1 - 3 = \Delta(G) - 4$ colors available. Hence, $u_2, u_3, \ldots, u_{k-1}$ can be colored. Next, we shall color $w_1, w_2, \ldots, w_m$. If $d(v^1) = 3$, then $|S(w_1)| \geq \Delta(G) + 1 - 5 \geq 2$, $|S(w_2)| \geq \Delta(G) + 1 - 5 \geq 2$. Thus we can color w_1, w_2. If $d(v^1) = 4$, then $|S(w_1)| \geq \Delta(G) + 1 - 4 \geq 3$, $|S(w_2)| \geq \Delta(G) + 1 - 4 \geq 3$, $|S(w_3)| \geq \Delta(G) + 1 - 4 \geq 3$. Thus we can color w_1, w_2, w_3. If $d(v^1) \geq 5$, then $|S(w_1)| \geq \Delta(G) + 1 - 4 \geq 3$, $|S(w_2)| \geq \Delta(G) + 1 - 3 \geq 4$, $|S(w_h)| \geq \Delta(G) + 1 - 2 \geq \Delta(G) - 1, 3 \leq h \leq m - 2$, $|S(w_{m-1})| \geq \Delta(G) + 1 - 3 \geq 4$, $|S(w_m)| \geq \Delta(G) + 1 - 4 \geq 3$. Thus we can color w_1, w_m, w_{m-1}, w_2 in order. Now we have $m - 4 \leq \Delta(G) - 5$ vertices uncolored and each vertex has $\Delta(G) - 1 - 4 = \Delta(G) - 5$ colors available. Hence, G has an injective $(\Delta(G) + 1)$-coloring.

By the analysis above, we have proven Theorem 1.2.

References

1. Borodin, O.V., Ivanova, A.O.: List injective colorings of planar graphs. Discrete Math. **311**(2–3), 154–165 (2011)
2. Borodin, O.V., Ivanova, A.O.: Injective $(\Delta + 1)$-colorings of planar graphs with girth 6. Sib. Math. J. **52**(1), 23–29 (2011)
3. Bu, Y., Lu, K.: List injective coloring of planar graphs with girth 5, 6, 8. Discrete Appl. Math. **161**(1011), 1367–1377 (2013)
4. Bu, Y., Lu, K., Yang, S.: Two smaller upper bounds of list injective chromatic number. J. Comb. Optim. **29**(2), 373–388 (2013). https://doi.org/10.1007/s10878-013-9599-7
5. Chen, H.-Y., Wu, J.-L.: List injective coloring of planar graphs with girth $g \geq 6$. Discrete Math. **339**(12), 3043–3051 (2016)

6. Cranston, D.W., Kim, S.-J., Yu, G.: Injective colorings of graphs with low average degree. Algorithmica **60**(3), 553–568 (2011)
7. Wei, D., Wensong, L.: Injective coloring of planar graphs with girth 6. Discrete Math. **313**(12), 1302–1311 (2013)
8. Wei, D., Wensong, L.: Injective coloring of planar graphs with girth 5. Discrete Math. **315–316**, 120–127 (2014)
9. Alain, D., Geňa, H., André, R.: Some bounds on the injective chromatic number of graphs. Discrete Math. **310**(3), 585–590 (2010)
10. Hahn, G., Kratochvíl, J., Širáň, J., Sotteau, D.: On the injective chromatic number of graphs. Discrete Math. **256**(1–2), 179–192 (2002)
11. Hahn, G., Raspaud, A., Wang, W.: On the injective coloring of K_4-minor free graphs, preprint (2006)
12. Jansen, T.R., Toft, B.: Graph Colouring Problems, pp. 156–158. Wiley, New York (1995)
13. Li, R., Xu, B.: Injective choosability of planar graphs of girth five and six. Discrete Math. **312**(6), 1260–1265 (2012)
14. Lužar, B.: Planar graphs with largest injective chromatic numbers, Institute of Mathematics, Physics and Mechanics Jadranska 19, 1000 Ljubljana. Slovenia Preprint 48, 1110 (2010)
15. Borut, L., Riste, Š., Martin, T.: Injective colorings of planar graphs with few colors. Discrete Math. **309**(18), 5636–5649 (2009)

The List $L(2,1)$-Labeling of Planar Graphs with Large Girth

Zhu Haiyang[1], Zhu Junlei[2(✉)] ⓘ, Liu Ying[1], Wang Shuling[3], Huang Danjun[4] ⓘ,
and Miao Lianying[5]

[1] Department of Flight Support Command, Air Force Logistics College,
Xuzhou 221000, People's Republic of China
[2] College of Mathematics Physics and Information Engineering, Jiaxing University,
Jiaxing 314000, People's Republic of China
zhujl-001@163.com
[3] Department of Basic Courses, Air Force Logistics College,
Xuzhou 221000, People's Republic of China
[4] Department of Mathematics, Zhejiang Normal University,
Jinhua 321004, People's Republic of China
[5] College of Sciences, China University of Mining and Technology,
Xuzhou 221008, People's Republic of China

Abstract. A list assignment of a graph G is a function $L : V(G) \longrightarrow 2^N$ that assigns each vertex v a list $L(v)$ for all $v \in V(G)$. We say that G has an L-$L(2,1)$-labeling if there exists a function ϕ such that $\phi(v) \in L(v)$ for all $v \in V(G)$, $|\phi(u) - \phi(v)| \geq 2$ if $d(u,v) = 1$ and $|\phi(u) - \phi(v)| \geq 1$ if $d(u,v) = 2$. The list $L(2,1)$-labeling number of G, denoted by $\lambda^l_{2,1}(G)$, is the minimum k such that for every list assignment $L = \{L(v) : |L(v)| = k, \ v \in V(G)\}$, G has an L-$L(2,1)$-labeling. We prove that for planar graph G with maximum degree $\Delta(G)$ and girth $g(G)$, $\lambda^l_{2,1}(G) \leq \Delta(G) + 3$ holds if $\Delta(G) = 4$ and $g(G) \geq 19$ or $\Delta(G) = 3$ and $g(G) \geq 32$. Moreover, there exist planar graphs having $\lambda^l_{2,1}(G) = \Delta(G) + 3$ for arbitrarily large $\Delta(G)$.

Keywords: Planar graph · Girth · $L(2,1)$-labeling ·
List $L(2,1)$-labeling number

1 Introduction

All graphs considered in this paper are finite simple graphs. We denote the vertex set, edge set, face set, maximum degree, minimum degree by $V(G), E(G)$, $F(G), \Delta(G), \delta(G)$, respectively. For vertices x and y, let $d(x,y)$ be the distance between x and y. Two cycles are adjacent if they share at least one edge. Undefined terminologies and notations are referred to [3].

Supported by National Science Foundation of China under Grant Nos.11901243, 11771403 and Zhejiang Provincial Natural Science Foundation of China under Grant Nos. LQ19A010005, LY18A010014.

ⓒ Springer Nature Switzerland AG 2020
Z. Zhang et al. (Eds.): AAIM 2020, LNCS 12290, pp. 501–512, 2020.
https://doi.org/10.1007/978-3-030-57602-8_45

A k-$L(2,1)$-labeling of a graph G is a mapping $\phi\colon V(G) \to \{0,1,2,\ldots,k\}$ such that $|\phi(x) - \phi(y)| \geq 2$ if $d(x,y) = 1$ and $|\phi(x) - \phi(y)| \geq 1$ if $d(x,y) = 2$. The smallest integer k such that G has such a labeling is called the $L(2,1)$-labeling number of G, denoted by $\lambda_{2,1}(G)$.

Motivated by the channel assignment problem, Griggs and Yeh [8] investigated the $L(2,1)$-labeling problem of graphs. Using a greedy labelling, they proved that for every graph G with maximum degree Δ, $\lambda_{2,1}(G) \leq \Delta^2 + 2\Delta$. Furthermore, for graphs with diameter 2, the upper bound can be decreased to $\Delta^2 + \Delta$. In the proof of case $\Delta \geq \frac{|V|-1}{2}$, the condition $diam(G) = 2$ was not used. Therefore, they proposed the following conjecture.

Conjecture 1. For any graph G with maximum degree $\Delta \geq 2$, $\lambda_{2,1}(G) \leq \Delta^2$.

In 1996, Chang and Kuo [5] proposed an algorithm funded on the concept of 2-stable set of a graph to obtain an $L(2,1)$-labeling of a given graph G and proved that $\lambda_{2,1}(G) \leq \Delta^2 + \Delta$, which remained the best general upper bound for about a decade. Král' and Škrekovski [12] brought this upper bound down by 1. Gonçalves [7] decreased this bound by 1 again. However, there is still a gap from Δ^2. In 2012, Havet et al. [9] settled this conjecture for general graphs with sufficiently large maximum degree by probabilistic tools.

For planar graph G with maximum degree Δ, Jan van den Heuvel and Sean McGuinness [10] proved that $\lambda_{2,1}(G) \leq 2\Delta + 34$. Bella et al. [2] proved that $\lambda_{2,1}(G) \leq 32$ if $\Delta \leq 6$, $\lambda_{2,1}(G) \leq 25$ if $\Delta \leq 5$ and $\lambda_{2,1}(G) \leq 16$ if $\Delta \leq 4$. Borodin et al. [4] improved the upper bound to $\lambda_{2,1}(G) \leq \lceil \frac{9\Delta}{5} \rceil + 9$ if $\Delta \geq 47$. To the best of our knowledge, the best known upper bound $\lambda_{2,1}(G) \leq \lceil \frac{5\Delta}{3} \rceil + 95$ is due to Molloy and Salavatipour [13]. For planar graph with girth and maximum degree restriction or without some short cycles, more results can be seen in [6,15–21].

Let L be a list assignment of G, we say that G has an L-$L(2,1)$-labeling if there exists a function ϕ such that $\phi(v) \in L(v)$ for all $v \in V(G)$, $|\phi(u)-\phi(v)| \geq 2$ if $d(u,v) = 1$ and $|\phi(u) - \phi(v)| \geq 1$ if $d(u,v) = 2$. If $|L(v)| = k$ for all $v \in V(G)$, then L is a k-list assignment. The list $L(2,1)$ labeling number of G, denoted by $\lambda_{2,1}^l(G)$, is the minimum k such that for each k-assignment L, G has an L-$L(2,1)$-labeling. The list $L(2,1)$-labeling problem is a generalization of the $L(2,1)$-labeling problem, and it easy to see $\lambda_{2,1}^l(G) \geq \lambda_{2,1}(G) + 1 \geq \Delta(G) + 2$.

The bound of Van den Heuvel et al. [10] and Bella et al. [4] implies that if G is a planar graph with maximum degree $\Delta \neq 3$, then $\lambda_{2,1}^l(G) \leq \Delta^2 + 1$. In this paper, we considered the list $L(2,1)$-labeling of planar graphs with maximum degree and girth restrictions and proved that

Theorem 1.1. If G is a planar graph, then $\lambda_{2,1}^l(G) \leq \Delta(G) + 3$ in each of the following cases:

(i) $\Delta(G) = 4$ and $g(G) \geq 19$;
(ii) $\Delta(G) = 3$ and $g(G) \geq 32$.

Wang [14] showed that for any integer $t \geq 3$, there exists a tree T with $\Delta(T) = t$ such that $\lambda_{2,1}(T) = \Delta(T) + 2$ and $\lambda_{2,1}^l(T) = \Delta(T) + 3$. This implies that there are graphs with $\Delta(G) \geq 3$ having $\lambda_{2,1}^l(G) = \Delta(G) + 3$. The main tool used in our proofs in this article is the following famous Combinatorial Nullstellensatz which is due to Alon [1].

Lemma 1.1 (Alon [1]). Let F be an arbitrary field, and let $f = f(x_1, x_2, \cdots, x_n)$ be a polynomial in $F[x_1, x_2, \cdots, x_n]$. Suppose the degree $deg(f)$ of f equals $\sum_{i=1}^n k_i$, where each k_i is a nonnegative integer, and suppose the coefficient of $x_1^{k_1} x_2^{k_2} \cdots x_n^{k_n}$ in f is non-zero. Then if $L_1, L_2, \cdots, L_n$ are subsets of F with $|L_i| > k_i$, there are $s_1 \in L_1, s_2 \in L_2, \cdots, s_n \in L_n$ so that $f(s_1, s_2, \cdots, s_n) \neq 0$.

2   Preliminaries

Let L be a list assignment of G and $b = \min \bigcup_{v \in V(G)} L(v) > 1$. Then there exists a list assignment $L' = \{L'(v)| v \in V(G)\}$, where $L'(v) = \{n - (b-1)| n \in L(v)\}$ for all $v \in V(G)$ such that G has an L-$L(2,1)$-labeling if and only if G has an L'-$L(2,1)$-labeling. Obviously, $\min \bigcup_{v \in V(G)} L'(v) = 1$ and $|L'(v)| = |L(v)|$ for all $v \in V(G)$. In the following, we can assume that the list assignment $L: V(G) \longrightarrow 2^N$ of G satisfies that $\min \bigcup_{v \in V(G)} L(v) = 1$. A k-path is a path $P_{k+2} = v_1 v_2 \cdots v_{k+2}$, where $d(v_1) \geq 3$, $d(v_2) = d(v_3) = \cdots = d(v_{k+1}) = 2$ and $d(v_{k+2}) \geq 3$. If $n \geq k$ or $n \leq k$, then n-path is also called k^+-path or k^--path.

Lemma 2.1. Let L be a list assignment of P_k. If L satisfies one of the following conditions, then P_k has a L-$L(2,1)$-labeling.

(1) $k = 2, |L(v_1)| = 2$ and $|L(v_2)| = 3$;
(2) $k = 2, |L(v_1)| = |L(v_2)| = 2$ and $L(v_1) \neq L(v_2)$ or $L(v_1) = L(v_2) = \{\beta, \gamma\}$, where $|\beta - \gamma| \geq 2$;
(3) $k = 3, |L(v_1)| = 2, |L(v_2)| = 4$ and $|L(v_3)| = 3$;
(4) $k = 4, |L(v_1)| = 2, |L(v_2)| = 5, |L(v_3)| = 5$ and $|L(v_4)| = 2$.

Proof.

(1) If $1 \in L(v_1)$, then we define an L-$L(2,1)$-labeling $\sigma: \sigma(v_1) = 1, \sigma(v_2) \in L(v_2) \backslash \{1, 2\}$. Otherwise, $1 \notin L(v_1)$ and $1 \in L(v_2)$. Then we can define an L-$L(2,1)$-labeling $\sigma: \sigma(v_2) = 1, \sigma(v_1) \in L(v_1) \backslash \{2\}$.
(2) If $L(v_1) = L(v_2) = \{\beta, \gamma\}$, where $|\beta - \gamma| \geq 2$, then we define an L-$L(2,1)$-labeling $\sigma: \sigma(v_1) = \beta, \sigma(v_2) = \gamma$. Now we assume that $L(v_1) \neq L(v_2)$. W.l.o.g, let $1 \in L(v_1)$. If $L(v_2) \backslash \{1, 2\} \neq \emptyset$, then we define an L-$L(2,1)$-labeling $\sigma: \sigma(v_1) = 1, \sigma(v_2) \in L(v_2) \backslash \{1, 2\}$. Otherwise, $L(v_2) = \{1, 2\}$. Since $L(v_1) \neq L(v_2)$, we can define an L-$L(2,1)$-labeling $\sigma: \sigma(v_2) = 1, \sigma(v_1) \in L(v_1) \backslash \{1, 2\}$.
(3) If $1 \notin L(v_1) \cup L(v_3)$, then $1 \in L(v_2)$. Define an L-$L(2,1)$-labeling $\sigma: \sigma(v_2) = 1, \sigma(v_1) \in L(v_1) \backslash \{2\}, \sigma(v_3) \in L(v_3) \backslash \{2, \sigma(v_1)\}$.

Suppose now that $1 \notin L(v_1)$ and $1 \in L(v_3)$. Let $L'(v_2) = L(v_2) \backslash \{1, 2\}$ and $L'(v_1) = L(v_1)$. It follows easily that $|L'(v_1)| = 2, |L'(v_2)| \geq 2$. If $|L'(v_2)| \geq 3$ or $L'(v_1) \neq L'(v_2)$, then by Lemma 2.1(1)(2), $P_3 - v_3$ has an L'-$L(2,1)$-labeling and we can extend it to P_3 by labeling v_3 with 1. Otherwise, we can assume that $L(v_1) = \{a, b\}, L(v_2) = \{1, 2, a, b\}, L(v_3) = \{1, c, d\}$, where $3 \leq a < b, d \geq 3$. Define an L-$L(2,1)$-labeling σ of P_3 such that $\sigma(v_2) = 1, \sigma(v_3) = d, \sigma(v_1) \in L(v_1) \backslash \{d\}$.

Suppose finally that $1 \in L(v_1)$. Let $L'(v_2) = L(v_2) \backslash \{1, 2\}$ and $L'(v_3) = L(v_3) \backslash \{1\}$. It follows easily that $|L'(v_2)| \geq 2, |L'(v_3)| \geq 2$. If $\max \{|L'(v_2)|, |L'(v_3)|\} \geq 3$ or $L'(v_2) \neq L'(v_3)$, then by Lemma 2.1(1)(2), $P_3 - v_1$ has an L'-$L(2,1)$-labeling and we can extend it to P_3 by labeling v_1 with 1. Now assume that $L(v_2) = \{1, 2, a, b\}, L(v_3) = \{1, a, b\}, L(v_1) = \{1, k\}$, where $3 \leq a < b$. If $k = 2$, then we define an L-$L(2,1)$-labeling σ of P_3 such that $\sigma(v_1) = 2, \sigma(v_2) = b, \sigma(v_3) = 1$. If $k \geq 3$, then let $\sigma(v_1) = k, \sigma(v_2) = 1, \sigma(v_3) \in L(v_3) \backslash \{1, k\}$.

(4) Suppose that $1 \in L(v_1) \cup L(v_4)$. W.l.o.g, assume that $1 \in L(v_1)$. Let $L'(v_2) = L(v_2) \backslash \{1, 2\}, L'(v_3) = L(v_3) \backslash \{1\}, L'(v_4) = L(v_4)$. Note that $|L'(v_2)| \geq 3$, $|L'(v_3)| \geq 4$, and $|L'(v_4)| = 2$, by Lemma 2.1(3), $P_4 - v_1$ has an L'-$L(2,1)$-labeling and we can extend it to P_4 by labeling v_1 with 1.

Suppose now that $1 \notin L(v_1) \cup L(v_4)$ and $1 \in L(v_2) \cup L(v_3)$. W.l.o.g, assume that $1 \in L(v_2)$. Let $L'(v_3) = L(v_3) \backslash \{1, 2\}, L'(v_4) = L(v_4) \backslash \{1\}$. If $2 \notin L(v_1)$, since $|L'(v_3)| \geq 3$ and $|L'(v_4)| = 2$, then by Lemma 2.1(1), $P_4 - \{v_1, v_2\}$ has an L'-$L(2,1)$-labeling ϕ and we can extend it to P_4 by labeling v_2 with 1 and v_1 with a label in $L(v_1) \backslash \{\phi(v_3)\}$. Otherwise, $L(v_1) = \{2, \alpha\}$, where $\alpha \geq 3$. Let $L'(v_3) = L(v_3) \backslash \{1, 2, \alpha\}, L'(v_4) = L(v_4)$. Note that $|L'(v_3)| \geq 2$ and $|L'(v_4)| = 2$. If $|L'(v_3)| \geq 3$ or $L'(v_3) \neq L'(v_4)$ or $L'(v_3) = L'(v_4) = \{\beta, \gamma\}$ with $|\beta - \gamma| \geq 2$, then by Lemma 2.1(1)(2), $P_4 - \{v_1, v_2\}$ has an L'-$L(2,1)$-labeling and we can extend it to P_4 by labeling v_2 with 1, v_1 with α.

Finally, assume that $L'(v_3) = L'(v_4) = \{\beta, \beta+1\}, L(v_3) = \{1, 2, \alpha, \beta, \beta+1\}$ where $\beta \geq 3$. Let $L''(v_2) = L(v_2) \backslash \{1, 2\}$ and $L''(v_1) = L(v_1)$. Note that $|L''(v_2)| \geq 3$ and $|L''(v_1)| = 2$, then by Lemma 2.1(1), $P_4 - \{v_3, v_4\}$ has an L'-$L(2,1)$-labeling and we can extend it to P_4 by labeling v_3 with 1, v_4 with a label in $\{\beta, \beta+1\} \backslash \{\sigma(v_2)\}$.

Lemma 2.2. Let L_i be a list assignment of tree T_i, $1 \leq i \leq 3$. If L_i satisfies one of the following conditions, then T_i has an L_i-$L(2,1)$-labeling.

(1) T_1 is a tree with seven vertices, where $|L_1(v_1)| = 7, |L_1(v_2)| = |L_1(v_3)| = 6$, $|L_1(v_4)| = 5, |L_1(v_5)| = 3$ and $|L_1(v_6)| = |L_1(v_7)| = 2$;
(2) T_2 is a tree with ten vertices, where $|L_2(v_1)| = |L_2(v_5)| = |L_2(v_7)| = 6$, $|L_2(v_2)| = |L_2(v_8)| = |L_2(v_9)| = 3, |L_2(v_3)| = |L_2(v_4)| = 7, |L_2(v_6)| = 5$ and $|L_2(v_{10})| = 2$;
(3) $T_3 = v_1 v_2 v_3 v_4 v_5 v_6$ is a path with six vertices, where $|L_3(v_1)| = |L_3(v_3)| = 3$, $|L_3(v_2)| = 5, |L_3(v_4)| = |L_3(v_5)| = 6$ and $|L_3(v_6)| = 2$.

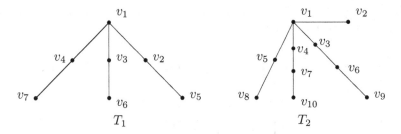

$$T_1 \qquad\qquad\qquad\qquad T_2$$

Proof.

(1) For $1 \leq i \leq 7$, let $L_1(v_i)$ be the label set assigned to vertex v_i and assign variables x_i to v_i. Define a polynomial $g = g(x_1, x_2, \cdots, x_7) \in R[x_1, x_2, \cdots, x_7]$ as follows:

$$g(x_1, x_2, \cdots, x_7) = \prod_{i=2}^{4}(x_1 - x_i)(x_1 - x_i + 1)(x_1 - x_i - 1)\prod_{l=2}^{4}(x_l - x_{l+3})$$

$$(x_l - x_{l+3} + 1)(x_l - x_{l+3} - 1)\prod_{2 \leq k < j \leq 4}(x_k - x_j)\prod_{m=5}^{7}(x_1 - x_m).$$

Delete the constant 1 and -1, we can obtain a homogeneous polynomial $f = f(x_1, x_2, \cdots, x_7)$ as follows:

$$f(x_1, x_2, \cdots, x_7) = \prod_{i=2}^{4}(x_1 - x_i)^3\prod_{l=2}^{4}(x_l - x_{l+3})^3\prod_{2 \leq k < j \leq 4}(x_k - x_j)\prod_{m=5}^{7}(x_1 - x_m).$$

Note that $deg(g) = deg(f) = 24$, the coefficient of the monomial with degree $deg(g)$ in g is non-zero if and only if the coefficient of the same monomial in f is non-zero. By means of a program in MATLAB (see the Appendix in [11]), we can calculate easily

$$\frac{\partial^{24} f}{\partial x_1^3 \partial x_2^5 \partial x_3^4 \partial x_4^4 \partial x_5^3 \partial x_6^3 \partial x_7^2} = 6628884480,$$

and

$$\frac{\partial^{24} f}{\partial x_1^6 \partial x_2^5 \partial x_3^5 \partial x_4^4 \partial x_5^2 \partial x_6^1 \partial x_7^1} = 74649600000,$$

which implies obviously that the coefficient of the monomial $x_1^3 x_2^5 x_3^4 x_4^4 x_5^3 x_6^3 x_7^2$ and $x_1^6 x_2^5 x_3^5 x_4^4 x_5^2 x_6 x_7$ in f are non-zero. Since $|L_1(v_1)| = 7 > 6, |L_1(v_2)| = |L_1(v_3)| = 6 > 5, |L_1(v_4)| = 5 > 4, |L_1(v_5)| = 3 > 2, |L_1(v_6)| = |L_1(v_7)| = 2 > 1$, by Lemma 1.1, there exist $s_1 \in L_1(v_1), s_2 \in L_1(v_2), \cdots, s_7 \in L_1(v_7)$ such that $f(s_1, s_2, \cdots, s_7) \neq 0$. Thus, T_1 has a L_1-$L(2, 1)$-labeling.

(2) For $1 \leq i \leq 10$, Let $L_2(v_i)$ be the label set assigned to vertex v_i and assign variables x_i to v_i. Define

$$f(x_1, x_2, \cdots, x_{10}) = \prod_{i=2}^{5}(x_1 - x_i)^3 \prod_{l=3}^{7}(x_l - x_{l+3})^3 \prod_{2 \leq k < j \leq 5}(x_k - x_j)$$

$$\prod_{m=6}^{8}(x_1 - x_m)(x_3 - x_9)(x_4 - x_{10})$$

Since

$$\frac{\partial^{38} f}{\partial x_1^5 \partial x_2^2 \partial x_3^6 \partial x_4^6 \partial x_5^5 \partial x_6^4 \partial x_7^5 \partial x_8^2 \partial x_9^2 \partial x_{10}^1} = -278628139008000000,$$

the coefficient of the monomial $x_1^5 x_2^2 x_3^6 x_4^6 x_5^5 x_6^4 x_7^5 x_8^2 x_9^2 x_{10}$ in f is non-zero. Since $|L_2(v_1)| = |L_2(v_5)| = |L_2(v_7)| = 6 > 5, |L_2(v_2)| = |L_2(v_8)| = |L_2(v_9)| = 3 > 2, |L_2(v_3)| = |L_2(v_4)| = 7 > 6, |L_2(v_6)| = 5 > 4, |L_2(v_{10})| = 2 > 1$, by Lemma 1.1, there exist $s_1 \in L_2(v_1), s_2 \in L_2(v_2), \cdots, s_{10} \in L_2(v_{10})$ such that $f(s_1, s_2, \cdots, s_{10}) \neq 0$. Thus, T_2 has an L_2-$L(2,1)$-labeling.

(3) For $1 \leq i \leq 6$, Let $L_3(v_i)$ be the label set assigned to vertex v_i and assign variables x_i to v_i. Define

$$f(x_1, x_2, \cdots, x_6) = \prod_{i=1}^{5}(x_i - x_{i+1})^3 \prod_{j=1}^{4}(x_j - x_{j+2}).$$

Since

$$\frac{\partial^{19} f}{\partial x_1^2 \partial x_2^4 \partial x_3^2 \partial x_4^5 \partial x_5^5 \partial x_6^1} = 149299200,$$

the coefficient of the monomial $x_1^2 x_2^4 x_3^2 x_4^5 x_5^5 x_6$ in f is non-zero. Since $|L_3(v_1)| = |L_3(v_3)| = 3 > 2, |L_3(v_2)| = 5 > 3, |L_3(v_4)| = |L_3(v_5)| = 6 > 5, |L_3(v_6)| = 2 > 1$, by Lemma 1.1, there exist $s_1 \in L_3(v_1), s_2 \in L_3(v_2), \cdots, s_6 \in L_3(v_6)$ such that $f(s_1, s_2, \cdots, s_6) \neq 0$. Thus, T_3 has an L_3-$L(2,1)$-labeling.

3   Proof of Theorem 1.1

Let G be a minimal counterexample with the smallest number of vertices and edges. Then G has a $(\Delta(G) + 3)$- assignment L such that G has no L-$L(2,1)$-labeling. However, for every proper graph $H \subset G$, H has an L-$L(2,1)$-labeling. Obviously, G is connected and $\delta(G) \geq 2$. We will show that G does not exist, which contradicts the assumption.

Using the relation $\sum_{v \in V} d(v) = \sum_{f \in F} d(f) = 2|E|$, Euler's formula can be rewritten as: $\sum_{v \in V}(d(v) - 4) + \sum_{f \in F}(d(f) - 4) = -8$. We define a charge function w by

$w(v) = d(v) - 4$ for all $v \in V \cup F$. We design appropriate rules and redistribute charge accordingly, preserving their sum, produce a new non-negative charge function $w'(x) \geq 0$ for all $x \in V \cup F$. This leads to the following obvious contradiction, hence demonstrates that no such a counterexample can exist.

Since G is connected, we define a weight function w by $w(v) = \frac{n-2}{2}d(v) - n$ for $v \in V$ and $w(f) = d(f) - n$ for $f \in F$. By Euler's formula $|V| - |E| + |F| = 2$ and formula $\sum_{v \in V} d(v) = 2|E| = \sum_{f \in F} d(f)$, we can derive

$$\sum_{v \in V(G)} \left(\frac{n-2}{2}d(v) - n\right) + \sum_{f \in F(G)} (d(f) - n) = -2n.$$

To prove the non-existence of G, we first give some structural properties of G, then we design appropriate discharging rules and redistribute weights accordingly. Once the discharging is finished, a new weight function w' is produced. During the process, the total sum of weights is kept fixed. It follows that

$$\sum_{x \in V \cup F} w'(x) = \sum_{x \in V \cup F} w(x) = -2n.$$

However, we will show that after the discharging is complete, the new weight function $w'(x) \geq 0$ for all $x \in V \cup F$. This leads to the following obvious contradiction

$$0 \leq \sum_{x \in V \cup F} w'(x) = \sum_{x \in V \cup F} w(x) = -2n < 0.$$

3.1   Structure and Properties of a Counterexample to Theorem 1.1.

For $f \in F(G)$, we use $\partial(f)$ to denote the boundary walk of f and write $f = v_1 v_2 \ldots v_n$ if $v_1, v_2, \ldots, v_n$ are the vertices of $\partial(f)$ in the clockwise order. If the degree of a vertex (or face) x is k, at least k, or at most k, then x is called a k-vertex (k-face), k^+-vertex, or k^--vertex, respectively. A $(d_1, d_2, \cdots, d_i^+, \ldots, d_k^-)$-vertex is a k-vertex incident with k different paths, where the 1th, ith, and kth of them is a d_1-path, d_i^+-path and d_k^--path, respectively. Let $F(x)$ be the set of forbidden labels cannot be used for x. Let $\tau(x \to y)$ be the charge transferred from x to y. Let $F(x)$ be the set of forbidden labels cannot be used for x.

Lemma 3.1. *G has no k-path in each of the following cases:*

(1) $\Delta(G) = 4$ and $k \geq 5$;
(2) $\Delta(G) = 3$ and $k \geq 6$.

Proof.(1) Assume that G has a k-path $P_{k+2} = v_1 v_2 \cdots v_{k+2}$, where $k \geq 5$. By the minimality of G, $G - v_3 v_4$ has a L-$L(2,1)$-labeling. Remove the labels on vertices v_3, v_4 and v_5. Note that $|L(v_i) \backslash F(v_i)| \geq 3$ for $i = 3, 5$ and $|L(v_4) \backslash F(v_4)| \geq 5$, we can relabel v_3, v_4 and v_5 by Lemma 2.1(3), a contradiction.

(2) Assume that G has a k-path $P_{k+2} = v_1v_2 \cdots v_{k+2}$, where $k \geq 6$. By the minimality of G, $G - v_3v_4$ has a L-$L(2,1)$-labeling. Remove the labels on vertices v_3, v_4, v_5 and v_6. Note that $|L(v_i)\backslash F(v_i)| \geq 2$ for $i = 3, 6$ and $|L(v_j)\backslash F(v_j)| \geq 5$ for $j = 4, 5$, we can relabel v_3, v_4, v_5 and v_6 by Lemma 2.1(4), a contradiction.

Lemma 3.2. *Let G be a graph with $\Delta(G) = 4$. If $v_1v_2v_3v_4v_5v_6$ is a 4-path, then $d(v_1) = d(v_6) = 4$.*

Proof. Assume that $d(v_1) = 3$. By the minimality of G, $G - v_3v_4$ has a L-$L(2,1)$-labeling. Erase the labels on v_2, v_3, v_4. Note that $|L(v_2)\backslash F(v_2)| \geq 2$, $|L(v_3)\backslash F(v_3)| \geq 5$, and $|L(v_4)\backslash F(v_4)| \geq 3$, we can relabel v_2, v_3, v_4 by Lemma 2.1(3), a contradiction. Similarly, $d(v_6) = 4$.

Lemma 3.3. *Let G be a graph with $\Delta(G) = 4$, then G does not contain one of the following configurations, where u is a k-vertex with $N(u) = \{x_1, x_2, \cdots, x_k\}$*

(1) 3-vertex u incident with one 3-path $ux_1y_1z_1w_1$ and two 2^+-paths $ux_2y_2z_2 \cdots$, $ux_3y_3z_3 \cdots$, where $\max\{d(z_2), d(z_3)\} \leq 3$.
(2) 3-vertex u incident with two 3-paths $ux_1y_1z_1w_1$, $ux_2y_2z_2w_2$ and one 1^+-path $ux_3y_3 \cdots$ where $d(w_1) \leq 3$.
(3) 4-vertex u incident with one 2^+-path $ux_1y_1 \ldots$, one 4-path $ux_2y_2z_2w_2t_2$ and two 3^+-paths $ux_3y_3z_3w_3 \cdots$, $ux_4y_4z_4w_4 \cdots$, where $d(w_3) \leq 3$.

Proof.(1) Assume G contains such a 3-vertex u. Let $Y = \{u, x_1, x_2, x_3, y_1, y_2, y_3\}$. Since $g(G) \geq 13$, $G[Y]$ is a tree. By the minimality of G, $G - ux_1$ has a L-$L(2,1)$-labelling. Erase the labels on vertices of $V(G[Y])$. For each vertex $x \in V(G[Y])$, let $L'(x) = L(x)\backslash F(x)$. Note that $|L'(u)| = 7$, $|L'(x_i)| \geq 6$ for $i = 1, 2, 3$, $|L'(y_1)| \geq 3$, $|L'(y_j)| \geq 2$ for $j = 2, 3$, by Lemma 2.2(1), we can relabel the vertices of $G[Y]$ to get an L-$L(2,1)$-labeling of G, a contradiction.
(2) Assume G contains such a 3-vertex. Let $Y = \{u, x_1, x_2, y_1, y_2, z_1\}$. Since $g(G) \geq 13$, $G[Y]$ is a tree. By the minimality of G, $G - ux_1$ has an L-$L(2,1)$-labelling. Erase the labels on vertices of $V(G[Y])$. For each vertex $x \in V(G[Y])$, let $L'(x) = L(x)\backslash F(x)$. Note that $|L'(u)| \geq 3$, $|L'(x_1)| \geq 6$, $|L'(x_2)| \geq 5$, $|L'(y_1)| \geq 6$, $|L'(y_2)| \geq 3$, $|L'(z_1)| \geq 2$, by Lemma 2.2(3), we can relabel the vertices of $G[Y]$ to get an L-$L(2,1)$-labeling of G, a contradiction.
(3) Assume G contains such a 4-vertex. Let $Y = \{u, x_1, x_2, x_3, x_4, y_2, y_3, y_4, z_2, z_3\}$. Since $g(G) \geq 13$, $G[Y]$ is a tree. By the minimality of G, $G - ux_1$ has a L-$L(2,1)$-labelling. Erase the labels on vertices of $V(G[Y])$. For each vertex $x \in V(G[Y])$, let $L'(x) = L(x)\backslash F(x)$. Note that $|L'(u)| \geq 6$, $|L'(x_1)| \geq 3$, $|L'(x_2)| = |L'(x_3)| = 7$, $|L'(x_4)| \geq 6$, $|L'(y_2)| \geq 6$, $|L'(y_3)| \geq 6$, $|L'(y_4)| \geq 3$, $|L'(z_2)| \geq 3$, $|L'(z_3)| \geq 2$, by Lemma 2.2(2), we can relabel the vertices of $G[Y]$ to get an L-$L(2,1)$-labeling of G, a contradiction.

Lemma 3.4. *Let G be a graph with $\Delta(G) = 3$, then G has no $(5, 5, 4^+)$-vertex.*

Proof. Assume G contains a $(5, 5, 4^+)$-vertex u which is incident with two 5-paths $ux_1y_1z_1w_1t_1v_1$, $ux_2y_2z_2w_2t_2v_2$ and one 4^+-path $ux_3y_3z_3w_3t_3\cdots$. By the minimality of G, $G - ux_1$ has a L-$L(2, 1)$-labelling f. Erase the labels on vertices of $u, x_i, y_i, z_i, w_1, w_2$, where $1 \leq i \leq 3$. Note that $|L(u)| = |L(x_1)| = 6$, then there exists a label $\alpha \in L(u)$ such that $|L(x_1) \backslash \{\alpha \pm 1, \alpha\}| \geq 4$. Label u by α. Now, we have that $|L(x_3) \backslash \{\alpha \pm 1, \alpha\}| \geq 3$, $|L(y_3) \backslash \{\alpha, f(w_3)\}| \geq 4$ and $|L(z_3) \backslash \{f(w_3) \pm 1, f(w_3), f(t_3)\}| \geq 2$. By Lemma 2.1(3), we can relabel x_3, y_3, z_3 to get a labeling f_1. After u, x_3, y_3, z_3 are labeled, $|L(x_2) \backslash \{\alpha \pm 1, \alpha, f_1(x_3)\}| \geq 2$, $|L(y_2) \backslash \{\alpha\}| \geq 5$, $|L(z_2) \backslash \{f_1(t_2)\}| \geq 5$ and $|L(w_2) \backslash \{f_1(t_2) \pm 1, f_1(t_2), f_1(v_2)\}| \geq 2$. By Lemma 2.1(4), we can relabel x_2, y_2, z_2, w_2 to get a labeling f_2. Now, we have that $|L(x_1) \backslash \{\alpha \pm 1, \alpha, f_2(x_3), f_2(x_2)\}| \geq 2$, $|L(y_1) \backslash \{\alpha\}| \geq 5$, $||L(z_1) \backslash \{f_2(t_1)\}| \geq 5$, $|L(w_1) \backslash \{f_2(t_1) \pm 1, f_2(t_1), f_2(v_1)\}| \geq 2$. By Lemma 2.1(4), we can relabel x_1, y_1, z_1, w_1. Thus, we have an L-$L(2, 1)$-labeling of G, a contradiction.

3.2   Case $\Delta(G) = 4$, $g(G) \geq 18$

Let $n = 18$ and define the discharging rules as follows.

Discharging Rules

(M1). If $v_1v_2v_3$ is a 1-path, then $\tau(v_1 \rightarrow v_2) = \tau(v_3 \rightarrow v_2) = 1$.

(M2a). If $v_1v_2v_3v_4$ is a 2-path and $d(v_1) = d(v_4) = 3$, then $\tau(v_1 \rightarrow v_2) = \tau(v_4 \rightarrow v_3) = 2$;

(M2b). If $v_1v_2v_3v_4$ is a 2-path and $d(v_1) = 3$, $d(v_4) = 4$, then $\tau(v_1 \rightarrow v_2) = \tau(v_4 \rightarrow v_2) = 1$;

(M2c). If $v_1v_2v_3v_4v_5$ is a 3-path and $d(v_1) = 3$, then $\tau(v_1 \rightarrow v_2) = 2$.

(M3a). If $v_1v_2v_3v_4v_5$ is a 3-path and $d(v_1) = d(v_5) = 3$, then $\tau(v_1 \rightarrow v_3) = \tau(v_5 \rightarrow v_3) = 1$;

(M3b). If $v_1v_2v_3v_4v_5$ is a 3-path and $d(v_1) = 3, d(v_5) = 4$, then $\tau(v_1 \rightarrow v_3) = \frac{1}{2}, \tau(v_5 \rightarrow v_3) = \frac{3}{2}$;

(M3c). If $v_1v_2v_3v_4v_5$ is a 3-path and $d(v_1) = d(v_5) = 4$, then $\tau(v_1 \rightarrow v_3) = \tau(v_5 \rightarrow v_3) = 1$.

(M4a). If $v_1v_2v_3v_4\cdots$ is a 2^+-path and $d(v_1) = 4$, then $\tau(v_1 \rightarrow v_2) = 2$;

(M4b). If $v_1v_2v_3v_4v_5v_6$ is a 4-path and $d(v_1) = 4$, then $\tau(v_1 \rightarrow v_3) = 2$.
    It is easy to see that $w'(f) = w(f) = d(f) - 18 \geq 0$ for $f \in F(G)$.

Checking $w'(v) \geq 0$ for each $v \in V(G)$

Case 1. $d(v) = 2$.
    Here, $w(v) = -2$. By Lemma 3.1(1), G does not contain 5^+-path. If $v_1v_2v_3v_4v_5v_6$ is a 4-path, then by Lemma 3.2, $d(v_1) = d(v_6) = 4$. By (M4a), $\tau(v_1 \rightarrow v_2) = \tau(v_6 \rightarrow v_5) = 2$. By (M4b), $\tau(v_1 \rightarrow v_3) = \tau(v_6 \rightarrow v_4) = 2$. Thus, $\sum_{i=2}^{5} w'(v_i) \geq -2 \times 4 + 2 \times 4 = 0$. If $v_1v_2v_3v_4v_5$ is a 3-path, then v_3 gets 2 from v_1 together with v_5 by (M3a), (M3b), (M3c), $\tau(v_1 \rightarrow v_2) = \tau(v_5 \rightarrow v_4) = 2$

by (M2c) and (M4a). Thus, $\sum_{i=2}^{4} w'(v_i) \geq -2 \times 3 + 2 + 2 \times 2 = 0$. If $v_1 v_2 v_3 v_4$ is a 2-path, then by (M2a), (M2b) and (M4a), $w'(v_2) + w'(v_3) \geq \min\{-2 \times 2 + 2 \times 2, -2 \times 2 + 2 + 1 + 1\} = 0$. If $v_1 v_2 v_3$ is a 1-path, then by (M1), $w'(v_2) \geq -2 + 1 \times 2 = 0$.

Case 2. $d(v) = 3$.

Here, $w(v) = \frac{18-2}{2} \times 3 - 18 = 6$. By Lemma 3.1(1), G does not contain 5^+-path. By Lemma 3.2, v is not incident with a 4-path. By (M1), v gives out 1 along each incident 1-path. By (M2a) and (M2b), v gives out at most 2 along each incident 2-path. By (M2c), (M3a) and (M3b), v gives out at most $2+1 = 3$ along each incident 3-path. By Lemma 3.3(1), v is incident with at most two 3-paths.

Subcase 2.1. v is incident with exactly two 3-paths.

If v is incident with two 3-paths $vx_1 y_1 z_1 w_1$, $vx_2 y_2 z_2 w_2$ and one 2-path $vx_3 y_3 z_3$, then $d(z_3) = 4$ by Lemma 3.3(1) and $d(w_1) = d(w_2) = 4$ by Lemma 3.3(2). Thus, v gives out $2 + \frac{1}{2} = \frac{5}{2}$ along every incident 3-path by (M2c) and (M3b) and 1 along $vx_3 y_3 z_3$ by (M2b). Hence, $w'(v) \geq 6 - \frac{5}{2} \times 2 - 1 = 0$. If v is incident with two 3-paths $vx_1 y_1 z_1 w_1$ and $vx_2 y_2 z_2 w_2$, then Lemma 3.3(2), $d(w_1) = d(w_2) = 4$. Thus, by (M2c) and (M3b), v gives out at most $2 + \frac{1}{2} = \frac{5}{2}$ along every incident 3-path. Hence, $w'(v) \geq 6 - \frac{5}{2} \times 2 - 1 = 0$. If v is a $(3,3,0)$-vertex, then $w'(v) \geq 6 - 3 \times 2 = 0$.

Subcase 2.2. v is incident with exactly one 3-path.

If v is incident with two 2-paths $vx_2 y_2 z_2$ and $vx_3 y_3 z_3$, then by Lemma 3.3(1), $\max\{d(z_2), d(z_3)\} = 4$. Thus, by (M2a) and (M2b), v gives out at most $2 + 1 = 3$ along $vx_2 y_2 z_2$ together with $vx_3 y_3 z_3$. Hence, $w'(v) \geq 6 - 3 - 3 = 0$. Otherwise, then $w'(v) \geq 6 - 3 - 2 - 1 = 0$.

Subcase 2.3. v is not incident with any 3-path.

Here, $w'(v) \geq 6 - 2 \times 3 = 0$.

Case 3. $d(v) = 4$.

Here, $w(v) = \frac{18-2}{2} \times 4 - 18 = 14$. By Lemma 3.1(1), G does not contain 5^+-path. By (M1), v gives out 1 along each incident 1-path. By (M2b) and (M4a), v gives out at most $1 + 2 = 3$ along each incident 2-path. By (M3b), (M3c) and (M4a), v gives out $\frac{3}{2} + 2 = \frac{7}{2}$ or $1 + 2 = 3$ along each incident 3-path. By (M4a) and (M4b), v gives out $2 + 2 = 4$ along each incident 4-path. By Lemma 3.3(3), v is incident with at most three 4-paths.

Subcase 3.1. v is incident with exactly three 4-paths.

By Lemma 3.3(3), v is a $(4,4,4,1^-)$-vertex. Thus, $w'(v) \geq 14 - 4 \times 3 - 1 = 1$.

Subcase 3.2. v is incident with exactly two 4-paths.

By Lemma 3.3(3), v is not a $(4,4,3,3)$-vertex or $(4,4,3,2)$-vertex. Thus, $w'(v) \geq \min\{14 - 4 \times 2 - \frac{7}{2} - 1, 14 - 4 \times 2 - 3 \times 2\} = 0$.

Subcase 3.3. v is incident with exactly one 4-path.

If v is incident with three 3-paths $vx_i y_i z_i w_i$ for $2 \leq i \leq 4$, then by Lemma 3.3(3), $d(w_i) = 4$. Thus, by (M3c) and (M4a), v gives out $1 + 2 = 3$

along each incident 3-path. Hence, $w'(v) \geq 14 - 4 - 3 \times 3 = 1$. Otherwise, $w'(v) \geq 14 - 4 - \frac{7}{2} \times 2 - 3 = 0$.

Subcase 3.4. v is not incident with any 4-path.

Here, $w'(v) \geq 14 - \frac{7}{2} \times 4 = 0$.

By the analysis above, we have proven that for each $x \in V(G) \cup F(G)$, $w'(v) \geq 0$.

3.3   Case $\Delta(G) = 3$, $g(G) \geq 32$

Let $n = 32$ and define the discharging rules as follows

Discharging Rules

(N1). Each 3-vertex gives 2 to each adjacent 2-vertex.

(N2). If $v_1 v_2 v_3 v_4 v_5$ is a 3-path, then $\tau(v_1 \rightarrow v_3) = \tau(v_5 \rightarrow v_3) = 1$.

(N3). If $v_1 v_2 v_3 v_4 v_5 v_6$ is a 4-path, then $\tau(v_1 \rightarrow v_3) = \tau(v_6 \rightarrow v_4) = 2$.

(N4). If $P = v_1 v_2 v_3 v_4 v_5 v_6 v_7$ is a 5-path, then $\tau(v_1 \rightarrow v_3) = \tau(v_7 \rightarrow v_5) = 2$ and $\tau(v_1 \rightarrow v_4) = \tau(v_5 \rightarrow v_4) = 1$.

It is easy to see that $w'(f) = w(f) = d(f) - 32 \geq 0$ for $f \in F(G)$.

Checking $w'(v) \geq 0$ for each $v \in V(G)$

Case 1. $d(v) = 2$.

Here, $w(v) = -2$. By Lemma 3.1(2), G does not contain 6^+-path. If $v_1 v_2 v_3$ is a 1-path, then by (N1), $w'(v_2) = -2 + 2 = 0$. If $v_1 v_2 v_3 v_4$ is a 2-path, then by (N1), $w'(v_2) + w'(v_3) = -2 \times 2 + 2 \times 2 = 0$. If $v_1 v_2 v_3 v_4 v_5$ is a 3-path, then by (N1) and (N2), $\sum_{i=2}^{4} = -2 \times 3 + 2 \times 2 + 1 + 1 = 0$. If $v_1 v_2 v_3 v_4 v_5 v_6$ is a 4-path, then by (N1) and (N3), $\sum_{i=2}^{5} = -2 \times 4 + 2 \times 4 = 0$. If $v_1 v_2 v_3 v_4 v_5 v_6 v_7$ is a 5-path, then by (N1) and (N4), $\sum_{i=2}^{6} = -2 \times 5 + 2 \times 4 + 1 \times 2 = 0$.

Case 2. $d(v) = 3$.

Here, $w(v) = \frac{32-2}{2} \times 3 - 32 = 13$. By (N1) and (N4), v gives out $2 + 2 + 1 = 5$ along every incident 5-path. By (N1) and (N3), v gives out $2 + 2 = 4$ along every incident 4-path. By (N1) and (N2), v gives out $2 + 1 = 3$ along every incident 3-path. By (N1), v gives out 2 along every incident 2-path or 1-path. By Lemma 3.1(2), G does not contain 6^+-path. By Lemma 3.4, if v is incident with two 5-paths, then v is a $(5, 5, 3^-)$-vertex and thus $w'(v) \geq 13 - 5 \times 2 - 3 = 0$. Otherwise, $w'(v) \geq 13 - 5 - 4 \times 2 = 0$.

By the analysis above, we have proven that for each $x \in V(G) \cup F(G)$, $w'(v) \geq 0$.

References

1. Alon, N.: Combinatorial nullstellensatz. Combin. Probab. Comput. **8**, 7–29 (1999)
2. Bella, P., Král, D., Mohar, B., Quittnerová, K., et al.: Labeling planar graphs with a condition at distance two. Eur. J. Comb. **28**, 2201–2239 (2007)

3. Bondy, J.A., Murty, U.S.R.: Graph Theory with Applications. American Elsevier (1976)
4. Borodin, O., Broersma, H.J., Glebov, A., van den Heuvel, J.: Stars and bunches in planar graphs. Part II: general planar graphs and colourings. CDAM Res. Rep. **05**, 2002 (2002)
5. Chang, G.J., Kuo, D.: The $L(2,1)$-labeling problem on graphs. SIAM J. Discrete Math. **9**, 309–316 (1996)
6. Dvořák, Z., Kŕal, D., Nejedlý, P., Škrekovski, R.: Coloring squares of planar graphs with no short cycles, Preprint series (2005)
7. Gonçalves, D.: On the $L(p,1)$-labelling of graphs. Discrete Math. **308**(8), 1405–1414 (2008)
8. Griggs, J.R., Yeh, R.K.: Labeling graphs with a condition at distance 2. SIAM J. Discrete Math. **5**, 586–595 (1992)
9. Havet, F., Reed, B., Sereni, J.S.: Griggs and Yeh's conjecture and $L(p,1)$-labellings. SIAM J. Discrete Math. **26**(1), 145–168 (2012)
10. Van den Heuvel, J., McGuinness, S.: Coloring of the square of a planar graph. J. Graph Theory **42**, 110–124 (2003)
11. Huo, J.J., Wang, W.F., Xu, C.D.: Neighbor sum distinguishing index of subcubic graphs. Graphs Comb. **33**, 419–431 (2017)
12. Král, D., Škrekovski, R.: A theorem about channel assignment problem. SIAM J. Discrete Math. **16**(3), 426–437 (2003)
13. Molloy, M., Salavatipour, M.R.: A bound on the chromatic number of the square of a planar graph. J. Comb. Theory Ser. B **94**(2), 189–213 (2005)
14. Wang, W.F.: The L(2,1)-labelling of trees. Discrete Appl. Math. **154**, 598–603 (2006)
15. Wang, W.F., Cai, L.Z.: Labelling planar graphs without 4-cycles with a conditionon distance two. Discrete Appl. Math. **156**, 2241–2249 (2008)
16. Wang, W.F., Lih, K.W.: Labeling planar graphs with conditions on girth and distance two. SIAM J. Discrete Math. **17**, 264–275 (2003)
17. Zhu, H.Y., Gu, Y., Sheng, J.J., Lv, X.Z.: List 2-distance $\Delta + 3$-coloring of planar graphs without 4, 5-cycles. J. Comb. Optim. **36**(4), 1411–1424 (2018)
18. Zhu, H.Y., Hou, L.F., Chen, W., Lv, X.Z.: The $L(p,q)$-labelling of planar graphs without 4-cycles. Discrete Appl. Math. **162**, 355–363 (2014)
19. Zhu, H.Y., Miao, L.Y., Chen, S., Lv, X.Z., Song, W.Y.: The list $L(2,1)$-labeling of planar graphs. Discrete Math. **341**, 2211–2219 (2018)
20. Zhu, H.Y., Lv, X.Z., Wang, C.Q., Chen, M.: Labelling planar graphs without 4,5-cycles with a condition on distance two. Discrete Math. **26**, 52–64 (2012)
21. Zhu, J.L., Bu, Y.H., Pardalos, M.P., Du, H.W., Wang, H.J., Liu, B.: Optimal channel assignment and $L(p,1)$-labeling. J. Global Optim. **72**, 539–552 (2018)

The Frequency of the Optimal Hamiltonian Cycle Computed with Frequency Quadrilaterals for Traveling Salesman Problem

Yong Wang[✉][iD] and Zunpu Han[iD]

North China Electric Power University, Changping 102206, BJ, China
yongwang@ncepu.edu.cn

Abstract. Traveling salesman problem is generally represented as a complete graph. The complete graph is converted into a frequency graph where the frequency of each edge is computed with a number of frequency quadrilaterals. The frequency consistency is introduced for deriving the lower frequency bound for an edge in the optimal Hamiltonian cycle (OHC). The frequency of the OHC is compared with that of the other Hamiltonian cycles. Considering the frequency of edges, the OHC is better cultivated. Finally, the experimental results are provided for demonstrating the frequency of the OHC edges.

Keywords: Traveling salesman problem · Frequency of optimal Hamiltonian cycle · Frequency quadrilateral · Frequency consistency

1 Introduction

Traveling salesman problem (TSP) is one of the well-known combinatorial optimization problems. Given a set of n points $\{1, 2, \cdots, n\}$, a distance function $d(u, v) > 0$ is defined on each pair of distinct points $u, v \in \{1, 2, \cdots, n\}$ and $u \neq v$. A Hamiltonian cycle (HC) visiting each of the points once is denoted by $(v_1, v_2, \cdots, v_n)$, where $v_i \in \{1, 2, \cdots, n\}$ and $1 \leq i \leq n$. The HC distance is computed as $d(HC) = d(v_1, v_n) + \sum_{i=1}^{n-1} (d(v_i, v_{i+1}))$. The objective of TSP is to compute the shortest cycle, namely the optimal Hamiltonian cycle (OHC). In this paper, the symmetrical TSP is considered. That is, for any pair of points u and v, $d(u, v) = d(v, u)$ holds. TSP has been extensively studied in operations research and computer science in order to find the OHC within an acceptable computation time [1].

TSP is usually represented as the complete graph K_n. The search space increases exponentially in proportion to the size of TSP. Karp [2] has shown that TSP is NP-complete, which means that no exact polynomial-time algorithms

Supported by the Fundamental Research Funds for the Central Universities (Nos.2018ZD09 and 2018MS039).

Z. Zhang et al. (Eds.): AAIM 2020, LNCS 12290, pp. 513–524, 2020.
https://doi.org/10.1007/978-3-030-57602-8_46

exist for TSP, unless $P = NP$. The computation time of current exact algorithms is generally $O(a^n)$ [3,4], where $a > 1$. By incorporating additional constraints, the branch-and-bound [5] and cutting-plane [6] methods can resolve the TSP on thousands of points. In 2006, a VLSI problem containing 85,900 points was resolved by a networked computer system with 128 nodes [7]. Recently, Cook reported that the OHC through the 109,399 stars was found [8]. The experiments indicated that it was difficult to reduce the computation time of the exact algorithms for a large-scale TSP.

Numerous approximation algorithms and heuristics are available for various TSP on K_n. For the metric TSP, the best approximation ratio is $\frac{3}{2}$ owing to Christofides [1]. Moreover, Arora [9] designed the polynomial time approximation schemes for the constant-dimensional Euclidean TSP. As the approximation approaches the optimal solution more closely, the computation time required will increase. Experiments illustrated that the cycle improvement heuristics [10] were efficient for determining a satisfactory solution. The results also demonstrated that the edge distances were insufficient to be taken as the heuristic information for finding the OHC.

The OHC search space is so huge in K_n that the algorithms usually require a significant amount of time to resolve a big-scale TSP. In order to reduce the OHC search space, we convert the K_n into a frequency graph by means of frequency quadrilaterals. In a frequency quadrilateral, the frequency of an edge is the number of the optimal four-vertex paths containing this edge in the corresponding quadrilateral, see [11]. As all the frequencies of an edge in a number of frequency quadrilaterals are added together, the (total) frequency of this edge is obtained. Experiments have demonstrated that the frequency of an OHC edge is significantly higher than that of most of the other edges. Thus, when the TSP is resolved based on the frequency of edges, the edges of low frequency can be neglected in the search of the OHC. Wang and Remmel [11] proposed a binomial distribution model for cultivating the OHC based on frequency quadrilaterals. However, the probability model that they assumed for the OHC edges was not proven [19]. This is the reason why we present this paper. The probability model will be proven and the lower frequency bound will be derived for the OHC edges. Moreover, the frequency of the OHC will be compared with that of the other HCs. According to the frequency of edges, the OHC is better cultivated.

The remainder of this paper is organized as follows. In Sect. 2, we shall review the frequency quadrilaterals and the frequency of edges computed with the frequency quadrilaterals. In Sect. 3, we will discuss the frequency consistency for an edge in different frequency quadrilaterals. It is found that, if an edge (A, B) has a frequency $f(A, B) \in \{1, 3, 5\}$ in the first frequency quadrilateral $ABCD$ in K_n, it will maintain the same frequency with a probability $\frac{2}{3}$ in another frequency quadrilateral $ABCE$ or $ABDE$ and $E \neq \{A, B, C, D\}$. In Sect. 4, we will study the combinatorics of frequency quadrilaterals for the OHC edges. When we select N frequency quadrilaterals containing an edge to compute its frequency, the lower frequency bound will be derived according to the frequency consistency. In the 5th section, we shall discuss the total frequency of the OHC edges, and it is found that the OHC frequency is the maximum among those of all the HCs. In Sect. 6, we shall give certain experimental results for eliminating the non-OHC edges before the TSP is resolved. Finally, conclusions of this work are drawn and possibilities of the future research are proposed.

2   The Frequency Quadrilaterals

Given K_n on a set of n points $\{1, 2, \cdots, n\}$, it contains $\binom{n}{4}$ quadrilaterals $ABCD$, where $A, B, C, D \in \{1, 2, \cdots, n\}$. As we assume a vertex set of $\{1, 2, \cdots, n\}$, there is a total order on the elements in $\{A, B, C, D\}$ induced by the natural ordering on $\{1, 2, \cdots, n\}$, which we assume to be $A < B < C < D$. Each quadrilateral $ABCD$ contains six optimal four-vertex paths with two given endpoints [11]. For example, for the two endpoints A and C, there are two four-vertex paths (A, B, D, C) and (A, D, B, C). The shorter path (A, B, D, C) or (A, D, B, C) is taken as the optimal four-vertex path for A and C. Since there are six pairs of endpoints, there are six optimal four-vertex paths. The frequency quadrilateral is computed with the six optimal four-vertex paths in $ABCD$. The frequency of an edge is the number of the optimal four-vertex paths containing this edge.

Wang and Remmel [11] presented the six frequency quadrilaterals for $ABCD$ according to the orders of the three distance sums $d(A, B) + d(C, D)$, $d(A, C) + d(B, D)$, and $d(A, D) + d(B, C)$, where each distance sum is the addition of the distances of two vertex-disjoint edges in $ABCD$. The six frequency quadrilaterals are illustrated in Fig. 1. Under each frequency quadrilateral, the inequality containing the three distance sums is illustrated. It mentions that the frequency quadrilaterals will change if the three distance sums are equal or two of them are equal in $ABCD$. Thus, the following probability model does not hold for all TSP. This model is only useful for the TSP containing the quadrilaterals in which the three distance sums are not equal.

It can be seen that the frequency of an edge $e = \{(A, B), (A, C), (A, D), (B, C), (B, D), (C, D)\}$ is 1, 3, and 5 in each of the frequency quadrilaterals $ABCD$, frequency $ABCD$ for short. Moreover, e has each of the frequencies 1, 3, and 5 twice in the six frequency $ABCDs$. Given $ABCD$ in K_n, the corresponding frequency quadrilateral may be one of the six frequency quadrilaterals illustrated in Fig. 1. Therefore, they assumed that the probability that e has one frequency of 1, 3, and 5 in a frequency $ABCD$ is $p_1(e) = p_3(e) = p_5(e) = \frac{1}{3}$. Here, $p_i(e)$ means the probability that e has the frequency $i \in \{1, 3, 5\}$ in a frequency quadrilateral containing e.

For an OHC edge $e_o = (A, B)$, Wang and Remmel [11] constructed the $n - 3$ quadrilaterals with the vertex-disjoint edges (A, B) and (C, D) in the OHC. As (A, B) and (C, D) are in the OHC, the distance inequality $d(A, B) + d(C, D) < d(A, C) + d(B, D)$ holds. It can be seen that the inequality holds for the three cases (1), (2), and (3) in Fig. 1. The frequency of $e_o = (A, B)$ in the three frequency quadrilaterals (1), (2), and (3) in Fig. 1 is 5, 5, and 3, respectively. Thus, the probability set for e_o becomes $p_5(e_o) = \frac{2}{3}$, $p_3(e_o) = \frac{1}{3}$, and $p_1(e_o) = 0$

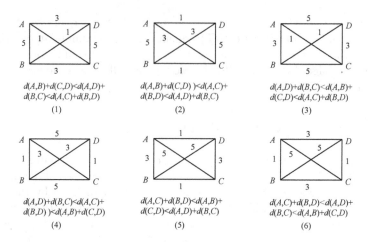

Fig. 1. Six frequency quadrilaterals for a quadrilateral $ABCD$.

according to the three frequency quadrilaterals. It should be noted that each edge is included in $\binom{n-2}{2}$ quadrilaterals in K_n. In addition, for the other $\binom{n-3}{2}$ frequency quadrilaterals containing e_o, they assumed the probability $p_1(e_o) = p_3(e_o) = p_5(e_o) = \frac{1}{3}$. Thus, the probability model (1) [18] is built for e_o.

$$p_5(e_o) = \frac{1}{3} + \frac{2}{3(n-2)} \text{ and}$$

$$p_3(e_o) = \frac{1}{3} \text{ and}$$

$$p_1(e_o) = \frac{1}{3} - \frac{2}{3(n-2)}. \tag{1}$$

The frequency of an edge is computed with the frequency quadrilaterals. An edge e is contained in $\binom{n-2}{2}$ quadrilaterals in K_n. e is also included in the same number of frequency quadrilaterals. One can select N frequency quadrilaterals containing e for computing its total frequency $F(e)$. In the ith $(1 \le i \le N)$ frequency quadrilateral, we note that the frequency of e is $f_i(e) \in \{1, 3, 5\}$. Thus, the total frequency is computed as $F(e) = \sum_{i=1}^{N} f_i(e)$. All of the edges e and their frequencies $F(e)$ form the frequency graph which has the same topological structure as the weighted graph K_n.

When we select N frequency quadrilaterals containing e to compute $F(e)$, the expected frequency is $3N$ as e has the probability of $p_1(e) = p_3(e) = p_5(e) = \frac{1}{3}$. For an OHC edge, the expected frequency is $F(e_o) = 3N + \frac{8N}{3(n-2)}$ according to the probability model (1). It says the frequency of an OHC edge will be bigger than that of a common edge. Although this property has been verified by experiments [11], it is not proven in the previous papers. In the next section, we introduce the frequency consistency which demonstrates the probability that an edge maintains the same frequency in different frequency quadrilaterals. According to the frequency consistency, the probability model (1) will be proven.

3   The Frequency Consistency

In this section, we focus on the frequency consistency in order to determine the probability that an edge (A, B) preserves the same frequency $f(A, B) \in \{1, 3, 5\}$ in different but related frequency quadrilaterals. Here, $f(A, B)$ denotes the frequency of (A, B) in a frequency quadrilateral containing (A, B). Lemma 1 is stated as follows.

Lemma 1. *Given a frequency quadrilateral $ABCD$ in K_n on n points $\{1, 2, \cdots, n\}$, if the frequency $f(A, B) > f(A, C)$ (or $f(A, B) < f(A, C)$) exists, the inequality $f(A, B) > f(A, C)$ (or $f(A, B) < f(A, C)$) holds with the probability $\frac{2}{3}$ in another frequency quadrilateral $ABCE$, where $E \in \{1, 2, \cdots, n\}$ and $E \neq \{A, B, C, D\}$.*

We take this situation as the frequency consistency for (A, B) in two different but related frequency quadrilaterals.

Proof. As $f(A, B) > f(A, C)$ in the frequency $ABCD$, the distance inequality $d(A, B) + d(C, D) < d(A, C) + d(B, D)$ holds in $ABCD$ (see Fig. 1). When vertex D is replaced by another vertex $E \neq \{A, B, C, D\}$, we obtain another quadrilateral $ABCE$. We are interested in the probability of the event $f(A, B) > f(A, C)$ in the new frequency $ABCE$, which is denoted by p. In order to aid in understanding the frequency consistency, we construct the third quadrilateral $BCDE$, so that three quadrilaterals $ABCD$, $ABCE$, and $BCDE$ exist. Because $f(C, D) > f(B, D)$ in the frequency $ABCD$, the event $f(C, D) > f(B, D)$ has the same probability p in the frequency $BCDE$ as that for $f(A, B) > f(A, C)$ in the frequency $ABCE$.

In $BCDE$, the two events $d(B, D) + d(C, E) < d(B, E) + d(C, D)$ and $d(B, D) + d(C, E) > d(B, E) + d(C, D)$ will occur. If $d(B, D) + d(C, E) < d(B, E) + d(C, D)$, we obtain $f(B, D) > f(C, D)$, and this event occurs with the probability $1 - p$, as we assume that the event $f(C, D) > f(B, D)$ has the probability p. Plus the distance inequality $d(A, B) + d(C, D) < d(A, C) + d(B, D)$ according to $ABCD$, the inequality $d(A, B) + d(C, E) < d(A, C) + d(B, E)$ is derived for $ABCE$. In this case, the event $f(A, B) > f(A, C)$ occurs with the probability $1 - p$ in the frequency $ABCE$.

Otherwise, if $d(B, D) + d(C, E) > d(B, E) + d(C, D)$ in $BCDE$, $f(B, D) < f(C, D)$ is derived and this event occurs with the probability p. According to the known inequality $d(A, B) + d(C, D) < d(A, C) + d(B, D)$ (which has a probability of 1), the order of $d(A, B) + d(C, E)$ and $d(A, C) + d(B, E)$ cannot be determined. In this situation, we divide this case into two new cases, namely $d(A, B) + d(C, E) < d(A, C) + d(B, E)$ and $d(A, B) + d(C, E) > d(A, C) + d(B, E)$. In the best case, $d(A, B) + d(C, E) < d(A, C) + d(B, E)$ occurs with the probability p. The total probability that $f(A, B) > f(A, C)$ in the frequency $ABCE$ is

$1-p+p=1$. In the worst case, $d(A,B)+d(C,E) < d(A,C)+d(B,E)$ occurs with the probability zero. The total probability of $f(A,B) > f(A,C)$ in the frequency $ABCE$ becomes $1-p+0 = 1-p$. Since we assume $f(A,B) > f(A,C)$ has the probability p in the frequency $ABCE$, one can derive the equality $p = 1-p$. Thus, the minimum probability that $f(A,B) > f(A,C)$ in the frequency $ABCE$ is $\frac{1}{2}$. On average, we assume each of the two new cases occurs with a probability of $\frac{p}{2}$. Thus, each of the two events $f(A,B) > f(A,C)$ and $f(A,B) < f(A,C)$ occur with the probability $\frac{p}{2}$ under the condition $f(B,D) < f(C,D)$, respectively.

A total of three cases of the results $f(A,B)$ vs $f(A,C)$ exist, according to the orders of the $d(A,B) + d(C,E)$ and $d(A,C) + d(B,E)$ under the pre-conditions $d(A,B) + d(C,D) < d(A,C) + d(B,D)$ and $d(B,D) + d(C,E) < d(B,E) + d(C,D)$ or $d(B,D) + d(C,E) > d(B,E) + d(C,D)$. The three cases are illustrated in Fig. 2. The case $d(A,B) + d(C,E) < d(A,C) + d(B,E)$ occurs with a total probability of $1 - \frac{p}{2}$. This means that the event $f(A,B) > f(A,C)$ in the frequency $ABCE$ occurs with the probability $1 - \frac{p}{2}$. As we assume that $f(A,B) > f(A,C)$ has the probability p, $1 - \frac{p}{2} = p$ holds and $p = \frac{2}{3}$.   □

It can be seen that, if $f(A,B) > f(A,C)$ in the first frequency $ABCD$, $f(A,B) > f(A,C)$ will occur with the probability $\frac{2}{3}$ in another frequency $ABCE$, where $E \neq \{A,B,C,D\}$. It mentions that $ABCD$ and $ABCE$ contains the two edges (A,B) and (A,C). This phenomenon is taken as the frequency consistency for an edge in two different but related frequency quadrilaterals.

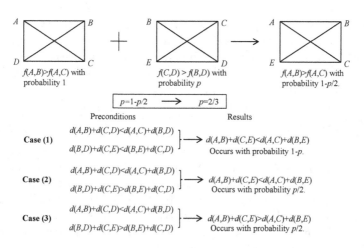

Fig. 2. Illustration of frequency consistency for two adjacent edges (A,B) and (A,C) in two frequency quadrilaterals $ABCD$ and $ABCE$.

4   The Lower Frequency Bound for an *OHC* Edge

In this section, we will estimate the minimum frequency of an *OHC* edge computed with the frequency quadrilaterals based on frequency consistency.

According to the known frequency $f(e_o) \in \{1, 3, 5\}$ of $e_o \in OHC$ in an appropriate number of frequency quadrilaterals, the expected number of frequency quadrilaterals where e_o has the frequency of 1, 3, and 5 in K_n can be evaluated according to the frequency consistency. Then, the probability that e_o has the frequency of 1, 3, and 5 in K_n can be computed. When we select N frequency quadrilaterals containing e_o to compute $F(e_o)$, the expected value of $F(e_o)$ will be estimated.

Given the OHC in K_n in Fig. 3, edge $(A, B) \in OHC$ is included in the $n - 3$ frequency $ABCD$s, where $f(A, B) = 5$ and 3 since $d(A, B) + d(C, D) < d(A, C) + d(B, D)$. Moreover, there are $\frac{2}{3}(n - 3)$ frequency $ABCD$s where $f(A, B) = 5$ and $\frac{1}{3}(n - 3)$ frequency $ABCD$s where $f(A, B) = 3$ according to $d(A, B) + d(C, D) < d(A, C) + d(B, D)$, see Fig. 1. Using the $n - 3$ frequency $ABCD$s, we will construct the other frequency quadrilaterals containing (A, B), and compute the expected frequency of $F(A, B)$ based on the frequency consistency.

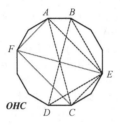

Fig. 3. $ABCD$ containing two non-adjacent edges (A, B) and (C, D) in OHC.

Theorem 1. *As the frequency $F(e_o)$ of an edge $e_o \in OHC$ is computed with N random frequency quadrilaterals, the inequality $F(e_o) \geq \frac{29N}{9}$ holds.*

Proof. Given $ABCD$ in Fig. 3, we replace vertex D (or C) with another vertex $E \neq \{A, B, C, D\}$. As E moves along the OHC, we obtain $n - 4$ quadrilaterals $ABCE$s (or $ABDE$s) containing (A, B). Since $f(A, B) > f(A, C)$ and $f(A, B) > f(B, D)$ in the frequency $ABCD$, $f(A, B) > f(A, C)$ (or $f(A, B) > f(B, D)$) occurs with the probability $\frac{2}{3}$ in each frequency $ABCE$ (or $ABDE$).

Moreover, there are $n - 3$ edges (C, D)s for (A, B) in the OHC. According to the symmetry of the changing quadrilaterals $ABCE$ and $ABDE$ (or vertices C, D and E), the number of the quadrilaterals $ABCE$s (or $ABDE$s) is $\frac{(n-3)(n-4)}{2}$. Based on the frequency consistency and plus the $n - 3$ quadrilaterals $ABCD$s, there will be $\frac{(n-3)(n-1)}{3}$ frequency quadrilaterals where $f(A, B) > f(A, C)$ (or $f(A, B) > f(B, D)$). In each of these frequency quadrilaterals, (A, B) has the frequency 3 and 5 with the probability $\frac{1}{3}$ and $\frac{2}{3}$, respectively.

Since $f(A, B) = 3$ or 5 in each of the $n - 3$ frequency $ABCD$s, it will maintain the same frequency in the generated frequency $ABCE$s with the probability $\frac{2}{3}$. According to the six frequency quadrilaterals in Fig. 1 and frequency consistency, the number of the frequency quadrilaterals in case of $f(A, B) = 5$ is

$\frac{4}{9}\binom{n-3}{2} + \frac{2(n-3)}{3} = \frac{2(n-3)(n-1)}{9}$. The number of frequency quadrilaterals in case of $f(A,B) = 3$ is $\frac{2}{9}\binom{n-3}{2} + \frac{n-3}{3} = \frac{(n-3)(n-1)}{9}$. In each of the residual $\frac{1}{3}\binom{n-3}{2}$ frequency quadrilaterals containing (A,B), $f(A,B) = 1$ occurs in the worst case. In this case, the probability $p_5(e_o) = \frac{4}{9} + \frac{4}{9(n-2)}$, $p_3(e_o) = \frac{2}{9} + \frac{2}{9(n-2)}$ and $p_1(e_o) = \frac{1}{3} - \frac{2}{3(n-2)}$ can be derived. As the frequency consistency is considered, $p_3(e_o)$ or $p_5(e_o)$ is not equal to that in formula (1). As we choose N frequency quadrilaterals containing (A,B) to compute $F(A,B)$, a lower frequency bound is derived as $\frac{29N}{9} + \frac{28N}{9(n-2)}$ which is bigger than $\frac{29N}{9}$ as n goes infinity. □

This lower frequency bound is better than that given in paper [11]. In average case, (A,B) will have each of the frequencies 1, 3 and 5 with the probability $\frac{1}{3}$ in the residual $\frac{1}{3}\binom{n-3}{2}$ frequency quadrilaterals. In this case, $p_5(e_o) = \frac{5}{9} + \frac{2}{9(n-2)}$, $p_3(e_o) = \frac{1}{3}$ and $p_1(e_o) = \frac{1}{9} - \frac{2}{9(n-2)}$ are derived. It can be seen that $p_5(e_o)$ is significantly larger than $p_1(e_o)$. Therefore, e_o will be contained in many more frequency quadrilaterals where $f(e_o) = 5$ than those where $f(e_o) = 1$ in K_n. When N frequency quadrilaterals containing e_o is used to compute the total frequency, $F(e_o) \geq \frac{35}{9}N \approx 3.89N$.

It should be mentioned that $p_5(e_o) \to 1$ as n is sufficiently large [19]. The frequency $F(e_o) \to 5N$ means e_o will have the frequency 5 in nearly all of the frequency quadrilaterals containing e_o. For medium and large-size TSP, the expected frequency $3.89N$ can be taken as the frequency threshold to eliminate the edges with $F(e) < 3.89N$. Thus, a graph with a smaller number of edges can be generated, and the OHC search space will be reduced.

5   The Frequency of the Optimal Hamiltonian Cycle

Given a frequency graph on n vertices $\{v_i\}$ and $v_i \in \{1, 2, \cdots, n\}$, the frequency of an edge (v_i, v_j) is $F(v_i, v_j)$. The frequency of an HC is the sum of frequency of the n sequential edges (v_i, v_{i+1}) in the HC, that is $F(HC) = F(v_1, v_n) + \sum_{i=1}^{n-1} F(v_i, v_{i+1})$. For the OHC, the following theorem is given.

Theorem 2. *If there is only one OHC in K_n and each quadrilateral contains six optimal four-vertex paths of given endpoints, the frequency of the OHC is the maximum among those of the HCs.*

Proof. Given an edge $(A,B) \in OHC$ in Fig. 3, it has $n - 3$ non-adjacent edges $(C,D) \in OHC$. In $ABCD$, $d(A,B) + d(C,D) < d(A,C) + d(B,D)$ holds since there is only one OHC. One sees there are three corresponding frequency quadrilaterals (1), (2) and (3) in Fig. 1. In the three frequency quadrilaterals, $f(A,B) = f(C,D) > f(A,D) = f(B,C)$ and $f(A,B) = f(C,D) > f(A,C) = f(B,D)$ occur two and three times, respectively. In addition, $f(A,D) = f(B,C) > f(A,C) = f(B,D)$ occurs two times. As $d(A,B) + d(C,D) < d(A,C) + d(B,D)$ in $ABCD$, we compute the probability of $f(A,B) = f(C,D) > f(A,D) = f(B,C)$, $f(A,B) = f(C,D) > f(A,C) = f(B,D)$ and $f(A,D) = f(B,C) > f(A,C) = f(B,D)$ according to the three frequency $ABCD$s and note them as

$p(f(A,B) = f(C,D) > f(A,C) = f(B,D)) = 1$ and $p(f(A,B) = f(C,D) > f(A,D) = f(B,C)) = p(f(A,D) = f(B,C) > f(A,C) = f(B,D)) = \frac{2}{3}$.

Firstly, we keep A, B, C and replace vertex D with another vertex E. Based on the frequency consistency, $f(A,B) > f(A,C)$ occurs with the probability $\frac{2}{3}$ in the frequency $ABCE$. It says $d(A,B) + d(C,E) < d(A,C) + d(B,E)$ holds with the probability $\frac{2}{3}$ in $ABCE$. If $f(A,B) > f(A,C)$ in the frequency $ABCE$, $f(A,B) > f(A,E)$ and $f(A,E) > f(A,C)$ occur with the probability $\frac{2}{3}$, see the frequency quadrilaterals in Fig. 1. On the other hand, $f(A,B) < f(A,E)$ and $f(A,E) < f(A,C)$ occur with the probability $\frac{1}{3}$ in the frequency $ABCE$. As $f(A,B) < f(A,C)$ holds in the frequency $ABCE$, the probability of $f(A,B) < f(A,E)$ and $f(A,E) < f(A,C)$ is $\frac{2}{3}$ whereas the probability of $f(A,B) > f(A,E)$ and $f(A,E) > f(A,C)$ becomes $\frac{1}{3}$. In summary, the probability that $f(A,B) > f(A,E)$ and $f(A,E) > f(A,C)$ in the frequency $ABCE$ is computed as $\frac{2}{3} \times \frac{2}{3} + \frac{1}{3} \times \frac{1}{3} = \frac{5}{9}$.

Secondly, we keep A, B, E and replace vertex C with another vertex F and a quadrilateral $ABEF$ is obtained, see Fig. 3. Based on the frequency consistency, one can derive the probability of $f(A,B) > f(A,E)$ is $\frac{14}{27}$ in the frequency $ABEF$. As E and F change, there are $\binom{n-3}{2}$ frequency quadrilaterals where $f(A,B) > f(A,E)$ with the probability $\frac{14}{27}$. Simultaneously, one can substitute B with another vertex F in $ABCE$ and construct $\binom{n-3}{2}$ frequency quadrilaterals $ACEF$ in which $f(A,E) > f(A,C)$ occurs with the probability $\frac{14}{27}$. Thus, there are $\binom{n-3}{2}$ pairs of frequency $ABEF$ and $ACEF$ where $f(A,B) > f(A,E)$ and $f(A,E) > f(A,C)$ occur with the probability $\frac{14}{27}$, respectively. Based on these pairs of frequency $ABEF$ and $ACEF$, $f(A,B) > f(A,C)$ holds with the probability $\frac{14}{27}$. Plus the $n - 3$ frequency quadrilaterals $ABCD$s where $f(A,B) > f(A,C)$, $f(A,B) > f(A,C)$ occurs $\frac{14}{27}\binom{n-3}{2} + n - 3$ times according to the frequency quadrilaterals containing them.

$f(A,B)$ and $f(A,C)$ take the same frequencies 1, 3, 5 under the conditions $f(A,B) > f(A,C)$ and $f(A,C) > f(A,B)$, respectively. In this case, $F(A,B) > F(A,C)$ holds since the probability of $f(A,B) > f(A,C)$ is bigger than $\frac{14}{27}$ according to every frequency pairs $f(A,B)$ vs $f(A,C)$. It says the frequency of an OHC edge is bigger than that of a common edge. As the OHC becomes an HC by exchanging (two or more) edges, the frequency of the non-OHC edges is smaller than that of the replaced OHC edges. Thus, the frequency of the OHC is the maximum. □

6   Experimental Results

In this section, we conduct experiments for certain real-world TSP instances in order to demonstrate the frequency of the OHC edges. We verify the following viewpoint: the OHC is better cultivated in the frequency graph than in the weighted graph. In other words, the frequencies of the OHC edges are significantly higher than those of most other edges based on frequency quadrilaterals.

The TSP instances are downloaded from $TSPLIB$ [12]. In order to show the difference between the OHC edge frequency and that of the other edges, the

OHC of these TSP instances is computed using the Concorde package [13]. We use all of the frequency quadrilaterals to compute the frequency graph for each TSP instance, and record the minimum frequency $F_{min}(e_o) = f_{min}\binom{n-2}{2}$ of the OHC edge e_o. In each frequency graph, the number $N_{<f_{min}}$ of edges having the frequency below $F_{min}(e_o)$ is also recorded. As this computation consumes $O(n^4)$ time, we only did experiments for certain small and medium TSP instances. Even for medium TSP instances, the value of F_{min} and $N_{<f_{min}}$ will be very large. For the sake of convenience, we compute the values $f_{min} = \frac{F_{min}(e_o)}{\binom{n-2}{2}}$ and $\frac{N_{<f_{min}}}{\binom{n}{2}}$ for comparison. The experimental results are provided in Table 1.

For the various types of TSP instances, it can be seen that f_{min} is larger than $\frac{29}{9}$ for most of these small and medium TSP instances. The case of $f_{min} < \frac{29}{9}$ exists only for two small TSP instances ($n \leq 26$). In addition, all the $f_{min} > 3$ except for the small TSP instance ulysses16. Therefore, for most TSP instances, $\frac{29N}{9}$ can be used as the frequency threshold to eliminate the edges with a smaller frequency. The elimination of useless edges will speed up the algorithms to find the OHC for TSP, see the experiments in [14,15].

The value $\frac{N_{<f_{min}}}{\binom{n}{2}}$ illustrates the percentage of edges with the frequency below $F_{min}(e_o)$. The percentage is over 0.7 for most TSP instances, and even over 0.8 for most medium TSP instances. This means the frequency of the OHC edges is much higher than that of most of the other edges. Comparing to the weighted graph, the OHC edges are better cultivated in the frequency graph.

According to the frequency of edges, we can eliminate the useless edges in order to reduce the computation time of the algorithms for TSP. Taillard and Helsgaun [17] have done intensive experiments for resolving large scale of TSP instances. They argued that it was essential to reduce K_n of large TSP instances to sparse graphs for treatment. In early stage, Jonker and Volgenant [14] recognized certain useless edges based on $2 - opt$ rule for the symmetric TSP. The edge recognition was used to speed up the branch-and-bound method for TSP. The experiments for several $Euclidean\ TSP$ instances showed the computation time of branch-and-bound method was reduced to half. In 2014, Hougardy and Schroeder [15] eliminated the useless edges according to the $3 - opt$, and computed a sparse graph for TSP. Thereafter, they resolved the TSP instances on sparse graphs using the Concorde. Their experimental results demonstrated that the Concorde computation time was reduced by 11 times for certain large TSP instances. We also designed an algorithm to reduce the number of edges that must be considered based on frequency quadrilaterals [16]. In the algorithm, we firstly select N frequency quadrilaterals containing an edge to compute its average frequency. Following this, the edges with a small frequency below a given threshold are eliminated. The two steps are repeated until the algorithm reaches the terminal conditions. The experiments conducted on the TSP instances illustrated that our algorithm outperforms the edges elimination method [15].

Table 1. The minimum frequency of the OHC edge (n is the TSP scale).

TSP	n	f_{min}	$\frac{N_{<f_{min}}}{\binom{n}{2}}$	TSP	n	f_{min}	$\frac{N_{<f_{min}}}{\binom{n}{2}}$	TSP	n	f_{min}	$\frac{N_{<f_{min}}}{\binom{n}{2}}$
ulysses16	16	2.9560	0.4667	gr120	120	3.3372	0.6975	att532	532	3.9791	0.8851
gr17	17	3.2667	0.6875	bier127	127	3.9098	0.8651	ail535	535	3.9190	0.8801
gr21	21	3.4912	0.7500	ch130	130	3.7072	0.8062	rat575	575	3.8862	0.8693
gr24	24	3.2510	0.6522	ch150	150	3.6142	0.7852	p654	654	3.4890	0.7810
fri26	26	3.0072	0.5600	kroA150	150	3.7042	0.8188	d657	657	3.8669	0.8506
bays29	29	3.3418	0.7143	u159	159	3.6963	0.8228	gr666	666	4.0929	0.9128
att48	48	3.4946	0.7447	si175	175	4.0371	0.8793	u724	724	3.9748	0.8838
gr48	48	3.5140	0.7447	d198	198	3.5973	0.8071	rat783	783	3.9977	0.8926
eil51	51	3.7397	0.8200	kroB200	200	3.6288	0.7990	pr1002	1002	3.9109	0.8601
berlin52	52	3.7281	0.8235	gr202	202	4.0113	0.8955	u1060	1060	3.9109	0.8772
brazil58	58	3.6766	0.8070	tsp225	225	3.9140	0.8616	vm1084	1084	3.8889	0.8578
st70	70	3.2950	0.6667	gr229	229	3.3836	0.7061	pcb1173	1173	4.0065	0.8857
pr76	76	3.3983	0.7200	pr264	264	3.4532	0.7414	rl1323	1323	4.1215	0.9092
gr96	96	3.6639	0.8105	a280	280	3.6408	0.7993	u1432	1432	4.0468	0.8882
rat99	99	3.3862	0.7143	lin318	318	3.7391	0.8170	fl1577	1577	3.7533	0.8280
kroA100	100	3.3652	0.7071	rd400	400	3.8671	0.8471	d1655	1655	3.6385	0.7872
kroD100	100	3.3644	0.7071	gr431	431	3.6926	0.8163	vm1748	1748	4.0973	0.9027
eil101	101	3.7499	0.8200	pcb442	442	3.8624	0.8503	rl1889	1889	3.9141	0.8549
pr107	107	3.5366	0.7736	d493	493	3.8885	0.8679				

7   Conclusions

The frequency graph computed with frequency quadrilaterals is another proper model to represent TSP. When the frequency of an edge is computed with N frequency quadrilaterals, the frequency of the OHC edges will be higher than $\frac{29N}{9}$. The time to compute the frequency of all edges is $O(Nn^2)$. Moreover, the OHC has the maximum frequency among those of all HCs. The OHC is therefore better cultivated in the frequency graph. The edges with a high frequency are preferred as the candidate OHC edges. The lower frequency bound for the OHC edges is used to filter out the other edges with a smaller frequency, so that the OHC search space will be reduced.

We complete this paper with two questions for future research. The first question is the evaluation of the complexity of the exact or approximation algorithms used to resolve the TSP according to the frequency of edges. The second question is the exploration of the performance of the frequency graph computed with the i-vertex frequency subgraphs for TSP, where $i > 4$. In theory, the frequency graph computed with frequency subgraphs containing additional vertices will exhibit superior properties for resolving the TSP. We will pursue these questions in future research.

References

1. Johnson, D.S., McGeoch, L.A.: The Traveling Salesman Problem and Its Variations. Combinatorial Optimization. Springer, New York (2007)

2. Karp, R.: On the computational complexity of combinatorial problems. Networks (USA) **5**(1), 45–68 (1975)
3. Held, M., Karp, R.: A dynamic programming approach to sequencing problems. J. Soc. Ind. Appl. Math. **10**(1), 196–210 (1962)
4. Bellman, R.: Dynamic programming treatment of the traveling salesman problem. J. ACM **9**(1), 61–63 (1962)
5. de Klerk, E., Dobre, C.: A comparison of lower bounds for the symmetric circulant traveling salesman problem. Discrete Appl. Math. **159**(16), 1815–1826 (2011)
6. Levine, M.S.: Finding the right cutting planes for the TSP. J. Exp. Algorithmics (JEA) **5**(6), 1–16 (2000)
7. Applegate, D., et al.: Certification of an optimal TSP tour through 85900 cities. Oper. Res. Lett. **37**(1), 11–15 (2009)
8. Cook, W.: The traveling salesman problem: postcards from the edge of impossibility (Plenary talk). In: The 30th European Conference on Operational Research, Dublin, Ireland, 23–26 June 2019
9. Arora, S.: Polynomial time approximation schemes for Euclidean traveling salesman and other geometric problems. J. ACM **45**(5), 753–782 (1998)
10. Helsgaun, K.: An effective implementation of the Lin-Kernighan traveling salesman heuristic. Eur. J. Oper. Res. **126**(1), 106–130 (2000)
11. Wang, Y., Remmel, J.B.: A binomial distribution model for the traveling salesman problem based on frequency quadrilaterals. J. Graph Algorithms Appl. **20**(2), 411–434 (2016)
12. Reinelt, G.: http://comopt.ifi.uni-heidelberg.de/software/TSPLIB95/. Accessed 30 Apr 2020
13. Mittelmann, H.: http://NEOSServerforConcorde.neos-server.org/neos/solvers/co:concorde/TSP.html. Accessed 5 May 2020
14. Jonker, R., Volgenant, T.: Nonoptimal edges for the symmetric traveling salesman problem. Oper. Res. **32**(4), 837–846 (1984)
15. Hougardy, S., Schroeder, R.T.: Edge elimination in TSP instances. In: Kratsch, D., Todinca, I. (eds.) WG 2014. LNCS, vol. 8747, pp. 275–286. Springer, Cham (2014). https://doi.org/10.1007/978-3-319-12340-0_23
16. Wang, Y.: An improved method to compute sparse graphs for traveling salesman problem. Int. J. Ind. Manuf. Eng. **12**(3), 92–100 (2018)
17. Taillard, É.D., Helsgaun, K.: POPMUSIC for the traveling salesman problem. Eur. J. Oper. Res. **272**(2), 420–429 (2019)
18. Wang, Y., Remmel, J.B.: An iterative algorithm to eliminate edges for traveling salesman problem based on a new binomial distribution. Appl. Intell. **48**(11), 4470–4484 (2018). https://doi.org/10.1007/s10489-018-1222-2
19. Wang, Y.: Sufficient and necessary conditions for an edge in the optimal Hamiltonian cycle based on frequency quadrilaterals. J. Optim. Theory Appl. **181**(2), 671–683 (2019)

Single Bounded Parallel-Batch Machine Scheduling with an Unavailability Constraint and Job Delivery

Jing Fan[1], C. T. Ng[2], T. C. E. Cheng[2], and Hui Shi[3]($\boxtimes$)

[1] Shanghai Polytechnic University, Shanghai 201209, China
[2] Logistics Research Centre, Department of Logistics and Maritime Studies,
The Hong Kong Polytechnic University, Hong Kong SAR, China
[3] Shanghai General Hospital, School of Medicine, Shanghai Jiaotong University,
Shanghai 200080, China
sspu_fj@163.com

Abstract. We consider a scheduling problem where a manufacturer processes a set of jobs for a customer and delivers the completed jobs to the customer. The job sizes and processing times are given. The objective is to minimize the maximum delivery time to the customer. In the production stage, one machine with an unavailability period is used to process the jobs. The machine has a fixed capacity and the jobs are processed in batches under the condition that the total size of the jobs in a batch cannot exceed the machine capacity. The processing time of a batch is the maximum processing time of the jobs contained in the batch. In addition, each batch is non-resumable, i.e., if the processing of a batch cannot be completed before the unavailability period, the batch needs to be processed anew after the unavailability interval. In the distribution stage, the manufacturer assigns a vehicle with a fixed capacity to deliver the completed jobs. The total size of the completed jobs in one delivery cannot exceed the vehicle capacity. We first consider the case where the jobs have the same size and arbitrary processing times, for which we provide a 3/2-approximation algorithm and show that the worst-case ratio is tight. We then consider the case where the jobs have the same processing time and arbitrary sizes, for which we provide a 5/3-approximation algorithm and show that the worst-case ratio is tight.

Keywords: Parallel-batch · Production and delivery · Unavailability constraint · Approximation algorithm

This research was supported in part by the National Natural Science Foundation of China (No.11601316). Fan was also supported in part by the key discipline "Applied Mathematics" of Shanghai Polytechnic University (No. XXKPY1604), Research Center of Resource Recycling Science and Engineering, and Gaoyuan Discipline of Shanghai − Environmental Science and Engineering (Resource Recycling Science and Engineering) of Shanghai Polytechnic University. Cheng was also supported in part by The Hong Kong Polytechnic University under the Fung Yiu King - Wing Hang Bank Endowed Professorship in Business Administration.

© Springer Nature Switzerland AG 2020
Z. Zhang et al. (Eds.): AAIM 2020, LNCS 12290, pp. 525–536, 2020.
https://doi.org/10.1007/978-3-030-57602-8_47

1 Introduction

Although production scheduling integrated with logistics is more difficult to deal with than production scheduling alone from the theoretical research perspective, there is an abundance of literature on the former because such problems are more practical in the realistic manufacturing environment.

Potts [1] was probably the first researcher that considered scheduling with job delivery. Hall and Potts [2] studied integrated scheduling that involves a supplier, a manufacturer, and a customer. For integrated scheduling with one machine and one customer, Chen and Vairaktarakis [3] presented a polynomial-time dynamic programming solution algorithm to minimize the maximum arrival time, under the condition that the last completed job delivered to the customer and the total distribution cost do not exceed given bounds. Different from [3] in which the number of vehicles has no limit, Lee and Chen [4] considered the problem with a limited number of vehicles and showed that it is polynomially solvable by dynamic programming. Studying the same problem in [4] where the jobs have different sizes, Chang and Lee [5] showed that the problem is strongly NP-hard and provided a heuristic with a worst-case performance bound of 5/3. He et al. [6] and Zhong et al. [7] studied the same problem in [5], and proposed improved approximation algorithms with worst-case performance bounds of 53/35 and 3/2, respectively.

However, due to the occurrence of breakdowns or the necessity for maintenance and repair, the machines may become unavailable during the production stage. There are many studies on integrated scheduling under the constraint of machine unavailability. When job processing is interrupted by machine unavailability, the interrupted job is often assumed to be non-resumable, which means that the job needs to processed anew as defined in [8]. For the problem with delivery using a capacitated vehicle to minimize the maximum arrival time, Wang and Cheng [9] showed that it is NP-hard, and proposed a 3/2-approximation algorithm and showed that the worst-case ratio is tight. More details on this research stream can be found in Chen [10] and Ma et al. [11].

In most models of scheduling with delivery, the machine processes one job at a time. However, it is noted that batch-processing machines with limited capacities are also widely used in real production. When a bounded parallel-batch machine is used for production, the jobs are processed in batches under the condition that the total size of the jobs in a batch cannot exceed the machine capacity, and the processing time of a batch is equal to the longest processing time of the jobs in it. Li et al. [12] and Gong et al. [13] considered several parallel-batch machine scheduling problems with job delivery. Lu and Yuan [14] considered unbounded parallel-batch scheduling with job delivery to minimize the makespan. They provided a polynomial-time algorithm to solve the case where the jobs have identical sizes, and a heuristic with a worst-case performance ratio of 7/4 for the case where the jobs have non-identical sizes. Cheng et al. [15] considered integrated scheduling of production and distribution on parallel batch-processing machines. They presented a $(2-1/m)$-approximation algorithm for the case where the jobs have the same size and arbitrary processing times, and provided a 13/7-approximation algorithm for the case where the jobs have the same processing time and arbitrary sizes.

In this paper we consider scheduling with job delivery for a customer on a bounded parallel-batch machine with a machine unavailability period, where a capacitated vehicle is used to deliver the completed jobs to the customer. The objective is to minimize the time of the last completed job delivered to the customer. We consider two cases of the problem corresponding to different conditions on the job processing times and sizes, and design approximation algorithms for them.

2 Description of the Problem

In our problem, there is a manufacturer that processes jobs on a bounded parallel-batch machine and delivers the completed jobs to a customer. Given a set of n jobs $J = \{J_1, J_2, \cdots, J_n\}$, where job J_j has the processing time p_j and size s_j for $j = 1, 2, ..., n$. In the production stage, the machine has a capacity B, i.e., it can simultaneously process at most B jobs as a batch. The processing time of a batch is the maximum processing time of the jobs contained in the batch. Due to reasons such as maintenance, breakdown etc, the machine has an unavailability period $[t_1, t_2]$. Let I be the length of the unavailability period, i.e., $I = t_2 - t_1$. The processing of batches is non-resumable, i.e., if there is at least one job in a batch that is interrupted by the unavailable period $[t_1, t_2]$, the whole batch needs to be processed anew after t_2. In the delivery stage, there is a vehicle with a capacity c to deliver the completed jobs to the customer. The transport time between the machine and the customer is T. Let D_j be the delivery time of job J_j, i.e., the arrival time of the batch containing job J_j to the customer. The objective is to minimize the maximum delivery time of all the jobs, denoted by D_{max}.

Chen [10] proposed a five-field notation to denote an integrated scheduling problem as $\alpha|\beta|\pi|\delta|\gamma$, where α represents the facility configuration of the manufacturer; β represents the production constraints; π represents the vehicle configuration and is often denoted by (v_1, v_2), where v_1 represents the number of vehicles and v_2 represents the vehicle capacity; δ represents the number of customers; and γ represents the scheduling objective to be minimized. Using the above notation, we denote the two cases of the problem under study as follows:

$$(\mathcal{P}1) : 1, h_1|nr - a, p - batch, s_j = 1, p_j|V(1, c)|1|D_{max},$$
$$(\mathcal{P}2) : 1, h_1|nr - a, p - batch, s_j, p_j = 1|V(1, c)|1|D_{max}.$$

For problem $(\mathcal{P}1)$, each job has a unit size but an arbitrary processing time. On the contrary, for problem $(\mathcal{P}2)$, each job has an arbitrary size but a unit processing time.

We organize the rest of the paper as follows: In Sect. 3 we show that problem $(\mathcal{P}1)$ is NP-hard and propose an approximation algorithm for it. In Sect. 4 we prove that problem $(\mathcal{P}2)$ is strongly NP-hard and present an approximation algorithm for it.

3   Algorithm for $(\mathcal{P}1) : 1, h_1 | nr - a, p - batch, s_j = 1,$ $p_j | V(1, c) | 1 | D_{max}$

In this section we study problem $(\mathcal{P}1)$, where the jobs have the same size, i.e., $s_j = 1$. Similar to [15], we assume that the manufacturer uses appropriate equipment to improve the efficiency of the supply chain. Specifically, we let $c = \mu B$, where $\mu \geq 2$ and μ is a positive integer.

We first analyze the computational complexity of $(\mathcal{P}1)$.

Theorem 1. $(\mathcal{P}1) : 1, h_1 | nr - a, p - batch, s_j = 1, p_j | V(1, c) | 1 | D_{max}$ is NP-hard.

Proof. Consider the special case $(\mathcal{P}1')$, where $B = 1$, i.e., each batch contains at most one job. It is clear that $(\mathcal{P}1')$ is equivalent to $1, h_1 | nr - a | V(1, c) | 1 | D'_{max}$, where D'_{max} is the maximum of the return time D'_j of the vehicle after delivering the completed jobs to the customer. Obviously, $D'_j = D_j + T$ for job J_j. Given that Wang and Cheng [9] have shown that problem $(\mathcal{P}1')$ is NP-hard, we obtain the conclusion.

Next, we derive some properties of the optimal solution for problem $(\mathcal{P}1)$.

Lemma 1. *There exists an optimal schedule σ^* possessing the following properties:*

(1) Let N^* to be the number of batches, then $N^* = \lceil n/B \rceil$;
(2) The batches are processed consecutively before and after the unavailability period;
(3) The batch that becomes available earlier is delivered earlier.
(4) The first delivery includes b^* batches, and each of the last a^* deliveries includes μ batches, where a^* and b^* are two positive integers satisfying $N^* = a^*\mu + b^*$ and $0 < b^* \leq \mu$, respectively.

The lemma can be proved similar to the proof in [9]. Because every job has the same size 1, we construct the following algorithm including the same number of batches as the optimal schedule.

Algorithm A1

Step 1: Sequence all the jobs in non-increasing order of their processing times.

Step 2: Create the first batch H_{N^*} and put the first B jobs in H_{N^*}. Then create batch H_{N^*-1} and put the next B jobs in it. Repeat the assignment until there are y jobs left, where $0 < y \leq B$. Put them in batch H_1. The obtained batch set is $\{H_1, H_2, ..., H_{N^*}\}$. The batches are in non-decreasing order of their batch processing times.

Step 3: Regard batch H_j as job $\widetilde{J}_j$ for $j = 1, 2, ..., N^*$, whose processing time is the maximum processing time of the jobs in H_j and the size is 1. Take the obtained job set $\{\widetilde{J}_1, \widetilde{J}_2, ..., \widetilde{J}_N^*\}$ as the job set of problem $(\mathcal{P}1')$:

$1, h_1|nr - a|V(1,c)|1|D'_{max}$. Use the approximation algorithm proposed in Wang and Cheng [9] to obtain the schedule σ.

Furthermore, we can obtain the following lemma.

Lemma 2. *Sorting all the batches of σ and σ^* in non-decreasing order, we have $P(H_j) \leq P(H_j^*)$ for $j = 1, ..., N^*$, where $P(H_j)$ and $P(H_j^*)$ are the processing times of the j-th batch in schedules σ and σ^*, respectively,*

The proof is similar to Cheng et al. [15]. Given that the worst-case ratio of the approximation algorithm Wang and Cheng [9] proposed for $(\mathcal{P}1')$ is $3/2$, we derive the worst-case ratio of algorithm $A1$ for $(\mathcal{P}1)$ as follows:

Theorem 2. *Solving $(\mathcal{P}1)$, algorithm $A1$ has the worst-case ratio of $3/2$, which is tight.*

Proof. It suffices to prove that the objective value produced by the optimal schedule for $(\mathcal{P}1')$, denoted by $\sigma^*(\mathcal{P}1')$, is not greater than that for $(\mathcal{P}1)$, denoted by $\sigma^*(\mathcal{P}1)$. Otherwise, we construct a new schedule $\hat{\sigma}(\mathcal{P}1)$ by replacing the corresponding batches of $\sigma^*(\mathcal{P}1)$ with the batches of $\sigma^*(\mathcal{P}1')$ to process in non-decreasing order of the batches in the two schedules, and delivering the batches as $\sigma^*(\mathcal{P}1)$. Because the completion times of the batches in $\hat{\sigma}(\mathcal{P}1)$ are no later than those in $\sigma^*(\mathcal{P}1)$ by Lemma 2, so are the delivery times. Hence, the new schedule $\hat{\sigma}(\mathcal{P}1)$ is no worse than $\sigma^*(\mathcal{P}1)$.

Next, consider the following instance: $n = 6$, $B = 2$, $\mu = 2$, $[t_1, t_2] = [2, 2+\epsilon]$, $p_1 = p_2 = \epsilon$, $p_3 = p_4 = 1$, and $p_5 = p_6 = 1$. The delivery time is $T = \epsilon$. The schedule produced by algorithm $A1$ is as follows: The first delivery including J_1 and J_2 is delivered at ϵ, and the second delivery including J_3, J_4, J_5, and J_6 is delivered at $3 + \epsilon$. Hence, $D_{max} = 3 + 2\epsilon$. However, in the optimal schedule, there are two deliveries: the first delivery consisting of J_3 and J_4, and the second delivery consisting of the others. The optimal objective function value is $D_{max}^* = 2 + 3\epsilon$. So, we have $\frac{D_{max}}{D_{max}^*} = \frac{3}{2}$ if ϵ is sufficiently small.

4   Algorithm for $(\mathcal{P}2)$: $1, h_1|nr - a, p - batch, s_j, p_j = 1|V(1,c)|1|D_{max}$

In $(\mathcal{P}2)$: $1, h_1|nr - a, p - batch, s_j, p_j = 1|V(1,c)|1|D_{max}$, all the jobs have a unit processing time but arbitrary sizes. First we analyze the computational complexity of $(\mathcal{P}2)$.

Theorem 3. $(\mathcal{P}2)$: $1, h_1|nr - a, p - batch, s_j, p_j = 1|V(1,c)|1|D_{max}$ *is strongly NP-hard.*

Proof. Consider the special case $(\mathcal{P}2')$, where $t_1 = t_2$ and $T = 0$, i.e., there is no unavailability interval on the machine and no delivery is needed. Since each job has a processing time of 1, every batch has a processing time of 1. Hence, $(\mathcal{P}2')$ is equivalent to minimizing the number of batches, i.e., the bin-packing problem, which is a well-known strongly NP-hard problem. Therefore, $(\mathcal{P}2)$, as well as $(\mathcal{P}2')$, is strongly NP-hard.

Obviously, the optimal schedule for ($\mathcal{P}2$) possesses properties (2)–(4) in Lemma 1.

In this section we use the same notation and the corresponding meanings, such as N and N^*, as those used in Sect. 3. Next, we propose the following approximation algorithm for ($\mathcal{P}2$).

Algorithm A2

Step 1: Sort the jobs in non-increasing order of their sizes. Re-label them as job $J_1, J_2, \cdots, J_n$.

Step 2: Use the First Fit Decreasing (FFD) rule to assign the jobs into batches. Create the first empty batch H_1 and put job J_1 in it. Check the following jobs one by one as to whether it can be put in the batch. If so, put the job in H_1 and delete it from the job list. If not, go on to check the next job. When all the jobs have been assigned, create the second batch H_2 and assign the remaining jobs in the job list. Repeat the assignment until there is no job in the job list. The obtained batch set is $\{H_1, ..., H_N\}$.

Step 3: Assign the batches in an arbitrary order to the machine for processing.

Step 4: Deliver the first completed b batches in D_1. For the following batches, deliver μ batches immediately in each delivery. If the vehicle is available when the μ batches are completed, deliver them immediately. If the vehicle is not available at the time, wait until the vehicle returns to the manufacturer and deliver the batches. When the last μ batches are delivered to the customer, production and distribution are finished, and the obtained deliveries are $D_1, ..., D_{a+1}$.

To analyze the performance of algorithm A2, recall that for the bin-packing problem, the number of bins obtained by FFD is no more than the sum of 6/9 and 11/9 times of the optimal number of bins. In algorithm $A2$, Steps 1 and 2 assign the jobs to batches by the FFD rule, so we have the following results on N and a.

Lemma 3. ([16]) $N \leq \frac{11}{9}N^* + \frac{6}{9}$, where N^* is the optimal number of batches.

Lemma 4. $a^* \leq a < \frac{11}{9}a^* + \frac{14}{9}$.

Proof. It is clear that $a^* \leq a$.

By Lemma 3, we have $N \leq \frac{11}{9}N^* + \frac{6}{9} \leq \frac{11}{9}a^*\mu + \frac{11b^*+6}{9}$ and $\frac{N}{\mu} \leq \frac{11}{9}a^* + \frac{11b^*+6}{9\mu}$. Since $0 < b^* \leq \mu$ and $\mu \geq 2$,

$$a \leq \frac{N}{\mu} \leq \frac{11}{9}a^* + \frac{11b^*+6}{9\mu} \leq \frac{11}{9}a^* + \frac{11}{9} + \frac{6}{9\mu} \leq \frac{11}{9}a^* + \frac{14}{9}.$$

But if $a = \frac{11}{9}a^* + \frac{14}{9}$, by $N \leq \frac{11}{9}N^* + \frac{6}{9}$, $N = a\mu + b$, and $N^* = a^*\mu + b^*$, we have $\frac{11}{9}a^*\mu + \frac{14}{9}\mu + b \leq \frac{11}{9}(a^*\mu + b^*) + \frac{6}{9}$, i.e.,

$$\frac{14}{9}\mu < \frac{11}{9}b^* + \frac{6}{9} \leq \frac{11}{9}\mu + \frac{6}{9}.$$

So, we deduce that $\mu < 2$, which contradicts the assumption $\mu \geq 2$.

As a result, we can easily obtain the maximal value of a when $a^* \leq 4$ in Table 1.

Table 1. The maximal values of a when $a^* \leq 4$.

a^*	1	2	3	4
Maximal value of a	2	3	5	6

For convenience, we use L and L^* to denote the numbers of deliveries in schedules π and π^*, respectively, which means $L = a + 1$ and $L^* = a^* + 1$. Meanwhile, we use $C(D_j)$ and $C(D_j^*)$ to denote the completion times of deliveries D_j and D_j^* in π and π^*, and D_{max} and D_{max}^* to denote the objective values of π and π^*, respectively. Because every job has a unit processing time, we can easily obtain the following relationships between π and π^*.

Lemma 5. *(1) $\lambda = \lambda^*$, where λ and λ^* are the numbers of batches completed before the unavailability interval $[t_1, t_2]$ in π and π^*, respectively;*
(2) $\delta \leq \delta^*$, where δ and δ^* are the idle times on the machine before $[t_1, t_2]$ in π and π^*, respectively.

The results are obvious and we omit the proof.

Lemma 6. $k - 1 \leq l \leq k + 1$, *where l and k are the first deliveries completed after the unavailability interval $[t_1, t_2]$ in π and π^*, respectively.*

Proof. Because k is the first delivery completed after t_2, there are at most $k\mu$ batches completed in the total k deliveries in π^*. We prove the result by contradiction.

If $l \geq k + 2$, there are at least $(k + 1)$ deliveries before t_1 in π, i.e., there are at least $k\mu$ batches completed before t_1. So it is a contradiction.

If $l \leq k - 2$, there are $(k - 1)$ deliveries before t_1 in π^*, i.e., there are no fewer than $(k - 2)\mu$ batches completed before t_1, which is a contradiction to the situation that at most $(k - 2)\mu$ batches completed after t_2 in π.

In the following we analyze the worst-case ratios of algorithm $A2$ for $(\mathcal{P}2)$ according to $a^* \geq 5$ and $a^* \leq 4$, respectively.

Lemma 7. *When $a^* \geq 5$, the worst-case ratio of algorithm $A2$ is $\frac{5}{3}$.*

Proof. We prove the result in three cases.

Case 1: $C(D_1^*) \geq t_2$, which means $k = 1$ and $l = 1$ or 2.

- If $D^*_{max} = C(D^*_{L^*}) + T = N^* + I + \delta^* + T$, then we deduce that $\mu \geq 2T$. Moreover, $D_{max} = C(D_L) + T = N + I + \delta + T$, so

$$\frac{D_{max} - D^*_{max}}{D^*_{max}} \leq \frac{N - N^*}{N^*} = \frac{N}{N^*} - 1.$$

Since $a^* \geq 5$ and $\mu \geq 2$, $N^* \geq 10$. Similar to [15], we can find positive integers $\alpha \geq 1$ and $1 \leq \beta \leq 9$ such that $N^* = 9\alpha + \beta$, and we can obtain an upper bound on $\frac{N}{N^*}$ from Table 2 as follows:

$$\frac{N}{N^*} \leq \frac{9\alpha + 2}{11\alpha + 3} \leq \frac{9}{11} + \frac{5}{9(9\alpha + 2)} \leq \frac{14}{11} < \frac{5}{3}. \tag{1}$$

Table 2. The upper bound on $\frac{N}{N^*}$.

N^*	$9\alpha+1$	$9\alpha+2$	$9\alpha+3$	$9\alpha+4$	$9\alpha+5$	$9\alpha+6$	$9\alpha+7$	$9\alpha+8$	$9\alpha+9$
Maximal value of N	$11\alpha+1$	$11\alpha+3$	$11\alpha+4$	$11\alpha+5$	$11\alpha+6$	$11\alpha+8$	$11\alpha+9$	$11\alpha+10$	$11\alpha+11$
Upper bound on $\frac{N}{N^*}$	$\frac{11}{9}$	$\frac{14}{11}$	$\frac{5}{4}$	$\frac{16}{13}$	$\frac{11}{9}$	$\frac{19}{15}$	$\frac{5}{4}$	$\frac{21}{17}$	$\frac{11}{9}$

- If $D^*_{max} = C(D^*_1) + (2a^* + 1)T = b^* + I + \delta^* + (2a^* + 1)T$, then $\mu < 2T$ and the objective value of schedule π is

$$D_{max} = \begin{cases} C(D_1) + (2a + 1)T, & \text{for } l = 1 \text{ or } 2, \\ C(D_l) + (2(a - l) + 3)T, & \text{for } l = 2. \end{cases}$$

For the first case, $D_{max} \leq b + I + \delta + (2a + 1)T$. By Lemma 5 and $b - b^* \leq \mu$, we have

$$\frac{D_{max} - D^*_{max}}{D^*_{max}} \leq \frac{(b - b^*) + 2(a - a^*)T}{C(D^*_1) + (2a^* + 1)T} < \frac{a - a^* + 1}{a^* + \frac{1}{2}}.$$

Moreover, because of $a^* \geq 5$ and Lemma 4, we have

$$\frac{a - a^* + 1}{a^* + \frac{1}{2}} \leq \frac{2}{9} + \frac{22}{9(a^* + \frac{1}{2})} \leq \frac{2}{3}. \tag{2}$$

For the second case, we have $b < b^*$; otherwise, $C(D_1) > t_2$, which contradicts $l = 2$. Hence,

$$\begin{aligned} \frac{D_{max} - D^*_{max}}{D^*_{max}} &\leq \frac{(b - b^*) + (l - 1)\mu + 2(a - a^* - l + 1)T}{C(D^*_1) + (2a^* + 1)T} \\ &= \frac{(b - b^*) + 2(a - a^*)T + \mu - 2T}{C(D^*_1) + (2a^* + 1)T} \\ &\leq \frac{a - a^*}{a^* + \frac{1}{2}} < \frac{2}{3}. \end{aligned}$$

Case 2: $C(D^*_k) > t_2$ for $1 < k < L^*$, where D^*_k is the first delivery completed after the unavailability interval $[t_1, t_2]$ in π^*.

- If $D^*_{max} = C(D^*_{L^*}) + T = N^* + I + \delta^* + T$ and $D_{max} = C(D_L) + T = N + I + \delta + T$, we obtain the same result as (1).
- If $D^*_{max} = C(D^*_1) + (2a^* + 1)T = b^* + I + \delta^* + (2a^* + 1)T$, then $\mu < 2T$ and $2(k-1)T > (k-1)\mu + I + \delta^*$. Therefore, the objective vale of schedule π is

$$
D_{max} = \begin{cases} C(D_1) + (2a+1)T, & \text{for } l = k-1, \ k \text{ or } k+1, \\ C(D_l) + (2(a-l)+3)T, & \text{for } l = k-1. \end{cases}
$$

For the first case, we obtain the same result as (2). For the second case, we have

$$
\frac{D_{max} - D^*_{max}}{D^*_{max}} \le \frac{(b - b^*) + (l-1)\mu + I + \delta + 2(a - a^* - l + 1)T}{b^* + (2a^* + 1)T}.
$$

Because of $b - b^* \le \mu$ and $l = k-1$, $(b-b^*) + (l-1)\mu + I + \delta + 2(a - a^* - l + 1)T \le 2(a - a^* + 1)T$. Moreover, $\frac{D_{max} - D^*_{max}}{D^*_{max}} \le \frac{2}{3}$.

- If $D^*_{max} = C(D^*_k) + (2(a^* - k) + 3)T = b^* + (k-1)\mu + I + \delta^* + (2(a^* - k) + 3)T$, then $\mu < 2T$ and $2(k-1)T \le (k-1)\mu + I + \delta^*$. Hence, the objective value of schedule π is

$$
D_{max} = \begin{cases} C(D_1) + (2a+1)T, & \text{for } l = k+1, \\ C(D_l) + (2(a-l)+3)T, & \text{for } l = k-1, \ k \text{ or } k+1. \end{cases}
$$

For the first case, we have

$$
\begin{aligned}
\frac{D_{max} - D^*_{max}}{D^*_{max}} &\le \frac{(b - b^*) + 2(a - a^*)T - ((k-1)\mu + I + \delta^* + (2 - 2k)T)}{b^* + (k-1)\mu + I + \delta^* + (2(a^* - k) + 3)T} \\
&\le \frac{(b - b^*) + 2(a - a^*)T}{b^* + 2(k-1)T + (2(a^* - k) + 3)T} \\
&\le \frac{2(a - a^* + 1)T}{2(k-1)T + (2(a^* - k) + 3)T} \\
&= \frac{a - a^* + 1}{a^* + \frac{1}{2}} \le \frac{2}{3}.
\end{aligned}
$$

For the second case, we have

$$
\begin{aligned}
\frac{D_{max} - D^*_{max}}{D^*_{max}} &\le \frac{(b - b^*) + 2(a - a^*)T + (l - k)(\mu - T))}{b^* + (k-1)\mu + I + \delta^* + (2(a^* - k) + 3)T} \\
&\le \frac{(b - b^*) + 2(a - a^*)T + (l - k)(\mu - T))}{(2a^* + 1)T}.
\end{aligned}
$$

When $l = k-1$ or k, we easily obtain $\frac{D_{max} - D^*_{max}}{D^*_{max}} \le \frac{a - a^* + 1}{a^* + \frac{1}{2}} \le \frac{2}{3}$. When $l = k+1$, we have $b \le b^*$; otherwise, $C(D_{l-1}) > t_2$, which contradicts the definition of l. Hence, we obtain the same result.

Case 3: $C(D^*_{L^*}) \ge t_2$, i.e., the last delivery $D^*_{L^*}$ is completed after the unavailability interval $[t_1, t_2]$ in π^*. Obviously, $L^* = a^* + 1$.

- If $D^*_{max} = C(D^*_{L^*}) + T = b^* + (L^* - 1)\mu + I + \delta^* + T = b^* + a^*\mu + I + \delta^* + T$, then we deduce that $2a^*T \le a^*\mu + I + \delta^*$. Moreover,

$$
D_{max} = \begin{cases} C(D_L) + T, & \text{for } l = L^* - 1, \ L^* \text{ or } L^* + 1, \\ C(D_l) + (2(a - l) + 3)T, & \text{for } l = L^* - 1, \ L^*, \\ C(D_1) + (2a + 1)T, & \text{for } l = L^* + 1. \end{cases}
$$

For the first case, we obtain the same result as (1). For the second case, we have $\mu < 2T$, $D_{max} = b + (l-1)\mu + I + \delta + (2(a-l)+3)T$, and

$$\frac{D_{max}-D^*_{max}}{D^*_{max}} \leq \frac{(b-b^*)+(l-L^*)\mu+2(a-l+1)T}{b^*+(L^*-1)\mu+I+\delta^*+T}$$
$$\leq \frac{(b-b^*)+(l-L^*)\mu+2(a-l+1)T}{2(L^*-\frac{1}{2})T}.$$

When $l = L^* - 1$, $(b-b^*)+(l-L^*)\mu+2(a-l+1)T \leq \mu - \mu + 2(a-a^*+1)T = 2(a-a^*+1)T$. When $l = L^*$, $(b-b^*)+(l-L^*)\mu+2(a-l+1)T \leq \mu + 0 + 2(a-a^*)T \leq 2(a-a^*+\frac{1}{2})T < 2(a-a^*+1)T$. Therefore,

$$\frac{D_{max}-D^*_{max}}{D^*_{max}} \leq \frac{a-a^*+1}{a^*+\frac{1}{2}} \leq \frac{2}{3}.$$

For the third case, since $\mu < 2T$ and $D_{max} = C(D_1)+(2a+1)T = b+(2a+1)T$, we have

$$\frac{D_{max}-D^*_{max}}{D^*_{max}} = \frac{(b-b^*)-(a^*\mu+I+\delta^*)+2aT}{b^*+a^*\mu+I+\delta^*+T}$$
$$\leq \frac{(b-b^*)+2(a-a^*)T}{2(a^*+\frac{1}{2})T}$$
$$\leq \frac{a-a^*+1}{a^*+\frac{1}{2}} \leq \frac{2}{3}.$$

- If $D^*_{max} = C(D_1^*)+(2a^*+1)T = b^*+(2a^*+1)T$, then $\mu < 2T$ and $2(L^*-1)T > (L^*-1)\mu+I+\delta^*$, i.e., $2a^*T > a^*\mu + I + \delta^*$. We obtain

$$D_{max} = \begin{cases} C(D_1) + (2a+1)T, & \text{for } l = L^*-1, \; L^* \text{ or } L^*+1, \\ C(D_l) + (2(a-l)+3)T, & \text{for } l = L^*-1. \end{cases}$$

For the first case, we achieve the same result as (2). For the second case, since $l = L^* - 1$, we have $2(L^*-2)T \leq (L^*-2)\mu+I+\delta^*$ and $D_{max} = b+(l-1)\mu+I+\delta+(2(a-l)+3)T \leq b+2(L^*-2)T+2(a-L^*+1)T+3T = b+(2a+1)T = C(D_1)+(2a+1)T$. Hence, we obtain the inequalities in (2).

Lemma 8. *When $a^* \leq 4$, the worst-case ratio of algorithm A2 is $\frac{5}{3}$.*

Proof. We prove the result in two cases.

Case 1: $b \leq b^*$. In fact, this inequality holds for $a^* \leq 4$ and the values of a satisfy Lemma 4 except $a^* = 3$ and $a = 4$. Now we show it by contradiction. If $b > b^*$, then $a\mu + b^* < a\mu + b \leq \frac{11}{9}(a^*\mu + b^*) + \frac{6}{9} = \frac{11}{9}a^*\mu + \frac{11}{9}b^* + \frac{6}{9}$, i.e., $(a - \frac{11}{9}a^*)\mu < \frac{2}{9}b^* + \frac{6}{9} \leq \frac{2}{9}\mu + \frac{6}{9}$. So

$$(a - \frac{11}{9}a^* - \frac{2}{9})\mu < \frac{6}{9}.$$

Using the corresponding data in Table 1, we obtain $\mu < 2$, contradicting $\mu \geq 2$.

Most parts of the remaining proof are similar to Lemma 7. Here we discuss two different situations.

The first situation is that $D^*_{max} = C(D^*_{L^*})+T = N^*+I+\delta^*+T = a^*\mu + b^*+I+\delta^*+T$ and $D_{max} = C(D_L)+T = N+I+\delta+T = a\mu+bI+\delta+T$. Hence,

$\frac{D_{max}-D^*_{max}}{D^*_{max}} = \frac{(b-b^*)+(a-a^*)\mu}{b^*+a^*\mu+I+\delta^*+T} \leq \frac{a-a^*}{a^*}$. For $a^* = 2,3,4$ and the corresponding maximal value of a in Table 1, we obtain $\frac{D_{max}-D^*_{max}}{D^*_{max}} \leq \frac{2}{3}$. For $a^* = 1$ and $a = 2$, we have $\frac{D_{max}-D^*_{max}}{D^*_{max}} \leq \frac{N-N^*}{N^*} = \frac{N}{N^*} - 1$. Given the upper bound on $\frac{N}{N^*}$ for $N^* \leq 9$ in Table 3, we have $\frac{N}{N^*} \leq \frac{3}{2} < \frac{5}{3}$.

Table 3. The upper bound on $\frac{N}{N^*}$.

N^*	1	2	3	4	5	6	7	8	9
Maximal value of N	1	3	4	5	6	8	9	10	11
Upper bound on $\frac{N}{N^*}$	1	$\frac{3}{2}$	$\frac{4}{3}$	$\frac{5}{4}$	$\frac{6}{5}$	$\frac{4}{3}$	$\frac{9}{7}$	$\frac{5}{4}$	$\frac{11}{9}$

The second situation corresponds to other combinations of D^*_{max} and D_{max}. We always obtain

$$\frac{D_{max}-D^*_{max}}{D^*_{max}} \leq \frac{(b-b^*)+2(a-a^*)T}{2(a^*+\frac{1}{2})T} \leq \frac{a-a^*}{a^*+\frac{1}{2}} = \frac{a+\frac{1}{2}}{a^*+\frac{1}{2}} - 1.$$

The upper bound on $\frac{a+\frac{1}{2}}{a^*+\frac{1}{2}}$ is $\frac{5}{3}$, which we deduce from Table 4.

Table 4. The upper bound on $\frac{a+\frac{1}{2}}{a^*+\frac{1}{2}}$ when $a^* \leq 4$.

a^*	1	2	3	4
Upper bound on $\frac{a+\frac{1}{2}}{a^*+\frac{1}{2}}$	$\frac{5}{3}$	$\frac{7}{5}$	$\frac{11}{7}$	$\frac{13}{9}$

Case 2: $b > b^*$ for $a^* = 3$ and $a = 4$. Note that $k - 1 \leq l \leq k$, i.e., $l \neq k + 1$. Most parts of the remaining proof are similar to Lemma 7. Here we discuss two different situations.

The first situation is that $D^*_{max} = C(D^*_{L^*}) + T = a^*\mu + b^* + I + \delta^* + T$ and $D_{max} = C(D_L) + T = a\mu + b + I + \delta + T$. Since $b - b^* \leq \mu$, $\frac{D_{max}-D^*_{max}}{D^*_{max}} = \frac{(b-b^*)+(a-a^*)\mu}{b^*+a^*\mu+I+\delta^*+T} \leq \frac{a-a^*+1}{a^*} = \frac{2}{3}$.

The second situation is that $D^*_{max} = C(D^*_1) + (2a^* + 1)T = b^* + (2a^* + 1)T$ and $D_{max} = C(D_1) + (2a + 1)T = b + I + \delta + (2a + 1)T$ for $k = 2$ and $l = 1$. Given that $\mu + I + \delta \leq 2T$ in this situation, $\frac{D_{max}-D^*_{max}}{D^*_{max}} = \frac{(b-b^*)+I+\delta+2(a-a^*)T}{b^*+(2a^*+1)T} \leq \frac{\mu+I+\delta+2(a-a^*)T}{b^*+(2a^*+1)T} < \frac{4}{7} < \frac{2}{3}$.

From Lemma 7 and Lemma 8, we derive the performance of $A2$ as follows:

Theorem 4. *Solving* $(\mathcal{P}2) : 1, h_1|nr - a, p - batch, s_j, p_j = 1|V(1,v)|1|D_{max}$, *algorithm A2 has the worst-case ratio of* $\frac{5}{3}$, *which is tight.*

Proof. It is obvious that the worst-case ratio of algorithm $A1$ is $\frac{5}{3}$ by Lemma 7 and Lemma 8.

Next, consider the following instance: $n = 12$, $B = 7$, $\mu = 2$, $[t_1, t_2] = [1, 1+\epsilon]$, $s_1 = s_2 = s_3 = s_4 = 3$, and $s_5 = s_6 = \ldots = s_{12} = 2$. The delivery time is $T > \frac{2+\epsilon}{2}$. The schedule produced by algorithm $A2$ is as follows: the first delivery including J_1 and J_2 is delivered at time 1; the second delivery including J_3, ..., J_7 and the third delivery including J_8, ..., J_{12} are delivered at $1 + 2T$ and $1 + 4T$, respectively. Hence, $D_{max} = 1 + 5T$. However, in the optimal schedule, there are two deliveries: the first delivery consisting of J_1, J_2, J_5, J_6, J_7, and J_8, and the second delivery consisting of the remaining jobs. The optimal objective value is $D_{max}^* = 2 + 3T$. When $T \to +\infty$, $\frac{D_{max}}{D_{max}^*} = \frac{1+5T}{2+3T} \to \frac{5}{3}$.

References

1. Potts, C.N.: Analysis of a heuristic for one machine sequencing with release dates and delivery times. Oper. Res. **28**, 1436–1441 (1980)
2. Hall, N.G., Potts, C.N.: Supply chain scheduling: batching and delivery. Oper. Res. **51**(4), 566–584 (2003)
3. Chen, Z.L., Vairaktarakis, G.L.: Integrated scheduling of production and distribution operations. Manag. Sci. **51**(4), 614–628 (2005)
4. Lee, C.Y., Chen, Z.L.: Machine scheduling with transportation considerations. J. Sched. **4**, 3–24 (2001)
5. Chang, Y.C., Lee, C.Y.: Machine scheduling with job delivery coordination. Eur. J. Oper. Res. **158**, 470–487 (2004)
6. He, Y., Zhong, W.Y., Gu, H.K.: Improved algorithms for two single machine scheduling problems. Theor. Comput. Sci. **363**, 257–265 (2006)
7. Zhong, W., Dosa, G., Tan, Z.Y.: On the machine scheduling problem with job delivery coordination. Eur. J. Oper. Res. **182**(3), 1057–1072 (2007)
8. Lee, C.Y.: Machine scheduling with an availability constraints. J. Global Optim. **9**, 363–382 (1996)
9. Wang, X., Cheng, T.C.E.: Machine scheduling with an availability constraint and job delivery coordination. Naval Res. Log. **54**, 11–20 (2007)
10. Chen, Z.L.: Integrated production and outbound distribution scheduling: review and extensions. Oper. Res. **58**, 130–148 (2010)
11. Ma, Y., Chu, C.B., Zuo, C.R.: A survey of scheduling with deterministic machine availability constraints. Comput. Ind. Eng. **58**, 199–211 (2010)
12. Li, S.S., Yuan, J.J., Fan, B.Q.: Unbounded parallel-batch scheduling with family jobs and delivery coordination. Inf. Process. Lett. **111**(12), 575–582 (2011)
13. Gong, H., Chen, D.H., Xu, K.: Parallel-batch scheduling and transportation coordination with waiting time constraint. Sci World J. **15**, 1–8 (2014)
14. Lu, L.F., Yuan, J.J.: Unbounded parallel batch scheduling with job delivery to minimize makespan. Oper. Res. Lett. **36**, 477–480 (2008)
15. Cheng, B., Pei, J., Li, K., Pardalos, P.M.: Integrated scheduling of production and distribution for manufacturers with parallel batching facilities. Optim. Lett. **12**(7), 1609–1623 (2017). https://doi.org/10.1007/s11590-017-1201-2
16. Dosa, G., Tan, Z., Tuza, Z., Yan, Y., Lányi, C.S.: Improved bounds for batch scheduling with nonidentical job sizes. Naval Res. Log. **61**(5), 351–358 (2014)

A Batch Scheduling Problem
of Automatic Drug Dispensing System
in Outpatient Pharmacy

Lili Liu[1] and Chunyu Fu[2]

[1] College of Arts and Sciences, Shanghai Polytechnic University, Shanghai 201209,
China
03902707@163.com
[2] Shanghai General Hospital, School of Medicine, Shanghai Jiaotong University,
Shanghai 200080, China
fuchunyu2020@163.com

Abstract. This paper studies the batch scheduling problem with incompatible job families which can be applied to the automatic drug dispensing of outpatient pharmacies. We prove that the problem is strongly NP-hard even if the processing time and the weight of each job are same, and we propose a pseudo-polynomial time algorithm for the special case where the jobs of each family have a common due date.

Keywords: Drug dispensing · Batch scheduling · Incompatible job families

1 Introduction

1.1 Automatic Drug Dispensing in Outpatient Pharmacies

An automatic drug dispenser is an automated drug delivery device used for dispensing drugs in the pharmacy.

The automatic drug dispensing devices in pharmacy popular in Europe and America are mainly available in four modes: (1) Vertical carousel, with low cost, but in company with low automation level and speed; (2) Miniload, capable of handling the packages in different shapes, but unsuitable for the quick-moving drugs; (3) Commissioning system, applicable for a small amount of drugs in different types, but needing the drug barcode technology and being characterized by low speed; (4) Gravity, with low cost, but unable to handle the package of irregular shape or the fragile drugs.

The demand for automatic drug dispenser in China is much more stringent than that in Europe and America. The reasons are as follows. Reason I:

Supported by the National Natural Science Foundation of China (No. 11601316) and discipline "Applied Mathematics" of Shanghai Polytechnic University (No. XXKPY1604).

Z. Zhang et al. (Eds.): AAIM 2020, LNCS 12290, pp. 537–543, 2020.
https://doi.org/10.1007/978-3-030-57602-8_48

the outpatient pharmacy of the Chinese hospitals handles a huge amount of prescriptions per day. A large-sized tertiary hospital handles more than 6,000 prescriptions per day, and in the peak hours, the hourly quantity is up to 1,500 and even 2,000 prescriptions. Reason II: the pharmacy of the hospital has the limited space. Most hospitals are situated in the commercial center of the cities and cannot be expanded, while the pharmacy of the traditional hospital is usually very small. Reason III: the drug package in the pharmacy of the hospital varies. Except the square box for drugs, the hospitals have the packages, such as cylindrical, glass bottle and triangular packages. While the unified package and barcode for drugs lack in China, and it is very regretted that no automatic drug dispensing mode popular in Europe and America can address the automation demand in the pharmacy of the Chinese hospitals. We need to improve and customize the design.

The automatic drug dispensing devices run online in Shanghai General Hospital have basically met the following characteristics.

- Automatic drug dispensing speed can meet the demand of the patients in the peak hours.
- Simple and quick process for automatic drug prescription and billing to the device.
- Capable of handling the drugs in different packages and sizes.
- Capable of handling the drugs in fragile package, for example, glass bottle and so on.
- The carousel can store sufficient types of drugs.
- The software system can fully interface with the HIS system of the hospital.
- The operating interface of the software is concise and practical.
- There is reasonable emergency response scheme in case of any emergency event, such as power interruption.
- The machine is convenient for installation, and does not have special requirements for the civil construction, and may be dismantled and then reinstalled.
- The hardware system occupies small floor area.
- The machine has low fault rate.

In recent years, a lot of research has been developed to apply scheduling theory and methodology to solve health care management problems. Liu et al. [1] model the drug dispensing system of outpatient pharmacies as two-person cooperative games. Fan and Lu [2] study the supply chain scheduling with periodic working time in the warehouse of a hospital. Wang et al. [3] address prioritized surgery scheduling for a single operating room with surgeon tiredness and fixed off-duty period. Yang et al. [4] design a surgical scheduling method for an operation room with surgeons preferences considered. Zhang et al. [5] study patients scheduling problems with deferred deteriorated functions. Zhang et al. [6] explore a two-stage medical supply chain scheduling problem with an assignable common due date and shelf life. Li and Chai [7] discuss an on-line scheduling problem arising from medical laboratory.

This paper discusses the drug dispensing system from the point of view of batch scheduling with incompatible job families. For a given period of time, there are many electronic prescriptions that arrive in the drug dispensing system. They have different processing times, different weights, and different due dates. The drug dispensing system can process several electronic prescriptions at the same time. For a given scheduling objective function, we try to optimize it by assigning the electronic prescriptions into different batches and ordering these batches.

1.2   Batch Scheduling with Incompatible Job Families

Batch scheduling is originally motivated by the burn-in operation of the final testing stage in the very large-scale integrated circuit manufacturing, in which a machine can process several jobs at the same time. The processing time of a batch is equal to the largest processing time among all the jobs in the batch. If the jobs with different processing times cannot be processed at the same time, batch scheduling with incompatible job families is brought about. Generally, jobs having identical processing times belong to the same family.

We first briefly describe some notations that will be used in this paper. There are n jobs that have to be processed on a single batch processing machine. These jobs have t different processing times, that is, there are totally t job families. For $i = 1, \cdots, t$, the processing time of family i is denoted by p_i, and the number of jobs of family i by n_i. For $i = 1, \cdots, t; j = 1, \cdots, n_i$, job J_{ij} of family i is associated with a weight w_{ij}, the importance of job J_{ij} compared to the other jobs, and a due date d_{ij}, the time that job J_{ij} should be finished (completion of a job after its due date is allowed, but a penalty is incurred). A batch processing machine can process up to B jobs at the same time and B is called the machine capacity. Given a schedule, for job $J_{ij}(i = 1, ..., t; j = 1, ...n_i)$, we use C_{ij} to denote its completion time, and U_{ij} its unit penalty which is defined as $U_{ij} = 1$ if $C_{ij} > d_{ij}$ and zero otherwise. This paper considers the objective of the total weighted number of tardy jobs $\sum w_j U_j$, which indicates the efficiency of medication dispensing system and the patient satisfaction.

For the problem of scheduling jobs with incompatible job families on a single batch processing machine, Uzsoy [8] first considers batch scheduling with incompatible job families. He provides polynomial time algorithms or heuristic algorithms for several problems. Mehta and Uzsoy [9] prove that the problem of minimizing total tardiness is strongly NP-hard, and a dynamic programming algorithm and several heuristic algorithms are presented. Jolai [10] shows that the problem of minimizing the number of tardy jobs is NP-hard with respect to id-encoding and presents a dynamic programming algorithm for this problem. Perez et al. [11] develop several heuristic algorithms for the strongly NP-hard problem of minimizing total weighted tardiness.

This paper is organized as follows. In Sect. 2 we show that the problem of scheduling jobs with incompatible job families on a single batch processing machine to minimize the total weighted number of tardy jobs is strongly NP-hard even if the processing time and the weight of each job are same. In Sect. 3 we propose a pseudo-polynomial time dynamic programming algorithm for the special case where the jobs of each family have a common due date.

2   The Strong NP-hardness of a Special Case

Liu et al. [12] prove that the problem of scheduling jobs with incompatible job families on a single batch processing machine to minimize the total weighted number of tardy jobs is strongly NP-hard. In this section we show that it is strongly NP-hard even if the processing time and the weight of each job are same.

3-Partition: Given $3m$ positive integers $\{a_1, \ldots, a_{3m}\}$ and an integer A such that $\sum_{i=1}^{3m} a_i = mA$ and $\frac{A}{4} < a_i < \frac{A}{2}$ for $i = 1, \ldots, 3m$, is there a partition of the index set $\{1, \ldots, 3m\}$ into $A_1, A_2, \ldots, A_m$ such that $\sum_{i \in A_j} a_i = A$ for $j = 1, \ldots, m$?

Theorem 1. *The problem of minimizing the total weighted number of tardy jobs with incompatible job families on a single batch processing machine is strongly NP-hard even if the weight and the processing time of each job are same.*

Proof. Given a 3-Partition instance, we construct the following scheduling instance with $n = 3m^2 + m$ jobs and the machine capacity is $B = m$. Let $L = \frac{m(m+1)}{2}A$. For $j = 1, \ldots, m$, there is a job L_j with processing time $p_{L_j} = jL$, weight $w_{L_j} = jL$, and due date $d_{L_j} = jA + \frac{j(j+1)}{2}L$. For $i = 1, \ldots, 3m$ and $j = 1, \ldots, m$, there is a job S_{ij} with processing time $p_{S_{ij}} = a_i$, weight $w_{S_{ij}} = a_i$, and due date $d_{S_{ij}} = jA + \frac{j(j-1)}{2}L$.

Clearly, the scheduling instance can be constructed in polynomial time and there are totally $4m$ different job families in this instance. Set the threshold value of the total weighted number of tardy jobs as $\frac{m(m-1)}{2}A$. We will show that there is a solution for the scheduling instance such that $\sum w_j U_j \leq \frac{m(m-1)}{2}A$ if and only if there is a solution for the 3-Partition instance.

Suppose that there is a solution $A_1, \ldots, A_m$ for the 3-Partition instance. For $j = 1, \ldots, m$, assign job L_j to batch B_{L_j}. For $i = 1, \ldots, 3m$, assign jobs $S_{i1}, S_{i2}, \cdots, S_{im}$ to batch B_{S_i}. Construct the following schedule:

$$\sigma = \{\{B_{S_i} : i \in A_1\}, B_{L_1}, \{B_{S_i} : i \in A_2\}, B_{L_2}, \ldots, \{B_{S_i} : i \in A_m\}, B_{L_m}\}.$$

In schedule σ, for $i = 1, \cdots, 3m; j = 1, \ldots, m$, batches $B_{S_i}(i \in A_j)$ finish at time $jA + \frac{j(j-1)}{2}L$, and batch B_{L_j} finishes at time $jA + \frac{j(j+1)}{2}L$. Hence, the jobs in batches $B_{L_j}(j = 1, \cdots, m)$ are on time since $d_{L_j} = jA + \frac{j(j+1)}{2}L$, the jobs S_{ij} in batches $B_{S_i}(l \leq j)$ are on time since $d_{S_{ij}} = jA + \frac{j(j-1)}{2}L$, and only the jobs S_{ij} in batches $B_{S_i}(l > j)$ are tardy. We have,

$$\sum w_j U_j = \sum_{i \in A_2} a_i + 2 \sum_{i \in A_3} a_i + \ldots + (m-1) \sum_{i \in A_m} a_i$$
$$= A + 2A + \ldots + (m-1)A$$
$$= \frac{m(m-1)}{2}A.$$

On the other hand, if the scheduling instance has a solution σ such that $\sum w_j U_j \leq \frac{m(m-1)}{2}A$. For $j = 1, \cdots, m$, job L_j must be assigned to one batch and processed in increasing order of their processing times since $w_{L_j} = jL > \frac{m(m-1)}{2} A(j = 1, \cdots, m)$ and each of them must be completed on time, denote the batch containing job L_j as B_{L_j}. Without loss of generality, for $i = 1, \cdots, 3m$, we assume that the m jobs with processing time $p_{S_{ij}} (j = 1, \cdots, m)$ are assigned to one batch in σ. Otherwise, if there are jobs of the same family are assigned to different batches, move jobs with the same processing time from the subsequent batch(es) to the precedent batch and the completion times of all the jobs do not increase. Therefore, the total weighted number of tardy jobs does not increase. For $i = 1, \cdots, 3m$, following the above procedure, we get a full batch with processing time a_i, and this batch is denoted by B_{S_i}. For $j = 1, \ldots, m$, let TP_j be the total processing time of the batches with processing times $p_{S_{ij}} (i = 1, \cdots, 3m)$ processed after batch B_{L_j}. Since the job in batch B_{L_j} cannot be tardy, we have $TP_j \geq (m-j)A$. According to the definition of instance, we know that jobs $S_{il} (i = 1, \ldots, 3m; l = 1, \ldots, j)$ processed after batch B_{L_j} must be tardy. Since $p_{S_{ij}} = w_{S_{ij}} = a_i$, it follows that

$$\sum w_j U_j \geq \sum_{j=1}^{m} TP_j \geq \sum_{j=1}^{m} (m-j)A.$$

Suppose that there is $1 \leq j \leq m$ such that $TP_j > (m-j)A$. Then

$$\sum w_j U_j > \sum_{j=1}^{m} (m-j)A = m^2 A - \frac{m(m+1)}{2} A$$
$$= \frac{m(m-1)}{2} A.$$

It's a contradiction. Thus, for $j = 1, \ldots, m$, we have $TP_j = (m-j)A$. This implies that the sum of processing times of batches processed before batch $B_{L_j}(j = 1, \ldots, m)$ is exactly A. For $i = 1, \cdots, 3m$, since $\frac{A}{4} < a_i < \frac{A}{2}$, there are exactly three batches of B_{S_i} are scheduled before batch B_{L_j}. Denote the batch indices set of batches B_{S_i} processed before B_{L_j} as A_j, which is a solution of the 3-Partition instance.

3   A Pseudo-polynomial Time Solvable Case

From the last section, we know that even if the jobs of each family have a same weight, the problem under consideration is strongly NP-hard. In this section, we consider a special case where the jobs of each family have a common due date. It is at least binary NP-hard since the classical problem of scheduling jobs with a common due date on a single machine to minimize the total weighted number of tardy jobs is NP-hard [13]. We develop a dynamic programming algorithm for the special case of incompatible batch scheduling problem where the jobs of each family have a common due date, which runs in pseudo-polynomial time and indicates that this special case cannot be strongly NP-hard unless P = NP.

Lemma 1. *If the jobs of each family have a common due date, there exists an optimal schedule in which the batches of each family are full except possibly the last one.*

Proof. Consider an optimal schedule containing two partial batches of same family, move jobs from the subsequent batch to the precedent batch until the precedent batch is full or the subsequent batch is empty. The completion times of the moved jobs decrease and the completion times of the other jobs do not increase, hence the schedule remains optimal. Repeating this procedure for the jobs of each family, we get an optimal schedule in which the batches of each family are full except possibly the last one.

We say that a schedule is in the batch LWF-order if for any two batches P and Q, batch P is processed before batch Q and there does not exist two jobs J_k, J_l such that $J_k \in P, J_l \in Q$ and their weights $w_{J_k} < w_{J_l}$.

Lemma 2. *If the jobs of each family have a common due date, there exists an optimal schedule in which the batches of each family are in batch LWF-order.*

Proof. Consider an optimal schedule in which there are two batches P and Q of same family, where P is processed before Q and there exist two jobs J_k, J_l such that $J_k \in P, J_l \in Q$ and $w_{J_k} < w_{J_l}$. Exchange jobs J_k and J_l by moving J_k to Q and J_l to P. Since the two jobs belong to the same family, they have the same processing times and same due dates. The completion times of the batches after the exchange will not increase. Hence, the total weighted number of tardy jobs will not increase, too.

For $i = 1, \cdots, t$, denote the common due date of jobs in family i as d_i, the weights as $w_{i1}, w_{i2}, \cdots, w_{i,n_i}$. Based on Lemma 1 and Lemma 2, first order the jobs of family i in non-increasing order of their weights, then assign adjacent B jobs as a batch from the beginning until all the jobs of family i have been assigned. The number of batches of family i is $\lceil \frac{n_i}{B} \rceil$, where $\lceil r \rceil$ denotes the smallest integer larger than or equal to r. Denote the batches of family i as $B_{i1}, B_{i2}, \cdots, B_{i,\lceil \frac{n_i}{B} \rceil}$. For $i = 1, \cdots, t; j = 1, \cdots, \lceil \frac{n_i}{B} \rceil$, regard batch B_{ij} as a job with processing p_i, due date d_i and weight W_{ij}, where W_{ij} is the sum of weights of jobs in batch B_{ij}. Then we obtain a single machine scheduling problem to minimize the total weighted number of tardy jobs. Apply the dynamic programming algorithm presented by Brucker [14] to find an optimal schedule for all the batches. The time complexity of this dynamic programming algorithm is $O(n \sum_{i=1}^{t} \lceil \frac{n_i}{B} \rceil p_i)$, which is pseudo-polynomial time.

4   Conclusion

In this paper, we study the batch scheduling problem with incompatible job families on a single batch processing machine, which can be applied to the drug dispensing system at outpatient pharmacies. We first prove that the problem is strongly NP-hard even if the processing time and the weight of each job are same, we also develop a pseudo-polynomial time algorithm for the special case where the jobs of each family have a common due date.

To develop efficient heuristic algorithms for the problem under consideration and to investigate its special case are very challenging topics for future research.

References

1. Liu, L., Tang, G., Fan, B., Wang, X.: Two-person cooperative games on scheduling problems in outpatient pharmacy dispensing process. J. Comb. Optim. **30**(4), 938–948 (2015). https://doi.org/10.1007/s10878-015-9854-1
2. Fan, J., Lu, X.: Supply chain scheduling problem in the hospital with periodic working time on a single machine. J. Comb. Optim. **30**(4), 892–905 (2015). https://doi.org/10.1007/s10878-015-9857-y
3. Wang, D., Liu, F., Yin, Y., Wang, J., Wang, Y.: Prioritized surgery scheduling in face of surgeon tiredness and fixed off-duty period. J. Comb. Optim. **30**(4), 967–981 (2015). https://doi.org/10.1007/s10878-015-9846-1
4. Yang, Y., Shen, B., Gao, W., Liu, Y., Zhong, L.: A surgical scheduling method considering surgeons preferences. J. Comb. Optim. **30**, 1016–1026 (2015)
5. Zhang, X., Wang, H., Wang, X.: Patients scheduling problems with deferred deteriorated functions. J. Comb. Optim. **30**(4), 1027–1041 (2015). https://doi.org/10.1007/s10878-015-9852-3
6. Zhang, L., Zhang, Y., Bai, Q.: Two-stage medical supply chain scheduling with an assignable common due window and shelf life. J. Comb. Optim. **37**(1), 319–329 (2017). https://doi.org/10.1007/s10878-017-0228-8
7. Li, W., Chai, X.: The medical laboratory scheduling for weighted flow-time. J. Comb. Optim. **37**(1), 83–94 (2017). https://doi.org/10.1007/s10878-017-0211-4
8. Uzsoy, R.: Scheduling batch processing machines with incompatible job families. Int. J. Prod. Res. **33**, 2685–2708 (1995)
9. Mehta, S., Uzsoy, R.: Minimizing total tardiness on a batch processing machine with incompatible job families. IIE Trans. **30**, 165–178 (1998)
10. Jolai, F.: Minimizing number of tardy jobs on a batch processing machine with incompatible job families. Eur. J. Oper. Res. **162**, 184–190 (2005)
11. Perez, I., Fowler, J., Carlyle, W.: Minimizing total weighted tardiness on a single batch process machine with incompatible job families. Comput. Oper. Res. **32**, 327–341 (2005)
12. Liu, L., Ng, C., Cheng, T.: On the complexity of bi-criteria scheduling on a single batch processing machine. J. Sched. **13**, 629–638 (2010)
13. Karp, R.: Reducibility among combinatorial problems. In: Miller, R.E., Thatcher, J.W. (eds.) Complexity of Computer Computaitons, pp. 85–103. Plenum Press, New York (1972)
14. Brucker, P.: Theory of Scheduling. Springer, Berlin (1995)

A Two-Stage Medical Expenses Estimation Model for Inpatients During Diagnosis Process Under Artificial Intelligence Environment

He Huang[1], Baizhou Shi[1], Yuelan Zhu[2], and Wei Gao[3](✉)

[1] Business School, University of Shanghai for Science and Technology, Shanghai 200093, China
[2] Shanghai General Hospital of Nanjing Medical University, Nanjing 211166, China
[3] Shanghai General Hospital, Shanghai Jiaotong University, Shanghai 200080, China
gaowei1108@hotmail.com

Abstract. Under artificial intelligence (AI) environment, medical staff and patients benefit tremendously from sophisticated AI algorithms in various health-care activities. In order to enhance diagnosis performance for coronary heart disease (CHD) inpatients with surgery (e.g. stent), this research presented a two-stage medical expenses estimation model. Two intelligent modules were integrated into this model, SVM-based module and SOM-based module, and they were developed to estimate total medical expenses and detailed medical expenses. The model was also compared with classic AI techniques, back propagation neural networks and random forests. The data from a real hospital was introduced. For the target disease CHD, several attributes were extracted for inputs/outputs. Based on experimental results, the proposed model not only achieved excellent performance for total medical expenses estimation, but also a powerful tool for detail expenses estimation. The related managerial insights would assist medical staff and patients in reliable decision-making.

Keywords: Artificial intelligence algorithm · Healthcare management · Support vector machine (SVM) · Self-organizing maps (SOM) · Medical expenses estimation

1 Introduction

The healthcare industry is one of the largest and rapidest developing industries in the world. Improving the performance of healthcare management is an important but challengeable task, especially after the global health incident, 2019-nCoV (novel coronavirus). Nowadays, artificial intelligence (AI) is bringing a paradigm shift to healthcare management with the outstanding capabilities of data acquisition and analysis [1]. As a result, a large amount of healthcare activities have experienced a proliferation of innovations under AI environment. Among a variety of healthcare activities, an excellent medical expenses estimation system would benefit medical staff and patients for accurate and reliable decisions [2], especially in disease diagnosis process. For instance, medical staff would provide appropriate therapeutic regimen and equipment for patients, and

© Springer Nature Switzerland AG 2020
Z. Zhang et al. (Eds.): AAIM 2020, LNCS 12290, pp. 544–556, 2020.
https://doi.org/10.1007/978-3-030-57602-8_49

patients would participate in those decisions based on their financial conditions. In fact, this kind of superiority is more distinct for inpatients with major surgery, like coronary heart disease (CHD) patients. CHD, or coronary artery disease, is a narrowing of the blood vessels that supply blood and oxygen to the heart. Usually, most CHD patients are the elderly rather than young people, and a proper surgery is required (e.g. stent). Based on these situations, this study concentrates more on CHD inpatients over 60 years old. However, there are many factors affecting CHD patents' medical expenses, and the relationship between these factors and medical expenses is not clear with a non-linear characteristic. Not only that, there are tremendous history data in healthcare system. The raw data is chaotic and disorder with various information. Therefore, a novel AI-based model with multi-function for target disease is on demand. As a result, this study attempts to investigate the issue of medical expenses estimation from two perspectives, total medical expenses estimation and detailed medical expenses estimation. Considering the capability and performance of different AI algorithms, support vector machine (SVM) [3] and self-organizing maps (SOM) [4] are employed in this study. The former is well-known for perdition, and the latter is a powerful clustering tool.

2 Literature Review

2.1 Medical Expenses Issues in Healthcare Management

The increasing costs/expenses in healthcare domain is one of the world's most important problems [5]. The analysis of medical expenses could help healthcare manager to control the cost of healthcare organizations; Meanwhile, the patients would make appropriate financial decisions based on sufficient medical expenses information.

Kim & Park (2019) focused on high-cost healthcare users [6]. In this study, various methods and data sources were introduced to build high-cost user prediction model. Contrarily, Zhou, Zhou, & Li (2016) proposed an AI algorithm that was able to select a low-cost subset of informative features and achieved better performance than other state-of-art feature selection methods in medical diagnosis [7]. Cao, Ewing, & Thompson (2012) concentrated on another aspect of medical expenses, medical cost inflation rates [8]. They found that AI algorithms resulted in better prediction performance. Among these previous studies, it is demonstrated that AI-based algorithms are adequate for clustering and prediction in medical expenses issues. However, in terms of CHD expenses, further research is required for optimization of input/output variables as well as sample selection, and the comparison of different AI-based algorithms is necessary. From the perspective of statistics, Liu, Deng, & Wang (2019) estimated the medical costs in disease diagnosis to a terminal event [9]. The combined scheme of both inverse probability of censoring weighting (IPCW) technique and longitudinal quantile regression model was used to develop a novel procedure to the estimation of cumulative quantile function (CQF) based on history process with time-dependent covariates and right censored time-to-event variable. In addition, Khazbak et al. (2016) introduced the novel concept of cost-effective mobile healthcare which leverages the multiple wireless interfaces onboard most mobile phones today, and their proposed method and algorithm achieved cost saving [10]. The cost of medical equipment is another area in medical expenses, and Khalaf, Djouani, Hamam, & Alayli (2015) investigated the failure-cost model for medical equipment [11].

Based on these related papers, it is found that there are three main techniques involved to discuss medical expenses: AI algorithms, statistical methods, and operational research models. A certain number of papers demonstrated that AI algorithms outperform other techniques. With such observation, AI algorithms are introduced in this study. In addition, CHD mainly happens in the elderly, and factors affecting expenses differ from other diseases. Hence, it is essential to optimize input/output variables as well as sample selection to develop a new model for CHD diagnosis process. Identifying appropriate AI algorithm is also required based on the comparison of different algorithms. As a result, this study attempts to investigate the issue of medical expenses estimation for inpatients during diagnosis process. Different from previous papers, the proposed two-stage model is integrated of two intelligent modules, total expense estimation module, and detailed expense estimation module.

2.2 Artificial Intelligence Algorithms in Healthcare Management

Nowadays, AI is gradually changing healthcare practice with numerous breakthroughs in algorithm [12]. In summary, three main healthcare fields have been explored, basic biomedical research, translational research, and clinical practice [13]. By observation of these papers, it is found that SVM and SOM are two powerful algorithms to solve healthcare management issues.

SVM is one of the pattern recognition approaches, and it constructs a hyperplane or set of hyperplanes used for classification and regression. Huang, Gao, & Ye (2019) proposed a data-driven model with SVM. The model intelligently predicted patients' conditions by 12 assay indexes for the diagnosis of cough variant asthma (CVA) [14]. Gao, Bao, & Zhou (2019) also used SVM to assist medical staff with CVA diagnosis [15]. Differently, Zhu, Liu, Lu, & Li (2016) discussed pre-diagnosis with privacy considera-tion, and they presented an efficient and privacy-preserving online medical pre-diagnosis framework by using SVM [16]. Recently, Nawaz et al. (2017) investigated the impact of social media on healthcare management based on SVM [17]. In these previous papers, SVM technique were widely used in diagnosis process to determine whether patients were attacked with diseases. However, in disease diagnosis process, expenses infor-mation is also crucial for decision-making of medical staff and patients. Motivated by this reality, SVM algorithm is employed to estimate medical expenses of CHD during diagnosis process in this study.

SOM is an effective tool for the visualization of high-dimensional data. It is actu-ally a center-based clustering algorithm that preserves density and topology of the data distribution, and the data is mapped to a topological grid of cluster centers [4]. To solve outlier detection in healthcare, Elmougy, Hossain, Tolba, Alhamid, & Muhammad (2019) proposed a new parameter based growing SOM ensemble [18]. By introducing the clus-tering function of SOM, Orjuela-Cañón, Mendoza, García, & Vela (2018) developed a technique to help medical staff make decisions about management of subjects under suspicious of tuberculosis, with limited infrastructure and data [19]. Malek et al. (2018) investigated the healthcare issue of pediatric fracture healing time of the lower limb, SOM was then applied for analysis of the relationship between the selected variables with fracture healing time [20]. According to these previous studies, it is demonstrated that SOM is equipped with the effective capability of clustering based on various features. As a result, it motivates us to classify patients into different groups based on their historical medical expenses data. Additionally, individual characteristics are distinctive for each cluster. In consequence, a reliable database would be created by utilizing these clusters. Obviously, further analysis with such database would affect healthcare management policies, and a wealth of managerial insights could be acquired.

3 Proposed Two-Stage Model

In this section, a two-stage medical expenses estimation model is developed. Aiming at estimating both total medical expenses and detailed medical expenses, the model is integrated of two modules, SVM-based module and SOM-based module.

In the first stage, the module is used for regression prediction with the basic concept of SVM [14]. Suppose that the data set is $D = \{(x_1, y_1), (x_2, y_2), (x_3, y_3), \cdots, (x_m, y_m)\}$. Here, x indicates the main attributes affecting total CHD medical expenses, and y indicates total expenses. The detailed explanation of these variables is given is Sect. 4. The module concept is to find a function $f(x)$ that deviates from y by a value no greater than ε for each training point x, and at the same time is as flat as possible, Fig. 1. Similar to the classification problem, the purpose here is also to minimize error, individualizing the hyperplane which maximizes the margin. Notice that, part of the error is tolerated with epsilon.

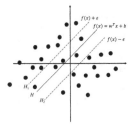

Fig. 1. The basic concept of support vector machine (SVM) regression

The problem of this module is initially expressed as follow,

$$min\frac{1}{2}\|w\|^2 + C\sum_{i=1}^{m} l_\in(f(x_i) - y_i) \tag{1}$$

$$l_\in(z) = \begin{cases} 0, & |z| < \varepsilon \\ |z| - \in, & otherwise \end{cases} \tag{2}$$

The constant C is the box constraint, and it is a positive numeric value. The function of C is to control the penalty imposed on observations that lie outside the epsilon margin as well as help to prevent overfitting. The slack variables ξ_i and ξ_i^* are then introduced based on the "soft margin" concept in classification problem. Including slack variables leads to the objective function as Eq. (3).

$$
\begin{cases}
min \frac{1}{2}\|w\|^2 + C \sum_{i=1}^{m} \left(\xi_i + \xi_i^*\right) \\
s.t. f(x_i) - y_i \le \varepsilon + \xi_i \\
y_i - f(x_i) \le \varepsilon + \xi_i^* \\
\xi_i \ge 0, \xi_i^* \ge 0
\end{cases}
\tag{3}
$$

The Lagrange function is then developed with Lagrange multipliers $\mu_i \ge 0$, $\mu_i^* \ge 0$, $\alpha_i \ge 0$, $\alpha_i^* \ge 0$.

$$
L = \frac{1}{2}\|w\|^2 + C \sum_{i=1}^{m}\left(\xi_i + \xi_i^*\right) - \sum_{i=1}^{m}\mu_i\xi_i - \sum_{i=1}^{m}\mu_i^*\xi_i^* + \sum_{i=1}^{m}\alpha_i(f(x_i) - y_i - \varepsilon - \xi_i) + \sum_{i=1}^{m}\alpha_i^*\left(y_i - f(x_i) - \varepsilon - \xi_i^*\right)
\tag{4}
$$

The corresponding dual problem is given in Eq. (5).

$$
\begin{cases}
max \sum_{i=1}^{m} y_i\left(\alpha_i^* - \alpha_i\right) - \varepsilon\left(\alpha_i^* + \alpha_i\right) - \frac{1}{2}\sum_{i=1}^{m}\sum_{j=1}^{m}(\alpha_i^* - \alpha_i)\left(\alpha_j^* - \alpha_j\right)x_i^T x_j \\
s.t. \sum_{i=1}^{m}(\alpha_i^* - \alpha_i) = 0 \\
0 \le \alpha_i^*, \alpha_i \le C
\end{cases}
\tag{5}
$$

Meanwhile, the problem is constrained by Karush-Kuhn-Tucker (KKT) complementarity conditions, see Eq. (6), to obtain optimal solutions. The problem is finally expressed as Eq. (7) based on the results of b and w.

$$
\begin{cases}
\alpha_i(f(x_i) - y_i - \varepsilon - \xi_i) = 0 \\
\alpha_i^*\left(y_i - f(x_i) - \varepsilon - \xi_i^*\right) = 0 \\
\alpha_i\alpha_i^* = 0, \xi_i\xi_i^* = 0 \\
(C - \alpha_i)\xi_i = 0, \left(C - \alpha_i^*\right)\xi_i^* = 0
\end{cases}
\tag{6}
$$

$$
f(x) = \sum_{i=1}^{m}\left(\alpha_i^* - \alpha_i\right)\cdot\langle x_i, x\rangle + b
\tag{7}
$$

However, some regression problems cannot adequately be described using a linear model. With this consideration, the nonlinear kernel function is employed that transforms data into a higher dimensional feature space to make it possible to perform the linear separation. Therefore, the problem is expressed as Eq. (8) and (9) in the nonlinear situation, with KKT complementarity conditions.

$$
f(x) = \sum_{i=1}^{m}\left(\alpha_i^* - \alpha_i\right)\cdot\langle\varphi(x_i), \varphi(x)\rangle + b
\tag{8}
$$

$$
f(x) = \sum_{i=1}^{m}\left(\alpha_i^* - \alpha_i\right)\cdot K(x_i, x) + b
\tag{9}
$$

In fact, there are various types of kernel function, such as Polynomial kernel function and Gaussian kernel function.

$$
\text{Polynomial: } k\left(x_i, x_j\right) = \left(x_i, x_j\right)^d
\tag{10}
$$

$$\text{Gaussian: } k\left(x_i, x_j\right) = exp\left(-\frac{\left\|x_i - x_j\right\|^2}{2\sigma^2}\right) \tag{11}$$

Based on the theory of SOM [21], the SOM-based module of the proposed model is developed for clustering. In the second stage, the data of detailed medical expenses is represented by an n-dimensional vector X in the input layer (Fig. 2). $X = [X_1, X_2, X_3, ..., X_n]$, n is the number of attributes (medical expenses items), also the number of input neurons. It is also assumed that there are j neurons in the output layer (or the competition layer), and j is related to the maximum number of final clusters.

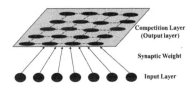

Fig. 2. The basic concept of self-organizing maps (SOM)

The neurons in the input layer are directly connected to the neurons in the output layer without hidden layer. Therefore, each output neuron is connected to the input vector X via an n-dimensional weight vector $W = [W_1, W_2, W_3, ..., W_n]$. Additionally, neurons in the output layer are arranged in form of a topological architecture, and the two-dimensional architecture is introduced. Firstly, the module randomly initializes weight vectors. Meanwhile, the learning rate (α^t) and radius of neighbourhood (R^t) are assigned the starting values. Based on the weight vectors, a best matching unit (BMU) is selected for input vectors. In other words, the module calculates the Euclidean distance (EUD), donated as d_j, between input neurons and output neuron j by Eq. (12). Here, W_{ij} indicates the ith weight of the jth output neuron.

$$d_j = \sqrt{\sum_i (X_i - W_{ij})^2} \tag{12}$$

According to EUD between input vector and output neuron's weight vector, the best matching neuron is defined as winner, represented by $j*$, that achieves the minimum distance among d_j. It means that the winner is the neuron whose weight vector is most similar to the input vector. (C_x, C_y) is the topological coordinates of the winner.

$$d_{j*} = \min\{d_j\} \tag{13}$$

The module then updates the weight W between the input layer and the output layer. This updating towards the input sample is related to learning rate (α^t) and neighborhood size. The learning rate gradually decreases with the iteration step index t. With the weight vector $W_{ij}(t)$ of the winning neuron at iteration t, the updated weight vector $W_{ij}(t + 1)$ at iteration $t + 1$ is computed by Eq. (14).

$$W_{ij}(t + 1) = W_{ij}(t) + \alpha^t (X_i(t) - W_{ij}(t)) * R_c^j \tag{14}$$

$$R_c^j = f\left(R^t, r_j\right) \tag{15}$$

$$r_j = \sqrt{\left(X_j - C_x\right)^2 + \left(Y_j - C_y\right)^2} \tag{16}$$

In Eq. (14), R_c^j represents the neighborhood coefficient. In Eq. (15), R^t is the neighborhood radius and r_j is the neighborhood distance. In Eq. (16), (X_j, Y_j) is the topological coordinates of the jth output neuron. The above process would continue until convergence (i.e. no noticeable changes in the weights) or the pre-defined number of training cycles is satisfied. Meanwhile, the learning rate and the neighborhood radius would be decreased in each iteration by Eq. (17). Here, α_c and R_c are the corresponding decreasing coefficients for the learning rate and the neighborhood radius, respectively.

$$\begin{cases} \alpha^t = \alpha^{t-1} * \alpha_c \\ R^t = R^{t-1} * R_c \end{cases} \tag{17}$$

4 Description of Data and Variables

The experimental platform is Intel Core i5-5200U CPU, 4 GB RAM, Windows 10 (64 bit). The data is from a real hospital in Shanghai, China. With the cooperation of medical staff in cardiology department, the data from 2017 to 2019 is obtained.

4.1 Data and Variables in the First Stage

In the first stage, the total expenses of CHD are estimated by the SVM-based module. In order to acquire effective and reasonable variables, preliminary work is required, including several interviews with related medical staff and analysis of previous papers. As a result, 8 attributes were extracted for this stage. They are explained as follows, and the descriptive statistics analysis is given in Table 1.

Table 1. The descriptive statistics analysis for the numerical attributes (before normalization)

Attribute	Maximum	Minimum	Mean	Median
Total expenses	99958.00	20663.26	57441.51	53927.69
Age	98.00	60.00	72.07	70.00
Hospital length of stay (HLOS)	35.00	1.00	6.38	5.00
Quantity of surgical equipment (QOSE)	4.00	1.00	1.38	1.00
Level of surgical equipment (LOSE)	81672.00	4763.00	21953.18	17200.00

1) Total Expenses. It indicates the expenses patients would afford during CHD treatment. The proposed model attempts to intelligently estimate such expenses in diagnosis process. It is the single output variable.

2) Gender. Male and female. The number 1 and 0 are used to distinguish them.
3) Age. Age is a numerical variable. It is noticed that most CHD patients are the elderly. Hence, this study only collected the data of patients over 60 years old.
4) Hospital length of stay (HLOS). HLOS is a numerical variable. Usually, medical staff provide CHD patients the possible HLOS information based on their conditions during diagnosis process.
5) Month. The month when patient is admitted to hospital. Month is the indicator of season, it therefor affects patient's condition (especially for the elderly) and expenses (e.g. hospitalization expenses). There are 12 months in a year, and the number 1 and 0 are used to distinguish them.
6) Type of surgery (TOS). For CHD patients, stent is a common surgery type. Additionally, CHD is the main factor resulting in arrhythmia (e.g. bradycardia), and pacemaker is used for such patients. Therefore, two surgery types are involved in this study, stent and pacemaker. Obviously, TOS is a crucial factor affecting medical expenses, and number 1 and 0 are used to distinguish them.
7) Quantity of surgical equipment (QOSE). It is a numerical variable. In this research, stent and pacemaker are the main equipment for CHD surgery, and the quantities of them are different based on diagnosis.
8) Level of surgical equipment (LOSE). This study also considers LOSE. In fact, there are different levels of surgical equipment in the market. By discussing with medical staff, patients could choose the appropriate one based on their financial conditions. Here, the unit price is used to denote LOSE, and it is a numerical variable.

In terms of experiment samples, there are 2100 samples in total after removing the samples with incomplete data and outliers. Besides, according to the proposed model, two datasets are required, training data and testing data. Therefore, all samples were randomly divided into two groups. Specifically, training dataset with 1600 samples and testing dataset with 500 samples. Additionally, all data is normalized to avoid the effect of dimensions and units, see Eq. (18). Where, D_{new} is the new data after normalization. D is the original data. D_{min} is the minimum value of the corresponding variable, and D_{max} is maximum value of the corresponding variable.

$$D_{new} = \frac{D - D_{min}}{D_{max} - D_{min}} \tag{18}$$

4.2 Data and Variables in the Second Stage

The total medical expenses of CHD are estimated in the first stage, and detailed expenses are analyzed in the second stage by SOM-based module. With the discussion of medical staff, 6 main items are involved in the total medical expenses of CHD. Specifically, 1) Hospitalization fee, 2) Treatment fee, 3) Examination fee, 4) Assay fee, 5) Material fee and 6) Medicine fee. More importantly, it is found that material fee accounts for a large proportion of total expenses. Obviously, it has a significant influence on total expenses with maximum weight. Taking this into consideration, the clustering process by SOM was implemented in two steps. All samples were clustered with material fee item first, resulting in primary clusters. In each primary cluster, the samples were then clustered with other five items, resulting in secondary clusters. Consequently, total cluster quantity is the number of primary clusters multiplies the number of secondary clusters. Based on these precise clusters, more accurate and reliable decision-making insights for medical expenses would be acquired.

5 Experiments and Results

5.1 Experiments and Results Analysis for the First Stage

In this section, total medical expenses are estimated by SVM-based module, Eq. (1) to Eq. (11). In order to verify the performance of the proposed model, another two AI algorithms were employed to make a comparison, back propagation neural networks (BPNN) and random forests (RF). Besides, RMSE (root mean square error) was introduced to evaluate the estimation performance, Eq. (19). Here, D_r is the real data, D_e is the estimated data, m is the sample number. In terms of RMSE value, this study computed both the normalization value and inverse-normalization value.

$$RMSE = \sqrt{\frac{1}{m} \sum_{i=1}^{m} (D_r - D_e)^2} \tag{19}$$

(1) **Total medical expenses estimation by SVM-based module**
    As described above, the inputs of SVM-based module include 7 attributes; the output of SVM-based module is the data of Total Expenses. The key parameters in this process are given in Table 2, and the results are shown in Table 3.

Table 2. The key parameters in SVM-based module

Key parameter	Value or Type
The set type of SVM	nu-SVR
The set type of kernel function	radial basis function
The optimal objective value of the dual SVM problem	-123.480819
The bias term in the decision function	-0.378418
The number of support vectors and bounded support vectors	827 and 773
The number of iterations	2136

(2) **Total medical expenses estimation by BPNN and RF**
    BPNN and RF were introduced to make a comparison with SVM-based module. The same input data and output data that used in SVM-based module experiment is employed for BPNN and RF experiments. The key parameters in BPNN and RF are given in Table 4, and the results are shown in Table 3.

Table 3. The estimation results by different approaches

	SVM-based Module	BPNN	RF
RMSE (normalization)	0.0459	0.0583	0.0665
RMSE (inverse-normalization)	3636.9459	4624.8852	5273.7238

Table 4. The key parameters in BPNN and RF

Key parameters in BPNN		Key parameters in RF	
Input neurons	20	The number of mtry	6
Hidden layers	1	The number of trees	500
Hidden neurons	4		
Output neurons	1		
Optimizer	adam		
Learning rate	0.001		
Loss function	mse		
Dropout (hidden layer)	0.1		
Activation function (hidden layer)	relu		
Activation function (output layer)	sigmoid		
Validation split	0.1		
Batch size	50		
Epochs	10,000		

Based on the results obtained above, some conclusions are generated,

1) By comparing the estimation performance of SVM-based module, BPNN and RF, it is found that SVM-based module outperforms the other two approaches, with both RMSE (normalization) and RMSE (inverse-normalization) value.
2) Due to the inverse-normalization RMSE value is more appropriate to reflect real deviation of total medical expenses, then it is compared with the average value and median value of total medical expenses among all samples (see Table 1). The inverse-normalization RMSE value of SVM-based module (3636.9459) only accounts for 6.33% of average total medical expenses (57441.51) and 6.74% of median value (53927.69). It is demonstrated that the SVM-based module of the proposed model achieved satisfied results and it is adaptive for medical expenses estimation.

5.2 Experiments and Results Analysis for the Second Stage

In this section, detailed medical expenses are analyzed by SOM-based module, see Eq. (12) to Eq. (17). As discussed above, a certain number of clusters would be generated based on detailed medical expenses. Besides, this process includes two steps, primary cluster step and secondary cluster step. As a result, 5 primary clusters were generated based on material fee, and 4 secondary clusters were generated for each primary cluster based on the other 5 expenses items, hospitalization fee, treatment fee, examination fee, assay fee and medicine fee. Consequently, the total cluster quantity is 20 that is 5 (primary clusters) multiplies 4 (secondary clusters). The key parameters are shown in Table 5, and Table 6 illustrates partial results of this module. The values in Table 6 are cluster centers with corresponding attributes, that means sample values in each cluster are close to their centers. For instance, the material fee values of each sample are all close to 38956.53 in primary cluster 1, and the hospitalization fee values of each sample are all close to 632.68 in secondary cluster 20. It could be seen that expenses characteristic of each cluster differ greatly. Meanwhile, detailed medical expenses information is associated with personal and medical information of inpatients in each cluster, that creating a database of medical expenses and patients' information. Specifically, with patients' information, the corresponding cluster would be identified with expenses characteristic. In fact, the availability of accurate and reliable databases is important to provide the right tools for better healthcare decisions [22], especially in diagnosis process. With increasing inpatient quantity, this database would be more and more reliable and accurate. Definitely, it will help medical staff and inpatients make better estimation decisions during diagnosis process under artificial intelligence environment.

Table 5. The key parameters in SOM-based module

Parameter	Value (Primary cluster step)	Value (Secondary cluster step)
Input neurons	1	5
Output neurons	1×5	1×4
Neighborhood	3	1
Learning rate	0.1	0.1
Neighborhood reduction	0.99	0.99
Learning rate reduction	0.99	0.99
The minimum of learning rate	0.001	0.001
Epochs	10	10

Table 6. Results of SOM-based module (partial)

	NO. of secondary cluster	Material fee	Hospitalization fee	Treatment fee	Examination fee	Assay fee	Medicine fee	Quantity
Primary cluster 1	1	38956.53	2603.83	3250.88	3304.82	5022.54	3821.26	104
	2	38956.53	1383.94	2570.47	2310.50	3807.64	1936.20	231
	3	38956.53	482.91	2097.20	2269.58	3181.67	1460.71	344
	4	38956.53	292.43	2035.66	1888.72	2540.70	1248.91	482
......	......	......	......	......	......	......	......	......
Primary cluster 5	17	65875.12	3629.18	7034.76	5532.20	7155.37	12286.37	34
	18	65875.12	1803.93	4417.25	3229.24	4547.50	3877.01	123
	19	65875.12	941.47	3401.73	2340.72	3351.80	1980.45	158
	20	65875.12	632.68	2605.05	2064.43	2774.09	1235.52	404

6   Conclusions

Focusing on CHD inpatients with surgery, this study developed a two-stage medical expenses estimation model for diagnosis process. In the first stage, the total medical expenses were estimated by SVM-based module. The experiment results indicated that the proposed model achieved satisfied perdition performance. In the second stage, the SOM-based module of the proposed model was used to analyze detailed medical expenses. The clustering process successfully divided patients into primary clusters and secondary clusters. In these clusters, detailed medical expenses information combined with patients' information would generate a databased to provide reliable estimation decisions. With the application of the proposed model to CHD, it is demonstrated that this model could be a user-friendly estimation tool for medical staff and patients.

Acknowledgements. This research is supported by three projects: Clinic management and optimization project of Shanghai Shenkang hospital development center (SHDC12017623); Hospital management and development project of SJTU (CHDI-2019-B-14); National natural science foundation of China (71801150).

References

1. Wong, B.N., Ho, G.T.S., Tsui, E.: Development of an intelligent e-healthcare system for the domestic care industry. Ind. Manag. Data Syst. **117**(7), 1426–1445 (2017)
2. Kuo, R.J., Cheng, W.C., Lien, W.C., et al.: A medical cost estimation with fuzzy neural network of acute hepatitis patients in emergency room. Comput. Methods Programs Biomed. **122**(1), 40–46 (2015)
3. Bai, Y.Q., Han, X., Chen, T., et al.: Quadratic kernel-free least squares support vector machine for target diseases classification. J. Comb. Optim. **30**(4), 850–870 (2015)
4. Creput, J.C., Hajjam, A., Koukam, A., et al.: Self-organizing maps in population based metaheuristic to the dynamic vehicle routing problem. J. Comb. Optim. **24**(4), 437–458 (2012)
5. Bertsimas, D., Bjarnadottir, M.V., Kane, M.A., et al.: Algorithmic prediction of health-care costs. Oper. Res. **56**(6), 1382–1392 (2008)

6. Kim, Y.J., Park, H.: Improving prediction of high-cost health care users with medical check-up data. Big Data **7**(3), 163–175 (2019)
7. Zhou, Q.F., Zhou, H., Li, T.: Cost-sensitive feature selection using random forest: selecting low-cost subsets of informative features. Knowl.-Based Syst. **95**, 1–11 (2016)
8. Cao, Q., Ewing, B.T., Thompson, M.A.: Forecasting medical cost inflation rates: a model comparison approach. Decis. Support Syst. **53**(1), 154–160 (2012)
9. Liu, X.F., Deng, D.L., Wang, D.H.: Estimating the quantile medical cost under time-dependent covariates and right censored time-to-event variable based on a state process. Stat. Methods Med. Res. (2019)
10. Khazbak, Y., Izz, M., ElBatt, T.: Cost-effective data transfer for mobile health care. IEEE Syst. J. **11**(4), 2663–2674 (2016)
11. Khalaf, A., Djouani, K., Hamam, Y., et al.: Maintenance strategies and failure-cost model for medical equipment. Qual. Reliab. Eng. Int. **31**(6), 935–947 (2015)
12. Oliveira, T., Novais, P., Neves, J.: Development and implementation of clinical guidelines: an artificial intelligence perspective. Artif. Intell. Rev. **42**(4), 999–1027 (2014)
13. Yu, K.H., Beam, A.L., Kohane, I.S.: Artificial intelligence in healthcare. Nat. Biomed. Eng. **2**(10), 719–731 (2018)
14. Huang, H., Gao, W., Ye, C.M.: An intelligent data-driven model for disease diagnosis based on machine learning theory. J. Comb. Optim. (2019)
15. Gao, W., Bao, W.P., Zhou, X.: Analysis of cough detection index based on decision tree and support vector machine. J. Comb. Optim. **37**(1), 375–384 (2019)
16. Zhu, H., Liu, X.X., Lu, R.X., et al.: Efficient and privacy-preserving online medical prediagnosis framework using nonlinear SVM. IEEE J. Biomed. Health Inform. **21**(3), 838–850 (2016)
17. Nawaz, M.S., Bilal, M., Lali, M.I., et al.: Effectiveness of social media data in healthcare communication. J. Med. Imaging Health Inform. **7**(6), 1365–1371 (2017)
18. Elmougy, S., Hossain, M.S., Tolba, A.S., et al.: A parameter based growing ensemble of self-organizing maps for outlier detection in healthcare. Cluster Comput. **22**(1), 2437–2460 (2019)
19. Orjuela-Canon, A.D., Mendoza, J.E.C., Garcia, C.E.A., et al.: Tuberculosis diagnosis support analysis for precarious health information systems. Comput. Methods Programs Biomed. **157**, 11–17 (2018)
20. Malek, S., Gunalan, R., Kedija, S.Y., et al.: Random forest and self organizing maps application for analysis of pediatric fracture healing time of the lower limb. Neurocomputing **272**, 55–62 (2018)
21. Chang, P.C., Lai, C.Y.: A hybrid system combining self-organizing maps with case-based reasoning in wholesaler's new-release book forecasting. Expert Syst. Appl. **29**(1), 183–192 (2005)
22. Cesari, U., De Pietro, G., Marciano, E., et al.: A new database of healthy and pathological voices. Comput. Electr. Eng. **68**, 310–321 (2018)

The Early-Warning Model of Evaluation and Prevention for Venous Thromboembolism in Gynecological Tumor Surgical Patients Based on WSOM

Shi Yin[1,2] and Jian Chang[3(✉)]

[1] Research Center of Resource Recycling Science and Engineering,
Shanghai Polytechnic University, Shanghai 201209, China
[2] School of Economics and Management, Shanghai Polytechnic
University, Shanghai 201209, China
[3] Nursing Department, Shanghai General Hospital, Shanghai 200080, China
changjiancn@163.com

Abstract. The pathologic analysis of gynecological tumor patients is relatively complex, and the patients without preventive measures after surgery are the high-risk population of VTE. Based on the inpatient data of patients admitted to a hospital in Shanghai and undergoing gynecological tumor surgery, this paper directly verified whether patients in the model construction group had VTE according to the VTE early warning model. Based on direct verification, according to the coagulation report, medical history and hormone level of the patients in the model construction group, the WSOM algorithm model was used to construct the VTE early warning model for gynecological tumor patients after surgery. The study found that based on the coagulation report, three new indicators of FSH, LH and E2 of sex hormone levels in patients with gynecological tumors after surgery were needed to be added to the VTE early warning model after surgery. Meanwhile, the physical and drug preventive measures of VTE for gynecological tumor patients after surgery were proposed.

Keywords: Gynecologic tumor · Venous thromboembolism · Early warning model

1 Introduction

Venous thromboembolism (VTE) includes deep vein thrombosis (DVT) and pulmonary embolism (PE), is a common vascular disease after gynecologic tumor surgery. Deep vein thrombosis (DVT) is a condition in which blood clots in a deep vein that partially or completely block the lumen of the vein, causing an obstruction to venous return. Pulmonary embolism is a disease caused by pulmonary embolism that blocks blood flow. In patients without preventive measures after gynecological surgery in China, the incidence of DVT is as high as 9.2%–40.0%, and the incidence of PE in DVT is as high as 46.0%. The incidence of VTE is higher in patients with malignant tumors, with

© Springer Nature Switzerland AG 2020
Z. Zhang et al. (Eds.): AAIM 2020, LNCS 12290, pp. 557–568, 2020.
https://doi.org/10.1007/978-3-030-57602-8_50

approximately 20.0% of new cases occurring in cancer patients. By improving the self-organizing competitive network algorithm for the selection of initial value connection weights, the project research team established an early warning model based on a total of 14 factors before and after VTE by deeply mining the relevant pathological data before and after VTE. Pathological analysis of gynecological tumor patients is more complicated than that of male tumor patients. The focus of this study is whether the 14 indicators of coagulation report can give early warning of gynecological tumor VTE and what measures should be taken for the early warning and protection of female VTE.

Patients with gynecological tumor usually have less activities before the preparation of surgery and need a long period of bed rest after the surgery, which leads to the slowing of venous blood flow, which is an indirect cause of the occurrence of VTE symptoms. Intraoperative anesthesia and surgical trauma caused by the patient's age, weight, tumor and other factors, leading to hypercoagulable state or thrombosis is the direct cause of VTE symptoms.

The establishment of VTE after venous thromboembolism is the primary factor for the life safety of patients after the perioperative operation of gynecological tumors, which has attracted extensive attention from medical institutions and medical staff, but its evaluation and corresponding prevention guidelines have not been formed. This paper refers to the diagnosis and treatment guidelines, based on the existing Chinese experience in the prevention and treatment of gynecological tumors after surgery, based on the WSOM algorithm for gynecological tumors after surgery VTE early warning model and protection guidelines.

2 VTE Warning Model Based on WSOM Algorithm

2.1 WSOM Algorithm

SOM algorithm is to change the parameters and structure of the network by automatically finding the rules and essential properties of the memory in the sample. The goal of SOM is to represent all the points in the high-dimensional space by the points in the low-dimensional (usually two-dimensional or three-dimensional) target space, and to keep the distance and proximity between the points as much as possible. The typical SOM structure is input layer and competition layer. The SOM input layer is used to receive information from the outside world and transmit the input information to the competition layer, playing the role of "observation". The competition layer is responsible for "analyzing and comparing" input information, looking for patterns and categorizing them.

The competitive learning steps are as follows:

(1) Vector normalization

The current input mode vector X in the self-organizing competition network and the internal star weight vector $w_j (j = 1, 2, \cdots, m)$ corresponding to each neuron in the competition layer are all normalized, $\overset{\Lambda}{X}$ and $\overset{\Lambda}{w}_j$ are obtained.

$$\overset{\Lambda}{X} = \frac{X}{\|X\|} \quad \overset{\Lambda}{w} = \frac{w_j}{\|w_j\|}$$

(2) Identify winning neurons

A similar comparison was made between $\overset{\Lambda}{X}$ and all neurons in the competitive layer with respect to the internal star weight vector and $\overset{\Lambda}{w}_j(j = 1, 2, \cdots, m)$. The most similar neuron wins, and the weight vector is $\overset{\Lambda}{w}_{j*}$.

(3) Network output and weight vector adjustment

According to the WTA learning rule, the winning neuron outputs "1" and the rest 0, which are:

$$y_i(t + 1) = \begin{cases} 1, j = j* \\ 0, j \neq j* \end{cases}$$

Only the winning neuron has the right to adjust its weight vector $\hat{w}_{j*}$, and its weight vector learning adjustment is as follows:

$$\begin{cases} w_{j*}(t + 1) = \hat{w}_{j*}(t) + \Delta w_{j*} = \hat{w}_{j*}(t) + \alpha\left(\hat{X} - \hat{w}_{j*}\right) \\ w_j(t + 1) = w_j(t) \quad j \neq j* \end{cases}$$

$0 < \alpha \leq 1$ is the learning efficiency, which generally decreases with the multidimensional progress of learning, that is, the degree of adjustment becomes smaller and smaller and tends to the clustering center.

(4) renormalization

After the normalized weight vector is adjusted, the resulting new vector is no longer a unit vector. Therefore, the adjusted learning vector should be normalized again, and the learning rate should be reduced to 0.

2.2 Improving Algorithm

The advantage of SOM is that it can strengthen the adjacent relation to the center of mass of the cluster, which is conducive to the interpretation and observability of clustering effect. White SOM also has the problem that the user must choose parameters, neighborhood function, network type and number of centroid, and lacks specific objective function. And although it usually converges in practice, SOM does not guarantee convergence. Based on this, the training speed was improved by initial weight during the research. The specific steps are as follows:

Set data matrix $X = [x_1, x_2, \cdots, x_n]^T$, where $x_i = (x_{i1}, x_{i2}, \cdots, x_{ip}), i = 1, 2, \cdots, n$.

Step1. Calculate the Euclidean distance between two data:

$$d(x_i, x_j) = \left((x_{i1} - x_{j1})^2 + (x_{i2} - x_{j2})^2 + \cdots (x_{ip} - x_{jp})^2\right)^{1/2}$$

Step 2. Calculate the average distance between sample points:

$$\bar{d} = \frac{\sum d(x_i, x_j)}{C_n^2}$$

Step 3. Calculate the density parameters of data object points, which are recorded as $density(x_i, \bar{d})$. That is, with the point x_i as the center, the region of radius $\bar{d}$ is called the neighborhood of the point, and the number of points in the region is called the density parameter of point x_i based on distance $\bar{d}$.

$$density(x_i, \bar{d}) = \sum_{j=1}^{n} u(\bar{d} - d(x_i, x_j))$$

where $u(z) = \begin{cases} 1, z \geq 0 \\ 0, \text{other} \end{cases}$. Calculate the density parameters of all objects and form the set $D = density(x_i)$.

Step 4. Find the maximum density parameter value D_i in D and check the number of the maximum density parameter $\max I$.

(i) If $\max I = 1$, the density parameter value D_i corresponds to sample x_i in the sample set S, which is the first weight value w_1.

(ii) If $\max I > 1$, fine the maximum density parameter value D_i of D, and all samples x_i, $i \in n$ corresponding to sample set S, and calculate the following,

$$sum_i = \sum_{j=1}^{n} d(x_i, x_j)$$

where $d(x_i, x_j) \leq \bar{d}(S), j = 1, 2, \cdots n$, and comprises the set of SUM. $sum_i = \min(SUM)$, the 'ith' sample in S, is the first weight, called w_1.

Step5. Calculate the distance from other sample points in sample set S to w_1, and select the point with the greatest distance as the second weight w_2.

Step6. Calculate the distance between other sample points in sample set S to w_1 and w_2, which are denoted as d_{i1} and d_{i2}. If $d_i = max\{min(d_i, d_{i2})\}, i = 1, \cdots, n$ and $d_i > \frac{1}{2}d_{12}$, the corresponding data point is the third weight w_3. d_{12} is the distance between w_1 and w_2.

Step 7. If w_3 exists, then calculate the distance between other sample points in sample set S to w_1, w_2 and w_3, denoted as d_{i1}, d_{i2} and d_{i3}.

If $d_j = max\{min(d_{i1}, d_{i2}, d_{i3})\}, i = 1, \cdots, n$ and $d_j > \frac{1}{2}d_{12}$, the corresponding data point is the fourth weight w_4. d_{12} is the distance between w_1 and w_2.

Step 8. Followed the above steps until the distance is no more than d_{12}.

In this way, a series of network initial weights $W_j = [w_{j1}, w_{j2}, \cdots w_{jm}]^T$, $j = 1, 2, \cdots, k$ is obtained. Following the algorithm steps to identify the winning neuron according to the formula, determine the strengthening center of the winning neuron, and update the weight vector of the neurons in the topological neighborhood of the winning neuron on the grid.

2.3   General Warning Indicators

The general early warning model based on the WSOM algorithm as shown in Table 1, which is to give early warning of VTE to patients according to the abnormal changes of preoperative and postoperative coagulation report indicators. The experimental results in the hospital of case selection show that the accuracy is over 95%. In order to further explore postoperative VTE early warning, this paper carried out postoperative VTE early warning research for gynecological tumor patients based on the original research results.

Table 1. VTE warning indicators based on WSOM algorithm.

Before surgery	D- dimer (mg/L)	FDP (mg/L)	PT (S)	APTT (S)	TT (S)	FIB (g/L)	AT3 (%)
After surgery	D- dimer (mg/L)	FDP (mg/L)	PT (S)	APTT (S)	TT (S)	FIB (g/L)	AT3 (%)

According to the research on the composition characteristics and changing trend of inpatients with gynecological tumors since 1995 in Central South Hospital in Wuhan, the number of inpatients with various gynecological tumors shows an increasing trend, and the first four are cervical tumor, ovarian tumor, endometrial cancer and malignant trophoblastic tumor. Their study further analyzed the characteristics of gynecological tumor patients and male tumor patients, and further supported the study of selecting gynecological tumor patients for postoperative VTE early warning. In the meanwhile, VTE early warning after tumor surgery based on computer learning will have a broader application prospect.

3 Clinical Data and Research Methods

3.1 Clinical Data

According to the domestic gynecological tumor patients VTE general clinical diagnostic criteria, DVT can be diagnosed by diffuse pain appeared in both lower limbs after the operation, or D-dimer > 500 ηg/L, color doppler ultrasound examination of blood vessels in both lower extremities revealed strong echo in veins, and the veins could not retract after being pressurized. The 132 samples were collected from a hospital in Shanghai, involving uterine body malignant tumor, ovarian benign tumor, uterine fibroid (adenomyosis), gestation-related diseases, genital prolapse, genital polyps, cervical malignant tumor, and other non-inflammatory uterine diseases. In addition to the indicators obtained from each case sample, case history and hormone level sample values of clinical observation of gynecological tumor diseases were collected.

3.2 Data Processing

The collected sample values were preprocessed, and finally 89 available samples were obtained, of which 10 were recurrent samples after surgery and 79 were non-recurrent samples, as shown in Table 2.

72 postoperative cases were randomly selected as the model construction group, among which 7 cases had VTE, 65 cases had no VTE, and the remaining 17 cases were the model verification group. The age of the patients in the model construction group

Table 2. Database summary table.

Data type		Data volume(cases)
Overall count of clinical cases	Summary	89
	1 (no recurrence after surgery)	79
	0 (recurrence after surgery)	10

was (45 ± 10.1) years old, including 7 cases of uterine body malignant tumor, 10 cases of ovarian benign tumor, 15 cases of uterine fibroids (adenomyosis), 15 cases of gestation-related diseases, 5 cases of genital tract prolapse, 8 cases of genital tract polyps, 7 cases of cervical malignant tumor and 5 cases of other non-inflammatory uterine lesions. There were 17 cases in the control group, including 3 cases with VTE. The age of the patients in the model verification group was (46 ± 10.3) years old, including 1 case of uterine body malignant tumor, 3 cases of ovarian benign tumor, 2 cases of uterine fibroids (adenomyosis), 2 cases of pregnancy-related diseases, 1 case of genital prolapse, 3 cases of genital polyps, 2 cases of cervical malignant tumors and 3 cases of other non-inflammatory uterine lesions. There was no statistically significant difference in age and disease composition between the two groups ($p > 0.05$), suggesting comparability.

3.3 Application Test and Analysis of General Early Warning Model

All 89 samples were studied and evaluated using the VTE early warning model (as shown in Table 1), and the results were shown in Table 3.

Table 3. General VTE warning model validation.

Data type		Data volume (cases)	Exact clustering number (cases)	Clustering accuracy (%)
Overall count of clinical cases	Summary	89	74	83.15
	1 (no recurrence after surgery)	79	12	84.81
	0 (recurrence after surgery)	10	7	70.00

It can be seen from Table 4 that the general warning model was used for the test. Among the 89 samples, there were an average of 12 errors (type A errors) in the test of no recurrence cases, with the test accuracy of 84.81%, but the error tolerance rate of some recurrence samples was 30%.

In order to construct a VTE early warning model for gynecological tumors after surgery, the sample history and gonadal hormone concentrations were selected as additional indicators for the model, but PRL in the hormone level of cases was further

Table 4. Gonadal hormone concentrations in the samples after surgery.

Diagnosis	E2	LH	FSH	R	PRL	T
Cervical malignancy	100.51 ± 132.16	17.14 ± 14.77	18.91 ± 22.65	2.48 ± 4.21	22..48 ± 14.21	0.74 ± 2.39
Malignant Tumor of Ovary	160.15 ± 260.89	14.29 ± 12.27	27.67 ± 30.01	3.33 ± 6.72	23.01 ± 16.72	0.28 ± 0.27
Leiomyoma of uterus	140.51 ± 112.16	12.54 ± 13.79	14.78 ± 17.91	3.31 ± 5.56	21.31 ± 15.56	0.82 ± 2.99
Endometrial malignancy	142.51 ± 112.16	12.54 ± 13.72	14.25 ± 17.69	3.34 ± 5.53	22.24 ± 16.12	0.18 ± 0.13
Endometrial polyp	134.51 ± 116.16	13.54 ± 17.18	17.19 ± 20.18	3.31 ± 4.51	20.15 ± 17.15	0.5 ± 3.01
Genital prolapse	140.51 ± 112.16	12.84 ± 17.98	15.51 ± 20.76	3.15 ± 5.02	21.32 ± 15.43	0.27 ± 0.21
Pregnancy related illness	140.51 ± 112.16	13.54 ± 17.71	14.57 ± 23.21	3.16 ± 4.55	21.92 ± 11.35	0.24 ± 0.20
Other noninflammatory lesions	121.71 ± 125.16	10.54 ± 11.5	14.53 ± 21.89	2.19 ± 4.76	20.91 ± 14.21	0.28 ± 0.22

analyzed, and the PRL level of cases was within the normal range (see Table 4). Therefore, in the construction of the new early warning model, PRL levels of hormones are preferentially eliminated.

Seven indicators of coagulation report from the common warning indicators were listed as one group (14 indicators before and after operation), three indicators of medical history were grouped into one group, the remaining five indexes of gynecological hormone level (10 indicators before and after operation), these seven groups (25 indicators) were used for computer training and learning of WSOM algorithm to construct a VTE warning system after gynecological tumor surgery (Table 5).

Table 5. Grouping of sample indicators.

Indicator group	Group	Indicator						
1	Coagulation report	D-dimer (mg </L)	FDP (mg/L)	PT (S)	APTT (S)	TT (S)	FIB (g/L)	AT3 (%)
2	Medical history	Menopause	Reproductive history	Diseased site (uterus/ovary)				
3	Gynecological hormone level	LH						
4		R						
5		E2						
6		FSH						
7		T						

3.4   VTE Early Warning Model of Gynecology Was Constructed Based on WSOM Algorithm

WSOM algorithm is used to establish the VTE early warning model of gynecology, which consists of three steps: learning and training, index drop and model test.

72 groups of case data were randomly selected for model training from 89 groups of case samples, and the remaining 17 groups of case data were used as verification samples. Among the 72 data sets, there were 65 samples with no postoperative recurrence and 7 samples with postoperative recurrence. A matrix of size is established to represent a total of 72 groups of sample data, and each group of data includes 25 indicators. Using the NEWC function in Matlab software, the two categories are postoperative recurrence (denoted as 1) and postoperative non-recurrence (denoted as 0). To speed up learning, set the learning rate to 0,1. According to the research objective of postoperative early warning of gynecological VTE in this paper, the errors in the training results were divided into type A errors and type B errors. Meanwhile, the selection criteria of optimal clustering model were determined as follows:

(1)   High clustering accuracy.
(2)   Relative uniformity in the distribution of misclassified cases.
(3)   Under the same conditions, type A errors are easier to accept than type B errors.
(4)   If the maximum number of training rounds is reached, the algorithm is terminated, with the upper limit of training rounds set to 10,000.

We input all the 72×25 data to train the iteration. The upper limit of each training iteration was 100. After 10,000 training rounds, the classification results were still not ideal, which meant that the consideration of more influencing factors slowed the convergence rate of model optimization, making it difficult to identify the optimal classification model.

There is no doubt that using all 40 characteristic indicators as early-warning indicators of VTE can realize accurate early warning but based on empirical judgement at present it is unrealistic for clinical diagnosis and early-warning. Therefore, we adopted the 'cluster sampling' method to improve the value of the research results.

Based on the existing research results of the project, the coagulation report group was put into one group, the medical history was put into one group, and the five factors of hormone level were the single element group. A total of 7 groups were obtained, and the whole group sampling was conducted to improve the learning and training speed.

The number of indexes required by the model was optimized using the dimension-reduction learning and training method. According to the learning and training method outlined in the first step, all possible combinations of the index group are traversed, namely, $C_7^1, C_7^2, \ldots, C_7^7$, for a total of $2^7 - 1 = 131$ possible combinations.

To ensure the stability of the model, the learning data generated each time are not fixed. Sixty of the 72 groups of case data were randomly selected as learning data, and the remaining 12 groups were test data. Different learning data with the same indicator combination generated 100 different prediction models, and we took the average prediction accuracy of these 100 models, namely the 'average accuracy' referred to in Table 6.

Then, we selected 10 indicators (7 indicators in the preoperative and postoperative coagulation reports and 3 indicators in the hormone level group) in this model for whom the average accuracy was highest at 84.28%.

Table 6. Model reference factors.

Coagulation report	D- dimer (mg/L)	FDP (mg/L)	PT (S)	APTT (S)	TT (S)	FIB (g/L)	AT3 (%)
Hormone level	FSH	LH	E2				

According to the operation of WSOM algorithm, the final postoperative early warning model of gynecological tumor was obtained, as shown in Table 6.

The 10 constructed indicators were used to test whether VTE early warning after gynecological tumor surgery has application and promotion value. The model is tested twice with the remaining 17 indicators, and the test criteria are the same as the above. The results are shown in Table 7. Due to the randomness of the remaining indicators and two type A errors in the secondary test results of the 17 test samples, there were no type B errors, so the test accuracy was up to 88.24%. The VTE early warning model after gynecological tumor surgery based on WSOM algorithm has A strong application.

Table 7. Data classification results of secondary test.

Serial number	Actual classification	Forecast classification	Result	Serial number	Actual classification	Forecast classification	Result
1	1	1	T	10	1	1	T
2	1	1	T	11	1	1	T
3	1	0	F-A	12	1	1	T
4	1	1	T	13	1	1	T
5	1	1	T	14	1	1	T
6	1	1	T	15	0	0	T
7	1	0	F-A	16	0	0	T
8	1	1	T	17	0	0	T
9	1	1	T				

4   Model Analysis

4.1   Factor Analysis of VTE Prevention Model After Gynecological Tumor Surgery

The causes of VTE are increased blood solidness, slow blood flow and vein wall damage. Many studies have shown that patients with gynecological tumor surgery are at high risk of VTE. Gynecologic tumor surgery is mostly radical surgery, surgery can directly damage the local tissue and vascular wall, promote the activation of platelets. The tumor cells in the patient can express the pro-coagulant activity of the cells themselves, that is, excessive tissue factor, pro-coagulant protein and factor V receptor, and activate the coagulation process. Preoperative preparation of gynecological tumor patients requires fasting, blood further concentration, increased the chance of thrombosis. Stored plasma is commonly used intraoperatively and may increase the likelihood of thromboembolic events due to the high consistency of stored plasma. At the same time, intraoperative anesthesia causes the veins below the anesthesia plane to dilate and the blood flow rate to slow down, which increases the risk of intraoperative thrombosis. Although the patient recovers blood flow after the operation, it is difficult to repair the endothelial damage caused by the operation. Therefore, targeted early warning can create the best opportunity for prevention and provide scientific evidence for the most reasonable prevention measures for physicians.

In this study, indicators of coagulation report (d-dimer, FDP, PT, PTT, TT, FIB, AT3) and hormone levels of FSH, LH, E2 were established as an early warning of VTE after gynecological tumor, which can more accurately reflect the detection indicators of t-PA and PAI in body fiber capacity. At the same time, the coagulation reporting index can detect the postoperative vWFF, DD, and FDP levels of tumor patients, and the appropriate prevention and treatment measures can help patients improve the ability of fibrinolytic system correction and avoid the occurrence of postoperative VTE. The hormone levels of FSH, LH and E2 in patients with gynecological tumors after surgery can be dynamically detected and help the attending physician to analyze PAP, PAI, PLI, PLG and PLM, which is conducive to the evaluation of the pre-thrombotic status of patients after surgery and targeted drug prevention, so as to reduce thrombosis and improve the therapeutic effect of embolism.

4.2   Prevention of Gynecologic Tumors After Surgery

Gynecological tumor surgery is the most difficult operation in gynecological surgery, compared with the postoperative complications of other benign diseases, gynecological tumor surgery postoperative complications will be more. At present, most gynecological tumor surgery is minimally invasive surgery, the common postoperative complications are mechanical injury and thermal injury. In the process of puncture, there may also be vascular injury, especially great vessels injury. In the process of surgical injury, there may also be vascular injury, intestinal injury, urinary system injury, etc. These complications are insidious and fatal, which are often serious and even life-threatening in the later stage. Therefore, appropriate nursing intervention before surgery for gynecological tumor patients can improve the effect of targeted surgical treatment, reduce the incidence of complications, improve the satisfaction of patients with nursing, and relieve the patients' bad mood.

Clinical prevention of VTE after gynecological tumor surgery is divided into physical prevention and drug prevention. And the preventive measures according to the early warning model of the early warning report to choose the immediate preventive measures. Physical prevention mainly adopts GCS, IPC and VFP, which can reduce the formation of DVT by preventing blood stasis in the calf. Intermittent air pressure treatment can reduce the incidence of VTE disease. Drug prophylaxis was mainly administered with low-dose unfractionated heparin (LDUH) and low molecular weight heparin (LMWH), it is one of the most widely studied methods of thromboprophylaxis. Several control studies have shown that low-dose heparin is most effective in preventing DVT, and subcutaneous injection is recommended to start 2 h before surgery and continue every 8 to 12 h after surgery. LDUH can greatly reduce the incidence of VTE in patients with gynecologic tumors after surgery. Low molecular weight heparin has the same efficacy as LDUH, with better bioavailability, less adverse reactions, less antithrombotic activity, lower bleeding risk and lower dose of heparin. It is easy to use in clinical practice, but costs more than LDUH.

The preventive measures for gynecological tumor patients after surgery are generally to the patient's complete freedom of movement. The American college of obstetricians and gynecologists (ACOG) recommends that gynecologic surgeons take precautions until they leave the hospital. For extremely high-risk patients, including those who underwent radical tumor resection, aged over 60 years old, and had a previous history of VTE, continued preventive measures are recommended for 2 to 4 weeks after discharge. At the same time, special attention should be paid to the mental health care of patients in the recovery process, to strengthen the effective communication with their families, to obtain their support and coordination, and to guide patients. In general, it is recommended that scientific assessment and individualization of preventive measures be taken.

5   Conclusion

Based on the post-operative early warning model of gynecological tumor proposed by WSOM clustering algorithm, this paper screened out the characteristic variables that can be passed and interpreted by all books with higher target variables, and constructed the VTE early warning model after gynecological tumor surgery. At the same time, based on the factor analysis of VTE early warning model after gynecological tumor surgery, physical prevention and drug preventive measures were proposed. In clinical practice, its prevention is conducted in grades. How to determine the grade of postoperative VTE in gynecological tumor patients will be the direction of further study in this paper. The research will further optimize the algorithm to enrich the case samples and improve the transformation and application of the results.

Acknowledgments. This work was supported in part by Fund Project of SSPU (EGD20XQD17), Gaoyuan Discipline of Shanghai–Environmental Science and Engineering (Resource Recycling

Science and Engineering), Discipline of Management Science and Engineering of Shanghai Polytechnic University (Grant No. XXKPY1606). Scientific Research Project in Hospital Management Construction of Shanghai Jiaotong University (CHDI-2018-A-16) and Scientific, and Technological Key Project of Shanghai Songjiang District Science and Technology Commission (2017SJKJGG32).

References

1. Zhang, Z., Lang, J.: Deep vein thrombosis (DVT) and pulmonary embolism (PE) after gynecological surgery are important surgical complications. Chinese J. Obstetrics Gynecol. **52**(10), 654–666 (2017)
2. Li, X.: An Empirical Study of Traditional Chinese and Western Medicine in the Prevention of Deep Vein Thrombosis in Gynecological Laparoscopy. ChongQing Medical University, Chongqing (2015)
3. Peedicayil, A., Weaver, A., Li, X., Carey, E., Cliby, W., Mariani, A.: Incidence and timing of venous thromboembolism after surgery for gynecological cancer. Gynecol. Oncol. **121**(1), 64–69 (2011)
4. Yin, S., Chang, J., Pan, H., Mao, H., Wang, M.: Early warning of venous thromboembolism after surgery based on self-organizing competitive network. J. Comb. Optim. https://doi.org/10.1007/s10878-019-00504-z
5. Graul, A., Latif, N., Zhang, X., Dean, T., Morgan, M., Giuntoli, R.: Incidence of venous thromboembolism by type of gynecologic malignancy and surgical modality in the national surgical quality improvement program. Int. J. Gynecol. Cancer **27**(3), 581–587 (2017)
6. Worley, M.J., Rauh-Hain, J.A., Sandberg, E.M., Muto, M.G.: Venous thromboembolism prophylaxis for laparoscopic surgery: a survey of members of the society of gynecologic oncology. Int. J. Gynecol. Cancer **23**(1), 208–215 (2012)
7. Morimoto, A., Ueda, Y., Yokoi, T., Tokizawa, Y., Yoshino, K., Fujita, M.: Perioperative venous thromboembolism in patients with gynecological malignancies: a lesson from four years of recent clinical experience. Anticancer Res. **34**(7), 3589–3595 (2014)
8. Jorgensen, E.M., Li, A., Modest, A.M., Leung, K., Hur, H.C.: Incidence of venous thromboembolism after different modes of gynecologic surgery. Obstetrical Gynecol. Survey **74**(4), 207–208 (2019)
9. Mittal, V., Ahuja, S., Vejella, S.S., Stempel, J.M., Leighton, J.C.: Trends and outcomes of venous thromboembolism in hospitalized patients with ovarian cancer: results from nationwide inpatient sample database 2003 to 2011. Int. J. Gynecol. Cancer **28**(8) (2018)

The Coordinated Decisions of Service Supply Chain Between Tumor Healthcare Alliance and Patients Under Government Intervention

Gengjun Gao[1], Zhen Wu[1], and Shuyun Wang[2(✉)]

[1] Logistics Research Center, Shanghai Maritime University, Shanghai 201306, China
[2] Shanghai First Maternity and Infant Hospital, Shanghai 201204, China
sywangsh@163.com

Abstract. Malignant tumor is a major disease that seriously threatens the health of residents, as one of the vital measures to promote treatment of the tumor, healthcare alliance is widely advocated and adopted. However, conflicting interests between the involved stakeholders in alliance hinder the implementation of healthcare alliance effectively. Based on government intervention, this paper investigated the supply chain of healthcare alliance-patient services at two levels: the tumor healthcare alliance and tumor patients, comprehensive hospitals (CH) and member hospitals (MH) that are low utilized (e.g. Primary Hospitals) within the alliance and respectively constructed the evolutionary game model between the alliance and patients, and between CH and MH to analyze the main influencing factors of the evolutionary strategies and study the optimal coordination mechanisms of the healthcare alliances to achieve the balance of the supply chain of alliance-patient services. Finally, numerical experiments are conducted to further verify the validity and rationality of the model. The results show that: "widen different reimbursement policy", "increasing government subsidies", "enhancing the reputational impact of the different strategies of the healthcare alliance" and "reducing the CH's costs of supporting MH" are effective measures to encourage patients to follow first diagnosis at MH and the CH support the MH to strengthen the ability. In addition, "increasing government subsidies" and "reducing the costs of cooperation" can promote both sides to collaborate with each other. Moreover, the initial strategy probability of stakeholders would affect the evolutionary trajectories. The conclusions provide managerial insights of much practical value for the operation of tumor healthcare alliance.

Keywords: Tumor healthcare alliance · Government intervention · Coordinated decisions · Evolutionary game theory

1 Introduction

As the population ages, the incidence and mortality of malignant tumor in China are on the rise, and malignant tumor is the second leading common cause of death, according to the latest data, patients with malignant tumors accounted for 23.91 percent of all deaths [1], such as lung, gastrointestinal, colorectal and esophageal cancers [2, 3].

© Springer Nature Switzerland AG 2020
Z. Zhang et al. (Eds.): AAIM 2020, LNCS 12290, pp. 569–580, 2020.
https://doi.org/10.1007/978-3-030-57602-8_51

As the high-quality services provider, the comprehensive hospitals are crowded and over utilized frequently [4]. Another salient issue is the long treatment cycle, involving chemotherapy, surgery, and radiation therapy [2], incurring high costs in high-level hospitals. Hence, there exists the salient problem of "difficult and expensive medical treatment" for residents.

As an important initiative of the new medical reform, the healthcare alliance has been effective in vertically integrating the medical resources of different levels of medical institutions, but there are still some barriers in the actual operation of the healthcare alliance service supply chain. Firstly, the overwhelming majority of the patients with malignant tumor prone to go directly to specialist CH for the first diagnosis. Consequently, the CH are frequently over utilized while the MH often remain idle [5]. Besides, hospitals within the alliance are all self-interest behaviors, simultaneously, patients are the source of income, which results in inefficient of the referral system [6]. In essence, the peers in alliance stay independent, every peer pursues maximum profits [7]. Then we thereby argue that the lack of interest coordination mechanisms, which prevents the formation of a stable service supply chain. Accordingly, the government should give full play to its function of propelling new medical reform [8]. By exploring the conflict of interest and coordination mechanism between the healthcare alliance and patients, we expect to realize the sinking of superior resources and achieve the standardize and homogenize tumor treatment, with a view to better serve patients.

2 Literature Review

As the problem of uncoordinated operation of the services supply chain of the healthcare alliance-patient has become more prominent, many scholars have strived to explore the causes of the problem and the solutions from different perspectives. Representative studies on the analysis and coordination of health consortium stakeholders include: Chen et al. [9] explored game matrixes among various stakeholders of the healthcare alliance, and discussed some suggestions on the positive operation of healthcare alliance. Lei [10] pointed that the lack of incentive and restraint mechanisms of basic medical care insurance is the major factor in the inefficient operation of two-way referral and put forwards suggestions to improve the process of referral. In addition, scholars have conducted abundant reviews to show that government intervention has played a critical role in the coordinated operation of the alliance: Lin et al. [11] analyzed the operation mode of the compact tumor healthcare alliance under the intervention of the government, and put forward constructive opinions on the possible risks in the later stage of operation. Xu et al. [12] studied the characteristics of the model of the government-hospital co-building healthcare alliance, and analyzed the benefits of this model through the experimental data. Game theory is potently applied to solve the problem of cooperation and coordination among multiple independent stakeholders, there are some typical researches: Gao [7] proposed a two-stage game-theoretic approach to study the operations of the healthcare alliance and further analyze the revenue sharing in the alliance. Gan et al. [13] constructed a three-stage dynamic game model of medical insurance funds, patients, and higher and lower hospitals, which showed that medical insurance payment levels should be used rationally to improve the hierarchical diagnosis and treatment system.

To sum up, domestic and foreign scholars have conducted many beneficial explorations of the healthcare alliance-patient service supply chain. However, most of the solutions to the alliance-patient problems lack the perspective of the evolution of events; The healthcare alliance-patient service supply chain involves multiple stakeholders, and the above literature does not discuss the balance of interests from the internal and external of the alliance. To fill the knowledge gaps, this paper focuses on the conflicts of interest in the healthcare alliance-patient service supply chain. Under the premise of government intervention, this paper will strive to explore the evolutionary processes of the healthcare alliance and tumor patients, the CH and MH of the alliance, and discuss the stable strategies for them, aiming to provide decision-making references and policy recommendations for the efficient operation of the healthcare alliance-patient service supply chain under the new medical reform.

3 Evolutionary Stability Analysis of Tumor Healthcare Alliance and Patients

In the service supply chain composed of tumor healthcare alliance and patients, hospitals within the alliance provide corresponding treatment plans and medical services to patients according to the patient's physical condition and pecuniary condition. Subsequently, the patient chooses whether to follow the provided plan for the first diagnosis and two-way referral behaviors [14]. Owing to the inadequate competence of the MH, the patients are reluctant to MH for the first diagnosis. On the other hand, the healthcare alliance is unwilling to spend costs to improve primary medical services. To investigate the strategies for propelling the service supply chain being operated smoothly, we build a healthcare alliance-patient evolutionary game model to constantly adjust their respective strategies for optimizing the dynamic evolution equilibrium.

3.1 A Subsection Sample

Related Hypothesis. In this section, the decision-making behaviors of the participants in the service supply chain constitute a game system L. Consider two players in the game: tumor healthcare alliance and patients, meanwhile, both of them are bounded rationality.

The healthcare alliance has two alternative strategies: "helping the MH to strengthen the ability of the malignant tumor treatment" (strategy S_1) with the probability $x(0 \leq x \leq 1)$; "not helping the MH to strengthen ability of the malignant tumor treatment" (S_2) with the probability $1 - x$; In the meanwhile, the patients also have two pure strategies: "going to MH for first diagnosis" (strategy P_1) with the probability $y(0 \leq y \leq 1)$; "not going to MH for first diagnosis" (P_2) with the probability $1 - y$.

Based on the above conflicts of healthcare alliance and patients, relevant parameters of the game models are defined:

(1) For healthcare alliance: If the patients choose strategy P_1, the medical alliance obtains the payoff I_1. Under the circumstances, if the healthcare alliance chooses S_1, the healthcare alliance incurs costs C to promote the sinking of medical resources and improve the MH's medical capacity. Beyond that, they will obtain the extra

payoff I_0, and the government compensation $G(G < C)$ and increasing in reputation M_1. By contrast, If the healthcare alliance chooses strategy S_2, the healthcare alliance's reputational loss is qualified as M_2. If patients choose strategy P_2, the healthcare alliance obtains the payoff I_2. Under the circumstances, if the healthcare alliance chooses strategy S_1, the healthcare alliance incurs costs C and obtain government compensation G.

(2) For patients: On account of the limited treatment level, the patients with malignant tumor who adopt strategy P_1 obtain the physical recovery payoff R_1 and incur the lower time and economic cost Q_1. Certainly, it contains the cost of patients with misdiagnosis, illness delay and so on. In addition, the government and the medical insurance institutions compensate them for their total expenses at a rate of μ percent ($0 < \mu < 1$). Under the circumstances, if the healthcare alliance chooses to strategy S_1, the patients will get relatively higher-quality treatment leading extra payoff R_0 and save the costs Q_0 incurred by mistreatment; The patients with malignant tumor who adopt strategy P_2 obtain the physical recovery payoff R_2 and incur the more costs $Q_2(R_1 < R_2; Q_1 < Q_2)$. In addition, the government and the medical institutions compensate them for their total expenses at a rate of v percent.($0 < v < 1$, $v < \mu$). And we suppose that when the patients do not comply with the primary diagnosis, which strategy the healthcare alliance chooses, it will not affect the patients' medical service benefits.

Then, evolutionary game payoff matrix of tumor healthcare alliance and patients under the government intervention can be obtained. Table 1 shows its payoff matrix:

Table 1. The evolution game model between healthcare alliance and patients

Patients	Healthcare alliance			
	S_1		S_2	
P_1	$R_1 + R_0 - (1-\mu)(Q_1 - Q_0)$	$I_1 + I_0 + M_1 + G - C$	$R_1 - (1-\mu)Q_1$	$I_1 - M_2$
P_2	$R_2 - (1-v)Q_2$	$I_2 + G - C$	$R_2 - (1-v)Q_2$	I_2

Modeling. According to the hypothesis and the matrix (1), it is supposed that U_{S1} and U_{S2} denote the expected payoff for the healthcare alliance on adopting strategy S_1 and S_2 respectively, and $\overline{U}_S$ represents the average excepted payoff of the healthcare alliance, all of which are described as follows:

$$U_{S1} = y(I_1 + I_0 + M_1 + G - C) + (1 - y)(I_2 + G - C) \tag{1}$$

$$U_{S2} = y(I_1 - M_2) + (1 - y)I_2 \tag{2}$$

$$\overline{U}_S = xU_{S1} + (1 - x)U_{S2} \tag{3}$$

Based on the Malthusian dynamic equation, from (1) and (3), the constructed replication dynamic equation of healthcare alliance is $F(x)$:

$$F(x) = \frac{dx}{dt} = x(U_{S1} - \overline{U_S})$$
$$= x(1 - x)[G - C + (I_0 + M_1 + M_2)y] \quad (4)$$

Similarly, the constructed replication dynamic equation of patients is $F(y)$:

$$F(y) = \frac{dy}{dt} = y(U_{P1} - \overline{U_P})$$
$$= y(1 - y)\{R_1 - (1 - \mu)Q_1 - [R_2 - (1 - v)Q_2] + [R_0 + (1 - \mu)Q_0]x\} \quad (5)$$

From (4) and (5), we can build a two-dimensional nonlinear dynamic system L as follows:

$$\begin{cases} F(x) = x(1 - x)[G - C + (I_0 + M_1 + M_2)y] \\ F(y) = y(1 - y)\{R_1 - (1 - \mu)Q_1 - [R_2 - (1 - v)Q_2] + [R_0 + (1 - \mu)Q_0]x\} \end{cases} \quad (6)$$

3.2 The Strategy Evolution of the System L

The Healthcare Alliance's Stable Strategy. Let $F(x) = \frac{dx}{dt} = 0$, we can obtain $x^* = 0x^* = 1$ or $y^* = \frac{C-G}{I_0 + M_1 + M_2}$. From the properties of the evolutionary stable strategy and the stability theorem of differential equation, when, $\frac{dF(x^*)}{dx}|_{x=x^*}$ and then x^* is an ESS. Then, the first-order derivative with respect to $F(x)$, leading to

$$F'(x) = (1 - 2x)[G - C + (I_0 + M_1 + M_2)y] \quad (7)$$

Only when $F'(x) < 0$, the evolutionary Stability Strategy (ESS) will be realized.

When $y = y^*$, $F(x)$ always remains 0, at this point, all x are stable condition; When $y \neq y^*$, two evolutionary stable strategy equilibrium points, respectively, but whether the point becomes an evolutionarily stable strategy depends on the plus or minus of $G - C + (I_0 + M_1 + M_2)y$. Furthermore, there exists three cases:

(1) When $y = y^*$, $F(x) = 0$, all x are stable;
(2) When $y < y^*$, $F'(0) < 0$, therefore $x^* = 0$ is an evolutionary stable strategy;
(3) When $y > y^*$, $F'(1) < 0$, therefore $x^* = 1$ is an evolutionary stable strategy;

The Patients' Stable Strategy. Let $F(y) = \frac{dy}{dt} = 0$, we can obtain $y^* = 0 y^* = 1$ or $x^* = \frac{R_2 - (1-v)Q_2 - [R_1 - (1-\mu)Q_1]}{[R_0 + (1-\mu)Q_0]}$ and after similar analysis, we found:

(1) When $x = x^*$, $F(y) = 0$, all y are stable;
(2) When $x < x^*$, $F'(0) < 0$, therefore $y^* = 0$ is an evolutionary stable strategy;
(3) When $x > x^*$, $F'(1) < 0$, therefore $y^* = 1$ is an evolutionary stable strategy;

3.3 Analysis of the Influence of Parameters on Evolutionary Equilibrium

From the proceeding analysis, we obtained five equalization points: $(0, 0)$, $(0, 1)$, $(1, 0)$, $(1, 1)$ and (x^*, y^*). And it can be known from the above hypothesis: $0 < x^* < 1$ $0 < y^* < 1$. The evolution analysis of portfolio strategy of healthcare alliance and patients with malignant tumor under the government intervention is shown in Fig. 1.

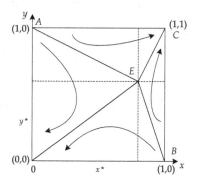

Fig. 1. The phase diagrams of the system L

As depicted in Fig. 1, $(0, 0)$ and $(1, 1)$ are evolutionary stable strategies, namely $\{S_1, P_1\}$ and $\{S_2, P_2\}$. By enlarging the quadrilateral S_{AEBC}, the evolutionary game is more prone to converge to Pareto optimal equilibrium $\{S_1, P_1\}$.It can be seen that the area of S_{AEBC} is negatively related to the value of x^* and y^*. By analyzing the factors influencing the value of x^* and y^*, we found that:

(i) $\frac{\partial x^*}{\partial \mu} < 0$, $\frac{\partial x^*}{\partial v} > 0$, $\frac{\partial x^*}{\partial (R_2 - Q_2)} > 0$, $\frac{\partial x^*}{\partial (R_1 - Q_1)} < 0$, $\frac{\partial x^*}{\partial R_0} < 0$;

(ii) $\frac{\partial y^*}{\partial C} > 0$, $\frac{\partial y^*}{\partial G} < 0$, $\frac{\partial y^*}{\partial (M_1 + M_2)} < 0$, $\frac{\partial y^*}{\partial I_0} < 0$;

To obtain the minimum value of x^* and y^*: (1) The government ought to improve the incentive mechanism and increase financial subsidies to alleviate the cost of the healthcare alliance's assistance to primary hospitals; (2) Medical insurance institutions should formulate relevant differential payment policies within the alliance, and increase the proportion of reimbursement under the strategy of patients going to MH for the first diagnosis and not going to MH for the first diagnosis; (3) The publicity should be intensified to enhance the influence of the strategy of choosing whether to support the MH of alliance on their reputation, so as to effectively influence the strategic choice of the alliance; (4) Formulate effective guidance measures to improve the players' spillover benefits of optimal decisions; (5) Besides, the ultimate ESS depends not only on the saddle point (x^*, y^*), but also on the initial proportion chosen by the participants. The ESS converges to Pareto optimal equilibrium $\{S_1, P_1\}$ with a greatest probability only if the initial proportion of peers' choices is in the upper right of $O(x^*, y^*)$, where $x > x^*, y > y^*$. Hence, it is needful to develop effective measures to increase the initial possibilities of the participants.

4 Evolutionary Stability Analysis of CH and MH in the Healthcare Alliance

Within the healthcare alliance, hospitals at different levels propose referral suggestions to the tumor patients according to their conditions. Some prime factors that hinder the implementation of two-way referral: (1) The CH lack the incentive to divert patients to lower-level hospitals for healing treatment, because it may sacrifice their revenues. Besides, the MH loath to easily access transferred patients due to the lack of high-quality medical service resources, contributing to the failure of the implementation of standardized treatment plans. Therefore, the evolutionary game model between CH and MH is established to explore a cooperative mechanism to form a community of interests between CH and MH, like establishing the green referral channel.

4.1 Hypothesis and Model

Related Hypothesis. In this section, we divide the healthcare alliance into two levels: CH and MH. And the decision-making behaviors of the players constitute a game system L'. Meanwhile, both of them are bounded rationality.

The CH have two pure strategies: "cooperation (strategy H_1)" and "non-cooperation (H_2)" with the probability $\gamma (0 \leq \gamma \leq 1)$ and $1 - \gamma$. Meanwhile, the MH can also play two pure strategies: "cooperation (strategy J_1)" and "non-cooperation (J_2)" with the probability $\eta (0 \leq \eta \leq 1)$ and $1 - \eta$.

Based on the above conflicts of CH and MH, relevant parameters of the game model are defined:

(1) For CH:D is the CH's payoff when the cooperation strategy between the CH and MH is not reached;D_0 is the CH's spillover payoff when the cooperation strategy is reached;F is the CH's input the cost for cooperating with MH, the cooperation cost includes the information platform construction, green channel construction in patients with malignant tumor, and so on; MH invest part of their cooperation cost βL into the CH as incentive cost $(0 < \beta < 1)$, since the CH play an dominating role in the medical alliance, and it is more inclined to maintain the exclusive right to its own resources;G_1 is the financial subsidies for the MH who adopt strategy $J_1(G_1 < F)$.

(2) For MH:M is the MH's payoff when the cooperation strategy between the CH and MH is not reached;M_0 is the MH's spillover payoff when the cooperation strategy between the CH and MH is reached;L is the MH's input the cost for cooperating with CH; G_2 is the financial subsidies for the MH who adopt strategy $H_1(G_2 < L)$.

From the preceding definitions, evolutionary game payoff matrix of CH and MH under the government intervention can be obtained. Table 2 shows its payoff matrix (2):

Modeling. It is supposed that U_{H1} and U_{H2} signify the expected payoff for the CH on adopting strategy H_1 and H_2, and $\overline{U}_H$ represents the average excepted payoff:

$$U_{H1} = \eta(D + D_0 + G_1 + \beta L - F) + (1 - \eta)(D + G_1 - F) \qquad (8)$$

Table 2. The evolutionary game model between MH and CH

MH	CH			
	H_1		H_2	
J_1	$M + M_0 + G_2 - L$	$D + D_0 + G_1 + \beta L - F$	$M + G_2 - L$	$D + \beta L$
J_2	M	$D + G_1 - F$	M	D

$$U_{H2} = \eta(D + \beta L) + (1 - \eta)D \tag{9}$$

$$\overline{U}_H = \gamma U_{H1} + (1 - \gamma)U_{H2} \tag{10}$$

From (8) and (10), the constructed replication dynamic equation of CH is $F(\gamma)$:

$$F(\gamma) = \frac{d\gamma}{dt} = \gamma(1 - \gamma)(G_1 - F + D_0\eta) \tag{11}$$

Similarly, the constructed replication dynamic equation of patients is $F(\eta)$:

$$F(\eta) = \frac{d\eta}{dt} = \eta(1 - \eta)(G_2 - L + M_0\gamma) \tag{12}$$

Furthermore, we can build a two-dimensional nonlinear dynamic system L' as follows:

$$\begin{cases} F(\gamma) = \gamma(1 - \gamma)(G_1 - F + D_0\eta) \\ F(\eta) = \eta(1 - \eta)(G_2 - L + M_0\gamma) \end{cases} \tag{13}$$

$$U_{H1} = \eta(D + D_0 + G_1 + \beta L - F) + (1 - \eta)(D + G_1 - F) \tag{8}$$

$$U_{H2} = \eta(D + \beta L) + (1 - \eta)D \tag{9}$$

$$\overline{U}_H = \gamma U_{H1} + (1 - \gamma)U_{H2} \tag{10}$$

From (8) and (10), the constructed replication dynamic equation of CH is $F(\gamma)$:

$$F(\gamma) = \frac{d\gamma}{dt} = \gamma(1 - \gamma)(G_1 - F + D_0\eta) \tag{11}$$

Similarly, the constructed replication dynamic equation of patients is $F(\eta)$:

$$F(\eta) = \frac{d\eta}{dt} = \eta(1 - \eta)(G_2 - L + M_0\gamma) \tag{12}$$

Furthermore, we can build a two-dimensional nonlinear dynamic system L' as follows:

$$\begin{cases} F(\gamma) = \gamma(1 - \gamma)(G_1 - F + D_0\eta) \\ F(\eta) = \eta(1 - \eta)(G_2 - L + M_0\gamma) \end{cases} \tag{13}$$

4.2 The Strategy Evolution of the System L'

The CH' Stable Strategy. Let $F(\gamma) = \frac{d\gamma}{dt} = 0$, we can obtain $\gamma^* = 0$; $\gamma^* = 1$ and $\eta^* = \frac{F-G_1}{D_0}$. From the properties of the evolutionary stable strategy and the stability theorem of differential equation, when, $\frac{dF(\gamma^*)}{d\gamma}|_{\gamma=\gamma^*}$ and then γ^* is an ESS.

The first-order derivative with respect to $F(\gamma)$, leading to

$$F'(\gamma) = (1 - 2\gamma)(G_1 - F + D_0\eta) \tag{14}$$

When $\eta = \eta^*$, $F(\gamma)$ always remains 0, at this point, all γ are stable condition; When $\eta \neq \eta^*$, two evolutionary stable strategy equilibrium points, respectively, but whether the point becomes an evolutionarily stable strategy depends on the plus or minus of $(G_1 - F + D_0\eta)$. Furthermore, there exists three cases:

(1) When $\eta = \eta^*$, $F(\gamma) = 0$, all γ are stable;
(2) When $\eta < \eta^*$, $F'(0) < 0$, therefore $\gamma^* = 0$ is an evolutionary stable strategy;
(3) When $\eta > \eta^*$, $F'(1) < 0$, therefore $\gamma^* = 1$ is an evolutionary stable strategy;

The MH's Stable Strategy. Let $F(\eta) = \frac{d\eta}{dt} = 0$, we can obtain $\eta^* = 0$ $\eta^* = 1$ or $\gamma^* = \frac{L-G_2}{M_0}$. And after the similar analysis, we found:

(1) When $\gamma = \gamma^*$, $F(\eta) = 0$, all η are stable;
(2) When $\gamma < \gamma^*$, $F'(0) < 0$, therefore $\eta^* = 0$ is an evolutionary stable strategy;
(3) When $\gamma > \gamma^*$, $F'(1) < 0$, therefore $\eta^* = 1$ is an evolutionary stable strategy;

4.3 Analysis of the Influence of Parameters on Evolutionary Equilibrium

From the proceeding analysis, we obtained five equalization points: $(0, 0)$, $(0, 1)$, $(1, 0)$, $(1, 1)$ and (γ^*, η^*). And it can be known from the above hypothesis: $0 < \gamma^* < 1$ $0 < \eta^* < 1$. The evolution analysis of portfolio strategy of CH and MH under the government intervention is shown in Fig. 2.

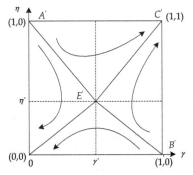

Fig. 2. The phase diagrams of the system L'

As depicted in Fig. 2, $\{H_1, J_1\}$ and $\{H_2, J_2\}$ are evolutionary stable strategies. Similar to the analysis in Sect. 3.3, we found that:

(i) $\frac{\partial \gamma^*}{\partial G_2} < 0$, $\frac{\partial \gamma^*}{\partial L} > 0$, $\frac{\partial \gamma^*}{\partial M_0} < 0$; (ii) $\frac{\partial \eta^*}{\partial F} > 0$, $\frac{\partial \eta^*}{\partial G_1} < 0$, $\frac{\partial \eta^*}{\partial D_0} < 0$;

To obtain the minimum value of γ^* and η^*: (1) The government should optimize the policies to reduce the cooperation costs of different levels of hospitals within the healthcare alliance; (2) In addition, it is also recommended to reduce the cooperation costs by simplifying the referral process, scale effect and other measures to promote collaboration and cooperation between each other; (3) Formulate effective guidance measures to improve the players' spillover benefits of optimal decisions; (4) Additionally, the inital proportion of strategies of participants exercises a great influence on the evolution trajectories. The higher the initial proportion, the greater probability the game converges to Pareto optimal equilibrium.

5   Numerical Simulations

In this section, a simulation of the model is conducted through MATLAB for the purpose of verifying the accuracy and effectiveness of the above results of the model and intuitively observe the dynamic evolutionary trajectories of the evolutionary system. The parameters of the system L are assigned as follows: $G = 14$; $C = 20$; $I_0 = 5M_1 = 7$; $M_2 = 2; Q_0 = 1$; $Q_1 = 5$; $Q_2 = 5$; $\mu = 0.7$; $\nu = 0.5$; $R_0 = 3$; $R_1 = 8$; $R_2 = 10$. Simultaneously, the parameters of system L' are assigned as follows: $G_1 = 9$; $G_2 = 6$; $D_0 = 12$; $M_0 = 10$; $F = 14$; $L = 13$.

The trend of the dynamic evolutionary are displayed in Fig. 3 and Fig. 4, it thus appears that the evolutionary trajectory will eventually tend to the ESS $(0, 0)$ and $(1, 1)$, it means that $\{S_1, P_1\}$ and $\{S_2, P_2\}$, $\{H_1, J_1\}$ and $\{H_2, J_2\}$ are evolutionary stable strategies, which accords with the above model analysis.

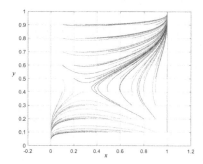

Fig. 3. The dynamic of the system L

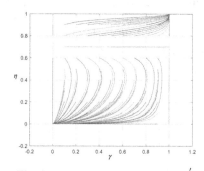

Fig. 4. The dynamic of the system L'

In order to simulate and verify the influences of the initial selective probability of the healthcare alliance and tumor patients on the ESS, we assume that the initial strategy of x and y are $(0.2, 0.2)$ and $(0.8, 0.8)$ respectively. Subsequently, the detailed results are displayed in Fig. 5, it can be easily found that the initial probabilities of the two peers are all 0.8, the ultimate ESS will tend to $(1, 1)$, namely $\{S_1, P_1\}$. Similar to the above assumption and analysis, Fig. 6 also prove that when the initial probabilities are all 0.8, the ESS will tend to $\{H_1, J_1\}$. Accordingly, it is necessary to formulate effective measures to increase the possibility of participants choosing the optimal strategy, which has important guiding significance for the coordinated operation of the tumor healthcare alliance.

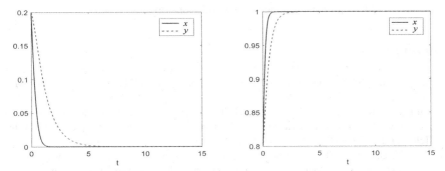

Fig. 5. Dynamic evolutionary paths under various initial probabilities of the system L

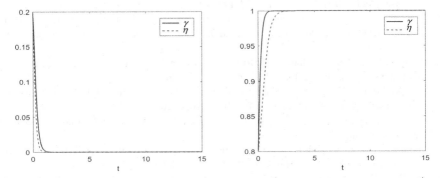

Fig. 6. Dynamic evolutionary paths under various initial probabilities of the system L'

6 Conclusion

The construction of the medical alliance of tumor specialists is a long-term work, which requires the government to play a dominating role, deepening the medical insurance policy, increasing the investment in primary hospitals, coordinating the interests of all stakeholders and making high-quality medical resources can better serve the people. To address this challenge, our paper concentrated on the issue of aligning the self-interest of the stakeholders related to the healthcare alliance. In particular, this paper constructs the evolutionary game model of the healthcare alliance and the patients with malignant tumor, CH and MH under government intervention. And then it analyzes the evolutionary game equilibrium strategy and the factors affecting the optimal strategy. By exploring reasonable incentive and compensation mechanisms, coordinating the interests of the game participants, the construction of healthcare alliance can be effectively promoted. What's more, the tumor defense and treatment mode of "primary hospital first diagnosis, two-way referral, triage treatment, and linkage up and down" can be realized. Finally, the healthcare alliance can better benefit the people, which is also conducive to forming a long-term stable healthcare alliance-patient relationship.

References

1. http://www.360doc.com/content/20/0113/17/19913717_885989219.shtml. Accessed 28 May 2020
2. Billingsley, K.G., Schwartz, D.L., Lentz, S.: The development of a tele-medical cancer center within the Veterans Affairs Health Care System: a report of preliminary clinical results. Telemed. J. e-Health **8**(1), 123–130 (2002)
3. Gao, Z., Gai, A.L., Zhou, J.: Current situation of cancer prevention and two-way referral under the new rural cooperative medical system in a pilot region of Hubei Province. China Cancer **126**(5), 327–332 (2017)
4. Li, N., Kong, N., Li, Q., Jiang, Z.: Evaluation of reverse referral partnership in a tiered hospital system–a queuing-based approach. Int. J. Prod. Res. **55**(19), 5647–5663 (2017)
5. Li, N., Pan, J., Xie, X.: Operational decision making for a referral coordination alliance- when should patients be referred and where should they be referred to? Omega-Int. J. Manag. Sci. **96**, 1–7 (2019)
6. Wang, J., Li, Z., Shi, J., Chang, A.: Hospital referral and capacity strategies in the two-tier healthcare systems. Omega-Int. J. Manag. Sci., 1–56 (2020)
7. Gao, X., Wen, J., Song, J.: Simulation study of revenue sharing in healthcare alliances. In: 2018 Winter Simulation Conference (WSC), pp. 2680–2689 (2018)
8. Gao, X., Wen, J., Song, J.: Capacity allocation and revenue sharing in healthcare alliances. Flex. Serv. Manuf. J. **2**, 1–23 (2019)
9. Chen, L.L., Yu, C.Y., Wei, L.: Analysis of operating mode of regional medical union based on game theory. China Med. Herald **13**(21), 131–133 (2016)
10. Lei, G.H.: Construction of incentive and restraint mechanism of dual-referral: from the age of basic medical care insurance. Chin. Gen. Pract. **16**(6A), 1829–1832 (2013)
11. Lin, Y.F., Zhang, Y.R., Qiu, T.L.: Practice and reflection on close-type medical alliance of oncology department. China Cancer **28**(8), 592–595 (2019)
12. Xu, X.X., Li, Y.M., Lin, Q.Q.: Experience and reflection on the construction of medical alliances in a government and hospital partnership model. Chin. J. Hosp. Admin. **35**(11), 903–906 (2019)
13. Gan, Y.Q., You, M.X., Hu, K.: Reimbursement gap, patient behavior and medical expenses-A three-stage dynamic game analyses. Syst. Eng.-Theory Pract. **34**(11), 2794–2983 (2014)
14. Fuertes, J.N., Toporovsky, A., Reyes, M.: The physician-patient working alliance: theory, research, and future possibilities. Patient Educ. Couns. **100**(4), 610–615 (2017)

Medical Data Compression and Sharing Technology Based on Blockchain

Yi Du[1] and Hua Yu[2]

[1] College of Engineering, Shanghai Polytechnic University, No. 2360 Jinghai Road,
Pudong District, Shanghai 230020, People's Republic of China
[2] Information Center, Shanghai General Hospital, No. 100 Haining Road, Shanghai 200080,
People's Republic of China
h.yu@shgh.cn

Abstract. Cloud service provides distributed storage spaces and avoids computing bottlenecks of centralized database storage, but it cannot provide secure data storage and sharing. How to store and share medical data efficiently and reliably is an important issue to ensure the safety of medical data in the process of medical data management of multi-regional hospitals. The public access of blockchains and the non-modifiable characteristics of the stored data make it an effective implementation scheme for medical data sharing. Meanwhile, with the continuous transactions in blockchains, the data of the blockchains are bound to be larger and larger, which leads to serious issues for the use and storage of data. Based on the LZW (Lemple-Ziv-Welch) algorithm, this paper presents a lossless compression technology for Chinese text compression with a compression storage and sharing scheme for medical data using blockchains to provide safer and more efficient access services for medical data.

Keywords: Blockchain · Medical data · Text compression · Data storage · Data sharing

1 Preface

Medical information is the valuable information from patients. However, in the current medical system of Chinese domestic hospitals, most information cannot be shared by all the hospitals, so the patients sometimes should apply for a new medical card to record the medical information of themselves, making the previous medical information of the patients useless or easily wrong. Although most hospitals use paper-based medical records, these records are very easy to be damaged or lost, which is a very unreliable way for medical information recording. On the other hand, with the development of cloud computing, distributed databases can realize medical information sharing, but may cause losses to patients because of the leakage of patients' information. Therefore, medical staff and patients need a system that can share medical information among hospitals and ensure the safety and reliability of stored medical data. As a distributed database system with multiple independent nodes, a blockchain is the ideal way to realize this system at present due to the advantages of decentralization, no trust, strong tamper resistance, etc.

© Springer Nature Switzerland AG 2020
Z. Zhang et al. (Eds.): AAIM 2020, LNCS 12290, pp. 581–592, 2020.
https://doi.org/10.1007/978-3-030-57602-8_52

A blockchain is expected to solve the common problems of low data security or poor sharing in the existing medical system. At the same time, with the continuous expansion of transactions in a blockchain, the amount of data it stores is also increasing. In order to improve the efficiency of a blockchain, this paper proposes the following solutions based on blockchains and text lossless compression technology:

1. Hospitals build a blockchain to jointly manage and maintain the stable operation of the blockchain, ensuring data security and reducing the cost of data protection;
2. The LZW (Lemple-Ziv-Welch) algorithm is used to compress the original medical data, compress and store the medical text data and realize the effective compression of medical information to relieve the storage pressure;
3. Based on the idea of text data compression given in this paper, the realization of medical data management and sharing under the blockchain is discussed, which provides a new idea for the effective management of medical data under the blockchain.

Section 2 of this paper introduces the related work of text compression algorithm and blockchain technology. Section 3 introduces the blockchain, text lossless compression and other related technologies involved in this method. Section 4 introduces the specific design of the method, and finally the work is summarized in Sect. 5.

2 Research Progress

2.1 Blockchain

In recent years, many researchers and institutions at home and abroad use blockchain technology to explore and practice in the fields of data protection and sharing [1]. In 2013, Araoz et al. realized the authenticity protection of electronic files by storing hash values in fields in blockchain transactions [2]. Based on the blockchain, Vaughan et al. proposed a general file protection framework, which computes the file hash values and builds the Merkel tree to reduce the cost of data protection [3, 4]. In 2016, Azaria and others constructed a decentralized medical data access and authority management system by using smart contracts to realize patients' ownership of their medical data and enable them to independently share and manage medical records [5]. In November of the same year, Weide Cai and others put forward the development method of application system based on blockchain, including the design model of account chains and transaction chains, as well as the application principle of parallel code execution model on the chain [6].

In China, ant financial services carried out tracking management of donation and donation flow based on blockchain technology in 2017, improving the transparency, traceability and nonmodifiable characteristics of its system and data [7]. In 2018, Baidu applied the blockchain technology to the data protection of Baidu Encyclopedia, recording the historical version of each update of the encyclopedia entry, the author, editing time and other information on the blockchain, achieving the purpose of data protection and storage [8]. In October 2017, Zhang Ning et al. realized a solution framework of personal privacy data protection by using blockchain, database, asymmetric encryption and other technologies, initially realizing the protection of personal privacy in the Internet car rental scene [9].

In the field of medical treatment, American scholar Kevin Peterson et al. used blockchain technology to realize medical information sharing, and proposed a novel consensus mechanism, which uses the accuracy of semantics as proof to mine and generate blocks [10]. Researchers such as Ekblaw from MIT Media Laboratory in the United States, used Ethereum as the platform to write smart contracts to realize a distributed information management system "MedRec", realizing the information security of sensitive medical information through identity authentication, encryption, sharing and distributed storage of data [11]. At the same time, a proof of concept mechanism was proposed to ensure the normal operation and maintenance of the system. Researchers at the University of California, San Diego, proposed a blockchain-based privacy protection framework for decentralized medical information "ModelChain", which analyzes the data in the context of the private chain with artificial intelligence, and can design the information certificate without explicitly displaying the patient's medical information The algorithm is used to determine the processing order, which provides a good solution to protect the sensitive data of patients by combining the related technology of artificial intelligence [12].

In China, Xia Qi and others from the University of Electronic Science and technology proposed MeDShare scheme, which uses blockchain to realize medical data sharing, and completes the verification, audit and information sharing control of medical data [13]. In the scheme, smart contract is also introduced to track data information and realize data traceability. If there is an illegal transaction, smart contract can automatically revoke the access rights of illegal users to improve the security of the system. In 2018, Daiying Dong and Xueming Wang proposed to let the nodes in the hospital alliance service group register the public key on the network, and then add the public key to the header of each item of transaction data. The client uses the API interface provided by Web3 to interact with the node, and encrypts the user's private key once through the user's password, so as to ensure that patients can query cases through identity documents [14]. Yanhui Ren from Xi'an University of Electronic Science and technology proposed a scheme of medical information privacy protection and sharing based on blockchain. On the basis of ensuring the dense storage of archive data, the scheme enables users to achieve fine-grained access control for each record, and verifies the feasibility and performance overhead of the scheme [15].

2.2 Data Compression Algorithm

In the early 19th century, researchers replaced the characters commonly used in text using the code named MC (Morse Codex's), that is, using shorter characters to encode commonly used characters to achieve the compression effect. Then, the S-F (Shannon fan) coding algorithm appeared. According to the probability of symbol occurrence, the algorithm uses shorter coding symbols to replace the original symbols. In 1952, David A. Huffman proposed Huffman coding, which was monopolized from 1960s to 1980s [16]. Although Huffman coding has some advantages in data compression compared with the compression algorithm before 1952, it also has a fatal weakness, that is, when the algorithm compresses the data, it needs to scan the original data twice [17]. In

1977, Abraham Lempel and Jacob Ziv used dynamic dictionaries to replace the repeated strings in the text with shorter symbols, which greatly improved the compression ratio. In 1978, they published the improved LZ78 algorithm, whose idea is to generate a static dictionary set based on the input data. The algorithm is more stable and effective.

According to the requirements of application background, there are more and more improved algorithms based on LZ77 and LZ78 compression algorithms [18], among which LZW algorithm is the most popular. LZW algorithm is an optimized version of LZ78 algorithm proposed by Terry Welch in 1984 [19]. In addition to inheriting the fast compression and decompression characteristics of LZ78 algorithm, the compression ratio of this algorithm is higher. LZW can dynamically generate a dictionary set to save the processed historical data when compressing and decompressing the data. When "prefix, character" cannot correspond to each other in the algorithm dictionary, the prefix is encoded and output to realize data compression. For medical data, this paper presents a dictionary-based LZW Chinese compression method in the process of blockchain management and sharing to reduce the increasing pressure of transaction data storage and improve the transaction efficiency in blockchains.

3 Related Technologies

3.1 Main Idea of Blockchain

Blockchain technology can safely store bitcoin transactions or other data, and ensure the security of these data or information to prevent tampering and forgery. Different from the common relational database and the non-relational database, the core of blockchain is decentralization. By using encryption algorithms such as digital signatures, hash algorithms and distributed consensus algorithms, the stored data are very difficult to be tampered with, destroyed or erased from the database operation log. Under the premise that peer-to-peer nodes do not need trust, the decentralized point-to-point transaction, coordination and cooperation are realized to solve the problems of high cost, low efficiency and insecure data storage in the centralized database application system.

Blockchain system is divided into six layers: the data layer, the network layer, the consensus layer, the incentive layer, the contract layer and the application layer. Among them, the data layer encapsulates the underlying block of a blockchain, hash function, data encryption and time stamp; the network layer includes point-to-point technology, propagation mechanism and verification mechanism; the consensus layer encapsulates various consensus algorithms of network nodes; the incentive layer mainly includes the distribution mechanism; the contract layer mainly encapsulates various scripts, algorithms and smart contracts, which is the basis of the programmable characteristics of the blockchain; the application layer is the application of the blockchain in various scenarios. Smart contract is the core element of blockchain, which is triggered by events and runs on the blockchain data ledger. It is used to realize the interaction between a blockchain and contract application.

Each node in the blockchain can encapsulate the transaction data into a data block with time stamp, and link to the main block to form the latest block, and then synchronize the block information into the whole blockchain network. The data block includes block head and block body. The block header encapsulates the address of the previous block and a series of hash values, which are used to link the front block and the back block; the block body mainly contains the main information of the block. The composition of a blockchain is shown in Fig. 1.

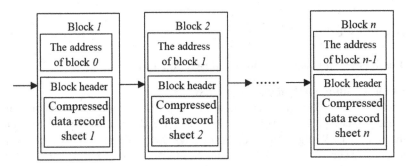

Fig. 1. The composition of a blockchain

3.2 LZW Algorithm

LZW algorithm is improved by Terry A. Welch on the bases of LZ78 and LZ78. It is a compression algorithm based on dictionary. The idea of dictionary compression algorithm is to store the strings appearing in the data as entries in the dictionary, and code these entries to replace the relatively long strings in the data with the encoding of entries. Its characteristic is that the dictionary does not need to be stored in the compressed file with the compressed data.

LZW algorithm initializes the dictionary at the beginning of compression. After initialization, the dictionary contains 256 single characters. After the compression starts, read the strings and match the entries in the dictionary one by one in order. If the match is successful, continue to read in the next character to form a new string and continue to match until the match fails. Number the string and add it to the dictionary as a new entry. Output the last matching to the number corresponding to the string. In this way, the number of entries in the dictionary will automatically increase as the data are compressed. Then the matching probability of strings in the data will be increased to achieve the purpose of data compression.

LZW algorithm is independent of the statistical characteristics of probability with compressed data. Therefore, this algorithm can be used in real-time data compression, which is very important because sometimes it is impossible to know the probability and statistical characteristics of each character in the compressed data in advance. For the data source with a high repetition rate, LZW algorithm can get a satisfactory compression rate.

LZW algorithm is a compression algorithm based on dictionary. The dictionary is built dynamically in the process of data compression. In the process of compressed data transmission and storage, the dictionary does not need to be transferred and stored together with the compressed data. When the data is decompressed, the dictionary can be generated dynamically in the process of decompressing. The dictionary constructed during compression is exactly the same with the dictionary constructed during decompressing. The capacity of the dictionary is not large, and the compression and decompression speed is high.

4 Method Design

4.1 Lossless Compression of Chinese Based on LZW

(1) **Chinese coding method based on LZW**

The core of LZW compression algorithm is a conversion table maintained in the process of compression, which is a dictionary. Because the basic processing unit of the general LZW algorithm is bytes for English characters, if it is directly applied to Chinese characters, the hidden semantic information in Chinese data coding will be lost artificially. Therefore, in this paper, the original LZW algorithm is improved to make it more suitable for the actual application scenarios.

In the improved algorithm given in this paper, GB2312 standard is used to obtain the code value of Chinese characters. In order to avoid too large initial dictionary, only common Chinese characters are added to the basic code set of the dictionary in advance, and space is reserved for undefined code words that may be encountered in the compression process. Through this dictionary, the longer Chinese string in the input data can be converted into shorter encoding as entry to achieve the compression goal instead of the relatively longer strings in the input data. The process of algorithm compression is shown in Table 1. In the process of compression, according to certain rules, the algorithm adds the first Chinese string encountered in the encoder to the dictionary, and assigns a unique flag value called code value to the added string.

Table 1. Chinese coding process based on LZW

Step 1	Initialize dictionary
Step 2	Input a Chinese character c, prefix string $P=c$
Step 3	Encoding conversion:
(1)	While c is not an end character do
(2)	If $P+c$ is in the dictionary
(3)	Then $P=P+c$
(4)	Else
(5)	Find the code of P in the dictionary;
(6)	Add P and c to the dictionary;
(7)	$P=c;$
(8)	Output the encoding of P

(2) **Chinese decoding method based on LZW**

Compared with the compression process, the restoration process of LZW algorithm is shown in Table 2. The key to the restoration process is that the initialized dictionary must be consistent with the compression program, and the dictionary maintained during the restoration process is almost synchronized with the compression process. In a blockchain, this process is mainly used for data users to decode the acquired data information.

Table 2. Chinese reduction and decoding process based on LZW

Step 1	Initialize dictionary
Step 2	Input the first encoding num and assign it to the reserved string O
Step 3	Output O
Step 3	Decoding conversion:
(1)	While num is not an end character do
(2)	Find the string N corresponding to the encoding num in the dictionary;
(3)	If N is Null
(4)	Then $N=O+N$;
(5)	Output N
(6)	Add $O+N$ to the dictionary;
(7)	$O=N$

4.2   Sharing and Acquisition of Medical Compressed Data

(1) **Blockchain model**

The compressed blockchain model proposed in this paper mainly includes three entities: a data owner, a data manager and a data demander. The details are as follows:

1. Data owner: the owner of data, responsible for the collection and provision of data. Here it mainly refers to patients or scientific research institutions or hospitals authorized by patients. The data owner needs to ensure the authenticity and reliability of data.
2. Data manager: responsible for compressed data storage and publishing. The data manager encrypts the compressed medical data and saves them on the cloud server. Only authorized users can download them.
3. Data demanders: entities that need to retrieve and use medical data. In the paper, they mainly refer to scientific research institutions or hospitals. Data demanders can query the medical data they are interested in from the blockchain network, record the hash values and download the complete data from the cloud server.

(2) **Blockchain sharing process of medical data**

The sharing of medical data refers to the sharing of safe and reliable medical data within the owner of the medical blockchain or outside the blockchain by using the demanders through intelligent contracts and hybrid encryption mechanisms, and can ensure the safe and efficient storage of data in the blockchain. It includes:

1. Building a data sharing model: in this stage, build a medical data sharing model on each sharing node, which includes data processing module;
2. Data processing stage: in this stage, the data owners participating in the sharing use the data processing module to collect, compress and store the data under their jurisdiction, make classification marks on the data as the sharing labels, and sign the data information using the private key;
3. Data communication stage: in this stage, the data users participating in data sharing perform node initialization configuration, and generate index data blocks containing the unique identity of the node in each node of the blockchain;
4. Data acquisition: the data demander applies for obtaining encrypted data information, and obtains the decryption key by sending his/her own identity; the data sharing module checks the authenticity of the shared record through the smart contract to determine whether the shared information matches successfully;
5. Data operation: each node in the area chain determines whether to allow this operation. If the number of nodes allowed for this sharing request is less than half of the total number of nodes, the data sharing request will be rejected. Otherwise, the operation will be allowed, and the operation will be time-stamped to record, generating a data operation block.

5 Performance Analyses

The mixed compression method including Chinese and English characters is implemented using Java. The experimental environment is given as follows:

- CPU: Intel (R) Cor e(TM) i7-5600U CPU @ 2.60 GHz;
- Memory: 12 GB;
- Operating system: Windows 7 professional (64 bit);

In order to test the effectiveness of the improved LZW algorithm, five groups of medical data files were randomly selected as experimental data to test the compression effect. The code length of the Chinese dictionary is 16 bit and the number of entries is 64K. Due to the fact that both Chinese and English characters are included in the conventional medical data information, the compression effects of Chinese medical data, English medical data and mixed medical data were tested and compared in the experiment. In the compression process of Chinese and English medical data, the experiment is completed by separating Chinese characters and English characters, and then compressing them respectively. The experimental results record the data compression ratio (data size after compression/data size before compression) and the compression time. The experimental results of compression performance are shown in Figs. 2, 3 and 4, respectively. Through the comparison of the three figures, it is not difficult to find that although the amount of mixed data changes, the compression rate of the algorithm remains constant, which continues the better compression performance of LZW compression algorithm.

Tables 3, 4 and 5 show the time taken to compress different data files. It can be seen that the compression time increases with the increase of data volume, but the compression time is also within the acceptable range for users.

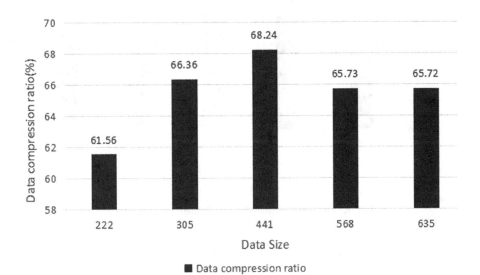

Fig. 2. The compression effect of mixed medical data

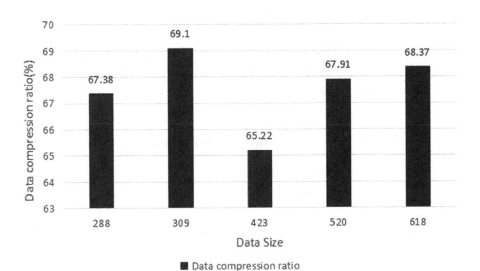

Fig. 3. The compression effect of Chinese medical data

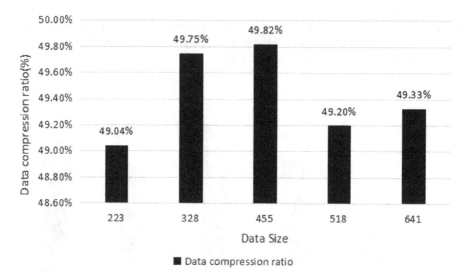

Fig. 4. The compression effect of English medical data

Table 3. The compression effect and time of mixed medical data

File size	Mixed medical data	
	Compression ratio (%)	Compression time (ms)
222	61.56	136
305	66.36	181
441	68.24	253
568	65.73	320
635	65.72	337

Table 4. The compression effect and time of Chinese medical data

File size	Chinese medical data	
	Compression ratio (%)	Compression time (ms)
288	67.38	123
309	69.10	162
423	65.22	242
520	67.91	297
618	68.37	318

Table 5. The compression effect and time of English medical data

File size	English medical data	
	Compression ratio (%)	Compression time (ms)
223	49.04	117
328	49.75	156
455	49.82	235
518	49.2	308
641	49.33	347

6 Conclusions

At present, the sharing of medical data among medical related institutions is always a hot research issue, so it is of great significance to ensure the privacy of medical data and realize the sharing of electronic medical compressed records based on blockchains. In this paper, based on the characteristics of blockchains, such as decentralization and non-modifiable characteristics, and in view of the increasing data volume with blockchain transactions, a medical data compression and sharing scheme based on blockchain is proposed. In this paper, a mixed LZW algorithm is proposed. In this experiment, some medical data information is randomly extracted, and the effectiveness of the algorithm is analyzed from the aspects of compression effect and compression time. The corresponding blockchain compressed data model and sharing process are given. It provides a technical idea for realizing the safe and efficient sharing of medical data among authorized users.

Acknowledgments. This work is supported by the National Natural Science Foundation of China (Nos. 41672114, 41702148).

References

1. Zhao, Z.: Research and Design of Digital Archive Management System Based on Blockchain. University of Science and Technology of China, Hefei (2018)
2. Proof of existence–An online service to prove the existence of documents (2018). https://docs.proofofexistence.com/
3. Merkle, R.C.: A digital signature based on a conventional encryption function. In: Pomerance, C. (ed.) CRYPTO 1987. LNCS, vol. 293, pp. 369–378. Springer, Heidelberg (1988). https://doi.org/10.1007/3-540-48184-2_32
4. Verify a chainpoint proof directly using Bitcoin (2017). https://runkit.com/tierion-/verify-a-chainpoint-proof-directly-using-bitcoin
5. Azaria, A., Ekblaw, A., Vieira, T., et al.: MedRec: using blockchain for medical data access and permission management. In: Proceedings of the International Conference on Open and Big Data, pp. 25 – 30. IEEE (2016)
6. Tsai, W.T., Yu, L., Wang, R., Liu, N., Deng, E.Y.: Blockchain application development techniques. J. Softw. **28**(6), 1474–1487 (2017)

7. Blockchain + public welfare, concept or trend (2017). http://www.xinhuanet.com/gongyi/2016-12/21/c_129414848.htm
8. Baidu's 'Wikipedia' now logs revisions on a blockchain (2018). https://www.coin-desk.com/-baidus-wikipedia-now-logs-revisionson-a-blockchain
9. Zhang, N., Zhong, S.: Mechanism of personal privacy protection based on blockchain. J. Comput. Appl. **37**(10), 2787–2793 (2017)
10. Peterson, K., Deeduvanu, R., Kanjamala, P., et al.: A blockchain-based approach to health information exchange networks. In: Proceedings of the NIST Workshop Blockchain Healthcare, vol. 1, pp. 1–10 (2016)
11. Ekblaw, A., Azaria, A., Halamka, J.D., et al.: A case study for blockchain in healthcare: "MedRec" prototype for electronic health records and medical research data. In: Proceedings of IEEE Open & Big Data Conference, vol. 13, pp. 1–13 (2016)
12. Kuo, T.T., Ohno-Machado, L.: ModelChain: Decentralized Privacy-Preserving Healthcare Predicting Modeling Framework on Private Blockchain Networks. arXiv preprint arXiv:1802.01746 (2018)
13. Xia, Q., Sifah, E.B., Asamoah, K.O., et al.: MeDShare: trust-less medical data sharing among cloud service providers via blockchain. IEEE Access **5**, 14757–14767 (2017)
14. Dong, D., Wang, X.: Research on electronic medical record sharing model based on blockchain. Comput. Technol. Dev. **05**, 1–4 (2019)
15. Ren, Y.: A privacy protection and sharing scheme of medical information based on blockchain. Xi'an University of Electronic Science and technology (2018)
16. Huffman, D.A.: Technique for the construction of minimum-redundancy codes. Proc. IRE **40**(9), 1098–1101 (1952)
17. Anas Almarri, A., Al Yami, B., et al.: Toward a better compression for DNA sequences using Huffman encoding. J. Comput. Biol. **24**(4), 280–288 (2017)
18. Ziv, J., Lempel, A.: Compression of individual sequences via variable-rate coding. IEEE Trans. Inf. Theory **24**(5), 530–536 (1978)
19. Welch, T.A.: A technique for high-performance data compression. Computer **17**(6), 8–19 (1984)

Model Establishment and Algorithm Research of Tumor Marker Combination Prediction for Colorectal Cancer

Bin Li[2], Tengfei Li[1], Xinye Zhou[2], Chen Huang[1(⊠)], and Guochun Tang[2]

[1] Shanghi General Hospital, Shanghai Jiaotong University, Shanghai 200080, China
richard-hc@hotmail.com
[2] Shanghai Polytechnic University, Shanghai 2001209, China
libin@sspu.edu.cn

Abstract. Colorectal Cancer (CRC) is one of the most common malignant tumors of digestive tract in the world; the incidence of CRC is increasing year by year. With the development of early screening, the improvement of surgical technique and the application of new treatment methods such as targeted therapy and immunotherapy, the mortality rate of patients with colorectal cancer has decreased obviously, but the long-term therapeutic effect remains suboptimal. Early detection, early diagnosis and early treatment are still the main treatment strategies for CRC. Currently, effective tumor markers are the primary basis for analysis and reference for clinical decision making in CRC treatment. Due to the large variety and quantity of data, the sensitivity and specificity between tumor marker data are difficult to determine, which brings some inconvenience to treatment. We use two-layer neural network's self-organizing-mapping (SOM) model and unsupervised learning algorithms to map high-dimensional data to a two-dimensional topology, allowing the relationships between the data to be visually represented, which can help doctors better judge the prognosis of CRC and monitor the recurrence and metastasis of CRC from the relationship between the data.

Keywords: Colorectal cancer · Tumor marker · SOM · Data clustering · Statistics

1 Introduction

Colorectal Cancer (CRC) is one of the most common malignant tumors of digestive tract in the world, the incidence of colorectal cancer is increasing year by year. The latest cancer statistics for the United States in 2020 show that CRC accounts for 9 percent of all cancers in men and 8 percent in women. It ranks 4th and 3rd in morbidity and mortality among all malignancies [1]. Cancer Statistics from the National Cancer Center of China in 2019 show that the morbidity and mortality of colorectal cancer ranked 3rd and 5th among all malignant tumors respectively, posing a serious threat to the lives and health of the country's residents [2]. With the development of early screening, the improvement of surgical technique and the application of new treatment methods such

© Springer Nature Switzerland AG 2020
Z. Zhang et al. (Eds.): AAIM 2020, LNCS 12290, pp. 593–603, 2020.
https://doi.org/10.1007/978-3-030-57602-8_53

as targeted therapy and immunotherapy, the mortality of patients with colorectal cancer has decreased obviously, but the long-term effect remains suboptimal [3]. Our group has long been devoted to the basic and clinical research of colorectal cancer, putting forward new opinions on the mechanism of the occurrence and development of colorectal cancer, the progress of minimally invasive surgical treatment of colorectal cancer, the principles of management of obstructive colorectal cancer, the differences between left and right colorectal cancer and the principles of management [4–7]. At present, early detection, early diagnosis and early treatment are still the main treatment strategies for CRC. With the development of early screening, the improvement of surgical technique, and the application of targeted therapy and immunotherapy, the mortality rate of patients has been decreasing, but the long-term effect remains suboptimal.

At present, screening of effective and sensitive tumor marker in the existing diagnostic methods is of great significance for CRC evaluation, prognosis and recurrence monitoring. Common tumor markers for CRC include CEA, CA199, CA724, CA125, and CA50. Recent studies by Gao et Al have shown that in CRC serum tumor markers the sensitivity ranges of single markers from high to low are CEA, CA724, CA19-9 and CA125. The patients who are positive for preoperative serum CEA, CA199 and CA724 are more likely to have lymph node metastasis; CA125-positive patients are more likely to have vascular infiltration; CEA-positive and CA125-positive patients are more likely to have nerve infiltration. In addition, positive CA199, CA724 and CA125 are associated with poorly differentiated tumors, while the levels of CEA, CA199, CA72-4 and CA125 are positively correlated with staging of lymph node metastasis [8]. Therefore, the combined serum markers can be used not only in the diagnosis of colorectal cancer, but also in the evaluation of tumor status, guiding treatment, evaluation of efficacy and prognosis. However, the sensitivity and specificity of each single index in the first-visit patients are relatively low, so it is necessary to conduct optimized combination and matching of these tumor markers to improve the sensitivity and specificity of diagnosis.

In this study, by evaluating the co-expression levels of CRC five serum tumor markers and based on different weights, we constructed a mathematical model for CRC five serum tumor markers, patient's pathologic staging and prognosis. It is used to simulate clinical prognosis judgment and monitor recurrence in CRC patients.

Because of the complexity of the correspondence between tumor marker level and CRC prognostic status, it is difficult to address the issue with conventional approaches. This study combines medical, mathematical, information, and computational science methods to optimize the combination of CRC tumor marker in order to provide early more effective information of disease evaluation, prognosis judgment and relapse monitoring for clinical treatment of CRC.

2   Basic Principles and Methods

In clinical tests, a tumor marker is usually positive in a variety of tumors, and a tumor can also be positive in a variety of tumor marker, so the optimized combination of sensitive and specific markers is a very practical and meaningful issue in clinical work. Some literature evaluated the association between seven tumor markers and colorectal cancer and constructed an optimal combination for identifying CRC. Considering that the sensitivity and specificity of single-item test in the first-time patients are relatively low, the combined test of different methods can significantly improve the diagnostic efficiency and is worthy to be popularized. We intend to integrate the co-expression levels of CEA, CA199, CA50, CA125 and CA724 and construct a model of co-expression levels of five markers, CRC pathologic staging and prognosis based on different weights, which is used to evaluate the relationship between five serum tumor markers and pathological staging as well as prognosis in patients with CRC. It is expected to provide more effective information of disease evaluation, prognosis judgment and relapse monitoring for clinical treatment of CRC.

Recently, the department of gastroenterology at Shanghai General Hospital collected data on tumor marker, the clinicopathologic features and prognosis of 365 CRC patients. But the sensitivity and specificity of any single item of tumor marker data is not visually apparent, and the information provided to medical staff is incomplete. In order to facilitate the analysis of clinical data, we only made a two-dimensional data sample table for the marker data set obtained during treatment for each CRC patient, listing five measurement data indexes and two related state indexes. The actual data set contains values for 365 patients, and this paper presents a sample of data for only five patients, for example, as shown in Table 1. To save space in this article, only five sample data are presented, all of which are drawn from a total database sample of 365 CRC patients. In Table 1, the first column of symbols, such as I, IIA, IIB, IIIC, represents the developmental stages of CRC status; the second column represents the patient's postoperative survival; and the remaining columns represent the measurements of the corresponding tumor markers.

Table 1. For example: positive level of CRC markers in 5 samples.

Staging	Days	CEA	CEA199	CA50	CA125	CA724
I	998	3.31	0.6	0.01	15.19	1.27
IIA	287	2.49	4.28	2.98	11.47	1.68
IIA	287	1.45	5.71	6.35	3.4	2.15
IIB	312	5.97	7.4	6.93	5.31	0.57
IIIC	298	215	76.88	33.64	25.56	3.52

Doctors at Shanghai General Hospital analyzed data collected from 365 CRC patients to determine the sensitivity and specificity of the tumor markers. The data analysis of sensitivity and specificity basically belongs to clustering analysis. For example, for a data sample {2, 4, 6, 16, 18, 20, 50} clustering analysis, generally a well-known k-means clustering method will be used [15]. But when the sample size is n and the cluster number k and the spatial dimension d are fixed values, all the Euclidean Distance must be calculated repeatedly. The computational complexity at this point is $O(n^{dk+1} \log n)$. It is still possible to converge to a local optimal solution using the k-means algorithm. This is essentially an NP-hard problem when trying to analyze the relationship between different vector samples as the number of patient samples increases [16].

Artificial Neural Networks (ANNs) are powerful and efficient data analysis tools. An important property of neural networks is their ability to learn from their sample dataset and, through training, to improve classification performance in some statistical sense [9].

In order to present more information, we use Kohonen SOM model and unsupervised learning algorithm in artificial neural network. The feature of SOM algorithm is that it can map the input vector data in the high-dimensional space to the low-dimensional space. In the process of feature mapping, the network is self-learning the features of the input vector data, and then re-clustering the input data according to the similarity between the vector data. SOM method can be used to map data to two-dimensional space, and the relationship between data types can be visualized, which can provide useful information for CRC treatment. It is a self-learning training process for finding clusters in data. This algorithm is mainly for their interactive ensemble clustering monitoring [14].

3 Neural Network Model and Algorithm

There are two main types of artificial neural network model and algorithm. One is multilayer neural network, which adopts supervised learning algorithm. The other is two-layer neural network, which uses unsupervised learning algorithm. The Kohonen Self-Organizing Map (SOM) is an unsupervised neural network. The SOM clustering should be fast, robust and intuitive method in fact.

Kohonen SOM network topological structure consists of two layers of neurons: input layer and output layer. See Fig. 1. The principal goal of the self-organizing map (SOM) is to transform an input signal pattern of arbitrary dimension into a two-dimensional discrete map. Each neuron in the input vector is fully connected with each neuron in the output layer through the weight vector. This network represents a feedforward structure with a two-dimensional computation layer consisting of neurons arranged in rows and columns. The neurons in the output layer are arranged according to two-dimensional lattice points, which can be either rectangular or hexagonal.

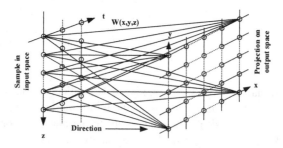

Fig. 1. Two-dimensional SOM topology

Definition. A real number set is Euclidean n-dimensional space, denoted by $\mathbb{R}^n$, as the set

$$\mathbb{R}^n := \{(x_1, x_2, \ldots, x_n) \mid x_i \in \mathbb{R} \, for \, i = 1, 2, \ldots, n\}.$$

Definition. A function $f : X \rightarrow Y$ is a relation between X and Y, such that for each $x \in X$, there is a unique y such that $(x, y) \in f$. We write this as $y = f(x)$. Functions are also called maps. A function can be injective and surjective. It is bijective if both.

Definition. A vector is called an ordered finite data set of numbers. The data set of all real number is written as $\mathbb{R}$, and the data set of all real number n-vectors is expressed $\mathbb{R}^n$. A collection of n-vectors in a d-dimensional Euclidean space, that are the columns of matrix $X \in \mathbb{R}^{d \times n}$, $X = \{(x_1, x_2, \ldots, x_n) \mid x_i \in \mathbb{R}^d \text{ for } i = 1, 2, \ldots, n\}$, then the Euclidean Distance of vector is the squared distance between x_i and x_j, that is given as $d_{i,j} = d(x_i, x_j) = \left\| x_i - x_j \right\|^2$, where $\| \, . \, \|$ denotes the Euclidean norm.

Definition. A topological mapping is said to be topology preserving, if it maps neighboring intervals from the topology space in $\mathbb{R}^n$ to the other 2D space nearby cells.

Algorithm description [9]:

Let n-dimensional vectors $X_1, X_2, \cdots, Xm$ are needed to applied in the input space, the components of these input vectors are $x^k = \{x_1^k, x_2^k, \cdots, x_n^k\}$, $k = 1, 2, \cdots m$, where m is the number of the sample data vectors, n-dimensional number is equal to the same number of components of the vectors. At one time learning, an input vector $X_p \in \{X_1, X_2, \cdots, Xm\}$ is applied to SOM. On map of the two-dimension output space, this neuron position (i, j) corresponds to the weight vector $W(i, j)$, where $W(i, j) = \{w_1(i, j), w_2(i, j), \cdots, w_n(i, j)\}$, $i, j = 1, \cdots, N$, N is map size. The learning process is to find a weight vector matrix $W(i, j)$, in order to minimal Euclidean distance $\left\| X_p - W(i, j) \right\|$. The neuron c with the minimal Euclidean distance to X_p is called a winner. According to the ability of topology protection, an input vector x^k is projected to the two-dimensional space.

When clustering data in the two-dimensional output space, by calculating the Euclidean distance D_p between vectors, compared with other neurons in the lattice, the weight vector $w_w(i, j)$ of the neuron c winning in the competition algorithm is the best matches to the input vector x^k, formula (1). The t th iterative training process of the first neuron is expressed as formula (2), where t is the number of iterations, $h_b(t)$ is the neighborhood function of the neuron c, and the simplest is the definition of the bubble, such as formula (3), where $\alpha(t)$ is the learning rate and the value range $0 < \alpha(t) < 1$. In addition, a Gaussian neighborhood function is defined as formula (4), σ is standard deviation. Last projection to output space [11]. When the SOM algorithm has converged on a solution, the feature map shows important statistical characteristics of the input space. The final result has significantly reduced the uncertainty of clustering data accuracy. The computational complexity is linearly with the number of data samples, and is quadrative with number of grid [12].

$$D_p(x^k, w_w(i, j)) = \left\| x^k - w_w(i, j) \right\| = \min_{1 \leq i, j \leq N}^{p} \left\| x^k - w(i, j) \right\| \tag{1}$$

$$w_{t+1}(i, j) = w_t(i, j) + h_b(t)\left[x^k(t) - w_t(i, j) \right] \tag{2}$$

$$\begin{cases} h_b(t) = \alpha(t), & \text{if } b \in R \\ h_b(t) = 0, & \text{if } b \notin R \end{cases} \tag{3}$$

$$h_b(t) = \alpha(t) \exp\left(-\frac{D_p(x^k, w_w(i, j))}{2\sigma^2(t)} \right) \tag{4}$$

Algorithm 1:

Input: Data Matrix: data set containing n objects. (N,N) grid of 2-dimensional MAP.

Output: Obtained a set of data clustering.

Begin:

  Initialization of the SOM weights with random number. learning rate: $0 < \alpha(t) < 1$

repeat: Draw a column vector X of Data Matrix .

  for all k , such that (eq.1) $D_p(\mathrm{x}^k, w_w(\mathrm{i,j})) = \left\| \mathrm{x}^k - w_w(\mathrm{i,j}) \right\| = \min_{1 \le i, j \le N} \left\| \mathrm{x}^k - w(\mathrm{i,j}) \right\|^{p}$ do

   determine the winning neurons c in the lattice and neighborhood(eq.3) and (eq.4).

  Update the weights of the SOM using (eq.2).

  end

 until all column data of Data Matrix.

   $w_{t+1}(\mathrm{i,j}) = =$new weight.

---

4 Experiments on Marker Data

Self-Organizing Map is here tried to provide doctors a supplementary tool to visualize easily the evolution of the medical data measured about the CRC patients. Taking everything into consideration in the experiments, the number of grid on 2-dimensional is about $\sqrt{n/2}$ for a data set of n points. For 500 to 1000 samples data, the map size of 4×4 or 5×5 rectangle can be taken.

The original data recorded the measurement of the markers of CRC patients during the clinical treatment. In order to find the rules and relationships between these markers, we need to process the original data in advance. After desensitization of data, we first distinguish the type of patients according to the two states of death and living, and two tables were obtained. Then make a sample table according to the CRC stage of each patient and the measurement values of five marker types. At last, the SOM learning classification of five kinds of markers was carried out [10]. See Fig. 2.

Because there are many types of data, this paper aims at CRC data experiment, mainly using the data of IIB stage of the living for comparison. The simulation results of these data partition experiments are based on MATLAB software. Their results are shown in the following figures. Figure 3, Fig. 4, Fig. 5, Fig. 6 and Fig. 7 are preliminary clustering results about relation of between CEA, CA50, CA724, CA199, CA125 and survival weeks number (instead of days) respectively with 365 CRC patients in IIB stage. The horizontal ordinate is survival weeks and the Y-axis ordinate is special tumor marker amplitudes. These green dots denote the whole tumor marker data of 365 patients. The black spot are centroids of clustering data. For areas where the green dots are denser, the number of patients is higher. The sparse green dots indicate a smaller number of patients. We circle the centroids represented by the corresponding black spot, and these samples will help doctors treat most patients further.

Figure 8, Fig. 9 and Fig. 10 are the data clustering of three tumor markers IIB-CEA, IIB-CA199, IIB-CA50 of CRC patients in the stage of state IIB. The figure (a) reflects the topology distance relationship between the sample values. The similarity features

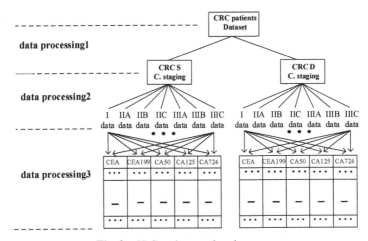

Fig. 2. CRC patient marker data type

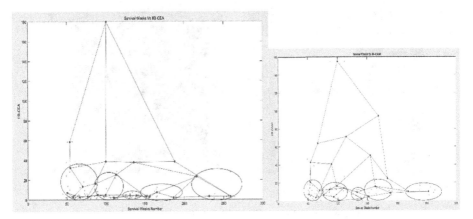

Fig. 3. CEA vs. s.w   **Fig. 4.** CA50 vs. s.w

values of the input sample data are projected to nearby grids on map. The number in the figure represents the number of times the similarity sample values occurs within a cluster. The difference between the samples is expressed by the size of the Euclidean distance. Because the sample data is extracted according to the tumor development stage of the patient, the sum of the total number of values in the small grid of figure(a) graph may be less than or equal to 365. The figure (b) is the size of the weight value of the sample. The darker the color, the greater the value.

Figure 11 and Fig. 12 are the classification of IIB-CA125 and IIB-CA724 in the stage of IIB. Figure (a) reflects the relationship between the three sample values and the number of samples in the data set. Figure (b) shows the weight value of the sample. The darker the color, the greater the value.

The experiment of data clustering using SOM Algorithm may lead to different locations of cluster centroids and different locations of neighbor samples. The main reasons

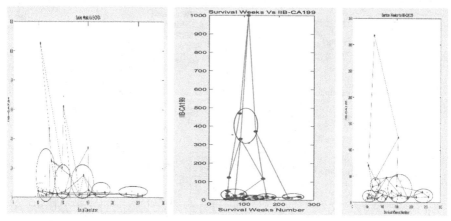

Fig. 5. CA724 vs. s.w   **Fig. 6.** CA199 vs. s.w   **Fig. 7.** CA125 vs. s.w

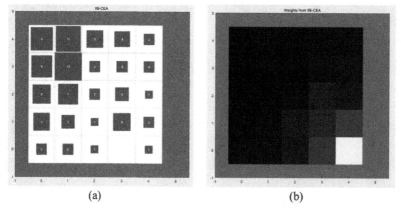

(a)   (b)

Fig. 8. (a) IIB stage CEA. (b) IIB stage CEA weight.

for this change are: First, the initial random weight will affect the last position of the centroid. Second, the final result depends on the training sample data set. Because of topology preservation, statistically speaking, a pair of adjacent data in a dataset or two specific adjacent data will be projected to the same or adjacent centroids in a two-dimensional graph with a high probability [13].

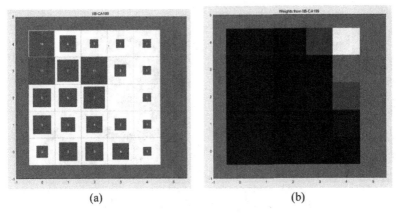

Fig. 9. (a) IIB stage CA199 (b) IIB stage CA199 weight

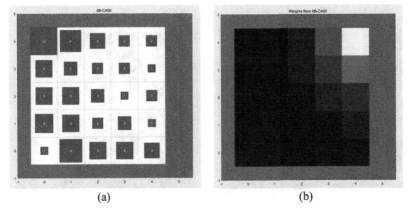

Fig. 10. (a) IIB stage-CA50 (b) IIB stage-CA50 weight

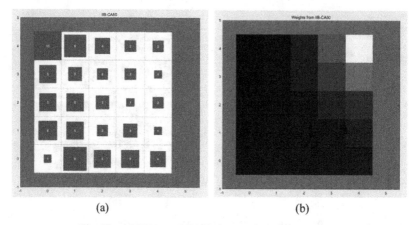

Fig. 11. (a) IIB stage-CA50 (b) IIB stage-CA50 weight

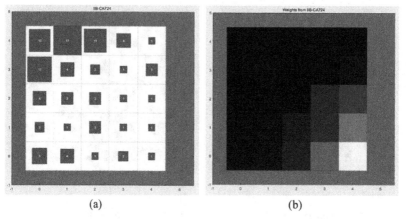

Fig. 12. (a) IIB stage-CA724 (b) IIB stage-CA724 weight

5   Conclusion

With the development of diagnosis and treatment technology, the existing examination tools can provide physicians with more and more information but it can also be more complex, so multiple diagnosis and treatment methods emerge. The self-organizing maps have possibilities not only to cluster, but also to visualize multidimensional marker date. With the rapid development of artificial intelligence technology (AI), AI can improve the ability of image recognition and natural language extraction [11]. As a difficult part in cancer research, CRC therapy is expected to be combined with artificial intelligence to bring more breakthroughs in the diagnosis and treatment of colorectal cancer. In addition, large amounts of tumor marker are numerical data that can be presented in an intuitive manner and are of practical importance for physicians to interpret and analyze pathological stages, judge prognosis and monitor recurrence in CRC patients. With Kohonen SOM network, input data can be clustered and analyzed, and the relationship between the data can be visualized by two-dimensional topological graph. It is vital to find out if the tumor marker data has cluster structure. This approach of medical care data visualization provides doctors with good data insight and complete information.

References

1. Siegel, R.L., Miller, K.D., Jemal, A.: Cancer statistics, 2020. A Cancer Journal for Clinicians, CA (2020)
2. Zheng, R., et al.: Analysis on the prevalence of malignant tumors in China in 2015. Chin. Jo. Oncol. **41**(1), 19–28 (2019)
3. Dekker, E., Rex, D.K.: Advances in CRC prevention: screening and surveillance. Gastroenterology **154**(7), 1970–1984 (2018)
4. Huang, C., Zhang, J.: The progression of surgical treatment strategies for left colon cancer with obstruction. Chin. J. Colorectal Dis. **6**(02), 94–97 (2017)
5. Fu, Z., Huang, C.: "Left, right": research progress in pathogenesis of colon cancer based on primary site and clinical treatment decision. Chin. J. Colorectal Dis. (Electron. Ed.) **8**(06), 546–552 (2019)
6. Rong, Z., Huang, C.: Research progress of colorectal cancer with schistosomiasis. J. Mod. Oncol. **27**(02), 335–339 (2019)
7. Huang, C., Rong, Z.: New progress in diagnosis and treatment of colorectal cancer. J. Bengbu Med. Coll. **43**(10), 1293–1298 (2018)
8. Gao, Y., et al.: Evaluation of serum CEA, CA19-9, CA72-4, CA125 and ferritin as diagnostic markers and factors of clinical parameters for colorectal cancer. Sci. Rep. **8**(1), 2732 (2018)
9. Haykin, S.: Neural Networks and Learning Machines, 3rd edn. Pearson, London (2009)
10. Giabbanelli, P.J., Mago, V.K., Papageoriou, E.I.: Advanced Data Analytics in Health. Springer, Heidelberg (2018). https://doi.org/10.1007/978-3-319-77911-9
11. Wu, H., Yu, Z., Huang, C.: Application and prospect of artificial intelligence technology in colorectal cancer. Chin. J. Colorectal Dis. (Electron. Ed.) **8**(02), 115–119 (2019)
12. Vesanto, J., Alhoniemi, E.: Clustering of the self-organizing map. IEEE Trans Neural Networks **11**(3), 586–600 (2000)
13. de Bodt, E., Cottrell, M., Verleysen, M.: Statistical tool to assess the reliability of self-organizing maps (1998)
14. Zhao, X., Cao, F., Liang, J.: A sequential ensemble clustering generation algorithm for mixed data. Appl. Math. Comput. **335**, 264–277 (2018)
15. Boyd, S.: Introduction to Applied Linear Algebra Vectors, Matrices, and Squares. Cambridge University Press, Cambridge (2018)
16. Strang, G.: Linear Algebra and Learning from Data. Wellesley-Cambridge Press, Cambridge (2019)

Safety Evaluation and Lean Disposal of Clinical Waste in Outpatient and Emergency Department of Large Hospitals

Huidan Lin[1] and Huijing Wu[2(✉)]

[1] Economy and Management School, Shanghai Polytechnic University, Shanghai 201209, China
hdlin@sspu.edu.cn
[2] Shanghai General Hospital, Shanghai 201620, China
vinogarden@126.com

Abstract. Large hospitals produce a large amount of medical waste in outpatient and emergency departments. The safety of medical waste is not only a huge problem faced by large hospitals themselves. Improper disposal will cross infection and expand the epidemic. The rational lean disposal of outpatient and emergency medical waste in large hospitals is an important measure, which is helpful to alleviate the problem. In this paper, we use fault tree analysis (FTA) of safety evaluation for medical waste logistics process in large hospitals with the new crown pneumonia for instance. Then we set up the safety evaluation model of this infectious medical waste and master waste streams of hazards and potential risks of large hospitals through calculation and analysis for the "discovery" not in time, "the lack of contingency plans and disposal technology". The results show that it is of great significance to adopt lean management method of outpatient and emergency treatment to solve these two hazard sources for the safety of infectious medical waste control in outpatient and emergency treatment.

Keywords: Medical waste · Logistics safety evaluation · FTA · Lean management

1 Introduction

In December 2019, continuous influenza and related disease surveillance was conducted in Wuhan, Hubei province, and multiple cases of viral pneumonia were found, all of which were diagnosed as viral pneumonia/pulmonary infection and were infectious. On January 30, 2020, the World Health Organization declared the outbreak of pneumonia caused by a novel coronavirus to be a public health emergency of international concern.

Supported by the Research Center of Resource Recycling Science and Engineering, Shanghai Polytechnic University Gao-yuan Discipline of Shanghai-Environmental Science and Engineering (Resource Recycling Science and EngineeringA30DB191202) and by Study on Optimization of Hospital Drug Supply Chain Based on SPD Model (EGD19XQD12) of Youth Teachers Training and Research Project of State Revenue in 2019 of Shanghai Polytechnic University.

© Springer Nature Switzerland AG 2020
Z. Zhang et al. (Eds.): AAIM 2020, LNCS 12290, pp. 604–614, 2020.
https://doi.org/10.1007/978-3-030-57602-8_54

By February 29, 2020, a total of 79,394 cases had been confirmed in China, with 2,838 deaths and 39,090 cures. A sudden outbreak of a new type of coronavirus pneumonia has roiled Wuhan, roiled the country and disrupted global supply chains. Facing the outbreak, from Wuhan and national medical emergency guarantee system of supply chain operation, there is confusion and regional segmentation, emergency response lag, the mismatch between supply and demand mismatches, the problem such as low efficiency and so on. So many patients with diagnosis of medical waste is also extremely dangerous, generated by the medical staff occupational health risk is great, if inappropriate treatment, easy to cause cross infection, expand the outbreak. In the early stage of the epidemic, due to the lack of human resources, the disposal, collection and transportation of medical waste are not reasonable, and there are also a variety of irregularities, which may lead to further cross-infection. Effective organization and rapid disposal of hospital medical waste, scientific exploration of its safe operation of hazards can provide better services for the new coronary pneumonia epidemic front line.

Medical waste has great risk [1], in the control over transboundary movement of hazardous wastes of the Basel convention, it mentions "from the hospital and clinic medical service, medical center of clinical waste" as a "category" which should strengthen the control of the waste in Y1 group, the risk characteristics of Grade 6.2, belong to infectious material [2]. In the process of classified collection, transportation, loading, unloading and storage, goods that are likely to cause damage to people, the environment, equipment and property and require special protection [3].

2   Methods for Logistics Risk Assessment of Infectious Medical Waste in Outpatient and Emergency Care of Large Hospitals

2.1   Safety Evaluation Method – Fault Tree Analysis (FTA)

Fault Tree Analysis (FTA) [4] is one of the important Analysis methods of safety system engineering. It can identify and evaluate the danger of various systems, which not only can analyze the direct cause of the accident, but also can deeply reveal the potential cause of the Accident.

According to the nature of the object system and the purpose of analysis, there are some differences in the analysis procedures, but the core analysis procedures are shown in Fig. 1 [5].

2.2   Safety Evaluation of Infectious Medical Waste in Outpatient and Emergency Department of Hospitals - a Case Study of Clinical Waste from New Coronavirus Pneumonia

According to the 2019 annual report on prevention and control of environmental pollution caused by solid waste released by the ministry of ecology and environment, 200 large and medium-sized cities in China released information on prevention and control of environmental pollution caused by solid waste in 2018. According to statistics, the output of general industrial solid waste in large and medium-sized cities released this time is 1.55 billion tons, the output of industrial hazardous waste is 46.43 million tons,

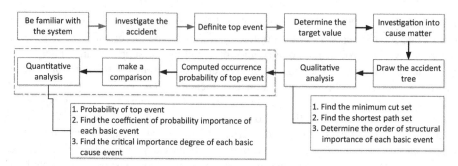

Fig. 1. Flow chart of accident tree analysis program

the output of medical waste is 817,000 tons, and the output of household garbage is 211.473 million tons. The amount of clinical waste produced is smaller than that of other wastes. The largest amount of medical waste was produced in Shanghai, which produced 55,000 tons. In the 2014–2018 report, the amount of medical waste produced in Shanghai ranked the first.

In addition, according to the latest dispatching data of the emergency office of the ministry of ecology and environment, by February 24, 2020, the national capacity of medical waste disposal is 5830.8 tons/day, which is 928.0 tons/day more than that before January 20. Since January 20, a total of 98,508.1 tons of clinical waste have been disposed of nationwide. On February 24, a total of 2,719.1 tons of medical waste were collected, of which 587.6 tons, or 21.6 percent, were from designated medical institutions. On that day, 2,749.8 tons of clinical waste was actually disposed of nationwide. All the clinical wastes from designated medical institutions in the new outbreak of pneumonia were transferred and disposed in a timely manner. In particular, the medical waste related to the epidemic situation basically achieved the "daily clear". Taking Shanghai as an example, all kinds of medical institutions produce about 130 tons of medical waste, among which 60–70 tons are generated by designated hospitals and medical institutions with fever outpatients.

New coronavirus pneumonia clinical waste contamination poses occupational health risks to public health and healthcare workers. With a greater understanding of the harm of new coronavirus pneumonia medical waste, on January 28, 2020, China CDC emphasized in the notice of the National Health Committee General Office regarding the new coronavirus infection pneumonia outbreak of medical wastes management of medical institutions during the period which made a clear emphasis on the new crown pneumonia of medical waste, collection, storage, transportation, strict management, the new coronavirus pneumonia medical waste contact staff must ensure the safety of the conditions. However, in practical operation, irregular operation is inevitable because of limited human resources, urgent time, heavy task and infection risk. People at risk of infection are mainly those in direct contact with clinical waste from new coronavirus pneumonia, such as hospital staff.

The United States ATSDR (the Agency for Toxic Substances and Disease Registry) has conducted a study on the impact of medical waste on public health, and has given Suggestions to the department of labor occupational safety and health management, that

Table 1. Number of risk factors of infectious medical waste pollution accidents

Serial no.	The name of the event	Serial no.	The name of the event	Serial no.	The name of the event
T	Infectious clinical waste contamination in hospital	X1	Untimely discovery	X17	Management rights and responsibilities are not clear
M1	Medical waste leakage, loss, pollution	X2	Lack of contingency plan and disposal tech	X18	Supervision and law enforcement are lax
M2	Not handled in time	X3	The accident was beyond control	X19	The rules and regulations are not sound
M3	Unsafe behavior of personnel	X4	The supporting equipment is not complete	X20	Lack of information system
M4	Management defect	X5	The signs are not standard	X21	unreasonable disposal operation
M5	Facility problem	X6	Quality not up to standard	X22	Lack of funds for waste disposal
M6	Operational flow problem	X7	Failing to disinfect and clean in time	X23	Management irregularities
M7	Poor professional awareness	X8	The network layout is not reasonable	X24	The management team is incompetent
M8	Wrong attitude	X9	Nonstandard operation	X25	Failing to implement the handover registration system
M9	Failure of remedial measures	X10	No protective equipment	X26	Irregular inspection
M10	Government departments fail to supervise	X11	No professional training	X27	Not finding the problem in time
M11	Dereliction of duty in hospital management	X12	Cognitive differences among different people	X28	The environmental pollution
M12	Equipment malfunction	X13	Think little	X29	Staff infection

(continued)

Table 1. (*continued*)

Serial no.	The name of the event	Serial no.	The name of the event	Serial no.	The name of the event
M13	The unsafe nature of clinical waste	X14	Insufficient protection awareness	X30	Infectious acute instrument injury
M14	infectious	X15	No regular physical examination	X31	Non-infective acute injury
M15	The injury	X16	The supervision team is not fully built		

is, measures to reduce injuries and infections of medical waste in the workplace. From this, we see that sharp instruments, blood and blood products, and the culture medium of pathogens are the most important medical wastes to harm the visual health.

2.2.1 Establish an Infectious Medical Waste Fault Tree Model

Through literature review and combined with the empirical evidence of new coronavirus pneumonia this paper analyzes the direct and indirect causes of infectious medical waste pollution accidents in outpatient and emergency hospitals. "Hospital outpatient and emergency infectious medical waste pollution" was taken as the top event, "medical waste leakage, loss, pollution" and "not treated in time" as the intermediate events. The accident risk factor number shows in the following Table 1 [6–9].

The Free FTA software was used to draw the accident tree. The accident tree of infectious clinical waste pollution is shown in Fig. 2.

2.2.2 Analysis and Calculation of Infectious Medical Waste Incident Tree in Outpatient and Emergency Department of Hospital

(1) Minimum cut set analysis

In this paper, the tree structure of infectious medical waste accident in hospital outpatient and emergency department was qualitatively analyzed.

$$T = M1*M2 = (M3+M4+M5+M13)*X1*M9$$

$$= (M6+M7+M8+M10+M11+X4+X5+X6+X7+X8+M12+M14+M15)*X1*(X2+X3)$$

$$= (X9+X10+X11+X12+X13+X14+X15+X16+X17+X18+X19+X20+X21+X22+X23+X24+X25+X4+$$
$$X5+X6+X7+X8+X26*X27+X28+X29+X30+X31)*X1*(X2+X3)$$

We see that there are 54 minimum cut sets of possible infection-injury accident trees for infectious medical waste, and the minimum cut set event combination shows in Table 2.

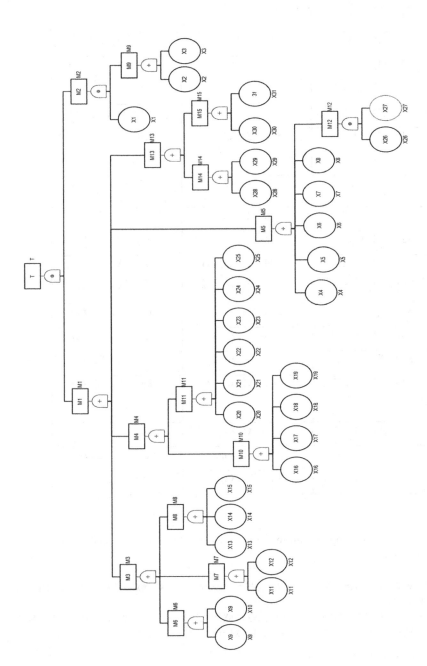

Fig. 2. FTA of Hospital outpatient and emergency infectious medical waste pollution

Table 2. Minimum cut set event combination table

K1={X1, X2, X10}	K2={X1, X2, X11}	K3={X1, X2, X12}
K4={X1, X2, X13}	K5={X1, X2, X14}	K6={X1, X2, X15}
K7={X1, X2, X16}	K8={X1, X2, X17}	K9={X1, X2, X18}
K10={X1, X2, X19}	K11={X1, X2, X20}	K12={X1, X2, X21}
K13={X1, X2, X22}	K14={X1, X2, X23}	K15={X1, X2, X24}
K16={X1, X2, X25}	K17={X1, X2, X26, X27}	K18={X1, X2, X28}
K19={X1, X2, X29}	K20={X1, X2, X30}	K21={X1, X2, X31}
K22={X1, X2, X4}	K23={X1, X2, X5}	K24={X1, X2, X6}
K25={X1, X2, X7}	K26={X1, X2, X8}	K27={X1, X2, X9}
K28={X1, X3, X10}	K29={X1, X3, X11}	K30={X1, X3, X12}
K31={X1, X3, X13}	K32={X1, X3, X14}	K33={X1, X3, X15}
K34={X1, X3, X16}	K35={X1, X3, X17}	K36={X1, X3, X18}
K37={X1, X3, X19}	K38={X1, X3, X20}	K39={X1, X3, X21}
K40={X1, X3, X22}	K41={X1, X3, X23}	K42={X1, X3, X24}
K43={X1, X3, X25}	K44={X1, X3, X26, X27}	K45={X1, X3, X28}
K46={X1, X3, X29}	K47={X1, X3, X30}	K48={X1, X3, X31}
k49={X1, X3, X4}	K50={X1, X3, X5}	K51={X1, X3, X6}
K52={X1, X3, X7}	K53={X1, X3, X8}	K54={X1, X3, X9}

The method of finding the minimum cut set is used to find the success tree of the accident tree.

$$T=M+M2=(M3*M4*M5*M13)*X1*M9=(M6*M7*M8*M10*M11*X4*X5*X6*X7*X8*M12*M14*M15)$$
$$+X1+(X2*X3)$$
$$=(X9*X10*X11*X12*X13*X14*X15*X16*X17*X18*X19*X20*X21*X22*X23*X24*X25*X4*X5*X6*X7*X8*(X26+X27)*X28*X29*X30*X31)+X1+(X2*X3)$$

(2) Minimum path set analysis

The Boolean algebra method can be used to obtain the four minimum cut sets of the successful tree of the infectious medical waste pollution accident tree. After the dual transformation, it can be obtained that there are four minimum path sets of the accident tree. The shortest path set event combination is shown in Table 3.

Table 3. Shortest path set event combination table

K1={X1}
K2={X2, X3}
K3={X4, X5, X6, X7, X8, X9, X10, X11, X12, X13, X14, X15, X16, X17, X18, X19, X20, X21, X22, X23, X24, X25, X26, X28, X29, X30, X31}
K4={X4, X5, X6, X7, X8, X9, X10, X11, X12, X13, X14, X15, X16, X17, X18, X19, X20, X21, X22, X23, X24, X25, X27, X28, X29, X30, X31}

(3) Structural significance analysis

Using the minimum cut set to discharge the order of structural importance:

$$I(X1) > I(X3) = I(X2) > I(X31) = I(X30) = I(X29) = I(X28) = I(X25) = I(X24) = I(X23) = I(X22)$$
$$= I(X21) = I(X20) = I(X19) = I(X18) = I(X17) = I(X16) = I(X15) = I(X14) = I(X13) = I(X12) = I(X11) = I(X10) = I(X9) = I(X8) = I(X7) = I(X6) = I(X5) = I(X4) > I(X27) = I(X26)$$

2.3 Analysis of Safety Risk Assessment Results of Infectious Medical Waste in Outpatient and Emergency Hospitals

Through the analysis of the possible pollution accident tree during the disposal of infectious medical waste in outpatient and emergency treatment, it is found that there are many basic events leading to pollution, and the minimum cut set is composed of ternary events and quaternary events, indicating that there are many ways leading to the contamination of infectious medical waste in outpatient and emergency treatment, and it is necessary to strengthen prevention.

In addition, from the above basic event structure of fault tree analysis of important degree, you can see that "found it in a timely manner to the structure of the highest importance", therefore we should strengthen this infectious medical waste regulation in the hospital, establish hospital infectious medical waste management information system, tracking of infectious medical waste, the infectious medical waste information transparency of the data network, always grasp the dynamic situation of infectious medical waste, put an end to disposal, discarded and selling waste. If any problem such as leakage and loss of infectious clinical waste is found, timely alert should be issued. "Lack of emergency plan and disposal technology" is the second important structure, so we should build a perfect emergency plan system, accident emergency plan. The emergency treatment should be equal to the safety treatment and the emergency treatment should be combined with the actual implementation. Actively explore the use of collaborative disposal facilities to carry out emergency treatment of infectious medical waste, rational construction of medical waste emergency treatment facilities.

Minimal path set is a minimum set of basic events which is not necessary to cause the top event happened. From the above set analysis it shows that in order to make outpatient emergency infectious medical waste accident in the hospital not occur there are four kinds of accident prevention programs, and generally less basic event Set corresponds to the path of the scheme for the optimal solution. Taking new coronavirus pneumonia outbreak as an example, if the infectious medical waste is improperly disposed, it will cause great impact and harm, and if the problem is timely found or regulated and effective emergency measures are taken, the pollution to the environment and the harm to the population can be reduced. Direct measures can be taken to solve the existing problems, such as strengthening the knowledge training of medical staff, strengthening and improving the management and maintenance of medical waste disposal facilities and other indirect measures that prevent the occurrence of accidents. Infectious medical waste disposal and contact belongs to high risk industry, we should build corresponding medical staff occupational health risk regulatory system and correctly handle new infectious pneumonia, reduce the probability of infection to the largest extent according to "the new coronavirus infection pneumonia outbreak of emergency disposal of medical waste management and technology guidelines (try out)".

3 Lean Disposal of Infectious Medical Waste in Outpatient and Emergency Treatment

Lean management is an efficient management method based on the lean production of Toyota. Lean management is the integration, enrichment and improvement of basic management, standard management, standardized management and fine management, and it pays more attention to management effect and management benefits [15]. On February 14, 2020, President Xi Jinping presided over the 12th meeting of the commission for deepening overall reform of the CPC central committee, which emphasized improving the systems and mechanisms for the prevention and control of major epidemics and the coordination mechanism for emergency response, and improving the national public health emergency management system. The types and scope of solid waste included in clinical waste management during the outbreak of new coronavirus infection pneumonia as well as the hygiene and epidemic prevention in the process of collection, storage, transportation and disposal shall be carried out in accordance with the relevant requirements of the competent sanitation and health authorities.

According to this infectious medical waste in the third section safety evaluation of the accident tree modeling calculation results of the hospital shows that the X1 "found" not in time, X2 "lack of contingency plans and disposal technology" is a major hazard of infectious medical waste disposal [10], we combine lean management method to explore feasible path of the infectious medical waste disposal for hospitals [11].

3.1 Establish Epidemic Prevention Instructions for Infectious Medical Waste in Outpatient and Emergency Hospitals

Take the prevention and control of new coronavirus pneumonia as an example. First, the prevention and control team of new coronavirus pneumonia was established, and the outpatient and emergency medical waste prevention and control team was established under the leadership of the medical waste disposal leader, and the responsibilities of the members of the outpatient and emergency medical waste prevention and control team were formulated. The personal information and travel status of outpatient and emergency medical personnel or cleaning personnel were investigated by the management of each outpatient and emergency department. Establish communication channels for the whole staff to timely report the situation of the epidemic. Arrange medical waste disposal management personnel to supervise, inspect the working status of medical waste disposal personnel every two hours, timely find problems and strengthen their health protection.

As facility management, the leader of the prevention and control team of medical waste disposal arranged the professional staff of medical waste disposal to conduct disinfection in every emergency room according to the hospital site. Logistics departments shall count the types and quantities of protective equipment for personnel and purchase them in time. They shall mainly check whether the protective equipment and packaging containers have the certificates and meet the national standards. In the hospital to set up abnormal personnel isolation area, equipped with the corresponding protective equipment. Strengthen the training of cleaning personnel to correctly use protective equipment, hand washing methods, mask wearing, etc.

As information dynamic management, the clinical waste disposal prevention and control team summarized the regional distribution of employees and the disposal of clinical waste, and implemented dynamic tracking. Management personnel at all levels should keep abreast of clinical waste trends.

3.2 Develop the Automatic Intelligent Conveying System for Medical Waste Treatment in Outpatient and Emergency Hospitals

There is a high risk of staff coming into contact with infectious waste in the clinical waste disposal link at the centralized disposal sites of hospitals. The emergence of scientific and technological information technology promotes the development of intelligentization. The use of automatic intelligent transportation system to conduct harmless treatment of infectious medical waste can greatly reduce the risk of human infection and realize the informatization, intelligentization and unmanned treatment of infectious medical waste. The process shall be strictly implemented in accordance with the relevant regulations issued by the local epidemic prevention and control team to ensure scientific, efficient and safe treatment of infectious medical waste.

The lean disposal and transportation process of infectious medical waste in outpatient and emergency treatment can be carried out according to the subordinate steps. Loading and unloading infectious medical waste in outpatient and emergency department, through the transport system in place for discharging operation, the whole barrel into the incinerator for incineration treatment, the barrel for thorough cleaning and disinfection.

From the drum feeding to cleaning and disinfection storage completely without manual operation, we can realize real-time monitoring of all aspects of the processing process.

The key is lifecycle management, leaving nothing to chance. For example, at a fixed time every morning, the medical waste collection and transportation vehicle starts from the solid waste company and goes to the major hospitals in the city to collect and transport medical waste. From classification, collector, transport into the incinerator to disposal of medical wastes from birth to destroy the whole process, clear display on the big screen of information platform of the whole flow of medical wastes, according to the number of medical waste of charge at the same time, large data analysis, analysis of the number of the hospital patients, surgery and prognosis, and then adjust the vehicle scheduling, ultimately guarantee to daily production daily cleaning of medical waste.

4 Conclusion

The new coronavirus pneumonia outbreak has been classified by WHO as "the highest level of global risk". Improper disposal of clinical waste can lead to cross infection and spread of the outbreak. Taking the new coronavirus pneumonia outbreak as an example, this paper constructs a safety evaluation model of infectious medical waste in outpatient and emergency department, and concludes that "untimely discovery" and "lack of emergency plan and disposal technology" are the main risk sources and hidden dangers through accident tree calculation and analysis. In this paper, it is of practical significance to adopt lean management method in outpatient and emergency department to control the two hazard sources for the safety management of medical waste in outpatient and emergency department.

References

1. Hu, B.: Research on the legal system of medical waste management in China. Guizhou University (2019)
2. Zhu, J.J.: Application of PDCA model in standardized management of clinical waste. J. Pract. Clin. Nurs. **4**(08), 154–157 (2019)
3. Liu, W., Du, W.T.: Improvement effect of 6 Sigma management mode on medical waste classification and treatment process in operating room. Nurs. Practice Res. **16**(22), 142–144 (2019)
4. Hong, F., Jing, L., Ling, L., Rong, Z.: Application of quality control circle in improving the safety management of pediatric medical waste. Contemp. Nurses **07**, 176–178 (2016)
5. Gao, Y.: A case study on the application of whole-process management in the centralized disposal of medical waste. Huazhong University of Science and Technology (2005)
6. Shen, S.S.: Research on safety risk evaluation system of medical waste logistics. Capital University of Economics and Business (2010)
7. Chen, Y., Shao, C.Y., Ding, Q.: Optimal Treatment Technology and Application of Medical Waste. Shanghai Science and Technology Press, Shanghai (2015)
8. Lin, H.D., Gao, G.G., Wang, H.Y.: Chemical Logistics and Transportation Management. Shanghai University of Finance and Economics Press, Shanghai (2015)
9. Marvin, R.: Risk Assessment Theory, Methods, and Applications. Wiley, Hoboken (2011)
10. Ministry of Ecology and Environment of China. National large and medium municipal solid waste in 2019. Environmental pollution prevention and control report (2019)
11. Tao, S.Q., Xu, J.: Toyota lean management – occupational health and safety (illustrated edition). People's Post and Telecommunications Press, Beijing (2015)

Effects of Taiji on Participants' Knees: A Behavioral-Modeling Approach

Jihong Yan[1], Liang Wu[1], Shuying Zhang[1], Hongguang Liang[2], and Qiang Lin[2(✉)]

[1] College of Computer and Information Engineering,
Shanghai Polytechnic University, Shanghai 201209, China
[2] Department of Tuina, Shanghai General Hospital,
Shanghai Jiao Tong University, Shanghai 200080, China
jhyanjihong@163.com

Abstract. More and more people like to practice taiji for different purposes. Taiji has been proposed as a treatment for the knee osteoarthritis. But there have always been different opinions that Taiji can cause knee injuries. The information on an individual participant is often incomplete. When complementary information is integrated, a better profile of a participant's knee can be built to explore the tension. This paper aims to investigate the effects of Taiji on knees. We introduce a method for buildting a mapping among Taiji participants and their knee healthcare. It consists of three key components: the first component identifies Taiji participants' unique behavioral patterns; the second component constructs features due to these behavioral patterns; and the third component exploys machine learning for effective computing. We formally define the problem and show that our method is effective. This study paves the way for analysis and mining knee osteoarthritis prevention and treatment. It also facilitates the creation of novel healthcare services.

Keywords: Taiji · Knees healthcare · Multiple logistic regression algorithm · Behavioral-modeling

1 Introduction

Taiji has a unique physical fitness effect, and is favored by more and more people. The number of people practicing Taiji is rapidly increasing. But Knee osteoarthritis is high ranking and causes disability [1]. It is convincing pointed out by Osteoarthritis Research Society International (OARSI), which reported "OA is also responsible for substantial health and societal costs, both directly and as a consequence of impaired work productivity and early retirement" [5]. Taiji is proposed as a potential option for the management of Knee osteoarthritis. It can reduce pain and improves physical function [3]. In recent years, the

Supported by Key Disciplines of Software Engineering of Shanghai Polytechnic University under Grant No. XXKZD1604.

different report mentions that "there are knee pain problems caused by Taiji exercises" [7]. In [8], it investigates and analysis the knee pain in middle-aged and elderly Taiji practitioners in Huai'an City.

In the practice of Taiji, the knees are mainly bent and squatting. The stress is loose waist, loose shoulders and elbows. Some data shows that the human knee flexes 30°, the knee joint bears the same pressure and weight; knee flexion of 60°, the knee joint pressure is 4 times the weight; knee flexion of 90°, the pressure is 6 times the weight. During the practice of Taiji, the human body is always in a squatting position. If some of the Taiji practitioners have incorrect movement techniques, they cannot "stand upright" or make mistakes such as "kneeling on knees", or they cannot "open in a round gear" so that the knees and toes are in the same direction, the knee injury will be more serious. In fact, the problem is not trival since there exist the following questions:

- Is knees pain caused by practicing Taiji?
- Does Taiji exercise benefit knees healthcare?

This paper proposes a behavioral-modeling approach to research the problem of effects of Taiji on participants' knees. In the first stage, we use mathematical statistics techniques to get the behavioral patterns. Then, we integrate the attributes to design the features to capture the information generated by these patterns. Finally, we model the problem of effects of Taiji on knee based multiple features. To match the model better, we design multiple logistic regression algorithm to optimize model matching parameter. In summary, the main contributions of this paper are fourfold:

- We propose and formalize the definition the problem of effects of Taiji on participants' knee.
- Based on the mathematical statistics technique, we model the Taiji participants behavioral patterns via generating some candidate pairs.
- Based on the feature engineering method, we propose an unsupervised learning framework to construct the features, and employ multiple logistic regression algorithm to infer the model parameters.
- We conduct extensive experiments on real data sets, those experiments can not only verify the rationality of our method, but also indicate that our method is effective.

The rest of the paper is organized as follows: we review the related work in Sect. 2. Methodology and technology is shown in Sect. 3. Modeling and computing are described in Sect. 4. In Sect. 5, we analyze experimental results. Finally, we conclude and discuss the future work in Sect. 6.

2   Related Work

To the best of our knowledge, the problem of effects of Taiji on knees is novel, there is however some related work about the research of Taiji in medical field and healthcare in informatical field.

2.1 Taiji in the Medical Field

At present, many international scientific research institutions have proved that Taiji has a very important therapeutic and health care role in many fields such as medicine, psychology and physiology. In the medical field, some studies have shown that Taiji may be an effective treatment for fibromyalgia, and it is worth long-term research (see [4]). Studies have also found that in the treatment of knee osteoarthritis, Taiji produces as good a therapeutic effect as standard physical therapy. It can relieve chronic pain in osteoarthritis and is beneficial for back pain and osteoporosis. In the field of psychology, research has found that peer-assisted Taiji can improve the mental health and social networks of lonely elderly people in society. In the area of health care, study has shown that practicing Taiji is associated with reducing mortality, just like walking and jogging.

A large number of studies have found that practicing does not cause knee pain. In many sports activities, Taiji has relatively little damage to the knee joint. The study [11] aims to examine the potential neuromuscular mechanisms underlying the benefits of Taiji and compare the effects of Taiji and balance and posture training on dynamic stability for knee osteoarthritis. There's been a lot of other medical research on knee, including Epidemiology on knee osteoarthritis [9], Obesity and knee osteoarthritis [2], Targeted therapy. In contrast, some studies have pointed out that if knee pain has been diagnosed or osteoarthritis of the knee has been diagnosed, the majority of people practicing Taiji will aggravate the symptoms, and a few people will get worse.

2.2 Healthcare in Informatical Field

With the development of artificial intelligence technology, more and more machine learning algorithms have also been applied to the field of medical health. Machine learning algorithms include traditional machine learning algorithms (such as clustering algorithms, classification algorithms, regression analysis algorithms, association rules, etc.) and deep learning algorithms (such as autoencoders, deep belief networks, volume neural networks, recurrent neural networks, etc.) (See [10]).

Because traditional machine learning algorithms can achieve relatively accurate results on small data sets by selecting appropriate feature values, research based on traditional machine learning algorithms has been more applied in the field of traditional Chinese medicine. The deep learning algorithm usually requires massive data to support the training and optimization of the model, so it has a wide range of applications in the field of western medicine. In addition to machine learning methods, common statistical methods have also been studied. For example: t-test, chi-square test, analysis of variance, Logistic statistical methods, etc. Aiming at the problem of traditional Chinese medicine pulse labeling, the pulse diagnosis signal data collected by the instrument is firstly subjected to double-tree complex wavelet transform, then features are extracted with Mel cepstrum coefficients, and finally clustered by fuzzy C-means clustering algorithm. The accuracy of pulse recognition by this method has reached

78.2%, to a certain extent, it has avoided the influence of doctors' subjective body on pulse marking. Guo Hong et al. [6] Used 560 outpatient cases of allergic rhinitis as a research object by Chinese doctors, and used association rule algorithm to explore the compatibility of allergic rhinitis drugs [?]. Used 1474 patients with chronic hepatitis B as the research object and used binary logistic regression analysis to explore the relationship between chronic hepatitis B symptoms and symptoms (see [12]). These existing works have presented a deep study on healthcare. However, these works simply apply in medicine research but not deliberate the traditional Chinese Taiji.

3   Methodology and Technology

Taiji participants often exhibit different behavioral patterns in the usual practice exercise. These patterns results can help identify and quanlify the effects of Taiji on knees. The behavior model contains behavioral patterns, features constructed to capture the important attributes pairs which can affect the participants' knees.

3.1   Bevavioral Patterns

Our methodology is outlined in Fig. 1. Different Taiji participants selects different practise behavior, they exhibit certain behavioral patterns.

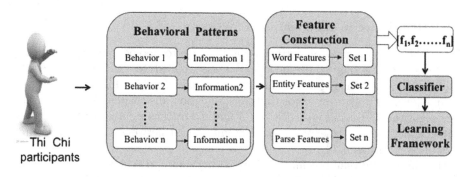

Fig. 1. The behavioral-modeling approach

In our paper, these patterns can be captured in terms of Taiji participants' knees attributes. Following the traditional maching learning and data mining research, we can learn a behavioral function by employing a supervised learning framework that utilized these features and prior information. Surpervised learning in behavioral patterns can be performed via either classification or regression. Depending on the learning framework, we can learn the probability that an Taiji participants own the specific attributes, generalizing our binary f function to a probabilistic model. This probability can help select the most likely behavioral attributes that due to the knee injury. The learning component is the most straightforward. Therefore, we next elaborate how to analyze behavioral patterns related to Taiji participants and how features can be constructed to capture attribute information redundancies due to these patterns.

3.2   Feature Construction

There may be a certain correlation between the factors that affect the position of knee pain and the health of the knee joint of Taiji participants. We calculate the relationship between the various factors and the knee pain.

- Correlation Analysis
  Because knee pain and knee function are classified and analyzed, and there are many influencing factors for these two variables, from the above correlation analysis, we can see that the correlation between the factors is weak, but the correlation cannot be ignored. The linear regression analysis cannot accurately describe the causal relationship between variables, so a multiple logistic regression analysis is used to fit the model, and its form is as follows:

$$\ln\left[\frac{P(y=j|x)}{P(y=J|x)}\right] = \alpha_j + \sum_{k=1}^{k}\beta_{jk}x_k(i \le j-1) \tag{1}$$

In the formula, J is the category, P is the probability of the principal component of the class j, and is the principal component of the class K; is the principal component coefficient of class K; it is a constant term. Where J, and the last category (I.e. the category J) is used as the reference category. And because the sum of the probabilities of each category is 1, that is

$$P(y=1|x)+P(y=2|x)+\ldots+P(y=J|x) = P(y=J|x)\left[1 + \sum_{i=t}^{J+t}e^{\alpha_2+\sum_{n-4}^{2}g_2=1}\right] = 1 \tag{2}$$

Therefore, the probabilistic formula of each category can be summarized in the following form:

$$P(y=f|x) = \frac{e^{a_n+\sum_{n=1}^{n}g_n z_n}}{1 + \sum_{n=1}^{n+1}e^{a_n+\sum_{n=1}^{n}g_{n-2}}} \tag{3}$$

- Decision Making
  Probabilities $P(t_i \in M|\gamma_i, \Theta)$ and $P(t_i \in U|\gamma_i, \Theta)$ can be calculated by the optimized model parameters. By selecting appropriate scoring function, the candidate feature is scored and the scoring function is defined as:

$$W_i = log(\frac{P(t_i \in M|\gamma_i, \hat{\Theta})}{P(t_i \in U|\gamma_i, \hat{\Theta})}) \propto \sum_{j=1}^{2}w_i^j \tag{4}$$

Among this,

$$w_i^j = (\Theta_{1,j}' S_{1,j}(\gamma_i^j) - z_{1,j}(\Theta_{1,j})) - (\Theta_{0,j}' S_{0,j}(\gamma_i^j) - z_{0,j}(\Theta_{0,j})) \qquad (5)$$

If $P(t_i \in M | \gamma_i, \Theta) > P(t_i \in U | \gamma_i, \Theta)$, which is equivalent to $W_i > 0$, then t_i is a candidate feature, otherwise the opposite. By choosing appropriate threshold $W_0 > 0$ to improve the accuracy of algorithm.

4   Modeling and Computing

4.1   Sample Data

In order to better understand the current situation of the effects of Taiji on knees in Shanghai, and to find a better solution based on the status quo and sports characteristics. The paper data is aimed at the in-depth participation of Taiji participants in Shanghai. Understand and use the questionnaire survey method and Irrgang Sports Ability Rating Scale. It is a scale designed for athletes to assess sports ability. It is divided into two categories of symptoms and exercise function. There are 10 items in total. The total score is 100 points. The higher the function, the better. The experimental data statistics of Taiji participants data are shown in Table 1.

4.2   Modeling

There are three main points of knee pain. The dependent variable is set as the knee pain which factors will affect the knee pain of Taiji participants. According to the results of the correlation analysis, we can do regression analysis on the relevant variables. Among them, all the factors other than the height factor have a significant contribution to the composition of the model. The logistic regression model can calculate the probability of where the knee of each Taiji exerciser will pain, and finally obtain the model prediction classification table shown in Table 2, where the model predicts the back side of the knee of the Taiji exerciser The accuracy rate of pain reached 89.80%, and the prediction of pain on the front side of the knee was low, and even the prediction of pain on the side of the knee was 0%. The overall prediction accuracy rate of the model was 52.40%, and the performance was average.

The performance of the model is average, and further stepwise regression analysis is done. It can be seen from the results of the stepwise regression analysis in the table that the new model retains the duration, gender, weight, and height as independent variables, and a new prediction model is obtained:

$$\begin{aligned} KneePain &= 0.074 * duration + 0.107 * gender \\ &+ 0.009 * weight - 0.007 * height + 1.882 \end{aligned} \qquad (6)$$

We take exercise frequency, duration of each exercise, posture, whether to warm up, knee pain position, age, knee joint function status as the characteristic value, and condition as the classification goal.

Table 1. Experimental data statistics of Taiji participants

Question	Answer option
Taiji routines you arecurrently exercising (multiple choices)	(1) Yang-style Taiji
	(2) Chen-style Taiji
	(3) Wu-style Taiji
	(4) Sun-style Taiji
	(5) Wu-style Taiji
	(6) 24-style simplified routine, 42-style competition routine
	(7) Daji sword, Taiji sword
	(8) Taiji pusher, vigorous
	(9) Other
Your main source of learning Taiji	(1) Taiji teachers in community parks
	(2) Taiji training organized by your school, unit, company
	(3) Signing up for Taiji classes in Taiji clubs and fitness clubs
	(4) Official Learn from Taiji Masters
	(5) Teach yourself from teaching CDs or online
What awards have you won in Thi Chi competitions ?	(1) International competitions
	(2)National competitions
	(3)Provincial and municipal competitions
	(4)Not yet won
Frequency of Taiji exercise	(1) Irregular exercise
	(2) 5-7 times per week
The duration of each time of your extreme climbing exercise	(1) Is less than half an hour
	(2) Half an hour to 1 hour
	(3) 1 hour to 2 hours
What are the main motivations for you to insist on Taiji? (Multiple choices)	(1) Can exercise anytime and anywhere
	(2) Can lose weight
	(3) Make the skin better
	(4) Can soothe emotions and release stress
	(5) Improve immunity
	(6) Improve sleep quality
	(7) Enhance muscle strength
	(8) Communicate and encourage among teammates
	(9) Gradually understand the profound cultural heritage
	(10) Other
Your stand posture for Taiji	(1) Low posture, knee flexion 60-70 degrees
	(2) Middle posture, knee flexion greater than 100 degrees
	(3) High, knee flexion 90-100 degrees
What is the warm-up exercise for the knee joint?	(1) No warm-up exercise
	(2) About 5 minutes
	(3) 15 minutes or more
The main positions of knee pain are	(1) Front of knee
	(2) Back of knee
	(3) Side of knee
How often do you experience knee pain after boxing?	(1) 5-15 minutes
	(2) 15-30 minutes
	(3) 30-60 minutes
	(4) After 1 hour
How do you deal with knee pain?	(1)No more Taiji after knee pain
	(2) After keep practicing Taiji for 1 month,knee pain is relieved
	(3) After keep practicing Taiji for 3 months,knee pain is relieved

Table 2. Parameter analysis of stepwise regression

Model	Coefficient	T-value	Sig
Constant	1.882	4.56	0
Duration	0.074	3.726	0
Sex	0.107	3.119	0.002
Weight	0.009	5.34	0
Height	−0.007	−2.617	0.009

4.3 Computing

Each factor of the model is significant to the model's composition. These factors can effectively predict the dependent variable. Import these effective variables into the model to get the following Table 3:

Table 3. Knee pain feature classification

Observation	Predictive value			
	1	2	3	Correlation
0	593	205	0	74.30%
1	304	1343	29	80.10%
2	19	173	48	20.00%
Percentage	33.80%	63.40%	2.80%	73.10%

It shows that the accuracy rate of the model in predicting that the knee joint of Taiji participants will have a slight impact has reached 80.10%, and the prediction that the knee joint of Taiji participants is very healthy is relatively low, but it The prediction of serious impact on the knee joint of Taiji participants is lower than 74.30%. The overall prediction accuracy of the model is 73.10%, which is better.

We can see in Fig. 2 the relationship of stand posture on knees, the duration on knees, the warm-up activity time and the location of knee pain can effect the knee condition separately. Among those surveyed, the duration of each exercise was between 30 min and 60 min, and the knee condition of these people is mild knee condition accounting for a large proportion.

About stand posture, there were many people who practiced low- and medium-strength Taiji. The knees of these people had the mild knee injury. Among the people surveyed in Taiji, the warm-up activity before each exercise was most likely to be around 5 min. These people had the largest percentage of mild knee conditions. Knee pain is the largest proportion of knee pain behind the knee, followed by the front of the knee, and these people have the largest percentage of mild knee conditions.

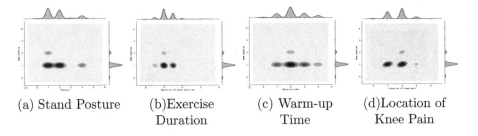

(a) Stand Posture   (b)Exercise Duration   (c) Warm-up Time   (d)Location of Knee Pain

Fig. 2. Exercise duration, stand posture, warm-up time and location of Taiji on knee

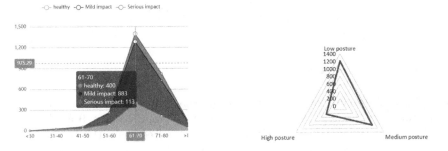

Fig. 3. Age distribution of Taiji partic-   **Fig. 4.** Stand posture score range
ipants

About the age, the correlation is shown in Fig. 3.

It can be seen from the figure that the knee condition of most tai chi practitioners is mildly affected, concentrated between 64–73 years old. Among the mildly affected people, the number of people aged 66 is the largest, with 171 in total. Among the severely affected people, the number of people aged 73 was the largest, with a total of 5 people.

The Knee condition statistical results can be shown in Fig. 5.

When assessing knee joint conditions, the status is divided into three levels of health, mild impact, and severe impact according to the score range. The proportion of mild impact is the largest.

Figure 4 shows the statistic results of exercise posture among these investigators. When assessing knee joint conditions, the status is divided into three levels of health, mild impact, and severe impact according to the score range. The proportion of mild impact is the largest.

Figure 6 is warm-up time statistics results of Taiji participants. It can be seen from the figure that the warm-up time of most Taijiquan practitioners is about 5 min, accounting for about 49% of the total number. Followed by warm-up time greater than 15 min, accounting for about 31% of the total number of people. It can be seen that in these three categories, there are far more people with mildly affected knee joints than the other two degrees. From this analysis, the warm-up time before exercise has little effect on the condition of the knee joint.

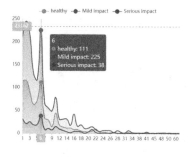

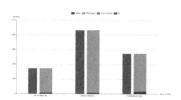

Fig. 5. Knee condition

Fig. 6. Warm-up times statistics of Taiji participants

5 Experimental Analysis

In this paper, chi-square test and feature importance calculation based on decision tree are used for feature selection. After data preprocessing, 9 attributes including gender, age, BMI, years of practice, exercise frequency, the duration of exercise, stand posture, warm-up time, and location of knee pain were selected as feature variables, and the condition of the knee joint as a classification variable. The results of the chi-square test are shown in Fig. 7.

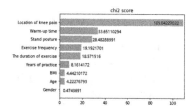

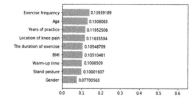

Fig. 7. The results of the chi-square test of Taiji participants

Fig. 8. The importance of features of Taiji participants

From the chi-square test results, it can be seen that the location of knee pain has the greatest correlation with the condition of the knee joint, while the correlation between BMI, age, gender and the knee joint condition is the smallest, especially gender. In the method of calculating feature importance based on decision tree, the calculation result is shown in Fig. 8. It can be seen from the figure that the feature importance calculated by the decision tree differs little among the features, but the importance of the "gender" feature is still the smallest. Combining the chi-square test results and the results of the feature importance calculated by the decision tree, the "gender" feature was finally selected to be removed. We collected 3352 real data samples. The survey content is mainly divided into three parts. The first part is the basic information of the interviewees, including age, gender, height, weight, etc. The second part is the information of the Taiji exercises of the respondents, including exercise frequency, stand posture, warm-up time, duration of each time of exercise, main source of learning Taiji, etc. The third part is a survey of the knee condition of the practitioner. The survey uses the Irrgang knee function scoring system. The content includes the degree of knee pain affecting the level of daily activity, the degree of friction in the knee affecting the level of daily activity, the degree of knee stiffness affecting the level of daily activity, etc. Based on this scoring system, we can obtain the overall knee joint condition of the practitioner.

6 Conclusion

In this paper, we proposed a behavioral-modeling approach to computing the effects of Taiji on participants' knees. To get the behavioral patterns, we choose the important attributes pairs by the mathematical statistics techniques. The results show that

Among the surveyed people who practice Taiji, the proportion of women is significantly larger than that of men. The age of Taiji practitioners is generally older, with the largest number of people between the ages of 60 and 70. Most of those who practice Taiji are low or medium. Few people practice in an elevated manner. Most people do warm-up exercises for about 5 min before practicing. The most painful part of the knee is the side of the knee, followed by the front of the knee and the back of the knee. The pain of the knee is related to the frequency of Taiji. Too low or too high the impact on the front, back, and sides of the knee. When assessing knee function, the function is divided into three levels of health, mild impact, and severe impact according to the range of scores. The proportion of mild impact is the largest, and the most affected in this population is between 50–70 years old.

To match the attributes pairs better, we design multiple logistic regression algorithm to optimize model matching parameter. Experiments show the effectiveness of our proposed method.

References

1. Bannuru, R.R., Osani, M., Vaysbrot, E., Arden, N., Bennell, K., Bierma-Zeinstra, S., Kraus, V., Lohmander, L., Abbott, J., Bhandari, M., et al.: Oarsi guidelines for the non-surgical management of knee, hip, and polyarticular osteoarthritis. Osteoarthritis and Cartilage **27**, 1578–1589 (2019)
2. Chan, W.N., Tsang, W.N.: The effect of Tai Chi training on the dual-tasking performance of stroke survivors: a randomized controlled trial. Clin. Rehabil. 026921551877787 (2018)
3. Chang, W.D., Chen, S., Lee, C.L., Lin, H.Y., Lai, P.T.: The effects of Tai Chi Chuan on improving mind-body health for knee osteoarthritis patients: a systematic review and meta-analysis. Evid.-Based Complement. Altern. Med. (2016)

4. Cheng, C.A., Chiu, Y.W., Wu, D., Kuan, Y.C., Chen, S.N., Tam, K.W.: Effectiveness of Tai Chi on fibromyalgia patients: a meta-analysis of randomized controlled trials. Complement. Ther. Med. **46**, 1–8 (2019)
5. Cross, M., et al.: The global burden of hip and knee osteoarthritis: estimates from the global burden of disease 2010 study. Ann. Rheum. Dis. **73**, 1323–1330 (2014)
6. Hong, G., Mp, L.: Mechanism of traditional chinese medicine in the treatment of allergic rhinitis. Chin. Med. J. **126**, 756–760 (2013)
7. Tao, K.: Study on rehabilitation training of middle-aged and elderly Tai Chi practitioners after knee joint movement injury p. 12. International Olympic Committee (2013)
8. Lei, S.: Investigation and analysis of knee pain in middle-aged and elderly Tai Chi practitioners in Huai'an city, Master's thesis. Yangzhou University (2013)
9. Niu, J., Zhao, X., Hu, H., Wang, J., Liu, Y., Lu, D.: Should acupuncture, biofeedback, massage, qi gong, relaxation therapy, device-guided breathing, yoga and tai chi be used to reduce blood pressure?: Recommendations based on high-quality systematic reviews. Complement. Ther. Med. **42**, 322–331 (2018)
10. Takamura, H., Okumura, M.: Text summarization model based on maximum coverage problem and its variant. In: Proceedings of the 12th Conference of the European Chapter of the Association for Computational Linguistics, pp. 781–789. Association for Computational Linguistics (2009)
11. Wang, X., et al.: Protocol: effects of Tai Chi on postural control during dual-task stair negotiation in knee osteoarthritis: a randomised controlled trial protocol. BMJ Open **10**, e033230 (2020)
12. Zhao, C., Li, G.Z., Wang, C., Niu, J.: Advances in patient classification for traditional Chinese medicine: a machine learning perspective. Evid.-Based Complement. Altern. Med. (2015)

Mixed Distribution of Relief Materials with the Consideration of Demand Matching Degree

Ling Gai[1], Ying Jin[1], and Binyuan Zhang[2(✉)]

[1] School of Management, Shanghai University, Shanghai 200444, China
{lgai,jinying612}@shu.edu.cn
[2] Renji Hospital Affiliated to Shanghai Jiaotong University, Shanghai 200127, China
Zhangbinyuan2020@163.com

Abstract. Timely and effective distribution of relief materials is one of the most important aspects when fighting with a natural or a man-made disaster. Due to the sudden and urgent nature of most disasters, it is hard to make the exact prediction and projection on the demand information. Meanwhile, timely delivery is also a problem. In this paper, we take the war against COVID-19 as an introductive example, analyze the process of material distribution hosted by a government department. We first show a fuzzy decision-making method to proper evaluate the demand degree of each hospital, then based on the result of it, we propose a mixed distribution model to improve the efficiency of delivery. Finally, we carry on a numerical experiment and compare results with the original way that adopted by the government.

Keywords: COVID-19 · Relief materials · COPRAS · Interval 2-tuple linguistic variable · Mixed distribution

1 Introduction

At the end of 2019, the emergence of 2019 novel coronavirus (COVID-19) in Wuhan, China, has caused a large global outbreak and a major public health issue. As of March 9, 2020, 80890 people have been confirmed being infected by COVID-19 in China. Wuhan, the region of epidemic, has 49448 people infected, accounting for 61% of total infected people in China and 73% of total infected people in Hubei province. As the main rescue force, hospitals need a variety of materials to support their rescue work. However, as the number of confirmed cases increased, the demand for materials is also increasing rapidly. The consumption of various materials in each hospital is very large and many hospitals are under the threat of shortage of materials in the early days of the epidemic.

In Wuhan, there is a government department- the Red Cross, being responsible for the distribution of donated relief materials. The relief materials can be mainly sorted into three types: daily necessities, drug medical material and protection medical material.

Research supported by NSFC (11201333).

Z. Zhang et al. (Eds.): AAIM 2020, LNCS 12290, pp. 627–638, 2020.
https://doi.org/10.1007/978-3-030-57602-8_56

However, the demand of these consumable materials is hard to predict or control, which makes its distribution being a severe problem in the reality.

Currently, the process of distribution hosted by government can be roughly divided into 2 stages. Firstly, each hospital reports their demand for materials, then, these materials are distributed from one single supple depot to each hospital. There are mainly three problems in the process of whole distribution:

a. Due to the situation of each hospital varies every day, government often fails to make an accurate prediction. Virtual high demand reported by hospitals and unfair allocation often happen.
b. The work efficiency of single supple depot is low.
c. In the stage of distribution, each hospital is served by one vehicle (belongs to the hospital or the supple depot). Most of the time vehicles are not fully loaded, which leads to an inefficient delivery. Furthermore, time and manpower are wasted during the whole process.

In order to solve the above problems, this paper puts forward an integrated method from two aspects. We first extend a multi-criteria decision-making method called COPRAS (COmplex PRoportional ASsessment) to evaluate the demand degree of each kind of materials for each hospital respectively. As we mentioned before, decision makers in government often fail to predict the demand of each hospital because the situation of each hospital is varied every day and cannot be counted precisely. Therefore, in this paper, a fuzzy approach is adopted instead of using precise number. We use interval 2-tuple linguistic variables as the evaluation language.

Then, based on the demand degree evaluated before, we prorate each kind of material to each hospital and take this result as the demand of each kind of material for each hospital. Finally, we propose a mixed distribution model to ensure that the relief materials are efficiently distributed.

The rest of this paper is organized as follows: Sect. 2 briefly reviews the literatures related to our problem. In Sect. 3, an extended COPRAS approach is proposed to evaluate the demand degree of each kind of material to each hospital. Section 4 models the distribution of materials and gives its mathematical formulation. Section 5 gives a numerical experiment for the whole process of distribution with 3 kinds of materials and 4 hospitals. Section 6 gives the conclusion of this paper.

2 Literature Review

We now review the literatures related to our problem. As a method of multi-criteria decision-making problem, COPRAS was first proposed by Zavadskas et al. [1] in 1994. They used this method to evaluate the building life cycles in order to select an optimal alternative. After that, many literatures have applied it to some multi-criteria decision-making problems. For instance, Mulliner et al. [2] used it to make an assessment of sustainable housing affordability. Kaklauskas et al. [3] used this method to choose the best option to design an efficient building refurbishment. Mulliner et al. [4] used the COPRAS method for the evaluation of sustainable housing affordability and compared it with other multi-criteria decision-making problem approaches. Pitchipoo et al. [5]

implemented the COPRAS decision-making model to find the optimal blind spot in heavy vehicles. Then, Peng et al. [6] extended the COPRAS method into Pythagorean fuzzy environment, which enriches the abundance of the COPRAS method. Zheng et al. [7] made a severity assessment of chronic obstructive pulmonary disease based on hesitant fuzzy linguistic COPRAS method. This paper combines the COPRAS method with interval 2-tuple linguistic to make an evaluation of each hospital's demand degree.

The literatures for distribution of materials can essentially be classified as two main situations: the demand value of each demand point is known and the demand value of each demand is a decision variable. The first situation is often described as vehicle routing problem (VRP) like reference [8–10] and the second is often described as inventory routing problem (IRP). Coelho [11] has made a comprehensive review for IRP, readers can search it if interested. Based on the number of depots, both of these two problems have been extended with VRP with multiple depots and IRP with multiple depots. For example, Zhen et al. [12] formulated the problem of the last mile distribution in electronic commerce as a multi-depot vehicle routing problem and solved some large-scale instances. Soeanu et al. [13] described the distribution in a supply chain management with the consideration of the risk of vehicle breakdown as a multi-depot vehicle routing problem. Multi-depot inventory was just proposed by Bertazzi et al. [14] in 2019, they described the problem of optimizing supply chain as a multi-depot inventory routing problem.

In summary, VRP tends to minimize the travel cost by determining the set of routes to deliver a given quantity to each customer in a single time period. IRP aims to minimize the routing cost only or the sum of inventory and routing costs over a time horizon by determining the quantity to each customer at each time period and the sets of routes at each time period [14]. This paper also describes the distribution of relief materials as a single-depot routing problem, but still has two differences with previous literatures: we additionally consider the consumption of vehicles and the coordination between vehicles.

3   Extended COPRAS for Evaluation of the Demand Degree

In the process of evaluation, in order to prevent the virtual high demand being reported by hospitals, the evaluation needs to be rated by experts from government. Meanwhile, experts from hospitals are also needed for improving the accuracy of the evaluation, because they often have a better understanding of the situation in hospitals. Therefore, the set of experts responsible for the evaluation should be composed of experts from hospitals and the government. It is usual that experts from different departments tend to use different linguistic term sets to express their judgments on criteria, and sometimes the experts may make a judgment between two linguistic terms for the uncertainty situation of each hospital. For example, there are 2 experts from different departments. Linguistic terms sets used by them and their ratings for a same criterion are respectively shown in Fig. 1 and Fig. 2.

Similar examples cannot be solved by normal fuzzy number like triangular fuzzy number, but it can be well handled by interval 2-tuple linguistic variables [15, 16]. Besides, compared to other fuzzy variables, interval 2-tuple linguistic variables have following advantages: (a) It improves the link between linguistic variables and numerical values. (b) The computational processes of dealing with interval 2-tuple linguistic

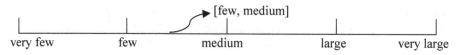

Fig. 1. The linguistic term set and rating of the first expert

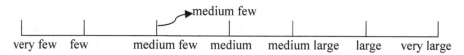

Fig. 2. The linguistic term set and rating of the second expert

variables can avoid loses of information [17]. Therefore, interval 2-tuple linguistic variable is a suitable fuzzy variable for our evaluation. More specific definitions of it are presented in Appendix 1.

The COPRAS is a method of multi-criteria decision-making problems. Compared with other multi criteria decision making methods like TOPSIS, TODIM, it is simple and efficiency, besides, it can process the information when both positive criteria and negative criteria exist [7]. And in our evaluation, criteria like the number of patients have a positive relationship with the demand degree; criteria like the materials inventory level have a negative relationship with the demand degree. Therefore, the COPRAS is an appropriate way for the evaluation of each hospital's demand degree. Finally, this paper uses the COPRAS method based on interval 2-tuple linguistic variable to evaluate each hospital's demand degree for different kind of materials respectively.

Suppose that the Evaluation has H decision makers $DM_h(h = 1, 2, .., H)$, P hospitals $A_p(p = 1, 2, .., P)$, Q criteria $C_q(q = 1, 2, \ldots, Q)$ and B kind of materials $(b = 1, 2, \ldots, B)$. Each decision maker DM_h is given a weight $\lambda_h > 0$ satisfying $\sum_{h=1}^{H} \lambda_h = 1$ to reflect the importance of each decision maker. Let $S = \{s_0, s_1, \ldots, s_g\}$ be the linguistic term set, $D_h = (d_{pq}^h)_{P \times Q}$ be the linguistic decision matrix of decision maker h, where d_{pq}^h is the linguistic information provided by DM_h on the assessment of criteria q for hospital p. Let $\omega_{hb} = \left(\omega_1^{hb}, \omega_2^{hb}, \ldots, \omega_Q^{hb}\right)$ be the linguistic weight vector given by the decision maker h, where ω_q^{hb} is the weight of criteria q under the evaluation for material b provided by DM_h. It is noteworthy that different decision makers can employ different linguistic term set. Based upon assumptions and notations, the procedure of interval 2-tuple linguistic COPRAS method for evaluation of material b's demand degree for each hospital can be defined as follows:

Step 1: Convert the linguistic decision matrix D_h into interval 2-tuple linguistic decision matrix $R_h = \left(\left[\left(r_{pq}^h, 0\right), \left(t_{pq}^h, 0\right)\right]\right)_{P \times Q}$, where $r_{pq}^h, t_{pq}^h \epsilon S$, and $r_{pq}^h \leq t_{pq}^h$.

Step 2: Convert the linguistic weight vector ω_{hb} into 2-tuple linguistic weight vector $w_{hb} = \left[\left(w_1^{hb}, 0\right), \left(w_2^{hb}, 0\right), \ldots \left(w_Q^{hb}, 0\right)\right]^T$, where $w_q^{hb} \epsilon S$.

Step 3: Convert each element in the above two interval 2-tuple linguistic decision matrix to its equivalent numerical value with the reverse function Δ^{-1} and the new matrix are separately written as R'_h and w'_h.

Step 4: Aggregate the all decision makers' ratings on each criterion to construct a collective interval 2-tuple linguistic decision matrix $R' = (r')_{P \times Q}$, where

$$r' = [\sum_{h=1}^{H} \lambda_h \Delta^{-1} \left(r_{pq}^h, 0 \right), \sum_{h=1}^{H} \lambda_h \Delta^{-1} \left(t_{pq}^h, 0 \right), p = 1, 2, \ldots, P, \ q = 1, 2, \ldots, Q \tag{1}$$

Step 5: Aggregate the all decision makers ratings on each criteria weights to construct a collective interval 2-tuple linguistic decision matrix $w' = \left[(w_{1b}, 0), (w_{2b}, 0), \ldots, (w_{Qb}, 0) \right]^T$, where

$$\left(w_{qb}, 0 \right) = (\sum_{h=1}^{H} \lambda_h \Delta^{-1} \left(w_q^{hb}, 0 \right)), q = 1, 2, \ldots, Q. \tag{2}$$

Step 6: Defuzzy the interval by the following equation:

$$\Delta^{-1} \left[\left(r_{pq}^h, 0 \right), \left(t_{pq}^h, 0 \right) \right] = [\beta_1, \beta_2] = \frac{\beta_1 + \beta_2}{2} \tag{3}$$

the final collective decision matrix is written as $R'' = \left[r_{pq}'' \right]_{P \times Q}$, the final collective weight vector is written as $w'' = \left[w_{1b}'', w_{2b}'', \ldots, w_{Qb}'' \right]^T$.

Step 7: Let $E = [e_{pq}]_{P \times Q}$ be the Normalization matrix of the decision-making, where

$$e_{pq} = \frac{w_q''}{\sum_{p=1}^{P} r_{pq}''} r_{pq}'' \tag{4}$$

Step 8: Calculate the sums of weighted normalized criteria for every hospital. The criteria are always composed of positive criteria and negative criteria, the higher the positive criteria's values are, the more demand degree of hospital is. Reversely, the higher the negative criteria's values are, the more demand degree of hospital is. The sums of positive and negative weighted normalized criteria are calculated by the following equation:

$$S_p^+ = \sum_{z_q} = +e_{pq} \tag{5}$$

$$S_p^- = \sum_{z_q} = -e_{pq} \tag{6}$$

where $z_q = \begin{cases} +, \ \text{if criteria } q \text{ is positive} \\ -, \ \text{if criteria } q \text{ is negative} \end{cases}$

Step 9: Calculate the relative significance Q_p of each hospital by the following equation:

$$Q_p = S_p^+ + \frac{S_{min}^- \sum_{p=1}^{P} S_p^-}{S_p^- \sum_{p=1}^{P} \frac{S_{min}^-}{S_p^-}} \tag{7}$$

Step 10: Calculate the normalization number $Q_p^{\sim}$ by the following equation,

$$Q_p^{\sim} = \frac{Q_p}{\sum_{p=1}^{P} Q_p} \tag{8}$$

This section proposes a COPRAS method based on 2-tuple linguistic variable for the evaluation of each hospital's demand degree for each kind of material. On the basis of it, we solve the problem unfair allocation and inefficient delivery in next section.

4 Mixed Distribution of the Relief Materials

The following subsections will introduce our model for the distribution of relief materials. In order to match the quantity of each kind of material allocated to each hospital with the demand degree of the hospital, we first prorate each kind of material to each hospital on the basis of the demand degree evaluated in Sect. 3 and take the results as the demand of each hospital. Then we propose a distribution model to improve the efficiency in the process of distribution. We make some assumptions for this distribution model as follows:

a) The definition of unit quantity refers to each undetachable package, so the number of each material means the number of undetachable package of each kind of material;
b) The volume of each package is same;
c) All the vehicles are homogenous and have a same capacity;
d) We consider the number of supple depot is single in our model. Actually, as we mentioned in introduction, the working efficiency of single supple depot is low when the number of relief materials is large. Therefore, multiple supple depots should be set up in face of a large number of materials. And for each supple depot, our model is suitable;
e) Each hospital can be served by multiple vehicles;
f) Vehicles must come back to the supple depot that they leave;
g) The distribution time of each vehicle refers to the total time required for each vehicle to start from the warehouse, visit some hospitals in a certain order, and finally return to the supple depot;
h) The objective is to minimize the sum of all vehicle distribution time, under the precondition that the demand of each hospital is satisfied.

4.1 Basic Notations

To avoid confusion with the notations in previous sections, we claim that the notations used in Sect. 3 are not suitable in Sect. 4 and 5.

Sets

$P = \{0\}$: The set of supple depot
$I = \{1, 2, \ldots, I\}$: The set of hospitals

$V = P \cup I$: The set of supple depots and hospitals
$R = \{1, 2, \ldots, r, \ldots, R\}$: The type of materials
$K = \{1, 2, \ldots, k, \ldots K\}$: The set of homogeneous vehicles
$E = \{(i, j)|i, j \in V, i \neq j\}$: The set of edges between i and j

Parameters

C: The capacity of each vehicle
w_{ir}: The demand degree of hospital i for material r calculated in Sect. 3
t_{ij}: The travel time between hospital i and j
B_r: The number of material r in supple depot

Decision Variables

q_{ir}^k: The number of material r to hospital i delivered from supple depot by vehicle k
z_{ir}^k: 1 if hospital i is served by vehicle k with material r, 0 otherwise
l_{ij}^k: 1 if hospital i is visited before hospital j by vehicle k, 0 otherwise
y_k : 1 if vehicle k is used, 0 otherwise
Q_{ir}: The demand of material r for hospital i

4.2   Mixed Distribution Model

Mathematical formulation of our model can be written as follows:

$$min \sum_{k=1}^{K} \sum_{i=1}^{V} \sum_{j=1}^{V} l_{ij}^k t_{ij} z_{ir}^{kp} \tag{9}$$

The objective function (9) minimizes the sum of all vehicle distribution time. Compared with traditional vehicle routing problem, our objective function takes vehicle consumption into account by using a binary variable y_k. In traditional vehicle routing problem, they often consume all the given vehicles and give a solution of vehicle routing, but don't have to consider the number of used vehicles. Moreover, the constraint that a demand point can only be visited once often lead to the result that each vehicle may finish its task with a relatively high no-load rate and waste of vehicle resources. During the background of epidemic, the government should not only give a solution of vehicle routing which aims to reduce the cost time for delivery, but also should reduce the movement of people to control the spread of the virus and this principle also works for those drivers of vehicles. Besides, hospitals have been filled with patients who get COVID-19 infected and all of them are badly in need of materials, tasks for material delivery are very heavy and it may result the number of vehicles is not enough. Hence, taking the consumption of vehicles into consideration is necessary.

Constraints of this model are presented in Eq. (10)–(15):

$$Q_{ir} = w_{ir} B_r \tag{10}$$

$$Q_{ir} = \sum_{k=1}^{K} q_{ir}^{k} \tag{11}$$

$$\sum_{r=1}^{R} \sum_{i=1}^{I} q_{ir}^{k} z_{ir}^{k} \leq C, \forall k \in K \tag{12}$$

$$\sum_{k=1}^{K} y_k \leq K \tag{13}$$

$$\sum_{i=1}^{I} z_{ir}^{k} \leq 1, \forall k \in K \tag{14}$$

$$\sum_{j=1}^{I} l_{ij}^{k} - l_{ji}^{k} = 0, \forall k \in K, i = 0 \tag{15}$$

Constraint (10) prorates each kind of material to each hospital. Constraint (11) shows the demand of each kind of material r for each hospital should be satisfied by a certain number of vehicles. Constraint (12) shows the capacity limitation of each vehicle. Constraint (13) shows the resource limitation of vehicles. Constraint (14) ensures a vehicle will visit a hospital not more than once. Constraint (15) ensures vehicles come back to the supple depot that they leave.

5   Numerical Experiment

This section examines the formulated problem using numerical example. Firstly, this paper evaluates the demand degree of three typical relief materials for four hospitals in Wuchang District, Wuhan, China. Then we give the result of mixed distribution model.

5.1   Evaluation of the Demand Degree

This paper chooses four hospitals in Wuchang District, Wuhan City, which are shortly named as A_1, A_2, A_3, A_4. An expert committee composed of four decision makers, DM_1, DM_2, DM_3, DM_4, has been formed to evaluate each hospital's demand degree for each kind of material. And DM_1, DM_2 are from hospitals, DM_3, DM_4 are from government. The evaluation is made on the basis of the following four criteria:

(1)   The number of hospitalizations who get COVID-19 infected every day.
(2)   The proportion of critical patients who get COVID-19 every day.
(3)   The overall protection level of hospital.
(4)   The inventory level of corresponding materials for COVID-19 every day.

The four decision makers employ different linguistic term sets to make evaluation. Then they give their assessments of four hospitals on each criterion and the weight of each criterion for different materials. It is noteworthy that for different kind of materials, the weight of each criterion given by each expert is also varied. Finally, we use the proposed ITL-COPRAS method to calculate the demand degree of hospitals for different materials. Specific steps of solution are shown in Appendix 2. The demand degree for materials of four hospitals is shown in Table 1.

Table 1. The demand degree of each kind of material for each hospital

Hospital, i	The demand degree of hospital i for material r, w_{ir}		
	1	2	3
1	0.280	0.260	0.267
2	0.189	0.142	0.161
3	0.229	0.186	0.203
4	0.302	0.413	0.369

5.2 Computational Study for Mixed Distribution Model

In this section, we virtually set the number of three kinds of materials in the supple depot and give a numerical experiment for this model. More relative parameters are specifically shown in Appendix 3.

The results of this instance are shown in Tables 2, 3, 4 and 5, the format of Tables 2, 3, 4 and 5 is as follows: Table 2 shows the demand of each kind of material for each hospital. Table 3 and Table 4 respectively show the results of distribution by using our model and the original way. Table 5 is an example of the situation of materials in vehicle 1.

Table 2. Demand of material r for hospital i, Q_{ir}

Hospital, i	The demand degree of hospital i for material r, w_{ir}			The demand of material r for hospital i, Q_{ir}		
	1	2	3	1	2	3
1	0.280	0.260	0.267	9	9	9
2	0.189	0.142	0.161	6	5	6
3	0.229	0.186	0.203	7	6	7
4	0.302	0.413	0.369	10	14	13

From Table 2, we can see that a hospital's demand degree is varied for different kinds of materials and hospitals with higher demand degree can get more materials in general. Take the allocation for hospital 4 as an example. Hospital 4 is a mobile cabin hospital which aims to isolate and give some simple medical treatment for those COVID-19 patients with mild clinical symptom. There are two main characteristics of this hospital: the number of patients is much larger than those general designated hospitals and the clinical symptom of these COVID-19 patients is much milder than those designated hospitals. So it is reasonable that the demand of hospital 4 for daily necessities and drug medical material are the highest. As for the demand of hospital 4 for protection medical

Table 3. The routes of each vehicle and distribution time by using our model

Vehicle, k	y_k	Route	Distribution time (/min)	Total distribution time (/min)
1	1	$0 \rightarrow 1 \rightarrow 2 \rightarrow 0$	18	
2	1	$0 \rightarrow 4 \rightarrow 0$	24	82
3	1	$0 \rightarrow 2 \rightarrow 3 \rightarrow 4 \rightarrow 0$	40	

Table 4. The routes of each vehicle and distribution time by using the original way

Vehicle, k	y_k	Route	Distribution time (/min)	Total distribution time (/min)
1	1	$0 \rightarrow 1 \rightarrow 0$	10	
2	1	$0 \rightarrow 2 \rightarrow 0$	14	
3	1	$0 \rightarrow 3 \rightarrow 0$	22	94
4	1	$0 \rightarrow 4 \rightarrow 0$	24	
5	1	$0 \rightarrow 4 \rightarrow 0$	24	

Table 5. The number of each kind of material to each hospital in vehicle 1 in our model

Materials, r	Hospitals, i			
	1	2	3	4
1	9	6	0	0
2	9	2	0	0
3	9	0	0	0

material is also the most, this phenomenon relates to the reason that the inventory level of this kind of material for newly-built hospital 4 is much lower than other hospitals.

Table 3 and Table 4 are the result of vehicle routing by using our model and original way respectively. According to the previous government's way of distribution which is shown in Table 3, each vehicle was assigned to deliver to one hospital at a time but without any programming. In our model, for the purpose of improving the efficiency of distribution, we project the driving path of used vehicles which aims to minimize the distribution time of all used vehicles. Obviously, the total cost time for distribution in our model is less than that of the original way.

From Table 5, we can see that all kinds of materials are mixed in vehicle 1 and we still give a solution about the task assignment of each vehicle for each hospital. Take the distribution for hospital 4 as an example, the demand of each kind of material for hospital 4 is satisfied by 2 vehicles, the combination of materials in each vehicle is different and the route of each vehicle is also different. In another word, the distribution of materials for hospital 4 is finished by the coordination of these vehicles with a variety number of materials. In traditional vehicle routing problem or inventory routing problem, the service for a hospital is usually satisfied by a single vehicle and can often lead to the waste of vehicles.

In general, compared with the previous methods adopted currently, our model ensures the fairness of material distribution and improves the efficiency. Moreover, we reduce the number of vehicles used.

6 Conclusions

This paper proposed an integrated method for the distribution of relief materials when facing at an emergency situation. Taking the COVID-19 rescue work as an introductory example, we first compute the demand degree of each hospital for each kind of relief material by COPRAS. Based on the degrees, make the decisions of distribution. Different from classical transportation problem and vehicle routing problem, we consider the consumption of vehicles into accountant, each hospital can be served by different vehicles, and each vehicle may just satisfy parts of the hospital's demand. The results suggest that our evaluation for the demand degree of each hospital for each kind of relief material is actually feasible and reasonable. For the distribution of various emergency relieves, our approach with mixed distribution is effective in reducing the use of vehicles.

References

1. Bergmann, F.M., Wagner, S.M., Winkenbach, M.: Integrating first-mile pickup and last-mile delivery on shared vehicle routes for efficient urban e-commerce distribution. Transp. Res. Part B: Methodol. **131**, 26–62 (2020)
2. Bertazzi, L., Coelho, L.C., Maio, A.D., Laganà, D.: A matheuristic algorithm for the multi-depot inventory routing problem. Transp. Res. Part E: Logist. Transp. Rev. **122**, 524–544 (2019)
3. Chen, C.T., Tai, W.S.: Measuring the intellectual capital performance based on 2-tuple fuzzy linguistic information. In: Proceedings of the 10th Annual Meeting of Asia Pacific Region of Decision Sciences Institute, Taiwan (2005)
4. Coelho, L.C., Cordeau, J.F., Laporte, G.: Thirty years of inventory routing. Transp. Sci. **48**(1), 1–19 (2014)
5. Herrera, F., Martínez, L.: A 2-tuple fuzzy linguistic representation model for computing with words. IEEE Trans. Fuzzy Syst. **8**(6), 746–752 (2000)
6. Hornstra, R.P., Silva, A., Roodbergen, K.J., Coelho, L.C.: The vehicle routing problem with simultaneous pickup and delivery and handling costs. Comput. Oper. Res. **115**. https://doi.org/10.1016/j.cor.2019.104858
7. Kaklauskas, A., Zavadskas, E.K., Raslanas, S.: Multivariant design and multiple criteria analysis of building refurbishments. Energy Build. **37**, 361–372 (2005)
8. Liu, G., Hu, J., Yang, Y., Xia, S., Lim, M.K.: Vehicle routing problem in cold chain logistics: a joint distribution model with carbon trading mechanisms. Resour. Conserv. Recycl. https://doi.org/10.1016/j.resconrec.2020.104715
9. Mulliner, E., Smallbone, K., Maliene, V.: An assessment of sustainable housing affordability using a multiple criteria. Omega **41**, 270–279 (2013)

10. Mulliner, E., Malys, N., Maliene, V.: Comparative analysis of MCDM methods for the assessment of sustainable housing affordability. Omega **59**, 146–156 (2016)
11. Peng, X., Selvachandran, G.: Pythagorean fuzzy set: state of the art and future directions. Artif. Intell. Rev. **52**(3), 1873–1927 (2017)
12. Pitchipoo, P., Vincent, D.S., Rajini, N., Rajakarunakaran, S.: COPRAS decision model to optimize blind spot in heavy vehicles: a comparative perspective. Procedia Eng. **97**, 1049–1059 (2014)
13. Soeanu, A., Ray, S., Berger, J., Boukhtouta, A., Debbabi, M.: Multi-depot vehicle routing problem with risk mitigation. Model and solution algorithm. Expert Syst. Appl. **145**, 113099 (2020)
14. You, X.Y., You, J.X., Liu, H.C., Zhen, L.: Group multi-criteria supplier selection using an extended VIKOR method with interval 2-tuple linguistic information. Expert Syst. Appl. **42**(4), 1906–1916 (2015)
15. Zavadskas, E.K., Kaklauskas, A., Sarka, V.: The new method of multi-criteria complex proportional assessment of projects. Technol. Econ. Dev. Econ. **1**(3), 131–139 (1994)
16. Zhang, H.: The multi-attribute group decision making method based on aggregation operators with interval-valued 2-tuple linguistic information. Math. Comput. Model. **56**, 27–35 (2012)
17. Zhen, L., Ma, C., Wang, K., Xiao, L., Zhang, W.: Multi-depot multi-trip vehicle routing problem with time windows and release dates. Transp. Res. Part E: Logist. Transp. Rev. **135**. https://doi.org/10.1016/j.tre.2020.101866

Author Index

Printed in the United States
By Bookmasters